Economical and Technical Considerations for Solar Tracking:

Methodologies and Opportunities for Energy Management

S. Soulayman
Higher Institute for Applied Sciences and Technology, Syria

A volume in the Advances in Environmental Engineering and Green Technologies (AEEGT) Book Series

IGI Global
DISSEMINATOR OF KNOWLEDGE

Published in the United States of America by
IGI Global
Engineering Science Reference (an imprint of IGI Global)
701 E. Chocolate Avenue
Hershey PA, USA 17033
Tel: 717-533-8845
Fax: 717-533-8661
E-mail: cust@igi-global.com
Web site: http://www.igi-global.com

Library of Congress Cataloging-in-Publication Data

Names: Soulayman, S., 1952- author.
Title: Economical and technical considerations for solar tracking : methodologies and opportunities for energy management / by S. Soulayman.
Description: Hershey, PA : Engineering Science Reference, [2018] | Includes bibliographical references.
Identifiers: LCCN 2017011949| ISBN 9781522529507 (h/c) | ISBN 9781522529514 (e-ISBN)
Subjects: LCSH: Solar collectors--Automatic control. | Automatic tracking. | Sun--Rising and setting. | Solar energy--Economic aspects.
Classification: LCC TJ812 .S685 2018 | DDC 621.47/2--dc23 LC record available at https://lccn.loc.gov/2017011949

This book is published in the IGI Global book series Advances in Environmental Engineering and Green Technologies (AEEGT) (ISSN: 2326-9162; eISSN: 2326-9170)

British Cataloguing in Publication Data
A Cataloguing in Publication record for this book is available from the British Library.

For electronic access to this publication, please contact: eresources@igi-global.com.

Advances in Environmental Engineering and Green Technologies (AEEGT) Book Series

ISSN:2326-9162
EISSN:2326-9170

Mission

Growing awareness and an increased focus on environmental issues such as climate change, energy use, and loss of non-renewable resources have brought about a greater need for research that provides potential solutions to these problems. Research in environmental science and engineering continues to play a vital role in uncovering new opportunities for a "green" future.

The **Advances in Environmental Engineering and Green Technologies (AEEGT)** book series is a mouthpiece for research in all aspects of environmental science, earth science, and green initiatives. This series supports the ongoing research in this field through publishing books that discuss topics within environmental engineering or that deal with the interdisciplinary field of green technologies.

Coverage

- Biofilters and Biofiltration
- Water Supply and Treatment
- Green Transportation
- Alternative Power Sources
- Waste Management
- Contaminated Site Remediation
- Air Quality
- Electric Vehicles
- Cleantech
- Radioactive Waste Treatment

IGI Global is currently accepting manuscripts for publication within this series. To submit a proposal for a volume in this series, please contact our Acquisition Editors at Acquisitions@igi-global.com or visit: http://www.igi-global.com/publish/.

The Advances in Environmental Engineering and Green Technologies (AEEGT) Book Series (ISSN 2326-9162) is published by IGI Global, 701 E. Chocolate Avenue, Hershey, PA 17033-1240, USA, www.igi-global.com. This series is composed of titles available for purchase individually; each title is edited to be contextually exclusive from any other title within the series. For pricing and ordering information please visit http://www.igi-global.com/book-series/advances-environmental-engineering-green-technologies/73679. Postmaster: Send all address changes to above address.

Titles in this Series

For a list of additional titles in this series, please visit:
https://www.igi-global.com/book-series/advances-environmental-engineering-green-technologies/73679

Computational Techniques for Modeling Atmospheric Processes
Vitaliy Prusov (University of Kyiv, Ukraine) and Anatoliy Doroshenko (National Academy of Sciences, Ukraine)
Information Science Reference • ©2018 • 460pp • H/C (ISBN: 9781522526360) • US $205.00

Advanced Nanomaterials for Water Engineering, Treatment, and Hydraulics
Tawfik A. Saleh (King Fahd University of Petroleum and Minerals, Saudi Arabia)
Engineering Science Reference • ©2017 • 384pp • H/C (ISBN: 9781522521365) • US $200.00

Handbook of Research on Inventive Bioremediation Techniques
Jatindra Nath Bhakta (University of Kalyani, India)
Engineering Science Reference • ©2017 • 624pp • H/C (ISBN: 9781522523253) • US $260.00

Handbook of Research on Entrepreneurial Development and Innovation Within Smart Cities
Luisa Cagica Carvalho (Universidade Aberta, Portugal & CEFAGE - Universidade de Évora, Portugal)
Information Science Reference • ©2017 • 661pp • H/C (ISBN: 9781522519782) • US $235.00

Applied Environmental Materials Science for Sustainability
Takaomi Kobayashi (Nagaoka University of Technology, Japan)
Information Science Reference • ©2017 • 416pp • H/C (ISBN: 9781522519713) • US $205.00

Environmental Issues Surrounding Human Overpopulation
Rajeev Pratap Singh (Banaras Hindu University, India) Anita Singh (University of Allahabad, India) and Vaibhav Srivastava (Banaras Hindu University, India)
Information Science Reference • ©2017 • 325pp • H/C (ISBN: 9781522516835) • US $200.00

Reconsidering the Impact of Climate Change on Global Water Supply, Use, and Management
Prakash Rao (Symbiosis International University, India) and Yogesh Patil (Symbiosis International University, India)
Information Science Reference • ©2017 • 430pp • H/C (ISBN: 9781522510468) • US $215.00

Environmental Sustainability and Climate Change Adaptation Strategies
Wayne Ganpat (The University of the West Indies, Trinidad and Tobago) and Wendy-Ann Isaac (The University of the West Indies, Trinidad and Tobago)
Information Science Reference • ©2017 • 406pp • H/C (ISBN: 9781522516071) • US $200.00

For a list of additional titles in this series, please visit:
https://www.igi-global.com/book-series/advances-environmental-engineering-green-technologies/73679

IGI Global
DISSEMINATOR OF KNOWLEDGE

701 East Chocolate Avenue, Hershey, PA 17033, USA
Tel: 717-533-8845 x100 • Fax: 717-533-8661
E-Mail: cust@igi-global.com • www.igi-global.com

Table of Contents

Preface

Solar energy is the Sun's nuclear fusion reactions within the continuous energy generated. Earth's orbit, the average solar radiation intensity is 1367kW/m^2. Circumference of the Earth's equator is 40000km, thus we can calculate the energy the Earth gets is up to 173,000 TW. At sea level on the standard peak intensity is 1kW/m^2, a point on the Earth's surface 24h of the annual average radiation intensity is 0.20kW/m^2, or roughly 102,000 TW of energy. Humans rely on solar energy to survive, including all other forms of renewable energy (except for geothermal resources). Although the total amount of solar energy resources is ten thousand times of the energy used by humans, but the solar energy density is low, and it is influenced by location, season, which is a major problem of development and utilization of solar energy.

The technical feasibility and economic viability of using solar energy depends on the amount of available solar radiation in the area where you intend to place solar heaters or solar panels. This is sometimes referred to as the available solar resource. Every part of Earth is provided with sunlight during at least one part of the year. The "part of the year" refers to the fact that the north and south polar caps are each in total darkness for a few months of the year. The amount of solar radiation available is one factor to take into account when considering using solar energy.

Day and night is due to the Earth's rotation generated, but the season is due to the Earth's rotation axis and the Earth's orbit around the sun's axis was 23°27' angle and generated. The Earth rotates around the "axis" which through its own north and south poles a circuit from west to east every day. Per revolution of the earth cause day and night, so the Earth's rotation per hour is 15°.In addition, the Earth goes through a small eccentricity elliptical orbit around the sun per circuit per year. The Earth's axis of rotation and revolution has always been 23.5° with the Earth orbit. The Earth's revolution remains unchanged when the direction of spin axis always points to the Earth's north pole. Therefore, the Earth's orbit at a different location when the solar radiation is projected onto the direction of the Earth is different, so it causes the formation of the Earth's seasons changes. Noon of each day, the sun's height is always the highest. In the tropical low-latitude regions (in the equatorial north and south latitude 23° 27' between the regions), solar radiation of each year, there are two vertical incidences at higher latitudes, the sun is always close to the equator direction. In the Arctic and Antarctic regions (in the northern and southern hemispheres are greater than 90° ~ 23° 27'), in winter the sun below the horizon for a long time.

Solar radiation on the horizontal surface is composed of two parts - direct solar radiation and diffuse radiation. Solar radiation goes through the atmosphere and reaches the ground, due to the atmosphere air molecules, water vapor and dust, such as solar radiation absorption, reflection and scattering, not only reduction of the intensity of solar radiation, but also to change the direction of solar radiation and spectral distribution of the radiation. Therefore, the actual solar radiation reaching the ground is usually caused by direct and diffusion of two parts. Direct solar radiation is the solar radiation directly coming from the sun and the direction of this radiation has not been changed; diffuse solar radiation is the reflection and scattering by the atmosphere changed after the direction of the solar radiation, which consists of three parts: the Sun around the scattering (surface of the Sun around the sky light), horizon circle scattering (horizon circle around the sky light or dark light), and other sky diffuse radiation.

In addition, the non-horizontal plane also receives the reflection of radiation from the ground. Direct sunlight, diffuse and reflected solar radiation shall be the sum of the total or global solar radiation. It can rely on the lens or reflector to focus on direct solar radiation. If the condenser rate is high, you can get high energy density, but loss of diffuse solar radiation. If the condenser rate is low it can also condense parts of the diffuse solar radiation. Diffuse solar radiation has a big range of variation, and when it's cloudless, the diffuse solar radiation is 10% of the total solar radiation. But when the sky is covered with dark clouds and the sun can't be seen, the total solar radiation is equal to the diffuse solar radiation. Therefore, poly-type collector is collecting the energy usually far higher than the non-poly-type collector. Reflected solar radiation is generally weak, but when there is snow-covered ground, the vertical reflection solar radiation can be up to 40% of the total solar radiation.

The objective of this book is to provide a platform to disseminate the knowledge regarding economical and technical considerations of long term and short term solar tracking to undergraduate and postgraduate students, learners, professional and designers with focusing on the methodologies and opportunities for energy management. In order to achieve the mentioned above goal it is important to build the knowledge base for treating the most effective parameters. Therefore, apparent Sun position in relation to the Earth's center as well as to the observer on the Earth's surface should be determined. Then, the maximum sunshine duration on the surface with different orientations should be calculated. The maximal solar irradiance on surfaces with different orientation should be provided as well as the different modes of solar tracking and the resulting maximal possible instantaneous energy gain should be given. The influence of the Earth's atmosphere on the different component of the received solar radiation is an important factor to be considered. Several models are proposed for treating this question. Therefore, it is reasonable to briefly comment these models and to suggest some of them for using within this book. Solar receiver optimum tilt angle is one of the most important factors the affects the gain of solar system in the long term tracking as well as in the azimuthal single axis tracking. Therefore, the question of optimum tilt finding should be treated in details with giving real and precise results for different users. Solar tracking is one of the main parts of the present book. Therefore, it is important to treat this question in a respectable manner. A general formula was proposed to describe the movement of the dual axis trackers. Another formula was also deduced for a single axis tracker. Some comments were provided on the effectiveness of dual solar tracking in relation to latitude tilted fixed PV panels as well as in relation to panels installed with an optimum early tilt angle from technical point of view as well as from economic point of view.

Thus, the present book has been written in eight chapters to study a basic knowledge of Sun's structure and its radiation.

Chapter 1 describes the development of equations to calculate the angle between a collector aperture normal and a central ray from the Sun. This development is done first for fixed and then for tracking collectors. These equations are then used to provide insight into solar angles measurement and geographic orientation. We defined first the Sun's position angles relative to Earth-center coordinates and then to coordinates at an arbitrary location on the Earth's surface. In the design of solar energy systems, it is most important to be able to predict the angle between the Sun's rays and a vector normal (perpendicular) to the aperture or surface of the collector. This angle is called the angle of incidence. Knowing this angle is of critical importance to the solar designer, since the maximum amount of solar radiation energy that could reach a collector is reduced by the cosine of this angle.

Chapter 2 describes an instrument (magnetic declination device) which could be used for determining the geographic north of the site where a solar system will be installed. The use of the magnetic declination device allows determine the azimuth angle of the solar collectors in the solar systems. This angle is required in choosing the best orientation of the solar collector as well as in the process of solar tracking.

Chapter 3 treats the topics that are based on extraterrestrial solar radiation. Different formulae for calculating the extraterrestrial solar radiation on surfaces with different orientations are provided.

This is background information for chapter IV which is concerned with effects of the atmosphere, radiation measurements and data manipulation. The main concepts of the direct geometric factor calculation using different modes of tracking are provided. The short term solar energy collection is also introduced. Chapter 4 describes methods (and gives comments on their applications) for the estimation of solar radiation information in the desired format from the data that are available. This includes estimation of average global solar radiation on the horizontal plane using sunshine hour duration based methods, ambient air temperature based methods, cloud cover based methods, satellite-based models and others. It includes also estimation of beam and diffuse solar radiation from total solar radiation on the horizontal plane as well as on the tilted surfaces.

In the Chapter 5, a general algorithm is proposed for treating the optimum tilt angles of solar receivers, β_{opt}, all over the world. The theoretical aspects that determine the optimal tilt angle, regarding to maximum solar energy collection, are examined. The computer program is used in determining the optimal tilt angle of any site at the Earth's surface between 66.45°S and 66.45°N. A regression analysis using site's latitude, solar declination angle and its corresponding optimal tilt angle is conducted to develop a mathematical model that allows the determination of the optimal tilt angle at which maximum solar radiation is collected using only the site's latitude and the day number of the year. A comparison with available experimental and theoretical results from other researchers is provided. A set of tables were provided in chapter V, where the daily, $\beta_{opt,d}$, monthly, $\beta_{opt,m}$, seasonally, $\beta_{opt,s}$, biannually, $\beta_{opt,b}$, and yearly, $\beta_{opt,y}$, optimum tilt angles are provided in the appendix of this chapter.

Chapter 6 treats the different aspects of short term tracking. In this chapter VI, a general formula for on-axis sun-tracking system has been derived using coordinate transformation method. The derived sun-tracking formula is the most general form of mathematical solution for various kinds of arbitrarily oriented on-axis sun tracker, where azimuth-elevation and tilt-roll tracking formulas are specific cases. Although the rotation angle is an intermediate value for determining the incidence angle, it has applications of its own for the control of tracker movement and for modeling the solar radiation available for a collector. For a motorized tracker with fixed gearing, the tracker rotation is directly proportional to the number of motor revolutions; consequently, the calculated rotation angle can be used to determine the number of motor revolutions to move the tracker to its optimum position. When modeling collector solar radiation, the rotation angle can also be used to account for non-optimum tracking that may occur when the optimum rotation angle exceeds the rotation limits of the tracker. Chapter VI sheds a light on the procedure of rotational angle determination.

In Chapter 7, the concept of energy gain is introduced. The energy gain is very useful in evaluating the performance of different types of tracking. This concept allows to evaluate the effectiveness of daily, weekly, fortnightly, monthly, seasonally, biannually and yearly adjustment of the solar receiver tilt angle in relation with the ideal instantaneous dual tracking, where the sun rays are kept permanently perpendicular to the flat-plate surface of the receiver. Thus, the incidence angle is kept to be zero all over the day. This evaluation determines the optimum tilt application over any period of consecutive days from technical point of view. A set of tables were provided in chapter VII, where the daily, monthly, seasonally, biannually and yearly energy gains in the case of long term tracking are provided in the appendix of this chapter.

Since the future of solar tracking depends on the cost of solar trackers and of the gained solar energy, Chapter 8 sheds a light on the principles of the economic analysis in general. These principles were applied on the solar thermal power system of commercial scale. It should be noted here that, energy price is one of the most influential factors in all case studies, yet it is among the most volatile indexes in the world market scale. Any deviation in energy price may significantly change the financial feasibility of any project. The comparison of dual axis tracking system with relation to horizontal, latitude tilted and yearly optimum tilted fixed solar systems of the same kind of PV solar panels from economic point of view is considered. It is proved that, dual axis tracking becomes more economic than latitude tilted fixed PV systems with increasing the PV panels' area. On the other hand, dual axis tracking is not economic in the sunny belt countries because of temperature effect on the performance of PV systems. Moreover, tracked and fixed PV systems of large scale are not economic with relation to traditional sources of energy. Anyway, the diffusion of photovoltaic systems is hindered until today by high investment costs. However, PV power generation is justified for special purposes. It is clearly demonstrated that, the small scale applications such as telecommunication systems, rural electrification, cathode protection and water lifting are economically feasible.

In order to make the book useful, the useful relationships in equations, graphical and tabular form were given wherever it is possible. The recommended standard nomenclature of the *Journal of Solar Energy* is used excepts for a few cases where additional symbols have benn needed for clarity. For example, G is used for solar irradiance (Wm^{-2}), I is used for integrated quantity over an hour (MJm^{-2}) and H (MJm^{-2}) is used for integrated quantity over a period of time (minimum one day and maximum one year). Therefore, we have daily, H_d, weekly, H_w, fortnightly, H_f, monthly, H_m, seasonally, H_s, biannually, H_b, and yearly, H_y, solar radiation. Moreover, the units and symbols follow, mainly, those suggested in Solar Energy Journal. S.I. units have been used throughout the book.

Numerous sources have been used in writing this book. The book of Duffie and Beckman (2013) is one of the most useful books with regard to the subjects of 1, 3, and 4 chapters of this book. The book of Tiwari (2010) is found to be helpful especially in preparing Chapter 8. The journals of Solar Energy, Renewable Energy, Energy and Energy Conversion and Management are very useful for all chapters of this book. They contain a variety of papers on different topics. The publications of NREL - the national laboratory of the U.S. Department of Energy- are also used. In addition to the above-mentioned sources, there exists a very large and growing body of literature in the form of reports to and by government agencies which contain useful information not readily available elsewhere.

This book has been aimed to provide a great insight in the subject particularly to the learning students and professionals doing self-study. In spite of my best efforts, some errors might have been crept in the text. I welcome the suggestions and comments, if any, from all readers for further improvement of the book in the next edition.

REFERENCES

Duffie, J. A., & Beckman, W. A. (2013). *Solar engineering of thermal processes* (3rd ed.). New York, NY: Wiley & Sons. doi:10.1002/9781118671603

Tiwari, G. N. (2010). *Solar energy fundamentals, design, modeling and applications* (7th ed.). New Delhi: Narosa Publishing House.

Acknowledgment

Individuals who have helped me with the preparation of this book are many. My graduate students and staff at the renewable energy laboratory at Higher Institute for Applied Sciences and Technology have provided me with ideas, useful information and reviews of parts of the manuscript. Their constructive comments have been invaluable.

The patience of my spouse, Mrs. Ahlam Hdewah in bearing with this lengthy project in good humor is greatly appreciated. My special thanks go out to my children, Haydar, Zainab and Fatemeh for willingly giving up their valuable time due to them, used in preparation of this book.

I acknowledge for moral support and encouragement extended by Mrs. A. Hdewah, Eng. I. Soulayman and Pr. A. Soulayman during the course.

I am thankful to Dr. K. Skeif for preparation of some of the figures on computer. Thanks are also due to Engineers M. Hammoud and A. Habbabeh for their help during preparation of excellent figures for the book.

It is of my pleasure to express my deep gratitude to my respected parents Sh. H. Soulayman and B. M. Daher for their blessing which helped me to reach my target.

Last but not least, I express my deep gratitude to my respected teachers V. I. Zubov and Y. P. Terletsky.

Full credit is due to the publishers for producing a nice print of the book. Special thanks are due to Ms. Kelsey Weitzel-Leishman, Editorial Assistant, Acquisitions CyberTech Publishing, Acquisitions Division, and Ms. Marianne Caesar, Development Editor, IGI Global, for their assistance.

Chapter 1
Solar and Collector Angles

ABSTRACT

The Sun position determination is required in several solar applications, within them is the Sun tracking. The Sun position is determined in this chapter with reference to the Earth's center and with reference to an observer on the Earth's surface. This procedure allows determining the possible relationships between different solar angles. The determination of the solar rays' incidence angle on the surface of different orientations is very important for determining sunshine duration on this surface as well as global solar radiation received by this surface. The obtained formulas could be used for determining the optimum tilt angle of solar receiver. Some procedures for measuring site latitude, solar elevation angle, solar zenith angle, hour angle and solar azimuth angle are presented. Some devices used in measuring sunshine duration are also described. The main systems of coordinates used in solar tracking are introduced. The provided information will be essential background for different types of Sun tracking.

INTRODUCTION

The Earth receives almost all its energy from the Sun's radiation. Sun also has the most dominating influence on the changing climate of various locations on Earth at different times of the year. From a fixed location on Earth, the Sun appears to move throughout the sky. The apparent motion of the Sun, caused by the rotation of the Earth about its axis, changes the angle at which the direct component of light will strike the tilted plane of a given orientations on the Earth' surface. The position of the Sun depends on the location of a point on Earth (observer), the time of day and the time of year. From our perspective on Earth, the Sun is always changing its position in the sky. It is pretty obvious that every day the Sun moves from the east to the west between Sunrise and Sunset, it also moves from north to south throughout the course of the year. When measuring the position of the Sun every day at solar noon, it would be at a different angle every day. The exact location of the Sun in the sky depends on the observer location on the Earth's surface, the day number in the year, and, of course, the time of the day. This effects the design decisions engineers make when they are installing photovoltaic (PV) panels. It is important for engineers to know where the Sun will be throughout the year so they can install PV panels

DOI: 10.4018/978-1-5225-2950-7.ch001

at the ideal angle to absorb the maximum amount of Sunlight during the course of a year. To improve PV panel efficiency, engineers also design creative ways so more Sunlight hits the surface of the panel.

In order to collect solar energy here on the Earth, it is important to know the angle between the Sun's rays and a collector surface (aperture). When a collector is not pointing (or more exactly, when the collector aperture normal is not pointing) directly at the Sun, some of the energy that could be collected is being lost. So, an algorithm for predicting relative Sun and collector positions for exact design conditions and locations is required.

In this chapter, we develop the equations to calculate the angle between a collector aperture normal and a central ray from the Sun. This development is done first for fixed and then for tracking collectors. These equations are then used to provide insight into solar angles measurement and geographic orientation. We defined first the Sun's position angles relative to Earth-center coordinates and then to coordinates at an arbitrary location on the Earth's surface. In the design of solar energy systems, it is most important to be able to predict the angle between the Sun's rays and a vector normal (perpendicular) to the aperture or surface of the collector. This angle is called the angle of incidence. Knowing this angle is of critical importance to the solar designer, since the maximum amount of solar radiation energy that could reach a collector is reduced by the cosine of this angle.

SUN POSITION

Before determining the Sun position in relation to the Earth's center or in relation to the Earth's surface, basing on different resources (Tian Pau Chang, 2009; Roberto Grena, 2008; Roberto Grena, 2012; Manuel Blanco-Muriel, Diego C. Alarcón-Padilla, Teodoro López-Moratalla and MartÍn Lara-Coira, 2001; Parkin, 2010) it is important to define some angles. The azimuth angle and the elevation angle at solar noon are the two key angles which are used to orient photovoltaic modules. However, to calculate the Sun's position throughout the day, both the elevation angle and the azimuth angle must be calculated throughout the day. In order to determine the Sun position in relation to Earth's center or in relation to an observer on the Earth's surface it is important to define two groups of angles (Kittler and Darula, 2013; Duffie and Beckman, 2013).

Earth-Sun Angles

These angles are:

Latitude, φ

This is the angle between a line that points from the center of the Earth to a location on the Earth's surface and a line that points from the center of the Earth to the equator. This can be easily found on a map. **φ** is positive in the Northern Hemisphere and negative in the Southern Hemisphere. **φ** varies between 90°S and 90°N or $-90^o \leq \varphi \leq 90^o$.

Declination, δ

This is the angle between the line that points to the Sun from the equator and the line that points straight out from the equator (at solar noon). North is positive and south is negative. This angle varies from 23.45° to -23.45° throughout the year, which is related to why we have seasons.

Hour Angle, ω

This is based on the Sun's angular displacement, east or west, of the local meridian (the line the local time zone is based on) due to the Earth rotation on its axis. The Earth rotates 15° per hour so at 11am, the hour angle is -15° and at 1pm it is 15°.

Local Meridian

The local meridian is closest longitude of 15° increments. If you live in Damascus, Syria, at a longitude of 36.2765° E, then your local meridian is 30° E (this is the closest number divisible by 15°).

Longitude

The longitude is the angle of your location on the Earth measured around the Equator, east or west from the prime meridian (0°).

Local Solar Time

Time based on the apparent angular motion of the Sun across the sky. The solar noon corresponds to the time when the Sun crosses the meridian of the observer.

Observer-Sun Angles

These angles are:

Solar Altitude (Elevation) Angle, α

This is the angle between the line that points to the Sun and the horizontal. It is the complement of the zenith angle. At Sunrise and Sunset this angle is 0°.

Zenith Angle, Θ_z

This is the angle between the line that points to the Sun and the vertical — basically, this is just where the Sun is in the sky. At sunrise and sunset this angle is 90°. This angle is just the angle of incidence of beam radiation on a horizontal plane.

Solar Azimuth Angle, γ_s

This is the angle between the line that points to the Sun and south. Angles to the east are negative. Angles to the west are positive. This angle is 0° at solar noon. It is probably close to -90° at sunrise and 90° at sunset, depending on the season. This angle is only measured in the horizontal plane; in other words, it neglects the height of the Sun. γ_s varies according to $-180^o \leq \gamma_s \leq 180^o$.

Sun Position in Relation to the Earth's Center

The Earth rotates about on a fixed plane that is tilted 23.5° with respect to its vertical axis around the Sun. The Earth needs 23hrs 56mins to complete one true rotation, or one sidereal period, around the Sun. The solar day, on the other hand, is the time needed for a point on Earth pointing towards a particular point on the Sun to complete one rotation and return to the same point. It is defined as the time taken for the Sun to move from the zenith on one day to the zenith of the next day, or from noon today to noon tomorrow. The length of a solar day can vary by up to 30 seconds during the year, and thus on the average is calculated to be 24hrs. So, a mean second is l/86,400 of the average time between one complete transit of the Sun, averaged over the entire year. In the course of the year, a solar day may differ to as much as 15mins. There are four reasons for this time difference (Jesperson and Fitz-Randolph, 1999):

1. The Earth's motion around the Sun is not perfect circle but is eccentric, so the Earth travels faster when it is nearer the Sun than when it is farther away;
2. The Earth's axis is tilted to the plane containing its orbit around the Sun and the Earth spins at an irregular rate around its axis of rotation;
3. The Sun's apparent motion is not parallel to the celestial equator;
4. The precession of the Earth's axis.

For simplicity, we averaged out that the Earth will complete one rotation every 24hrs (based on a solar day) and thus moves at a rate of 15° per hour (one full rotation is 360°). Because of this, the Sun appears to move proportionately at a constant speed across the sky. The Sun thus produces a daily solar arc, which is the apparent path of the Sun's motion across the sky. At different latitudes, the Sun will travel across the sky at different angles each day.

The Earth revolves around the Sun every 365.25 days in an elliptical orbit, with a mean Earth-Sun distance of 1.496×10^{11} m defined as one astronomical unit (1 AU). This plane of this orbit is called the ecliptic plane. The Earth's orbit reaches a maximum distance from the Sun, or aphelion, of 1.52×10^{11} m on about the third day of July. The minimum Earth-Sun distance, the perihelion, occurs on about January 2nd, when the Earth is 1.47×10^{11} m (91.3×10^6 miles) from the Sun.

The rotation of the Earth about its axis also causes the day and night phenomenon. The length of the day and night depends on the time of the year and the latitude of the location. For places in the northern hemisphere (NH), the shortest solar day occurs around December 21 (winter solstice) and the longest solar day occurs around June 21 (summer solstice) while the shortest solar day, in southern hemisphere (SH), occurs around June 21 and the longest solar day occurs around December 21.

In theory, during the time of the equinox, the length of the day should be equal to the length of the night. The Autumnal Equinox occurs when the Sun crosses the celestial equator during its apparent movement down from above it to below it. The intersection between the two planes is called Autumnal

Equinox. This usually happens around the 21st of September. When the Sun moves apparently up from below the celestial equator to above it, the intersection between the Sun trajectory and the celestial equator is called Spring (Vernal) Equinox. It usually happens around the 21st of March. During the equinoxes, all parts of the Earth experiences 12 hours of day and night and that is how equinox gets its name as equinox means equal night. At winter solstice (December), the North Pole is inclined directly away from the Sun. 3 months later, the Earth will reach the date point of the March equinox and that the Sun's declination will be 0°. 3 months later, the Earth will reach the date point of the summer solstice. At this point it will be at declination -23.5°. This cycle will carry on, creating the seasons that we experience on Earth.

The Earth is tilted 23.5°, so is the ecliptic, with respect to the celestial equator, therefore the Sun maximum angular distance from the celestial equator is 23.5°. At the summer solstice which occurs around 21st of June, the North Pole is pointing towards the Sun at an angle of 23.5°. Therefore the apparent declination of the Sun is positive 23.5° with respect to the celestial equator. At the Winter solstice which occurs around 21st December, the North Pole is pointing away from the Sun at an angle of 23.5°. Therefore the apparent declination of the Sun is negative 23.5° with respect to the celestial equator.

Seasons are caused by the Earth axis which is tilted by 23.5° with respect to the ecliptic and due to the fact that the axis is always pointed to the same direction. When the northern axis is pointing to the direction of the Sun, it will be winter in the southern hemisphere and summer in the northern hemisphere. Northern hemisphere will experience summer because the Sun's ray reached that part of the surface directly and more concentrated hence enabling that area to heat up more quickly. The southern hemisphere will receive the same amount of light ray at a more glancing angle, hence spreading out the light ray therefore is less concentrated and colder. The converse holds true when the Earth southern axis is pointing towards the Sun.

The Earth's rotation about its polar axis produces our days and nights; the tilt of this axis relative to the ecliptic plane produces our seasons as the Earth revolves about the Sun. The polar axis of the Earth is inclined to the ecliptic plane by 23.45 degrees, in approximately 24-hour cycles. The direction in which the polar axis points is fixed in space and is aligned with the North Star (Polaris) to within about 45 minutes of arc (13 mrad).

From the heliocentric point of view, the Earth rotates and revolves around the Sun in a counter clockwise direction. However, when we look at the Sun on Earth, it appears to be moving in a clockwise direction. This phenomenon is known as the apparent motion of the Sun.

If the origin of a set of coordinates is defined at the Earth's center C (see Figure 1), the y axis with a unit vector of $\vec{j}$ can be a line from the origin intersecting the equator at the point where the meridian of the observer at Q crosses. The x axis with a unit vector of $\vec{i}$ is perpendicular to the y axis and is also in the equatorial plane. The third orthogonal axis z with a unit vector of $\vec{k}$ may then be aligned with the Earth's axis of rotation. Then the unit direction vector $\vec{r}$ pointing to the Sun may be described in terms of its direction cosines r_i, r_j and r_k relative to the x, y, and z axes, respectively.

$$\vec{r} = r_i\vec{i} + r_j\vec{j} + r_k\vec{k} \tag{1}$$

where

Figure 1. Sun ray direction vector in Earth center coordinate system

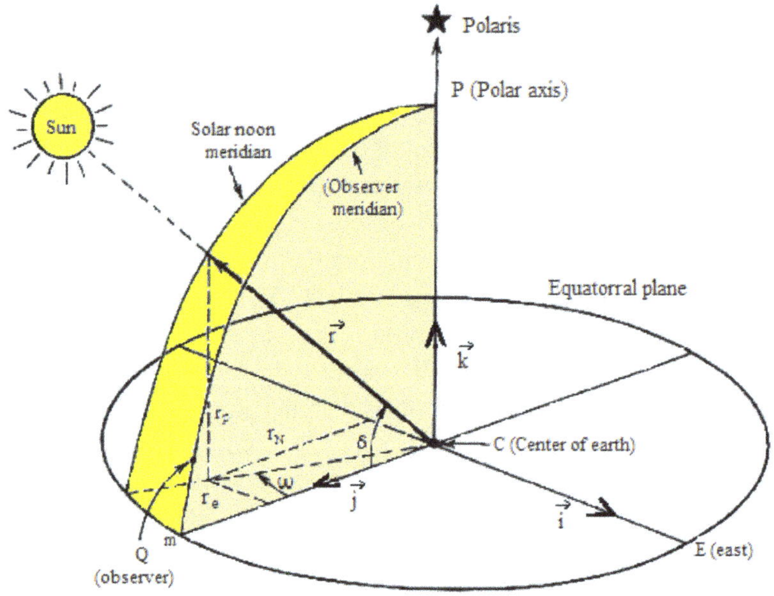

$$r_i = \cos(\delta)\sin(\omega) \tag{2}$$

$$r_j = \cos(\delta)\cos(\omega) \tag{3}$$

$$r_k = \sin(\delta) \tag{4}$$

According to Figure 1 the Earth's axis of rotation coincides with the Polaris (the North Star) position vector. The Polaris, is located almost exactly at North Celestial Pole - the point in the sky about which all the stars seen from the Northern Hemisphere rotate. The Polaris is up from the horizon exactly an angle equal to your latitude. So, if you live in Damascus at 33.51° latitude, the Polaris will be due north, up 33.51°.

ω is the hour angle and δ is the declination angle. The angle ω is based on the Sun's angular displacement, east or west, of the local meridian (the line the local time zone is based on). The Earth rotates 15° per hour so at 11am, the hour angle is -15° and at 1pm it is 15°. So, the hour angle ω in (°) is:

$$\omega = 15(LST - 12) \tag{5}$$

where LST is the local solar time. Hence, the hour angle converts the local solar time (LST) into the number of degrees which the Sun moves across the sky. By definition, the $\omega = 0^o$ at solar noon. Since the Earth rotates 15° per hour, each hour away from solar noon corresponds to an angular motion of the Sun in the sky of 15°. In the morning the hour angle is negative, in the afternoon the hour angle is positive. The daily variation of ω is demonstrated on Figure 2.

The angle δ is the angle between the line, drawn between the center of the Earth and the Sun, and the Earth's equatorial plane. At the time of year when the northern part of the Earth's rotational axis is inclined toward the Sun, the Earth's equatorial plane is inclined 23.45 degrees to the Earth-Sun line. At this time (about June 21), we observe that the noontime Sun is at its highest point in the sky and the declination angle $\delta = +23.45°$. This condition is known as the summer solstice, and it marks the beginning of summer in the Northern Hemisphere.

As the Earth continues its yearly orbit about the Sun, a point is reached about 3 months later where a line from the Earth to the Sun lies on the equatorial plane. At this point an observer on the equator would observe that the Sun was directly overhead at noontime. This condition is called an equinox since anywhere on the Earth, the time during which the Sun is visible (daytime) is exactly 12 hours and the time when it is not visible (nighttime) is 12 hours. There are two such conditions during a year; the autumnal equinox on about September 21, marking the start of the fall; and the vernal equinox on about March 21, marking the beginning of spring. At the equinoxes, the declination angle δ is zero.

The winter solstice occurs on about December 22 and marks the point where the equatorial plane is tilted relative to the Earth-Sun line such that the northern hemisphere is tilted away from the Sun. We say that the noontime Sun is at its "lowest point" in the sky, meaning that the declination angle is at its most negative value (i.e., $\delta = -23.45°$). By convention, winter declination angles are negative.

Figure 2. Hour angle variation during solar day

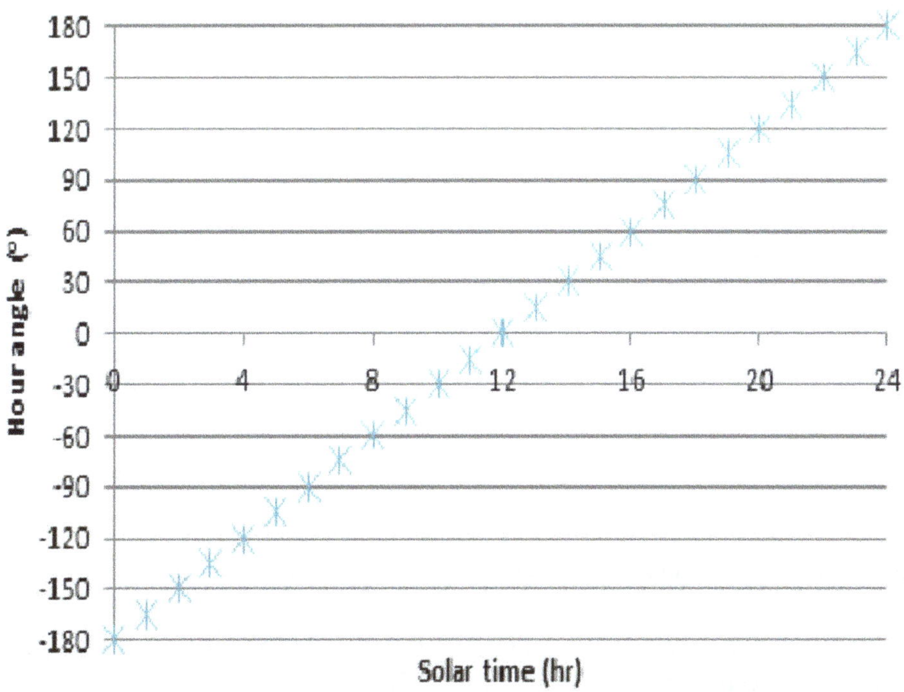

Accurate knowledge of the declination angle is important in navigation and astronomy. Very accurate values are published annually in tabulated form in an ephemeris. For most solar design purposes, however, an approximation accurate to within about 1 degree is adequate. In literature several formulas were proposed for calculating δ. The most used ones are those of (Cooper, 1969):

$$\delta = 23.458 \sin\left[\frac{2\pi\left(284 + n\right)}{365}\right] \tag{6}$$

of (Cooper, 1969):

$$\delta = \arcsin\left\{\sin\left(23.458\right)\sin\left[\frac{2\pi\left(n - 81\right)}{365}\right]\right\} \tag{7}$$

which is equivalent to that of (Smith and Wilson, 1976), of (Spencer, 1971):

$$\begin{aligned}\delta = 57.3\left[0.006918 - 0.399912\cos\left(B\right)\right] + 0.070257\sin\left(B\right)\\ -0.006758\cos\left(2B\right) + 0.000907\sin\left(2B\right) - 0.002697\cos\left(3B\right) + 0.00148\sin\left(3B\right)\end{aligned} \tag{8}$$

of (Soulayman and Shamiyah, 2000):

$$\delta = \arcsin\left\{0.39795\cos\left[\frac{2\pi\left(n - 173\right)}{365.25}\right]\right\} \tag{9}$$

of (Soulayman and Sabbagh, 2014):

$$\delta = 23.458\cos\left[\frac{2\pi\left(10.5 + n\right)}{365}\right] \tag{10}$$

where n being the number of days since January 1 and B is:

$$B = \frac{2\pi\left(n - 1\right)}{365} \tag{11}$$

However, four conditions should be satisfied by these formulas:

- Winter solstice occurs on about December 21 with δ = -23.45°.
- Summer solstice occurs on about June 21 with δ = 23.45°.

- Vernal equinox occurs on about March 21 with $\delta = 0^o$.
- Autumnal equinox occurs on about September 21 with $\delta = 0^o$.

When calculating δ using the Eqs (6) – (10) it was found that, the winter solstice occurs on December 21 according to Equations (6), (7), (9) and (10) while it occurs on January 1 according to equation (8); the summer solstice occurs on June 21 according to (6), (7) and (10) while it occurs on June 22 according to (9) and on July 2 according to (8); the vernal equinox occurs on about March 22 according to (6), (7) and (10) while it occurs on March 23 according to (9) and on April 1 according to (8) and the autumnal equinox occurs on about September 21 according to (6), (7), (9) and (10) while it occurs on October 3 according to (8). Table 1 demonstrates the ability of different formulas in satisfying the above mentioned conditions. Moreover, it is difficult to distinguish between the daily results of (6), (7), (9) and (10) while the results of (8) differ remarkably. Therefore, it is advised to one of the formulas (6), (7), (9) and (10). In this work, the declination angle δ will be calculated using (6). The annual variation of the declination angle is shown in Figure 3. Moreover, Table 2 has been prepared as an aid in rapid determination of values of n from calendar dates.

Sun Position in Relation to the Earth's Surface

When the Sun is observed from an arbitrary position on the Earth's surface, it is important to determine the Sun position relative to a coordinate system based at the point of observation, not at the center of the Earth (Budin and Budin, 1982). The conventional Earth-surface based coordinates are a vertical line (straight up) and a horizontal plane containing a north-south line and an east-west line. The position of the Sun relative to these coordinates can be described by two angles: the solar altitude angle and the solar zenith angle.

The solar altitude angle α is defined as the angle between the central ray from the Sun, and a horizontal plane containing the observer on the horizontal plane on the Earth's surface, as shown in Figure 4. The solar zenith angle θ_z, which is simply the complement of the solar altitude angle or:

$$\theta_z = \pi/2 - \alpha \tag{12}$$

The solar azimuth angle (γ_s) is the angle, measured clockwise on the horizontal plane, from the south-pointing coordinate axis to the projection of the Sun's central ray.

Since the Sun appears not as a point in the sky, but as a disc of finite size, all angles discussed hereafter are measured to the center of that disc, that is, relative to the "central ray" from the Sun.

Table 1. The ability of different formulas in calculating declination angle

Equation number	(6)	(7)	(8)	(9)	(10)
Winter solstice	-23.4578	-23.4578	-22.1319	-23.4492	-23.4491
Summer solstice	23.45776	23.45777	22.87411	23.4464	23.45
Vernal equinox	-0.00435	0	-3.93783	-0.29437	0.100918
Autumnal equinox	-0.19538	-0.19631	4.550123	0.098124	-0.30275

Figure 3. The declination angle as a function of day number

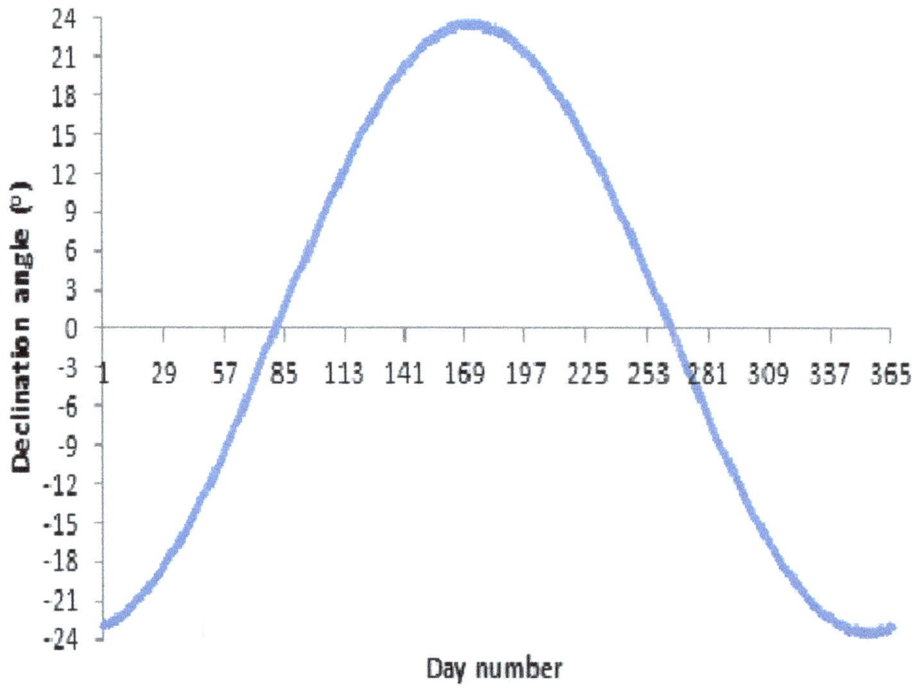

It is of the greatest importance in solar energy systems design, to be able to calculate the solar altitude and azimuth angles at any time for any location on the Earth. This can be done using the following three angles latitude φ, hour angle ω and declination δ. The latitude angle φ is one of the main angles that relate the position of the observer on the Earth's surface and the Earth center. The angle between a line, drawn from a point on the Earth's surface to the center of the Earth, and the Earth's equatorial plane, is φ. The intersection of the equatorial plane with the surface of the Earth forms the equator and is designated as 0 degrees latitude. The Earth's axis of rotation intersects the Earth's surface at 90° latitude (North Pole) and -90° latitude (South Pole). Any location on the surface of the Earth then can be defined by the intersection of a longitude angle and a latitude angle.

Other latitude angles of interest are the Tropic of Cancer (φ =+23.45°) and the Tropic of Capricorn (φ = - 23.45°). These represent the maximum tilts of the north and south poles toward the Sun. The other two latitudes of interest are the Arctic circle (φ = 66.55°) and Antarctic circle (φ = -66.5°) representing the intersection of a perpendicular to the Earth-Sun line when the south and north poles are at their maximum tilts toward the Sun. The tropics represent the highest latitudes where the Sun is directly overhead at solar noon, and the Arctic and Antarctic circles, the lowest latitudes where there are 24 hours of daylight or darkness. All of these events occur at either the summer or winter solstices. For this derivation, a Sun-pointing vector at the surface of the Earth should be defined and then mathematically translate it to the center of the Earth with a different coordinate system. Using Figure 4 as a guide, a unit direction vector $\vec{R}$ pointing toward the Sun from the observer location Q could be defined.

Table 2. Date-to-Day number conversion

Month	Day Number, *n*	Notes
January	d	
February	d + 31	
March	d + 59	Add 1 if leap year
April	d + 90	Add 1 if leap year
May	d + 120	Add 1 if leap year
June	d + 151	Add 1 if leap year
July	d + 181	Add 1 if leap year
August	d + 212	Add 1 if leap year
September	d + 243	Add 1 if leap year
October	d + 273	Add 1 if leap year
November	d + 304	Add 1 if leap year
December	d + 334	Add 1 if leap year
Days of Special Solar Interest		
Solar Event	Date	Day Number, *N*
Vernal equinox	March 21	80
Summer solstice	June 21	172
Autumnal equinox	September 21	264
Winter solstice	December 21	355
NOTES: 4. d is the day of the month 5. Leap years are 2000, 2004, 2008, 2012, 2016, 2020 etc. 6. Solstice and equinox dates may vary by a day or two. Also, add 1 to the solstice and equinox day number for leap years.		

$$\vec{R} = R_i\vec{i} + R_j\vec{j} + R_k\vec{k} \tag{13}$$

where $\vec{i}$, $\vec{j}$, and $\vec{k}$ are unit vectors along the *E, S,* and *Z* axes respectively. The direction cosines of $\vec{R}$ relative to the *E, S,* and *Z* axes are R_E, R_S and R_Z, respectively. These may be written in terms of solar altitude and azimuth as:

$$R_i = \cos\left(\alpha\right)\sin\left(\gamma_s\right) \tag{14}$$

$$R_j = \cos\left(\alpha\right)\cos\left(\gamma_s\right) \tag{15}$$

$$R_k = \sin\left(\alpha\right) \tag{16}$$

Figure 4. Sun position vector in Earth surface coordinate system

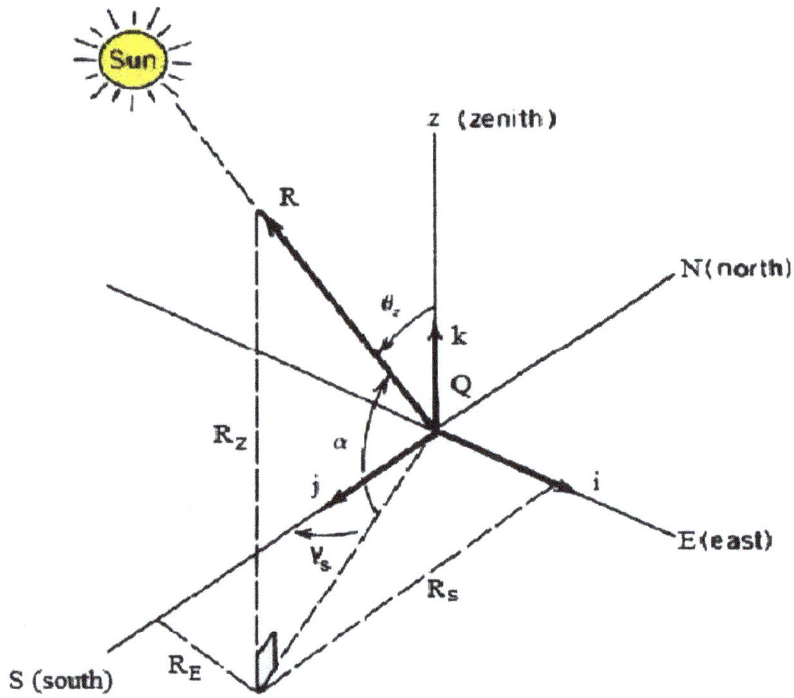

These two sets of coordinates are interrelated by a rotation about the E axis through the latitude angle and translation along the Earth radius QC. We will neglect the translation along the Earth's radius since this is about 1/23,525 of the distance from the Earth to the Sun, and thus the difference between the direction vectors $\vec{R}$ and $\vec{r}$ is negligible. The rotation from the E, m, P coordinates to the E, S, Z coordinates, about the E axis is depicted in Figure 5. Both sets of coordinates are summarized in Figure 6. Note that this rotation about the E axis is in the negative sense based on the right-hand rule. Accordingly, the following relations could be obtained:

$$R_i = r_i \tag{17}$$

$$R_j = r_j \sin(\varphi) - r_k \cos(\varphi) \tag{18}$$

$$R_k = r_k \sin(\varphi) + r_j \cos(\varphi) \tag{19}$$

Substituting Equations (2) – (4) and Equations (14) – (16) into Equations (17) – (19) for the direction cosine gives three important results:

Figure 5. The rotation from the E, m, P coordinates to the E, S, Z coordinates, about the E axis

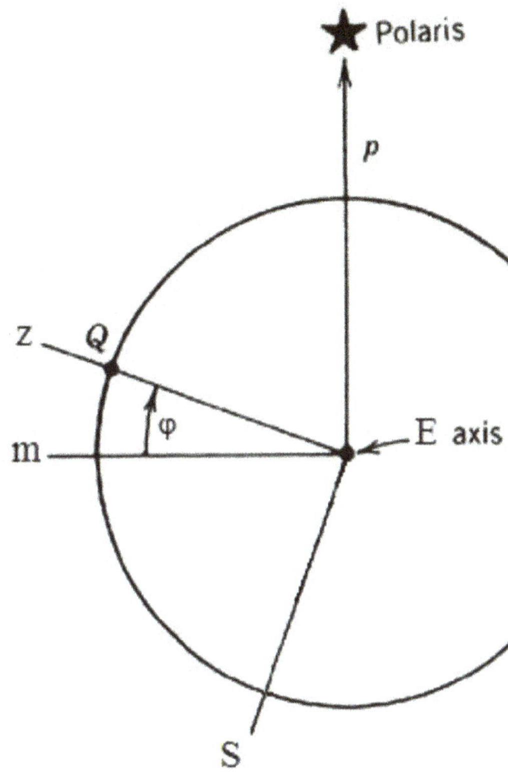

Figure 6. Sun position unit vector in Earth's center and surface coordinates

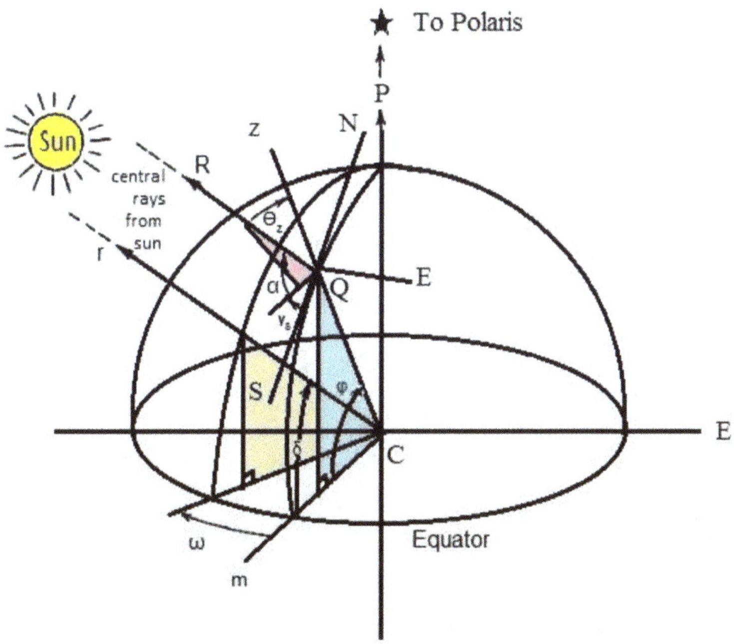

$$\sin(\alpha) = \sin(\delta)\sin(\varphi) + \cos(\delta)\cos(\varphi)\cos(\omega) \tag{20}$$

$$\sin(\gamma_s) = \frac{\cos(\delta)\sin(\omega)}{\cos(\alpha)} \tag{21}$$

$$\cos(\gamma_s) = \frac{-\sin(\delta)\cos(\varphi) + \sin(\varphi)\cos(\delta)\cos(\omega)}{\cos(\alpha)} \tag{22}$$

DEFINITIONS

Zenith Angle

Taking into consideration equation (12) the solar zenith angle θ_z, which is the angle of incidence on the horizontal surface (see Figure 4), could be determined from (20):

$$\cos(\theta_z) = \sin(\delta)\sin(\varphi) + \cos(\delta)\cos(\varphi)\cos(\omega) \tag{23}$$

Example 1

Calculate the solar zenith angle θ_z and solar altitude angle in Damascus $(\varphi = 33.51^o N)$ at 9:30 AM on February 13 and at 5:30 AM on July 1.

- **Solution**

On February 13 at 9:30 AM, the day number $n = 44$, $\delta = -14^o$ and $\omega = -37.5^o$. From equation (23),

$$\cos(\theta_z) = \sin(-14^o)\sin(33.51^o) + \cos(-14^o)\cos(33.51^o)\cos(-37.5^o)$$
$$= -0.24 * 0.55 + 0.83 * 0.97 * 0.79 = -0.13 + 0.42 = 0.29$$
$$\theta_z = 73.14^o$$

From equation (12),

$$\theta_z = 90^o - \alpha \rightarrow \alpha = 90^o - 73.14^o = 16.86^o$$

On July 1 at 5:30 AM, the day number $n = 182$, $\delta = 23.1^o$ and $\omega = -97.5^o$. From equation (23),

$$\cos\left(\theta_z\right) = \sin\left(23.1^o\right)\sin\left(33.51^o\right) + \cos\left(23.1^o\right)\cos\left(33.51^o\right)\cos\left(-97.5^o\right)$$
$$= 0.39 * 0.55 + +0.83 * 0.92 * \left(-0.13\right) = 0.21 - 0.10 = 0.11$$
$$\theta_z = 83.68^o$$

From equation (12),

$$\theta_z = 90^o - \alpha \rightarrow \alpha = 90^o - 83.68^o = 6.32^o$$

Sunshine Duration

Equation (23) could be used for determining the Sunrise ω_r and Sunset ω_s hour angles and the daylight length on the horizontal surface on the Earth's surface. The Sunrise and Sunset occur when $\theta_z = 90^o$. So, the use of (23) gives:

$$\omega_s = -\omega_r = \arccos\left[-\tan\left(\delta\right)\tan\left(\varphi\right)\right] \quad (24)$$

Total angle between Sunrise and Sunset is given by:

$$\omega_s - \omega_r = 2\arccos\left[-\tan\left(\delta\right)\tan\left(\varphi\right)\right] \quad (25)$$

Sunshine Duration on the Horizontal Surface

Since $15^o = 1$ hour, the daylight length on the horizontal surface on the Earth's surface S_0 in hours is:

$$S_0 = \frac{2\arccos\left[-\tan\left(\delta\right)\tan\left(\varphi\right)\right]}{15} \quad (26)$$

The variation of S_0 with latitude φ ($\varphi \leq 66.45^o$N) for different n (n[th] day of the year) in the Northern Hemisphere is given in Figure 7. It is clear from Figure 8 that the daylight length is 12 for March and September 21 whatever is the latitude. However, it is significantly varying with latitude for other days. The maximum total sunshine duration length is for June 21 and minimum for December 21 at latitude 66.45°.

For the arctic zone (66.45°N < $\varphi \leq$ 90°N) and (66.45°S < $\varphi \leq$ 90°S), it is reasonable to mention that equations (24) - (26) are applicable partially during the year as the absolute value of tan(δ)*tan(φ) is greater than 1 for a number of consecutive days. More precisely, the sunrise could not occur when:

$$\tan\left(\delta\right)\tan\left(\varphi\right) \leq -1 \quad (27)$$

Figure 7. Solar zenith and elevation angles

Figure 8. Daylight length variation with latitudes in NH for different n^{th} day of the year: Curves correspond to June 21 (upper), March and September 21 (middle) and December 21(lower)

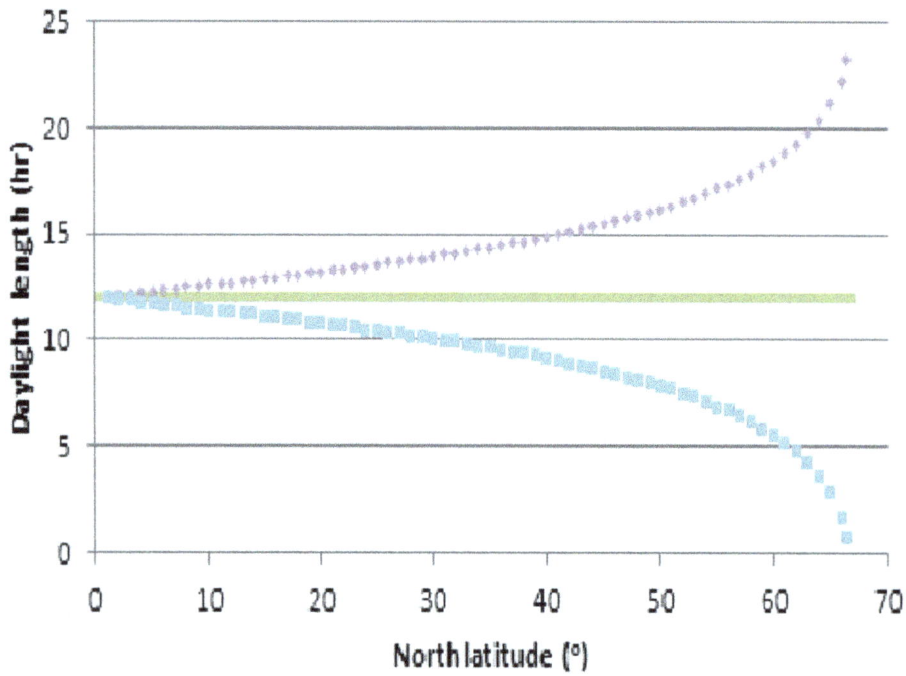

and the sunset could not occur when:

$$\tan(\delta)\tan(\varphi) \geq 1 \tag{28}$$

This means that, the daylight length could be either 0 hr over several days or months or 24 hr over several days or months in the arctic zone. The values of daylight length for different latitudes in the Northern Hemisphere are given in Table 3 while Table 4 gives these values for different latitudes in the Southern Hemisphere.

The greater the length of daylight will increase the amount of radiation received on the earth and hence add to the heating of the earth and contributes to hotter temperatures in the summer. Conversely,

Table 3. Daylight length during the characteristic periods in the Northern hemisphere

Latitude Angle φ (°)	Summer Solstice	Equinox	Winter Solstice
0 °N	12 hr	12 hr	12 hr
10°N	12 hr 35 min	12 hr	11 hr 25 min
20°N	13 hr 13 min	12 hr	10 hr 47 min
30 °N	13 hr 56 min	12 hr	10 hr 4 min
40 °N	14 hr 51 min	12 hr	9 hr 9 min
50 °N	16 hr 9 min	12 hr	7 hr 51 min
60 °N	18 hr 29 min	12 hr	5 hr 31 min
66.45°N	23 hr 15 min	12 hr	0 hr 45 min
70 °N	2 months	12 hr	0 hr
80 °N	4 months	12 hr	0 hr
90 °N	6 months	12 hr	0 hr

Table 4. Daylight length during the characteristic periods in the Southern hemisphere

Latitude Angle φ (°)	Summer Solstice	Equinox	Winter Solstice
0 °S	12 hr	12 hr	12 hr
10°S	11 hr 25 min	12 hr	12 hr 35 min
20 °S	10 hr 47 min	12 hr	13 hr 13 min
30 °S	10 hr 4 min	12 hr	13 hr 56 min
40 °S	9 hr 9 min	12 hr	14 hr 51 min
50 °S	7 hr 51 min	12 hr	16 hr 9 min
60 °S	5 hr 31 min	12 hr	18 hr 29 min
66.45°S	0 hr 45 min	12 hr	23 hr 15 min
70 °S	0 hr	12 hr	2 months
80 °S	0 hr	12 hr	4 months
90 °S	0 hr	12 hr	6 months

the shorter the length of daylight in the winter will decrease the amount of radiation received on the earth and contributes to the colder temperatures observed in the winter.

Equation (23) could be used also as a guide for determining the most probable orientation of solar collector on the Earth's surface. For this purpose it is recommended to study the zenith angle on noon, $\theta_{z,noon}$ dependence on latitude:

$$\theta_{z,noon} = |\varphi - \delta| \tag{29}$$

If imagining an axe of $\theta_{z,noon}$ changes as $\theta_{z,noon}$ is positive when solar rays incident from south direction and θnoon is negative when solar rays incident from north direction, then φ - δ determine the period where solar collector should be oriented to Equator or to opposite direction. This may be done by studying the dependence of φ - δ as function of day number in the year. Figures 9 and 10 show the daily variation of $\theta_{z,noon}$ for different latitudes.

It is seen from Figure 9 that outside the tropical zone, 23.45°S≤φ≥23.45°N, the solar rays incident from the south in the Northern Hemisphere and from the north in the Southern Hemisphere. Therefore, the solar collector should be oriented towards Equator when φ≥23.45°N and φ≥23.45°S while the solar collector should be oriented towards Equator when φ - δ is positive and to opposite direction when it is negative in the Northern Hemisphere. In the Southern Hemisphere, the solar collector should be oriented to Equator when φ - δ is negative and to opposite direction when it is positive (see Figure 10).

Figure 9. The daily variation of $\theta_{z,noon}$ for latitudes (60°N, 40°N, 20°N and 0° from up to down respectively)

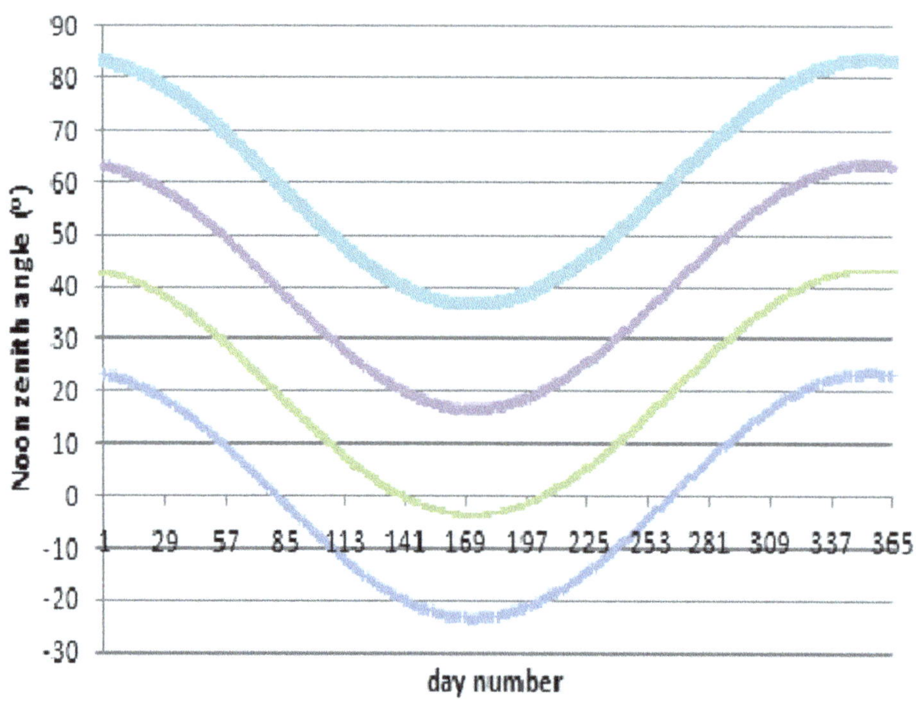

Figure 10. The daily variation of $\theta_{z,noon}$ for latitudes (60°S, 40°S, 20°S and 0° from up to down respectively)

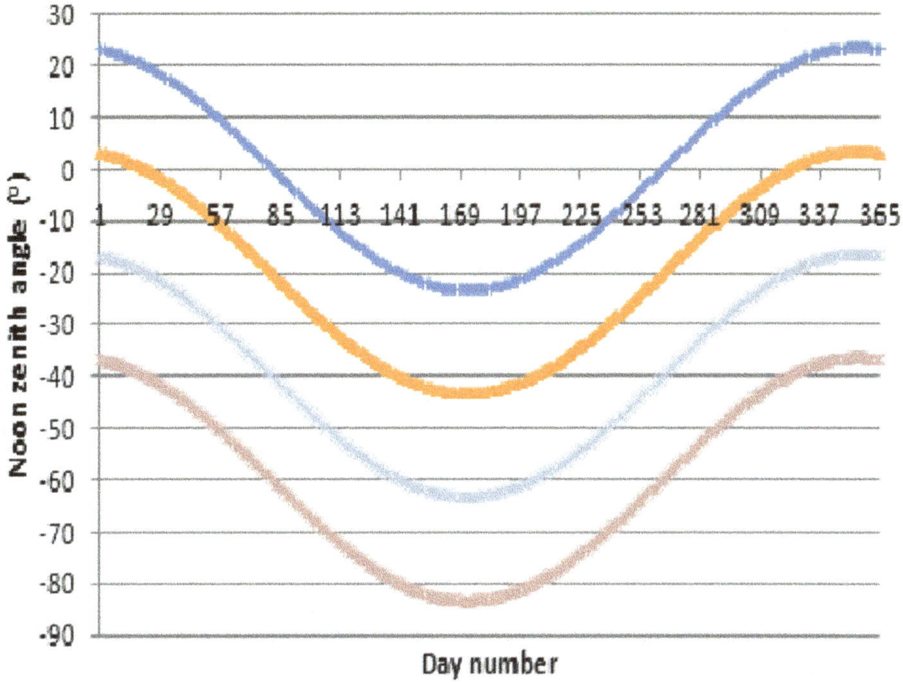

At the Tropic of Cancer on summer solstice, the sun is directly overhead and the elevation angle is 90°. In summer at latitudes between the equator and the Tropic of Cancer, the elevation angle at solar noon is greater than 90°, implying that the sunlight is coming from the north rather than from the south as in most of the northern hemisphere. Similarly, at latitudes between the equator and the Tropic of Capricorn, during some periods of the year, sunlight is incident from the south, rather than from the north. Table 5 shows the periods that sunlight comes from the Poles at solar noon for different tropical latitudes at Northern and Southern Hemispheres. Here it should be mentioned that at Equator the solar collector should be oriented toward the North Pole (NP) during half year (from 22/3 to 21/9) and another half year (from 22/9 to 21/3) should be oriented toward the South Pole (SP). It is seen from Table 5 that, for high latitudes, latitudes $\varphi = 23.5°$N and $\varphi = 23.5°$S, solar collectors should be oriented toward Equator permanently while, for tropical zone, Equator direction and opposite direction should be used depending on the application period. Moreover, for any other latitude at tropical region there are two periods for solar collector orientation: one is laid between 22/9 and 21/3 while the other is laid between 22/3 and 21/9. The duration of each period depends on the latitude value.

Solar Azimuth Angle

The knowledge of solar azimuth angle γ_s is important in observing the daily apparent Sun position trajectory. The azimuth angle is the compass direction from which the sunlight is coming. Outside the tropical zone and at solar noon, the sun is always directly south in the northern hemisphere and directly north in the southern hemisphere. The azimuth angle varies throughout the day. At the equinoxes, the sun

Table 5. Latitudes, corresponded duration (days) and required orientation (Soulayman,2015)

φ (°)	Equator Facing	North Pole Facing	South Pole Facing
23.45N	[1-365]	-	-
20N	[1- 140] + [204-365]	[141- 203]	-
15N	[1-121] + [223-365]	[121-222]	-
10N	[1-106] + [239-365]	[107-238]	-
5N	[1- 93] + [251-365]	[94-250]	-
0	-	[81-263]	[1- 80] + [264-365]
5S	[69-275]	-	[1-68] + [276-365]
10S	[56 - 288]	-	[1-55] + [289-365]
15S	[41 - 304]	-	[1-40] + [305-365]
20S	[22 - 322]	-	[1-21] + [323-365]
23.45S	[1 - 365]	-	-

rises directly east and sets directly west regardless of the latitude, thus making the azimuth angles -90° at sunrise and 90° at sunset. In general however, the azimuth angle varies with the latitude and time of year and the full equations (21) and (22) are suitable to calculate the sun's position throughout the day.

The equations (21) and (22) allow to determine the azimuth angle and to answer to question: In which quadrant the Sun will be in any instant during the day. Hereafter γ_s will be calculated for solstices and equinoxes at different latitudes in order to demonstrate the daily behavior of γ_s.

Winter Solstice Case

During winter in Northern Hemisphere the maximum sunshine duration S_0 is ≤ 12 hrs (see Table 3). So, the daily variations of γ_s should be studied during the period from 6 O'clock to 18 O'clock of solar time. The results are presented in Figure 11. It is seen from Figure 11 that the solar azimuth angle can have, during winter solstice, values in the range of -66.55° to +66.55° while this range extends, during the period of 22/9 to 21/3, to be of -90° to +90°.

Equinox Case

During equinox the maximum sunshine duration S_0 is 12 hrs (see Tables 3 and 4). So, the daily variations of γ_s should be studied during the period from 6 O'clock to 18 O'clock of solar time. The results are presented in Figure 12. According to results presented in Figure 12, γ_s will be between -90° and +90° whatever the latitude.

Summer Solstice Case

In summer solstice the maximum sunshine duration S_0 reaches its higher value which could reach 24 hours. Therefore, it is reasonable to study the daily variations of γ_s from 0 O'clock to 24 O'clock of solar time. Figure 13 shows the obtained results for different latitudes in the Northern Hemisphere.

Figure 11. The daily variation of γ_s for latitudes (15°N (x), 30°N(●), 45°N (-) and 60°N (♦)) in winter solstice

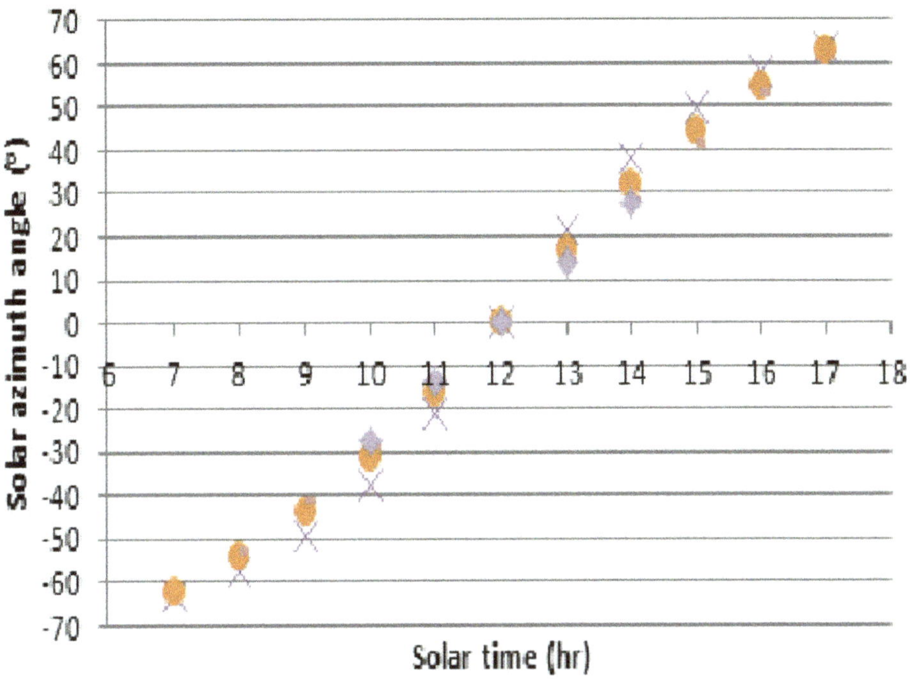

Figure 12. The daily variation of γ_s for latitudes (15°N (■), 30°N(x), 45°N (●) and 60°N (-)) in the equinox in the Northern hemisphere

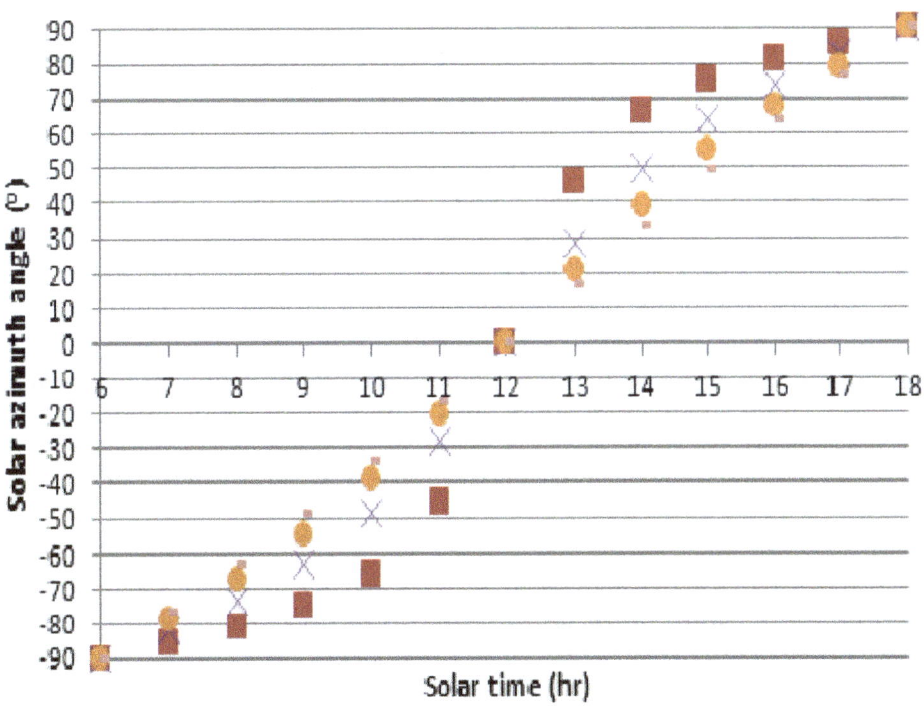

Figure 13. The daily variation of γ_s for latitudes (30°N (ж), 45°N (+) and 60°N (-)) in the summer solstice in the Northern hemisphere

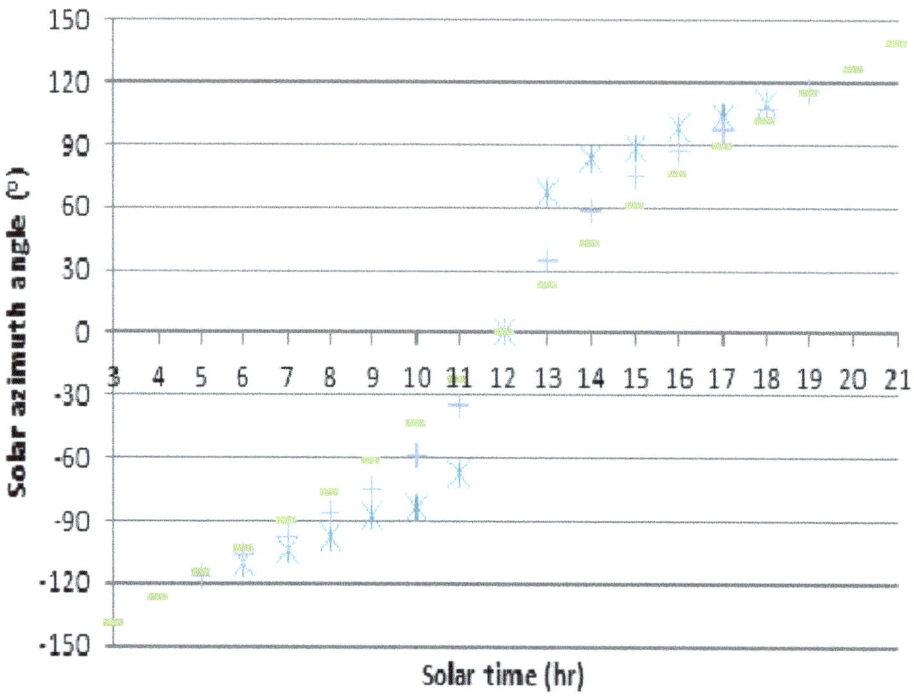

It is seen from Figure 13 that, outside the tropical zone, the solar azimuth angle γ_s can have values in the range of -180° to 180° in the summer solstice in the Northern Hemisphere. γ_s is less than -90° early morning after sunrise and greater than 90° late before sunset as the Sun is north of the East-West line. Moreover, the Sun shines from the north direction when γ_s is less than -90° or greater than 90°. Therefore, Equator facing vertical surfaces could not receive any portion of solar beam radiation during these periods.

Summer Solstice Case in the Tropical Zone

According to Table 5, at noon γ_s = -180° during a period containing the summer solstice for tropical latitudes in the Northern Hemisphere. The daily variations of γ_s for several latitudes of the tropical zone on summer solstice are shown in Figure 14.

For calculating the solar azimuth angle it is recommended (Duffie and Beckman, 2013) to use the formulation of Braun and Mitchell (1983). According to this formulation:

$$\gamma_s = C_1 C_2 \gamma' + \frac{180 C_3 \left(1 - C_1 C_2\right)}{2} \tag{30}$$

where γ' is determined by equation (21) and

Figure 14. The daily variation of γ_s for latitudes (5°N (+), 10°N (▲), 15°N (▲) and 20°N (ж)) in the summer solstice in the Northern hemisphere

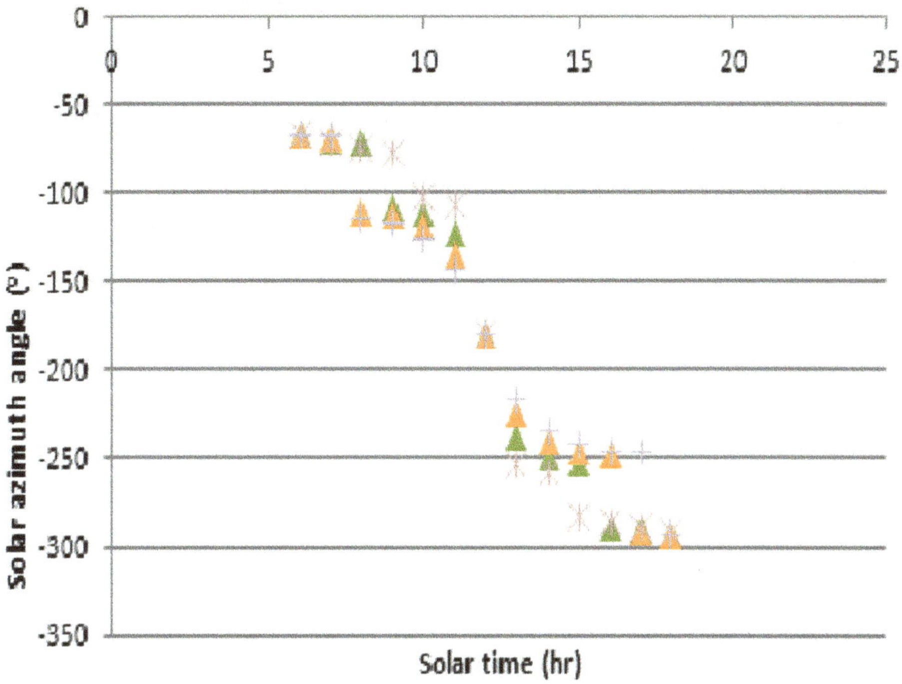

$$|\omega| \leq \arccos\left[\frac{\tan(\delta)}{\tan(\varphi)}\right] or - 1 if |\omega| > \arccos\left[\frac{\tan(\delta)}{\tan(\varphi)}\right] \tag{31}$$

$$C_2 = 1, if (\varphi - \delta) \geq 0 or - 1 if (\varphi - \delta) < 0 \tag{32}$$

$$C_3 = 1, if \omega \geq 0 or - 1 if \omega < 0 \tag{33}$$

The absolute value of tan(δ)/tan(φ) in the tropical zone is greater than one for several days in the year (the winter and summer solstices are within these days). So, the value of arcos[tan(δ)/tan(φ)] (see (30a)) can't be definitive. Therefore, the formulation of Braun and Mitchell (1983) is not applicable in the tropical zone.

Angle of Incidence on the Tilted Plane on the Earth's Surface

The orientation of the tilted plane on the horizontal Earth's surface could be described using surface tilt angle β and surface azimuth angle γ. The tilt angle β is the angle between the plane surface, under consideration, and the horizontal. It is taken to be positive for surface sloping towards south and negative for surfaces sloping towards north. Therefore, in Northern Hemisphere $\beta \geq 0^o$ for Equator facing surfaces and $\beta \leq 0^o$ for North Pole facing surfaces, while $\beta \geq 0^o$ for South Pole facing surfaces and $\beta \leq 0^o$ for Equator facing surfaces. The surface azimuth angle γ is the angle, that is between the line due south and the horizontal plane projection of the normal to the inclined plane. By convention, if the projection is east of south and positive if west of south for Northern Hemisphere and vice-versa for Southern Hemisphere. Finally, the angle of incidence θ_i on the tilted surface is the angle between solar rays on the tilted surface and the normal to the tilted surface.

In general, the unit vector of the normal to the inclined surface oriented with angle γ could be expressed using the system of coordinates related to the Earth's surface as follows:

$$\overrightarrow{I_N} = \sin\left(\gamma\right)\sin\left(\beta\right)\vec{i} + \cos\left(\gamma\right)\sin\left(\beta\right)\vec{j} + \cos\left(\beta\right)\vec{k} \tag{34}$$

Then the angle of incidence can be expressed as:

$$\cos\left(\theta_i\right) = \overrightarrow{s}\overrightarrow{I_N} = \cos\left(\theta_z\right)\cos\left(\beta\right) + \sin\left(\theta_z\right)\sin\left(\beta\right)\cos\left(\gamma_s - \gamma\right) \tag{35}$$

Substituting equations (22) and (23) into equation (35), $\cos(\theta_i)$ could be written as follows:

$$\begin{aligned}\cos\left(\theta_i\right) &= \sin\left(\delta\right)\sin\left(\varphi\right)\cos\left(\beta\right) - \sin\left(\delta\right)\cos\left(\varphi\right)\sin\left(\beta\right)\cos\left(\gamma\right) \\ &+ \cos\left(\delta\right)\cos\left(\beta\right)\cos\left(\varphi\right)\cos\left(\omega\right) + \cos\left(\delta\right)\sin\left(\varphi\right)\sin\left(\beta\right)\cos\left(\gamma\right)\cos\left(\omega\right) \\ &+ \cos\left(\delta\right)\sin\left(\beta\right)\sin\left(\gamma\right)\sin\left(\omega\right)\end{aligned} \tag{36}$$

Equation (23) can be obtained from equation (36) by taking $\beta=0^o$. Such case corresponds to the horizontal plane whatever is the plan azimuth angle γ. Moreover, three important cases could be considered when dealing with the equation (36):

1. Equator facing case. Accordingly, $\gamma = 0^o$ in Northern Hemisphere and $\gamma = 180^o$ in Southern Hemisphere. So, equation (36) takes the form:

$$\cos\left(\theta_i\right) = \sin\left(\delta\right)\sin\left(\varphi \pm \beta\right) + \cos\left(\delta\right)\cos\left(\varphi \pm \beta\right)\cos\left(\omega\right) \tag{37}$$

where '-' corresponds to the Northern Hemisphere and '+' corresponds to the Southern Hemisphere.

2. Pole facing case. Accordingly, $\gamma = 180^o$ in Northern Hemisphere and $\gamma = 0^o$ in Southern Hemisphere. So in equation (37), '+'corresponds to the Northern Hemisphere and '-'corresponds to the Southern Hemisphere.
3. Vertical surface facing due south, $\gamma = 0^o$, $\beta=90^o$. Then, equation (36) becomes:

$$\cos\left(\theta_i\right) = -\sin\left(\delta\right)\cos\left(\varphi\right) + \cos\left(\delta\right)\sin\left(\varphi\right)\cos\left(\omega\right) \tag{38}$$

Accordingly, Sun rises and sets on the tilted surface with different orientations differently from that on the horizontal plane.

Example 2

Calculate the solar the angle of incidence θ_i of beam radiation on a surface located at Damascus $\left(\varphi = 33.51^o N\right)$ at 10:30 AM solar time on February 13 if the surface is tilted 45° from the horizontal and pointed 15° west of south.

- Solution

On February 13 at 10:30 AM, the day number $n = 44$, $\delta = -14^o$ and $\omega = -22.5^o$. The surface is tilted 45° from the horizontal. This means that, $\beta = 45^o$. The surface is pointed 15° west of south. This means that, $\gamma = 15^o$. Then, from equation (33),

$$\begin{aligned}
&\cos\left(\theta_i\right) = \sin\left(-14^o\right)\sin\left(33.51^o\right)\cos\left(45^o\right) - \sin\left(-14^o\right)\cos\left(33.51^o\right)\sin\left(45^o\right)\cos\left(15^o\right) \\
&+\cos\left(-14^o\right)\cos\left(45^o\right)\cos\left(33.51^o\right)\cos\left(-22.5^o\right) + \cos\left(-14^o\right)\sin\left(33.51^o\right)\sin\left(45^o\right)\cos\left(15^o\right)\cos\left(-22.5^o\right) \\
&+\cos\left(-14^o\right)\sin\left(45^o\right)\sin\left(15^o\right)\sin\left(-22.5^o\right) = 0.85 \\
&\theta_i = 31.79^o
\end{aligned}$$

Sunshine Duration on a Tilted Surface

The determination of the sunrise and sunset on the tilted plane plays an important role on the long term tracking systems as these systems searches for an optimal orientation of solar collector during the period of tracking. Equation (36) can be used for this purpose. Let us, first, rewrite equation (33) in the form:

$$\cos\left(\theta_i\right) = A_1 + A_2\cos\left(\omega\right) + A_3\sin\left(\omega\right) \tag{39}$$

where

$$A_1 = \sin\left(\delta\right)\left[\sin\left(\varphi\right)\cos\left(\beta\right) - \cos\left(\varphi\right)\sin\left(\beta\right)\cos\left(\gamma\right)\right] \tag{40}$$

$$A_2 = \cos\left(\delta\right)\left[\cos\left(\varphi\right)\cos\left(\beta\right) + \sin\left(\varphi\right)\sin\left(\beta\right)\cos\left(\gamma\right)\right] \tag{41}$$

$$A_3 = \cos(\delta)\sin(\beta)\sin(\gamma) \quad (42)$$

Then, substituting θ_i by its value $\theta_i = 90^o$ in the equation (39) leads to the following equation:

$$0 = A_1 + A_2\cos(\omega) + A_3\sin(\omega) \quad (43)$$

The solution of (43), with taking into consideration that, Sun can't rise on the tilted surface before rising on the horizontal plane and can't setting on the tilted surface after setting on the horizontal plane, is:

$$\omega_{ss} = \min\left\{\arccos\left[-\tan(\delta)\tan(\varphi)\right], \arccos\left(-\frac{A_1}{A_4}\right) + \arcsin\left(\frac{A_3}{A_4}\right)\right\} \quad (44)$$

$$\omega_{sr} = \min\left\{-\arccos\left[-\tan(\delta)\tan(\varphi)\right], -\arccos\left(-\frac{A_1}{A_4}\right) + \arcsin\left(\frac{A_3}{A_4}\right)\right\} \quad (45)$$

ω_{ss} is the sunset hour angle on a tilted surface and ω_{sr} is the sunrise hour angle on a tilted surface and A_4 is a function of solar declination angle and collector angles:

$$A_4 = \sqrt{A_2^2 + A_3^2} \quad (46)$$

Consequently, the maximum sunshine duration on such a surface S' is:

$$S' = \frac{(\omega_{ss} - \omega_{sr})}{15} \quad (47)$$

Some interesting cases could be considered.

Equator Facing Case

In this case $\gamma = 0^o$ and the equations (40)-(42) and (46) become:

$$A_1 = \sin(\delta)\left[\sin(\varphi)\cos(\beta) - \cos(\varphi)\sin(\beta)\cos(\gamma)\right] = \sin(\delta)\sin(\varphi - \beta) \quad (40a)$$

$$A_2 = \cos(\delta)\left[\cos(\varphi)\cos(\beta) + \sin(\varphi)\sin(\beta)\cos(\gamma)\right] = \cos(\delta)\cos(\varphi - \beta) \quad (41a)$$

$$A_3 = 0 \tag{42a}$$

$$A_4 = \sqrt{A_2^2 + A_3^2} = A_2 \tag{46a}$$

Then, the equations (44), (45) and (47) take the form:

$$\omega_{ss} = \min\left\{\arccos\left[-\tan\left(\delta\right)\tan\left(\varphi\right)\right], \arccos\left[-\tan\left(\delta\right)\tan\left(\varphi - \beta\right)\right]\right\} \tag{44a}$$

$$\omega_{sr} = \max\left\{-\arccos\left[-\tan\left(\delta\right)\tan\left(\varphi\right)\right], -\arccos\left[-\tan\left(\delta\right)\tan\left(\varphi - \beta\right)\right]\right\} \tag{45a}$$

$$S' = \frac{\left(\omega_{ss} - \omega_{sr}\right)}{15} = \frac{2\omega_{ss}}{15} \tag{47a}$$

So, $\omega_{ss} = -\omega_{sr}$ and the sunset hour angle will depend on the tilt angle β, declination angle δ and latitude φ. When calculating the sunset hour angle for different latitudes and different tilts on the summer and winter solstices and substituting in equation (44a), the calculated results of maximum sunshine duration, are shown in Figures 15 and 16.

It is clear from Figure 15 that, the sunshine duration in the Northern Hemisphere increases with the latitude increase whatever the tilt angle is considered and tilting the Equator facing collector in the tropical zone (23.45°S <φ <23.45°N) by an angle greater than a value, that depends on the value of the latitude, prevents solar rays to be incident on it during summer solstice. The maximum tilt angle for φ = 15°N is 81.54°. This means that during summer sun rays incident from north direction in the Northern Hemisphere and from opposite direction in the Southern Hemisphere. Therefore, it is meaningful to present the sunshine duration in the tropical in the Northern and Southern Hemispheres separately (see Figure 17). Moreover, the sunshine duration decreases with the latitude increase during summer solstice. The decrease ratio differs from latitude to other. This result proves that during summer, the sunshine duration on the Equator facing tilt surface is $\leq S_0$ (see equation (26)). Figure 16 shows that, during winter solstice, the sunshine duration does not depend on the tilt angle. It depends on the latitude value only. When comparing the sunshine duration on the winter solstice in the Northern Hemisphere with the daylight length on the horizontal plane for the same latitude it was found that they are equal to each other. This result proves that during winter, the sun rises and sets on the Equator facing tilted collector at the same time at which the sun rises and sets on the horizontal plane. This result is repeated in the Southern Hemisphere but in the case of summer solstice. Figure 17 clearly demonstrates this result is on the example of latitudes 15°N and 15°S.

Finally, it should be mentioned here that outside the tropical region, during summer in the Northern Hemisphere, the sun can shine on the tilted North Pole facing surface after it rises on the horizontal plane and before its setting. The sunshine duration on such surface could be calculated basing on equations

Figure 15. Summer solstice maximum sunshine duration on the equator facing surface with different tilt at latitudes (φ = 15°N (■),30°N (♦),45°N (▲) and 60°N (x))

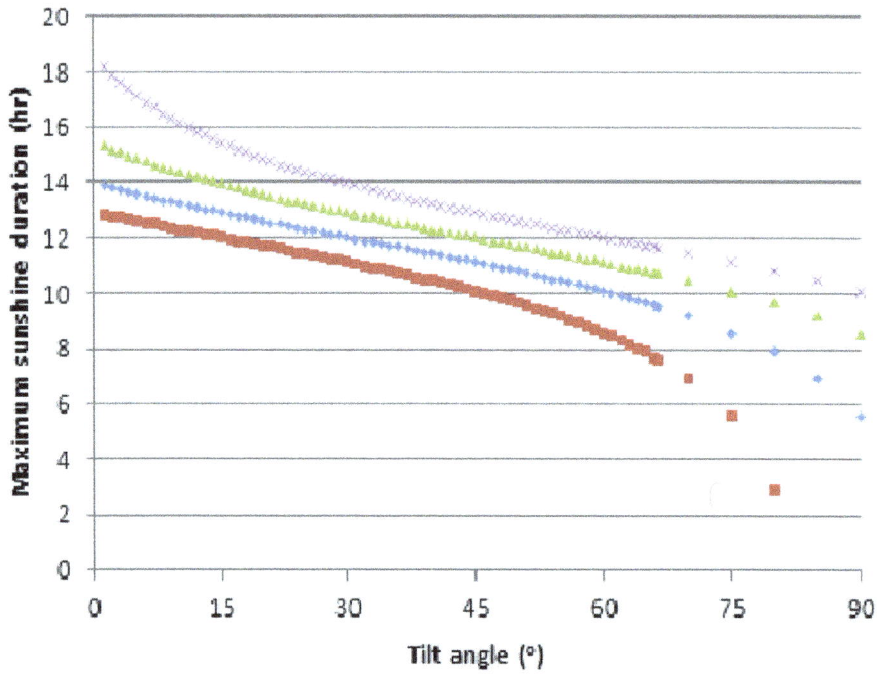

Figure 16. Winter solstice maximum sunshine duration on the equator facing surface with different tilt at latitudes (φ = 15°N (■),30°N (♦),45°N (▲) and 60°N (x))

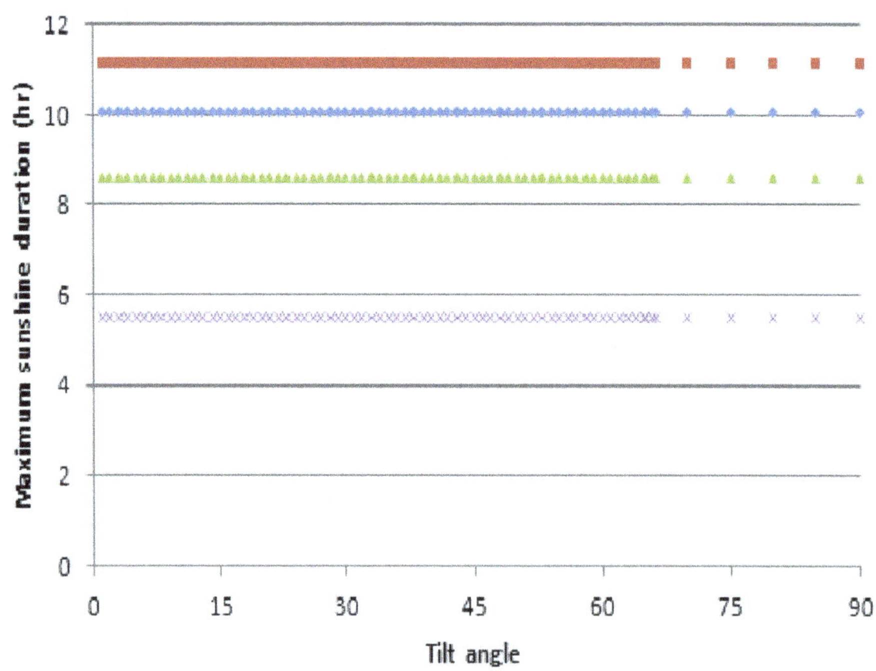

Figure 17. Summer solstice maximum sunshine duration on the equator facing surface (■) and North Pole facing (x) for latitude φ = 15°N and winter solstice maximum sunshine duration on the equator facing surface (■) and South Pole facing (x) for latitude φ = 15°S

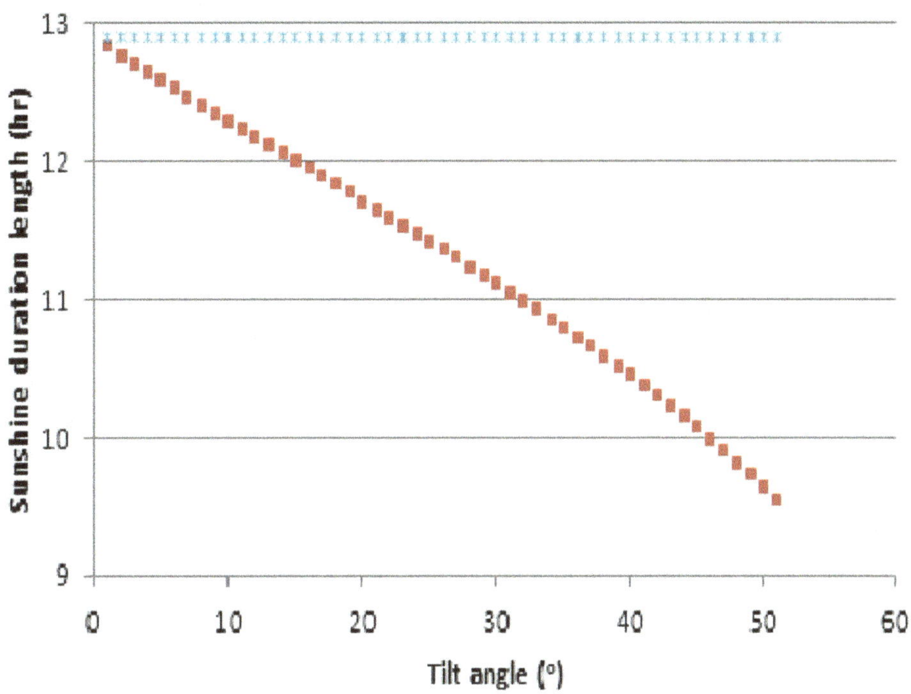

(41) and (42) but as this orientation could not be optimal for different economical feasible applications over a period of several months it will not considered with sufficient details.

Equator Facing Vertical Surface Case

Vertical surfaces are a part of building and the determination of solar radiation and sunshine duration on these vertical surfaces is very important for calculating the required energy for covering the solar building heating and cooling loads. The solar building is a building that includes as integral parts of its elements that admit, absorb, store and release solar energy and thus reduces the needs for auxiliary energy. Building walls could be constructed to have high heat capacity to store thermal energy and reduce temperature variations.

For latitudes greater than 23.45°N and 23.45°S the Sun shines at solar noon from the Equator direction all over the year. Therefore, it is important to calculate the sunshine on the Equator facing vertical surface. In this case, $\beta = \pi/2$ and $\gamma = 0°$. Thus, the equations (40a) and (41a) become:

$$A_1 = -\sin(\delta)\cos(\varphi) \tag{40b}$$

$$A_2 = \cos(\delta)\sin(\varphi) \tag{41b}$$

Hence the equations (44a) and (45a) become:

$$\omega_{ss} = \min\left\{\arccos\left[-\tan(\delta)\tan(\varphi)\right], \arccos\left[\frac{\tan(\delta)}{\tan(\varphi)}\right]\right\} \tag{44b}$$

$$\omega_{sr} = \max\left\{-\arccos\left[-\tan(\delta)\tan(\varphi)\right], -\arccos\left[\frac{\tan(\delta)}{\tan(\varphi)}\right]\right\} \tag{45b}$$

So, $\omega_{ss} = -\omega_{sr}$ and the sunshine duration on an Equator facing vertical surface S' could be determined using the equation (47a).

On winter solstice the sunshine duration on the Equator facing vertical surface is the same as that on the horizontal surface while it should be deduced from equations (41b) and (42b) on winter solstice. Figure 18 shows the sunshine duration on summer and winter solstices. It is seen from Figure 18 that, the Sun does not rise on the Equator facing vertical surface on summer solstice in the Northern part of the tropical zone. This means that the Sun shines on summer solstice in this part from North Pole direction.

Figure 18. Summer and winter solstice maximum sunshine duration on the equator facing vertical surface with different latitudes (summer (■),winter (♦)) in the Northern hemisphere

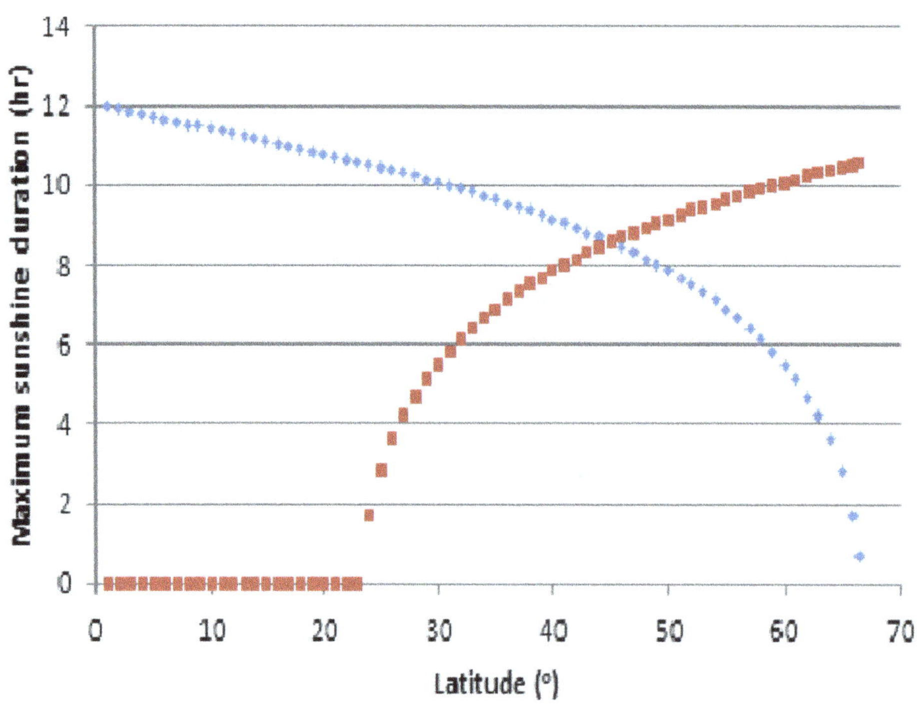

Latitude Tilted Equator Facing Surface

In this case, β = φ and γ =0°. Thus, the equations (40a) and (41a) become:

$$A_1 = \sin(\delta)\sin(\varphi - \beta) = 0 \tag{40c}$$

$$A_2 = \cos(\delta)\cos(\varphi - \beta) = \cos(\delta) \tag{41c}$$

Hence $A_4 = A_2$ and the equations (44a) and (45a) become:

$$\omega_{ss} = \min\left\{\arccos\left[-\tan(\delta)\tan(\varphi)\right], \pi/2\right\} \tag{44c}$$

$$\omega_{sr} = \max\left\{-\arccos\left[-\tan(\delta)\tan(\varphi)\right], -\pi/2\right\} \tag{45c}$$

So, $\omega_{ss} = -\omega_{sr}$ and the sunshine duration on an Equator facing latitude tilted surface S' could be determined using the equation (47a). When calculating *S'* for different latitudes, the results shown in Figure 19 are obtained.

Tilted Surface With Arbitrary Orientation

The importance of orientation of a solar absorber is to save energy while its importance of building is not only for saving energy but also to have a better house design, which not only gives comfortable living but also gives good health, prosperity and wealth to the house owners/occupiers and these families. There lies a co-relation between the rotational scenario of the planets and the house design and their different directions with respect of north. The building of any type and its construction meets the purpose if proper orientation has been given using suitable local building material. It increases not only its life span but also improves the condition of occupants. There are instances where buildings are not planned according to required local orientation were lost or deteriorated much faster than the buildings having built with proper studies of orientation.

The proper orientation means the proper knowledge of all the eight directions. It is a common knowledge that the direction from where the Sun arises is known as east and where it sets as west and when one faces the east direction, towards one's left is north and towards one's right is south. The corner where two directions meet obviously is more significant since it combines the forces emanating from both the directions. So, in all there are eight directions, north, northeast, east, east-south, south, southwest, west, west-north. Every direction has its own significance and has its own construction.

In general $\omega_{ss} \neq -\omega_{sr}$. Hence the sunshine duration *S'* on a surface of arbitrary orientation could be deduced from equations (44). The application of equation (44) on several examples on solstices and equinoxes gives a total idea about the variations of S'. These examples should cover the above mentioned proper orientations for building as well as those for solar collectors.

Figure 19. Summer and winter solstice maximum sunshine duration on the equator facing latitude tilted surface with different latitudes (summer (x), winter (▲)) in the Northern hemisphere

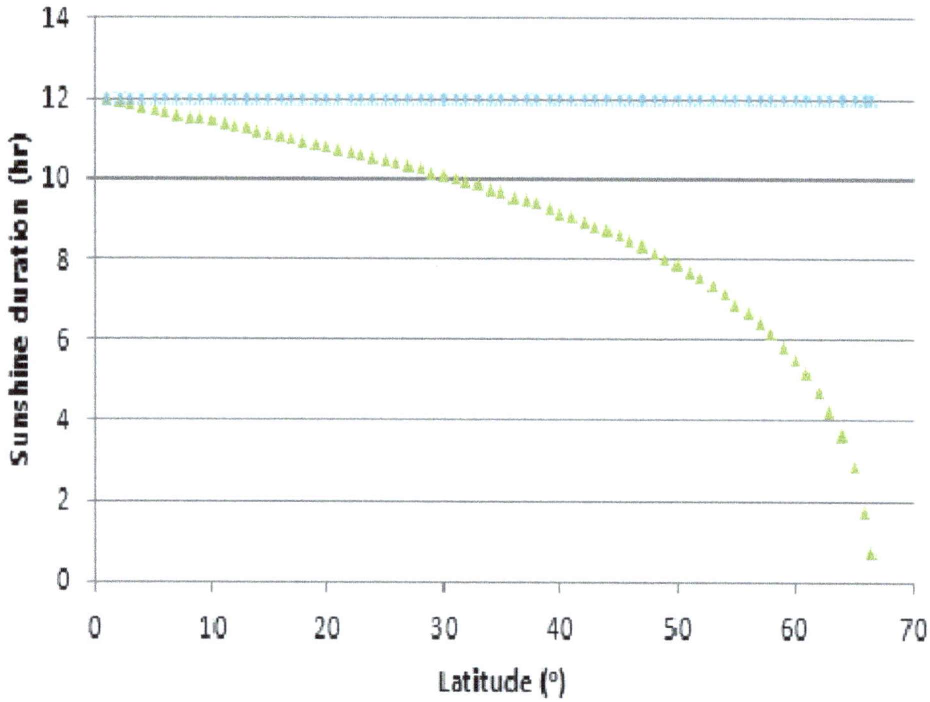

East and West Orientated Vertical Surface

In this case, $\beta = 90^{\circ}$ and $\gamma = \pm 90^{\circ}$: "+"is for west orientation while "-" is for east orientation. Thus, the equations (40)-(42) and (46) take the form:

$$A_1 = 0 \tag{40d}$$

$$A_2 = 0 \tag{41d}$$

$$A_3 = \cos\left(\delta\right) \tag{42d}$$

$$A_4 = A_3 \tag{46d}$$

Then, the equations (44) and (45) become:

$$\omega_{ss} = \min\left\{\arccos\left[-\tan\left(\delta\right)\tan\left(\varphi\right)\right], 0\right\} \tag{44d}$$

$$\omega_{sr} = \max\left\{-\arccos\left[-\tan\left(\delta\right)\tan\left(\varphi\right)\right], -\pi\right\} \tag{45d}$$

Then, as cos(δ)≠0 all over the year, equation (40) gives that ω_{ss}=0°. So, the sunset occurs on the solar noon on the vertical orientated towards east while the sunrise occurs on the moment of sunrise on the horizontal plane. So, the maximum sunshine duration variations, with latitude on the vertical surface orientated towards east for equinox and solstices, are given in Figure 20. In the case of vertical orientated towards west the sunrise occurs on the solar noon while the sunset occurs on the moment of sunset on the horizontal plane. The same results of Figure 20 will be obtained for west orientated vertical surface.

North Orientated Vertical Surface

In this case, β = 90° and γ =-180°. Thus, the equations (40)-(42) take the form:

$$A_1 = \sin\left(\delta\right)\cos\left(\varphi\right) \tag{40e}$$

Figure 20. Maximum sunshine duration on the East facing vertical surface with different latitudes (summer solstice (♦),winter solstice (■) and equinox (▲)) in the Northern hemisphere

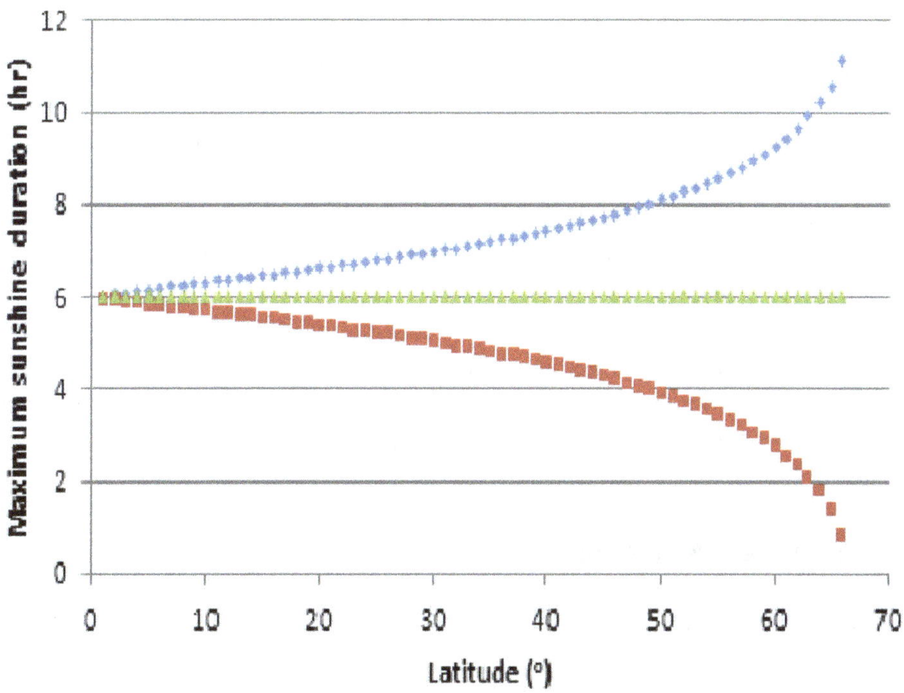

$$A_2 = -\cos(\delta)\sin(\varphi) \tag{41e}$$

$$A_3 = 0 \tag{42e}$$

So, $A_4 = A_2$ and the equations (44) and (45) become:

$$\omega_{ss} = \min\left\{\arccos\left[-\tan(\delta)\tan(\varphi)\right], \arccos\left[-\frac{\tan(\delta)}{\tan(\varphi)}\right]\right\} \tag{44e}$$

$$\omega_{sr} = \max\left\{-\arccos\left[-\tan(\delta)\tan(\varphi)\right], -\arccos\left[-\frac{\tan(\delta)}{\tan(\varphi)}\right]\right\} \tag{45e}$$

Out of the tropical zone, the Sun does not shine on the north orientated vertical surface, located in the Northern Hemisphere, during the period from 22/9 to 21/3 while it can rise and set on the both sides of the vertical wall during the period from 22/3 to 21/9. Hence, the maximum sunshine duration on the north orientated vertical surface is:

$$S' = \frac{\left(\omega_{sr1} - \omega_{sr}\right) + \left(\omega_{ss} - \omega_{ss1}\right)}{15} \tag{48}$$

where $\omega_{sr,1}$ and $\omega_{ss,1}$ are the sunrise and sunset hour angles on the south orientated surface of the vertical wall, ω_{sr} and ω_{ss} are the sunrise and sunset hour angles for the horizontal surface respectively. The values of the maximum sunset ($\omega_{ss,1}$) and minimum sunrise ($\omega_{sr,1}$) hour angles on the mentioned vertical wall with different latitudes on summer solstice in the Northern Hemisphere are given in the Figure 21. Here it should be mentioned that, for calculating the values of $\omega_{ss,1}$ and $\omega_{sr,1}$ on the same wall the hourly dependence of solar azimuth angle should be considered. The application of the equation (48) leads to results presented in the Figure 22.

According to the results given in Figure 22 it is possible to distinguish between mid-latitude and high-latitude zones. In the mid-latitude zone the maximum possible sunshine duration decreases with increasing the latitude value while it increases with increasing the latitude value in the high latitude zone.

The south orientated vertical surface was treated before when studying the Equator facing vertical surface.

East-South Orientated Vertical Surface

In the case of east-south orientated vertical surface, $\beta = 90^{o}$ and $\gamma = -45^{o}$. Thus, the equations (40)-(42) take the form:

Figure 21. Sunset hour angles ω_{ss}(ж) and $\omega_{ss,1}$ (x) and sunrise hour angles ω_{sr}(+) and $\omega_{sr,1}$ (●) on the horizontal surface and vertical surface orientated towards the Equator with different latitudes on summer solstice in the Northern Hemisphere

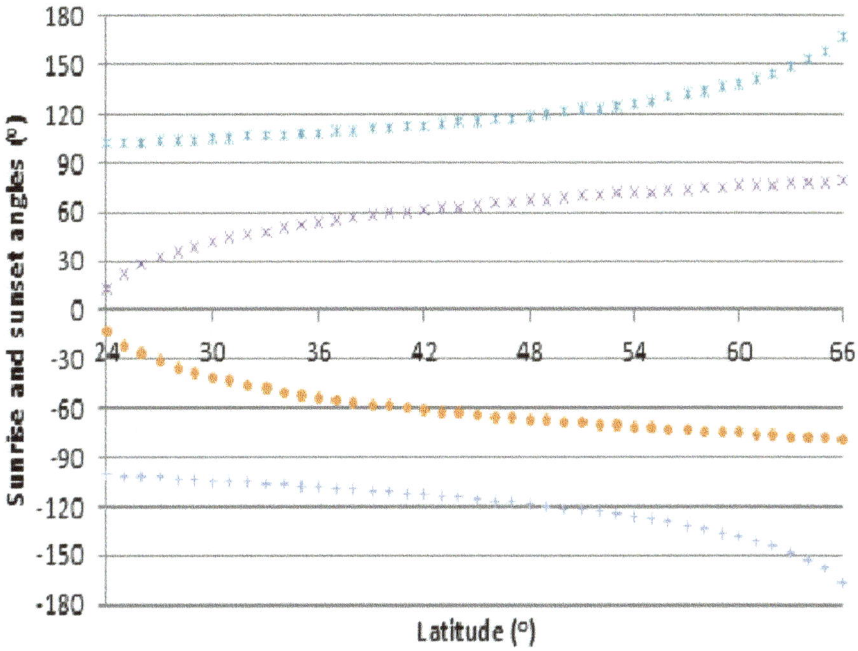

Figure 22. Maximum sunshine duration on a vertical surface orientated towards the North Pole with different latitudes on summer solstice in the Northern Hemisphere

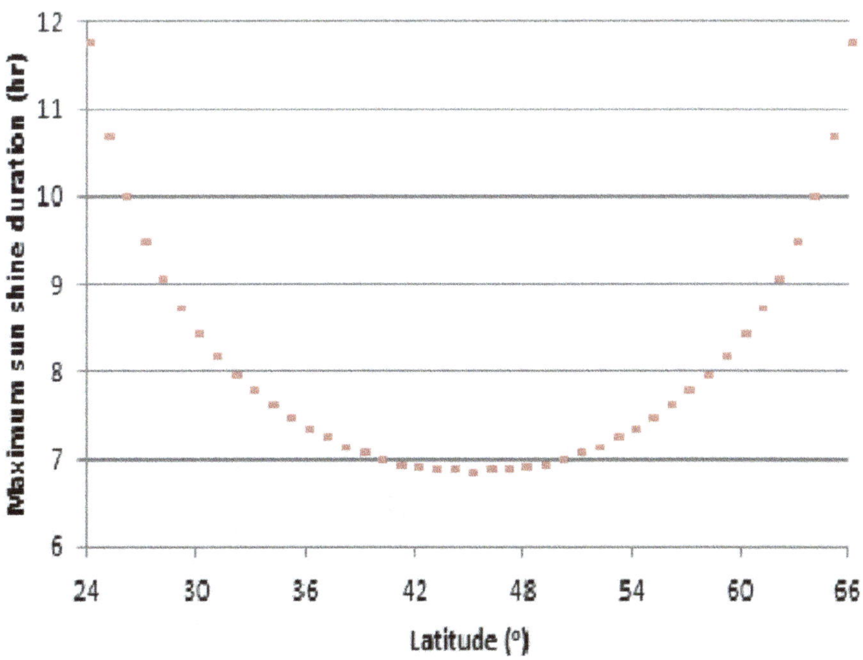

$$A_1 = -0.707 \sin(\delta) \cos(\varphi) \tag{40f}$$

$$A_2 = 0.707 \cos(\delta) \sin(\varphi) \tag{41f}$$

$$A_3 = -0.707 \sin(\delta) \tag{42f}$$

Therefore,

$$A_4 = 0.354\sqrt{3 - \cos(2\varphi)} \tag{46b}$$

Then the sunshine duration on winter and summer solstices on east-south orientated vertical surface can be calculated using equations (44), (45) and (47). The calculated results are given in Figure 23.

Figure 23. Maximum sunshine duration on the north facing vertical surface with different latitudes on winter (x) and summer (■) solstices in the Northern hemisphere

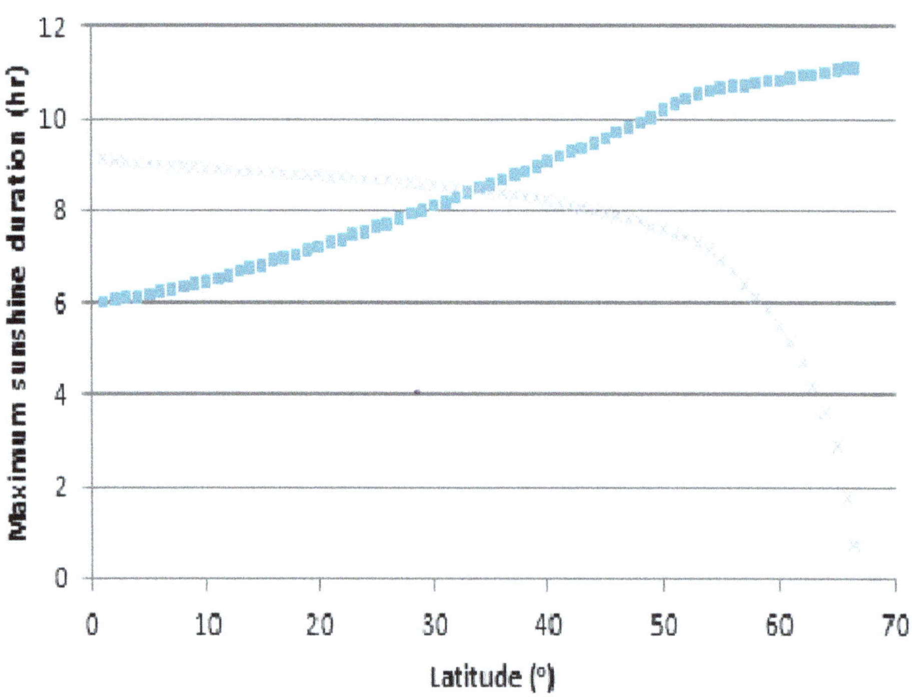

Southwest Orientated Vertical Surface

In the case of southwest orientated vertical surface, $\beta = 90^{\circ}$ and $\gamma = 45^{\circ}$. Thus, the equations (40) and (41) take the form of (40f) and (41f) while the equation (42) takes the form:

$$A_3 = 0.707 \sin(\delta) \tag{42g}$$

Therefore, A_4 could be calculated using equation (46b).

Then, using equations (44), (45) and (47) the sunshine duration on winter and summer solstices on southwest orientated vertical surface could be calculated. Here it should be mentioned that, even ω_{ss} and ω_{sr} in the case of southwest orientated vertical surface are different from those of southeast orientated vertical surface, the maximum sunshine duration is identical in the both cases.

Northeast Orientated Vertical Surface

In the case of northeast orientated vertical surface, $\beta = 90^{\circ}$ and $\gamma = -135^{\circ}$. Thus, the equations (40)-(42) take the form:

$$A_1 = 0.707 \sin(\delta) \cos(\varphi) \tag{40g}$$

$$A_2 = -0.707 \cos(\delta) \sin(\varphi) \tag{41g}$$

$$A_3 = -0.707 \cos(\delta) \tag{42g}$$

while the equation (46) takes the form of the equation (46b).

West-North Orientated Vertical Surface

In the case of west-north orientated vertical surface, $\beta = 90^{\circ}$ and $\gamma = 135^{\circ}$. Thus, the equations (40) and (41) take the form of the equations (40g) and (41g) while the equation (42) takes the form of the equation (42g) and the equation (43) takes the form of the equation (43b)

The calculation of the maximum sunshine duration using equations (41), (42) and (44) on winter and summer solstices on northwest orientated in the Northern Hemisphere leads to the same results presented in Figure 24 even ω_{ss} and ω_{sr} in the both cases are different but a simple relationships between them is observed.

Figure 24. Maximum sunshine duration on the northeast oriented vertical surface with different latitudes on winter (+) and summer (x) solstices in the Northern hemisphere

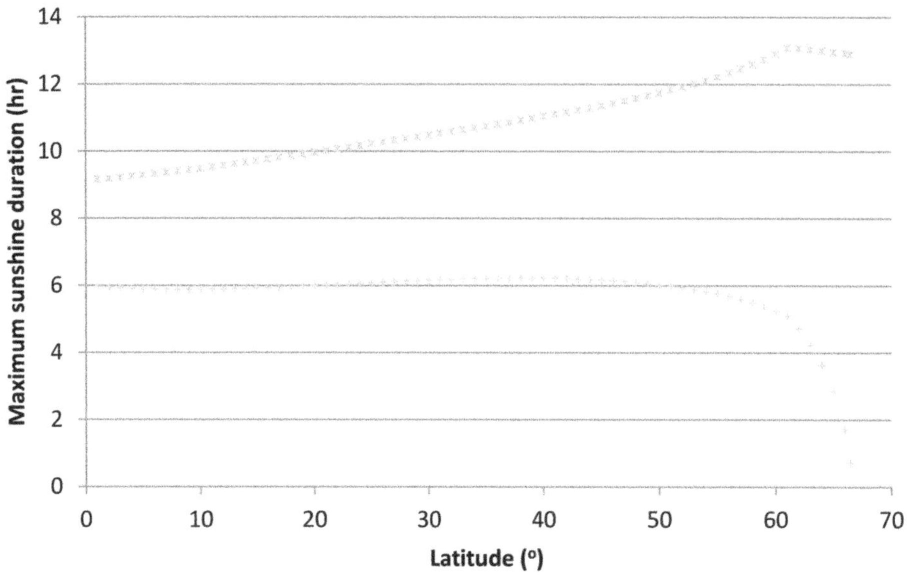

Southeast Orientated Tilted Surface

In the case of southeast orientated tilted surface, γ =-45°. Thus, the equations (40)-(42) take the form:

$$A_1 = \sin(\delta)\left[\sin(\varphi)\cos(\beta) - \left(\frac{\sqrt{2}}{2}\right)\cos(\varphi)\sin(\beta)\right] \quad (40h)$$

$$A_2 = \cos(\delta)\left[\cos(\varphi)\cos(\beta) + \left(\frac{\sqrt{2}}{2}\right)\sin(\varphi)\sin(\beta)\right] \quad (41h)$$

$$A_3 = -\left(\frac{\sqrt{2}}{2}\right)\cos(\delta)\sin(\beta) \quad (42h)$$

The calculated results of the sunshine duration with β variations on equinox and solstices for several latitudes in the Northern Hemisphere are given in the Figures 25 to 28.

According to equations (44), (45) and (47) and to the calculated results presented in Figures 25 to 28, the sunshine duration depends on the latitude φ, the declination angle δ and the surface tilt angle β.

Figure 25. Maximum sunshine duration on the southeast orientated tilted surface, located at latitude 15°N on equinox (x), winter (-) and summer (●) solstices

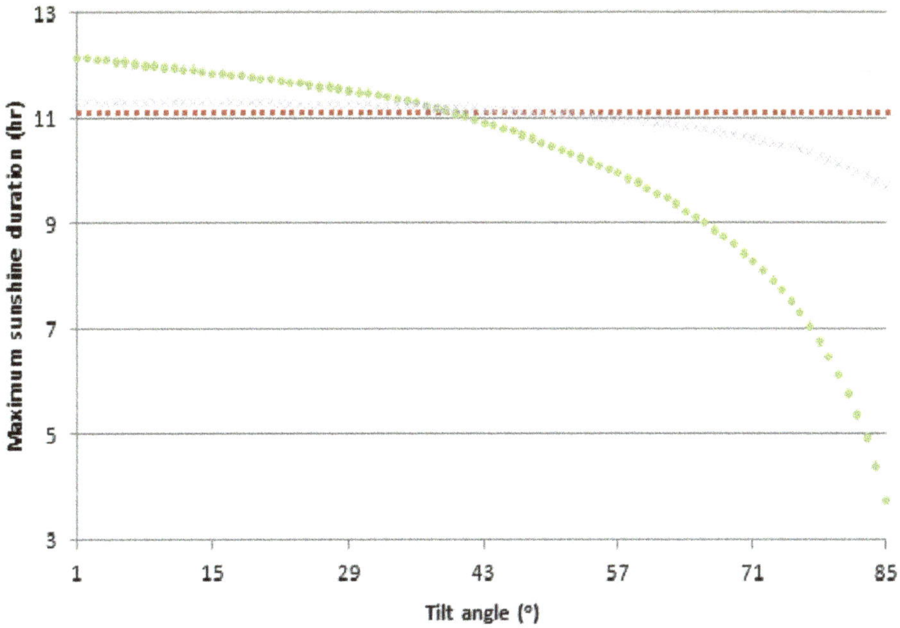

Figure 26. Maximum sunshine duration on the southeast orientated tilted surface, located at latitude 30°N on equinox (x), winter (-) and summer (●)solstices

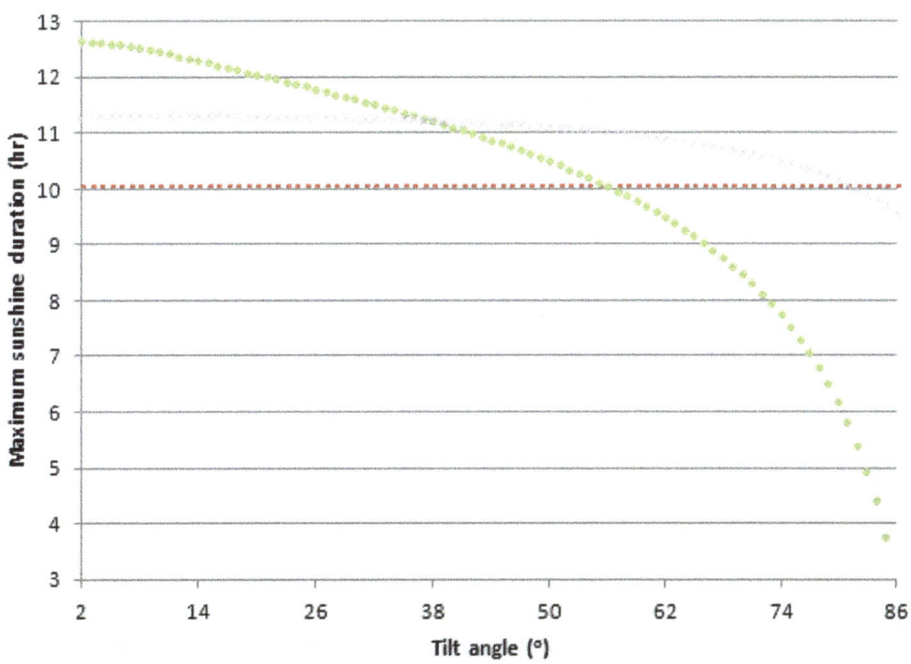

Figure 27. Maximum sunshine duration on the southeast orientated tilted surface, located at latitude 45°N on equinox (x), winter (-) and summer (●) solstices

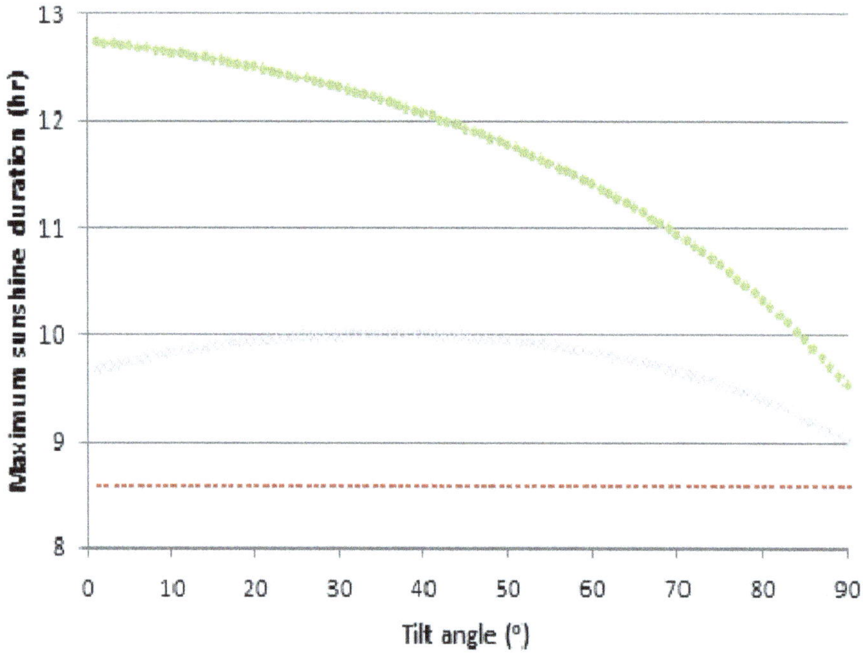

Figure 28. Maximum sunshine duration on the southeast orientated tilted surface, located at latitude 60°N on equinox (x), winter (-) and summer (●) solstices

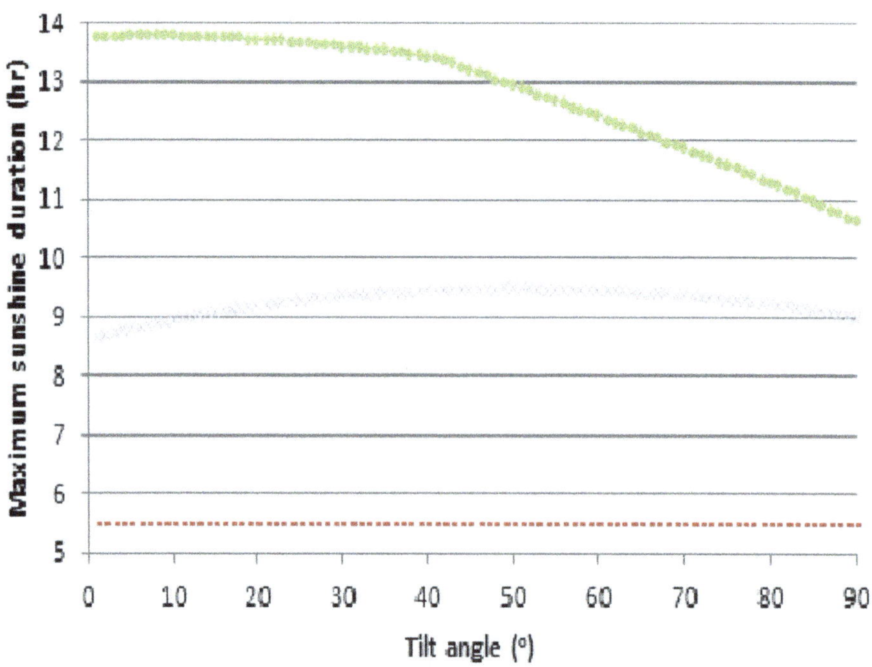

MEASUREMENTS OF SITE'S LATITUDE

When taking a vertical stick of length L_p and studying its shadow length L (see Figure 29) during the day it will be found that L varies with time during day and riches its minimum value on solar noon. Equation (23) becomes on solar noon (ω=0°):

$$\cos\left(\theta_{z,noon}\right) = \cos\left(\varphi - \delta\right) \tag{23a}$$

On the equinox δ=0° and:

$$\varphi = \theta_{z,noon} \tag{49}$$

Therefore, the accuracy of φ determination depends on the accuracy of θ_z determination which could be done by direct measurement using a protractor with an error of 0.5° or indirectly by measuring the shadow length of a mast of definite length. According to Figure 29:

$$\theta_z = \arctan\left(\frac{L}{L_p}\right) \tag{50}$$

Figure 29. The shadow length L of a vertical stick of length L_p

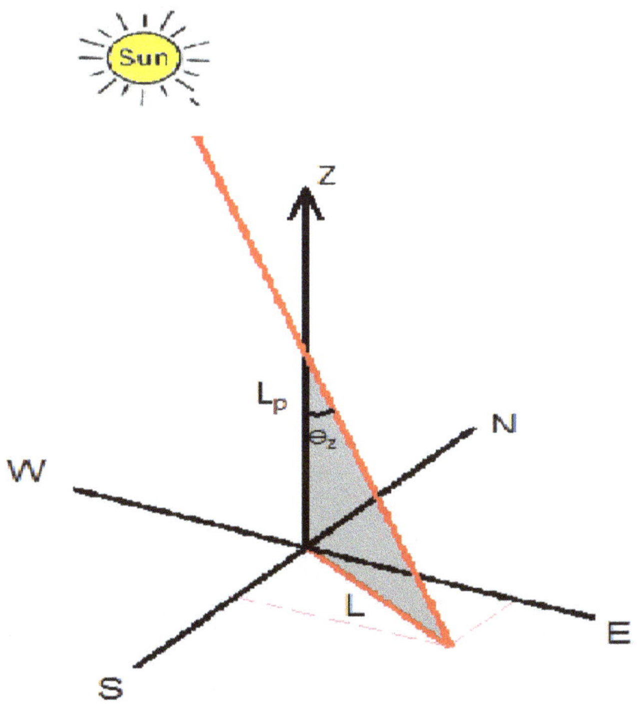

Since the normal ruler error committed in measuring the length of the mast and the length of its shadow is (± 1 mm). Therefore, and in order to reduce the sensitivity of the value of θ_z or φ error committed in the measurement, the appropriate length of the mast L_p, that lowers the error in measuring θ_z or φ, could be determined by studying the influence of its length on the measured value. The results are given in Figure 30. It is seen from Figure 30 that, with increasing the mast length the error decreases and the measured value approaches the precise value. Practically, the use of a stick of 1.5m in length leads to an acceptable result.

Unfortunately, equation (45) is only correct on 21 March and 21 September (the spring and autumn equinoxes). So, on two days every year, called the equinox, the Sun is positioned directly above Earth's Equator. On the equinox, the angle between a line that points to the Sun and a line that points straight up (vertical) exactly matches the latitude of the place you are standing. If you live on the equator, then in the very middle of the day (solar noon) the Sun will be directly above you, or at 0° from the vertical. If you live in Damascus, Syria, which is in the Northern Hemisphere at the latitude of 33.5138° N, then on the equinox the Sun will be 33.5138° to the south from the vertical (see Figure 29). However, it's possible to measure the latitude by comparing your position on the Earth with the position of the Sun on a clear day in the Northern or Southern Hemispheres when the Sun is easy to find. However, measurement of latitude isn't as straightforward as you might think. Accurate readings can only be taken at noon, when the Sun is at its highest in the sky.

The application of equation (46) could be considered by fixing either the length of the stick or the length of its shadow. These two possibilities will be treated below.

Figure 30. The measured latitude basing on measured vertical stick shadow at noon

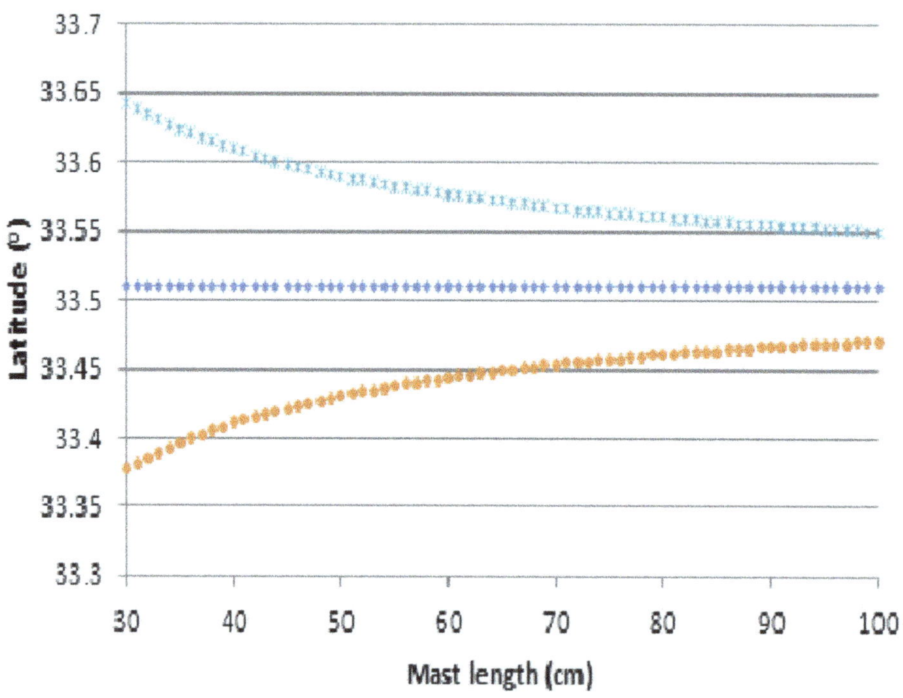

Site's Latitude Measurement Using Constant Stick's Length

In this case the sick's length will be constant all over the measuring period. For the purpose of measuring the latitude on equinox, it is important to do the following step by step:

1. Make a quadrant similar to the one shown in Figure 31. The aiming beam needs to be pivoted about its central point so that it can swing up and down. The protractor should be centered on the pivot, from which the plumb line should be suspended. There should be 4 nails at the hands of the cross, each only pounded far enough to keep it stable. There should be 2 nails at either end of the face of the cross and 2 nails at either end of the top of the hands of the cross (see Figure 31).
2. Using a compass, mark out a line on the ground that runs north to south (see Figure 32).
3. Set up the quadrant so that its aiming beam is parallel to this north-south line (see Figure 32).
4. Determine the noon by placing a stick at the southernmost end of the north-south line. Use a plumb line to make sure that the stick is vertical. When the shadow cast by the stick crosses the north-south line, it's noon (see Figure 33).
5. Align the sighting nails of the quadrant with the Sun. Once noon, line up the nails on the aiming beam with the Sun. Do not look directly at the Sun but rather use the shadows of the nails to find the correct position. Move the aiming beam up and down so that the two shadows from the nails move closer together and create a single shadow on the ground (see Figure 34).
6. Use the protractor to measure the small angle between the beam and the plumb line. Once the beam is aimed correctly, use the protractor to measure from the vertical plumb line to the portion of the aiming beam closest to it. The horizon should be kept at 90° when making the measurements.

Figure 31. Educational north direction finder quadrant

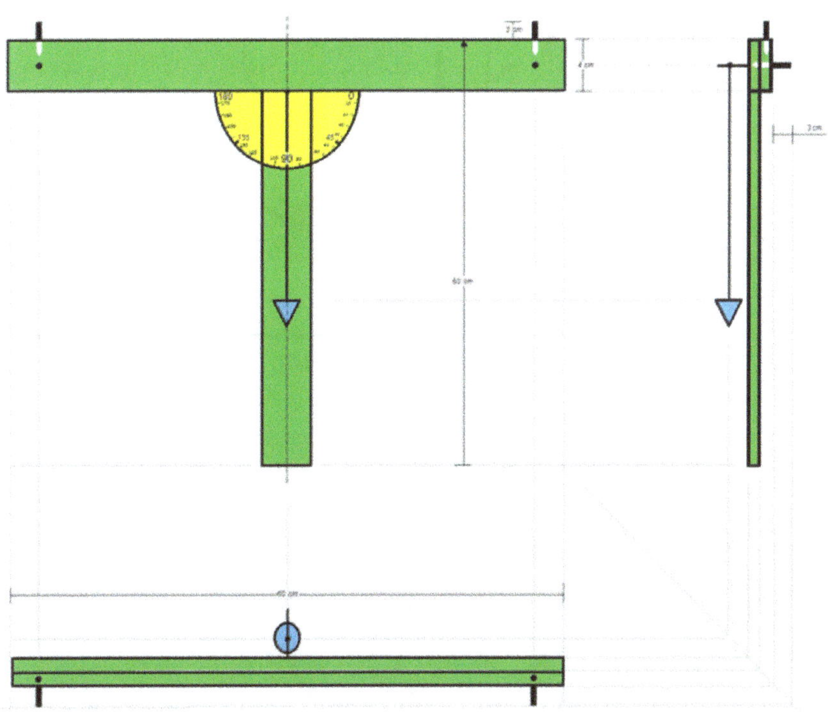

Figure 32. The quadrant with an aiming beam parallel to the north-south line

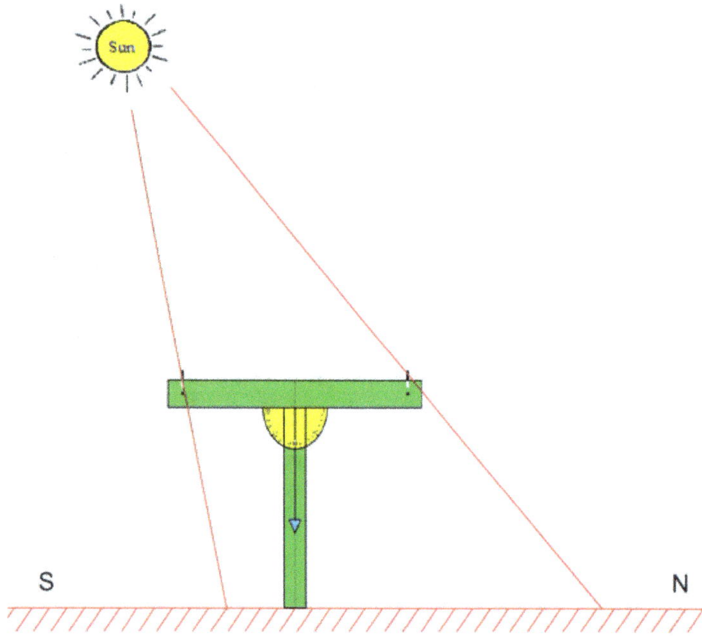

Figure 33. The stick shadow on the solar noon

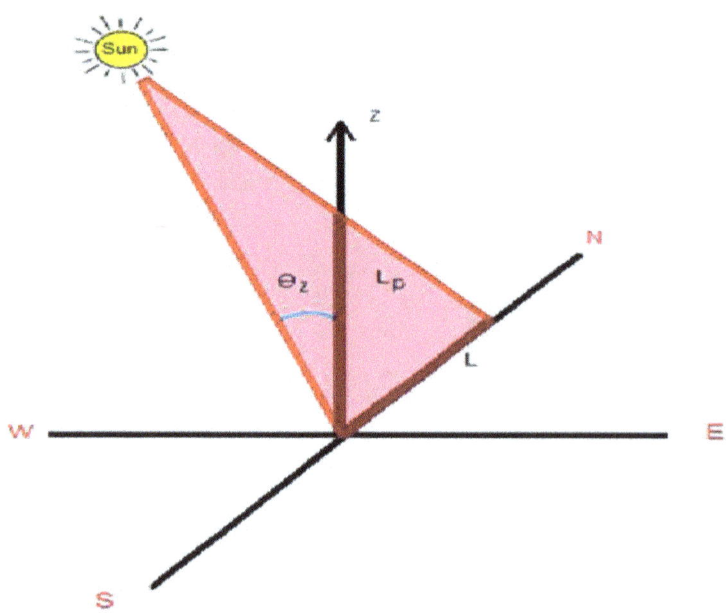

Figure 34. A single shadow for the two nails of the aiming beam on the ground

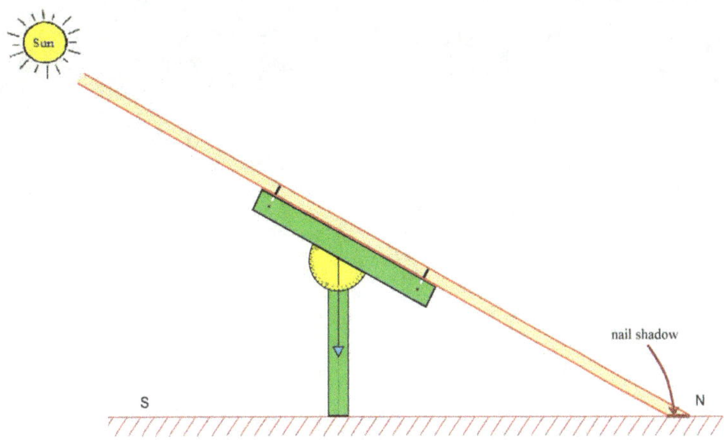

In order to verify the applicability of the mentioned method the designed and constructed educational north direction finder quadrant (see Figures 35 and 36) was used for measuring the solar zenith angle in Damascus (33.51°N) during August and September 2016 and calculating the latitude of Damascus basing on these measurements. The results are presented in Table 6.

The Sun's position on the equinox is the average location of the Sun throughout the year and is a great reference to use when designing a solar system for a specific location. But keep in mind, the Sun is always moving. In the summer, the Sun appears higher in the sky, which increases the duration of Sunlight seen in a day, and in the winter it appears lower, which decreases the length of Sunlight in a

Figure 35. A photo of the used right triangle for measuring site latitude

Figure 36. A photo of the used educational north direction finder quadrant

Table 6. The calculation of the Damascus latitude

Day Number	Measured Zenith Angle (°)	Calculated Declination Using Equation (6) (°)	Calculated Latitude (°)
236	22.8	10.69	33.49
237	23.1	10.33	33.43
240	24.4	9.23	33.63
241	24.7	8.86	33.56
242	25	8.48	33.48
243	25.4	8.10	33.50
244	25.8	7.72	33.52
247	27	6.57	33.57
248	27.2	6.18	33.38
249	27.7	5.79	33.49
250	28.1	5.40	33.50
251	28.5	5.01	33.51
261	32.5	1.01	33.51
262	33.1	0.61	33.71
263	33.3	0.20	33.50
264	33.5	-0.20	33.30
265	34.4	-0.61	33.80
268	35.5	-1.81	33.69
269	35.7	-2.22	33.48
		Average value for the calculated latitude	33.53

Note: The use of a compass makes the line marked out on the ground run north to south magnetic while the geographic orientations are required for precise measurements. The measurements are not completely semi-accurate, other than spring and autumn equinox, because of the way the Earth is tilted as it orbits around the Sun.

day. The Sun is highest in the sky on the summer solstice. To be more exact, it is 23.45° higher than on the equinox, or at 33.51-23.45 = 10.06° to the south of vertical.

The winter solstice is the day when the Sun appears lowest in the sky. On this day, the Sun is 23.45° lower than on the equinox, or at 33.51 + 23.45 = 56.96° to the south of vertical in Damascus. So, if you live in the northern hemisphere at latitude higher than 23.45°, then the Sun will never shine from the north. This means that the north side of your house would be a bad place for a solar panel (or a garden).

Site's Latitude Measurement Using Fixed Stick's Shadow Length

In this case the sick's shadow's length will be constant all over the measuring period. For the purpose of measuring the latitude on equinox, it is important to do the following step by step:

1. Make a right triangle similar to the one shown in Figure 36. The upper head of the right triangle able to be displaced up and down.
2. Using a compass, mark out a line on the ground that runs north to south.
3. Set up the right triangle so that its base is parallel to this north-south line and the other side based is parallel to the vertical.
4. Determine the noon by placing a stick at the southernmost end of the north-south line.
5. Move the triangle upper head until the hypotenuse's shadow coincides with the hypotenuse itself. This moment corresponds the solar noon.
6. Fix the triangle upper head.
7. Measure the vertical side based length.
8. Use the equation (46) for calculating the zenith angle. On equinox the measured angle is the site latitude (see equation (45).

In order to measure the latitude all over the year, two additional steps to the above mentioned ones (in a and b cases) should be done in accordance with equation (23a):

1. Take the value of the declination angle from the astronomical Almanac tables, (Meeus, 1998) or (Reda and Andreas, 2004).
2. Add the value of declination angle on the day of measurement to the measured value. The latitude is the obtained result.

So, at midwinter (21 December) you should deduct 23.45° from your reading, and at midsummer (21 June) add 23.45°. This is because of the way in which the 'tilted' Earth orbits the sun. Moreover, the above mentioned procedure is applicable for measuring φ, δ and θ_z.

The used length of the right triangle base was determined by studying the dependence of the measured latitude φ on the shadow length L (see Figure 37). It is seen from Figures 30 and 36 that the right triangle method decreases the measurement error (see Figure 38 which contains the results of the Figures 30 and 37)).

In order to make the results of measurements more accurate the geographic south-north direction should be determined precisely. This could be done either by compass correction tables or experimentally using a special device (Soulayman and Shamieah, 2000). This will be treated in detail in the chapter 2.

Figure 37. The measured latitude value dependence on the right triangle base's length

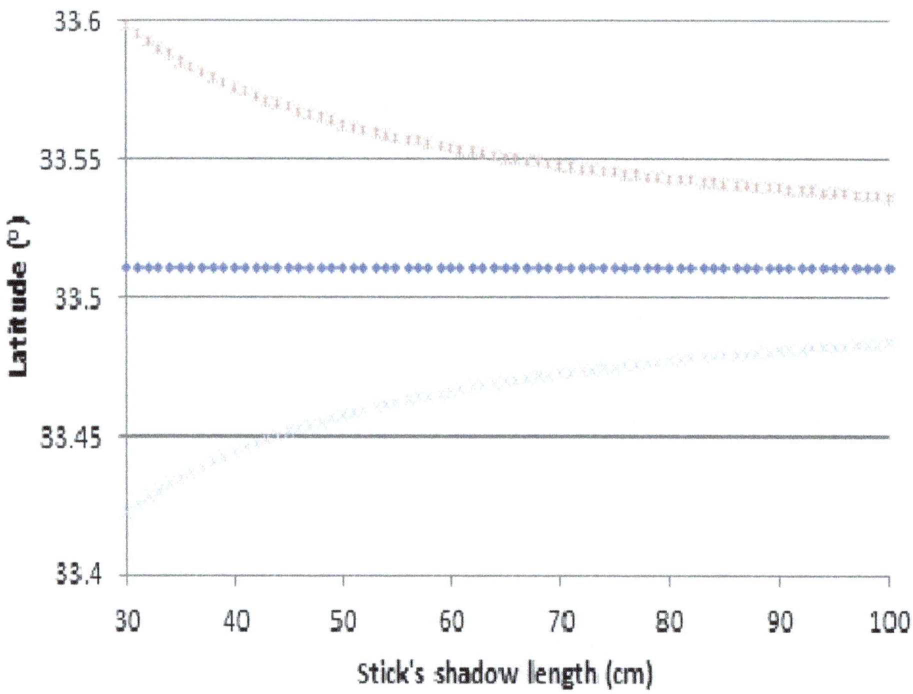

Figure 38. The measured latitude value dependence on the stick's (upper and lower curves) or stick's shadow's (other two curves) length. The middle line corresponds to the precise value.

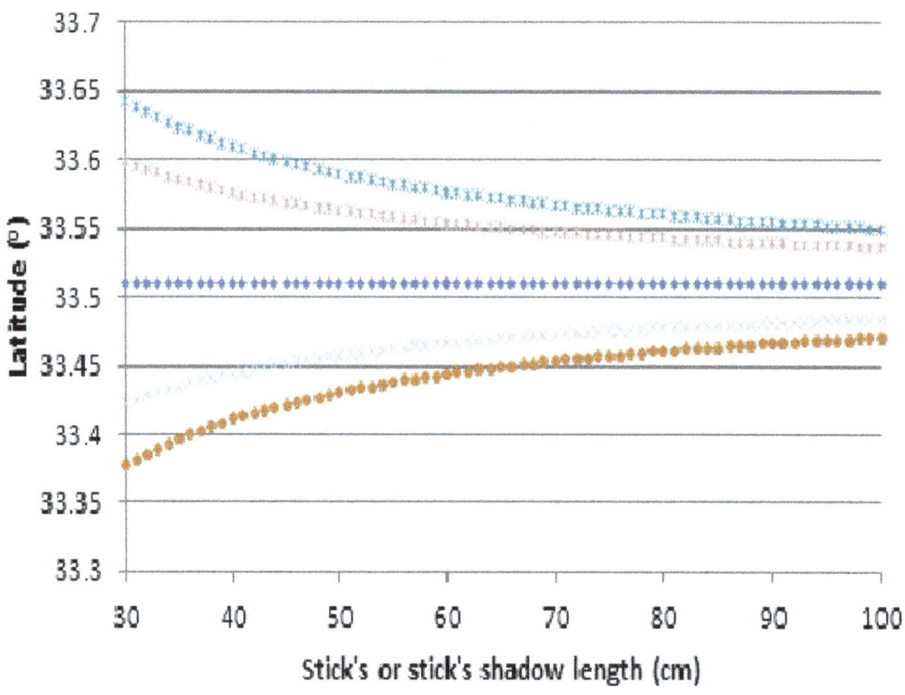

SOLAR AZIMUTH ANGLE MEASUREMENT

Hour Angle: Solar Azimuth Angle Measurement

Finding the locations of points often depends on angular measurements and directions of lines (Wolf and Ghilani, 2002). Determination of the directions of lines is crucial in many engineering applications. In order to determine the direction of a line, three requirements need to be met including a reference line, direction of angular measurement, and the value of the angle. The direction of a line is described by the horizontal angle between the reference line commonly known as meridian and the line of interest. If this angle is measured between the meridian's north or south directions, the angle is then referred to as the north or south azimuth.

There are two methods for determining the sun azimuth; the first is known as the hour angle method and the other is called the altitude method. The hour angle method uses the following formula (see equations (21) and (22)):

$$\tan\left(\gamma_s\right) = \frac{\sin\left(\omega\right)}{-\tan\left(\delta\right)\cos\left(\varphi\right) + \sin\left(\varphi\right)\cos\left(\omega\right)} \tag{51}$$

So, this method requires the determination of accurate time (hour angle ω), latitude φ and declination δ angles. Equation (47) was used by (Buckner, 1984) and (Ali, 2012) for measuring the solar azimuth angle clockwise from astronomic north. The obtained values were then normalized from 0° to 360° by adding algebraically a correction as follows. According to the proposed method, the hour angle varies from 0° (at solar noon) to 360°. So, when $0° \leq \omega \leq 180°$ a correction of 180° should added to γ_s if γ_s is positive otherwise the correction should be 360°. If $180° < \omega \leq 360°$ a correction of 0° should added to γ_s if γ_s is positive otherwise the correction should be 180°.

In the field, the horizontal angles from a line to the sun are obtained from direct and reverse (face left and face right or face I and face II pointing taken on the back- sight mark and the sun). It is suggested that repeating theodolites be used as directional instruments with one of two general measuring procedures being followed, which are: 1) A single foresight pointing on the sun for each pointing on the back-sight mark: here the sighting sequence is: direct on mark, direct on sun, reverse on sun, and reverse on mark, with times being recorded for each pointing on the sun. The two times and four horizontal circle readings constitute one set of data. An observation consists of one or more sets. A minimum of 3 sets is recommended. This procedure is similar to that of measuring an angle at a traverse station using a directional theodolite. This procedure is based on the assumption that pointing on the sun is of approximately the same accuracy as pointing on the back-sight mark. The single foresight procedure imparts itself to proper procedure for incrementing the horizontal circle and micrometer set- tings on the back-sight. 2) Multiple foresights pointing on the sun for each pointing on the back-sight mark: here, the sighting sequence is: direct on mark, several direct on the sun, an equal number reverse on sun, reverse on mark, with times being recorded for each pointing on the sun. A minimum of 6 pointing (3D and 3R) on the sun is recommended. The multiple times, multiple horizontal circle readings on the sun and the two horizontal circle readings on the back-sight mark constitute one observation. The multiple foresight procedure is based on the assumption that pointing on the sun is significantly less accurate than back-sight pointing.

Ali noted (Ali, 2012) that since a large difference usually exists between the vertical angle to the back-sight mark and the vertical angle to the sun, it is imperative that an equal number of both direct and reverse pointing be taken. This is even more important when using an objective lens filter. Also, the filter should not be removed or rotated between direct and reverse pointing on the sun. When reducing the data to compute the horizontal angles, the direct reading on the back-sight mark should always be subtracted from the direct foresight reading on the sun. Likewise, the reverse back-sight reading should always be subtracted from the reverse foresight reading, and add 360° if the resulting angle is negative. The vertical angle to the sun is usually larger than for typical surveying work. This increases the importance of accurately leveling the instrument. Because of this and other errors, it is recommended that observations not be made when the altitude of the sun is greater than approximately 45°.

Solar Azimuth Angle Measurement Using Elevation Angle

The second method is known as the altitude method which is based on the formula:

$$\gamma_s = arc\cos\left[\frac{\sin(\delta) - \sin(\varphi)\sin(\alpha)}{\cos(\varphi)\cos(\alpha)}\right] \tag{52}$$

which takes a simple form at solar sunrise or solar sunset as $\boldsymbol{\alpha} = 0o$:

$$\gamma_s = arc\cos\left[\frac{\sin(\delta)}{\cos(\varphi)}\right] \tag{53}$$

and it requires an accurate vertical angle.

On the other hand, Khavrus, and Shelevytsky (2012) defined the solar altitude angle time dependence using the following formula:

$$\alpha = \arcsin\left[\sin(\delta)\sin(\varphi) - \cos(\delta)\cos(\varphi)\cos\left(\frac{2\pi t}{1440}\right)\right] \tag{54}$$

where t is the solar time in minutes reckoned from midnight of the considered day in the year, $0 \le t < 1440$. At solar noon:

$$\alpha_{noon} = 90^o - \varphi + 23.45^o \sin\left(\frac{2\pi T}{365.25}\right) \tag{55}$$

where 23.45^o is the Earth's axial tilt (obliquity); T is the date counted from the spring equinox: T = 0 for the 20th of March, T = 1 for the 21st of March etc.

The derivation of equation (55) was based on the previously developed simple model describing the visible movement of the Sun on the celestial sphere as it can be viewed by a fixed observer situated at the latitude on the Earth (Khavrus and Shelevetsy, 2010). The geometrical idea of their model is based on a combination of four key components, namely:

1. A celestial sphere with radius equal to one;
2. A center of the celestial sphere that coincides with the observer;
3. A horizontal plane that passes through the observer;
4. A plane of the celestial equator that crosses the center of the sphere and is tilted relative to the horizontal plane at an angle of (90° -φ).

The intersection of the planes with the sphere gives the points of the geographical east and west, whereas intersection of the sphere with the plane of the celestial equator gives the path of the Sun during the days of the vernal and autumnal equinoxes. On other days of year, the movement of the Sun on the celestial sphere is defined by the plane, which is always parallel to the plane of the celestial equator with an angular distance determined by a declination angle δ.

It is seen from equation (55) that at the time of the winter solstice (T =274) the solar altitude during the whole day is lower than 66.55^o-φ while at the time of the summer solstice (T =92) the solar altitude during the whole day is lower than 113.45^o-φ. At equinox (T= 0 or T= 183) the solar altitude during the whole day is lower than 90°-φ.

Comments on Both Methods

When calculating the solar altitude angle using the equations (20) and (52) it was found that they lead to the same results. So, they can be used in azimuth angles using the equation (52). When comparing the results of equations (51) and (52) at the time of winter and summer solstices as well as at the time of equinox at the latitude of Damascus (33.5138° N) it was found that they give the same results, after correcting the results of equation (52) in order to be referenced to south direction ($\gamma_s = 0^o$ at south direction), at the time of winter solstice (see Figure 39). The deviations are remarkable at the time of summer solstice and equinox. Therefore, the results presented in Figures 40 to 42 are calculated using equation (52) in accordance with equations (21) and (22).

According to equation (55) the solar noon corresponds to *t*=720min. This means that:

$$\alpha_{noon} = \arcsin\left[\cos\left(\varphi - \delta\right)\right] \tag{56}$$

This implies that:

$$\alpha_{noon} = \frac{\pi}{2} - \left(\varphi - \delta\right) \tag{57}$$

This means that the daily dependence of the solar altitude angle α_{noon} is given by the daily dependence of the declination angle. So, the last term in equation (55) is but a δ. When calculating the declination angle using the last term of equation (55) and the equation (6) - the equation of (Cooper, 1969) – it was found that the maximum difference does exceed 1°. So, the last term of equation (55) is acceptable for

Figure 39. Daily azimuth angle variation at winter solstice (Damascus, 33.5138° N)

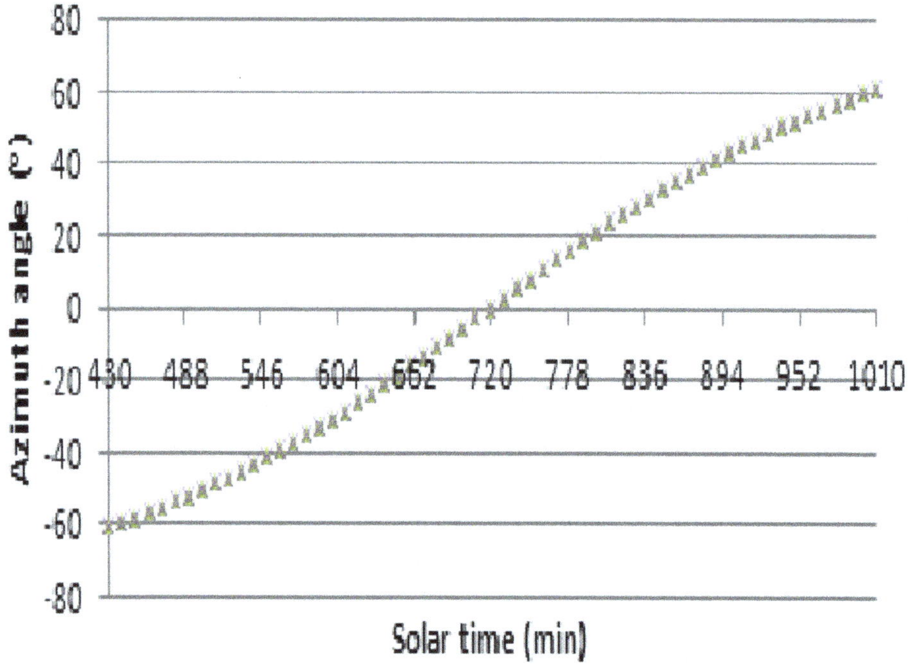

Figure 40. Daily azimuth angle variation at summer solstice (Damascus, 33.5138° N)

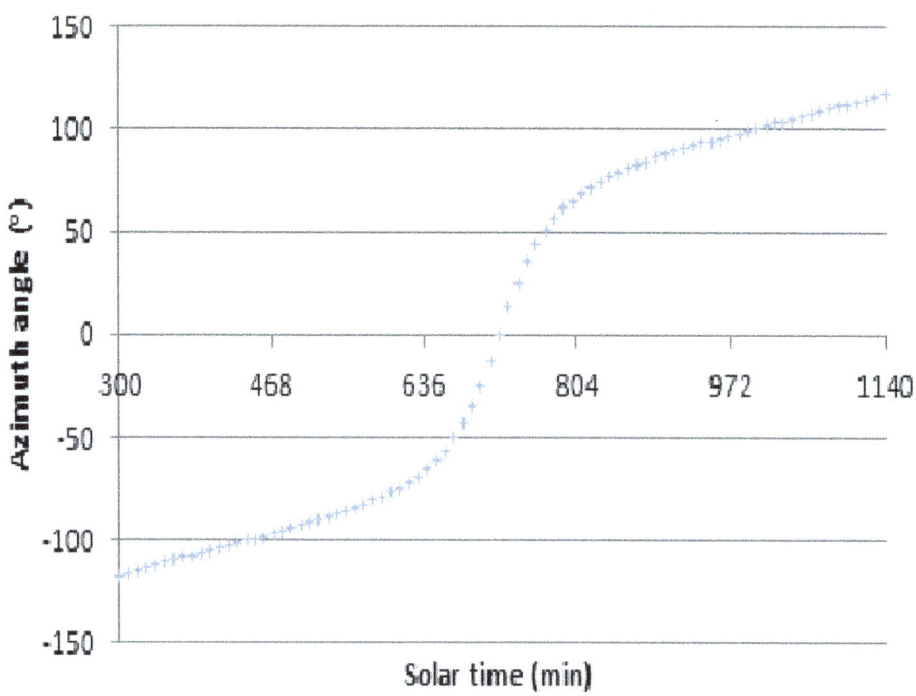

Figure 41. Daily azimuth angle variation at equinox (Damascus, 33.5138° N)

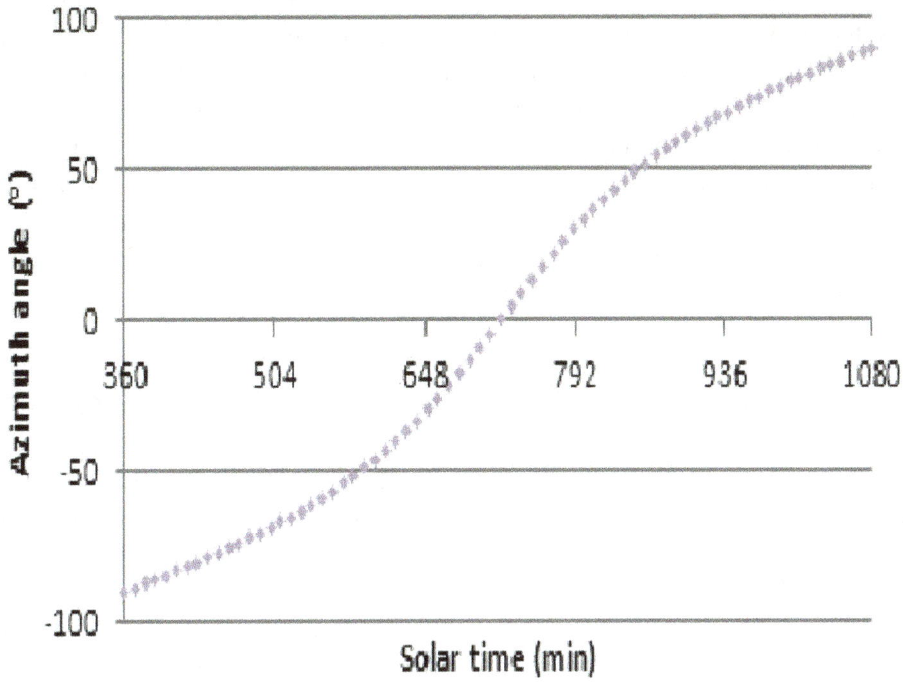

Figure 42. Daily azimuth angle variation at winter solstice (▲), summer solstice (x) and equinox (♦) at Damascus (33.5138° N)

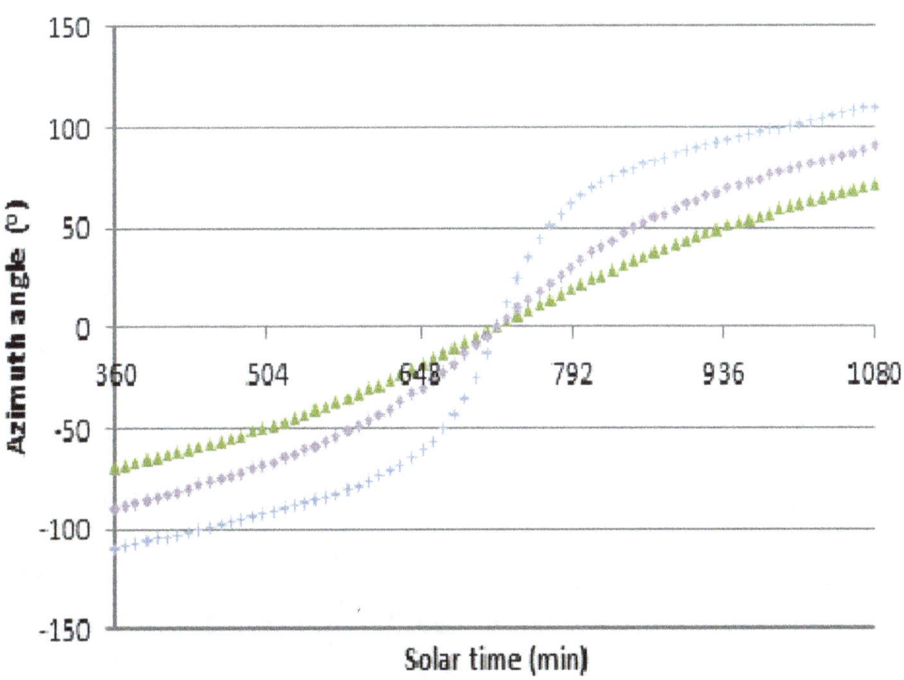

calculating the declination angle. On the other hand, the dependence of α_{noon} on latitude is linear with an inclination of -45°. The daily dependence of α_{noon} on latitude is shown in Figure 43.

Nowadays, the hour angle method is more popular because it is more accurate, can be performed at any time of day and is applicable to the sun, Polaris and other stars. In the past, the altitude method has been more popular primarily because of the difficulty of obtaining accurate time in the field. The development of time signals and accurate timepieces, particularly digital watches with split time features and time modules for calculators, has eliminated this obstacle to the extent that the hour angle is now preferred. The hour angle method is more accurate, faster, requires less training for proficiency, has fewer restrictions on time of day and geographic location, has more versatility (total-station instruments may be used), and is applicable to Polaris and other stars.

EDUCATIONAL DEVICE FOR MEASURING SOLAR ELEVATION ANGLE

The device basically consists of a mast mounted vertically on a flat circular disk perfectly horizontal, settled by mercurial bubble. Of course it is possible to manufacture this device in the form of a right angle. The vertical side should be greater than 80 cm in length and the other horizontal side should be longer than (160 cm) and thus must be wider in order to allow the mercurial bubble to be installed. The logical dimensions for the training of students to measure the solar angles and identify them are 100 cm for the mast and 100 cm of radius for the circular disk. The schematic view of the device is given in Figure 44.

Figure 43. Daily solar noon elevation angle variation for latitudes 25° (■), 35° (▲), 45° (x), 55° (ж) and 65° (●)

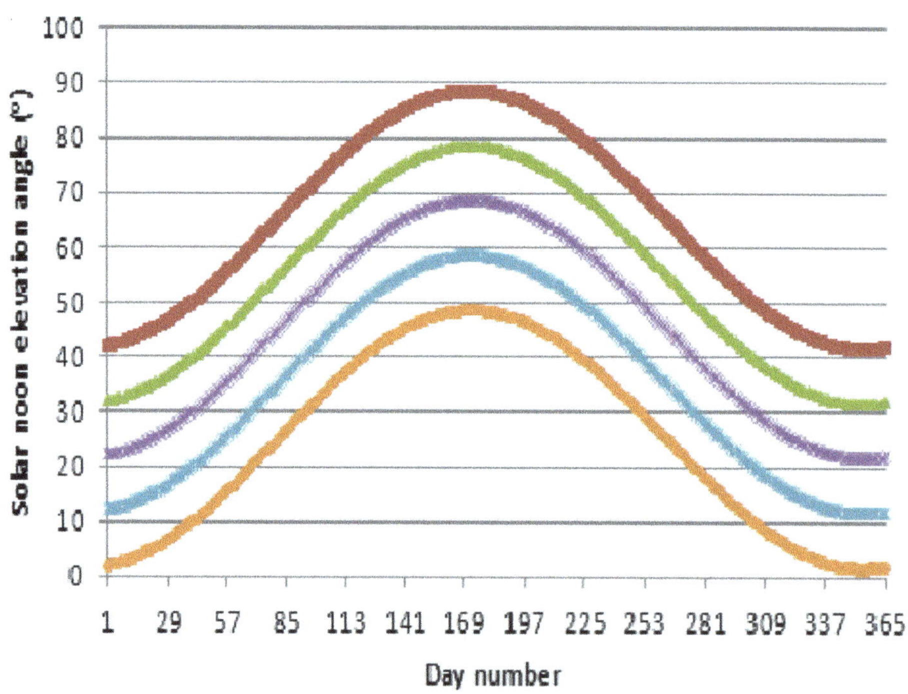

Figure 44. Solar angles recorder scheme

The upper face of the metallic circular disk is equipped with a mercurial bubble tool to guarantee the surface horizontality and it contains concentric circles and radii and numbered markers for different experiments regarding solar angles (see Figure 45). The disk is mounted on a base with screws for achieving the horizontality of the disk surface (see Figure 45). The device is also equipped with a compass to determine the magnetic north-south direction.

For measuring the solar azimuth angle the following steps should be considered:

Figure 45. A photo of the upper view of solar angles recorder

1. Make sure that the disk surface is horizontal using the mercurial bubble and screws on the disk base.
2. Make sure that the projection of the radius, passing near the number 12 engraved on the disk surface, passes through the mark on the ground which indicates the north direction.
3. Install the mast at the base of the three legs, taking advantage of the screw on the installed base of the mast.
4. Select the day number of your experiment in the calendar.
5. Observe the mast's shadow and select the point of its intersection with the external circuit or any inner circle on the disc points. This allows you to measure the mast's shadow length.
6. Select the moment of measuring the shadow's length and read the value of the solar hour angle on the intersected circle.
7. Calculate the declination angle using equation (6) and note the angle of latitude site (for Damascus, $\varphi = 33.5138°$ N).
8. Apply equations (21) and (22) for calculating the solar azimuth angle.
9. The daily variation of solar azimuth angle during any day in the calendar year could be obtained by repeating this procedure from sunrise to the sunset.

HOUR ANGLE MEASUREMENT

In conventional time keeping, regions of the Earth are divided into certain time zones. However, in these time zones, noon does not necessarily correspond to the time when the Sun is highest in the sky. Similarly, Sun rise is defined as the stage when the Sun rises in one part of the time zone. However, due to the distance covered in a single time zone, the time at which the Sun actually clears the horizon in one part of the time zone may be quite different to the "defined" Sun rise (or what is officially recognized as the time of Sun rise). Such conventions are necessary otherwise a house one block away from another would actually be different in time by several seconds. Solar time, on the other hand is unique to each particular longitude. For most solar design purposes, clock time is of little concern, and it is appropriate to present data in terms of solar time. Some situations, however, such as energy demand correlations, system performance correlations, determination of true south, and tracking algorithms require an accurate knowledge of the difference between solar time and the local clock time. On the other hand, the azimuth angle and the elevation angle are related to "solar time". Consequently, to calculate the Sun's position, first the local solar time is found and then the elevation and azimuth angles are calculated.

As mentioned before, equation (5) is used to converts the local solar time (LST) into the number of degrees (the hour angle ω) which the Sun moves across the sky. The local time (LT) usually varies from LST because of the eccentricity of the Earth's orbit and because of human adjustments such as time zones and daylight saving:

$$LT = LST - {T_c}/{60} \tag{58}$$

T_c being the time correction factor accounts for the variation of the LST within a given time zone due to the longitude variations within the time zone:

$$T_c = 4\left(Longitude - LSTM\right) + E \tag{59}$$

where *Longitude* is longitude of the site, LSTM is the local standard time meridian that is a reference meridian used for a particular time zone and is similar to the prime meridian, which is used for Greenwich mean time. The (LSTM) is calculated according to the equation:

$$LSTM = 15\Delta T_{GMT} \tag{60}$$

where ΔT_{GMT} is the difference of the local time (LT) from Greenwich mean time (GMT) in hours. The factor of 4 minutes (in equation (52)) comes from the fact that the Earth rotates 1° every 4 minutes and E (min.) is the equation of time (in minutes) which is an empirical equation that corrects for the eccentricity of the Earth's orbit and the Earth's axial tilt. The equation of time gives the difference between mean solar time and true solar time on a given date.

The level of accuracy required in determining the equation of time will depend on whether the designer is doing system performance or developing tracking equations. An approximation for calculating the equation of time in minutes is given by Woolf (1968) and claimed that is accurate to within about 30 seconds during daylight hours:

$$\begin{aligned} E = {} & 0.258\cos\left[\frac{2\pi\left(n-1\right)}{365.242}\right] - 7.416\sin\left[\frac{2\pi\left(n-1\right)}{365.242}\right] \\ & -3.648\cos\left[\frac{4\pi\left(n-1\right)}{365.242}\right] - 9.228\sin\left[\frac{4\pi\left(n-1\right)}{365.242}\right] \end{aligned} \tag{61}$$

or given by Spencer (1971):

$$E = 229.2\left\{\begin{aligned} & 0.000075 + 0.001868\cos\left[\frac{2\pi\left(n-1\right)}{365.242}\right] - 0.032077\sin\left[\frac{2\pi\left(n-1\right)}{365.242}\right] \\ & -0.014615\cos\left[\frac{4\pi\left(n-1\right)}{365.242}\right] - 0.04089\sin\left[\frac{4\pi\left(n-1\right)}{365.242}\right] \end{aligned}\right\} \tag{62}$$

The daily dependence of the equation of time is demonstrated on Figure 46 for these two methods. Since solar time is based on the Sun being due south at 12:00 noon on any specific day, the accumulated difference between mean solar time and true solar time can approach 17 minutes either ahead of or behind the mean, with an annual cycle.

To satisfy the control needs of concentrating collectors, a more accurate determination of the hour angle is often needed. An approximation of the equation of time claimed to have an average error of 0.63 seconds and a maximum absolute error of 2.0 seconds is proposed by Lamm (1981):

Figure 46. The daily dependence of the equation of time E in minutes using Woolf (1968) (blue) and Spencer (1971) (orange)

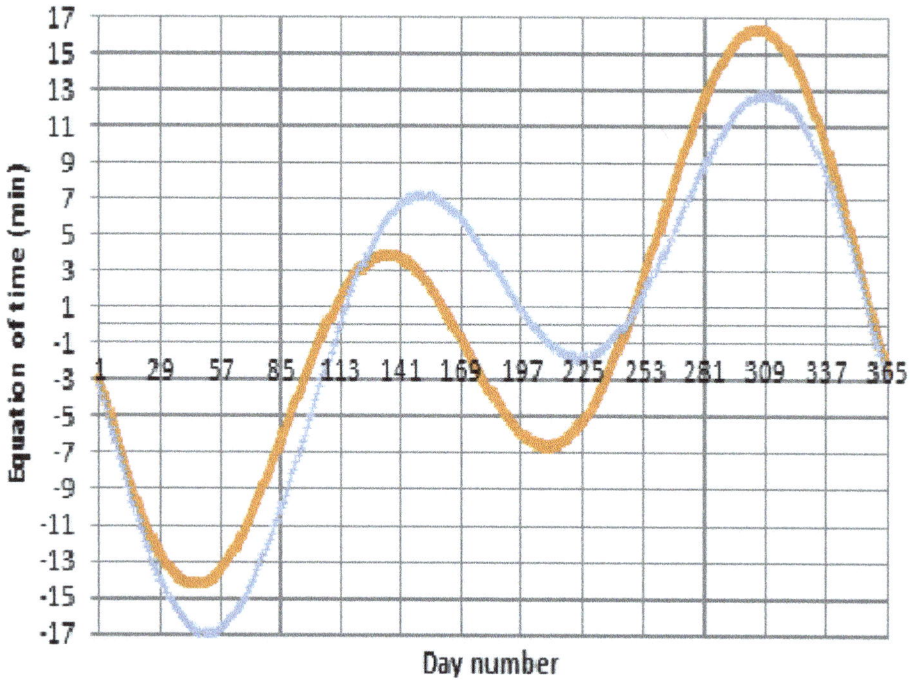

$$E = \sum_{k=0}^{k=5} \left[a_k \cos\left(\frac{2\pi k n_1}{365.25}\right) + b_k \sin\left(\frac{2\pi k n_1}{365.25}\right) \right] \tag{63}$$

Here n_l *is* the number of days into a leap year cycle with n_l = 1 being January 1 of each leap year, and n_l =1461 corresponding to December 31 of the 4th year of the leap year cycle. The coefficients A_k and B_k are given in the Table 7. Arguments for the cosine and sine functions are in degrees. The resulting value is in minutes and is positive when the apparent solar time is ahead of mean solar time and negative when the apparent solar time is behind the mean solar time.

The solar angels recorder shown in Figure 34 could be used for determining solar time and hour angle.

Table 7. Coefficients for Equation (63)

k	-10+4 x ak (hr)	-10+4 x bk (hr)
0	-2.0870	0
1	-92.869	222.9
2	522.58	1569.8
3	13.077	51.602
4	21.867	29.823
5	1.5100	2.3463

SUNSHINE DURATION RECORDER

In 2003, WMO defined sunshine duration as the period during which direct solar irradiance exceeds a threshold value of 120 watts per square meter (W/m^2). This definition, $I \geq 120\ Wm^{-2}$ where I being the direct solar radiation, was confirmed (WMO, 2008). This value is equivalent to the level of solar irradiance shortly after sunrise or shortly before sunset in cloud-free conditions. It was determined by comparing the sunshine duration recorded using a Campbell-Stokes sunshine recorder with the actual direct solar irradiance. So, the sunshine duration in a given period is as the sum of the time intervals for which the direct solar radiation exceeds the threshold of 120 Wm^{-2}. The instrument that is used for measuring the duration, in hours, of bright sunshine during the course of the day is called the sunshine recorder. In practice, four types of instruments are widely used for measuring sunshine duration: the Campbell-Stokes sunshine recorders, the Jordan sunshine recorders, the photoelectric sunshine recorders and the pyranometric method.

Campbell-Stokes Sunshine Recorder

The Campbell-Stokes sunshine recorder is known also as the burning card method. This recorder essentially consists of a glass sphere mounted concentrically in a section of a spherical brass bowl of approximately 10cm diameter with metallic grooves for holding the recorder cards (see Figure 47). The support is adjustable so that the axis of the sphere may be inclined to the angle of the local latitude. Outside the tropical zone the whole apparatus is set up to point south in the Northern Hemisphere and north in the Southern Hemisphere. The glass sphere is a lens that produces the Sun image on the card. So, Sun's rays are refracted and focused sharply on the record card beneath the glass sphere, leaving burnt marks on the card. During a day, a trace is burned whenever the Sun is shining. The length of the trace is the direct measure of the duration of bright sunshine. There are sets of metallic grooves for taking three sets of cards, straight cards for equinox, long curved for summer and short curved for winter. The errors of this recorder are mainly due to the dependence of burning initiation on card's temperature

Figure 47. The Campbell-Stokes sunshine recorder scheme: 1) Solid glass sphere, 2) Recorder cards holder, 3) The adjustable support, 4) The horizontal leveling base

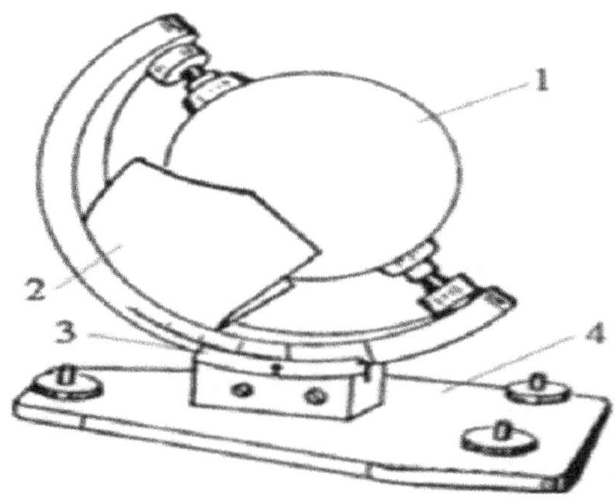

and humidity as well as to the over-burning effect, especially in case of broken clouds (Kerr & Tabony 2004; Ikeda, Aoshima & Miyake, 1986).

To obtain uniform results for observation of sunshine duration with a Campbell-Stokes sunshine recorder, the following points should be noted when reading records:

1. If the burn trace is distinct and rounded at the ends, subtract half of the curvature radius of the trace's ends from the trace length at both ends. Usually, this is equivalent to subtracting 0.1 hours from the length of each burn trace.
2. If the burn trace has a circular form, take the radius as its length. If there are multiple circular burns, count two or three as a sunshine duration of 0.1 hours, and four, five or six as 0.2 hours. Count sunshine duration this way in increments of 0.1 hours.
3. If the burn trace is narrow, or if the recording card is only slightly discolored, measure its entire length.
4. If a distinct burn trace diminishes in width by a third or more, subtract 0.1 hours from the entire length for each place of diminishing width. However, the subtraction should not exceed half the total length of the burn trace.

The advantages of Campbell-Stokes sunshine recorder are it has not moving parts and it doesn't require electric power while the disadvantages of this instrument are that the characteristics of the recording paper or photosensitized paper used in them affect measurement accuracy, differences between observers may arise in determining the occurrence of sunshine, and the recording paper must be replaced after sunset. The accepted version of Campbell-Stokes sunshine recorder by WMO is the IRSR grade with the technical specifications presented in the Table 8.

Jordan Sunshine Recorders

A Jordan sunshine recorder (see Jordan and Gasteb, 1886) lets in sunlight through a small hole in a cylinder or a semi-cylinder onto photosensitized paper set inside the cylinder on which traces are recorded. The instrument has its cylinder inclined to the relevant latitude and its ax set in the meridian direction.

Table 8. Campbell-Stokes recorder technical specifications

Glass Sphere	Spherical Segment	Record Cards
Shape: uniform	Material: gunmetal or equivalent durability	Material: good quality pasteboard not affected appreciably by moisture
Diameter: 10 cm	Radius: 73 mm	Width: Accurate to within 0.3 mm
Colour: Very pale or colourless	Additional specifications:	thickness: 0.4 ± 0.05 mm
Refractive index: 1.52± 0.02	- Central noon line engraved transversely	Moisture effect: Within 2 per cent
Focal length: 75 mm for sodium"D" light	- Adjustment for inclination of segment to horizontal	colour: Dark, homogeneous, no difference detected in diffuse daylight
	- Double base with provision for leveling and azimuth setting	graduations: Hour-lines in black

Photosensitized paper with a time scale printed on it is set in the cylinder in close contact with the inner surface. When direct solar radiation enters through the hole, the paper records the movement of the sun as a line. Sunshine duration is ascertained by measuring the length of time the paper was exposed to sunlight. The advantages and disadvantages of this recorder type are similar to those of Campbell-Stokes sunshine recorder type.

Photoelectric Sunshine Recorder

The photoelectric sunshine recorder incorporates several photovoltaic cells. The widely used one incorporates two identical selenium photovoltaic cells, one of which is shaded while the other is exposed to direct solar radiation. The duration of a critical radiation difference detected by the two cells is a measure of the bright sunshine duration.

Pyranometric Method

The pyranometric method implies measurement of global G and diffuse D solar irradiance; by subtraction, the direct beam solar irradiance is next derived, to be compared with the WMO threshold. The errors in the pyranometric method stem from the errors of measuring global and diffuse solar irradiance, which are amplified at higher zenith angles. Choosing a high quality pyranometer is of primary importance to reduce the results uncertainty level. The usage of shading rings has as a consequence the undervaluation of the incident diffuse solar energy. Corrections are required to diminish this negative effect. Last but not least, the sampling frequency is important. At a higher measurement rate the sunshine duration can be evaluated more precise. At least one sample per minute is required to properly capture the fast changes of the solar radiative regime.

ANGLES FOR TRACKING SURFACES

To date, a large number of solar installations, where the single-axis or dual-axis tracking is employed, is available in the world. The mode of solar tracking depends on the nature of solar application under consideration. Here, one can mention: horizontal north-south single-axis tracking, east-west single-axis tracking, polar axis tracking and dual-axis tracking. Some of these configurations are preferable to others depending on the considerations of cost and availability of resources. From energetic point of view, the effectiveness of each type of tracking was evaluated (Gordon and Wenger, 1991; Gordon, Kreider and Reeves, 1991). It was mentioned that horizontal north-south tracking can deliver about 90% of the yearly energy of two-axis tracking. This value decreases considerably in the case of east-west tracking at the same nominal and operating cost. Therefore, it should not be considered as it is not competitive with stationary collectors faced toward south and tilted by the angle of latitude to the horizontal plane as the later are the least expensive deployment option and they achieve about 75% of the yearly energy delivery of two-axes tracking. Polar axis tracking can achieve around 97% of the yearly collectible energy of two-axis trackers. This system offers the advantage of requiring one degree of freedom in the tracking mechanism. However, the tracking system represents a large part of the overall cost (Carroll, Schmidt and Bailor, 1990).

Nowadays, it is well known that solar tracking may be achieved using many different axis configurations. In addition to the above mentioned, all desired orientations may be obtained by means of any two perpendicular axis. This can be done also by means of suitable control electronics using a large variety of rotational axes or linear motion (Maish, 1990). Therefore, it seems reasonable to search a way for improving tracking structures.

The position of any celestial object can be determined using different systems of coordinates. The most usable ones are (Soulayman, 2009):

- The horizontal system of coordinates in which the reference point coincides with observer treated as material point.
- The equatorial system of coordinates where the center of the earth plays the role of the reference point.

So, any of these systems may be applied to determine the position of the sun during its apparent motion through the celestial dome. Moreover, the interaction between earth and sun is Newtonian. This leads to a planar motion. Therefore, two perpendicular axes are needed to determine the plane of motion. These two axes are generally preferred to be the azimuthal and elevation axes in the case of horizontal system. The elevation angle varies from 0° to 90° and the azimuth angle varies from 0° to 360°. This tracking can be realized using a schematic design shown in Figure 48 where the receiver turns around two axes: one of them is vertically fixed while the other one turns horizontally around the first axis. On the other hand, the reference axes in the equatorial system of coordinates are usually chosen to be parallel to the earth's orbit and spin axes (Davies, 1993). The earth spins around the axis of the equator. So, the first axis is parallel to the polar axis. This axis is called the equatorial axis of tracker. This axis should be tilted at earth's surface by an angle equal to the latitude of the tracker's site. The second axis is situated in the same plane and perpendicular to the orbital plane. So, the sun position can be determined using solar

Figure 48. The proposed scheme for implementation of solar tracking based on horizontal system of coordinates

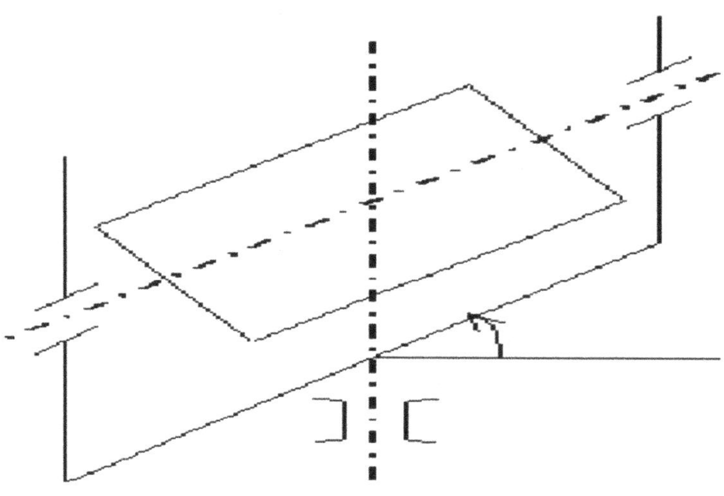

hour angle, which varies from 0 to 90 degrees, and solar declination angle, which varies from –23.45° to 23.45°. Tracking can be realized using a schematic design shown in Figure 49 where the polar axis is tilted by an angle of latitude to horizon and forms with the receiver an angle equals to solar declination. The receiver turns around the polar axis by an angular velocity equal 1 revolution per sidereal day.

When comparing values related to the mentioned systems of coordinates one finds the following:

- In contrast to horizontal system, tracking using equatorial system can be approximated to a tracking with one degree of freedom because daily change of declination angle is negligible.
- Daily variations of values related to equatorial system are smaller than those related to horizontal system (see Table 9 for more details).

Figure 49. The proposed scheme for implementation of solar tracking based on equatorial system of coordinates

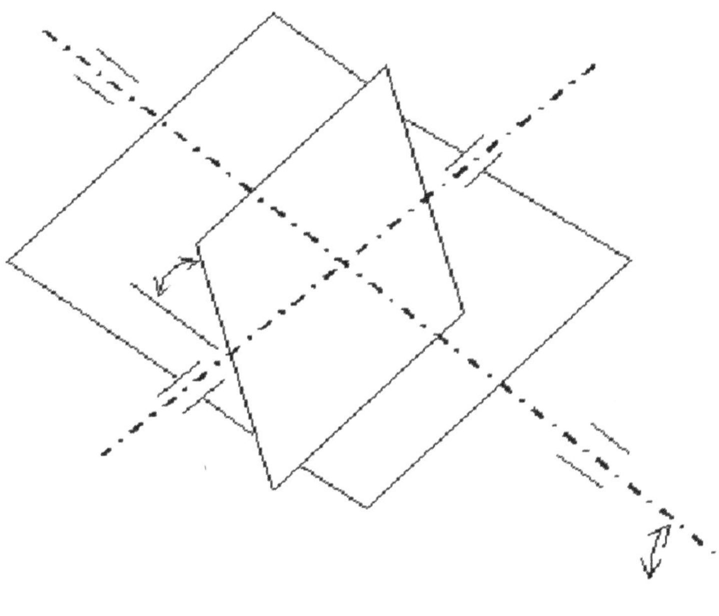

Table 9. Daylight changes of angles used in the equatorial and horizontal systems in Damascus (33.5138° N)

	Declination Angle δ (°)	Solar Elevation Angle α (°)	Solar Azimuth Angle γ_s(°)	Solar Hour Angle ω (°)
Equinox	0.41	56.5	180	180
Summer solstice	0.01	79.9	235.8	213.5
Winter solstice	0.01	33.0	122.9	146.6

CONCLUSION

In this chapter information for solar position determination, daylight length and the main important angles is introduced. Some procedures for measuring site latitude, solar elevation angle, solar zenith angle, hour angle and solar azimuth angle are presented. Some devices used in measuring sunshine duration are also described. The main systems of coordinates used in solar tracking are introduced. The provided information in this chapter will be essential background for other chapters.

REFERENCES

Ali, A. T. (2012). An error modeling framework for the Sun azimuth obtained at a location with the hour angle method. *Positioning*, *3*(02), 21–29. doi:10.4236/pos.2012.32004

Blanco-Muriel, M., Alarcón-Padilla, D. C., López-Moratalla, T., & Lara-Coira, M. Í.Manuel Blanco-Muriel. (2001). Computing the solar vector. *Solar Energy*, *70*(5), 431–441. doi:10.1016/S0038-092X(00)00156-0

Braun, J. E., & Michell, J. C. (1983). Solar geometry for fixed and tracking surfaces. *Solar Energy*, *31*(5), 439–444. doi:10.1016/0038-092X(83)90046-4

Buckner, R. (1984). *Astronomic and Grid Azimuth*. Landmark Enterprises, Rancho Cordova.

Budin, R., & Budin, L. (1982). A mathematical model for shading calculations. *Solar Energy*, *29*(4), 339–349. doi:10.1016/0038-092X(82)90249-3

Carroll, D., Schmidt, E., & Bailor, B. (1990). Production of the Alpha Solarco proof-of-concept array. *Proceedings of 21st IEEEPV Solar Energy Conference*, 2, 1136-1141. doi:10.1109/PVSC.1990.111793

Chang, T. P. (2009). The Suns apparent position and the optimal tilt angle of a solar collector in the northern hemisphere. *Solar Energy*, *83*(8), 1274–1284. doi:10.1016/j.solener.2009.02.009

Cooper, P. L. (1969). The absorption of radiation in solar stills. *Solar Energy*, *12*(3), 333–346. doi:10.1016/0038-092X(69)90047-4

Davies, P. A. (1993). Sun-tracking mechanism using equatorial and ecliptic axes. *Solar Energy*, *50*(6), 487–489. doi:10.1016/0038-092X(93)90110-A

Duffie, J. A., & Beckman, W. A. (2013). *Solar engineering of thermal processes* (3rd ed.). New York, NY: Wiley & Sons; doi:10.1002/9781118671603

Gordon, J. M., Kreider, J. F., & Reeves, P. (1991). Tracking and stationary flat plate solar collectors: Yearly collectible energy correlations for photovoltaic applications. *Solar Energy*, *47*(4), 245–252. doi:10.1016/0038-092X(91)90116-E

Gordon, J. M., & Wenger, H. J. (1991). Central-station solar photovoltaic systems: Field layout, tracker, and array geometry sensitivity studies. *Solar Energy*, *46*(4), 211–217. doi:10.1016/0038-092X(91)90065-5

Grena, R. (2008). An algorithm for the computation of the solar position. *Solar Energy*, *82*(5), 462–470. doi:10.1016/j.solener.2007.10.001

Grena, R. (2012). Five new algorithms for the computation of Sun position from 2010 to 2110. *Solar Energy*, *86*(5), 1323–1337. doi:10.1016/j.solener.2012.01.024

Ikeda, K., Aoshima, T., & Miyake, Y. (1986). Development of a new sunshine-duration meter. *Journal of the Meteorological Society of Japan*, *64*(6), 987–993. doi:10.2151/jmsj1965.64.6_987

Jesperson, J., & Fitz-Randolph, J. (1999). From Sundials to Atomic Clocks- Understanding time and frequency. US. Department of Commerce, National Institute of Standards and Technology, Monograph 155.

Jordan, J. E., & Gasteb, F. (1886). Notes as to the principle and working of Jordlys improved photographic sunshine recorder. *Quarterly Journal of the Royal Meteorological Society*, *12*(57), 21–25. doi:10.1002/qj.4970125703

Kerr, A., & Tabony, R. (2004). Comparison of sunshine recorded by Campbell–Stokes and automatic sensors. *Weather*, *59*(4), 90–95. doi:10.1256/wea.99.03

Khavrus, V., & Shelevytsky, I. (2010). Introduction to solar motion geometry on the basis of a simple model. *Phys. Educ.*, *45*(6), 641–53. Retrieved from www.iop.org/journals/physed

Khavrus, V., & Shelevytsky, I. (2012). Geometry and the physics of seasons. *Physics Education*, *47*(6), 681–692. doi:10.1088/0031-9120/47/6/680

Kittler, R., & Darula, S. (2013). Determination of time and Sun position system. *Solar Energy*, *93*, 72–79. doi:10.1016/j.solener.2013.03.021

Lamm, L. O. (1981). A new analytic expression for the equation of time. *Solar Energy*, *26*(5), 465. doi:10.1016/0038-092X(81)90229-2

Maish, A. B. (1990). Performance of a self-aligning photovoltaic array tracking controller. *Proceedings of 21st IEEEPV Solar Energy Conference*.

Meeus, J. (1998). *Astronomical Algorithms*. Richmond, VA: Willmann-Bell Inc.

Parkin, R. E. (2010). Solar angles revisited using a general vector approach. *Solar Energy*, *84*(6), 912–916. doi:10.1016/j.solener.2010.02.005

Reda, I., & Andreas, R. (2004). Solar position algorithm for solar radiation applications. *Solar Energy*, *76*(5), 577–589. doi:10.1016/j.solener.2003.12.003

Smith, F., & Wilson, C. B. (1976). The shading of ground by buildings. *Building and Environment*, *11*(3), 187–195. doi:10.1016/0360-1323(76)90005-6

Soulayman, S. (2009). An educational solar tracking system. In Proceedings of Solar Energy (IASTED, SOE 2009). Phuket, Thailand: Acta Press.

Soulayman, S. (2015). Comments on Optimum tilt angle for flat plate collectors all over the World – A declination dependence formula and comparisons of three solar radiation models by Stanciu, C., Stanciu, D. [Letter to the editor]. *Energy Conversion and Management*, *93*, 448–449. doi:10.1016/j.enconman.2015.01.005

Soulayman, S., & Sabbagh, W. (2014). An algorithm for determining optimum tilt angle and orientation for solar collectors. *Journal of Solar Energy Research Updates, 1*(1), 19–30. doi:10.15377/2410-2199.2014.01.01.3

Soulayman, S., & Shamiyah, S. (2000). Solar declination, geographic latitude and orientations measurements in Syria. *Proceedings of ENERGEX 2000*. Retrieved from http://trove.nla.gov.au/version/43244233

Spencer, J. W. (1971). Fourier series representation of the position of the Sun. *Search, 2*(5), 172.

WMO. (2008). *Guide to meteorological instruments and methods of observation*. World Meteorological Organization.

Wolf, P., & Ghilani, C. (2002). *Elementary surveying: An introduction to geomatics* (10th ed.). Upper Saddle River, NJ: Prentice Hall.

Woolf, H. M. (1968). *On the Computation of Solar Evaluation Angles and the Determination of Sunrise and Sunset Times*. National Aeronautics and Space Administration Report NASA TM-X-164, September.

Chapter 2
Geographic Orientations

ABSTRACT

The determination of the true geographic north is essential in many applications. There are different methods for doing that with different levels of complexity. In this chapter, the basic theoretical considerations, for determining the true north basing on shadow treatment, are described. The principles of using fence shadow for determining the true north all over the day are described. Basing on the above information, an instrument (magnetic declination device) is described in detail. This instrument could be used for determining the geographic north of the site where a solar system will be installed. The information provided in chapter 1 was used in this chapter for studying the shadow of different obstacles on the solar systems.

INTRODUCTION

The alignment of an inertial navigation system in a ground based device is one of the attractive problems. Clearly, the scope for applying motion to aid the process of alignment is very limited in such applications. Attention is focused now on the requirement of determining the orientation of a set of sensor axes with respect to the local geographic frame. For convenience, the local geographic axis set is often chosen to be the reference frame.

On the other hand, the site survey could be carried out to establish the geographic north line. Heading information would then be transferred to the aligning navigation system using theodolites and a prism attached to the aligning system. Although high accuracy can be obtained using this approach, it is both time consuming and labour intensive. The methods discussed in the following sections are usually more convenient to implement and avoid such problems.

One of the experimental works relevant to the problem of determining geographic directions is that of Bain (1961) which is based on the previously published work of the same author Bain (1956). In this work, an instrument capable of improved performance in conditions of multimode propagation Bain (1961) is used. The measurements were provided during the winter of 1954- 55, where a number of observations were carried out at the Radio Research Station in which phase differences were measured between the signals picked up at aerials spaced by several wave-lengths. The four aerials used in this

DOI: 10.4018/978-1-5225-2950-7.ch002

experiment were short unipolar; they were placed on the circumference of a circle of diameter 180 m so that one pair of aerials was aligned in the north-south direction and another in the east-west direction. Inside a hut at the center of the circle each of these pairs of aerials was connected to phase-measuring equipment of the type described by Ross, Bramley and Ashwell (1951). When the receivers were tuned to the desired station, the overall phase balance of the system was checked, and the phase differences between the signals picked up on the north-south and east-west pairs were recorded photographically at three-second intervals for a period not exceeding five minutes, thus giving about 70 values for each. The conditions of ionosphere were characteristic of the time in question and were never greatly disturbed. Observations were confined to the hours of daylight.

On the other hand, the National Oceanic and Atmospheric Administration (NOAA) maintains a Web site that gives up-to-date magnetic declination values for any site on the Earth's surface basing on two models: World Magnetic Model (WMM) and International Geomagnetic Reference Field (IGRF). For example, in Damascus, Syria, the angle of magnetic declination calculated using WMM model is 4° 46' E ± 0° 18' with a yearly change by 0° 5' E. The angle of magnetic declination calculated using IGRF model is 4° 45' E with a yearly change by 0° 5' E per year.

Accounting for declination is fairly straightforward using a compass. If the magnetic declination is towards east, then the declination should be subtracted from the magnetic north (0°; 360°) and magnetic south (180°) readings to get the true directional readings. Using the angle of Damascus magnetic declination of 4° 45' E, true north and south align with 355.25° (360° – 4.75°) and 175.25° (180° – 4.75°). To correct for a west declination, the declination should be added to (0°; 360°) and 180° to get the true directions.

However, to maximize the collection of the daily and seasonal solar energy possible, solar collectors (PV modules, solar absorbers, etc...) should be oriented geographically. In the Northern Hemisphere the optimum orientation for a solar receiver is the true south while the true north is the optimum orientation for a solar receiver in the Southern Hemisphere. True south or true north are where the sun will be at its highest during the day. Unfortunately simply using a compass needle aligned north-south isn't good enough. Therefore, before using compass to site an array of solar receivers, the declination error must be corrected for the considered site. In the Northern Hemisphere, a compass needle aligns itself along the magnetic north-south line. The solar receivers should be oriented to geographic south, so an account for magnetic declination—the angular difference between true and magnetic north— should be considered. The main cause for this discrepancy is the Earth's non-uniform, conductive, fluid outer core that consists mainly of iron and nickel. This layer pushes the compass needle off of true north. Depending upon the site's location on the planet, the "push" varies in strength and direction (Chisholm, 2007).

GEOGRAPHIC TRENDS FINDER

In addition to finding the direction of north you will need to know the latitude of the town or city where you intend to use your device in order to build it accurately. If you do not know the latitude of a particular part of the world you can figure it out by looking on a globe or a map that has the latitude lines marked. After that the geographic north could be found using different arrangements. One of the most popular arrangements is that based on shadow measurement. Within these arrangements it is useful to give some details on two different examples: an educational elementary way and more sophisticated way.

Theoretical Considerations on Determination of Geographic Orientations

Shadow dimensions are considered an important parameter in solar engineering applications, both in designing some devices that rely shadow as theoretical principle in its design, such as Sun tendency measuring device or latitude measuring device or geographical trends identifying device or studying the effect of shadow of trees or different physical buildings on the performance of solar collectors of all kinds (flat-plate and concentrated) and panels of photovoltaic cells or on the performance of the Sun tracking devices for measuring the intensity of solar radiation. The general procedure starts by studying the shadow of mast of length L_p installed on a tilted surface. The mast makes an angle ζ with the normal on the tilted surface. The mast's projection on the horizontal plane makes an angle (ψ) with the south direction. ψ is known as the mast's azimuth angle. The directions south – north and east-west are laid on the horizontal plane while the axis Z is directed to the Polaris.

Let us suppose that, the board, which hold the mast, makes an angle β with the horizon, and the projection of the board's side face makes an azimuth angle γ with the south direction as shown on Figure 1. To calculate the length of the mast's shadow we have to follow these steps:

1. Calculate the incident angle of solar rays on the horizontal and tilted surfaces.
2. Calculate ζ.
3. Calculate the azimuth angle of the mast's projection on the horizon and the azimuth angle of the projection of the tilted surface on the horizon.
4. Calculate the solar azimuth angle on the tilted surface.
5. Calculate the coordinates of the mast's shadow.

Figure 1. Mast installed on the tilted surface

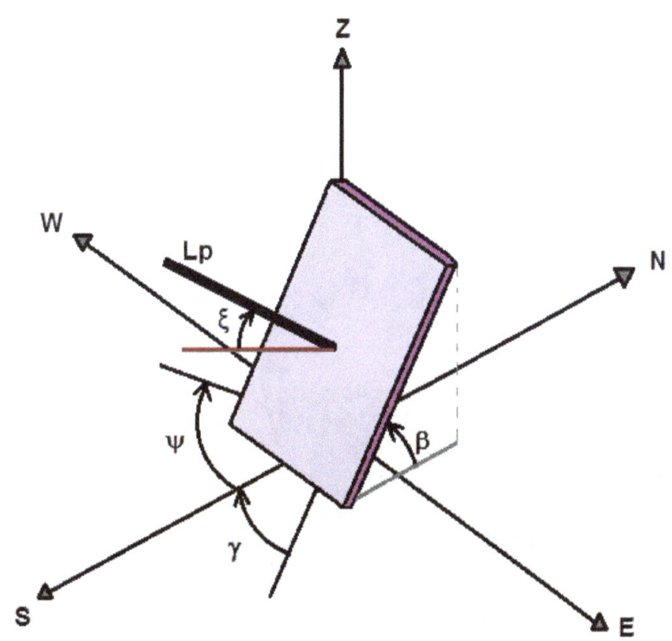

Let's stop on each of these steps separately.

The Incident Angle of Solar Rays on the Horizontal and Tilted Surfaces

Using vector algebra, definitions and the system of coordinates used in chapter 1 of this book for determining Sun's position; the important relations between different solar angles with incident solar rays angles could be obtained. Referring to the Figure 2 where the point C denotes the center of the Earth, it is possible to determine the position of any point P on the Earth's surface using its position vector CP. The position of the Sun could also be determined using the solar elevation angle α, the declination angle δ, the hour angle ω and the solar azimuth γ_s angle. After choosing the coordinate system in the studied case, the position vector of the Sun I_n and the normal vector on the tilted surface $V_{(CB)}$ which corresponds CP in the case of horizontal surface, the calculation of mentioned angle become very simple. The determination of Sun's position vector is useful for calculating the Sun's ray incidence angle on the tilted surface. For this purpose let's suppose that the plane XY corresponds to the equatorial plane of the earth. So axis Z coincides with the Polaris position vector. In this case the vector of Sun position is located in the XZ plane. Therefore the incident solar rays on the horizontal surface are also located in the XZ plane. The projection of position vector of the Sun on the XY-plane is CX as shown on Figure 3.

It is clear from Figure 3 that the point A represents the projection of point P on the equatorial plane of the earth XY, φ typifies the latitude of the point P on the Earth's surface and the angle XCA typifies the hour angle ω in the equatorial plane of the earth XY. The victor V typifies the normal line on the horizontal surface at the point P (position vector of point P). PN line indicates north direction and therefore PN located in the horizontal plane and perpendicular to CP. Let (a_1, b_1, c_1) the unit vector components of vector CP in XYZ coordinate system and (a_2, b_2, c_2) the unit vector components of vector I_n in the same system of coordinates, then:

Figure 2. Sun and site's position vectors in the equatorial system of coordinates

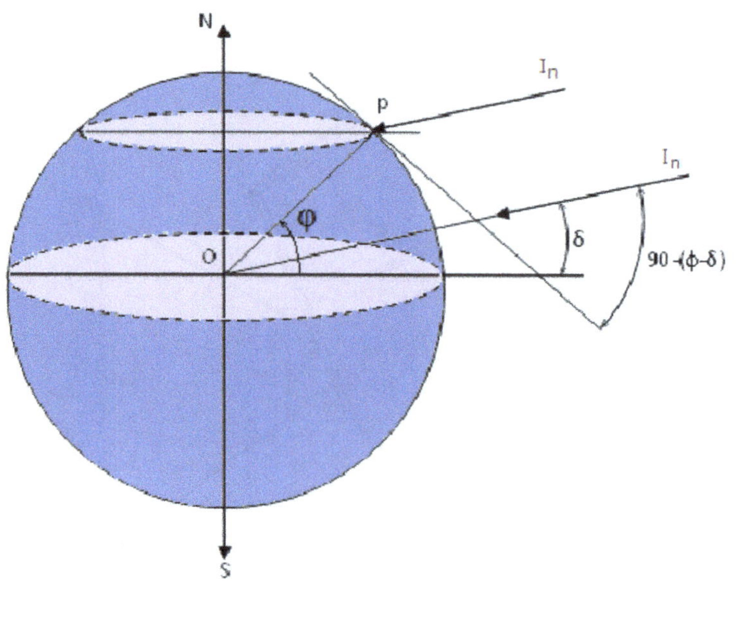

Figure 3. Equatorial and Earth's surface systems of coordinates

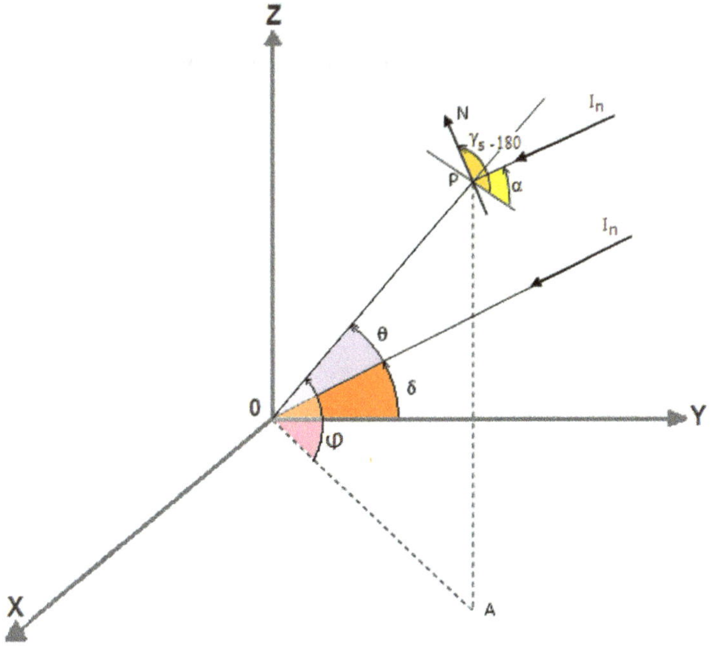

$$a_1 = \cos(\varphi)\cos(\omega) \tag{1}$$

$$b_1 = \cos(\varphi)\sin(\omega) \tag{2}$$

$$c_1 = \sin(\varphi) \tag{3}$$

$$a_2 = \cos(\delta) \tag{4}$$

$$b_2 = 0 \tag{5}$$

$$c_2 = \sin(\delta) \tag{6}$$

The scalar multiplication of vectors CP and I_n gives:

$$\cos\left(\theta_z\right) = a_1 a_2 + b_1 b_2 + c_1 c_2 \tag{7}$$

The substitution of (1) – (6) in (7) gives the equation (23) of chapter 1:

$$\cos\left(\theta_z\right) = \sin\left(\varphi\right)\sin\left(\delta\right) + \cos\left(\varphi\right)\cos\left(\delta\right)\cos\left(\omega\right)$$

So,

$$\theta_z = \arccos\left[\sin\left(\varphi\right)\sin\left(\delta\right) + \cos\left(\varphi\right)\cos\left(\delta\right)\cos\left(\omega\right)\right] \tag{8}$$

Here it should be noted that the solar elevation angle α is but the complementary angle with incidence solar angle θ on the horizontal surface ($\theta=\theta_z$).

Equation (8) could be used to calculate solar elevation angle at solar noon:

$$\alpha_{noon} = 90^o - \left|\varphi - \delta\right| \tag{9}$$

Equation (9) is very useful, especially in the tropical zone where the solar receiver could be orientated towards Earth's Poles, in determining the concept of long-term tracking.

The Incidence Angle on the Inclined Surface

The orientation of the inclined surface is determined by its azimuth angle, γ. Figure 4 shows the solar elevation and azimuth angles. Figure 5 shows the receiver tilt and azimuth angles while Figure 6 a shows the solar elevation, zenith and azimuth angles. Solar and tilted surface angles are shown in the Figure 7. With taking into consideration that γ can be change in the interval ($-360^o \le \gamma \le 360^o$), where negative values denote the orientation from south to the east and positive values denote the orientation from south to the west.

The tilt angle β can be change in the interval ($-90^o \le \beta \le 90^o$), where in the Northern Hemisphere, the negative values denote that the solar receiver is orientated towards the Northern Pole and the positive values denote that the solar receiver is orientated towards the Equator. In the Southern Hemisphere, the negative values denote that the solar receiver is orientated towards the Equator and the positive values denote that the solar receiver is orientated towards the Southern Pole.

Victor PV typifies the normal line on the inclined plane. Therefore the victor elevation angle will be the supplementary angle of angle (β), and its direction will be the plane direction itself. Therefore it typifies the components of the unit victors (l, m, n) of this victor in the coordinate system related to the horizontal plane on the Earth's surface as follows:

$$l = \cos\left(90^0 - \beta\right)\cos\left(\gamma\right) \tag{10}$$

Geographic Orientations

Figure 4. Solar elevation and azimuth angles

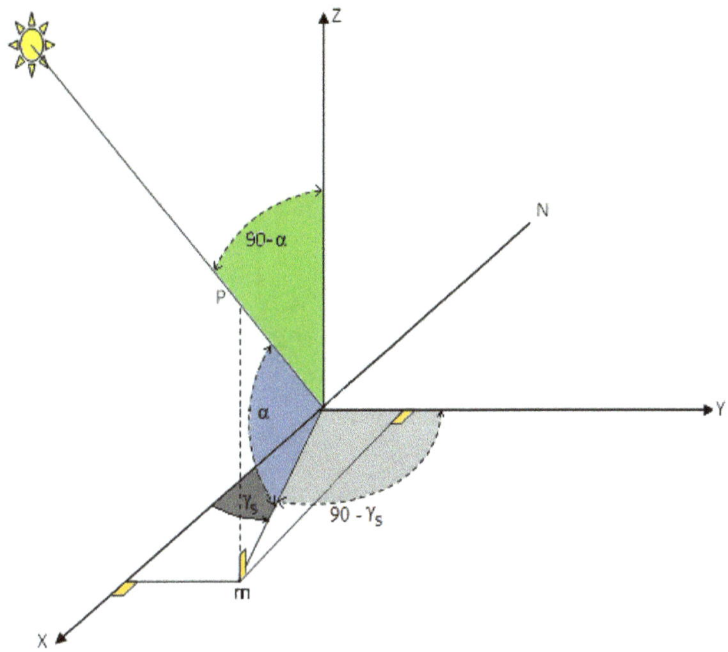

Figure 5. Receiver tilt and azimuth angles

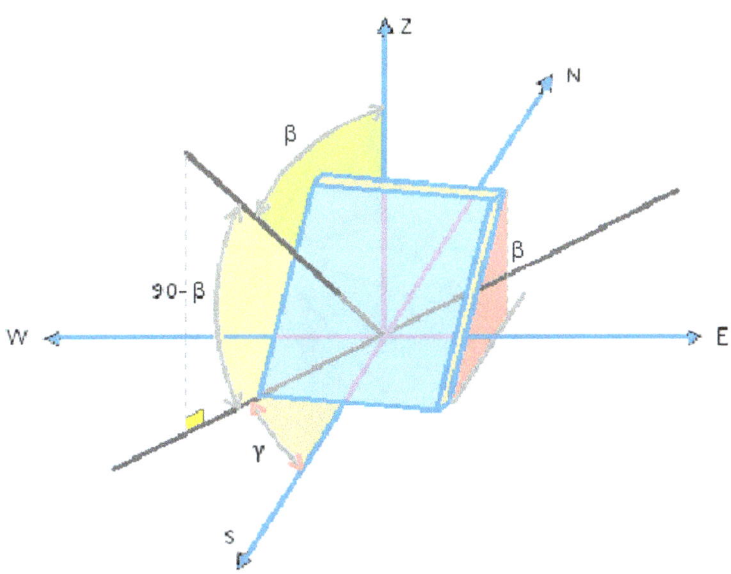

Figure 6. Solar angles in the ordinary geographic system of coordinates

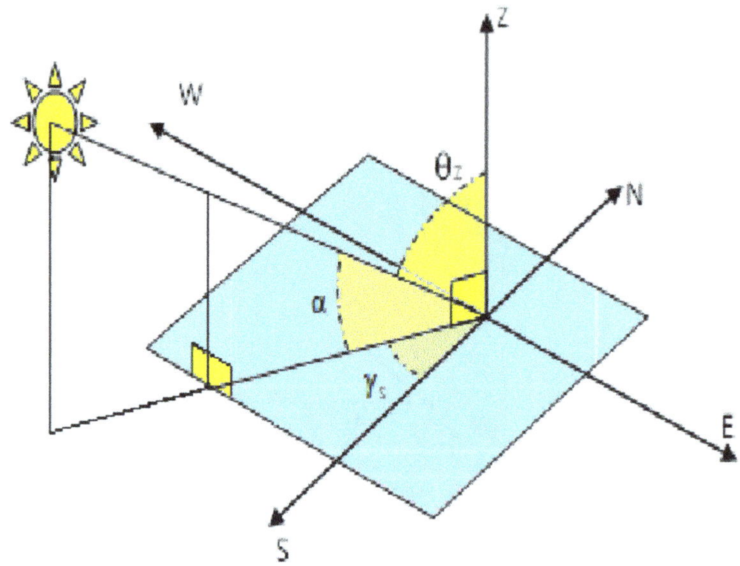

Figure 7. Solar and tilted surface angles

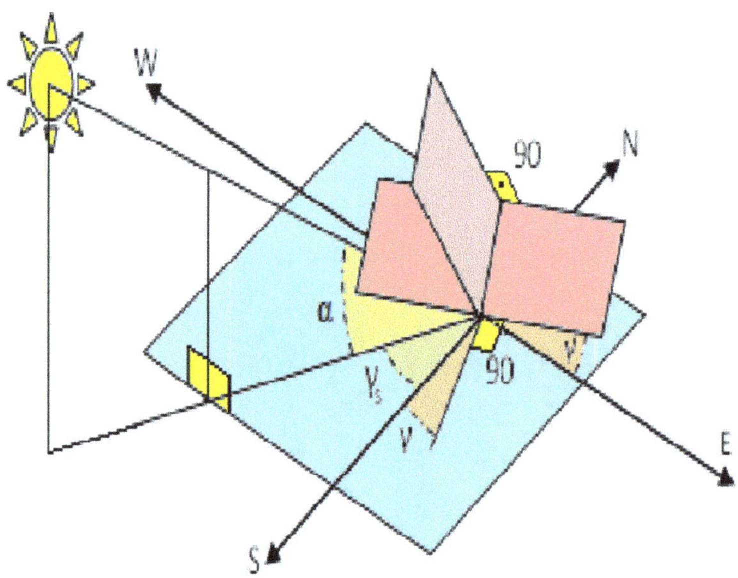

$$m = \cos\left(90^0 - \beta\right)\cos\left(90^0 + \gamma\right) = -\sin\left(\beta\right)\sin\left(\gamma\right) \tag{11}$$

$$n = \cos\left(\beta\right) \tag{12}$$

The components of unit victor of victor I_n are (l_1, m_1, n_1) which define the direction of direct solar ray are:

$$l_1 = \cos\left(\alpha\right)\cos\left(\gamma_s\right) \tag{13}$$

$$m_1 = \cos\left(\alpha\right)\cos\left(90^o - \gamma_s\right) = \cos\left(\alpha\right)\sin\left(\gamma_s\right) \tag{14}$$

$$n_1 = \cos\left(90^o - \alpha\right) = \sin\left(\alpha\right) \tag{15}$$

Thus solar incidence angle θ_i on the tilted surface is determined by the scalar multiplication of the unit vector of the normal line on the inclined plane (l, m, n) and the unit vector of the Sun's position vector (l_1, m_1, n_1):

$$\cos\left(\theta_i\right) = l.l_1 + m.m_1 + n.n_1 \tag{16}$$

The substitution of the values of l, m, n, l_1, m_1, n_1 according to equations (10)-(15) in equation (16) and rearranging the different terms of the obtained equation we get:

$$\cos\left(\theta_i\right) = \sin\left(\alpha\right)\cos\left(\beta\right) + \cos\left(\alpha\right)\sin\left(\beta\right)\cos\left(\gamma_s - \gamma\right) \tag{17}$$

The Angle Between the Mast and the Normal Line on Its Surface

The unit victor of the normal on the tilted surface is of the following components (see Figure 5):

$$p = \sin\left(\beta\right)\sin\left(\gamma\right) \tag{18}$$

$$q = \sin\left(\beta\right)\cos\left(\gamma\right) \tag{19}$$

$$r = \cos(\beta) \tag{20}$$

On the other hand, the unit victor of the victor that carries the mast has the following components (see Figure 1):

$$p_1 = \sin(\xi)\sin(\psi) \tag{21}$$

$$q_1 = \sin(\xi)\cos(\psi) \tag{22}$$

$$r_1 = \cos(\xi) \tag{23}$$

The scalar multiplication of the unit vectors of the mast victor and the victor of the normal on the inclined surface gives:

$$\cos(\zeta) = \sin(\xi)\sin(\psi)\sin(\beta)\sin(\gamma) + \sin(\xi)\cos(\psi)\sin(\beta)\cos(\gamma) + \cos(\psi)\cos(\beta) \tag{24}$$

So,

$$\begin{aligned}\zeta &= arc\cos\begin{bmatrix}\sin(\xi)\sin(\psi)\sin(\beta)\sin(\gamma) + \sin(\xi)\cos(\psi)\sin(\beta)\cos(\gamma) \\ + \cos(\psi)\cos(\beta)\end{bmatrix} \\ &= arc\cos\left\{\sin(\xi)\sin(\beta)\cos(\psi - \gamma) + \cos(\psi)\cos(\beta)\right\}\end{aligned} \tag{25}$$

The Azimuth Angle of Sun's Position γ_{st} on the Inclined Surface

This angle typifies the angle between the projection of Sun's position victor on the inclined surface and the projection of north-south direction on that surface. This angle could be given by the following equation:

$$\tan\left(\frac{\gamma_{st}}{2}\right) = \frac{\cos(\alpha)\sin(\gamma_s - \gamma)}{\sin(\theta_i) - \sin(\alpha)\sin(\beta) + \cos(\alpha)\cos(\beta)\cos(\gamma_s - \gamma)} \tag{26}$$

The Angle Between Mast and Normal Line Projections on the Horizontal Surface

This angle is determined by the following equation:

$$\tan\left(\frac{\alpha_1}{2}\right) = \frac{\sin(\xi)\sin(\psi-\gamma)}{\sin(\zeta)-\sin(\beta)\cos(\xi)+\cos(\beta)\sin(\xi)\cos(\psi-\gamma)} \tag{27}$$

Mast's Shadow's Coordinates on the Planar System of Coordinates

The mast's shadow coordinates on the planar system of coordinates are:

$$M = \sin(\zeta)\sin(\alpha) - \cos(\zeta)\sin(\gamma_{st})\tan(\theta_i) \tag{28}$$

$$N = \sin(\zeta)\cos(\alpha) - \cos(\zeta)\sin(\gamma_{st})\tan(\theta_i) \tag{29}$$

Thus, the length of the mast's shadow could be calculated using the following equation:

$$L = L_p\sqrt{M^2 + N^2} \tag{30}$$

This shadow makes an angle Ω with the surface inclination line:

$$\Omega = \arctan\left(\frac{M}{N}\right) \tag{31}$$

Finally, it is useful to shed a light on some cases of special importance.

Mast of Length L_p and Vertically Mounted on the Horizontal Surface

In this case, the length of the mast's shadow is:

$$L = {}^{L_p}\!\big/_{\tan(\alpha)} \tag{32}$$

and its direction is:

$$\Omega = \gamma_s + 180^o \tag{33}$$

Mast of Length L_p and Normally Mounted on a Tilted Surface With Angle β to the Horizontal Surface and Directed to the Equator

In this case, the length of the mast's shadow is:

$$L = L_p \tan\left(\theta_i\right) \tag{34}$$

and its direction is:

$$\Omega = \arctan\left[\frac{\sin\left(\zeta\right)\sin\left(\alpha\right) - \cos\left(\zeta\right)\sin\left(\gamma_{st}\right)\tan\left(\theta_i\right)}{\sin\left(\zeta\right)\cos\left(\alpha\right) - \cos\left(\zeta\right)\sin\left(\gamma_{st}\right)\tan\left(\theta_i\right)}\right] \tag{35}$$

Mast of Length L_p and Normally Mounted on a Tilted Surface With Angle β to the Horizontal Surface and has an Azimuth Angle γ

In this case, the length of the mast's shadow is:

$$L = L_p\sqrt{\sin^2\left(\beta\right) + \cos^2\left(\beta\right)\tan^2\left(\theta_i\right) + \sin\left(2\beta\right)\tan\left(\theta_i\right)} \tag{36}$$

and its direction is:

$$\Omega = \arctan\left[\frac{\cos\left(\beta\right)\sin\left(\gamma_{st}\right)\tan\left(\theta_i\right)}{\cos\left(\beta\right)\cos\left(\gamma_{st}\right)\tan\left(\theta_i\right) - \sin\left(\beta\right)}\right] \tag{37}$$

$$\gamma_{st} = \arctan\left[\frac{\cos\left(\alpha\right)\sin\left(\gamma_s - \gamma\right)}{\sin\left(\theta_i\right)}\right] \tag{38}$$

Shadow Width of an Inclined Wall

The general method for finding the width of inclined wall's shadow could be summarized as follows. Let's study a fence of H in height. The height of the fence is the normal distance from the base of the fence to its upper end as shown in the Figure 8. Let us assume that the fences base on the inclined surface makes with the sloped tilt line an angle χ on the inclined surface as shown on Figure 9. This means that χ is the fence base azimuth angle. Then the width of fence J, measured at its base, could be calculated using the following equation:

Figure 8. The shadow of a vertical fence

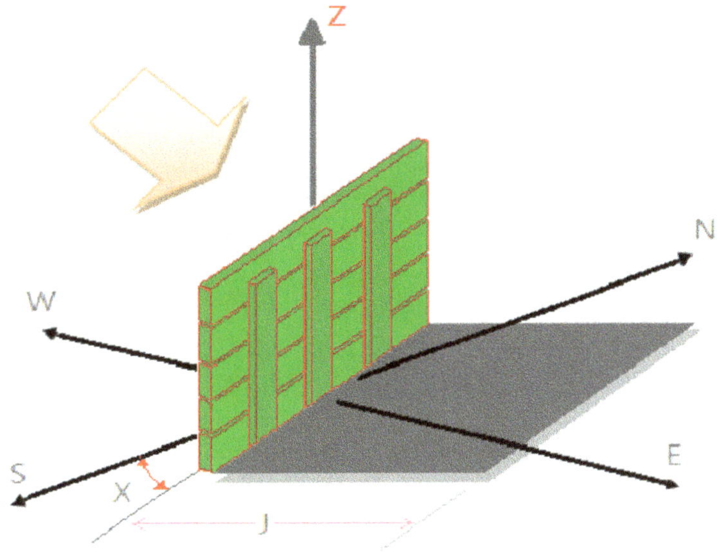

Figure 9. The shadow of an inclined fence

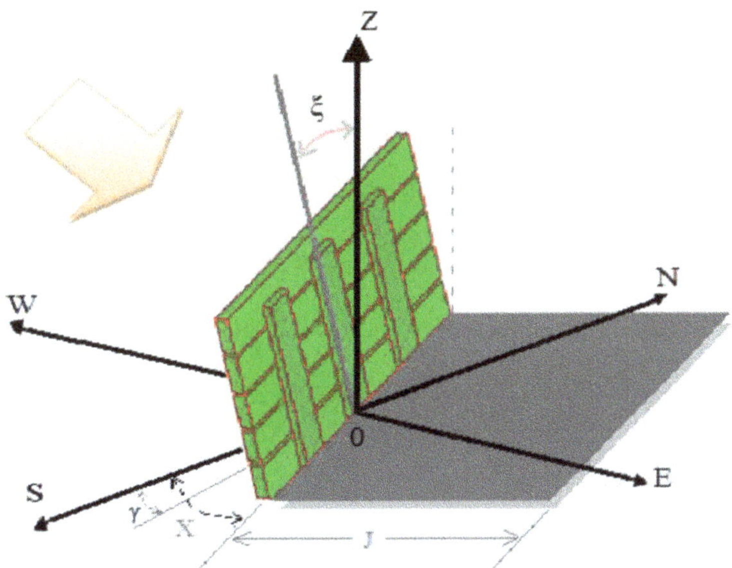

$$J = H\left[\left|\sin\left(\gamma_{st} - \chi\right)\cos\left(\xi\right)\tan\left(\theta_{i}\right)\right| \pm \sin\left(\xi\right)\right] \tag{39}$$

The signal "+" corresponds the case when the shadow is in the side of the base board while the signal "-" corresponds the case when the shadow is in the opposite side.

If the fence is vertical ($\xi = 0^0$), then the width of the shadow could be calculated using the following equation:

$$J = H\left|\sin\left(\gamma_{st} - \chi\right)\tan\left(\theta_i\right)\right| \tag{40}$$

Moreover, when the fence is directed along the sloped tilt line, and installed normally on the tilted plane, its azimuth angle will be zero, so, ($\xi = 0^0, \chi = 0^0$). In this case, the shadow width could be calculated as follows:

$$J = H\left|\sin\left(\gamma_{st}\right)\tan\left(\theta_i\right)\right| \tag{41}$$

Thus, using simple calculations and replacing γ_{st} by its value (see equation (38)), we get:

$$J = H\left|\frac{\cos\left(\gamma_s - \gamma\right)}{\cos\left(\theta_i\right)}\right| \tag{42}$$

Equation (42) can be used for determining the shadow length of the side wall of the thermal solar collector on its absorber plate.

On the other hand, in the case of installing the fence normally on the tilted surface with an azimuth angle of 90° then,($\xi = 0^0, \chi = 90^0$). In this case, the width of shadow could be calculated using the following equation:

$$J = H\left|\cos\left(\gamma_{st}\right)\tan\left(\theta_i\right)\right| \tag{43}$$

Replacing γ_{st} by its value (see equation (38)), the width of shadow could be calculated using the following equation:

$$J = H\left|\frac{\sqrt{\sin^2\left(\theta_i\right) + \cos^2\left(\alpha\right)\sin^2\left(\gamma_s - \gamma\right)}}{\cos\left(\theta_i\right)}\right| \tag{44}$$

The equation (44) allows us to determinate the shadows dimensions of the upper or lower walls of the thermal collector box on the absorber plate.

Finally, in the case of installing the fence normally on the horizontal surface we have: ($\beta = \xi = 0^0, \gamma_{st} = \gamma_s$). Then, the shadow's width could be calculated using the following equation:

$$J = H\left|\frac{\sin\left(\gamma_{st} - \chi\right)}{\tan\left(\theta_{i}\right)}\right| \tag{45}$$

Equation (45) allows determining the minimum distance between solar thermal collectors' rows as the optimum form of solar thermal collectors' installation is on a row along the west-east direction, where angle $\chi = 90^{\circ}$ while each collector is tilted by a suitable angle in relation to the horizontal surface and orientated towards the Equator. In this case, the solar thermal collectors' row could be treated as a fence inclined in relation to the horizontal surface. This means that, $\xi = 90^{0} - \beta, \gamma_{st} = \gamma_{s}, \theta_{i} = 90^{0} - \alpha$ and the shadow's width could be calculated using the following equation:

$$J = H\left[\left|\frac{\sin\left(\beta\right)\cos\left(\gamma_{s}\right)}{\tan\left(\theta_{i}\right)}\right| \pm \cos\left(\beta\right)\right] \tag{46}$$

It is clear from equation (36) that, the length of a mast's shadow, installed on an inclined surface, depends on the site's latitude, the angles that determine the inclined surface orientations and the hour angle. So, at solar noon $\left(\omega = 0^{o}\right)$ and the shadow's length at midday becomes hour angle independent. If the tilted surface has the latitude tilt angle and it is directed towards the Equator, the number of the variables, which determine the length of the mast's shadow, will be minimal. In this case, $\left(\gamma = 0^{o}\right)$ and $\left(\beta = \varphi\right)$ and the shadow's length could be calculated using the following equation:

$$L = L_{p} \tan\left\{\arccos\left[\cos\left(\delta\right)\cos\left(\omega\right)\right]\right\} \tag{47}$$

So, the ratio of the mast's shadow length to the mast's length depends on the declination and hour angles, and that will reduces the expected error of the measurement process.

Finally, it could be concluded that, the use of a single wedge does not determine precisely the direction all over the daylight. At solar noon, when the shadow's length is as short as possible the direction of the shadow corresponds to the axis of north-south direction $\left(\gamma = 0^{o}\right)$. So we should search for a way allows determine the geographical directions throughout the daylight.

An Elementary Educational Way for Finding Geographic Orientations

Plant a stick of 50cm length upright in the horizontal ground after sunrise and some hours (one hour is enough) before solar noon. Then a shadow of the stick will appears on the ground. Mark the tip of the stick's shadow by a stone. Wait solar noon that corresponds the shortest stick's shadow length. Mark the tip of the stick's shadow by another stone on solar noon. Wait again for the same period taken before solar noon when the first stone was put. Then use a third stone to mark the tip of the stick's shadow at this time. Draw a straight line from the stick to the second stone. The obtained line gives the south-north direction.

The second stone represents north. Draw a straight line between the first and third stones. The obtained line gives the west-east direction. The first stone represents west and the third stone represents east.

The geographic could be also determined using the same tools but with a sharp tool and a sting additionally. Tie a sharp tool to one end of the string, and the other end of the string to the pole, making sure it's just long enough to reach the first stone on the ground. Use the sharp tool that's attached to the pole to draw the circle in the ground around the stick. When the stick's shadow touches this circle again, mark the point where it connected with the third stone. Draw a straight line between the first and third stones. The obtained line gives the west-east direction. The first stone represents west and the third stone represents east. Draw a vertical line from the stick to the west-east direction and cross it with marking its end by the second stone. The obtained line gives the south-north direction. The second stone represents north.

An experimental arrangement for determining the geographic orientations were used in (Soulayman and Shamiyah, 1997; Soulayman and Shamiyah, 2000) and verified over three years. The results were of satisfactory accuracy. The error was less than 1°.

SOPHISTICATED WAY FOR FINDING GEOGRAPHIC ORIENTATIONS

This way is proposed to measure the declination of the magnetic north from the geographic north. Magnetic declination varies according to where you are located on the globe. The difference between the geographic north and the magnetic north is called the magnetic declination. Magnetic deviation is the error of a compass needle including nearby metallic objects. In order to point you in the right directions, users can compensate for magnetic declination by using charts of declination or local calibration.

The difference today is about 500 kilometers. But the Magnetic North Pole is actually moving several kilometers every year. This phenomenon is known as the Polar Shift Theory. Over the last 150 years, the magnetic pole has crept north over 1000 kilometers. Scientists suggest it migrates about 10 kilometers per year and can even flip from pole-to-pole. Lately, the speed has accelerated to about 40 kilometers per year and could reach Siberia in a few decades.

The magnetic field of the Earth, and of other planets that have magnetic fields, is generated by dynamo action in which convection of molten iron in the planetary core generates electric currents which in turn give rise to magnetic fields. In simulations of planetary dynamos, reversals often emerge spontaneously from the underlying dynamics. For example, Gary Glatzmaier and collaborator Paul Roberts of UCLA (Glatzmaiers and Roberts, 1995) ran a numerical model of the coupling between electromagnetism and fluid dynamics in the Earth's interior. Their simulation reproduced key features of the magnetic field over more than 40,000 years of simulated time and the computer-generated field reversed itself.

Shadow investigation is important in solar engineering applications, both in the design of some devices based on shadow behaviour, such as latitude measuring (see 1.4 in chapter 1 for more details) or to study the effect of shadow of trees, different physical obstacles and buildings on the performance of solar collectors field of all kinds (flat-plate collectors, concentrated collectors and Photovoltaic panels).

It was mentioned previously that the south-north direction could be determined at the time of solar noon, where the length of the vertical stick becomes as short as possible. At this time the solar azimuth angle $\gamma_s = 0^o$ and the shadow's length depends on the site's latitude. Therefore, it is desired to search a way for determining the geographical trends throughout the day and free of latitude dependence. This can be achieved by using a family of pegs in such manner that the length of the pegs increases from the end to the center. Thus the upper parts of the pegs form a semi-circle. These pegs should be mounted on an inclined surface. Obviously, the length of a single mast's shadow on the inclined surface depends on the sit's latitude, inclined surface angle and solar angle. At solar noon ($\omega = 0^{o}$) the shadow's length does depend on ω. When orienting the inclined surface towards the Equator and keeping its tilt angle to be equal to the site latitude the number of variables that determine the length of the imagination wedge becomes minimal and free of latitude and the shadow of the picket could be determined using the equation (47).

If orientating the axis X to east-west direction and axis Y to north – south direction, Table 1 gives the lengths and the coordinates of the pickets in relation to the central one. Then the shadows of the pegs pass from a single point throughout the day. This point is called the characteristic point. Then, it can easily be to study the moment of fantasy passage of the hallmark in terms of hour angle. It should be noted here that the hallmark lies on a straight perpendicular with the central peg and with the level of pickets in general. Table 2 gives the length of the picket that its shadow passes through the characteristic point and the solar time of its passage. Figure 10 gives the daily dependence of the characteristic point distance from the central picket.

The general scheme of the device for measuring the magnetic declination could be as it is shown on Figure 11 or as it is shown on Figure 12. The upper plate can move around a horizontal axis and the angle of rotation could be measured with a desired accuracy. As the magnetic compass is a part of the device the other device's part should be made of no magnetized materials. The base board of the device should be equipped with four adjusting no magnetized to provide its horizontality. When constructing the device for measuring the magnetic declination basing on both schemes it was found that the constructed device according to scheme presented on Figure 12 is less practical. Therefore, the scheme shown on the Figure 11 was used for constructing several devices.

The constructed device for measuring the magnetic declination is one of the shadow's use devices. The photography of this device is shown on Figure 13. This device is largely made of aluminum. It consists of a circular disk of aluminum of 30 cm in diameter and 1 mm thick, folded in the middle for forming a right angle. The disk is proved on a slab of aluminum so have half of the disk is located in the middle of the aluminum plate. And must be half a disk to be normal on the surface of the aluminum

Table 1. The coordinates and the lengths of the pickets.

Picket's Length	Coordinate X	Coordinate Y
3.75	-14.1	0
7.25	-12.92	0
10.25	-11.03	0
12.55	-8.52	0
14.01	-5.55	0
14.50	0	0
14.01	5.55	0
12.55	8.52	0
10.25	11.03	0
7.25	12.92	0
3.75	14.1	0

Table 2. The picket length and its shadow passage solar time through the characteristic point

Picket's Length	Hour Angle (°)
3.75	-75
7.25	-60
10.25	-45
12.55	-30
14.01	-15
14.50	0
14.01	15
12.55	30
10.25	45
7.25	60
3.75	75

Figure 10. The daily dependence of the characteristic point distance from the central picket

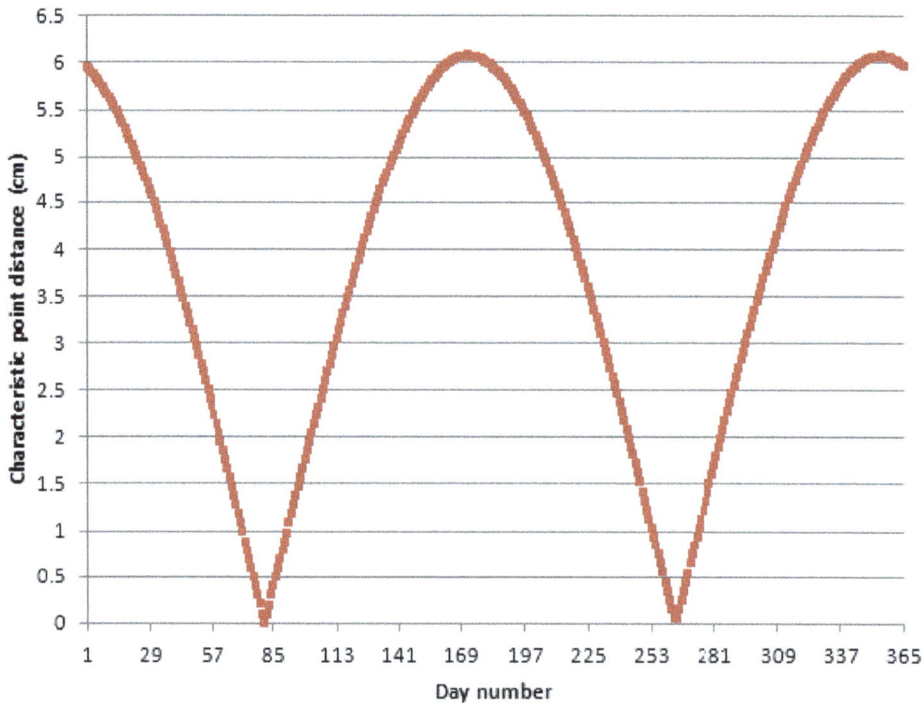

Figure 11. The first magnetic declination device scheme

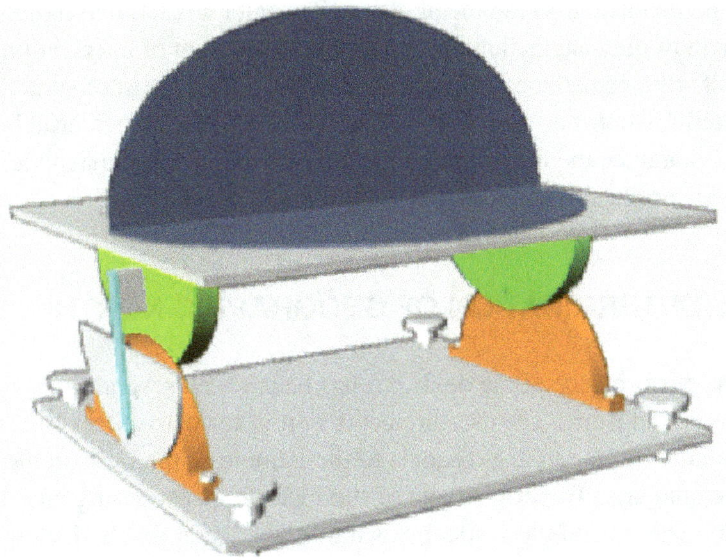

Figure 12. The second magnetic declination device scheme

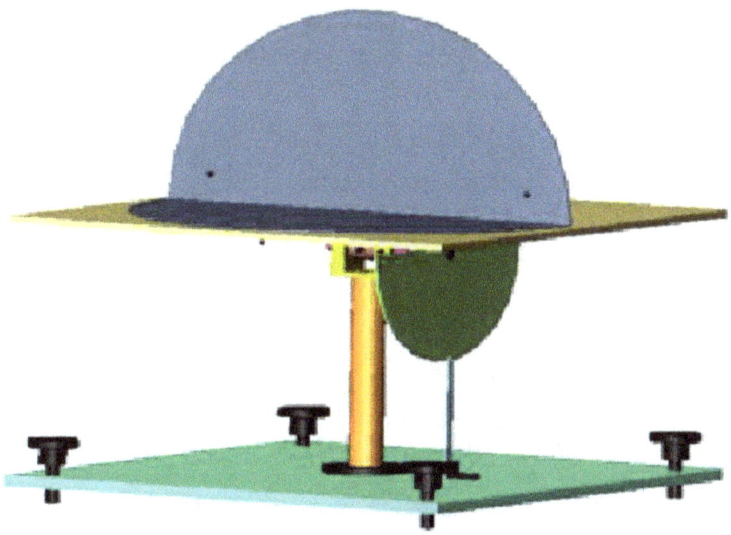

plate meticulously, branded disk axis of a half -ring and a clear black line and tight, and branded the rectum, mobile and perpendicular to the metal disc at its center a black line tightness also be allowed to determine the length of the imagination disk at noon and set point of intersection of the imagination metal disk with a half -ring centered easily according to the day of measurement. A similar device for geographic north determination was proposed by Harrison (1974). The mercurial bubble shown on the aluminum base is to guarantee the horizontality during measurement. A magnetic compass is used to determine the magnetic north.

EXPERIMENTAL DETERMINATION OF GEOGRAPHIC NORTH

In order to ensure the performance of the device, it has been tested several times during the period of 1997-1999 in Damascus and Homs. The measurements were taken place after tilting the aluminum plate which carries the circular disc at an angle equals to the latitude of the measurement area, with initial orientation using the compass. To perform any of the measurements during any day, the upper plate of the device should rotate around the vertical axis until the edge of disk's shadow intersects with the distinctive emblem which has been determined for each day of the year on the carrier aluminum plate. The distinctive emblem position can be determined basing on the direct calculations while the latitude value and declination angle could be measured directly or taken from the climatic tables issued by the World Meteorology Organization (WMO). During August and September 2016 the above mentioned measurements were repeated in Damascus. The obtained measuring data were compared with those of the National Oceanic and Atmospheric Administration (NOAA). Table 3 presents the measured and predicted results for August and September 2016.

Figure 13. A photo of magnetic declination device scheme

Table 3. The magnetic declination values during August and September 2016, Damascus

Date	Mean Measured Value (°), This Work	Predicted Value (°), World Magnetic Model (WMM)	Predicted Value (°), International Geomagnetic Reference Field (IGRF)
10/8	4°53'E ±0°18'	4° 46' E ± 0° 18'	4° 45' E
11/8	4° 48'E ±0°18'	4° 46' E ± 0° 18'	4° 45' E
15/8	4° 51'E ±0°18'	4° 47' E ± 0° 18'	4° 45' E
16/8	4° 52'E ±0°18'	4° 47' E ± 0° 18'	4° 45' E
17/8	4° 50'E ±0°18'	4° 47' E ± 0° 18'	4° 45' E
18/8	4° 52'E ±0°18'	4° 47' E ± 0° 18'	4° 46' E
22/8	4° 49'E ±0°18'	4° 47' E ± 0° 18'	4° 46' E
23/8	4° 47'E ±0°12'	4° 47' E ± 0° 18'	4° 46' E
24/8	4° 47'E ±0°12'	4° 47' E ± 0° 18'	4° 46' E
25/8	4° 47'E ±0°12'	4° 47' E ± 0° 18'	4° 46' E
28/8	4° 48'E ±0°12'	4° 47' E ± 0° 18'	4° 46' E
29/8	4° 47'E ±0°12'	4° 47' E ± 0° 18'	4° 46' E
30/8	4° 46'E ±0°12'	4° 47' E ± 0° 18'	4° 46' E
31/8	4° 47'E ±0°12'	4° 47' E ± 0° 18'	4° 46' E
1/9	4° 48'E ±0°12'	4° 47' E ± 0° 18'	4° 46' E
4/9	4° 45'E ±0°12'	4° 47' E ± 0° 18'	4° 46' E
5/9	4° 51'E ±0°12'	4° 47' E ± 0° 18'	4° 46' E
6/9	4° 45'E ±0°12'	4° 47' E ± 0° 18'	4° 46' E
7/9	4° 49'E ±0°6'	4° 47' E ± 0° 18'	4° 46' E
8/9	4° 51'E ±0°6'	4° 47' E ± 0° 18'	4° 46' E
18/9	4° 48'E ±0°6'	4° 47' E ± 0° 18'	4° 46' E
19/9	4° 47'E ±0°3'	4° 47' E ± 0° 18'	4° 46' E
20/9	4° 49'E ±0°1'	4° 47' E ± 0° 18'	4° 46' E
21/9	4° 50'E ±0°1'	4° 47' E ± 0° 18'	4° 46' E
22/9	4° 47'W ±0°3'	4° 47' E ± 0° 18'	4° 46' E
25/9	4° 46'W ±0°6'	4° 47' E ± 0° 18'	4° 46' E
26/9	4° 44'W ±0°7'	4° 47' E ± 0° 18'	4° 46' E

Then the compass indicates to the deviation between the direction of magnetic north and direction of the geographical north, the positive values indicate that the deviation happens towards the west, while negative values indicate that the deviation happens toward the east. Due to the fact that the magnetic compass does not allow measuring the angle with accuracy higher than an half of degree, the measurement accuracy is up to an half of degree too. Moreover, the accuracy of measurements varies during the day (see Figures 14 and 15). It should be noted that the deviation angle varies from a region to another and from a day to another.

Figure 14. The magnetic declination values during 18/8/2016 in Damascus

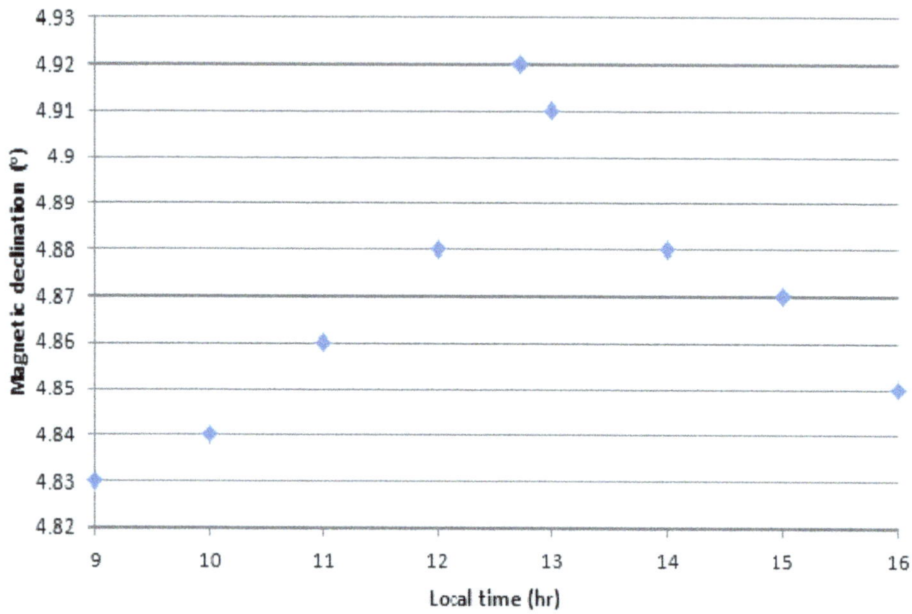

Figure 15. The magnetic declination values during 18/8/2016 in Damascus

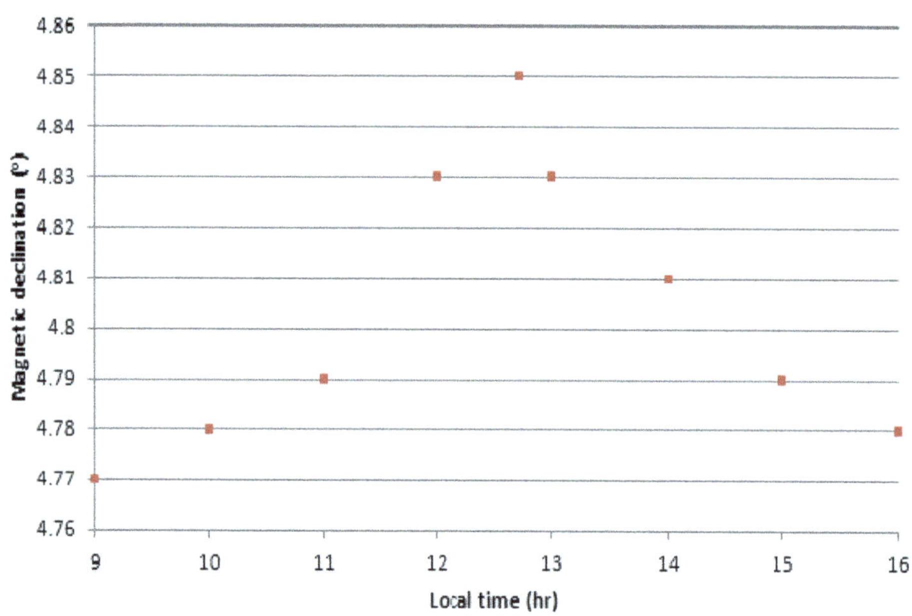

Impact of Installing Errors of the Device on the Accuracy

Possible errors of device installation can be classified to the three major errors:

1. Errors resulting of inclining the aluminum plate which carries the circular disk with the suitable angle. These errors also may be caused by either if the latitude angle is unknown or if the device is tilted with an angle doesn't equal to the latitude angle. Impact of this kind of errors can determined on the measurement accuracy while viewing the results of correct measurements along with the measurements accompanied by the mentioned errors.
2. Errors resulting of absence normality of the circular disk. These errors happen if the disk surface isn't horizontal or badly installed, or both of the tow problems. Experimentally we can determine the ambit of the impact of this type of errors on the measurement accuracy.
3. Errors resulting of absence of accuracy of distinctive emblem calculating. These errors are resulted by calculation errors, which maybe don't returned to the exploiter, or by rush reading. Also, experimentally we can determine the ambit of the impact of this type of errors on the measurement accuracy. Here it should be noted that these errors may happen because of the both reasons which mentioned in items a, and b.

It should be noted that the impact of these errors can be studied using three similar devices, one of them must be accurately installed while the others must be installed in accordance with the error types we want to study.

CONCLUSION

In this chapter we have described in detail an instrument (magnetic declination device) which could be used for determining the geographic north of the site where a solar system will be installed. The use of the magnetic declination device allows determine the azimuth angle of the solar collectors in the solar systems. This angle is required in choosing the best orientation of the solar collector as well as in the process of solar tracking. The information provided in chapter 1 was used in this chapter for studying the shadow of different obstacles on the solar systems.

REFERENCES

Bain, W. C. (1956). The theoretical design of direction-finding systems for high frequencies. *Proc. I.E.E., 103*(B), 113. doi:10.1049/pi-b-1.1956.0125

Bain, W. C. (1961). Phase difference observations at spaced aerials and their application to direction finding. *Journal of Research of the National Bureau of Standards. D, Radio Propagation, 65D*(3), 229–232. doi:10.6028/jres.065D.029

Chisholm, G. (2007). *Finding True South the Easy Way. Home Power Magazine, (120).*

Glatzmaiers, G. A., & Roberts, P. H. (1995, September 21). A three-dimensional self-consistent computer simulation of a geomagnetic field reversal. *Nature*, *377*(6546), 203–209. doi:10.1038/377203a0

Harrison, P. L (1974). A device for finding true north. *Solar Energy, 15*(4), 303-308. doi:10.1016/0038-092X(74)90020-6

Ross, W., Bramley, E. N. & Ashwell, C. E. (1951). A phase-comparison method of measuring the direction of arrival of ionospheric radio waves. *Journal of the Institution of Electrical Engineers, 1951*(8), 240 – 241. DOI: 10.1049/jiee-2.1951.0107

Soulayman, S., & Shamiyah, S. (1997). A simple method for measuring solar declination, geographic latitude and orientations. *Proceedings of National Seminar on renewable Energy for poverty Alleviation, NSREPA-97.*

Soulayman, S., & Shamiyah, S. (2000). Solar declination, geographic latitude and orientations measurements in Syria. *Proceedings of ENERGEX 2000*. Retrieved from http://trove.nla.gov.au/version/43244233

ADDITIONAL READING

Parkin, R. E. (2010). Solar angles revisited using a general vector approach. *Solar Energy*, *84*(6), 912–916. doi:10.1016/j.solener.2010.02.005

Chapter 3
Extraterrestrial Solar Radiation

ABSTRACT

Extraterrestrial solar radiation is the main source of terrestrial solar radiation components. Data on the spectrum of this radiation is available and a value of 1367 Wm^{-2} for the solar constant is accepted in solar literature. The knowledge of extraterrestrial solar radiation is essential for solar applications, within them is the Sun tracking. This radiation on horizontal surface is widely treated and a simple formula, for calculating it, is widely used. On Equator facing solar receivers, the appropriate equations for obtaining this radiation are also available, but the application of these equations by different authors was found to be not evident mainly on calculating the sunset hour angle on such surfaces. This question becomes problematic for some authors when treating surfaces of different orientations. The term of characteristic day is widely used in solar literature. The ambiguity of this term with regard to extraterrestrial solar radiation, declination angle and extraterrestrial solar radiation on a horizontal plane is described. In order to completely solve the above-mentioned problems, extraterrestrial solar radiation is calculated on surfaces of different orientations and the required relations are given for each case. The introduction of the energy gain on the basis of extraterrestrial solar radiation could be formulated mathematically very precisely. Therefore, this question is treated in details in this chapter. The difference between short term and long term Sun tracking is described also and the maximum possible energy gain of these two types of tracking is characterized.

INTRODUCTION

The sun, our ultimate source of energy, is just an average-sized star of average age, located in one of the spiral arms of the Milky Way galaxy. To astronomers, it is a main sequence star of spectral class G. This means that it has an apparent surface temperature around 6000K and is of average brightness. Other known main sequence stars have luminosities up to 1000 times greater and 1000 times less and temperatures ranging from 3000K to 16000K. At the center of the Sun it is presumed that hydrogen nuclei are combining to form helium nuclei in a thermonuclear fusion process where the excess binding energy is released into the body of the Sun. This energy is released at the rate of 3.83×10^{26} W.

DOI: 10.4018/978-1-5225-2950-7.ch003

Most of the electromagnetic radiation reaching the earth emanates from a spherical outer shell of hot dense gas called the photosphere. This region has a diameter of approximately 1.39×10^9 m and appears as a bright disc with some "limb darkening" (brighter near the center) since radiation coming to us from the outer edges comes from higher and cooler layers of gas. Observations of sunspot movement indicate that the Sun does not rotate uniformly. The region near its equator rotates with a period of about 27 days, whereas the polar regions rotate more slowly, with a period of about 32 days.

Beyond the photosphere are the chromosphere and the corona. These regions are characterized by low-density gases, higher temperature, and time wise variations in energy and diameter. Because of the low density and thus minimal energy emission from these regions, they are of little significance to Earth-based solar thermal applications. They do, however, produce uniform cyclic variations in the X-ray and ultraviolet (UV) components of the solar spectrum, having approximately 11-year periods, coincident with the sunspot cycles. Appendix A summarizes the important characteristics of the Sun.

The Sun sits at the center of the solar system and emits energy as electromagnetic radiation at an extremely large and relatively constant rate, 24 hours per day, 365 days of the year. The rate at which this energy is emitted is equivalent to the energy coming from a furnace at a temperature of about 6000 K. If we could harvest the energy coming from just 10 hectares of the surface of the Sun, we would have enough to supply the current energy demand of the world. However, there are three important reasons why this cannot be done: First, the Earth is displaced from the Sun, and since the Sun's energy spreads out like light from a candle, only a small fraction of the energy leaving an area of the Sun reaches an equal area on the Earth. Second, the Earth rotates about its polar axis, so that any collection device located on the Earth's surface can receive the Sun's radiant energy for only about one-half of each day. The third and least predictable factor is the atmosphere that surrounds the Earth's surface. At best, the Earth's atmosphere accounts for another 30 percent reduction in the Sun's energy. As is widely known, however, the weather conditions can stop all but a minimal amount of solar radiation from reaching the Earth's surface for many days in a row.

The rate at which solar energy reaches a unit area at the Earth is called the "solar irradiance" or "insolation". The units of measure for irradiance are Watts per square meter (W/m^2). Solar irradiance is an instantaneous measure of rate and can vary over time. The maximum solar irradiance value is used in system design to determine the peak rate of energy input into the system. If the storage is included in a system design, the designer also needs to know the variation of solar irradiance over time in order to optimize the system design. The designer of solar energy collection systems is also interested in knowing how much solar energy has fallen on a collector over a period of time such as a day, week or year. This summation is called solar radiation or irradiation. The units of measure for solar radiation are Joules per square meter (J/m^2) but often Watt-hours per square meter (Wh/m^2) are used. As will be described below, solar radiation is simply the integration or summation of solar irradiance over a time period.

The simplified picture of the Sun, its physical structure, temperature and density gradients indicate that the Sun, in fact, does not function as a blackbody radiator at a fixed temperature. Rather, the emitted solar radiation is the composite result of the several layers that emit and absorb radiation of various wavelengths. The photosphere is the source of most solar radiation and is essentially opaque, as gases, of which, are strongly ionized and able to absorb and emit a continuous spectrum of radiation. In addition to the total energy in the solar spectrum, it is useful to know the spectral distribution of the extraterrestrial solar radiation (ETR), which is the radiation that would be received in the absence of the atmosphere. The ETR refers to the beam of nearly parallel incident sunrays in a point at the top of Earth's atmosphere.

The ETR fluctuates about 6.9% during a year (from 1412.0 Wm^{-2} in January to 1321.0 Wm^{-2} in July) due to the Earth's varying distance from the Sun.

SOLAR SPECTRUM

The energy in solar irradiation comes in the form of electromagnetic waves of a wide spectrum. Longer wavelengths have less energy (for instance infrared) than shorter ones such as visible light or UV. The spectrum can be depicted in a Table or a graph, the spectral distribution, which shows the relative weights of individual wavelengths plotted over the wavelength variations. The diagram displays the spectrum of a Sun's ray just outside the entry into the Earth's atmosphere. The peak of the spectrum is within the visible spectrum, but there are still significant amounts of shorter and longer wavelengths present.

Figure 1 shows the spectral distribution of ETR at the mean Sun–Earth distance. The graph is plotted at low resolution with data from Gueymard and Myers (2008), which is detailed enough for many engineering applications, like forecasting of PV plants output. The extraterrestrial spectrum is also available online (NREL, 2012), along with other reference radiation spectra. It is seen from Figure 1 that the maximum spectral intensity occurs at about 478 nm wavelength (λ) in the green portion of the visible spectrum. About 6.4% of the total energy is contained in ultraviolet region (λ<380 nm). Another 45.6% is contained in the visible region (380 nm <λ<780 nm). So, the remaining 45.6% is contained in the infrared region (λ>780 nm).

According to the Planck's law of radiation, the energy E_λ emitted per unit area, per unit time and per unit wavelength interval at temperature T is given as:

Figure 1. Spectral extraterrestrial solar irradiance

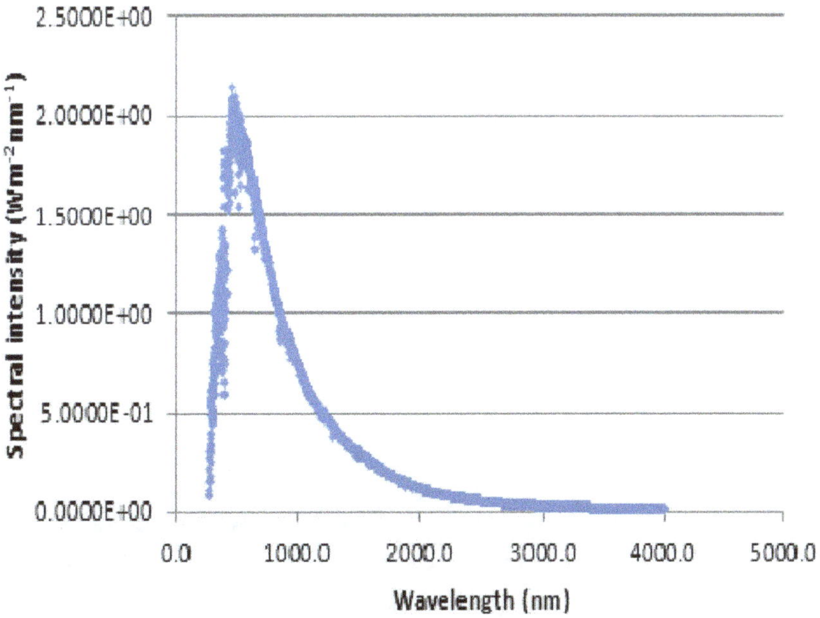

$$E_{\lambda} = \frac{C_1}{\lambda^5 \left[exp\left(\frac{C_2}{\lambda T} \right) - 1 \right]} \tag{1}$$

$$C_1 = 3.7405x10^{-16} Wm^2; \ C_2 = 1.43879x10^{-2} mK \tag{2}$$

The variation of E_{λ} with wavelength is shown in figures 2 to 4 for sources of temperatures 5777K (This temperature represents an approximation of the surface temperature of the Sun), 1000K and 288K (This temperature represents an approximation of the surface temperature of the Earth).

It is clear from Figures 5 and 6 that the wavelength of solar radiation emitted from the Sun at 5777K lies in the range of short wavelength while the radiation emitted from the Earth at 288 K lies in the range of long wavelength. Moreover, Figure 6 shows that the maximum of the spectral distribution shifts towards longer wavelengths with deceasing temperature. This displacement could be described using Wein's displacement law:

$$\lambda_{max} = \frac{2.8976x10^6}{T} nm \tag{3}$$

Figure 5 shows this displacement as a function of source temperature. The same information shown in Figure 5 could be obtained by studying the normalized spectral emissive power for different source's

Figure 2. Spectral distribution of blackbody radiation (T=5777K)

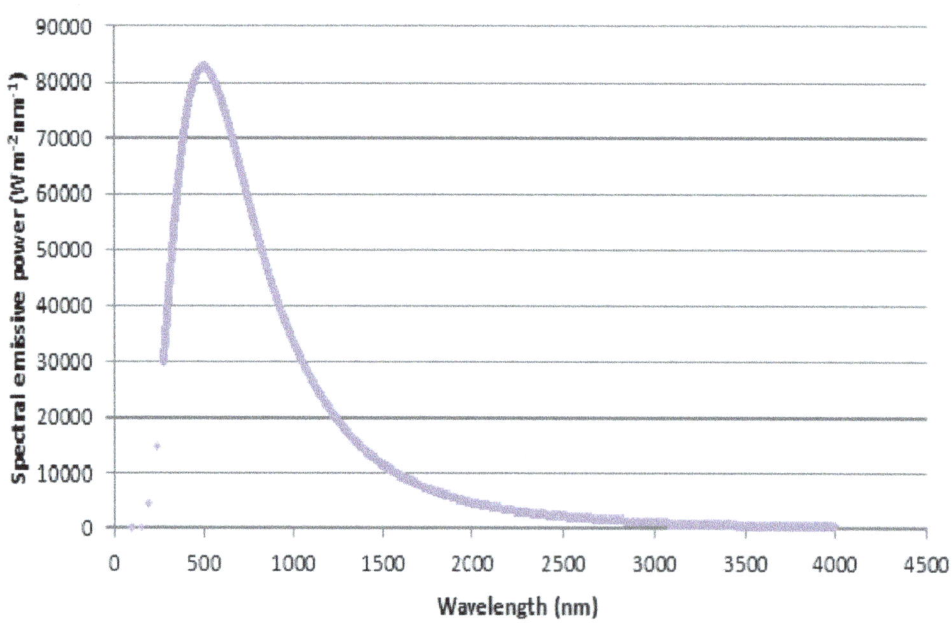

Figure 3. Spectral distribution of blackbody radiation (T=1000K)

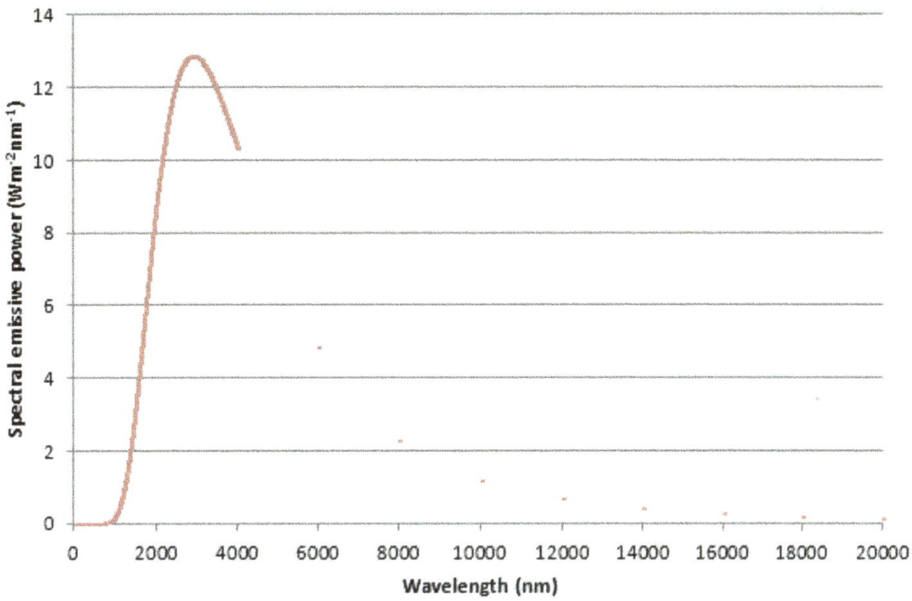

Figure 4. Spectral distribution of blackbody radiation (T=288K)

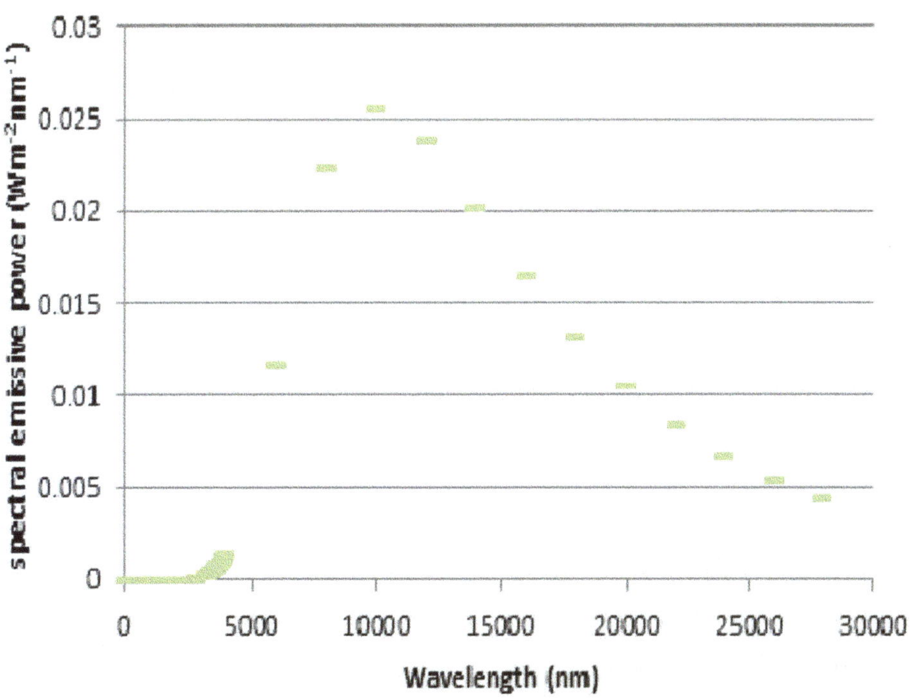

Figure 5. Wien's displacement for different source's temperature

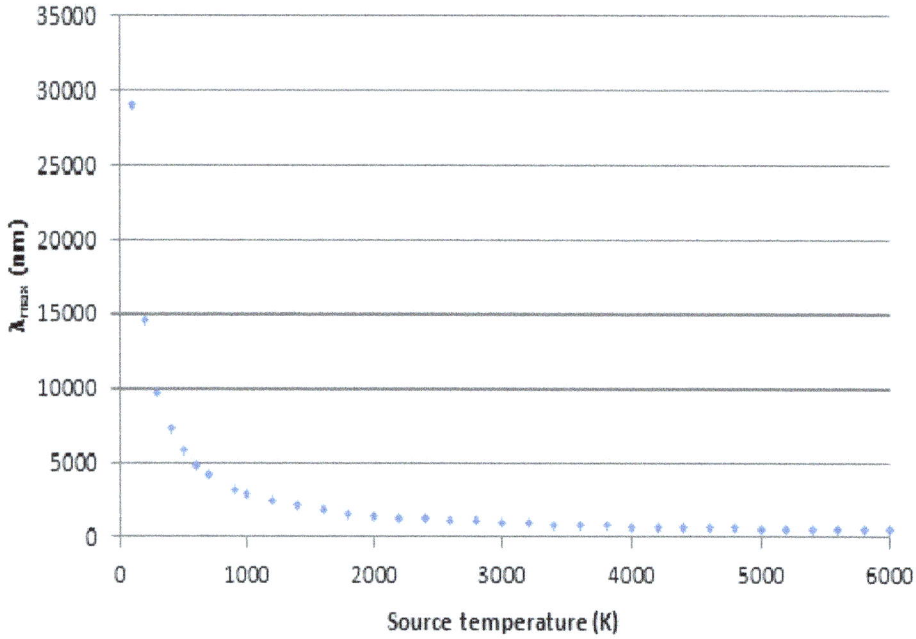

Figure 6. The ratio of the spectral emissive power to its maximum value at different source's temperature: 6000K (-), 5777 (-), 1000K (+) and 288K (●)

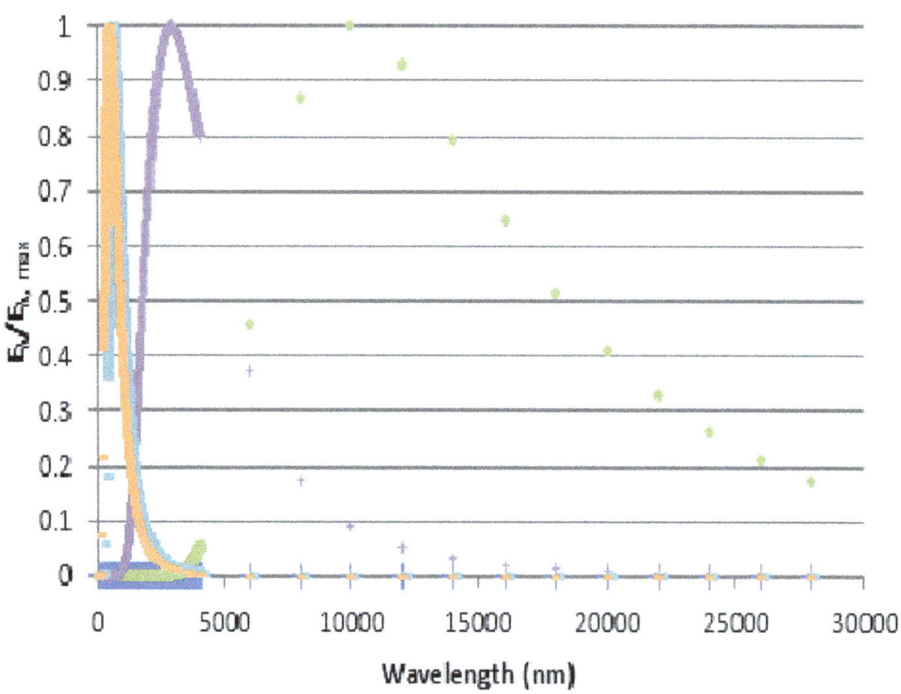

temperature. This can be done by plotting the ratio of the spectral emissive power to the maximum value at the same temperature. The results are shown in the Figure 6.

The Derivation of Wein's Displacement Law

Differentiating equation (1) with respect to the wavelength λ and equating the result to zero, λ_{max} which corresponds to the maximum of function E_λ could be determined:

$$\frac{\partial E_\lambda}{\partial \lambda} = \frac{\partial}{\partial \lambda}\left\{\frac{C_1}{\lambda^5\left[exp\left(\frac{C_2}{\lambda T}\right)-1\right]}\right\}$$

$$= C_1\left\{\frac{-5}{\lambda^6\left[exp\left(\frac{C_2}{\lambda T}\right)-1\right]} + \frac{exp\left(\frac{C_2}{\lambda T}\right)}{\lambda^5\left[exp\left(\frac{C_2}{\lambda T}\right)-1\right]^2}\left(\frac{C_2}{\lambda^2 T}\right)\right\} =$$

$$\frac{C_1\left\{-5\left[exp\left(\frac{C_2}{\lambda T}\right)-1\right]+\left(\frac{C_2}{\lambda T}\right)exp\left(\frac{C_2}{\lambda T}\right)\right\}}{\lambda^6\left[exp\left(\frac{C_2}{\lambda T}\right)-1\right]^2} = 0$$

$$\frac{C_1\left\{-5\left[1-exp\left(-\frac{C_2}{\lambda T}\right)\right]+\left(\frac{C_2}{\lambda T}\right)\right\}}{\lambda^6\left[exp\left(\frac{C_2}{\lambda T}\right)-1\right]^2} = 0$$

Since $C_2/(\lambda T)>>1$, then $exp[-C_2/(\lambda T)]\approx 0$ and:

$$\frac{C_1\left\{-5\left[1-exp\left(-\frac{C_2}{\lambda T}\right)\right]+\left(\frac{C_2}{\lambda T}\right)\right\}}{\lambda^6\left[exp\left(\frac{C_2}{\lambda T}\right)-1\right]^2} = 0$$

$$\left\{-5\left[1-exp\left(-\frac{C_2}{\lambda T}\right)\right]+\left(\frac{C_2}{\lambda T}\right)\right\} = 0 or \frac{C_2}{\lambda_{max} T} = 5$$

$$\lambda_{max} = \frac{C_2}{5T} = \frac{1.43879x10^7}{5T}nm = \frac{2.87758x10^6}{T}nm$$

The obtained result is very close to value given by equation (3).

Stefan-Boltzmann Equation

Planck's law gives the spectral distribution of radiation from a blackbody, but in engineering calculation the total energy is often of more interest. By integrating equation (10) over all wavelengths, the total energy emitted by a blackbody is:

$$E_b = \int_0^{\infty} E_\lambda d\lambda = \sigma T^4 \tag{4}$$

Equation (4) is known as the Stefan-Boltzmann equation where σ in equation (4) is the Stefan-Boltzmann constant and its value is equal to $5.6697 \times 10^{-8} Wm^{-2}K^{-4}$.

Radiation Table

The total emitted radiation from zero to any wavelength λ_1 from the Sun can be obtained from equation (1) as:

$$E_b = \int_0^{\lambda_1} E_\lambda d\lambda \tag{5}$$

If equation (5) is divided by the total energy emitted by a blackbody according to equation (4) the integral will be made to be only a function of λT as follows:

$$\frac{E_{0-\lambda T}}{\sigma T^4} = \frac{1}{\sigma T^4} \int_0^{\lambda T} E_\lambda d\lambda = \int_0^{\lambda T} \frac{C_1 d\left(\lambda T\right)}{\sigma\left(\lambda T\right)^5 \left[exp\left(\frac{C_2}{\lambda T}\right) - 1\right]} \tag{6}$$

The values of $f_{0\text{-}\lambda T}$ for different λT are given in the Table 1.

SOLAR CONSTANT

The integration of the extraterrestrial spectrum over all wavelengths defines the solar constant I_s. Thus, I_s represents the flux density of incoming solar radiation on a unitary surface perpendicular to the rays at the mean Sun–Earth distance.

The radiation intensity on the surface of the Sun is approximately 6.33×10^7 W/m^2. Since radiation spreads out as the distance squared, by the time it travels to the Earth (1.496×10^{11} m or 1 AU is the average Earth-Sun distance), the radiant energy falling on 1 m^2 of surface area is reduced to 1367 W.

Table 1. Values of $f_{0\text{-}\lambda T}$ for different λT for even increments of λT

λT (10^6nmK)	$f_{0\text{-}\lambda T}$(10^{-4})	λT (10^6nmK)	$f_{0\text{-}\lambda T}$(10^{-4})	λT (10^6nmK)	$f_{0\text{-}\lambda T}$(10^{-4})
1.0	3.0	4.5	5643.0	8.0	8562.0
1.1	9.0	4.6	5793.0	8.1	8601.0
1.2	21.0	4.7	5937.0	8.2	8639.0
1.3	43.0	4.8	6075.0	8.3	8676.0
1.4	77.0	4.9	6209.0	8.4	8711.0
1.5	128.0	5.0	6337.0	8.5	8745.0
1.6	197.0	5.1	6461.0	8.6	8778.0
1.7	285.0	5.2	6597.0	8.7	8810.0
1.8	393.0	5.3	6693.0	8.8	8841.0
1.9	521.0	5.4	6803.0	8.9	8871.0
2.0	667.0	5.5	6909.0	9.0	8899.0
2.1	830.0	5.6	7010.0	9.1	8927.0
2.2	1009.0	5.7	7107.0	9.2	8954.0
2.3	1200.0	5.8	7201.0	9.3	8980.0
2.4	1402.0	5.9	7291.0	9.4	9005.0
2.5	1613.0	6.0	7378.0	9.5	9030.0
2.6	1831.0	6.1	7461.0	9.6	9054.0
2.7	2053.0	6.2	7451.0	9.7	9076.0
2.8	2279.0	6.3	7618.0	9.8	9099.0
2.9	2506.0	6.4	7692.0	9.9	9120.0
3.0	2730.0	6.5	7763.0	10.0	9141.0
3.1	2958.0	6.6	7831.0	11.0	9318.0
3.2	3181.0	6.7	7897.0	12.0	9450.0
3.3	3401.0	6.8	7961.0	13.0	9550.0
3.4	3617.0	6.9	8022.0	14.0	9628.0
3.5	3829.0	7.0	8080.0	15.0	9689.0
3.6	4036.0	7.1	8137.0	16.0	9737.0
3.7	4238.0	7.2	8191.0	17.0	9776.0
3.8	4434.0	7.3	8244.0	18.0	9807.0
3.9	4624.0	7.4	8295.0	19.0	9833.0
4.0	4829.0	7.5	8343.0	20.0	9855.0
4.1	4987.0	7.6	8390.0	30.0	9952.0
4.2	5160.0	7.7	8436.0	40.0	9978.0
4.3	5327.0	7.8	8479.0	50.0	9988.0
4.4	5488.0	7.9	8521.0	∞	10000.0

Source: *(Duffie and Beckman, 2013)*

Other values for the solar constant are found in historical literature with the value 1353 W/m^2 appearing in many publications. It is now generally believed that most of the historical discrepancies have been due to instrument calibration error (White, 1977). Recent satellite and rocket data (Duncan, Willson, Kendall, Harrison and Hickey, 1982) and (Hickey, Alton, Griffin, Jacobwitz, Pellegrino, Maschhoff, Smith and Vonder Harr, 1982) have confirmed that the 1353 W/m^2 value was low. Also, these data confirm that there are daily and monthly variations, believed to be not over ±0.25 percent and changes over the 11-year sunspot cycle of by about 1 percent. Although none of these variations are of prime importance to the design of a solar energy system, studies are continuing to help explain the potential impact of these variations on our climate. It has been estimated that a drop of only 1 percent in the Sun's output of radiation would decrease the earth's mean global temperature by more than 1°C. The entire Earth would be covered with ice if the Sun's radiation decreased by only 6 percent.

The Sun radiance varies to some extent over short and long periods (Fröhlich, 1991). Therefore, the solar constant does not remain steady over time. There is a variation of about ± 1 Wm^{-2} around the mean solar constant during a typical Sun cycle of 11 years (Gueymard and Myers, 2008). The World Radiation Center (WRC) has adopted a value of solar constant 1367 Wm^{-2}. This value for was reported by Duncan, Willson, Kendall, Harrison and Hickey (1982) basing on measured data from three rocket flights. A near value was also proposed I_s = 1366.1 Wm^{-2} (Gueymard, 2004) basing on data collected over 25 years from terrestrial to space observations. On the other hand, basing on high-precision measurements of total solar irradiance, made by the active cavity radiometer irradiance monitor on the Solar Maximum Mission satellite, show the irradiance to have been variable throughout the first 153 days of observations. The corrected data resolve orbit-to-orbit variations with uncertainties as small as 0.001 percent. Irradiance fluctuations are typical of a band-limited noise spectrum, with high-frequency cutoff near 0.15 day^{-1} their amplitudes about the mean value of 1368.31 Wm^{-2}, approach ± 0.05 percent (Willson, Gulkis, Janssen, Hudson, and Chapman, 1981). Two large decreases in irradiance of up to 0.2 percent lasting about 1 week are highly correlated with the development of sunspot groups. The magnitude and time scale of the irradiance variability suggest that considerable energy storage occurs within the convection zone in solar active regions (Willson, Gulkis, Janssen, Hudson, and Chapman, 1981). Therefore, the value of I_s of 1367 Wm^{-2} will be used in this book. For more detailed information on the solar constant see Iqbal (1983). The different units, used in solar literature, for solar constants are listed in the appendix B.

The ETR is usually expressed in irradiance units (W/m^2) on a plane normal to the Sun's rays. It varies throughout the year because of the Earth's elliptical orbit, which results in the Earth-Sun distance varying during the year in a predictable way. Because the earth's orbit is slightly elliptical, the intensity of solar radiation received outside the earth's atmosphere varies as the square of the earth-sun distance. This effect can be represented empirically with the following equations:

$$G_s = I_s * \left(\frac{R_s}{R_{sa}}\right)^2 \tag{7}$$

where R_{sa} is the mean Sun-Earth distance and R_s is the actual Sun-Earth distance depending on the day of the year. Thus, the solar irradiance varies by ±3.4 percent with the maximum irradiance occurring at the perihelion i.e. earth closest to the sun (January 3-5) and the minimum at the aphelion (July 5).The equation (7) could be expressed as a function day number n in the year as follows (Spencer, 1971):

$$G_s = I_s * \left\{ \begin{array}{l} 1.00011 + 0.034221 cos\left[\frac{2\pi\left(n-1\right)}{365}\right] + 0.00128 \sin\left[\frac{2\pi\left(n-1\right)}{365}\right] \\ +0.000719 cos\left[\frac{2\pi\left(n-1\right)}{365}\right] +0.000077 * sin\left[\frac{2\pi\left(n-1\right)}{365}\right] \end{array} \right\} \quad (8)$$

Equation (8) was simplified by neglecting the last its two terms to be written in the form of:

$$G_s = I_s * \left\{1.00011 + 0.034221 cos\left[\frac{2\pi\left(n-1\right)}{365}\right] + 0.00128 \sin\left[\frac{2\pi\left(n-1\right)}{365}\right]\right\} \text{(8.a)}$$

And by neglecting the last term of (8.a) to have the form:

$$G_s = I_s * \left\{1.00011 + 0.034221 cos\left[\frac{2\pi\left(n-1\right)}{365}\right]\right\} \text{(8.b)}$$

However, the simple equation with accuracy adequate for most engineering calculations is given by (Duffie and Beckman, 2013):

$$G_s = I_s * \left[1 + 0.033 cos\left(\frac{2\pi n}{365}\right)\right] \text{(8.c)}$$

Figure 7 shows the daily variation of G_s basing on equations (8) and (8.a). The ignorance of the forth and fifth terms in equation (8) leads to a difference, in the calculated results, with a maximum value lower than $\pm$ 2 W/m^2. The results of equation (8.c) could differ from those of equation (8) by $\pm$ 4 W/m^2. So, equation (8.c) provides results that could differ from those of equation (8) by (0.27%) as maximum.

The ETR values, calculated using the equation (8), could be used for estimating the temperature of the Sun. For this purpose, the Sun's diameter and the Sun-Earth's distance should be given. As an example, let us suppose that the Sun-Earth's distance is that of the mean distance:

$$R_s = 1.5x10^{11}\, m$$

and the Sun's diameter is:

$$2r_s = 1.39x10^{9}\, m$$

Then, the amount of solar radiation emitted by the surface of the Sun is:

$$E = \sigma T_s^4 x 4\pi r_s^2 = G_s x 4\pi R_s^2$$

So,

$$T_s = \left[\frac{G_s x 4\pi R_S^2}{\sigma x 4\pi r_s^2}\right]^{0.25} \tag{9}$$

The daily variation of the approximate Sun's temperature is shown on Figure 8. The yearly mean Sun's temperature according to (9) is 5788.8K. The obtained approximate value of sun's temperature is very near to the blackbody temperature 5777K (Duffie & Beckman, 2013).

Finally, the apparent angular size of the solar disc may be calculated from the diameter of the photosphere and the earth-sun distance. At 1 AU, this is 9.3 mrad (0.553 degrees). As the earth-sun distance varies over the year, the apparent size of the sun varies by $\pm 1.7\%$.

EXTRATERRESTRIAL SOLAR RADIATION ON A HORIZONTAL SURFACE

Several types of radiation calculations are most conveniently done using radiation levels, that is, the radiation level to the theoretically possible radiation outside the Earth's atmosphere.

For these calculations, a method of calculating the extraterrestrial solar radiation is required. An instructional concept, and one often used in solar irradiance models, is that of the extraterrestrial solar irradiance falling on a horizontal surface. Consider a flat surface just outside the Earth's atmosphere

Figure 7. The daily variation of G_s basing on equations (8) (♦) and (8.a) (■)

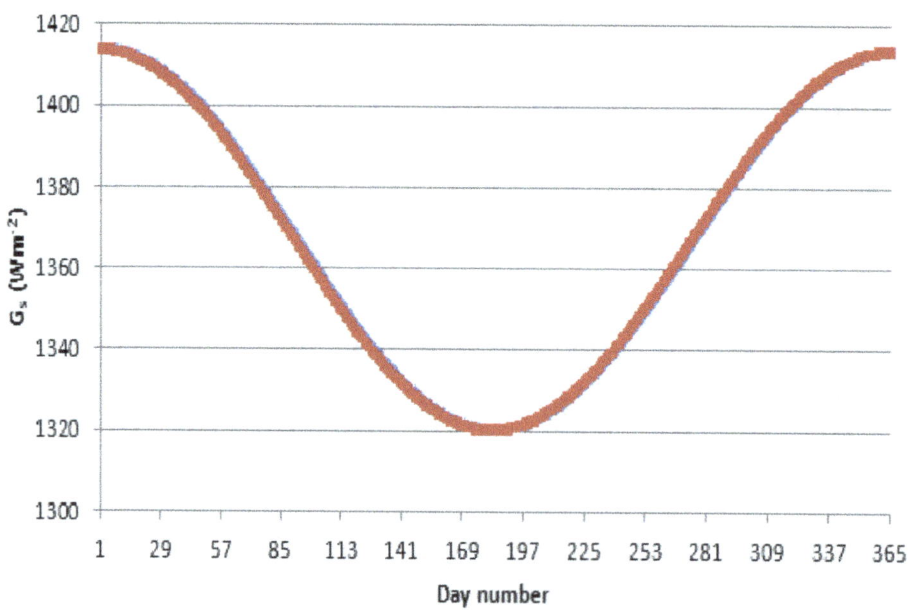

Figure 8. The daily variation of Sun's temperature T_s basing on equations (9)

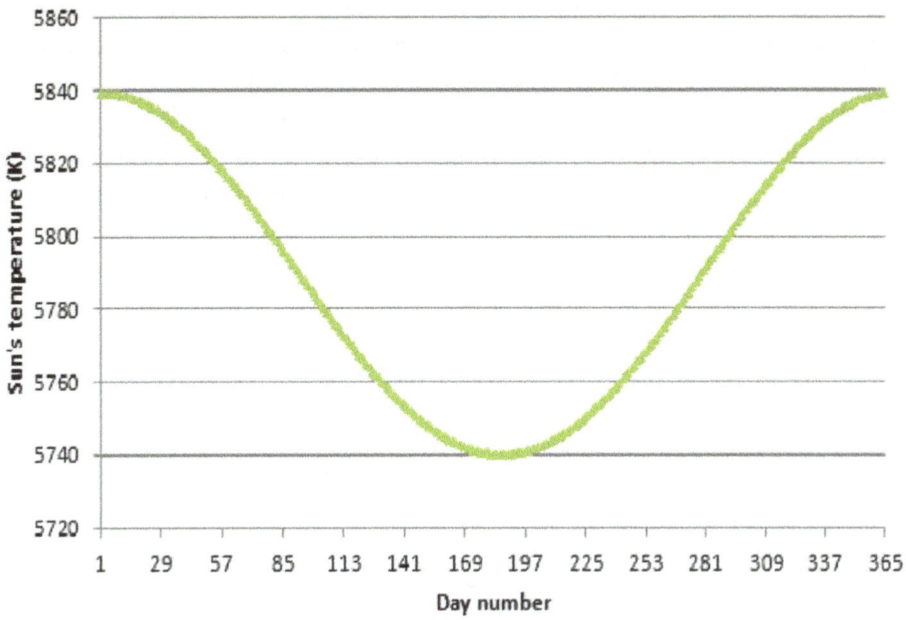

and parallel to the earth's surface below. When this surface faces the Sun (normal to a central ray), the solar irradiance falling on it, will be the maximum possible solar irradiance. If the surface is not normal to the Sun, the solar irradiance falling on it will be reduced by the cosine of the angle between the surface normal and a central ray from the Sun. Thus, at any point outside the Earth's atmosphere and at any time during the daylight period, the solar radiation incident on a surface of any orientation could be calculated using the following equation:

$$G_0 = G_s * cos\theta_i \tag{10}$$

where θ_i is the angle of solar rays' incidence (see equation (36) in chapter 1).

The reduction of radiation by the cosine of the angle between the solar radiation and a surface normal is called the cosine effect. Because of the cosine effect, the extraterrestrial solar irradiance on a horizontal varies cyclically as the Earth spins on its axis. The amount of solar radiation received on a horizontal surface outside the atmosphere forms an upper limit to the amount of radiation that will fall on a horizontal surface below the Earth's atmosphere. It also describes the cosine effect without the complication of air mass and cloud cover. The cosine effect is an extremely important concept in optimizing the orientation of solar collectors and is discussed extensively in Chapter 5.

In the case of horizontal surface $\theta = \theta_z$ and the equation (10) becomes:

$$G_0 = G_s * cos\theta_z = G_s * \left(cos\varphi cos\delta cos\omega + sin\varphi sin\delta\right) \tag{11}$$

Often it is of interest to determine the total amount of energy that has fallen on a surface over a full day outside the earth's atmosphere. The total amount of energy deposited on a surface over a period of time is found by integrating (or summing) solar irradiance over that period of time. This sum is called the solar radiation and has the units of energy per unit area (J/m²). Solar radiation is given the symbol H in this text. The daily extraterrestrial solar radiation on a horizontal surface $H_{o,h}$ may be calculated from the instantaneous values of extraterrestrial solar irradiance. Moreover, it is often necessary for calculation of daily solar radiation on horizontal plane to have the integrated daily extraterrestrial solar radiation on a horizontal surface, $H_{0,d}$ (Jm^{-2}). The integration of the equation (11) over daylight period gives:

$$H_{0,d} = \frac{24x3600xG_s}{\pi} * \left(cos\varphi cos\delta sin\omega_s + \frac{\pi * \omega_s}{180} sin\varphi sin\delta \right) \tag{12}$$

where ω_s is the sunset hour angle, in degrees, on the horizontal surface (see equation (24) of chapter 1) on that day.

Example 1

Calculate the $H_{0,d}$ on a horizontal surface in the absence of the atmosphere at latitude 33.51°N on June 21.

Solution

For these circumstances, n=172, G_s=1322.6W/m², δ=23.458° and φ=33.51°. Therefore, on a horizontal surface ω_{ss}=106.6° and

$$H_{0,d} = \frac{24x3600x1322.6}{\pi} * \begin{bmatrix} \cos\left(33.51^o\right)\cos\left(23.458^o\right) sin\left(106.6^o\right) \\ + \frac{\pi * 106.6^o}{180^o} \sin\left(33.51^o\right)\sin\left(23.458^o\right) \end{bmatrix} = 41.53 MJ / m^2$$

Example 2

Calculate the $H_{0,d}$ on a horizontal surface in the absence of the atmosphere at latitude 60°N on June 21.

Solution

For these circumstances, n=172, G_s=1322.6W/m², δ=23.458° and φ=60°. Therefore, on a horizontal surface ω_{ss}=138.1° and

$$H_{0,d} = \frac{24x3600x1322.6}{\pi} * \begin{bmatrix} \cos\left(60^o\right)\cos\left(23.458^o\right) sin\left(138.1^o\right) \\ + \frac{\pi * 138.1^o}{180^o} \sin\left(60^o\right)\sin\left(23.458^o\right) \end{bmatrix} = 41.37 MJ / m^2$$

In January 1, the $H_{0,d}$ values for 33.51°N and 60°N are 17.86 MJ/m²and 2.32 MJ/m² respectively. So, the daily amount of radiation received by a horizontal surface from the Sun at a particular place outside of the Earth atmosphere will depend upon:

1. The ETR, which depends upon:
   a. The energy output from the Sun,
   b. The distance from the Sun to the receiver surface.
2. The latitude and declination angles.
3. The angle at which the Sun's noon rays strike the receiver surface on the Vernal Equinox.
4. The angle at which the Sun's noon rays strike the receiver surface on the summer and winter Solstice.
5. The number of hours of daylight in a day.

For latitudes in the range 66.45°N to 66.45°S, the daily amount of extraterrestrial radiation, $H_{0,d}$, could be calculated using equation (12). The $H_{0,d}$ is plotted as a function of latitude for Northern and Southern Hemispheres at summer solstice, winter solstice and equinox in Figures 9 and 10.

It is seen from Figures 9 and 10 that, the dependence of the daily extraterrestrial solar radiation on a horizontal plane at equinox decreases with the latitude increase in both Hemispheres while its behavior at summer and winter solstices is opposite in both Hemispheres. On winter solstice in Northern hemisphere the daily extraterrestrial solar radiation decreases with the latitude increase and becomes nearly zero at the latitude 66.45° N. The same thing occurs in the Southern Hemisphere but on summer solstice. When the latitude becomes greater than 66.45° N, the observer reaches a site where the Sun does not rise for at least one day and does not set for approximately the same period during the year. This phenomenon

Figure 9. Daily extraterrestrial solar radiation on a horizontal plane at summer solstice (x), winter solstice (ж) and equinox (●) in the Northern hemisphere

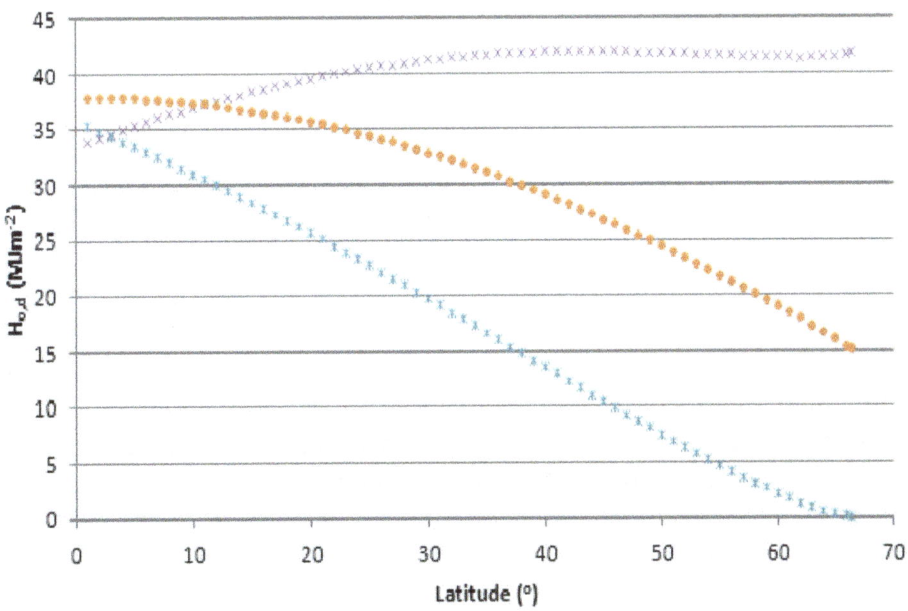

Figure 10. Daily extraterrestrial solar radiation on a horizontal plane at summer solstice (♦), winter solstice (■) and equinox (▲) in the Southern hemisphere

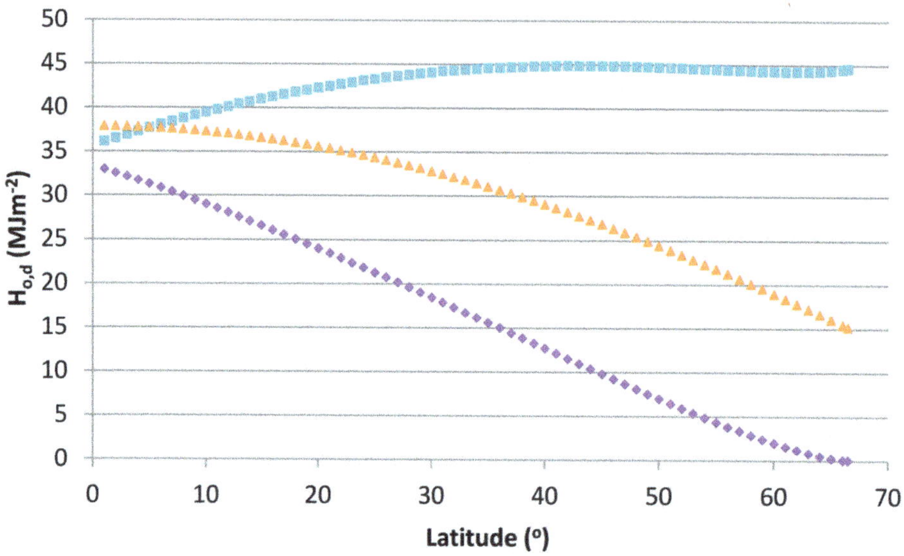

is clearly demonstrated in the Figures 11 and 12 which presents the values of $H_{0,d}$ all over the year for several latitudes in the Northern and Southern Hemispheres.

Note that the greatest amount of energy incident in one day occurs at the northernmost latitude or at the southernmost latitude. This high solar radiation value occurs during the summer in Northern Hemisphere and during winter in Southern Hemisphere, when the sun never sets. Also, note that at the Equator, the highest solar radiation occurs in the spring and fall (at the equinoxes), not in the summer as one might expect. A summer maximum occurs only at latitudes above 23.5°N and 23.5°S. The total amount of energy accumulated over a year could be calculated easily from data provided in the Table 2. As one might expect, the sum over a year of the daily solar radiation on a horizontal surface is highest at the Equator. Surfaces in the higher (polar) latitudes lose much of the available energy as a result of the cosine effect discussed above.

The monthly mean daily extraterrestrial radiation, $H_{0,m}$(Jm^{-2}), is a useful quantity in several calculations. For latitudes in the range 66.45°N to 66.45°S, this quantity could be calculated by summation of equation (12) over the days of each month in the year. The latitude dependence of $H_{0,m}$ values for latitudes 0°, 40°N, 66.45°N, 40°S and 66.45°S in the Northern and Southern Hemispheres are presented in Figures 13 and 14. The $H_{0,m}$ values for several latitudes in the Northern and Southern Hemispheres are given in the Table 2.

When analyzing the results of the Table 2 in order to find a relationship between the monthly values of extraterrestrial solar radiation on the horizontal surface and the latitude value, it was found that, the monthly average daily values of $H_{0,m}$(MJm^{-2}) dependence on latitude could be approximated, with a good accuracy, as a polynomial function of fourth order of the form:

$$H_{0,m} = a\varphi^4 + b\varphi^3 + c\varphi^2 + d\varphi + e \tag{13}$$

Figure 11. Daily extraterrestrial solar radiation on a horizontal plane for latitudes 0° (x), 10°N (●), 20° N (-), 30° N (♦), 40° N (▲), 50° N (ж), 60° N (+) and 66.45° N (-)

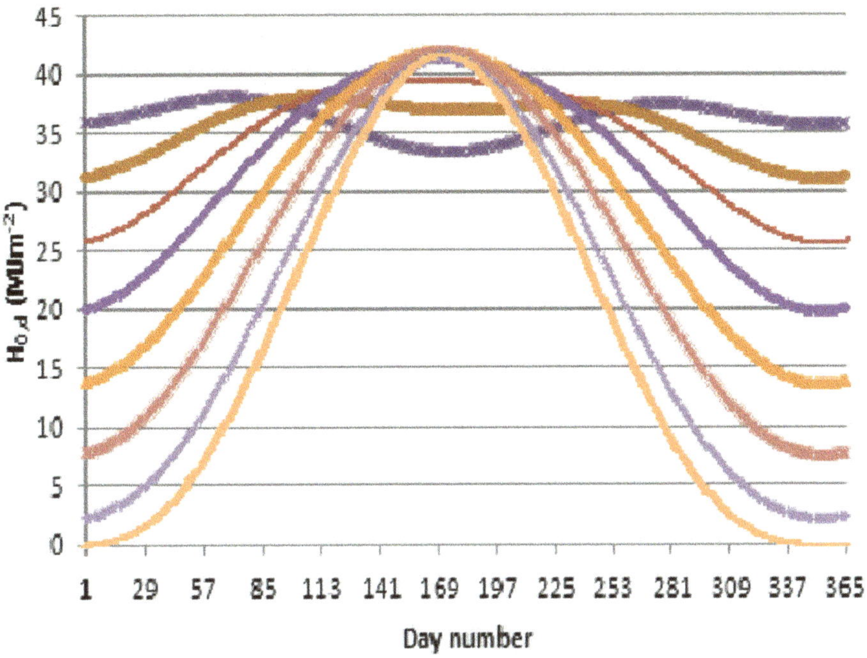

Figure 12. Daily extraterrestrial solar radiation on a horizontal plane for latitudes 0° (x), 10° S (●), 20° S (-), 30° S (♦), 40° S (▲), 50° S (ж), 60° S (+) and 66.45° S (-)

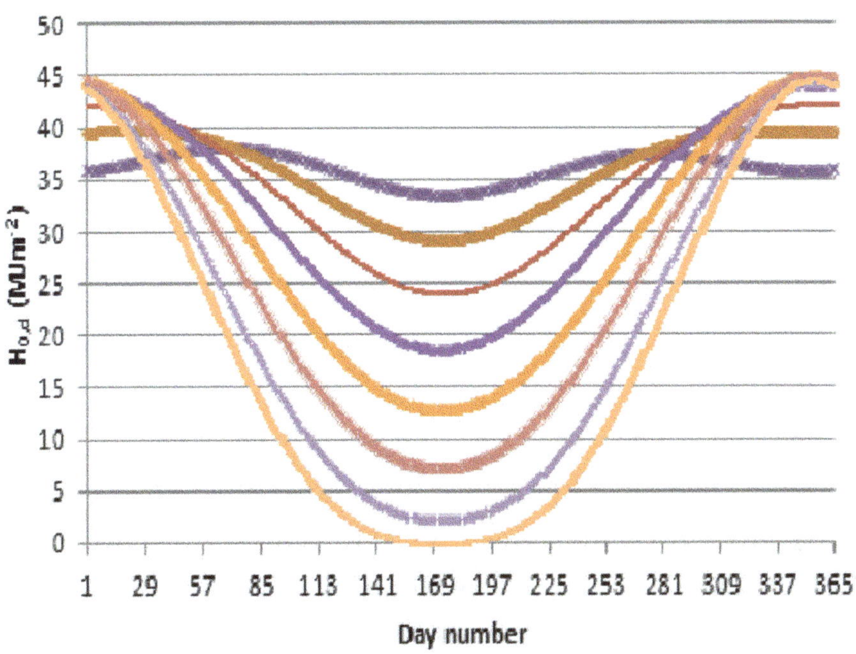

Figure 13. Daily average monthly extraterrestrial solar radiation on a horizontal plane for latitudes 0° (▲), 40° N (■) and 66.45° N (♦) In the Northern Hemisphere

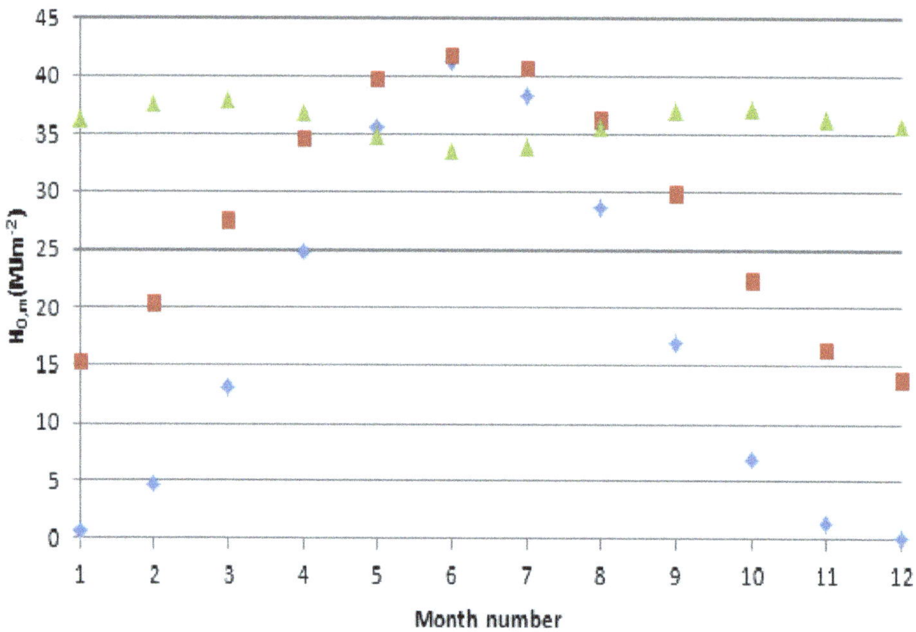

Figure 14. Daily average monthly extraterrestrial solar radiation on a horizontal plane for latitudes 0° (▲), 40° S (x) and 66.45° S (ж) In the Southern Hemisphere

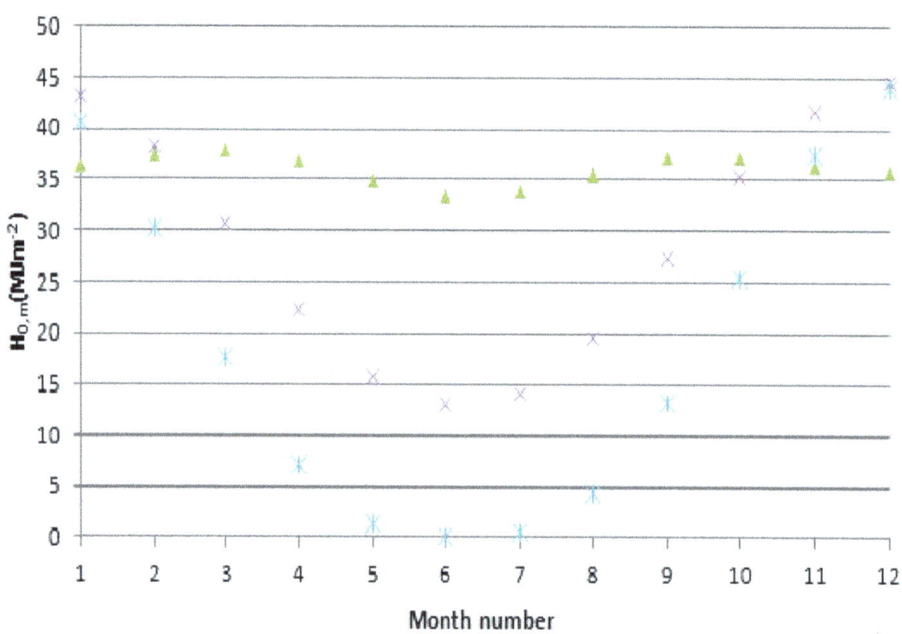

Table 2. Monthly average daily extraterrestrial solar radiation on a horizontal plane, MJ/m²

φ(°)	Jan.	Feb.	March	Apr.	May	June	July	Aug.	Sep.	Oct.	Nov.	Dec.
66.45	0.69	4.57	13.1	24.8	35.6	41.2	38.3	28.6	16.9	6.9	1.40	0.05
60	3.49	8.33	16.9	27.6	36.7	41.0	38.8	30.9	20.4	10.8	4.50	2.28
50	9.16	14.4	22.5	31.6	38.5	41.6	39.9	34.0	25.5	16.7	10.3	7.66
40	15.3	20.3	27.5	34.7	39.7	41.7	40.6	36.3	29.8	22.3	16.3	13.7
30	21.3	25.8	31.6	36.8	40.0	41.1	40.4	37.8	33.2	27.3	22.2	19.9
20	27.0	30.8	34.8	37.9	39.3	39.5	39.2	38.1	35.6	31.6	27.7	25.8
10	32.1	34.5	36.9	37.9	37.5	37.0	37.1	37.4	36.9	34.9	32.4	31.1
0	36.3	37.5	37.9	36.8	34.8	33.5	33.9	35.6	37.1	37.2	36.3	35.7
-10	39.6	39.4	37.8	34.6	31.1	29.2	29.9	32.8	36.2	38.4	39.3	39.4
-20	41.9	40.1	36.5	31.3	26. 6	24.2	25.2	29.1	34.2	38.5	41.1	42.2
-30	43.1	39.7	34.1	27.2	21.4	18.7	19.8	24.5	31.1	37.5	41.9	43.8
-40	43.2	38.3	30.7	22.3	15.8	12.9	14.1	19.4	27.2	35.4	41.7	44.5
-50	42.4	35.8	26.4	16.8	10.0	7.20	8.41	13.7	22.4	32.3	40.5	44.3
-60	41.1	32.5	21.3	10.9	4.46	2.14	3.13	7.96	16.9	28.4	38.6	43.7
-66.45	40.5	30.2	17.7	7.08	1.45	0.05	0.57	4.39	13.2	25.5	37.5	43.9

where a, b, c, d and e (MJm^{-2}) are constant for each month but their values are month's number dependent. The monthly values of a, b and c in addition to the correlation factor R^2 are given in the Table 3. The Table 3 could be used for determining the monthly average daily values of $H_{0,m}$ at any site with latitude being in the range 66.45°N to 66.45°S.

Table 3. The coefficients of the polynomial equation (13)

Month	a	b	c	d	e	R^2
January	$3x10^{-7}$	$2x10^{-5}$	-0.005	-0.3778	36.398	1
February	$2x10^{-7}$	$1x10^{-5}$	-0.0055	-0.2414	37.541	1
March	$1x10^{-7}$	$2x10^{-6}$	-0.0058	-0.0436	37.921	1
April	$2x10^{-7}$	$8x10^{-6}$	-0.0055	0.1672	36.738	1
May	$3x10^{-7}$	$-1x10^{-5}$	-0.0049	0.3228	34.839	1
June	$4x10^{-7}$	$-2x10^{-5}$	-0.0046	0.3893	33.617	1
July	$3x10^{-7}$	$-1x10^{-5}$	-0.0049	0.3228	34.839	1
August	$2x10^{-7}$	$-1x10^{-5}$	-0.0052	0.2298	35.63	1
September	$1x10^{-7}$	$2x10^{-6}$	-0.0056	0.0357	37.114	1
October	$2x10^{-7}$	$8x10^{-6}$	-0.0055	-0.177	37.207	1
November	$3x10^{-7}$	$2x10^{-5}$	-0.0051	-0.3428	36.375	1
December	$4x10^{-7}$	$2x10^{-5}$	-0.0049	-0.4161	35.833	0.9999

THE CHARACTERISTIC DAYS

In (Duffie and Beckman, 2013), some days in the year are recommended as average days for months with giving the declination angle δ values in these days. So, these days could be considered for δ only. These days are used in (Gunerhan and Hepbasli, 2007) as days representing the months of the year with providing the values of the declination angle, the optimum tilt angle, the extraterrestrial solar radiation on the tilted and horizontal surfaces as well as the measured total solar radiation. Therefore, they could be considered as characteristic days over all latitudes. In order to clarify this ambiguity it is useful to shed a light on these characteristic days in details.

The Characteristic Days for Declination Angle

When calculating the daily values of declination angle all over the year as well as its daily average monthly values, it will possible to determine for each month the characteristic day where the daily value of the declination angle δ_{cd} is approximately equal to the daily average monthly value δ_m: $\delta_{cd} \approx \delta_m$. The obtained results are presented in the Table 4 where the proposed values of (Duffie and Beckman, 2013) are reported also.

It is seen from the Table 4 that, the results of this work agree well with those of (Duffie and Beckman, 2013). The maximum difference in day number is 1 which is negligible. Even though, it is difficult to understand this difference as in this work the equation of (Cooper, 1969) is used for calculating the declination angle and it is mentioned in (Duffie and Beckman, 2013) that the declination angle can be found from the equation of (Cooper, 1969).

Table 4. The characteristic days for declination angle

Month	Day Number, n		Month's Day Number	Declination Angle, δ_{cd}(°)		Declination Angle, δ_m(°)
	This Work	a		This work	a	
January	17	17	17	-20.9241	-20.9	-20.8543
February	46	47	15	-13.2937	-13.0	-13.3298
March	75	75	16	-2.41856	-2.4	-2.38999
April	105	105	15	9.418105	9.4	9.496436
May	135	135	15	18.79833	18.8	18.81223
June	162	162	11	23.09379	23.1	23.08493
July	198	198	17	21.19092	21.2	21.10867
August	228	228	16	13.45955	13.5	13.30058
September	259	258	16	1.815332	2.2	1.994253
October	289	288	16	-9.96966	-9.6	-9.8519
November	319	318	15	-19.1543	-18.9	-19.057
December	345	344	11	-23.1284	-23.0	-23.1035

Source: (Duffie and Beckman, 2013).

The Characteristic Days for Extraterrestrial Solar Radiation

When calculating the daily values of extraterrestrial solar radiation intensity all over the year as well as its daily average monthly values using the equation (8), it will possible to determine for each month the characteristic day where the daily value of the ETR intensity $G_{s,cd}$ is approximately equal to the daily average monthly value $G_{s,m}$: $G_{s,cd} \approx G_{s,m}$. The obtained results are presented in the Table 5.

When comparing the characteristic day's numbers in Tables 4 and 5, it could be seen that the maximum difference riches 3 days.

The Characteristic Days for Extraterrestrial Solar Radiation on a Horizontal Surface H_0

For latitudes in the range 66.45°N to 66.45°S, the daily and monthly values of H_0 were calculated using the equation (12). Then, when comparing the daily monthly average values with the daily values it was found that there is a day in each month where $H_{0,d} \approx H_{0,m}$. These days are known as the characteristic days. Tables 6 to 9 show the number of these days in the year.

It is seen from Tables 6 to 9 that the characteristic days are latitude dependent. Therefore, the proposal of (Duffie and Beckman, 2013) is an approximation which should be evaluated before applying to any new case.

EXTRATERRESTRIAL SOLAR RADIATION ON THE EQUATOR FACING TILTED SURFACE

In the case of Equator facing surface, tilted by an angle β in relation to the horizontal plane, the incidence angle θi could be determined using the equation:

Table 5. The characteristic days for the extraterrestrial solar radiation intensity, G_s (W/m²)

Month	Year's Day Number, n	Month's Day Number	ETR, $G_{s,cd}$	ETR, $G_{s,m}$
January	19	19	1413.068	1413.128
February	46	15	1401.902	1401.849
March	75	16	1381.798	1381.653
April	105	15	1357.886	1357.633
May	135	15	1337.027	1336.811
June	165	14	1324.093	1324.167
July	200	19	1322.553	1322.581
August	229	17	1332.189	1332.601
September	259	16	1351.993	1351.818
October	289	16	1375.844	1375.772
November	319	15	1398.048	1398.022
December	348	14	1410.988	1410.988

Table 6. The characteristic days for the solar radiation H_o (MJ/m^2) at 0°N and 40°N

Month	φ = 0°				φ = 40°			
	Year's Day Number, n	**Month's Day Number**	$H_{o,cd}$	$H_{o,m}$	**Year's Day Number, n**	**Month's Day Number**	$H_{o,cd}$	$H_{o,m}$
January	17	17	36.31	36.30	17	17	15.24	15.29
February	45	14	37.49	37.47	46	15	20.36	20.33
March	80	21	37.89	37.89	75	16	27.48	27.48
April	106	16	36.78	36.77	110	20	35.69	34.68
May	135	15	34.81	34.78	135	15	39.72	39.71
June	163	12	33.49	33.50	161	10	41.74	41.74
July	198	17	33.90	33.92	198	17	40.63	40.59
August	228	16	35.63	35.62	229	17	36.27	36.35
September	257	14	37.10	37.09	259	16	29.67	29.75
October	291	18	37.22	37.22	289	16	22.25	22.31
November	319	15	36.32	36.32	319	15	16.28	16.34
December	342	8	35.72	35.72	344	10	13.77	13.75

Table 7. The characteristic days for the solar radiation H_o (MJ/m^2) at 50°N and 66.45°N

Month	φ = 50°				φ = 66.45°			
	Year's Day Number, n	**Month's Day Number**	$H_{o,cd}$	$H_{o,m}$	**Year's Day Number, n**	**Month's Day Number**	$H_{o,cd}$	$H_{o,m}$
January	17	17	9.102	9.160	19	19	0.724	0.693
February	46	15	14.41	14.41	46	15	4.488	4.567
March	75	16	22.51	22.54	75	16	12.93	13.07
April	105	15	31.50	31.55	105	15	24.60	24.76
May	134	14	38.50	38.50	134	14	35.48	35.56
June	161	10	41.54	41.56	162	11	41.20	41.17
July	198	17	39.98	39.93	198	17	38.33	38.28
August	228	16	34.10	33.98	228	16	28.72	28.63
September	259	16	25.32	25.46	258	15	16.97	16.85
October	289	16	16.60	16.71	288	15	6.951	6.894
November	318	14	10.35	10.28	318	14	1.367	1.398
December	344	10	7.682	7.661	352	18	0.0057	0.052

Table 8. The characteristic days for the solar radiation H_o (MJ/m^2) at 10°S and 40°S

Month	φ = 10°				φ = 40°			
	Year's Day Number, n	Month's Day Number	$H_{o,cd}$	$H_{o,m}$	Year's Day Number, n	Month's Day Number	$H_{o,cd}$	$H_{o,m}$
January	13	13	39.61	39.61	17	17	43.28	43.22
February	47	16	39.37	39.36	46	15	38.26	38.27
March	76	17	37.75	37.75	72	13	31.57	30.69
April	106	16	34.51	34.56	105	15	22.33	22.28
May	135	15	31.11	31.08	135	15	15.80	15.80
June	162	11	29.20	29.19	162	11	12.91	12.91
July	198	17	29.88	29.91	199	18	14.20	14.15
August	228	16	32.80	32.82	229	17	19.46	19.36
September	258	15	36.18	36.18	259	16	27.29	27.17
October	288	15	38.46	38.44	289	16	35.47	35.39
November	317	13	39.26	39.26	319	15	41.75	41.69
December	351	16	39.44	39.43	345	11	44.50	44.52

Table 9. The characteristic days for the solar radiation H_o (MJ/m^2) at 50°S and 66.45°S

Month	φ = 50°				φ = 66.45°			
	Year's Day Number, n	Month's Day Number	$H_{o,cd}$	$H_{o,m}$	Year's Day Number, n	Month's Day Number	$H_{o,cd}$	$H_{o,m}$
January	17	17	42.51	42.45	17	17	40.57	40.53
February	46	15	35.75	35.79	46	15	30.02	30.16
March	75	16	26.37	26.37	75	16	17.55	17.66
April	105	15	16.79	16.77	105	15	7.000	7.078
May	135	15	9.985	10.00	134	14	1.466	1.455
June	162	11	7.194	7.197	161	10	0.051	0.053
July	199	18	8.459	8.415	199	18	0.583	0.567
August	229	17	13.82	13.73	229	17	4.370	4.394
September	259	16	22.49	22.38	259	16	13.19	13.17
October	289	16	32.37	32.29	289	16	25.55	25.54
November	319	15	40.56	40.48	318	14	37.60	37.54
December	345	11	44.32	44.33	345	11	43.93	43.93

$$\cos\theta_i = \cos(\varphi-\beta)cos\delta cos\omega + \sin(\varphi-\beta)sin\delta \tag{14}$$

and the equation (10) becomes:

$$G_{0,\beta} = G_s * cos\theta = G_s * [cos(\varphi-\beta)cos\delta cos\omega + sin(\varphi-\beta)sin\delta] \tag{15}$$

It is often necessary for calculation of daily solar radiation on Equator facing tilted plane to have the integrated daily extraterrestrial solar radiation on the Equator facing tilted plane, $H_{0,d}$(Jm^{-2}). The integration of the equation (15) over daylight period gives:

$$H_{0,\beta d} = \frac{24x3600xG_s}{\pi} * \left[cos(\varphi-\beta)cos\delta sin\omega_{ss} + \frac{\pi * \omega_{ss}}{180^o} sin(\varphi-\beta)sin\delta\right] \tag{16}$$

where ω_{ss} is the sunset hour angle, in degrees, on the Equator facing tilted surface (see equation (41a) of chapter 1) on that day:

$$\omega_{ss} = min\{arccos[-tan(\delta)tan(\varphi)], arccos[-tan(\delta)tan(\varphi-\beta)]\} \tag{17}$$

Example 3

Calculate the $H_{0,d}$ on a surface, tilted by an angle 33.51° and orientated towards the Equator, in the absence of the atmosphere at latitude 33.51°N on June 21.

Solution

For these circumstances, n=172, G_s=1322.6W/m^2, δ=23.458° and φ=33.51°. Therefore, on a horizontal surface ω_{ss}=106.6° and on a surface, tilted by an angle 33.51° and orientated towards the Equator ω_{ss}=89.95°. Thus, according to the equation (17) ω_{ss}=89.95° and

$$H_{0,d} = \frac{24x3600x1322.6}{\pi} * \begin{bmatrix} \cos(0^o)\cos(23.458^o)sin(89.95^o) \\ + \frac{\pi * 89.95^o}{180^o} \sin(0^o)\sin(23.458^o) \end{bmatrix} = 9.93MJ / m^2$$

This means that the inclination of a solar receiver by an angle of the site latitude can lead to a decrease in the daily solar radiation received by a horizontal surface at the same site.

Equator Facing Tilted Surface in the Northern Hemisphere

For latitudes in the range 0° to 66.45°N, the daily amount of extraterrestrial radiation, $H_{0,d}$, could be calculated using equation (16). The $H_{0,d}$ is plotted as a function of the tilt and latitude for the Northern Hemispheres at summer solstice, winter solstice and equinox in Figures 15 to 20.

On the other hand, when calculating the daily extraterrestrial solar radiation on an Equator facing surface, tilted by an angle β in relation to the horizontal plane for latitudes 10°, 20°, 30°, 40°, 50°, 60° and 66.45° in the Northern Hemisphere, the results presented in the Figures 21 to 26 were obtained.

From Figures 21 to 26 it is seen that, in the Northern Hemisphere the daily extraterrestrial solar radiation on an Equator facing surface, tilted by an angle β in relation to the horizontal plane depends on the day number, latitude angle φ and tilt angle β. The dependence of $H_{0',d\,o}$n the latitude and day number is very similar to that of the daily extraterrestrial solar radiation on the horizontal plane $H_{0',d.}$ Concerning the dependence of $H_{0',\beta d\,on}$ the tilt angle, Figures 21 to 26 show that, there is a value of the tilt angle for each latitude which balances the values of the daily extraterrestrial solar radiation on an Equator facing surface all over the year. These values are very near to the latitude angle value. Table 10 sheds a light on the dependence of the average daily yearly values of $H_{0'},d_{on}$ β and φ. It was found that, in the tropical zone of the Northern Hemisphere, $0° \leq φ \leq 23.45°$, the Sun does rise on the Equator facing vertically mounted surface during two periods in the year. The length of each of these periods depends on the latitude value. For example, for the latitude of 10°N, one period started on 301st day number and ended on 58th day number while the second period started on 112th day number and ended on 237th day number. For the latitude of 20°N, one period started on 335th day number and ended on 23rd day number while the second

Figure 15. Daily extraterrestrial solar radiation on an Equator facing surface, tilted by an angle 15° in relation to the horizontal plane, at winter solstice (x), summer solstice (■) and equinox (●) in the Northern hemisphere

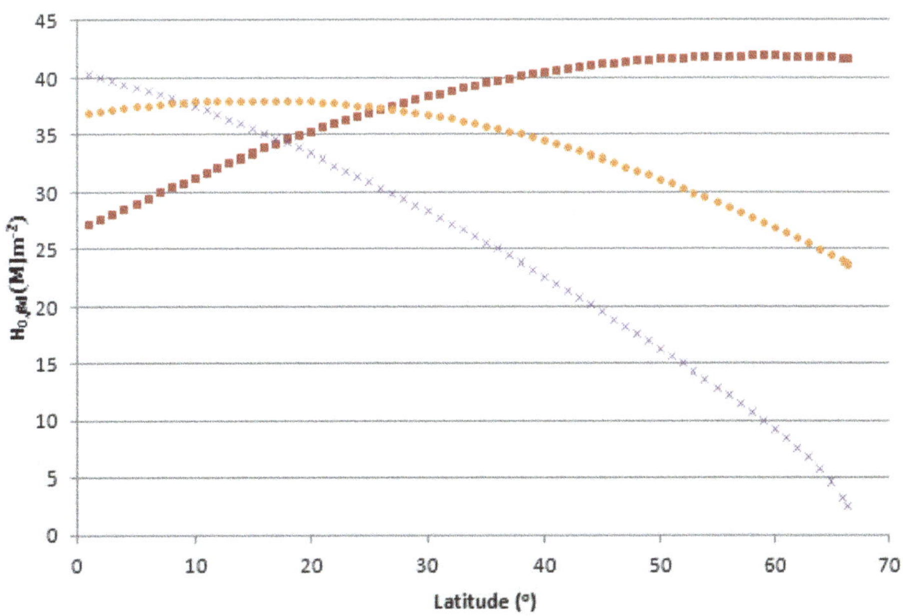

Figure 16. Daily extraterrestrial solar radiation on an Equator facing surface, tilted by an angle 30° in relation to the horizontal plane, at winter solstice (x), summer solstice (■) and equinox (●) in the Northern hemisphere

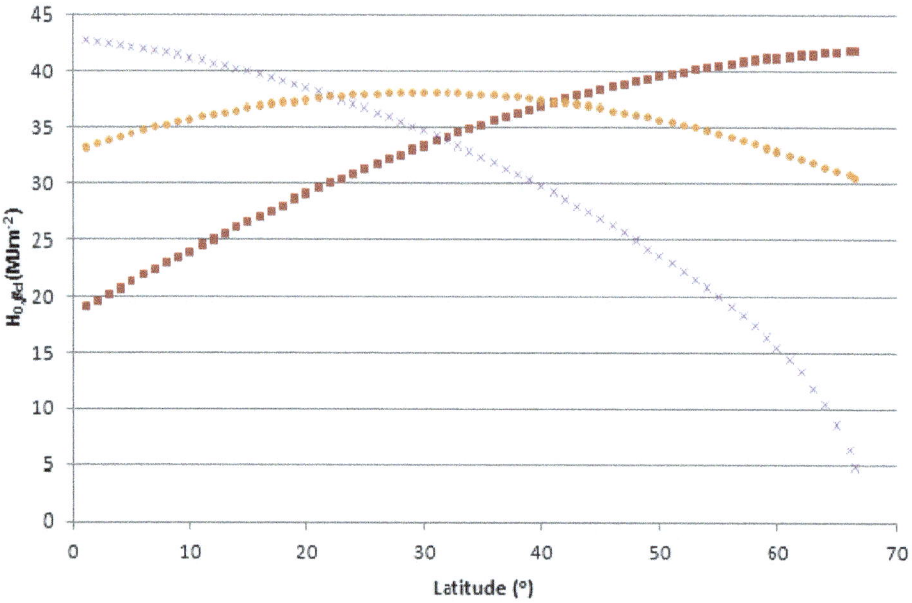

Figure 17. Daily extraterrestrial solar radiation on an Equator facing surface, tilted by an angle 45° in relation to the horizontal plane, at winter solstice (x), summer solstice (■) and equinox (●) in the Northern hemisphere

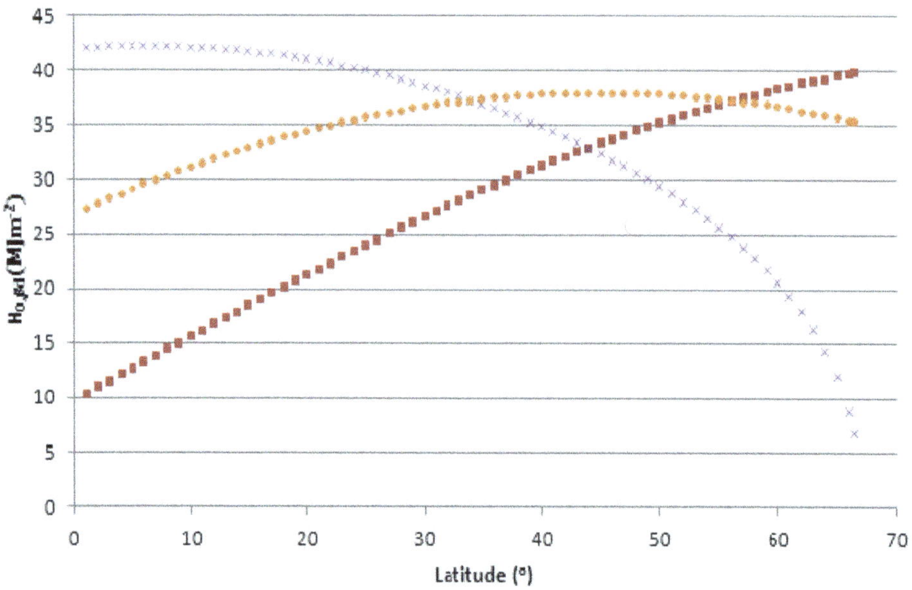

Figure 18. Daily extraterrestrial solar radiation on an Equator facing surface, tilted by an angle 60º in relation to the horizontal plane, at winter solstice (x), summer solstice (■) and equinox (●) in the Northern hemisphere

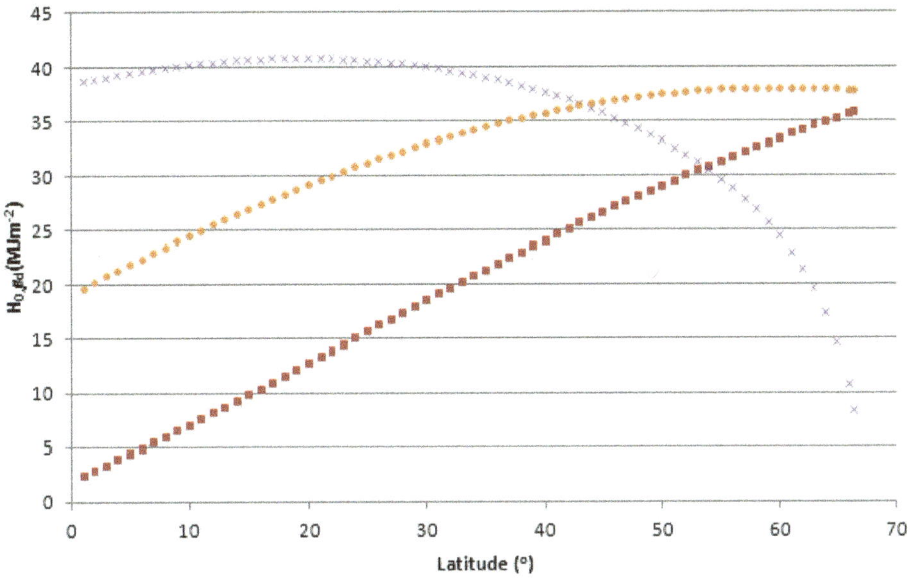

Figure 19. Daily extraterrestrial solar radiation on an Equator facing surface, tilted by an angle 75º in relation to the horizontal plane, at winter solstice (x), summer solstice (■) and equinox (●) in the Northern hemisphere

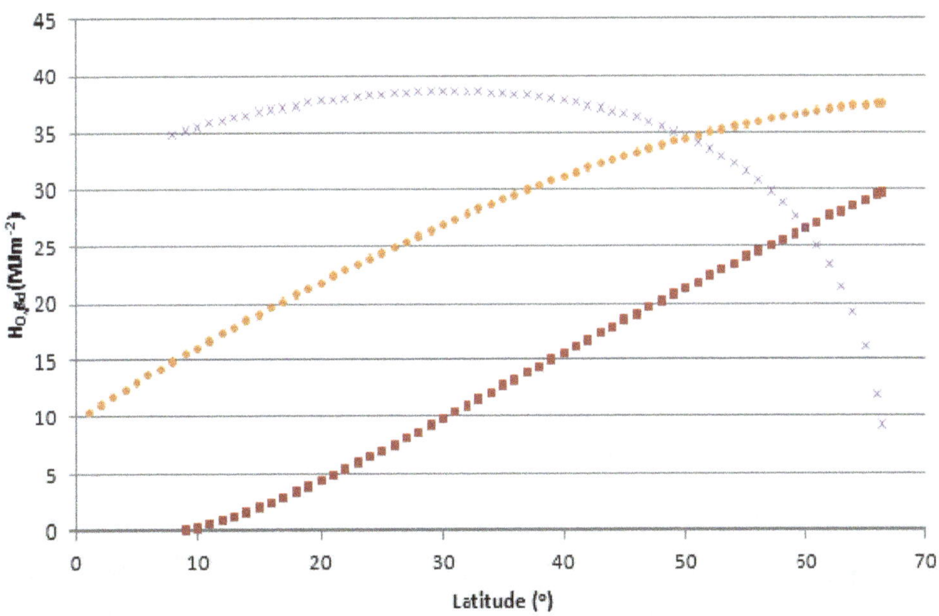

Figure 20. Daily extraterrestrial solar radiation on an Equator facing surface, tilted by an angle 90° in relation to the horizontal plane, at winter solstice (x), summer solstice (■) and equinox (●) in the Northern hemisphere

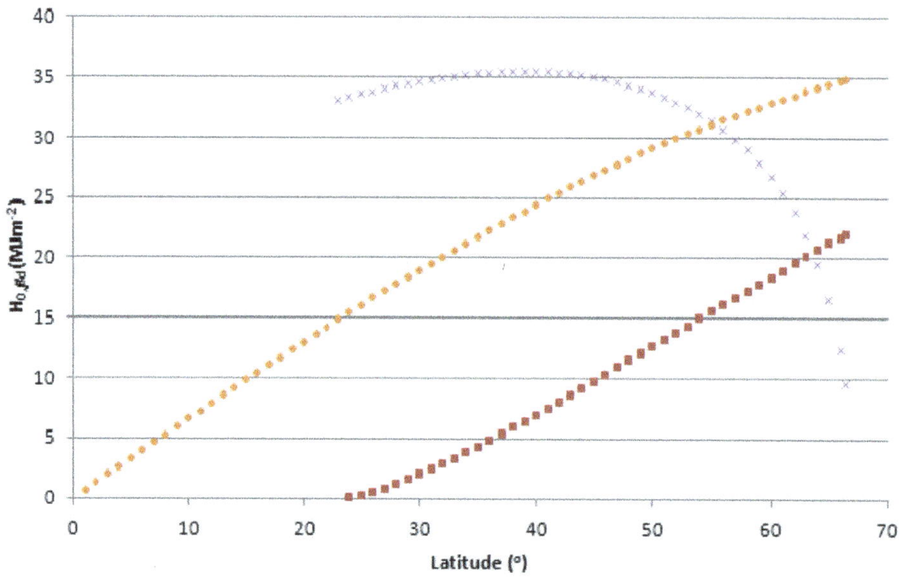

Figure 21. Daily extraterrestrial solar radiation on an Equator facing surface, tilted by an angle 15° in relation to the horizontal plane for latitudes 10°N(x), 20°N (●), 30°N (-), 40°N (♦), 50°N (▲), 60°N (ж) and 66.45°N (+) in the Northern hemisphere

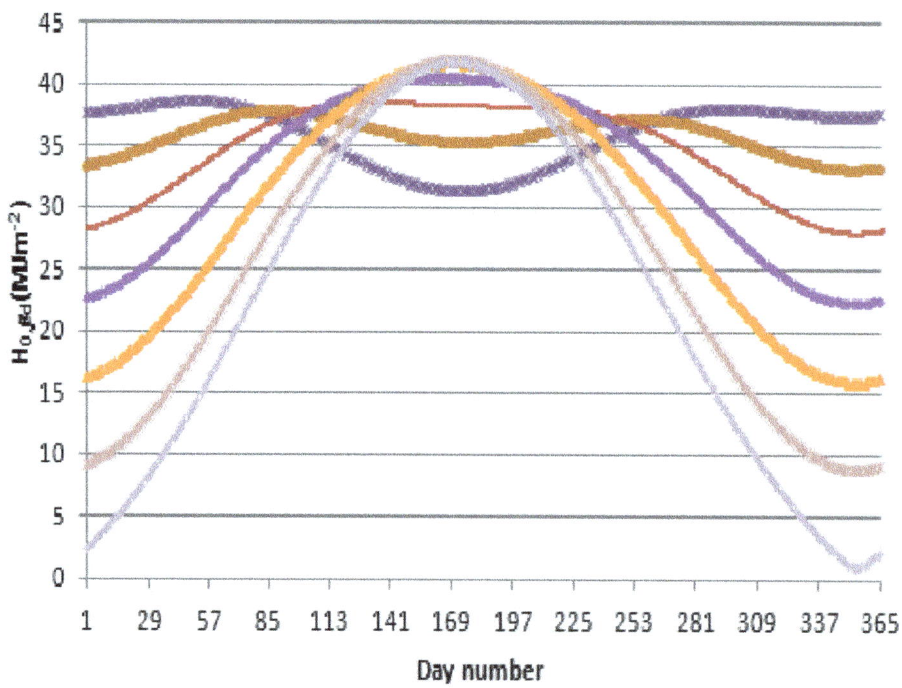

Figure 22. Daily extraterrestrial solar radiation on an Equator facing surface, tilted by an angle 30° in relation to the horizontal plane for latitudes 10°N(x), 20°N (●), 30°N (-), 40°N (♦), 50°N (▲), 60°N (ж) and 66.45°N (+) in the Northern hemisphere

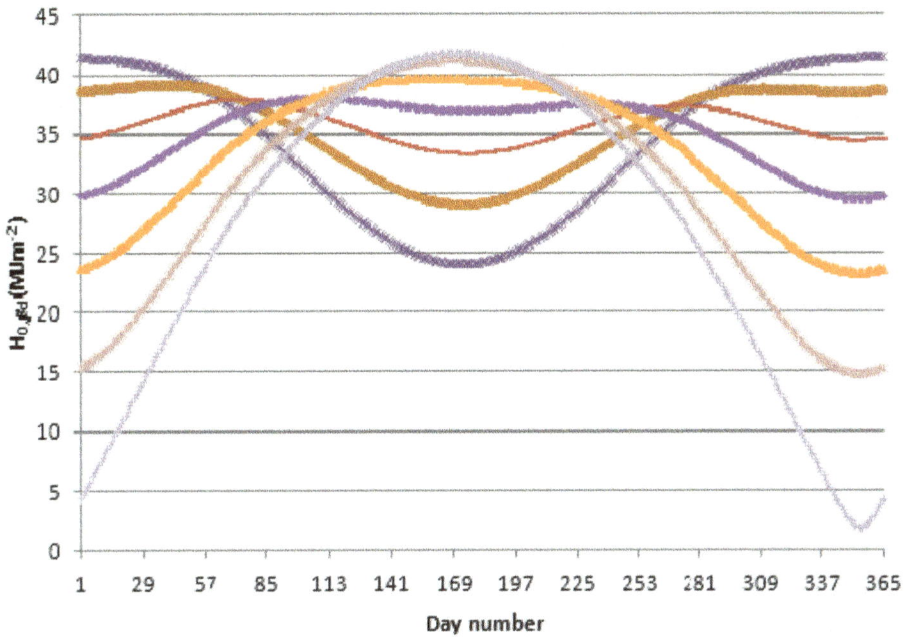

Figure 23. Daily extraterrestrial solar radiation on an Equator facing surface, tilted by an angle 45° in relation to the horizontal plane for latitudes 10°N(x), 20°N (●), 30°N (-), 40°N (♦), 50°N (▲), 60°N (ж) and 66.45°N (+) in the Northern hemisphere

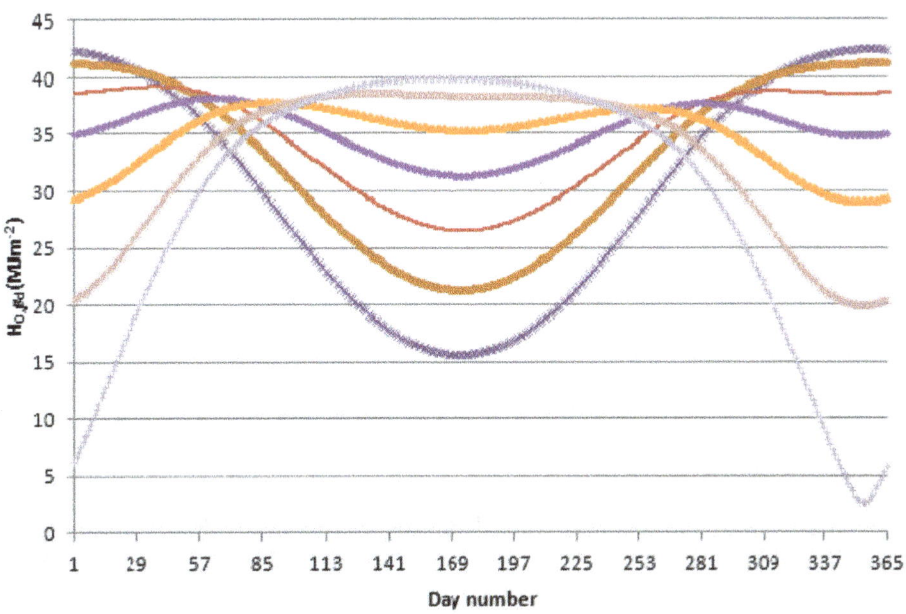

Figure 24. Daily extraterrestrial solar radiation on an Equator facing surface, tilted by an angle 60° in relation to the horizontal plane for latitudes 10°N(x), 20°N (●), 30°N (-), 40°N (♦), 50°N (▲), 60°N (ж) and 66.45°N (+) in the Northern hemisphere

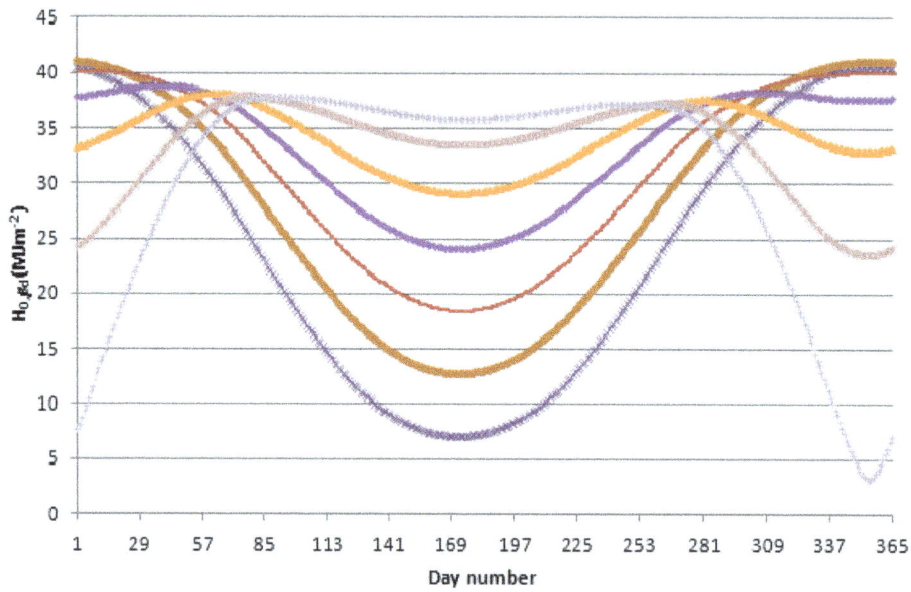

Figure 25. Daily extraterrestrial solar radiation on an Equator facing surface, tilted by an angle 75° in relation to the horizontal plane for latitudes 10°N(x), 20°N (●), 30°N (-), 40°N (♦), 50°N (▲), 60°N (ж) and 66.45°N (+) in the Northern hemisphere

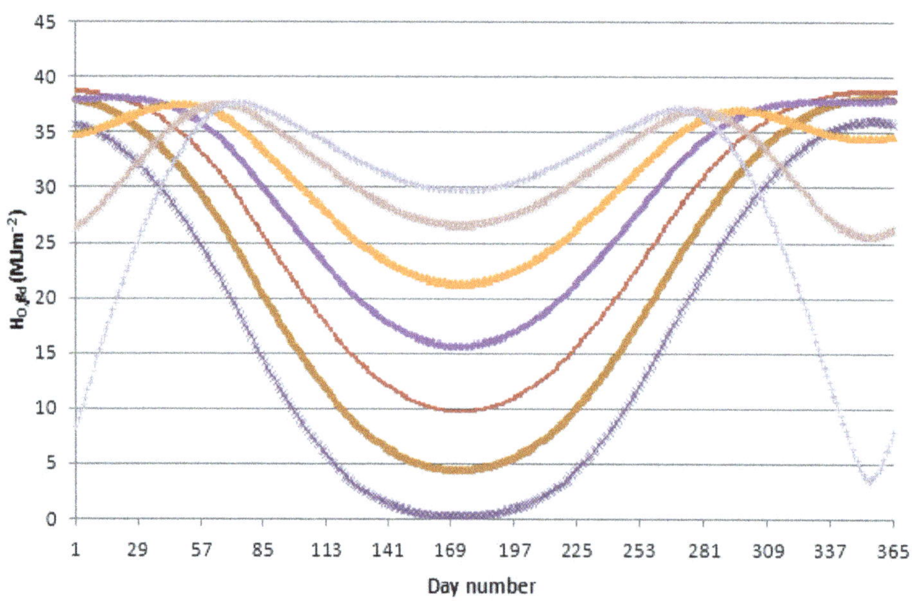

Figure 26. Daily extraterrestrial solar radiation on an Equator facing surface, tilted by an angle 90° in relation to the horizontal plane for latitudes 10°N(x), 20°N (●), 30°N (-), 40°N (♦), 50°N (▲), 60°N (ж) and 66.45°N (+) in the Northern hemisphere

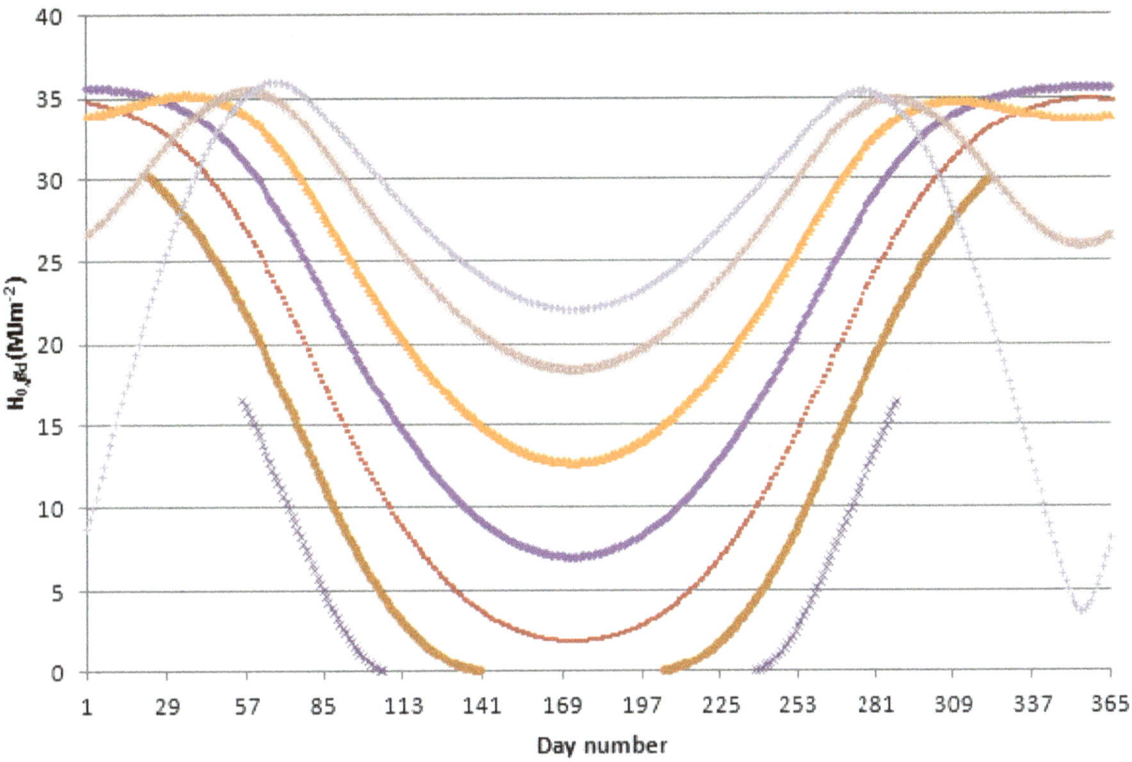

period started on 147th day number and ended on 212th day number. This result is clearly demonstrated in the Figure 26 and the low values, presented in the Table 7, confirm this result. Moreover, it is found in the first chapter of this book that, in the tropical zone the Equator facing surfaces are recommended for a period of several consecutive days while the North Pole facing surfaces are recommended for another period of several consecutive days during the year. This means that, the solar energy gain with respect to that received by a horizontal plane is greater than 1 for a range of tilt angle variation while it becomes lower than 1 for another range of tilt angle variation. These ranges are of latitude dependent values. This result is also observed in the mid-latitude zone (23.45° N ≤φ ≤43.45°N)

The monthly mean daily extraterrestrial radiation, $H_{0',m(J}$m-2), is a useful quantity in several calculations. For latitudes in the range 66.45°N to 66.45°S, this quantity could be calculated by summation of equation (16) over the days of each month in the year. The latitude dependence of $H_{0',m\ v}$alues for the latitudes of the Northern Hemisphere are presented in the Tables 11 to 16.

Table 10. The average daily yearly values of $H_{0,d}(MJm^{-2})$

β (o)	φ(°)						
	10N	20N	30N	40N	50N	60N	66.45N
0	35.47296	33.9322	31.48109	28.23104	24.36105	20.19514	17.73145
15	35.89163	35.81693	34.74349	32.70633	29.76865	25.97464	22.80702
30	33.97443	35.37088	35.76011	35.10375	33.3616	30.35104	26.95434
45	29.8756	32.62871	34.44844	35.22325	34.81469	32.83542	29.49302
60	23.95483	27.81395	30.9111	33.05242	34.00529	33.20407	30.15984
75	16.94376	21.37573	25.44358	28.7628	30.99283	31.4186	28.88232
90	2.001544	8.778078	18.60915	22.72707	26.01928	27.61395	25.74931

Table 11. The average daily monthly $H_{0,m}(Jm^{-2})$ at β=15o in the Northern hemisphere

Month	φ(°)						
	10N	20N	30N	40N	50N	60N	66.45N
Jan.	37.89503	34.08882	29.37786	23.89656	17.77511	10.96513	5.311697
Feb.	38.4792	36.03875	32.558	28.13991	22.90775	16.95719	12.6587
March	37.95902	37.52207	35.95051	33.29194	29.62647	25.06129	21.6935
April	35.79498	37.47638	38.05456	37.51551	35.88422	33.22811	31.02447
May	33.04107	36.28369	38.55068	39.78634	39.98507	39.20842	38.26499
June	31.44135	35.35086	38.37507	40.44213	41.53812	41.73456	41.48363
July	32.01924	35.60454	38.26569	39.93841	40.61213	40.35218	39.77537
Aug.	34.34102	36.64494	37.90026	38.07546	37.18112	35.2777	33.5719
Sep.	36.77132	37.12193	36.35008	34.47967	31.56865	27.70747	24.76799
Oct.	37.9228	36.14567	33.30156	29.47541	24.77772	19.31863	15.39139
Nov.	37.76204	34.30975	29.92372	24.73046	18.86022	12.32035	7.251005
Dec.	37.46496	33.26924	28.22268	22.46685	16.12638	9.049851	1.93645

Equator Facing Tilted Surface in the Southern Hemisphere

When calculating the daily extraterrestrial solar radiation on an Equator facing surface, tilted by an angle β in relation to the horizontal plane for latitudes 10°S, 20°S, 30°S, 40°S, 50°S, 60°S and 66.45°S in the Southern Hemisphere, the results presented in the Figures 27 to 32 were obtained.

Example 4

Calculate the $H_{0,d}$ on a surface, tilted by an angle 15° and orientated towards the Equator, in the absence of the atmosphere at latitudes 10°S to 66.45°S on June 21.

Table 12. The average daily monthly $H_{0,m}(Jm^{-2})$ at β=30o in the Northern hemisphere

Month	φ(°)						
	10N	20N	30N	40N	50N	60N	66.45N
Jan.	41.14707	38.84039	35.40791	30.8759	25.17857	17.69151	9.568
Feb.	39.82207	39.04478	37.10503	34.03221	29.8473	24.43129	19.88746
March	36.44827	37.72493	37.85813	36.84126	34.69933	31.47866	28.842
April	31.32582	34.55512	36.77124	37.90514	37.92407	36.83324	35.56043
May	26.57426	31.08312	34.77913	37.54368	39.29908	40.01337	39.93037
June	24.18883	29.19004	33.50037	36.9793	39.53068	41.10957	41.61918
July	25.15813	29.91247	33.92158	37.05577	39.22769	40.39826	40.62949
Aug.	29.08215	32.81629	35.62055	37.40658	38.12326	37.75952	36.96551
Sep.	34.17147	36.17483	37.0844	36.87204	35.54384	33.13904	31.04729
Oct.	38.35677	38.25616	37.00701	34.63077	31.1599	26.56854	22.83968
Nov.	40.5411	38.62326	35.57766	31.43386	26.15658	19.30226	12.60987
Dec.	41.26017	38.48988	34.61504	29.65428	23.49292	15.20398	3.688907

Table 13. The average daily monthly $H_{0,m}(Jm^{-2})$ at β=45o in the Northern hemisphere

Month	φ(°)						
	10N	20N	30N	40N	50N	60N	66.45N
Jan.	41.595	40.94505	39.02497	35.75111	30.86616	23.21225	13.17226
Feb.	38.45113	39.38997	39.12342	37.60528	34.75281	30.24043	25.76091
March	32.45508	35.35815	37.18696	37.88116	37.40894	35.75268	34.02735
April	24.81878	29.36236	33.06142	35.79498	37.47638	38.05456	37.83407
May	18.64523	24.06288	28.92135	33.04107	36.28369	38.55068	39.4673
June	15.80945	21.48493	26.76534	31.44135	35.35086	38.37507	39.82021
July	17.01714	22.55332	27.61875	32.01924	35.60454	38.26569	39.45972
Aug.	22.01895	26.90378	31.05852	34.34102	36.64494	37.90026	38.13722
Sep.	29.25637	32.7741	35.30256	36.76326	37.11025	36.32975	35.23301
Oct.	36.17678	37.75956	38.19048	37.42611	35.41859	32.00785	28.73149
Nov.	40.55735	40.30465	38.80704	35.9951	31.67041	24.96876	17.1094
Dec.	42.24358	41.08751	38.64843	34.82083	29.25847	20.32197	5.189971

Solution

For these circumstances, n=172, G_s=1322.6W/m^2 and δ=23.458°. Then, the application of the equations (16) and (17) leads to the following results:

φ=10°S ω_{ss}=85.60° $H_{o,d}$ = 35.03MJ/m^2

Table 14. The average daily monthly $H_{0,m}(Jm^{-2})$ at β=60o in the Northern hemisphere

Month	φ(°)						
	10N	20N	30N	40N	50N	60N	66.45N
Jan.	39.20829	40.25938	39.98255	38.18993	34.45027	27.15112	15.87885
Feb.	34.45981	37.0508	38.47561	38.61561	37.28997	33.98874	29.8788
March	26.25237	30.58339	33.98288	36.34071	37.57039	37.59155	36.89528
April	16.77204	22.27755	27.18724	31.32582	34.55512	36.77124	37.62935
May	10.00179	15.79646	21.41093	26.57426	31.08312	34.77913	36.67469
June	7.196696	12.90623	18.68129	24.18883	29.19004	33.50037	35.84649
July	8.414604	14.14852	19.82995	25.15813	29.91247	33.92158	36.04973
Aug.	13.73474	19.35768	24.54229	29.08215	32.81629	35.62055	36.89314
Sep.	22.36845	27.15499	31.12726	34.16033	36.15887	37.057	37.03155
Oct.	31.5314	34.6897	36.77134	37.67092	37.26356	35.26587	32.6653
Nov.	37.80969	39.23935	39.39179	38.10334	35.02595	28.93367	20.44295
Dec.	40.34815	40.88508	40.04801	37.61439	33.03009	24.05506	6.337347

Table 15. The average daily monthly $H_{0,m}(Jm^{-2})$ at β=75o in the Northern hemisphere

Month	φ(°)						
	10N	20N	30N	40N	50N	60N	66.45N
Jan.	34.14961	36.83009	38.21538	38.02618	35.68665	29.23967	17.50333
Feb.	28.12011	32.18667	35.20575	36.99435	37.28588	35.42077	31.96051
March	18.2656	23.72721	28.46481	32.32516	35.17273	36.8698	37.25007
April	7.930008	13.86416	19.58661	24.81878	29.36236	33.06142	34.94128
May	2.052331	7.158287	12.9006	18.64523	24.06288	28.92135	31.67267
June	0.343653	4.528987	10.01619	15.80945	21.48493	26.76534	29.86166
July	1.026969	5.668814	11.26419	17.01714	22.55332	27.61875	30.5435
Aug.	5.169971	10.84338	16.58572	22.01895	26.90378	31.05852	33.28303
Sep.	14.00311	19.7114	24.84819	29.24292	32.75494	35.26995	36.31775
Oct.	24.73722	29.2558	32.84629	35.34852	36.56907	36.12058	34.37301
Nov.	32.48535	35.49996	37.29205	37.61489	35.99454	30.92681	22.38335
Dec.	35.70306	37.89641	38.71837	37.8446	34.55077	26.14884	7.052844

φ=20°S ω_{ss}=80.90° $H_{o,d}$ = 31.04MJ/m²

φ=30°S ω_{ss}=75.52° $H_{o,d}$ = 26.26MJ/m²

φ=40°S ω_{ss}=68.75° $H_{o,d}$ = 20.83MJ/m²

φ=50°S ω_{ss}=59.01° $H_{o,d}$ = 14.86MJ/m²

Table 16. The average daily monthly $H_{0,m}(Jm^{-2})$ at β=90o in the Northern Hemisphere

Month	φ(°)						
	10N	20N	30N	40N	50N	60N	66.45N
Jan.	0	9.524594	33.84389	35.271	34.49104	29.3356	17.93498
Feb.	2.273641	25.12908	29.53668	32.85199	34.74082	34.43893	31.86415
March	9.057887	15.26147	21.01053	26.10894	30.37972	33.63676	35.06756
April	0.649499	5.035282	10.90007	16.77204	22.27755	27.18724	29.95431
May	0	0.399294	4.456556	10.00179	15.79646	21.41093	24.80676
June	0	0	2.146761	7.196696	12.90623	18.68129	22.28062
July	0	0.038974	3.134916	8.414604	14.14852	19.82995	23.32212
Aug.	0.041552	2.61334	7.962684	13.73474	19.35768	24.54229	27.55524
Sep.	4.917505	10.99446	16.90976	22.3536	27.13395	31.09167	33.14043
Oct.	7.026096	21.82816	26.68283	30.61717	33.38247	34.51374	33.73827
Nov.	0	17.00434	32.65092	34.56305	34.51015	30.81233	22.79836
Dec.	0	0	34.75015	35.49576	33.71687	26.46062	7.287701

φ=60°S ω_{ss}=41.25°$H_{o,d}$ = 8.19MJ/m^2

φ=66.45°S ω_{ss}=5.44°$H_{o,d}$ = 0.89MJ/m^2

Moreover, for latitudes in the range 0° to 66.45°S, the daily amount of extraterrestrial radiation, $H_{0,d}$, could be calculated using equation (16). The $H_{0,d}$ is plotted as a function of the tilt and latitude for the Northern Hemispheres at summer solstice, winter solstice and equinox in Figures 33 to 38.

From Figures 27 to 32 it is seen that, in the Southern Hemisphere the daily extraterrestrial solar radiation on an Equator facing surface, tilted by an angle β in relation to the horizontal plane depends on the day number, latitude angle φ and tilt angle β. The dependence of $H_{0',d\,o}$n the latitude and day number is very similar to that of the daily extraterrestrial solar radiation on the horizontal plane $H_{0',d.}$ Concerning the dependence of $H_{0',\beta d\,on}$ the tilt angle, Figures 27 to 32 show that there is a value of the tilt angle for each latitude which balances the values of the daily extraterrestrial solar radiation on an Equator facing surface all over the year. These values are very near to the latitude angle value. Table 17 sheds a light on the dependence of the average daily yearly values of H_0,d_{on} β and φ. It was found that, in the tropical zone of the Southern Hemisphere, 0°S≤φ≤ 23.45°S, 'he Sun does rise on the Equator facing vertically mounted surface during two periods in the year. The length of each of these periods depends on the latitude value. For example, for the latitude of 10°S, one period started on 290th day number and ended on 55th day number while the second period started on 107th day number and ended on 237th day number. For the latitude of 20°S, one period started on 323rd day number and ended on 21st day number while the second period started on 147th day number and ended on 204th day number. Moreover, it is found in the first chapter of this book that, in the tropical zone the Equator facing surfaces are recommended for a period of several consecutive days while the South Pole facing surfaces are recommended for another period of several consecutive days during the year. This means that, the solar energy gain with respect to that received by a horizontal plane is greater than 1 for a range of tilt angle variation while it becomes

Figure 27. Daily extraterrestrial solar radiation on an Equator facing surface, tilted by an angle 15° in relation to the horizontal plane for latitudes 10°S(x), 20°S (●), 30°S (-), 40°S (♦), 50°S (▲), 60°S (ж) and 66.45°S (+) in the Southern hemisphere

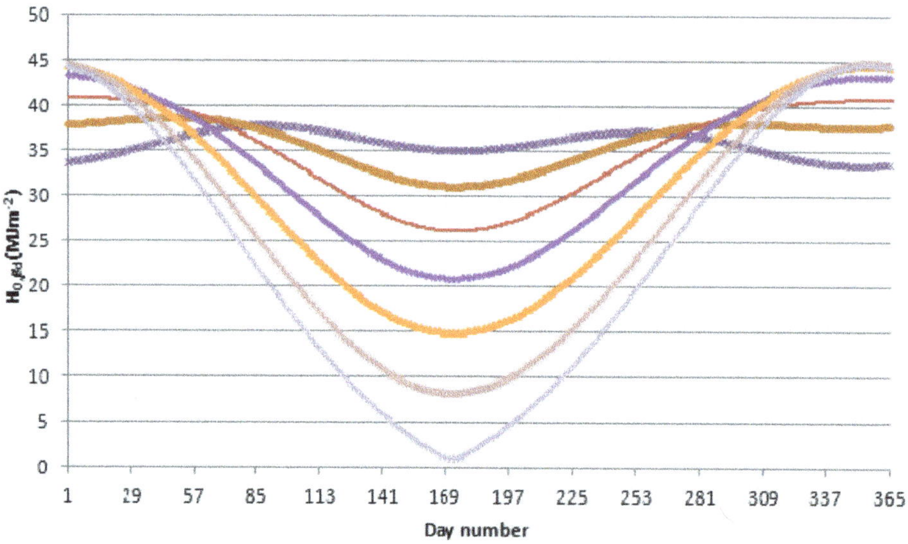

Figure 28. Daily extraterrestrial solar radiation on an Equator facing surface, tilted by an angle 30° in relation to the horizontal plane for latitudes 10°S(x), 20°S (●), 30°S (-), 40°S (♦), 50°S (▲), 60°S (ж) and 66.45°S (+) in the Southern hemisphere

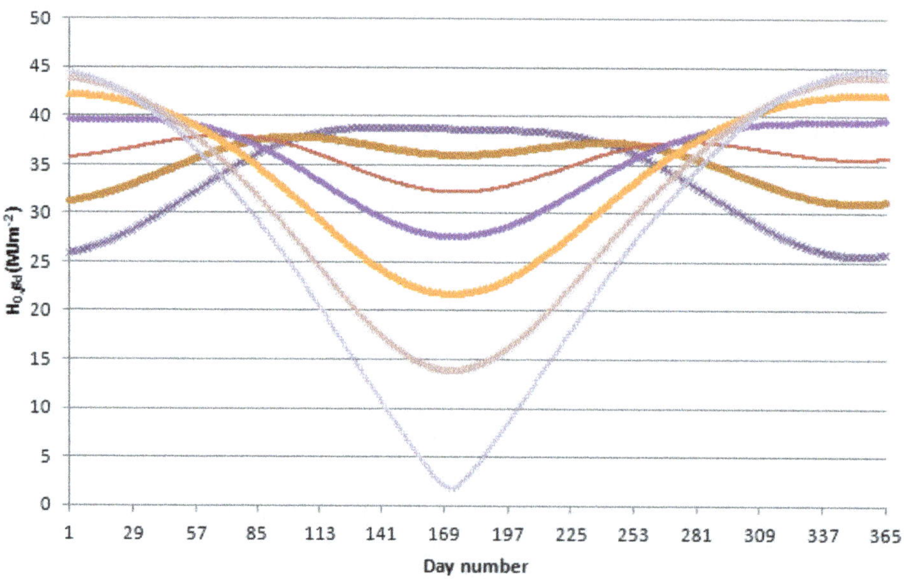

Figure 29. Daily extraterrestrial solar radiation on an Equator facing surface, tilted by an angle 45° in relation to the horizontal plane for latitudes 10°S(x), 20°S (●), 30°S (-), 40°S (♦), 50°S (▲), 60°S (ж) and 66.45°S (+) in the Southern hemisphere

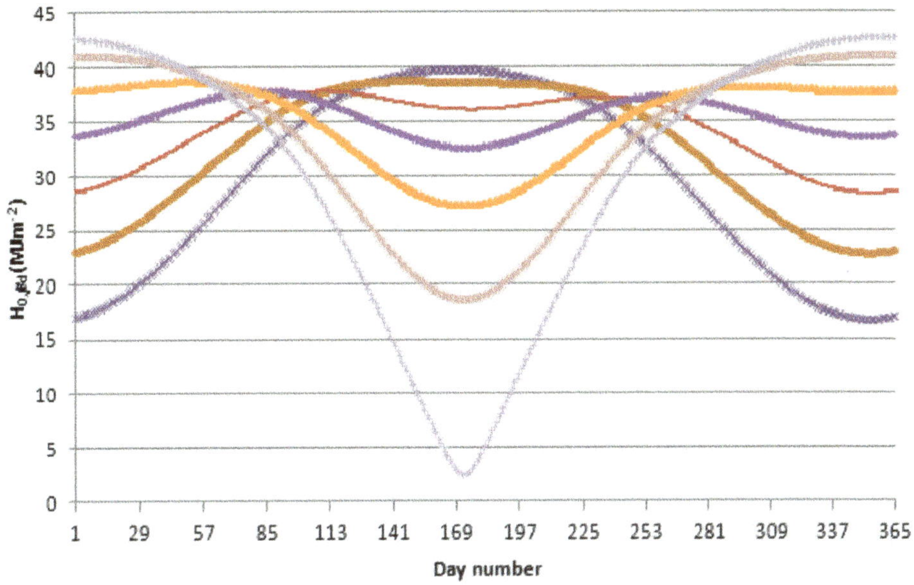

Figure 30. Daily extraterrestrial solar radiation on an Equator facing surface, tilted by an angle 60° in relation to the horizontal plane for latitudes 10°S(x), 20°S (●), 30°S (-), 40°S (♦), 50°S (▲), 60°S (ж) and 66.45°S (+) in the Southern hemisphere

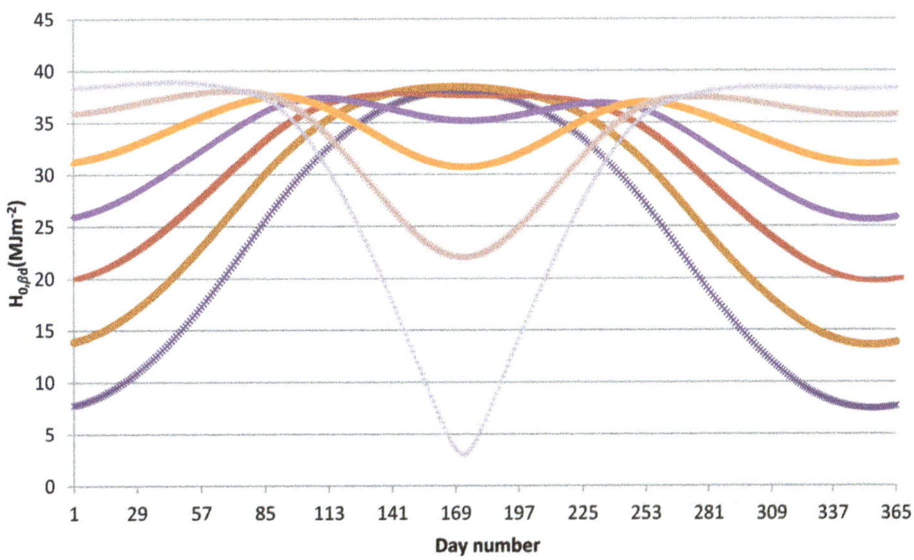

Figure 31. Daily extraterrestrial solar radiation on an Equator facing surface, tilted by an angle 75° in relation to the horizontal plane for latitudes 10°S(x), 20°S (●), 30°S (-), 40°S (♦), 50°S (▲), 60°S (ж) and 66.45°S (+) in the Southern hemisphere

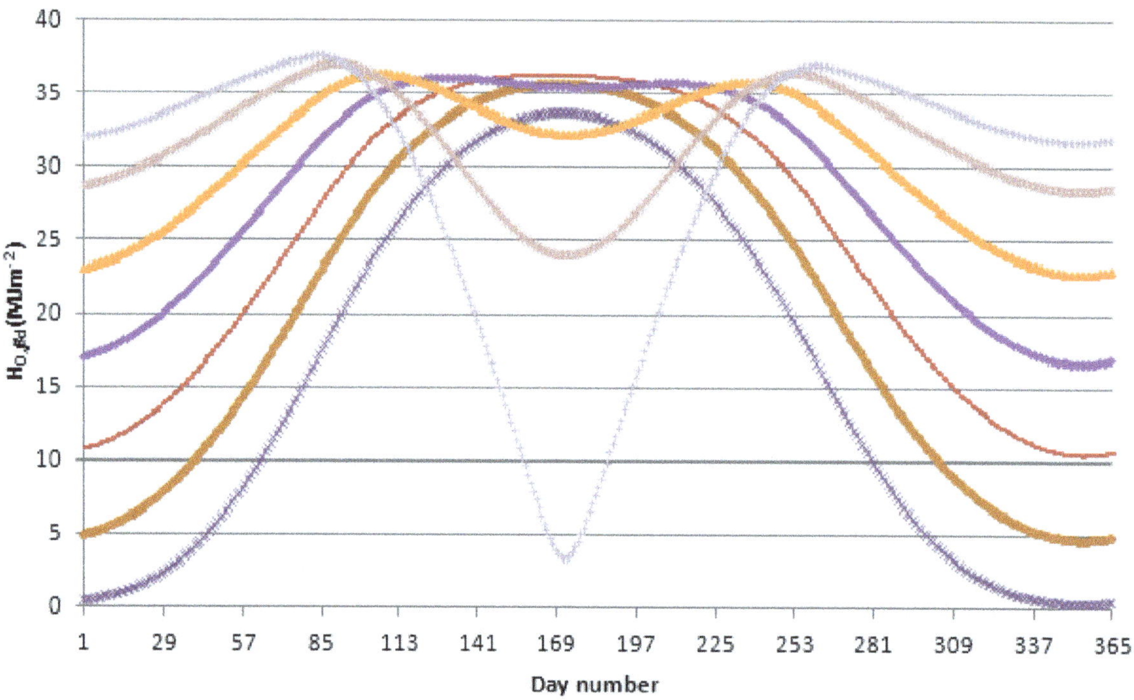

Figure 32. Daily extraterrestrial solar radiation on an Equator facing surface, tilted by an angle 90° in relation to the horizontal plane for latitudes 10°S(x), 20°S (●), 30°S (-), 40°S (♦), 50°S (▲), 60°S (ж) and 66.45°S (+) in the Southern hemisphere

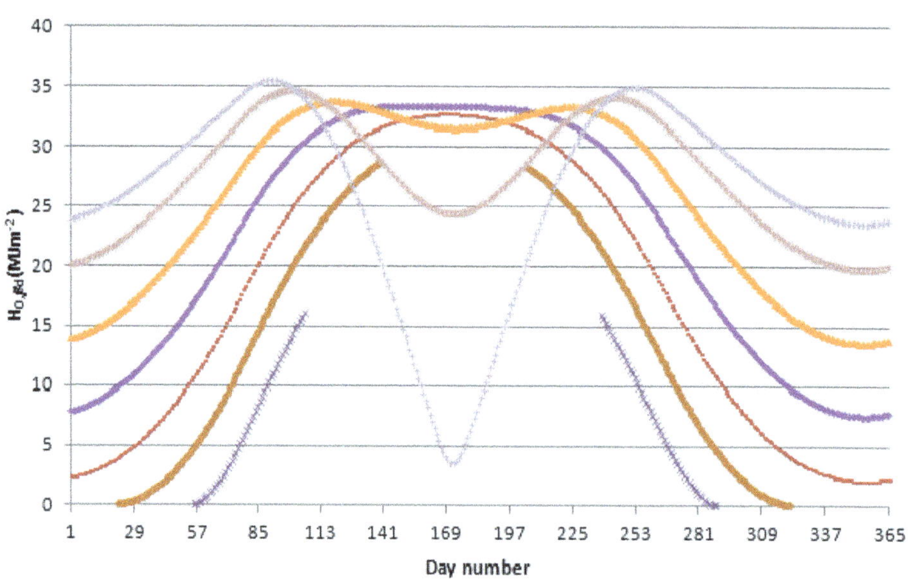

Figure 33. Daily extraterrestrial solar radiation on a tilted surface with an angle β=15o at summer solstice (▲), winter solstice (ж) and equinox (+) in the Southern hemisphere

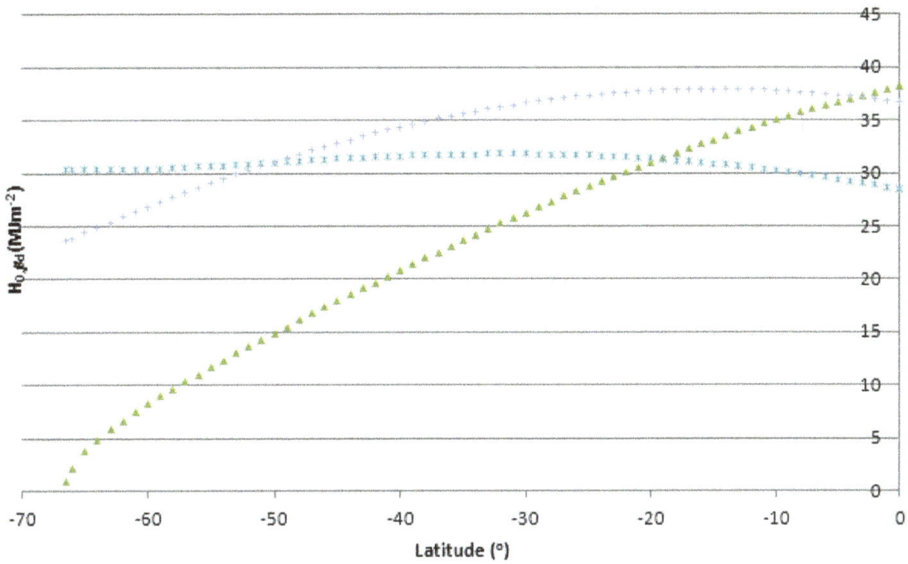

Figure 34. Daily extraterrestrial solar radiation on a tilted surface with an angle β=30o at summer solstice (▲), winter solstice (ж) and equinox (+) in the Southern hemisphere

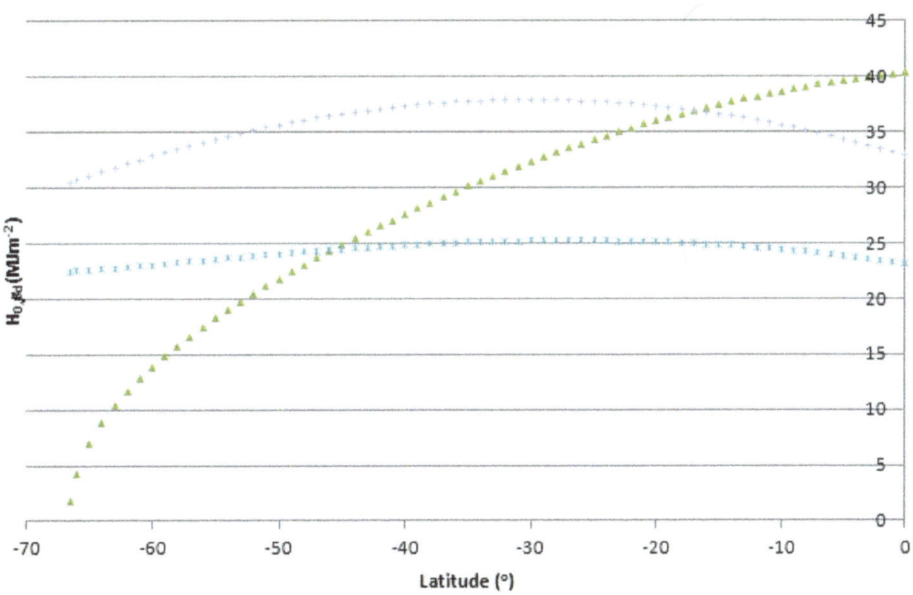

Figure 35. Daily extraterrestrial solar radiation on a tilted surface with an angle β=45o at summer solstice (▲), winter solstice (ж) and equinox (+) in the Southern hemisphere

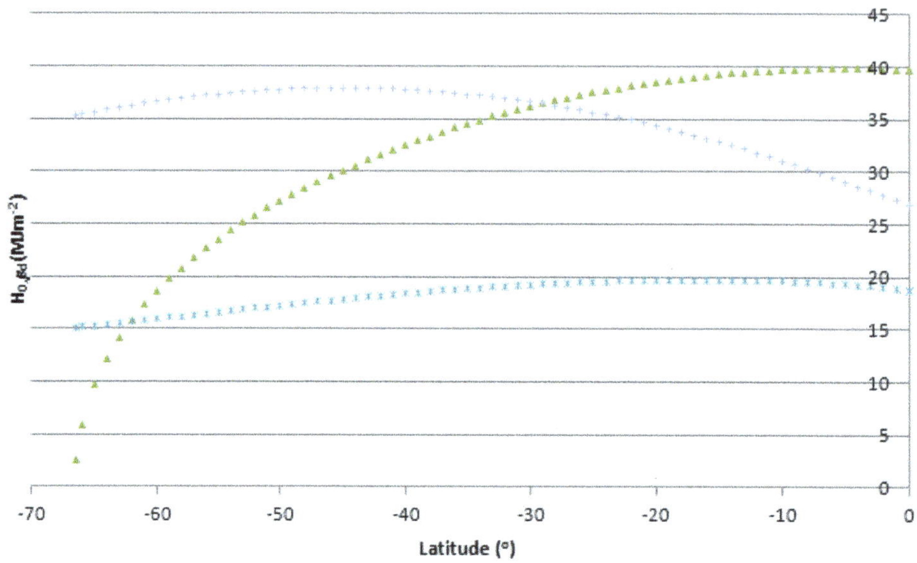

Figure 36. Daily extraterrestrial solar radiation on a tilted surface with an angle β=60o at summer solstice (▲), winter solstice (ж) and equinox (+) in the Southern hemisphere

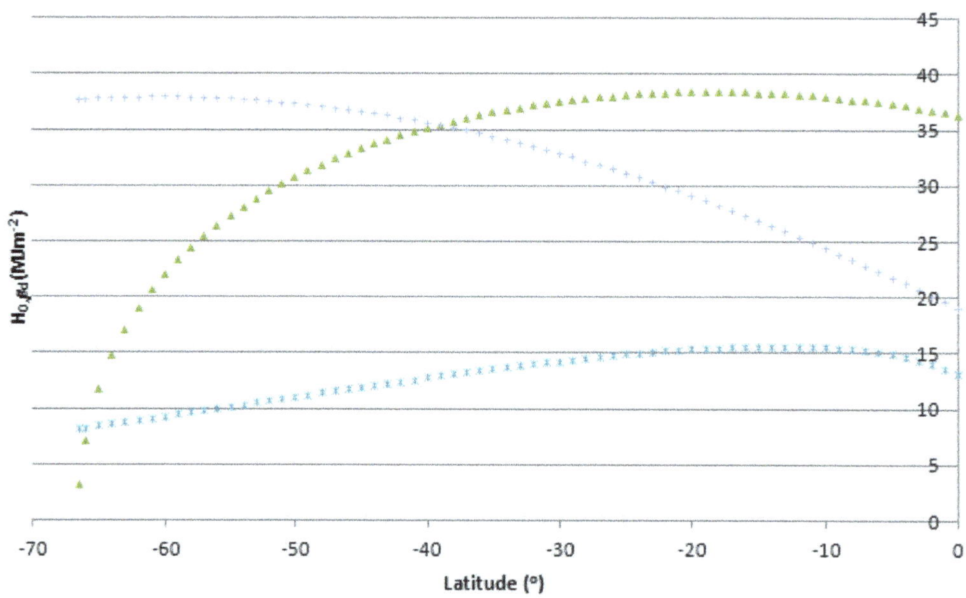

Figure 37. Daily extraterrestrial solar radiation on a tilted surface with an angle β=75o at summer solstice (▲), winter solstice (ж) and equinox (+) in the Southern hemisphere

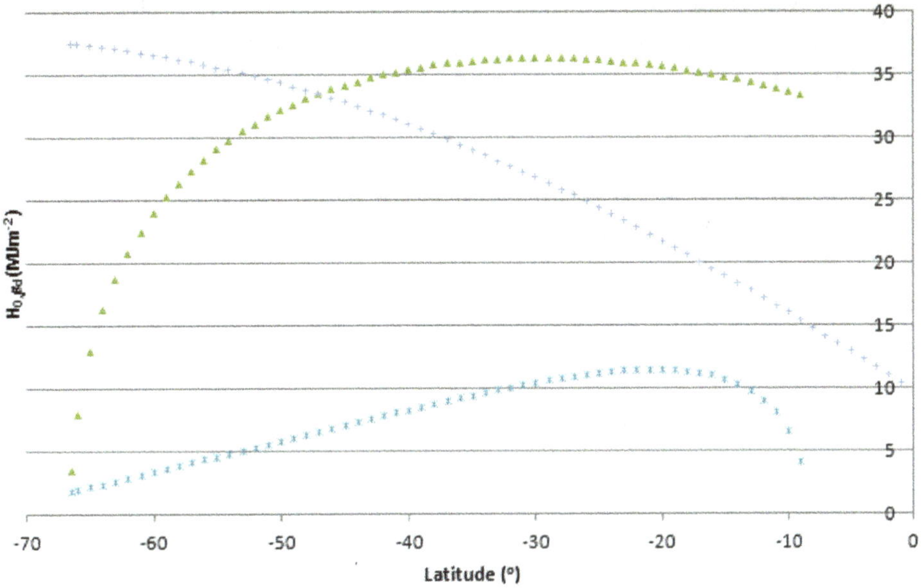

Figure 38. Daily extraterrestrial solar radiation on a tilted surface with an angle β=90o at summer solstice (▲), winter solstice (ж) and equinox (+) in the Southern hemisphere

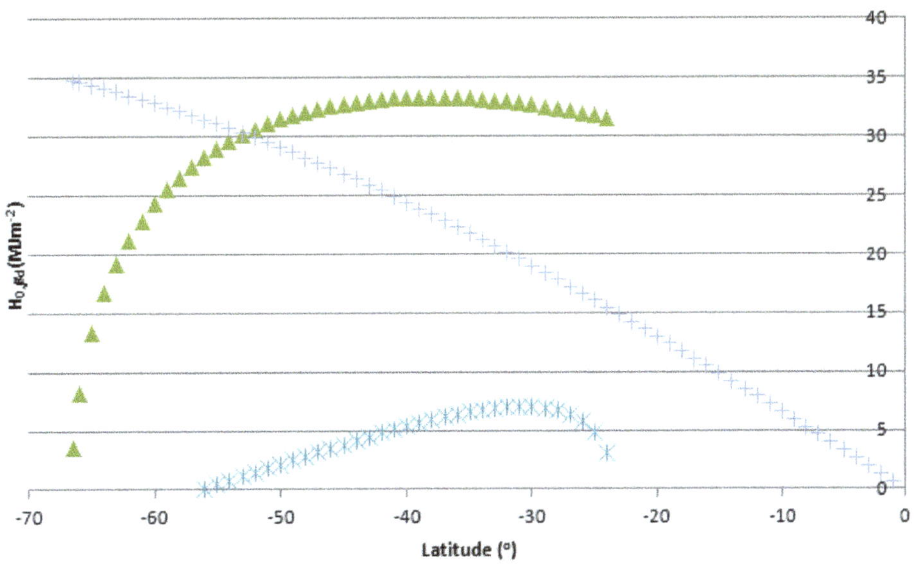

lower than 1 for another range of tilt angle variation. These ranges are of latitude dependent values. This result is also observed in the mid-latitude zone (23.45° S ≤φ ≤43.45°S). The dependence of the average daily yearly values of H_0,d($_{MJm}$-2) ^{on} the solar receiver surface tilt angle is plotted in the Figures 39 to 41 for different latitudes in the Southern Hemisphere. The Figure 39 is related to the tropical zone, the Figure 40 is related to the mid-latitude zone while the Figure 41 is related to the high-latitude zone.

EXTRATERRESTRIAL SOLAR RADIATION ON THE TILTED SURFACE WITH DIFFERENT ORIENTATIONS

The incidence angle of Sun's rays on the tilted surface with different orientations is determined by the equation (33) of the first chapter. This equation was also written in the first chapter as:

$$\cos\left(\theta_i\right) = A_1 + A_2\cos\left(\omega\right) + A_3\sin\left(\omega\right)$$

where

$$A_1 = \sin\left(\delta\right)\left[\sin\left(\varphi\right)\cos\left(\beta\right) - \sin\left(\beta\right)cos\left(\varphi\right)\cos\left(\gamma\right)\right]$$

$$A_2 = \cos\left(\delta\right)\left[\cos\left(\varphi\right)\cos\left(\beta\right) + \sin\left(\beta\right)sin\left(\varphi\right)\cos\left(\gamma\right)\right]$$

Figure 39. Yearly average daily extraterrestrial solar radiation on a tilted surface $H_{0,\beta d}$ variations with solar receiver surface tilt angle β for latitudes 10oS(♦)and 20oS (■)

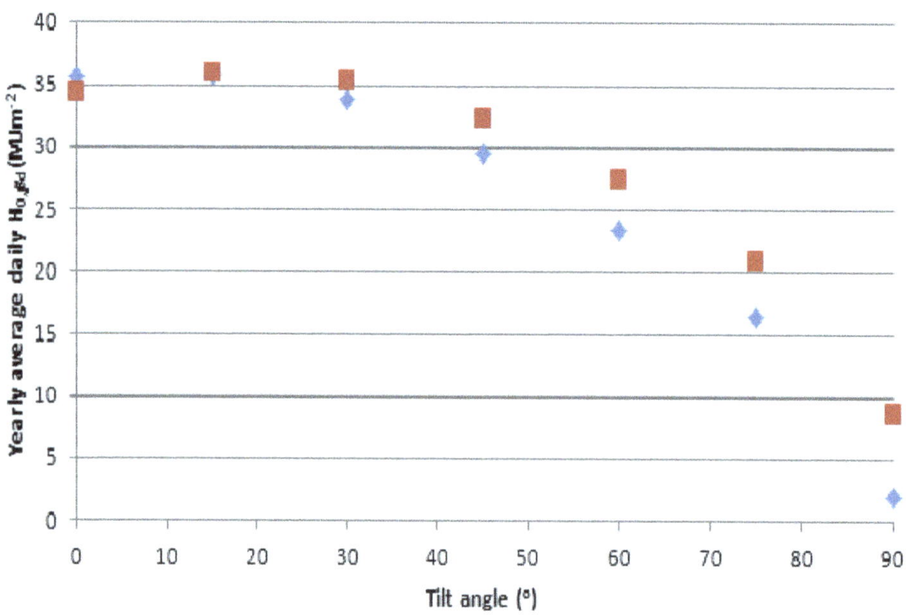

Figure 40. Yearly average daily extraterrestrial solar radiation on a tilted surface $H_{0,\beta d}$ variations with solar receiver surface tilt angle β for latitudes 30oS(♦)and 40oS(■)

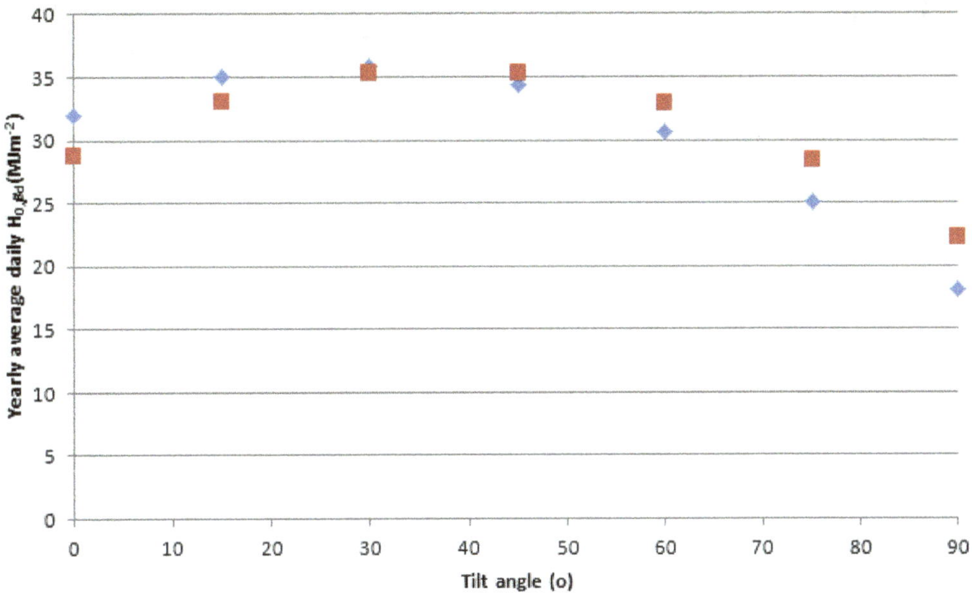

Figure 41. Yearly average daily extraterrestrial solar radiation on a tilted surface $H_{0,\beta d}$ variations with solar receiver surface tilt angle β for latitudes 50oS(♦), 20oS(■) and 66.45oS(▲)

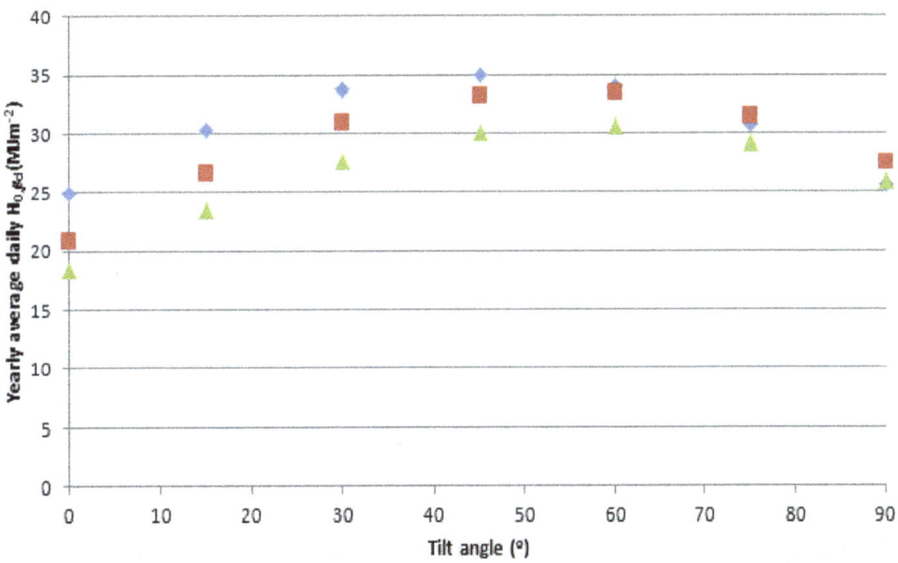

Table 17. The average daily yearly values of $H_{0,d}(MJm^{-2})$

β (o)	φ(°)						
	10S	20S	30S	40S	50S	60S	66.45S
0	35.60994	34.202	31.87552	28.73811	24.96534	20.87831	18.4546
15	35.82605	35.88918	34.95195	33.0457	30.23098	26.55388	23.47892
30	33.71716	35.24707	35.77551	35.2611	33.66297	30.80842	27.56632
45	29.45198	32.32393	34.27607	35.19453	34.94235	33.15009	30.01671
60	23.40623	27.3578	30.56981	32.846	33.95697	33.36168	30.56761
75	16.3396	20.81521	24.96693	28.40055	30.77848	31.41476	29.15285
90	1.977318	8.633665	18.05118	22.24619	25.66232	27.45601	25.87073

$$A_3 = \cos(\delta)\sin(\beta)sin(\gamma)$$

Accordingly, the extraterrestrial solar radiation on the tilted surface with different orientations could be written as follows (Soulayman, 1991):

$$H_{0,\beta d} = \frac{12x3600xG_s}{\pi} * \left[A_2\left(sin\omega_{ss} - sin\omega_{sr}\right) - A_3\left(cos\omega_{ss} - cos\omega_{sr}\right) + \frac{\pi * \left(\omega_{ss} - \omega_{sr}\right)}{180^o} A_1\right] \quad (18)$$

where

$$\omega_{ss} = min\left\{arccos\left[-tan(\delta)tan(\varphi)\right], arccos\left(-\frac{A_1}{A_4}\right) + arcsin\left(\frac{A_3}{A_4}\right)\right\} \quad (19)$$

$$\omega_{sr} = \max\left\{-\arccos\left[-\tan(\delta)\tan(\varphi)\right], -\arccos\left(-\frac{A_1}{A_4}\right) + \arcsin\left(\frac{A_3}{A_4}\right)\right\} \quad (20)$$

$$A_4 = \sqrt{A_2^2 + A_3^2}$$

The Equator facing case was treated previously. Some additional cases are useful to be treated also because of their wide area of applications. These cases are the North/South Pole facing case, the vertical east/west facing case and Equator facing case with ± 45° of receiver azimuth angle.

North Pole Vertical Surface Facing Case

In this case, β = 90° and γ =-180°. Thus, according to the equation (34) of the first chapter the incidence angle of Sun's rays on the North Pole facing tilted surface is:

$$\cos\left(\theta_i\right) = -\sin\left(\delta\right) * cos\left(\varphi\right) + \cos\left(\delta\right) * \sin\left(\varphi\right) * \cos\left(\omega\right) \tag{21}$$

and the sunrise and sunset hour angles satisfy the following equations (see the equations (41e) and (42e) of the first chapter):

$$\omega_{ss} = min\left\{arccos\left[-tan\left(\delta\right)tan\left(\varphi\right)\right], arccos\left[\frac{\tan\left(\delta\right)}{\tan\left(\varphi\right)}\right]\right\}$$

$$\omega_{sr} = \max\left\{-\arccos\left[-\tan\left(\delta\right)\tan\left(\varphi\right)\right], -\arccos\left[\frac{\tan\left(\delta\right)}{\tan\left(\varphi\right)}\right]\right\}$$

So, ω_{ss}=- ω_{sr} and the following relation:

$$-1 \leq \frac{\tan\left(\delta\right)}{\tan\left(\varphi\right)} \leq 1 \tag{22}$$

should be satisfied, elsewhere the Sun does not shine on the North Pole orientated vertical surface, located in the Northern Hemisphere. The relation (19) shows that it could not be applied in the tropical zone all over the year. On the other hand, outside of the tropical zone 0°≤φ≤23.45°, the Sun does not shine on the North Pole orientated vertical surface, located in the Northern Hemisphere, during the period from 22/9 to 21/3 while it can rise partially during the period from 22/3 to 21/9 in the mid-latitude zone 23.45°≤φ≤43.45°. In the high-latitude zone 43.45°<φ≤66.45° the Sun does not shine on the North Pole orientated vertical surface. The daily extraterrestrial solar radiation incident on a North Pole orientated vertical surface in the Northern Hemisphere could be calculated using the equation (18) which takes in this case the following form:

$$H_{0,\beta d} = \frac{24x3600xG_s}{\pi} * \left[\begin{array}{l}\cos\left(\delta\right) * sin\left(\varphi\right) * \left(sin\omega_s - sin\omega_{ss}\right) \\ -\frac{\pi * \left(\omega_{ss} - \omega_s\right)}{180^o}\sin\left(\delta\right) * cos\left(\varphi\right)\end{array}\right] \tag{18a}$$

where ω_s is the sunset our angle on the horizontal plane:

$$\omega_s = arccos\left[-tan\left(\delta\right)tan\left(\varphi\right)\right]$$

Example 5

Calculate the $H_{0,d}$ on a vertical surface, orientated towards the North Pole, in the absence of the atmosphere at latitudes 24°N to 59°N on June 21.

Solution

For these circumstances, n=172, G_s=1322.6W/m² and δ=23.458°. Then, the application of the equations (18a) leads to the following results:

φ=24°N $H_{o,d}$ = 30.64 MJ/m²

φ=30°N $H_{o,d}$ = 18.99 MJ/m²

φ=35°N $H_{o,d}$ = 14.80 MJ/m²

φ=40°N $H_{o,d}$ = 11.78 MJ/m²

φ=43.45°N $H_{o,d}$ = 10.50 MJ/m²

φ=50°N $H_{o,d}$ = 6.60 MJ/m²

φ=59°N $H_{o,d}$ = 0.16 MJ/m²

The daily extraterrestrial solar radiation incident on a North Pole orientated vertical surface in the Northern Hemisphere is calculated for several latitudes and plotted in the Figure 42. The results for the Southern Hemisphere are presented in the Figure 43, where it is seen that the Sun could rise partially on the South Pole orientated surface during the period from 22/9 to 21/3.

As it is seen from Figure 42 that, the Sun does not shine on the North Pole orientated vertical surface, located in the Northern Hemisphere, during the period from 22/9 to 21/3 while it can rise partially during the period from 22/3 to 21/9. On the other hand, Figure 43 that, the Sun shines partially on the South Pole orientated vertical surface, located in the Southern Hemisphere, during the period from 22/9 to 21/3. Moreover, for latitudes of 60°N and 60°S and higher (φ≥ 60°N, φ≥ 60°S) it is not observed any solar radiation on the North and South Poles orientated vertical surfaces.

Vertical Surface Orientated Towards East

In this case, β = 90° and γ =-90°. Thus, the incidence angle of Sun's rays on the east facing vertical surface is:

$$\cos\left(\theta_i\right) = -\cos\left(\delta\right) * \sin\left(\omega\right) \tag{20}$$

Figure 42. Daily extraterrestrial solar radiation $H_{0,90d}$ variations on a vertical surface orientated towards the North Pole for latitudes 24°N(x), 30°N (●), 35°N (-), 40°N (♦), 43.45°N (▲),50°N (ж) and 59°N (+)

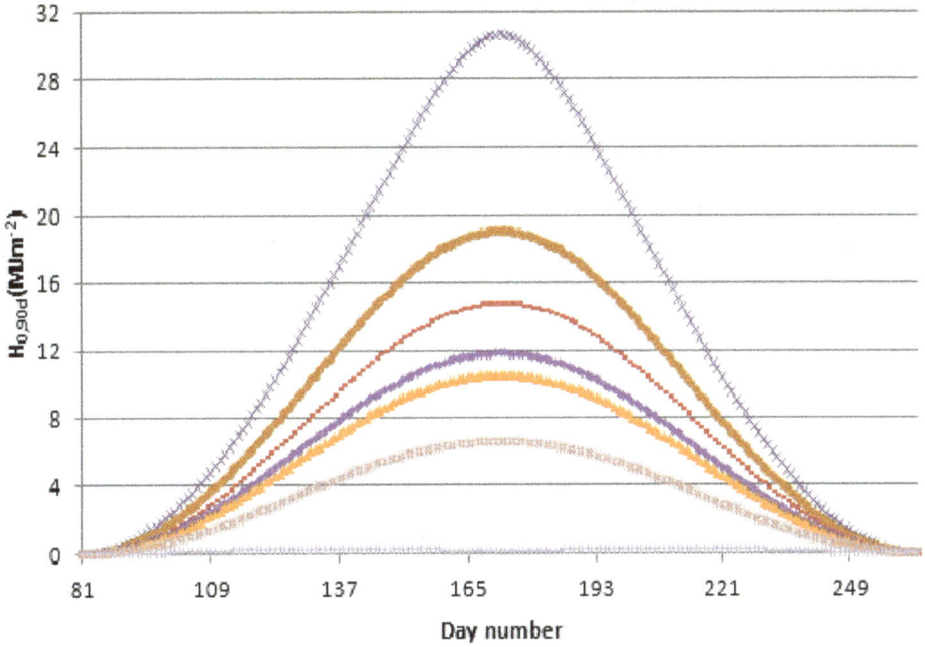

Figure 43. Daily extraterrestrial solar radiation $H_{0,90d}$ variations on a vertical surface orientated towards the South Pole for latitudes 24°S(▲), 30°S (●), 35°S (-), 40°S (▲), 43.45°S (●),50°S (-) and 59°S (▲)

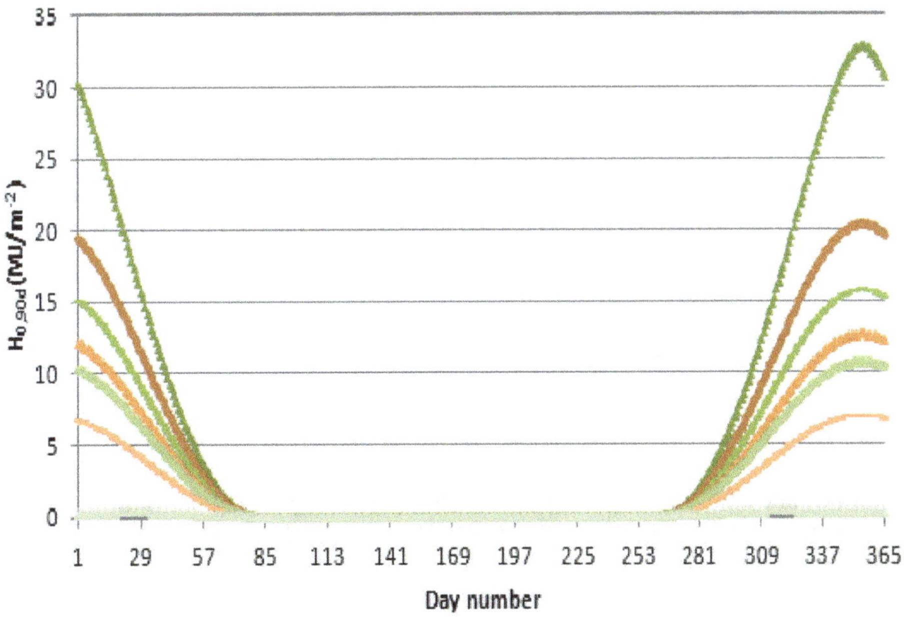

and the sunset and the sunrise hour angles satisfy the equation (41d) and (42d) of the first chapter:

$$\omega_{ss} = min\left\{arccos\left[-tan\left(\delta\right)tan\left(\varphi\right)\right],0\right\}$$

$$\omega_{sr} = \max\left\{-\arccos\left[-\tan\left(\delta\right)\tan\left(\varphi\right)\right],-\pi\right\}$$

The daily extraterrestrial solar radiation incident on on the east facing vertical surface in the Northern Hemisphere could be calculated using the equation (18) which takes in this case the following form:

$$H_{0,\beta d} = \frac{12x3600xG_s}{\pi} * \left[\sin\left(\delta\right) * sin\left(\omega_s\right)\right] \tag{18b}$$

Example 6

Calculate the $H_{0,d}$ on a vertical surface, orientated towards the east, in the absence of the atmosphere at latitudes 10°N to 66.45°N on June 21.

Solution

For these circumstances, n=172, G_s=1322.6W/m^2 and δ=23.458°. Then, the application of the equation (18b) leads to the following results:

φ=10°N ω_{sr}=-94.37° $H_{o,d}$ = 17.97 MJ/m^2

φ=20°N ω_{sr}=-99.06° $H_{o,d}$ = 19.32 MJ/m^2

φ=30°N ω_{sr}=-104.51° $H_{o,d}$ = 20.86 MJ/m^2

φ=40°N ω_{sr}=-111.33° $H_{o,d}$ = 22.76 MJ/m^2

φ=50°N ω_{sr}=-121.12° $H_{o,d}$ = 25.31 MJ/m^2

φ=60°N ω_{sr}=-138.71° $H_{o,d}$ = 29.22 MJ/m^2

φ=66.45°N ω_{sr}=-174.58° $H_{o,d}$ = 33.29 MJ/m^2

The daily extraterrestrial solar radiation incident on an east orientated vertical surface in the Northern Hemisphere is plotted in the Figure 44. The situation in the Southern Hemisphere becomes as shown in the Figure 45.

Figure 44. Daily extraterrestrial solar radiation $H_{0,90d}$ variations on a vertical surface orientated towards the east for latitudes 10°N(x), 20°N (●), 30°N (-), 40°N (♦), 50°N (▲), 60°N (ж) and 66.45°N (+)

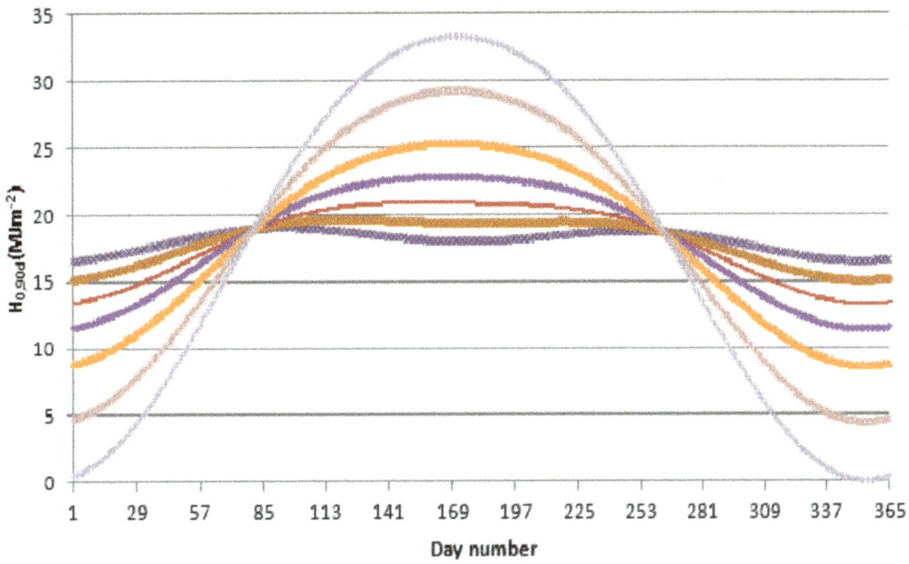

Figure 45. Daily extraterrestrial solar radiation $H_{0,90d}$ variations on a vertical surface orientated towards the east for latitudes 10°S(-), 20°S (■), 30°S (x), 40°S (●), 50°S (-), 60°S (♦) and 66.45°S (▲)

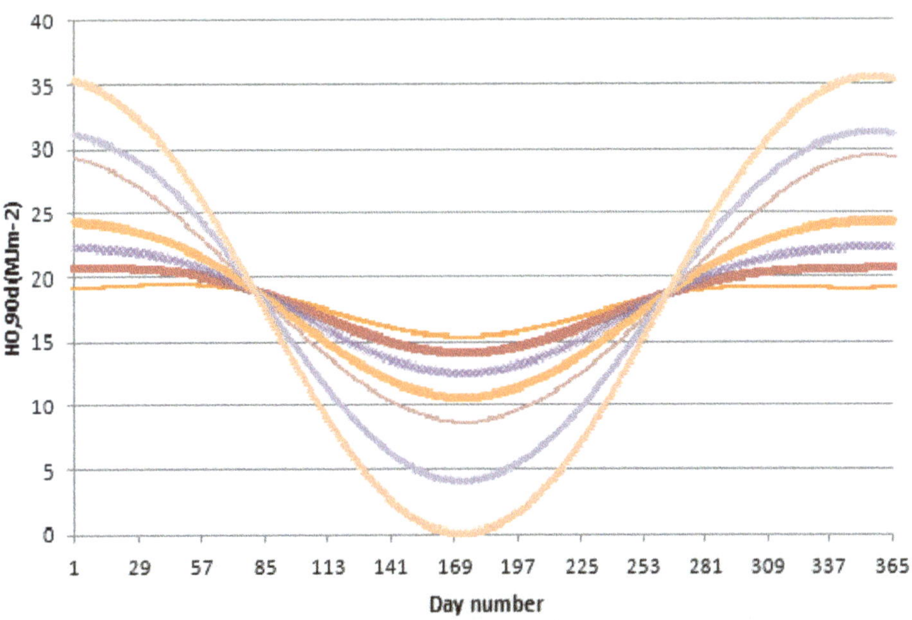

VERTICAL SURFACE ORIENTATED TOWARDS WEST

In this case, $\beta = 90^o$ and $\gamma = 90^o$. Thus, the incidence angle of Sun's rays on the east facing tilted surface is:

$$\cos\left(\theta_i\right) = \cos\left(\delta\right) * \sin\left(\omega\right) \tag{21}$$

and the sunset and the sunrise hour angles satisfy the equation (41d) and (42d) of the first chapter:

$$\omega_{ss} = min\left\{arccos\left[-tan\left(\delta\right)tan\left(\varphi\right)\right], \pi\right\}$$

$$\omega_{sr} = \max\left\{-\arccos\left[-\tan\left(\delta\right)\tan\left(\varphi\right)\right], 0\right\}$$

Example 7

Calculate the $H_{0,d}$ on a vertical surface, orientated towards the west, in the absence of the atmosphere at latitudes 10°N to 66.45°N on June 21.

Solution

For these circumstances, n=172, G_s=1322.6W/m^2 and δ=23.458°. Then, the application of the equation (18) leads to the following results:

φ=10°N ω_{ss}=94.37° $H_{o,d}$ = 17.97 MJ/m^2

φ=20°N ω_{ss}=99.06° $H_{o,d}$ = 19.32 MJ/m^2

φ=30°N ω_{ss}=104.51° $H_{o,d}$ = 20.86 MJ/m^2

φ=40°N ω_{ss}=111.33° $H_{o,d}$ = 22.76 MJ/m^2

φ=50°N ω_{ss}=121.12° $H_{o,d}$ = 25.31 MJ/m^2

φ=60°N ω_{ss}=138.71° $H_{o,d}$ = 29.22 MJ/m^2

φ=66.45°N ω_{ss}=174.58° $H_{o,d}$ = 33.29 MJ/m^2

The daily extraterrestrial solar radiation incident on the west orientated vertical surface in the Northern Hemisphere was found to be identical as shown in the Figures 46 and 47 in the Northern and Southern Hemispheres respectively.

Figure 46. Daily extraterrestrial solar radiation $H_{0,90d}$ variations on a vertical surface orientated towards the southeast for latitudes 10°N(-), 30°N(▲), 40°N (+), 50°N(♦), 60°N(ж) and 66.45°N (-)

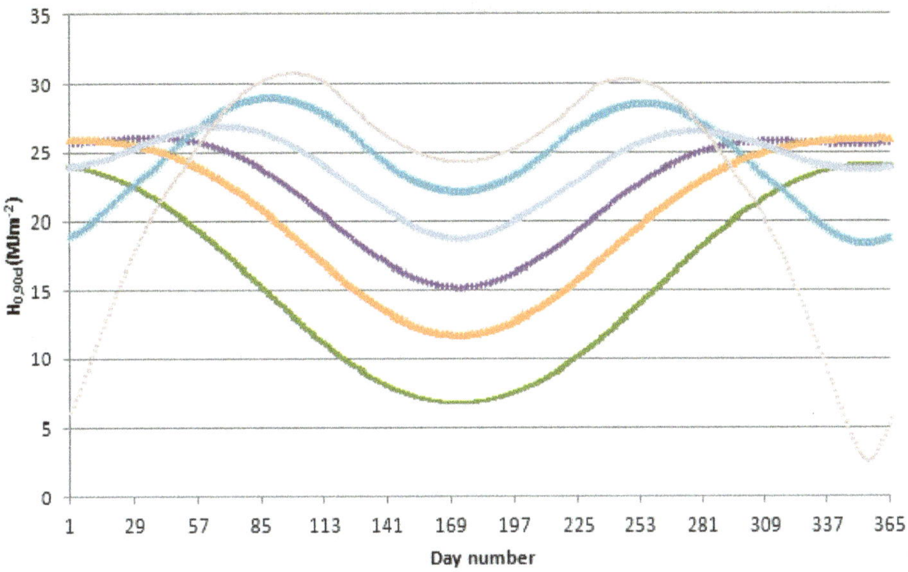

Figure 47. Daily extraterrestrial solar radiation $H_{0,90d}$ variations on a vertical surface orientated towards the southwest for latitudes 10°N(-), 30°N(▲), 40°N (+), 50°N(♦), 60°N(ж) and 66.45°N (-)

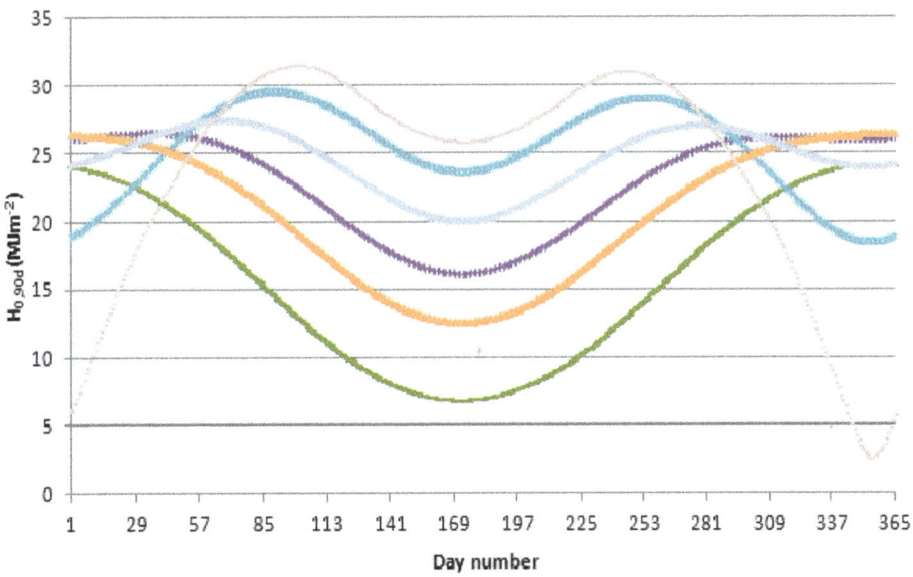

Southeast Orientated Vertical Surface

In the case of southeast orientated vertical surface, $\beta = 90^o$ and $\gamma = -45^o$. Thus, the incidence angle of Sun's rays on the east facing tilted surface is:

$$\cos\left(\Theta_i\right) = 0.707\left[-\sin\left(\delta\right)\cos\left(\varphi\right) - cos\left(\delta\right) * \sin\left(\omega\right) + \cos\left(\delta\right) sin\left(\varphi\right) cos\left(\omega\right)\right] \tag{22}$$

The daily extraterrestrial solar radiation incident on on the east facing vertical surface in the Northern Hemisphere could be calculated using the equation (18) which takes in this case the following form:

$$H_{0,\beta d} = \frac{0.707x12x3600xG_s}{\pi} * \left[\begin{array}{l} -\frac{\pi\left(\omega_{ss} - \omega_{sr}\right)}{180}\sin\delta cos\varphi \\ + \cos\delta sin\varphi\left(sin\omega_{ss} - sin\omega_{sr}\right) + \cos\delta\left(\cos\omega_{ss} - \cos\omega_{sr}\right) \end{array}\right] \tag{18c}$$

Then, the calculated results of the daily extraterrestrial solar radiation incident on the southeast orientated vertical surface, in the Northern are given in Figure 48.

Example 8

Calculate the $H_{0,d}$ on a vertical surface, orientated towards the southeast, in the absence of the atmosphere at latitudes 30°N to 60°N on June 21.

Figure 48. Daily extraterrestrial solar radiation $H_{0,90d}$ variations on a vertical surface orientated towards the southeast for latitudes 10°N(-), 30°N(▲), 40°N (+), 50°N(♦), 60°N(ж) and 66.45°N (-)

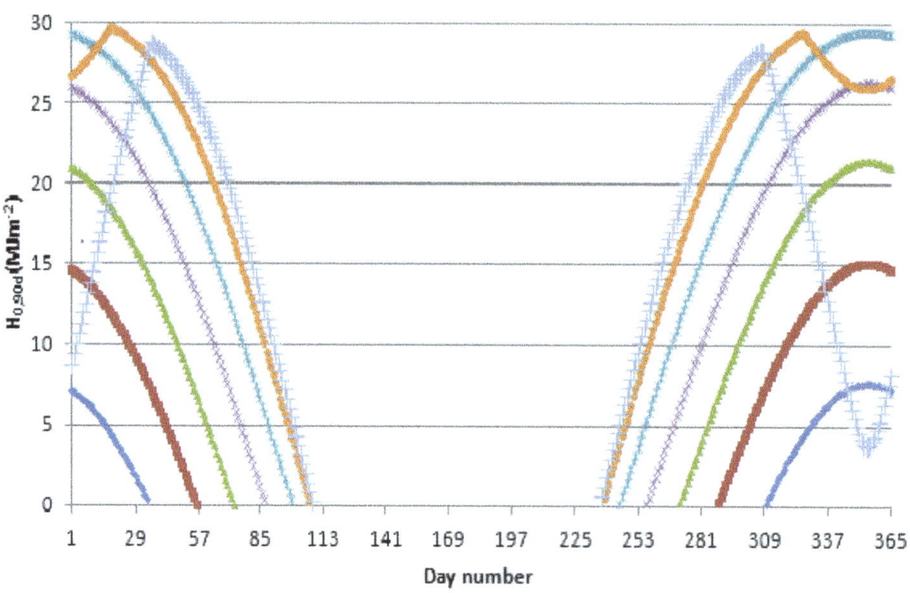

Solution

For these circumstances, n=172, G_s=1322.6W/m^2 and δ=23.458°. Then, the application of the equation (18.c) leads to the following results:

φ=30°N ω_{sr}=-104.51° ω_{ss}=26.87° $H_{o,d}$ = 11.69MJ/m^2

φ=40°N ω_{sr}=-111.32° ω_{ss}=37.13° $H_{o,d}$ = 15.18MJ/m^2

φ=50°N ω_{sr}=-115.97° ω_{ss}=45.49° $H_{o,d}$ = 18.74MJ/m^2

φ=60°N ω_{sr}=-115.05° ω_{ss}=51.91° $H_{o,d}$ = 22.11MJ/m^2

Southwest Orientated Vertical Surface

In the case of southwest orientated vertical surface, β = 90° and γ =45°. Thus, the incidence angle of Sun's rays on the southwest facing tilted surface is:

$$\cos\left(\theta_i\right) = 0.707\left[-\sin\left(\delta\right)\cos\left(\varphi\right) + cos\left(\delta\right) * \sin\left(\omega\right) + \cos\left(\delta\right) sin\left(\varphi\right) cos\left(\omega\right)\right] \tag{23}$$

The daily extraterrestrial solar radiation incident on the east facing vertical surface in the Northern Hemisphere could be calculated using the equation (18) which takes in this case the following form:

$$H_{0,\beta d} = \frac{0.707x12x3600xG_s}{\pi} * \left[\begin{array}{l} -\dfrac{\pi\left(\omega_{ss} - \omega_{sr}\right)}{180^o}\sin\delta cos\varphi \\ +\cos\delta sin\varphi\left(sin\omega_{ss} - sin\omega_{sr}\right) - \cos\delta\left(\cos\omega_{ss} - \cos\omega_{sr}\right) \end{array}\right] \tag{18d}$$

Then, the daily extraterrestrial solar radiation incident on the southwest orientated vertical surface could be calculated.

Example 9

Calculate the $H_{0,d}$ on a vertical surface, orientated towards the southwest, in the absence of the atmosphere at latitudes 30°N to 60°N on June 21.

Solution

For these circumstances, n=172, G_s=1322.6W/m^2 and δ=23.458°. Then, the application of the equation (18d) leads to the following results:

φ=30°N ω_{sr}=-104.51° ω_{ss}=26.87° $H_{o,d}$ = 12.47MJ/m^2

φ=40°N ω_{sr}=-111.32° ω_{ss}=37.13°$H_{o,d}$ = 16.07MJ/m²

φ=50°N ω_{sr}=-115.97° ω_{ss}=45.49°$H_{o,d}$ = 19.95MJ/m²

φ=60°N ω_{sr}=-115.05° ω_{ss}=51.91°$H_{o,d}$ = 23.59MJ/m²

When comparing the results presented in Figures 46 to 48 it was found that, the solar radiation on the southwest vertical surface is very similar to its values on the southeast vertical surface in the tropical, mid-latitude and high-latitude zones.

Northeast Orientated Vertical Surface

In the case of northeast orientated vertical surface, β = 90° and γ =-135°. Thus, the incidence angle of Sun's rays on the east facing tilted surface is:

$$\cos\left(\theta_i\right) = 0.707\left[\sin\left(\delta\right)\cos\left(\varphi\right) - cos\left(\delta\right) * \sin\left(\omega\right) - \cos\left(\delta\right) sin\left(\varphi\right) cos\left(\omega\right)\right] \tag{24}$$

The daily extraterrestrial solar radiation incident on the northeast facing vertical surface in the Northern Hemisphere could be calculated using the equation (18) which takes in this case the following form:

$$H_{0,\beta d} = \frac{0.707x12x3600xG_s}{\pi} * \left[\begin{array}{l} -\frac{\pi\left(\omega_{ss} - \omega_{sr}\right)}{180^0}\sin\delta cos\varphi + \\ \cos\delta sin\varphi\left(sin\omega_{ss} - sin\omega_{sr}\right) - \cos\delta\left(\cos\omega_{ss} - \cos\omega_{sr}\right) \end{array}\right] \tag{18e}$$

Example 10

Calculate the $H_{0,d}$ on a vertical surface, orientated towards the northeast, in the absence of the atmosphere at latitudes 30°N to 60°N on January 1.

Solution

For these circumstances, n=1, G_s=1414.9W/m² and δ=-23.02°. Then, the application of the equation (18e) leads to the following results:

φ=30°N $H_{o,d}$ = 20.98MJ/m²

φ=40°N $H_{o,d}$ = 25.91MJ/m²

φ=50°N $H_{o,d}$ = 29.24MJ/m²

φ=60°N $H_{o,d}$ = 26.66MJ/m²

Then, the daily extraterrestrial solar radiation incident on the northeast orientated vertical surface, in the Northern Hemisphere could be calculated. When doing these calculations it was found that, there is period in the year where the Sun does not shine on such a surface whatever is the latitude.

Northwest Orientated Vertical Surface

In the case of west-north orientated vertical surface, β = 90° and γ =135°. Thus, the incidence angle of Sun's rays on the east facing tilted surface is:

$$\cos\left(\theta_i\right) = 0.707\left[\sin\left(\delta\right)\cos\left(\varphi\right) + cos\left(\delta\right) * \sin\left(\omega\right) - \cos\left(\delta\right) sin\left(\varphi\right) cos\left(\omega\right)\right] \tag{25}$$

The daily extraterrestrial solar radiation incident on the northeast facing vertical surface in the Northern Hemisphere could be calculated using the equation (18) which takes in this case the following form:

$$H_{0,\beta d} = \frac{0.707x12x3600xG_s}{\pi} * \left[\begin{array}{l} -\frac{\pi\left(\omega_{ss} - \omega_{sr}\right)}{180^o}\sin\delta cos\varphi + \\ \cos\delta sin\varphi\left(sin\omega_{ss} - sin\omega_{sr}\right) - \cos\delta\left(\cos\omega_{ss} - \cos\omega_{sr}\right)\end{array}\right] \tag{18f}$$

Example 11

Calculate the $H_{0,d}$ on a vertical surface, orientated towards the northwest, in the absence of the atmosphere at latitudes 30°N to 60°N on January 1.

Solution

For these circumstances, n=1, G_s=1414.9W/m² and δ=-23.02°. Then, the application of the equation (18e) leads to the following results:

φ=30°N $H_{o,d}$ = 14.84MJ/m²

φ=40°N $H_{o,d}$ = 18.32MJ/m²

φ=50°N $H_{o,d}$ = 20.67MJ/m²

φ=60°N $H_{o,d}$ = 18.85MJ/m²

Then, the calculated results of the daily extraterrestrial solar radiation incident on the northwest orientated vertical surface, in the Northern Hemisphere are given in Figure 49 where it is clear that they are similar in behavior to those plotted in the Figure 49.

Figure 49. Daily extraterrestrial solar radiation $H_{0,90d}$ variations on a vertical surface orientated towards the northwest for latitudes 10°N(♦), 20°N(■), 30°N(▲), 40°N (x), 50°N(ж), 60°N(●) and 66.45°N (+)

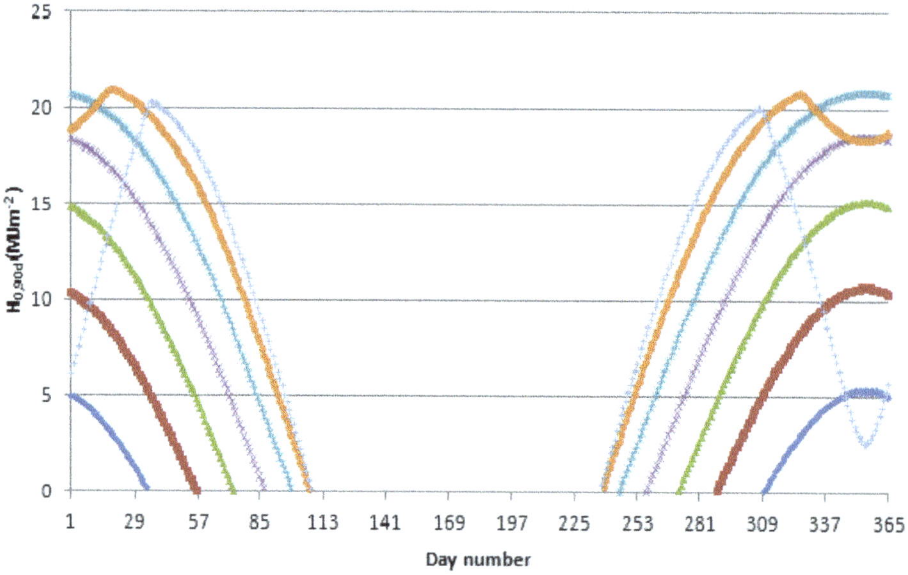

RATIO OF EXTRATERRESTRIAL SOLAR RADIATION ON A TILTED SURFACE TO THAT ON A HORIZONTAL SURFACE

The ratio of the extraterrestrial solar radiation on a tilted surface to that on a horizontal surface is known as the geometric factor *R*. This factor demonstrates the possible solar energy gain by tilting the solar receiver in relation to the horizontal surface and by orientating the solar receiver (with changing the azimuth angle of the solar receiver).

Short Term Ratio

The most commonly available data are the total solar radiation for hours or days on the horizontal plane, whereas the need is for direct and diffuse solar radiation on the a plane of the solar receiver. The geometric factor R_b, the ratio of the extraterrestrial solar radiation on the tilted surface to that on a horizontal plane at any time, can be calculated exactly by appropriate use of the equation (Duffie and Beckman, 2013):

$$R_b = \left(\frac{G_{0,\beta}}{G_0}\right) \tag{26}$$

where G_0 could be calculated using the equation (11) and:

$$G_{0,\beta} = G_s * cos\left(\theta_i\right) \tag{10a}$$

So, for tilted surfaces by an angle β in relation to the horizontal plane and orientated towards the Equator, the equation (26) could be written as follows:

$$R_b = \frac{G_{0,\beta}}{G_0} = \frac{\cos(\varphi - \beta)\cos(\delta)\cos(\omega) + \sin(\varphi - \beta)\sin(\delta)}{\cos(\varphi)\cos(\delta)\cos(\omega) + \sin(\varphi)\sin(\delta)} \tag{27}$$

Hottel and Woertz (1942) pointed out that the equation (27) provides a convenient method for calculating the geometric factor R_b. They also showed a graphical method for solving this equation. This graphical method has been revised by Whillier in 1975 (Duffie and Beckman, 2013). The results for R_b at equinox, summer and winter solstice on solar noon and 11 O'clock of solar time in the Northern Hemisphere (NH) are plotted in the Figures 50 to 55.

For a surface with slope 50° towards the Equator at latitude of Damascus 33.51°N, R_b is calculated during the year for various hour angles. The results are presented in Figures 56 and 57 where it is from this Figure that R_b becomes greater and greater with increasing hours from solar noon.

Here it should be noted that, the equation (27) could be applied to fixed solar receiver (β=constant) and to solar receivers moved in prescribed ways in order to track the Sun. In (Duffie and Beckman, 2013) some cases of solar tracking were noted. In the present work, it seems useful to shed more light on these cases.

In the case of receiver that rotated continuously about a horizontal east-west axis with continuous adjustment to minimize the angle of incidence and maximize the received solar radiation on it, the geometric factor R_b could be calculated using the following equation:

Figure 50. Latitude dependence of the geometric factor on solar noon at summer solstice for tilt angles 15°(♦), 30° (■), 45° (▲), 60° (x), 75° (ж) and 90° (●) at Northern Hemisphere (NH)

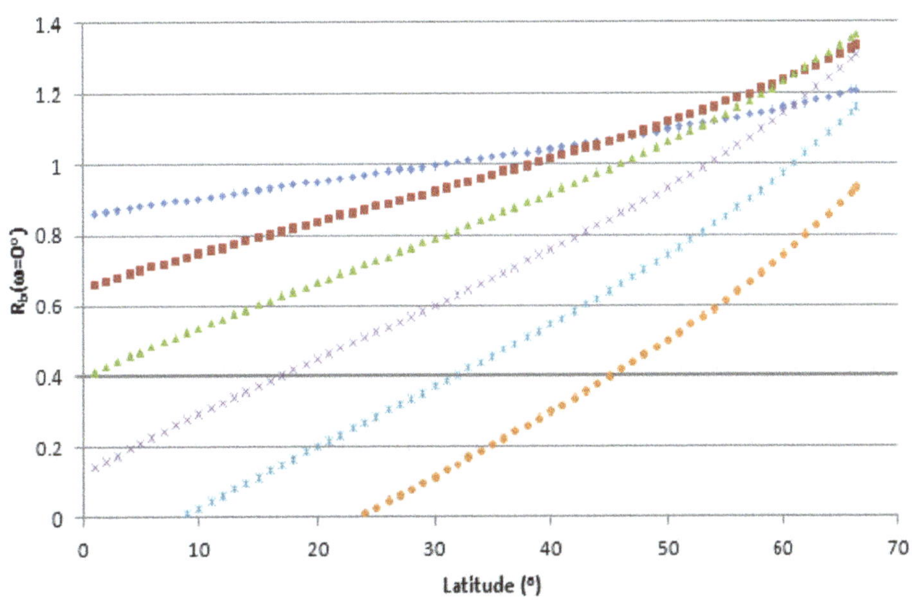

Figure 51. Latitude dependence of the geometric factor at 11 O'clock of solar time at summer solstice for tilt angles 15°(♦), 30° (■), 45° (▲), 60° (x), 75° (ж) and 90° (●) at NH

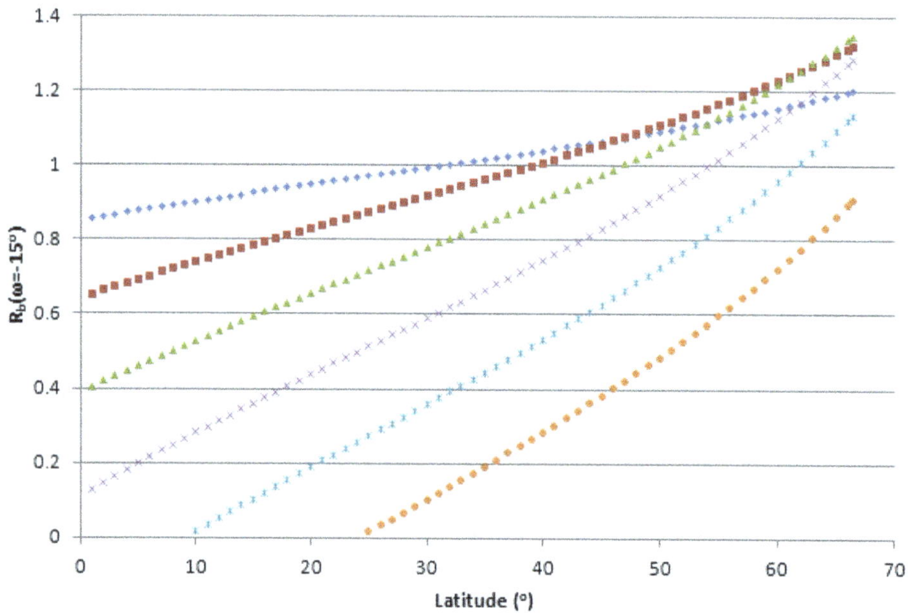

Figure 52. Latitude dependence of the geometric factor on solar noon at winter solstice for tilt angles 15°(♦), 30° (■), 45° (▲), 60° (x), 75° (ж) and 90° (●) at NH

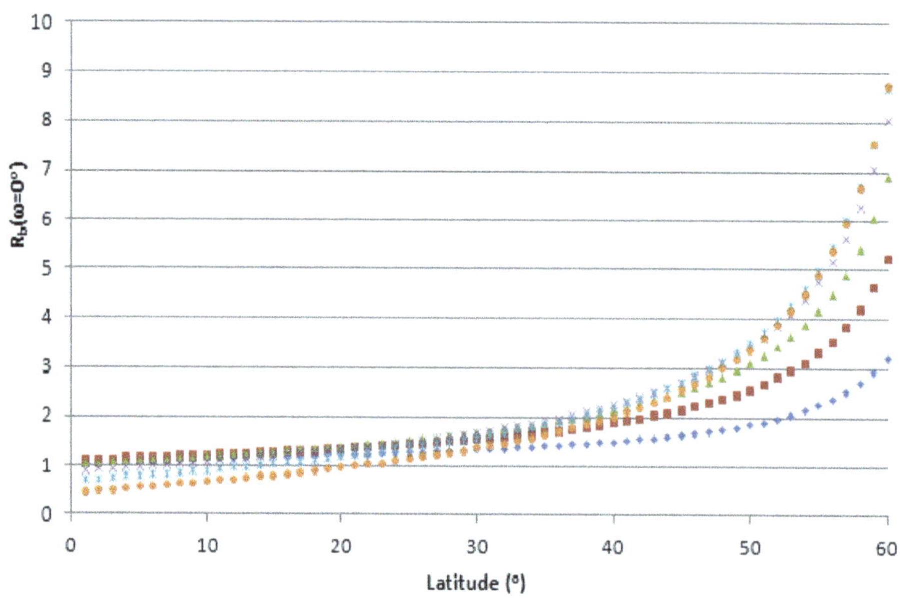

Figure 53. Latitude dependence of the geometric factor at 11 O'clock of solar time at winter solstice for tilt angles 15°(♦), 30° (■), 45° (▲), 60° (x), 75° (ж) and 90° (●) at NH

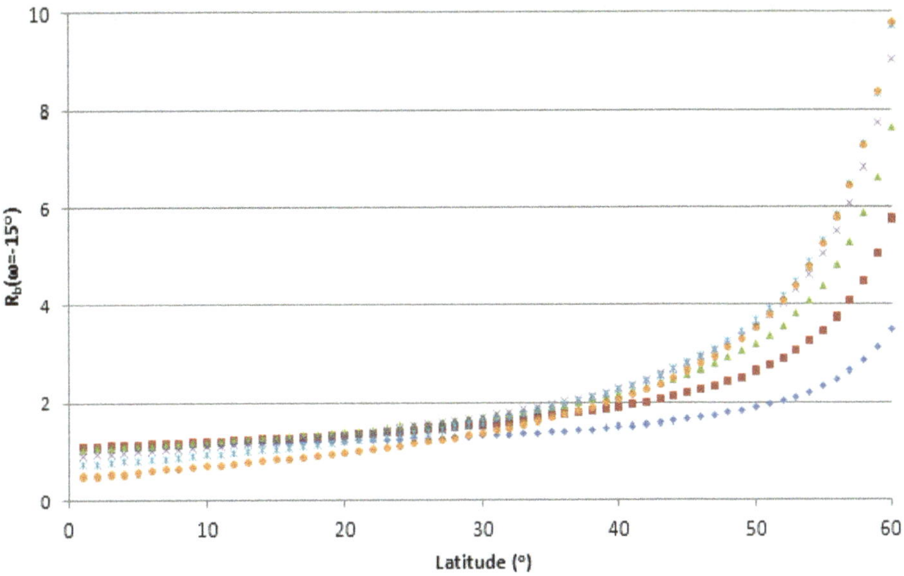

Figure 54. Latitude dependence of the geometric factor on solar noon at equinox for tilt angles 15°(♦), 30° (■), 45° (▲), 60° (x), 75° (ж) and 90° (●) at NH

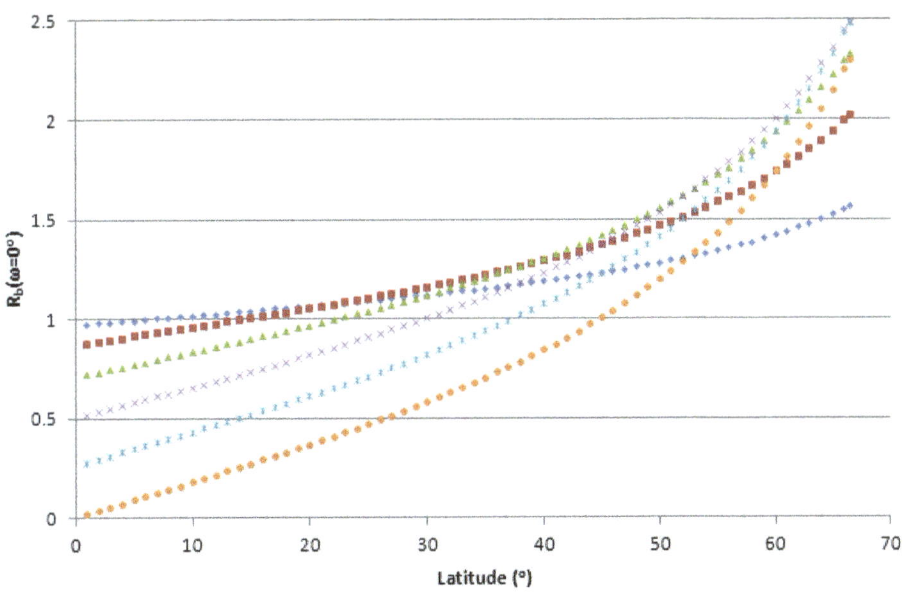

Figure 55. Latitude dependence of the geometric factor at 11 O'clock of solar time at equinox for tilt angles 15°(♦), 30° (■), 45° (▲), 60° (x), 75° (ж) and 90° (●) at NH

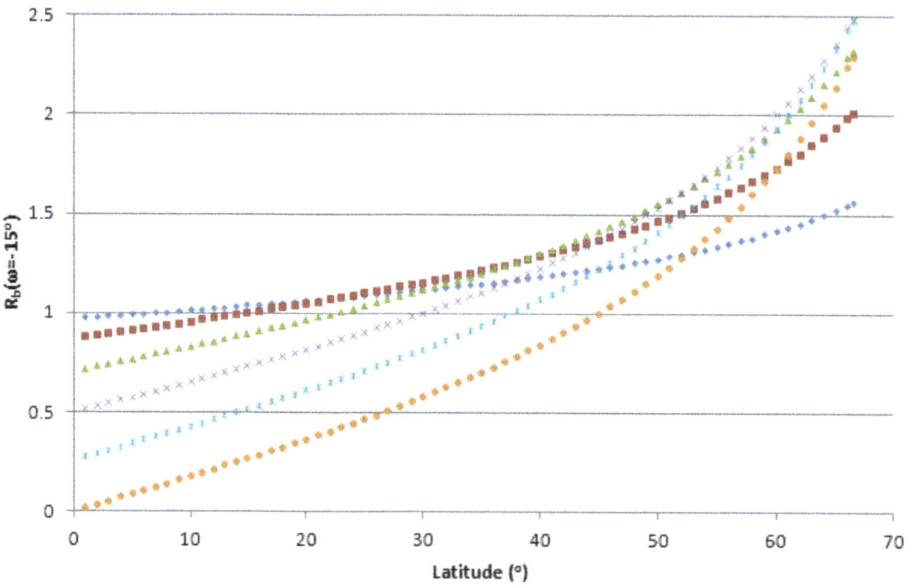

Figure 56. Daily variations of the geometric factor for various hours from solar noon for a surface with slope 50° towards the Equator at latitude of Damascus 33.51°. ± 0.5h(x), ± 1.5h (ж), ± 2.5h (●), ± 3.5h (+) and ± 4.5h (-)

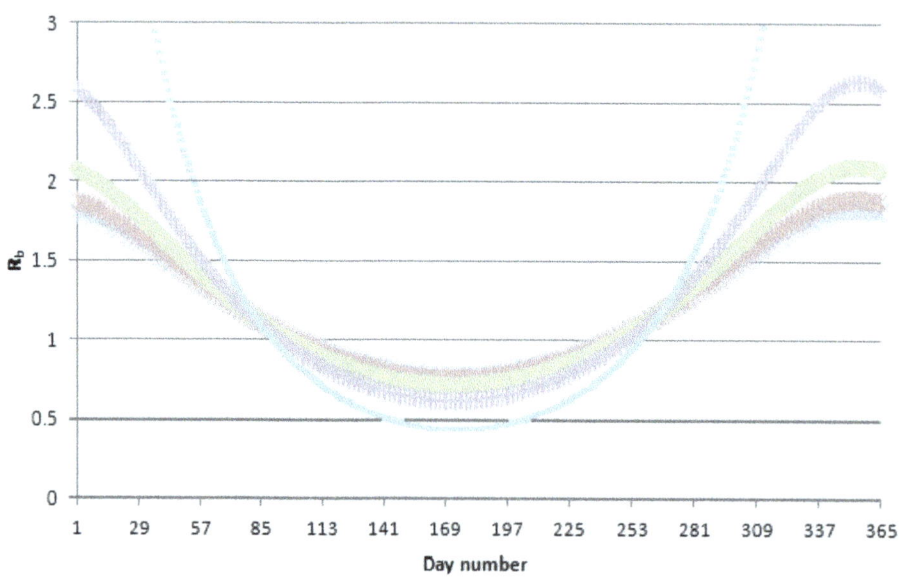

Figure 57. Daily variations of the geometric factor for various hours from solar noon for a surface with slope 50° towards the Equator at latitude of Damascus 33.51°. ± 0.5h(x), ± 1.5h (жс), ± 2.5h (●), ± 3.5h (+) and ± 4.5h (-)

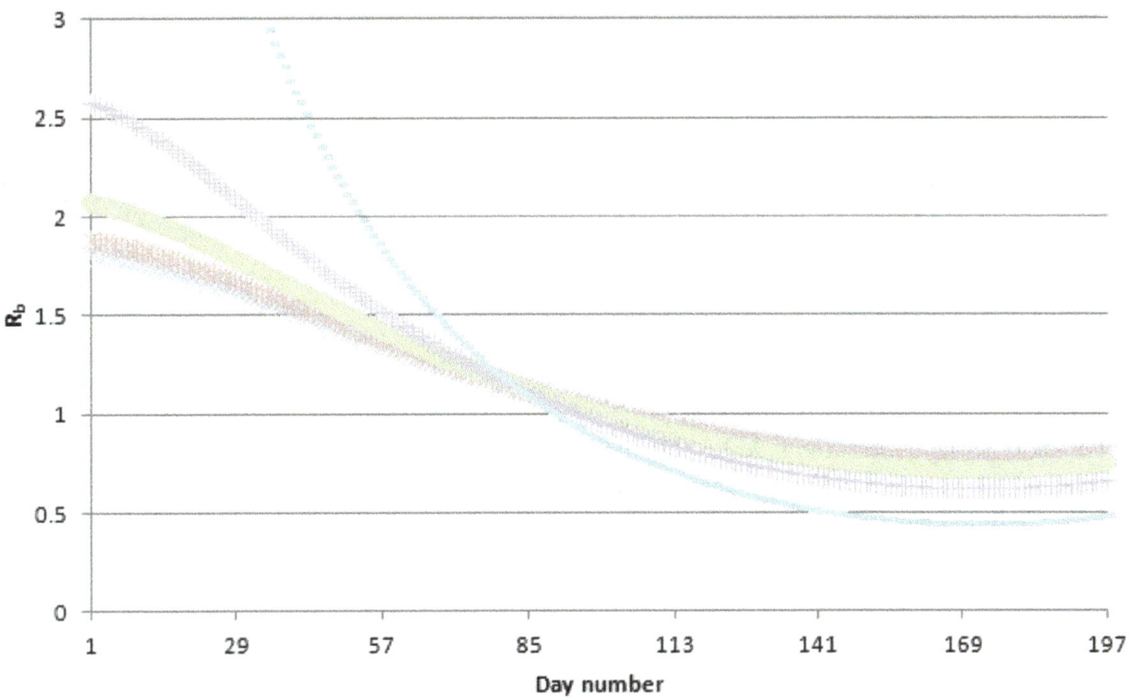

$$R_b = \frac{G_{0,\beta}}{G_0} = \frac{\sqrt{1 - cos^2\left(\delta\right) sin^2\left(\omega\right)}}{cos\left(\varphi\right) cos\left(\delta\right) cos\left(\omega\right) + sin\left(\varphi\right) sin\left(\delta\right)} \tag{28}$$

The surface azimuth angle for this mode of orientation will be 0° or 180°. If $\left|\gamma_s\right|$ <90° then γ=0° otherwise γ=180°. The tilt angle of the solar receiver satisfies the following equation:

$$tan\left(\beta\right) = tan\left(\theta_z\right)\left|cos\left(\gamma_s\right)\right| \tag{29}$$

In the case of receiver that rotated continuously about a horizontal north-south axis with continuous adjustment to minimize the angle of incidence and maximize the received solar radiation on it, the geometric factor R_b could be calculated using the following equation:

$$R_b = \frac{G_{0,\beta}}{G_0} = \frac{\sqrt{cos^2\left(\theta_z\right) - cos^2\left(\delta\right) sin^2\left(\omega\right)}}{cos\left(\varphi\right) cos\left(\delta\right) cos\left(\omega\right) + sin\left(\varphi\right) sin\left(\delta\right)} \tag{30}$$

The surface azimuth angle for this mode of orientation will be 90° or -90° depending on the sign of the solar azimuth angle. If γ_s>0° then γ=90° otherwise γ=-90°. The tilt angle of the solar receiver satisfies the following equation:

$$tan\left(\beta\right)=tan\left(\theta_z\right)\left|cos\left(\gamma-\gamma_s\right)\right| \tag{31}$$

In the case of receiver that rotated continuously about a horizontal east-west axis with a single daily adjustment so that the incident solar radiation is normal to the receiver at solar noon each day, the geometric factor R_b could be calculated using the following equation:

$$R_b=\frac{G_{0,\beta}}{G_0}=\frac{sin^2\left(\delta\right)+cos^2\left(\delta\right)cos\left(\omega\right)}{cos\left(\varphi\right)cos\left(\delta\right)cos\left(\omega\right)+sin\left(\varphi\right)sin\left(\delta\right)} \tag{32}$$

The daily surface azimuth angle for this mode of orientation will be 0° or 180° depending on the latitude and declination. If (φ-δ)>0° then γ = 0° otherwise γ = 180°. The tilt angle of the solar receiver satisfies the following equation:

$$\beta=\left|\varphi-\delta\right| \tag{33}$$

In the case of receiver with a fixed tilt angle and rotated continuously about a vertical axis, the angle of incidence is maximized when the receiver azimuth angle and solar azimuth angle are equal. In this case, the geometric factor R_b could be calculated using the following equation:

$$R_b=\frac{G_{0,\beta}}{G_0}=\frac{cos\left(\theta_z\right)cos\left(\beta\right)+sin\left(\theta_z\right)sin\left(\beta\right)}{cos\left(\varphi\right)cos\left(\delta\right)cos\left(\omega\right)+sin\left(\varphi\right)sin\left(\delta\right)} \tag{34}$$

The tilt angle of the solar receiver satisfies the following equation:

$$\beta=constant \tag{35}$$

In the case of receiver that rotated continuously about a north-south axis parallel to the Earth's axis with continuous adjustment to minimize the angle of incidence and maximize the received solar radiation on it, the geometric factor R_b could be calculated using the following equation:

$$R_b=\frac{G_{0,\beta}}{G_0}=\frac{cos\left(\delta\right)}{cos\left(\varphi\right)cos\left(\delta\right)cos\left(\omega\right)+sin\left(\varphi\right)sin\left(\delta\right)} \tag{36}$$

The tilt angle of the solar receiver varies continuously and it satisfies the following equation:

$$tan\left(\beta\right)=\frac{tan\left(\theta\right)}{cos\left(\gamma\right)} \tag{37}$$

In the case of receiver that continuously tracking about two axes with continuous adjustment to minimize the angle of incidence and maximize the received solar radiation on it, the geometric factor R_b could be calculated using the following equation:

$$R_b=\frac{G_{0,\beta}}{G_0}=\frac{1}{cos\left(\varphi\right)cos\left(\delta\right)cos\left(\omega\right)+sin\left(\varphi\right)sin\left(\delta\right)} \tag{38}$$

The tilt and azimuth angles of the solar receiver satisfy the following equations:

$$\beta=\alpha;\gamma=\gamma_s \tag{39}$$

The equation (28), (30), (32), (34), (36) and (38) could be divided into two categories with regard to tilt angle dependence: tilt dependent category (see equation (34)) and tilt independent category (see other equations). Therefore the application of each category separately is meaningful as it allows to have a comparison between the equations of each category and to evaluate the economic and energetic advantages and disadvantages of each mode of tracking.

Tilt Independent Category

When applying equation (28), (30), (32), (36) and (38) at latitude of Damascus 33.51°N for calculating R_b during the year for various hour angles. The results are presented in Figures 58 to 62 where it is seen from these figures that R_b becomes greater and greater with increasing hours from solar noon.

It is seen from Figure 58 that, for 4 hours from solar noon, the receiver that continuously tracking about two axes with continuous adjustment is the best mode of tracking. The second effective mode in this case is that when the receiver rotates continuously about a north-south axis parallel to the Earth's axis with continuous adjustment. The third effective mode in this case is that when the receiver rotates continuously about a horizontal north-south axis with continuous adjustment. The other two cases are practically identic in this case and they are the worst cases. The same situation is observed for 3 hours from solar noon (see Figure 59) but, during summer, the mode that when the receiver rotates continuously about a horizontal north-south axis becomes more effective that that when the receiver rotates continuously about a north-south axis parallel to the Earth's axis. For 2 hours from solar noon the mode that when the receiver rotates continuously about a horizontal north-south axis still good in summer but it becomes the worst case during other periods of the year (see Figure 60). From Figures 61 and 62 it is seen that the equation (30) is the worst all over the year for both cases.

It is clearly demonstrated from all above mentioned figures that, the tracking mode when the receiver is rotating continuously about a horizontal east-west axis with continuous adjustment to minimize the angle of incidence and maximize the received solar radiation (see equation (28)) and the tracking mode when the receiver is rotating continuously about a horizontal east-west axis with a single daily adjustment

Figure 58. Daily variations of the geometric factor in Damascus for 4 hours from solar noon using different modes of solar tracking, $R_{b(38)}$ (ж), $R_{b(36)}$ (x), $R_{b(30)}$ (▲), $R_{b(28)}$ (■) and $R_{b(32)}$ (♦). Note: that $R_{b(28)}$ means that R_b is calculated using the equation (28)

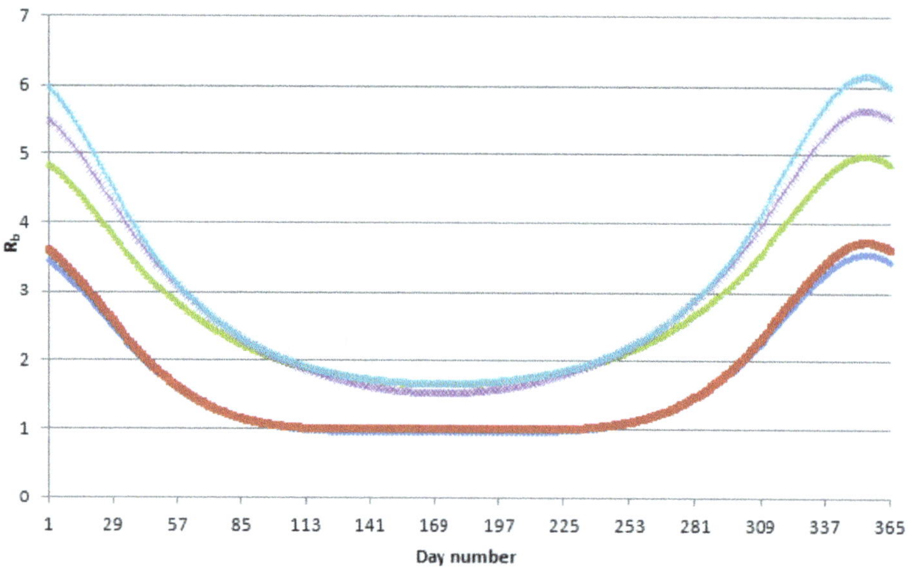

Figure 59. Daily variations of the geometric factor in Damascus for 3 hours from solar noon using different modes of solar tracking, $R_{b(38)}$ (ж), $R_{b(36)}$ (x), $R_{b(30)}$ (▲), $R_{b(28)}$ (■) and $R_{b(32)}$ (♦). Note: that $R_{b(28)}$ means that R_b is calculated using the equation (28)

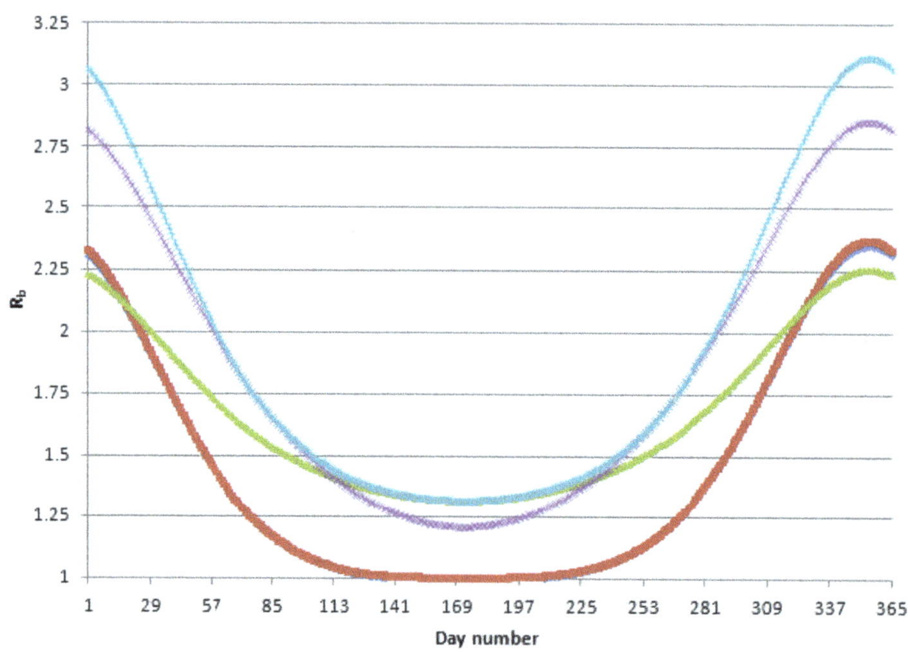

Figure 60. Daily variations of the geometric factor in Damascus for 2 hours from solar noon using different modes of solar tracking, $R_{b(38)}$ (ж), $R_{b(36)}$ (x), $R_{b(30)}$ (▲), $R_{b(28)}$ (■) and $R_{b(32)}$ (♦). Note: that $R_{b(28)}$ means that R_b is calculated using the equation (28)

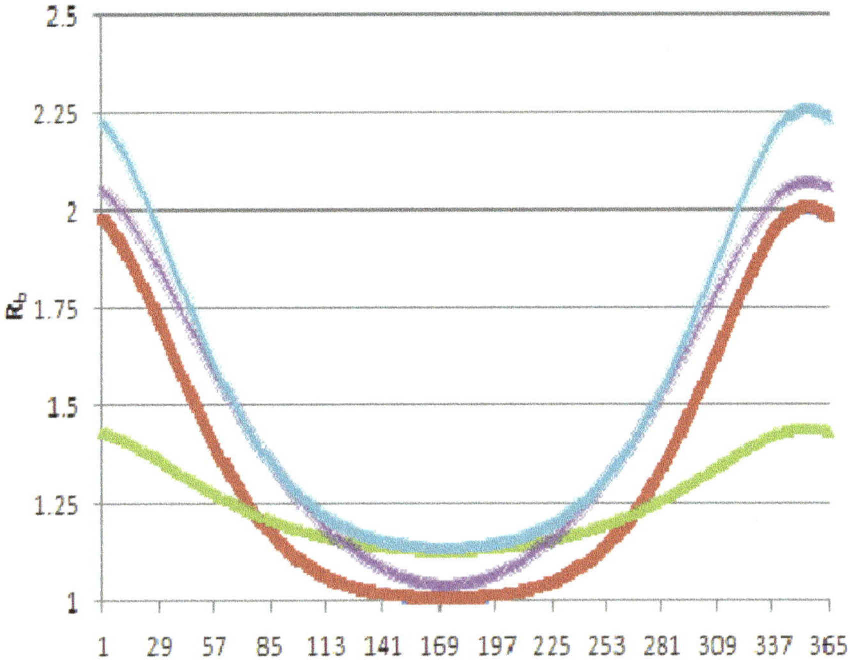

Figure 61. Daily variations of the geometric factor in Damascus for one hour from solar noon using different modes of solar tracking, $R_{b(38)}$ (ж), $R_{b(36)}$ (x), $R_{b(30)}$ (▲), $R_{b(28)}$ (■) and $R_{b(32)}$ (♦). Note: that $R_{b(28)}$ means that R_b is calculated using the equation (28)

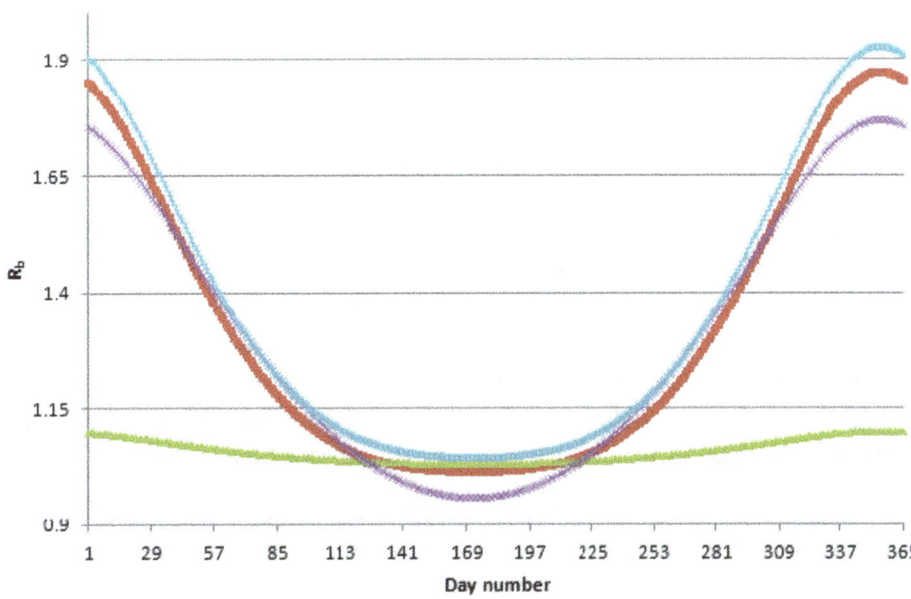

Figure 62. Daily variations of the geometric factor in Damascus at solar noon using different modes of solar tracking, $R_{b(38)}$ (ж), $R_{b(36)}$ (x), $R_{b(30)}$ (▲), $R_{b(28)}$ (■) and $R_{b(32)}$ (♦). Note: that $R_{b(28)}$ means that R_b is calculated using the equation (28)

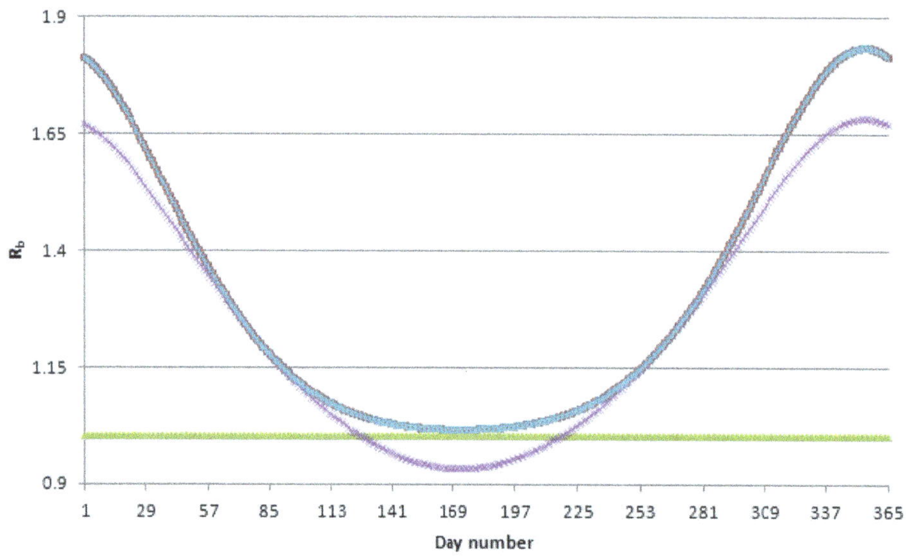

so that the incident solar radiation is normal to the receiver at solar noon each day (see equation (32)) are practically similar in effectiveness whatever is the latitude and whatever is the hour angle. So, it is sufficient to use one of these two modes for comparison with other modes. Moreover, the effectiveness of these two modes increases with decreasing the hours from the solar noon and on solar noon they become as effective as the tracking mode where the receiver is rotating continuously about two axes with continuous adjustment. On the other hand, when applying the tracking mode where the receiver rotates continuously about a horizontal north-south axis (see equation (30)), it was found that, the effectiveness of solar tracking decreases rapidly when approaching the solar noon. On solar noon the gain on solar energy collection was not observed, R_b=1 all over the year whatever the latitude is.

Finally, it is easy to conclude that, in the Northern Hemisphere the effectiveness of solar tracking modes is important during the period 22/9 to 21/3 while it could lead to the negative results, where R_b<1, during the period 22/3 to 21/9 (see the equation (36)). At the same time the effectiveness of solar tracking becomes higher and higher with increasing the site's latitude.

Moreover, in order to demonstrate the dependence of solar tracking on the latitude, the same equations were applied at several latitudes for various hour angles during a whole year. The results for the latitude 10°N, 40°N and 60°N are plotted in the Figures 63 to 67, 68 to 72 and 73 to 77 respectively.

It is seen from the above mentioned figures that, the mode of tracking, where the solar receiver continuously tracks about two axes with continuous adjustment to minimize the angle of incidence and maximize the received solar radiation on it, is the best mode. This is due to the fact that, this mode maintains the receiver surface normal to the Sun's rays regardless of the latitude.

Figure 63. Daily variations of the geometric factor at latitude 10°N for 4 hour from solar noon using different modes of solar tracking, $R_{b(38)}$ *(●),* $R_{b(36)}$ *(ЭC),* $R_{b(30)}$ *(▲),* $R_{b(28)}$ *(■) and* $R_{b(32)}$ *(x)*

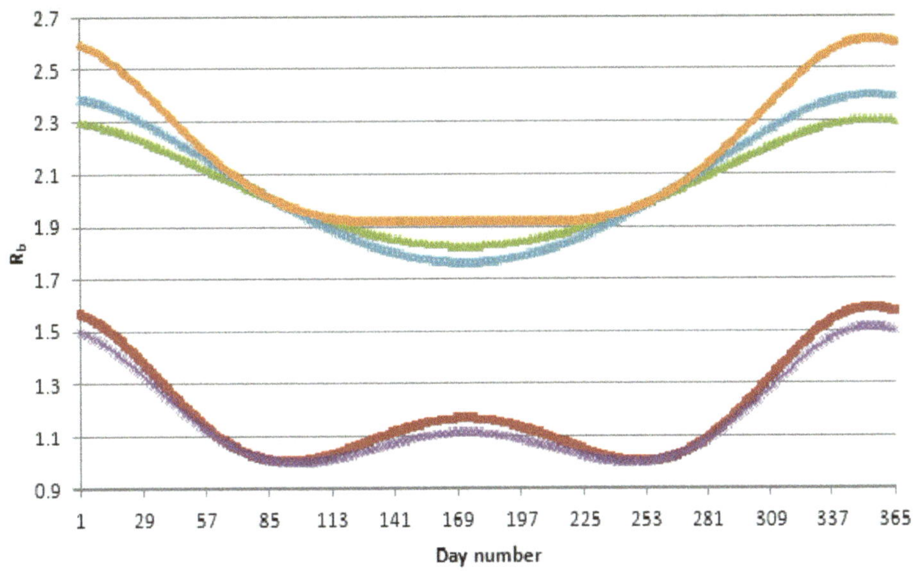

Figure 64. Daily variations of the geometric factor at latitude 10°N for 3 hour from solar noon using different modes of solar tracking, $R_{b(38)}$ *(●),* $R_{b(36)}$ *(ЭC),* $R_{b(30)}$ *(▲),* $R_{b(28)}$ *(■) and* $R_{b(32)}$ *(x)*

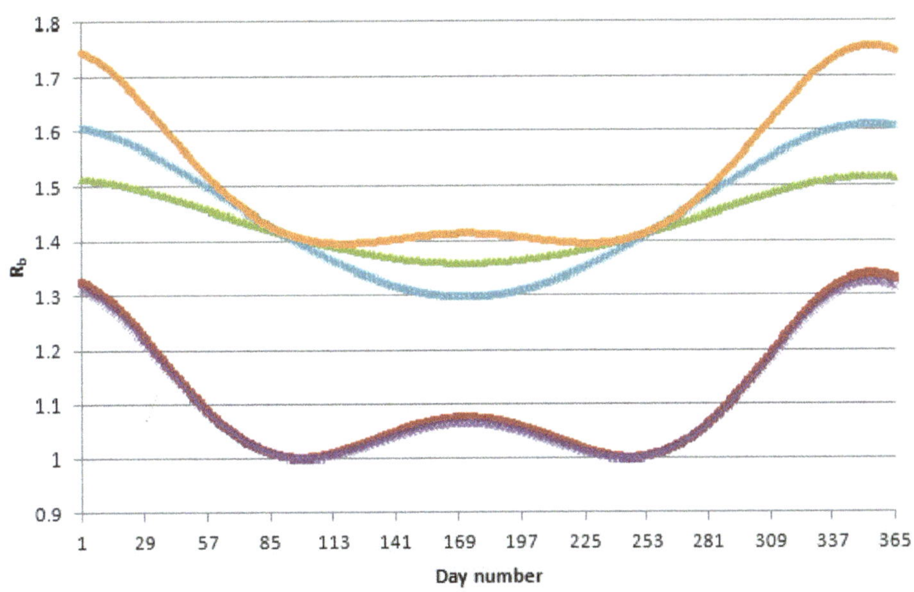

Figure 65. Daily variations of the geometric factor at latitude 10°N for 2 hour from solar noon using different modes of solar tracking, $R_{b(38)}$ (●), $R_{b(36)}$ (ж), $R_{b(30)}$ (▲), $R_{b(28)}$ (■) and $R_{b(32)}$ (x)

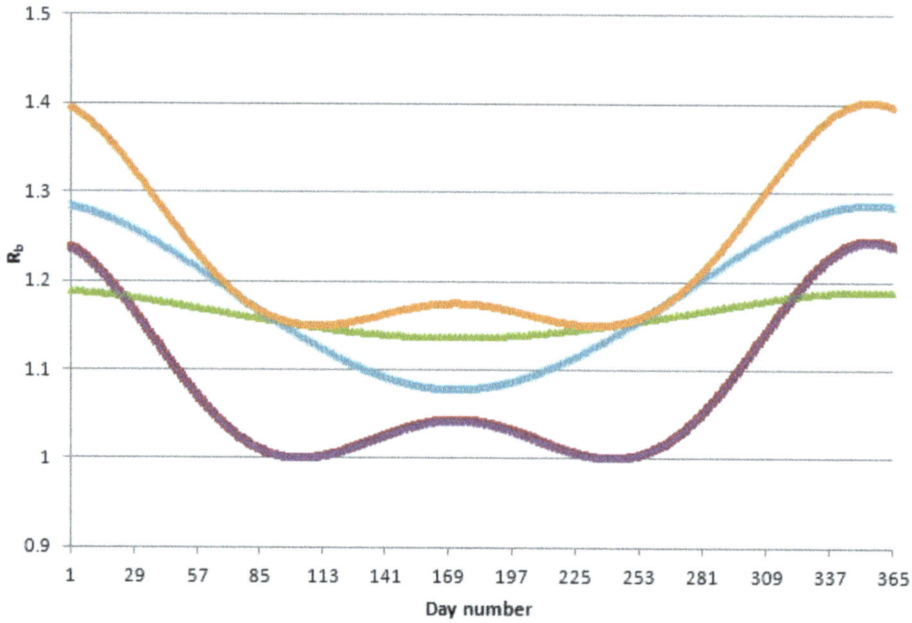

Figure 66. Daily variations of the geometric factor at latitude 10°N for 1 hour from solar noon using different modes of solar tracking, $R_{b(38)}$ (●), $R_{b(36)}$ (ж), $R_{b(30)}$ (▲), $R_{b(28)}$ (■) and $R_{b(32)}$ (x)

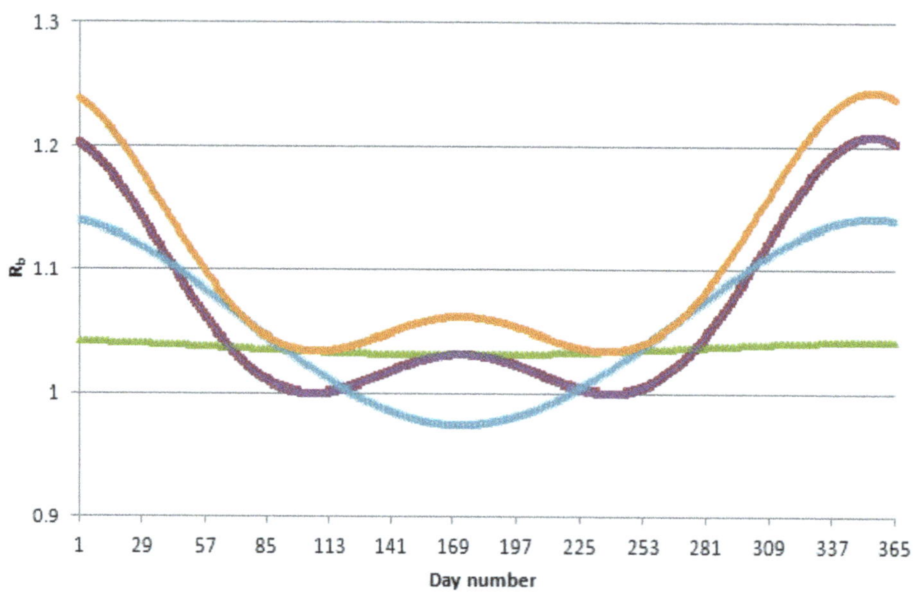

Figure 67. Daily variations of the geometric factor at latitude 10°N on solar noon using different modes of solar tracking, $R_{b(38)}$ (●), $R_{b(36)}$ (ж), $R_{b(30)}$ (▲), $R_{b(28)}$ (■) and $R_{b(32)}$ (x)

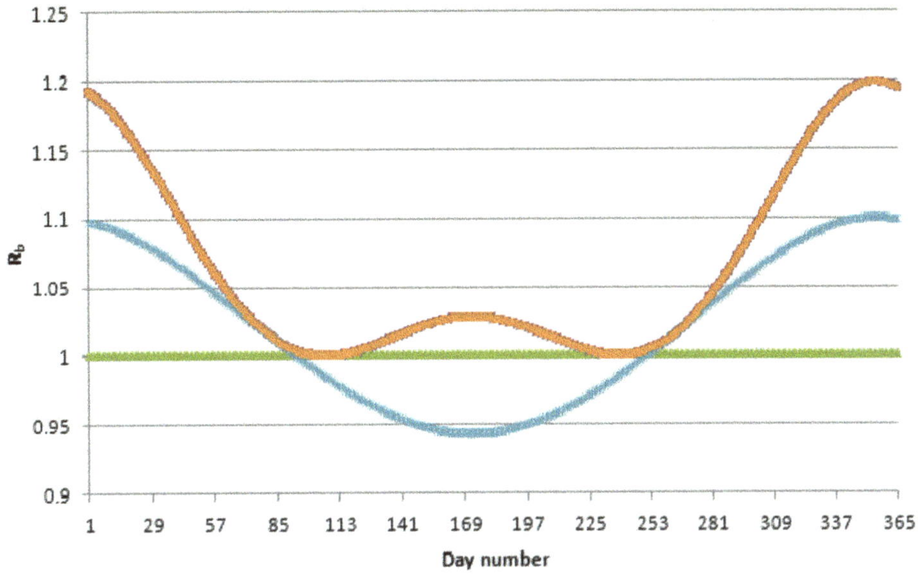

Figure 68. Daily variations of the geometric factor at latitude 40°N for 4 hour from solar noon using different modes of solar tracking, $R_{b(38)}$ (▲), $R_{b(36)}$ (■), $R_{b(30)}$ (-), $R_{b(28)}$ (-) and $R_{b(32)}$ (♦)

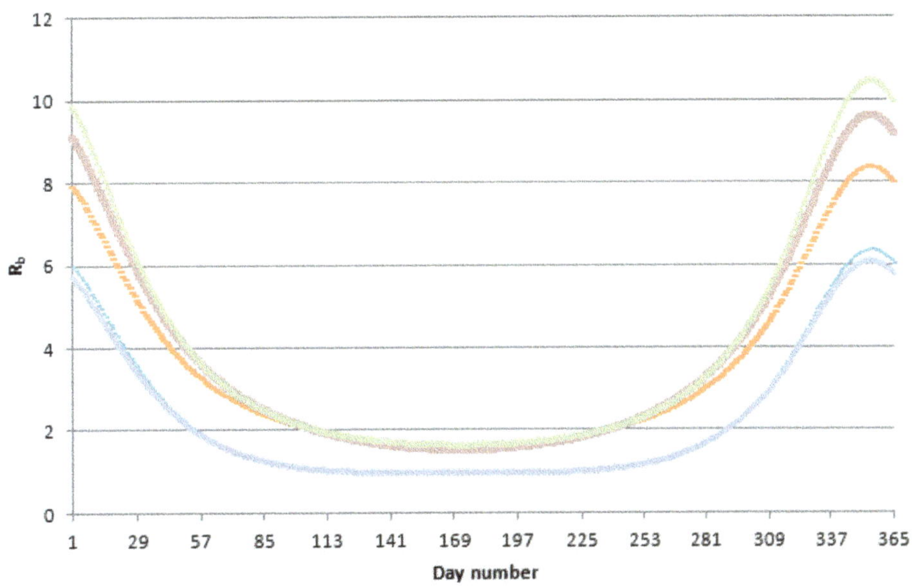

Figure 69. Daily variations of the geometric factor at latitude 40°N for 3 hour from solar noon using different modes of solar tracking, $R_{b(38)}$ (▲), $R_{b(36)}$ (■), $R_{b(30)}$ (-), $R_{b(28)}$ (-) *and* $R_{b(32)}$ (♦)

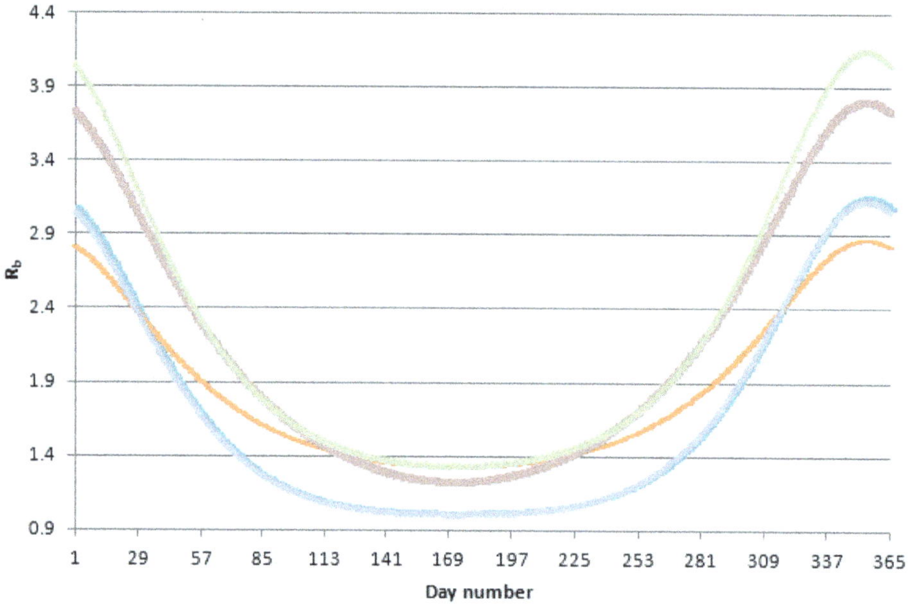

Figure 70. Daily variations of the geometric factor at latitude 40°N for 2 hour from solar noon using different modes of solar tracking, $R_{b(38)}$ (▲), $R_{b(36)}$ (■), $R_{b(30)}$ (-), $R_{b(28)}$ (-) *and* $R_{b(32)}$ (♦)

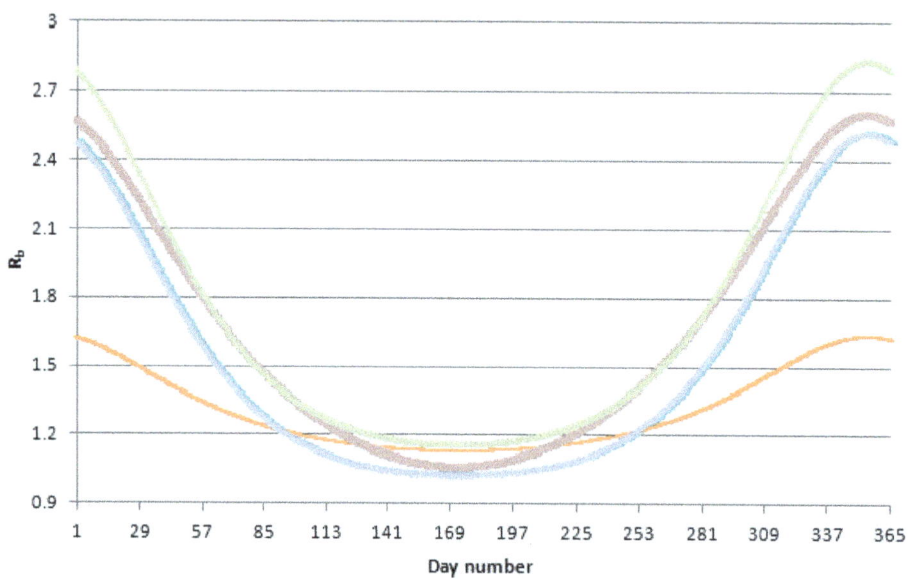

Figure 71. Daily variations of the geometric factor at latitude 40°N for 1 hour from solar noon using different modes of solar tracking, $R_{b(38)}$ (▲), $R_{b(36)}$ (■), $R_{b(30)}$ (-), $R_{b(28)}$ (-) *and* $R_{b(32)}$ (♦)

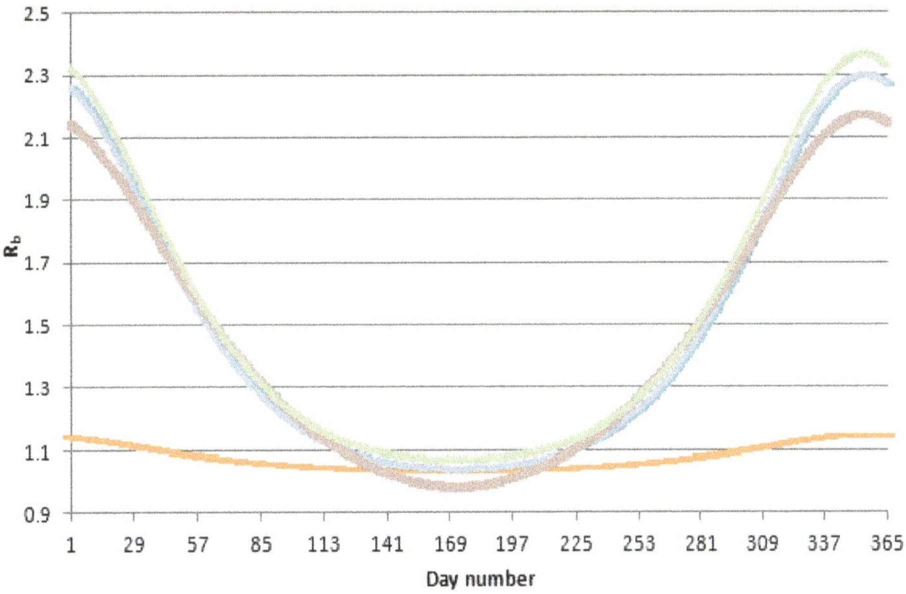

Figure 72. Daily variations of the geometric factor at latitude 40°N on solar noon using different modes of solar tracking, $R_{b(38)}$ (▲), $R_{b(36)}$ (■), $R_{b(30)}$ (-), $R_{b(28)}$ (-) *and* $R_{b(32)}$ (♦)

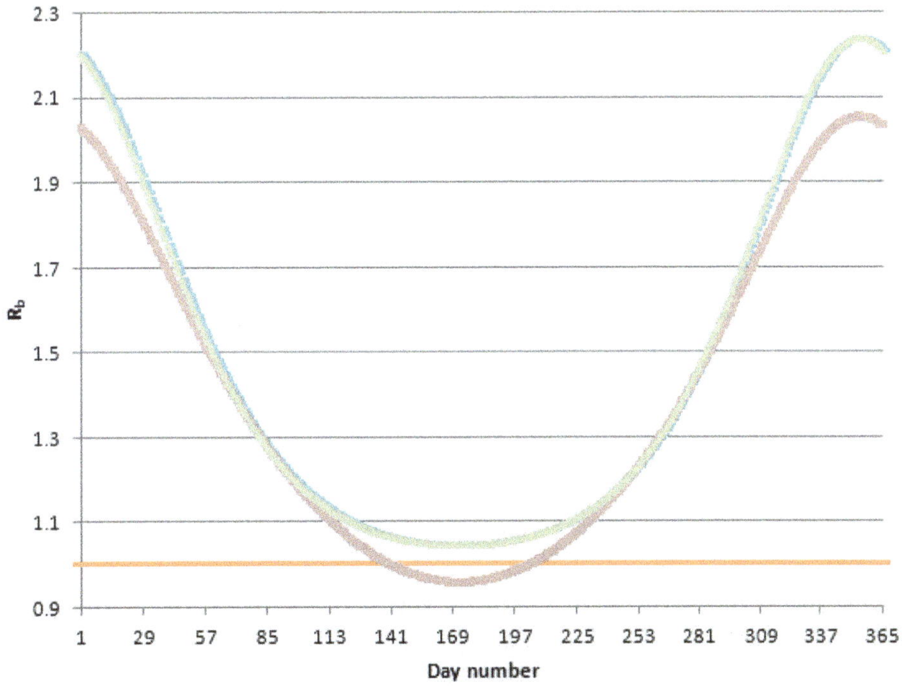

Figure 73. Daily variations of the geometric factor at latitude 60°N for 4 hour from solar noon using different modes of solar tracking, $R_{b(38)}$ (x), $R_{b(36)}$ (▲), $R_{b(30)}$ (♦), $R_{b(28)}$ (-) and $R_{b(32)}$ (■)

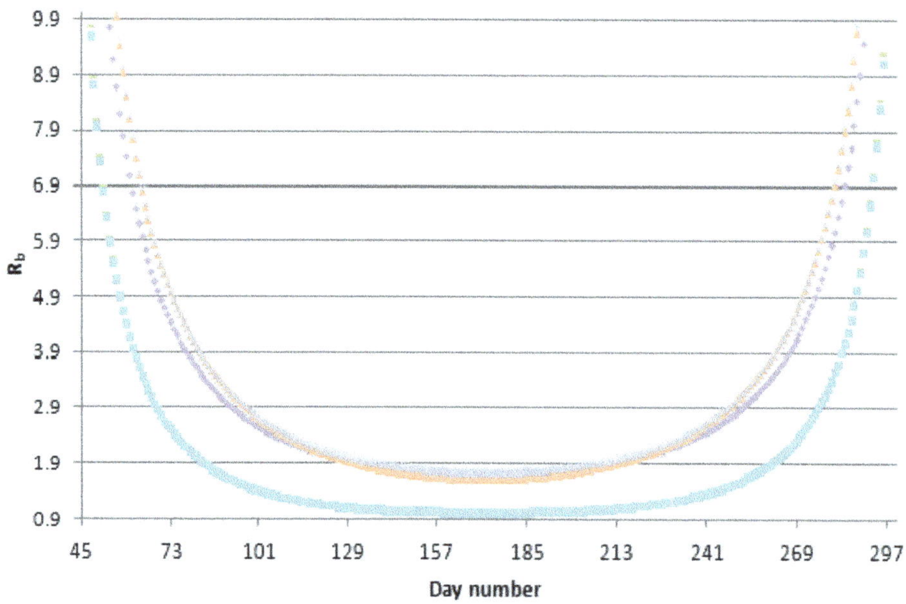

Figure 74. Daily variations of the geometric factor at latitude 60°N for 3 hour from solar noon using different modes of solar tracking, $R_{b(38)}$ (x), $R_{b(36)}$ (▲), $R_{b(30)}$ (♦), $R_{b(28)}$ (-) and $R_{b(32)}$ (■)

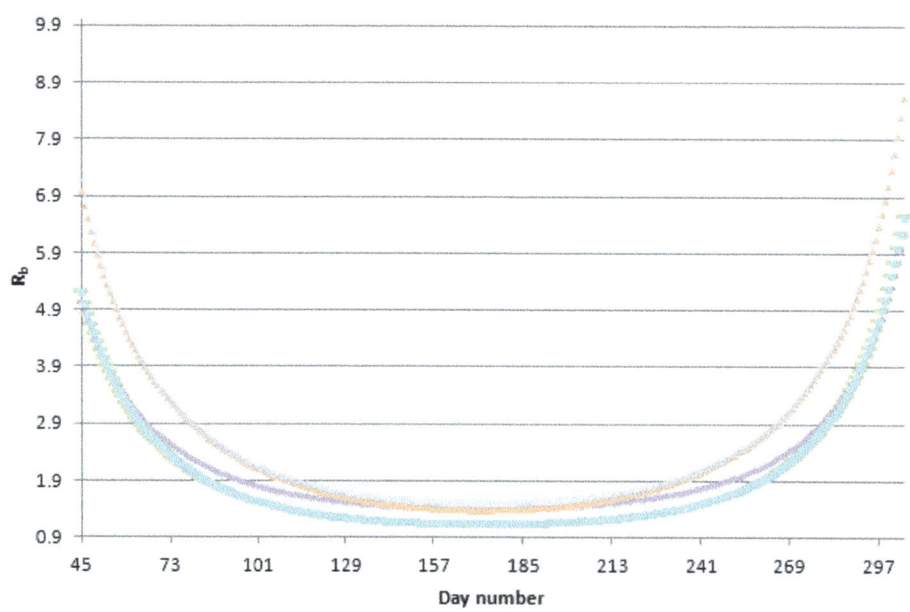

Figure 75. Daily variations of the geometric factor at latitude 60°N for 2 hour from solar noon using different modes of solar tracking, $R_{b(38)}$ *(x),* $R_{b(36)}$ *(▲),* $R_{b(30)}$ *(♦),* $R_{b(28)}$ *(-) and* $R_{b(32)}$ *(■)*

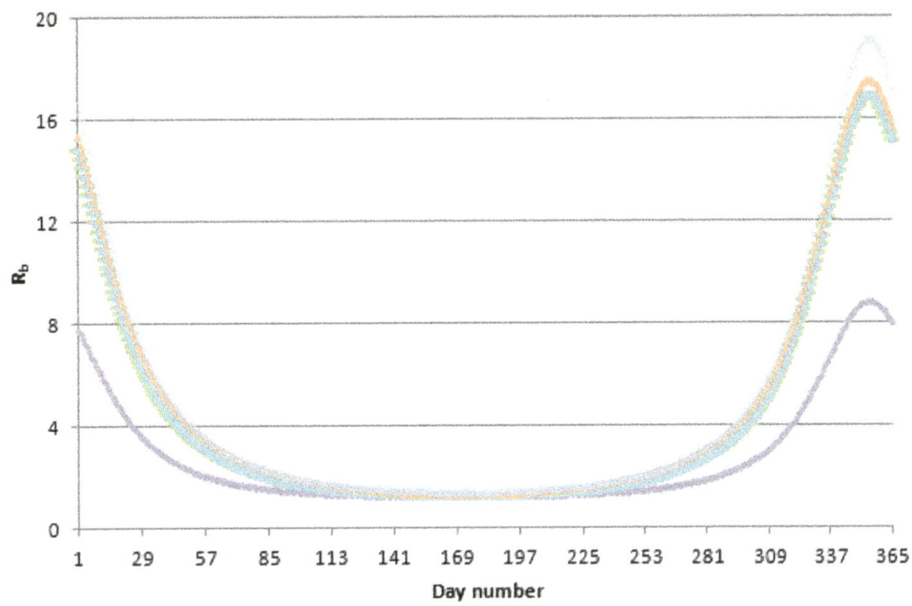

Figure 76. Daily variations of the geometric factor at latitude 60°N for 1 hour from solar noon using different modes of solar tracking, $R_{b(38)}$ *(x),* $R_{b(36)}$ *(▲),* $R_{b(30)}$ *(♦),* $R_{b(28)}$ *(-) and* $R_{b(32)}$ *(■)*

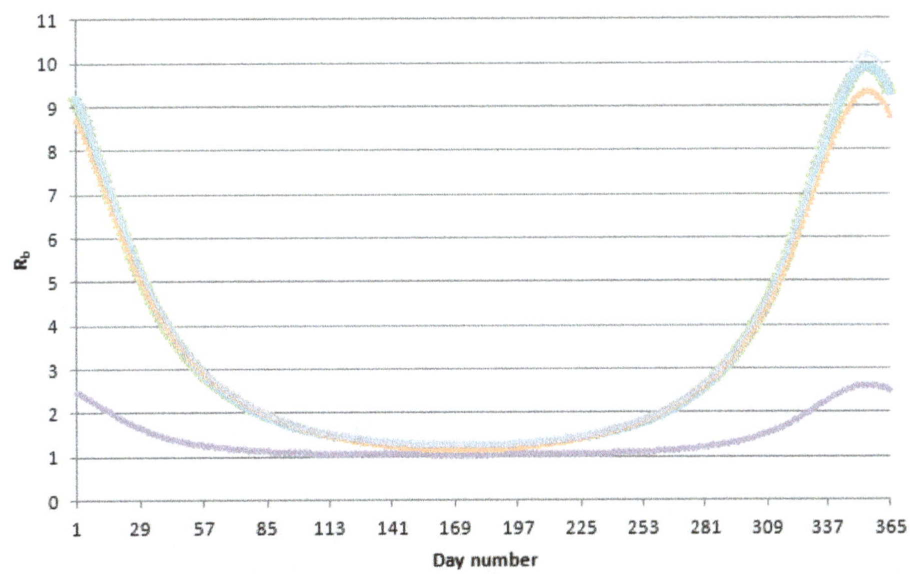

Figure 77. Daily variations of the geometric factor at latitude 60°N on solar noon using different modes of solar tracking, $R_{b(38)}$ (x), $R_{b(36)}$ (▲), $R_{b(30)}$ (♦), $R_{b(28)}$ (-) and $R_{b(32)}$ (■)

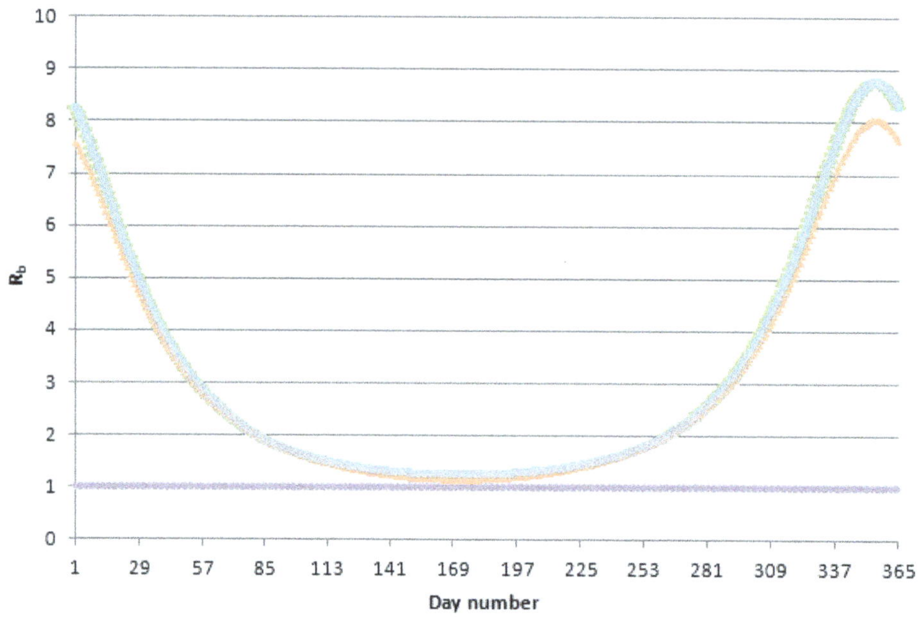

Tilt Dependent Category

The equation (34) is the sole mode of tracking that includes in this category. When applying the equation (34) at latitude of Damascus 33.51°N for calculating R_b during the year for various hour angles. The results are presented in Figures 78 to 82 where it is seen from these figures that R_b becomes greater and greater with increasing hours from solar noon. As daily values of R_b differ widely for each period of times from solar noon and for different tilt angles it is reasonable to search a suitable value of the tilt angle. This value may be daily, averaged weekly, fortnightly or monthly. Figures 78 to 82 demonstrate that averaged yearly daily slope does not lead to a suitable result.

When repeating the same calculations for different latitudes it was found that, for the same hour angle and the same tilt angle, the R_b could be greater than 1 and smaller than 1 depending on the latitude value. This means that, there is a tilt angle that makes the R_b as great as possible. The value of the tilt angle is latitude dependent. On the other hand, when applying the tracking mode where the receiver rotates continuously about a vertical axis (see equation (34)), it was found that, the effectiveness of solar tracking decreases rapidly when approaching the solar noon. On solar noon the gain on solar energy collection was not observed, $R_b<1$ for a long period during the year whatever the latitude is (see Figures 83 to 85).

Finally, it is easy to conclude that, in the Northern Hemisphere the effectiveness of solar tracking mode, basing on equation (34), is important during the period 22/9 to 21/3 while it could lead to the negative results, where $R_b<1$, during the period 22/3 to 21/9 (see Figure 78 to 82). At the same time the effectiveness of solar tracking becomes higher and higher with increasing the site's latitude. For Damascus, this is clearly demonstrated in the Figures 78 to 82. A similar behavior is observed for other latitudes in the Northern Hemisphere.

Figure 78. Daily variations of the R_b at latitude 33.51°N for 4 hours from solar noon using $R_{b(34)}$ mode of solar tracking at different slopes, 15°(■), 30°(▲), 45°(x), 60°(ж), 75°(●) and 90° (+)

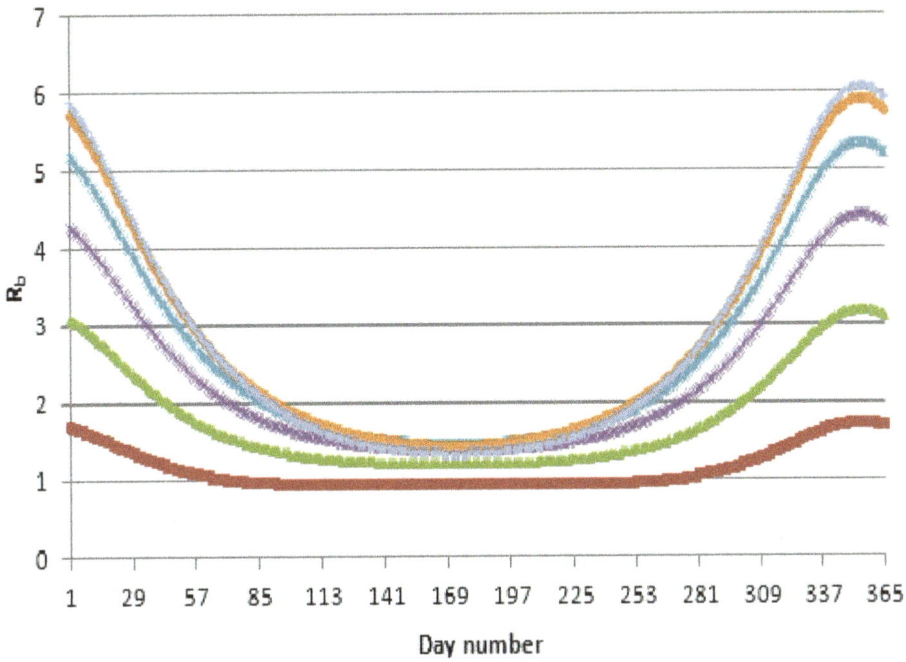

Figure 79. Daily variations of the R_b at latitude 33.51°N for 3 hours from solar noon using $R_{b(34)}$ mode of solar tracking at different slopes, 15°(■), 30°(▲), 45°(x), 60°(ж), 75°(●) and 90° (+)

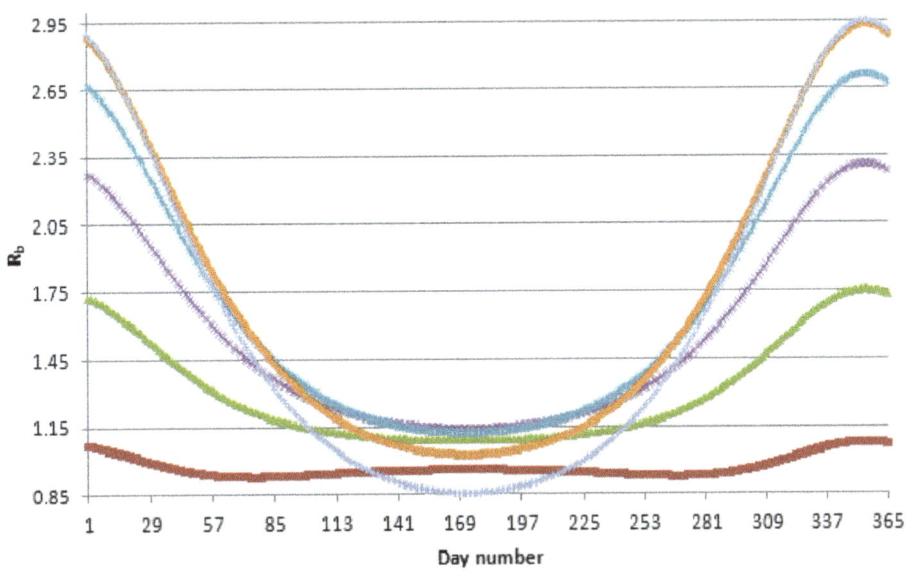

Figure 80. Daily variations of the R_b at latitude 33.51°N for 2 hours from solar noon using $R_{b(34)}$ mode of solar tracking at different slopes, 15°(■), 30°(▲), 45°(x), 60°(ж), 75°(●) and 90° (+)

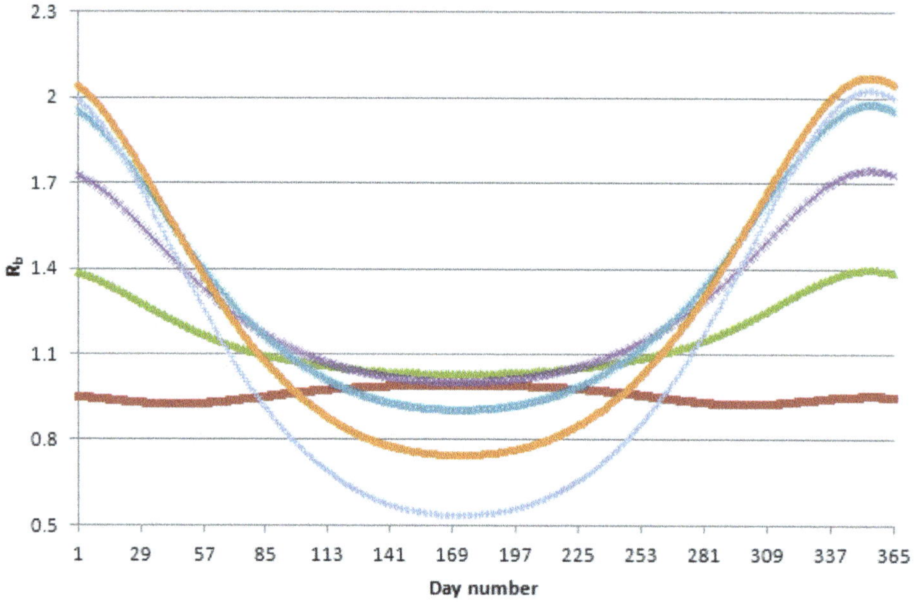

Figure 81. Daily variations of the R_b at latitude 33.51°N for 1 hours from solar noon using $R_{b(34)}$ mode of solar tracking at different slopes, 15°(■), 30°(▲), 45°(x), 60°(ж), 75°(●) and 90° (+)

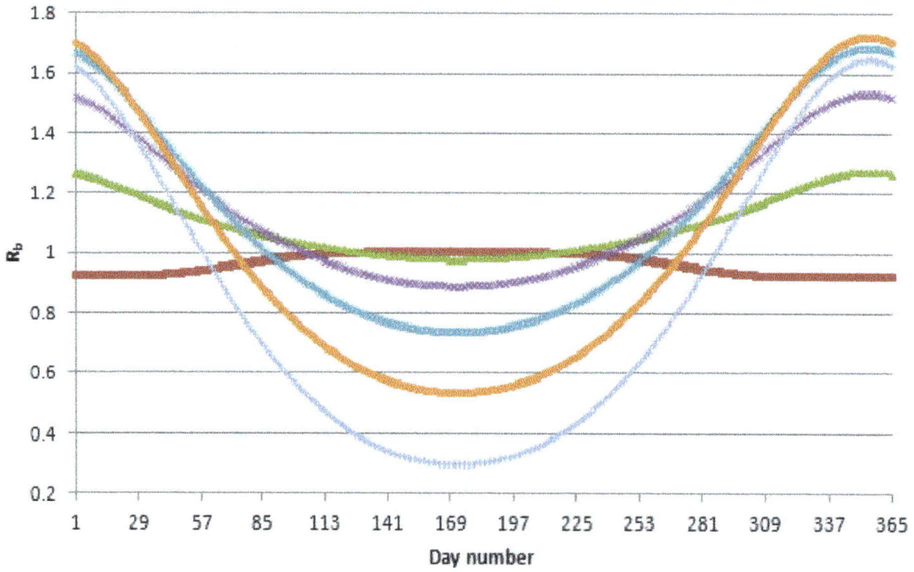

Figure 82. Daily variations of the R_b at latitude 33.51°N on solar noon using $R_{b(34)}$ mode of solar tracking at different slopes, 15°(■), 30°(▲), 45°(x), 60°(ж), 75°(●) and 90° (+)

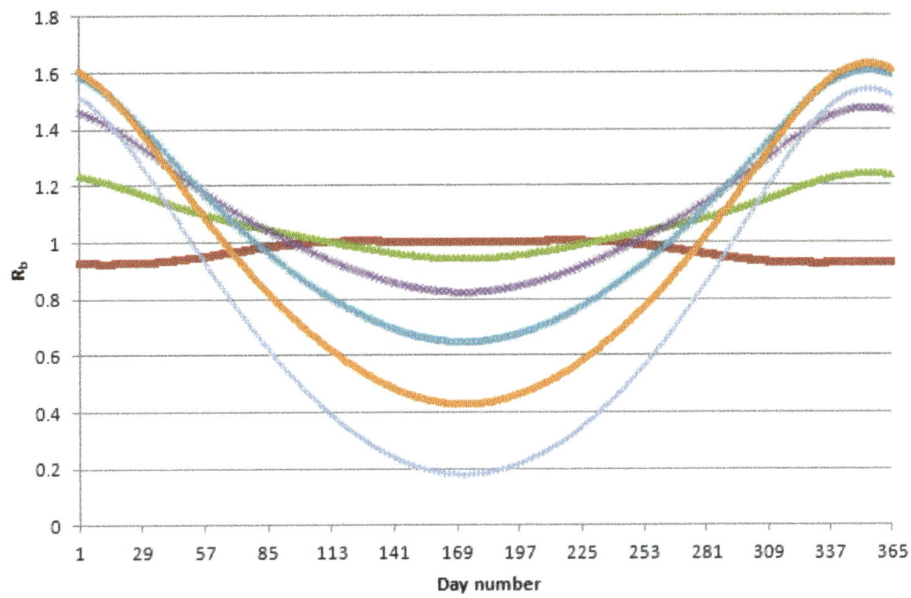

Figure 83. Daily variations of the R_b on solar noon using $R_{b(34)}$ mode of solar tracking at 15° of tilt angle for latitudes 10°N(ж), 30°N (●) and 45°N (+)

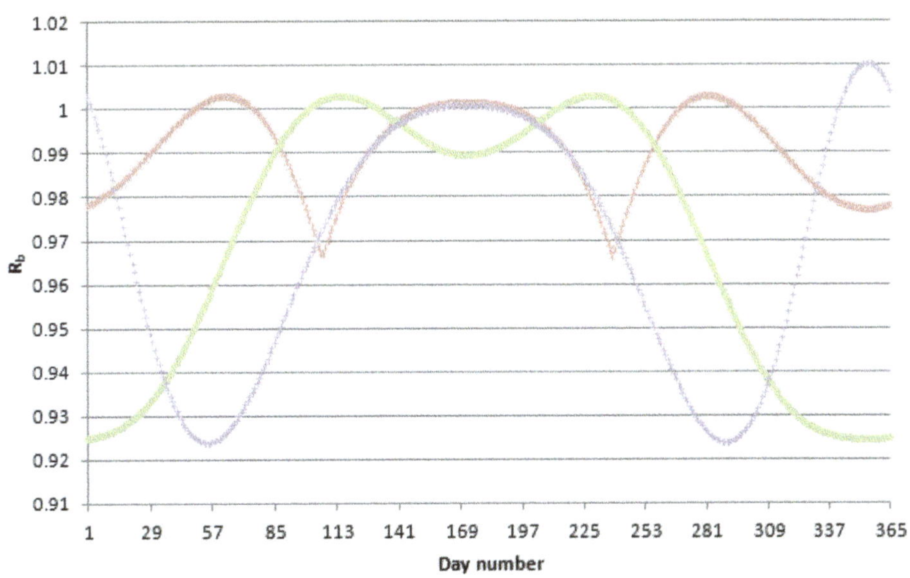

Figure 84. Daily variations of the R_b on solar noon using $R_{b(34)}$ mode of solar tracking at 30° of tilt angle for latitudes 10°N(ж), 30°N (●) and 45°N (+)

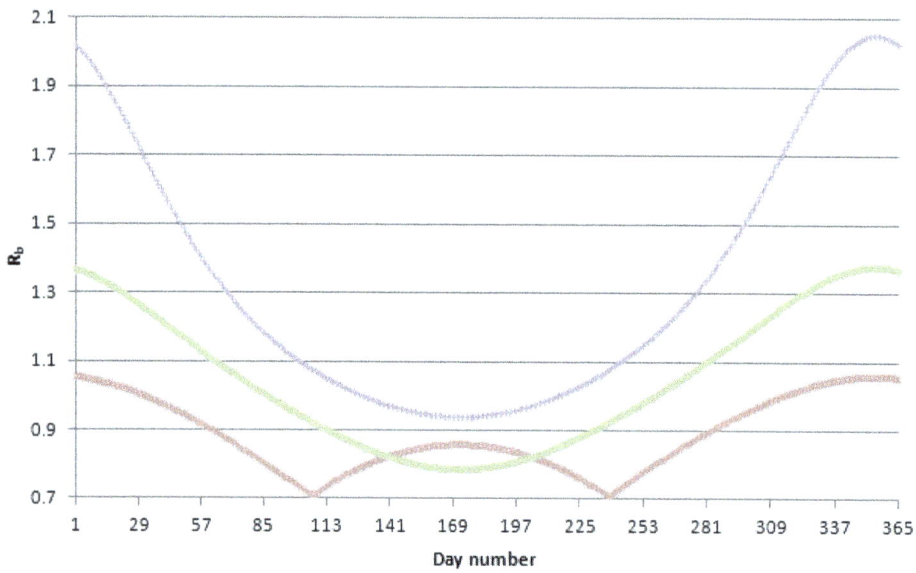

Figure 85. Daily variations of the R_b on solar noon using $R_{b(34)}$ mode of solar tracking at 45° of tilt angle for latitudes 10°N(ж), 30°N (●) and 45°N (+)

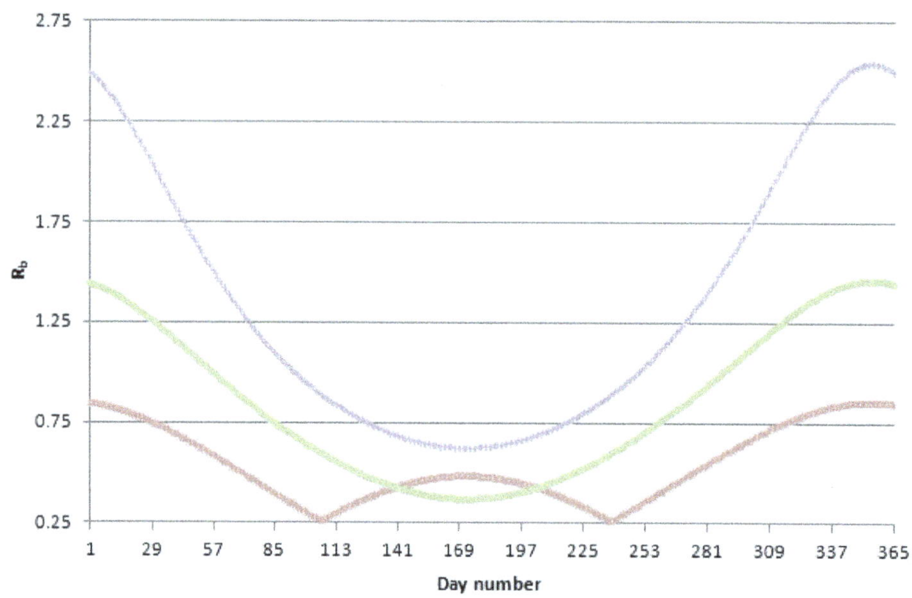

Long Term Ratio

The amount of solar energy that is converted in solar collector depends on tilt angle of collector with respect to horizontal surface and orientation of the collector. Although in the recent study of Beringer, Schilke, Lohse, and Seckmeyer (2011) the authors show that the tilt angle is nearly irrelevant that is the difference of plant yield is just 6% for tilt angles between 0° and 70°, these results are counterintuitive and as authors claim themselves the further investigations considering other locations and time periods are needed to clarify this issue. The tilt angle of a solar energy system was investigated for different applications: evacuated tube solar water heaters (see for example (Tang and Yang, 2014; Zhang, You, Xu, Wang, He and Zheng, 2014), hybrid power system (see for example Ismail, Moghavvemi and Mahlia, 2013a), PV generation ((see for example Yan, Saha, Meredith and Goodwin, 2013), solar cooker (see for example Sethi, V., Pal, D. and Sumathy, 2014), mirror-augmented PV system (see for example Fortunato, Torresi and Deramo, 2014), and many others.

The concept of energy gain is very useful in evaluating the tilt application over any period of consecutive days and in searching the optimum tilt angle and orientation. In this context, it is reasonable to introduce the following functions:

- Daily energy gain factor:

$$R = \left(\frac{H_{0,\beta d}}{H_{0,d}} \right) \tag{40}$$

- Weekly energy gain factor:

$$R_1 = \left(\frac{H_{0,\beta w}}{H_{0,w}} \right) \tag{41}$$

- Fortnightly energy gain factor:

$$R_2 = \left(\frac{H_{0,\beta f}}{H_{0,f}} \right) \tag{42}$$

- Monthly energy gain factor:

$$R_3 = \left(\frac{H_{0,\beta m}}{H_{0,m}} \right) \tag{43}$$

- Seasonally energy gain factor:

$$R_4 = \left(\frac{H_{0,\beta s}}{H_{0,s}}\right) \tag{44}$$

- Biannual energy gain factor:

$$R_5 = \left(\frac{H_{0,\beta b}}{H_{0,b}}\right) \tag{45}$$

- Yearly energy gain factor:

$$R_6 = \left(\frac{H_{0,\beta y}}{H_{0,y}}\right) \tag{46}$$

The division of the calendar year in days, weeks, fortnights and months is a common procedure and well definitive one but its division in 4 seasons and 2 half is different from one author to another. Different authors define seasons in different manner. Some authors mean by winter season the months: December to February (see for example (Sultan, Ali & Razaq, 2012; Skeiker, 2009; Ismail, Moghavvemi & Mahlia, 2013b; Khorasanizadeh, Mohammadi & Mostafaeipour, 2014)). Some authors define winter season from November to January (see for example (Yadav, 2013). Others understand by winter the first quarter of the year (see for example (Talebizadeh, Mehrabian & Abdolzadeh, 2011)) and others mention that winter season is the period started from 22 December to 21 March (see for example (Soulayman & Sabbagh, 2015)). Some authors (see for example (Moghadam, Tabrizi & Sharak, 2011)) used the original solar year with its division into 12 solar months according to information provided in the Table 18. The spring is the first solar season that includes Favardin, Ordibehesht and Khordad. This season covers the period from 20th March to 20th June. The summer season covers the period from 21st June to 21st September. The autumn season covers the period from 22nd September to 20th December while the winter season covers the period from 21st December to 19th March.

The first biannual year covers the period from 20th March to 21st September while the second biannual year covers the period from 22nd September to 19th March. For calculating the biannual energy gain it is advised here to consider that the solar year starts on 20/3 and ends on 19/3. So, the first half year covers the period from 20/3 to 21/9 while the second half year covers the period from 22/9 to 19/3.

Based on the functions in the equations (40) to (46), one can get reliable information regarding the solar receiver installation from practical and economic points of view. Furthermore, these functions will enable us in the chapter 5 to evaluate the validity of the two rules of thumbs which are widely used in solar energy applications. These rules are:

- The solar collector should be orientated towards Equator.
- The solar collector should have a latitude tilt angle value.

So, these seven functions should be considered and studied in detail over all latitudes in the Northern and Southern Hemispheres. Moreover, for purposes of solar process design and performance, it is often

Table 18. The solar year

Month Number	Name	The Period in the Usual Calendar Year
1	Favardin	From 20th March to 19th April
2	Ordibehesht	From 20th April to 20th May
3	Khordad	From 21st May to 20th June
4	Tir	From 21st June to 21st Julay
5	Mordad	From 22nd July to 21st August
6	Shahrivar	From 22nd August to 21st September
7	Mehr	From 22nd September to 21st October
8	Âbân	From 22nd October to 20th November
9	Âzar	From 21st November to 20th December
10	Dey	From 21st December to 19th January
11	Bahman	From 20th January to 18th February
12	Esfand	From 19th February to 19th March

necessary to calculate the hourly radiation on a tilted surface of a solar receiver from measurements or estimates of solar radiation on a horizontal plane.

For a surface with various slopes towards the Equator at various latitudes, the daily values of R were calculated during the year. The results are presented in Figures 86 to 91 where it is seen from these figures that R becomes greater and greater with increasing the latitude and the tilt angle. The results for the latitude 66.45°N are plotted in Figures 92 to 94 separately because of high R values.

As the general and detailed application of the equations (40) to (46) requires a large text area (see for example the Figures 86 to 94) and this question will considered when dealing with the long term tracking mode in chapter 5 it is reasonable to provide one example for illustration the upcoming results. Figure 95 shows the daily energy gain (in times of solar radiation on horizontal plane) of the Equator facing solar collector for L= 43.45 °N obtained for daily, monthly, seasonally and half-yearly solar collector tilt angle adjustments. It is seen from Figure 95 that: a) it is difficult to distinguish between the results obtained for daily and monthly adjustments which seems suggesting that the energy losses resulting from tilt angle adjustment on monthly basis is negligible; and b) even daily and monthly adjustments are slightly better than seasonally and half-yearly ones, all these four adjustments lead to comparable energy gain all over the year as the energy losses are not of practical importance. The daily yearly averages of R, R_4, R_5, R_6 are 1.7 and this means that it is reasonable, from practical and economical point of view, to adjust solar collector tilt twice a year (Soulayman and Hammoud, 2016).

When calculating the daily energy gain in the case of fixed installation over all the year with yearly optimum tilt angle, $\beta_{opt,y}$, and latitude tilted angle it was found that, even $\beta_{opt,y}$ is smaller than φ by few degrees (for φ= 43.45°N, $\beta_{opt,y}$ = 40.38°), the practical difference between the energy gain results is negligible (see Figure 96). Moreover, the daily yearly average of solar energy gain in both cases is 1.6. This means that the second rule of thumb is valid for mid-latitude zone.

Figure 86. Daily variations of the geometric factor R for various latitudes 10°N(ж), 20°N(-), 30°N(■), 40°N(ж), 50°N(-) and 60°N(-) for a surface with slope 15° towards the Equator

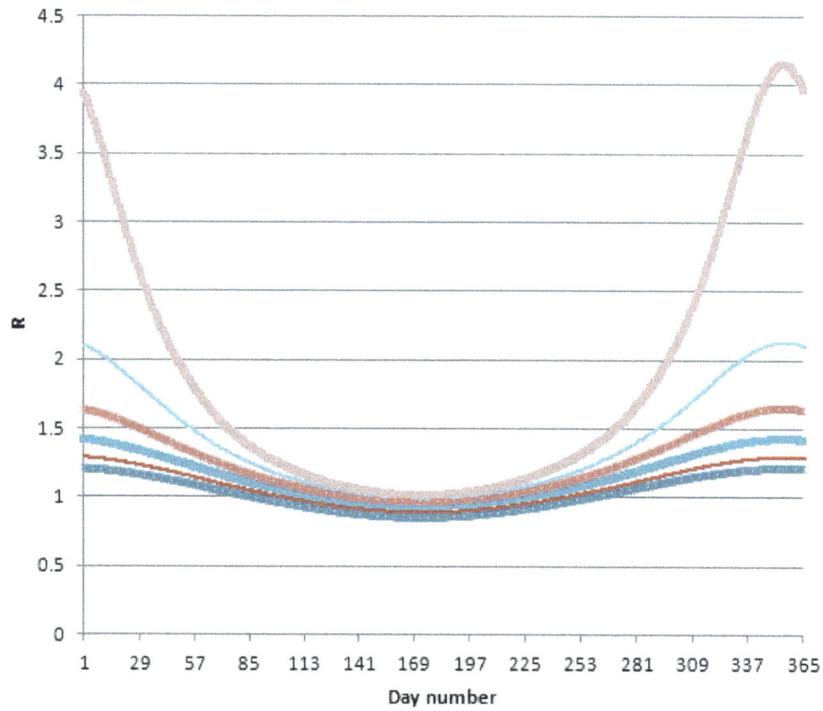

Figure 87. Daily variations of the geometric factor R for various latitudes 10°N(ж), 20°N(-), 30°N(■), 40°N(ж), 50°N(-) and 60°N(-) for a surface with slope 30° towards the Equator

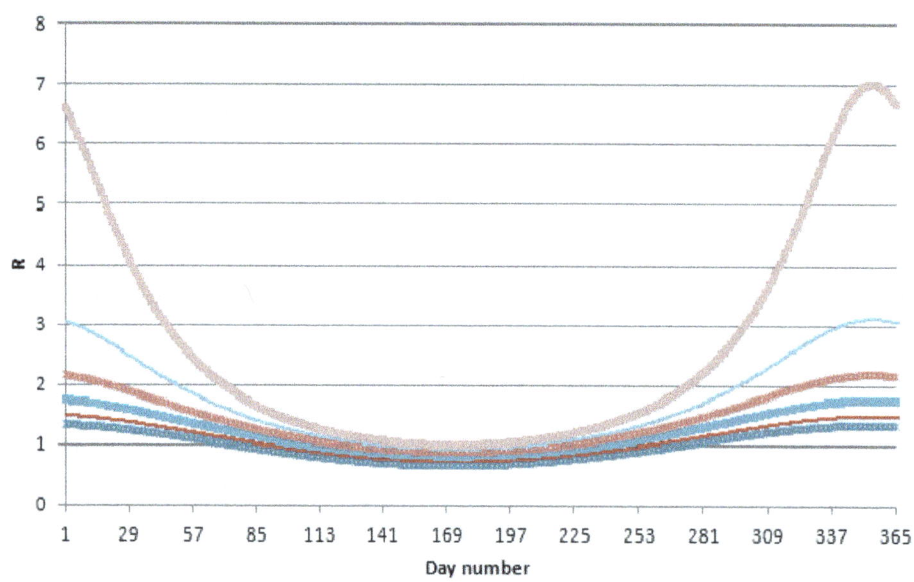

Figure 88. Daily variations of the geometric factor R for various latitudes 10°N(ж), 20°N(-), 30°N(■), 40°N(ж), 50°N(-) and 60°N(-) for a surface with slope 45° towards the Equator

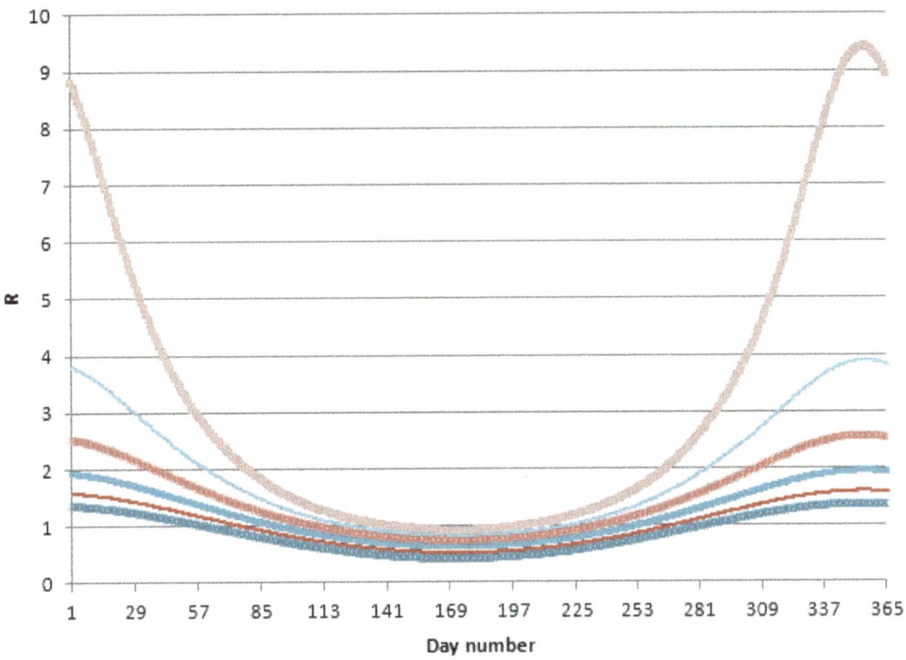

Figure 89. Daily variations of the geometric factor R for various latitudes 10°N(ж), 20°N(-), 30°N(■), 40°N(ж), 50°N(-) and 60°N(-) for a surface with slope 60° towards the Equator

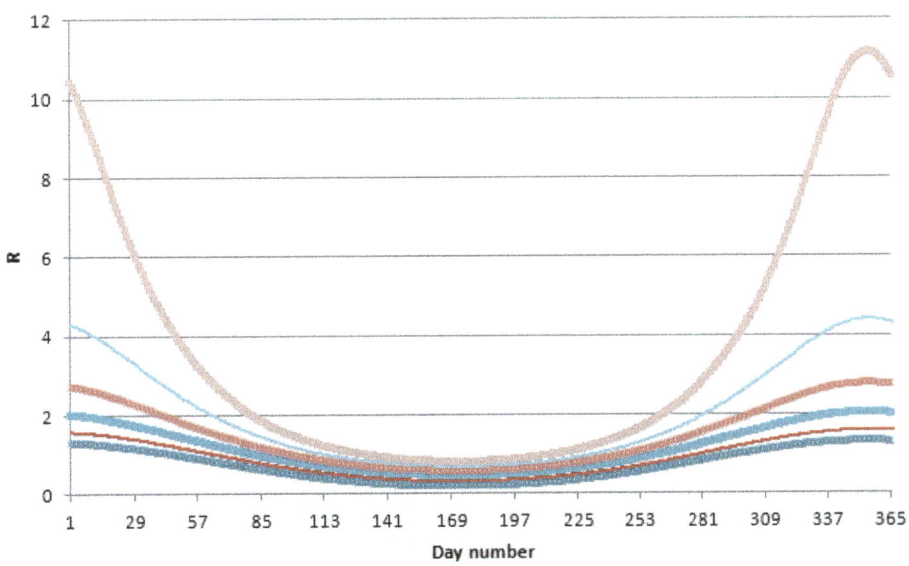

Figure 90. Daily variations of the geometric factor R for various latitudes 10°N(жс), 20°N(-), 30°N(■), 40°N(жс), 50°N(-) and 60°N(-) for a surface with slope 75° towards the Equator

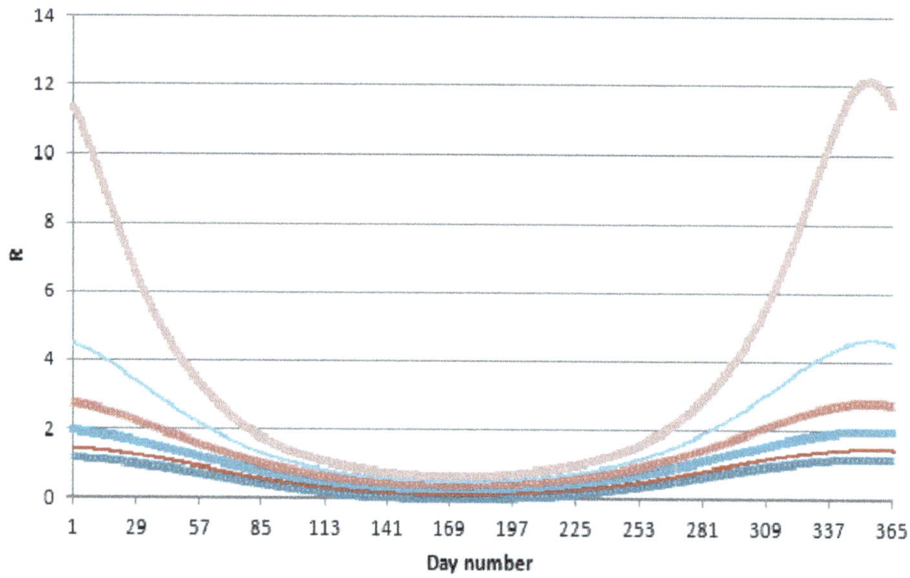

Figure 91. Daily variations of the geometric factor R for various latitudes 10°N(жс), 20°N(-), 30°N(■), 40°N(жс), 50°N(-) and 60°N(-) for a surface with slope 90° towards the Equator

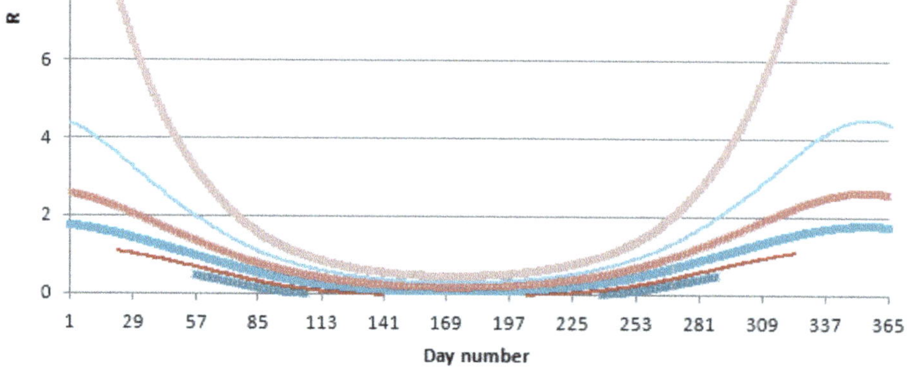

Figure 92. Daily variations of the geometric factor R for various slopes 15° (♦), 30° (■), 45° (▲), 60° (x), 75° (ж) and 90° (●) towards the Equator for the latitude 66.45°N during January

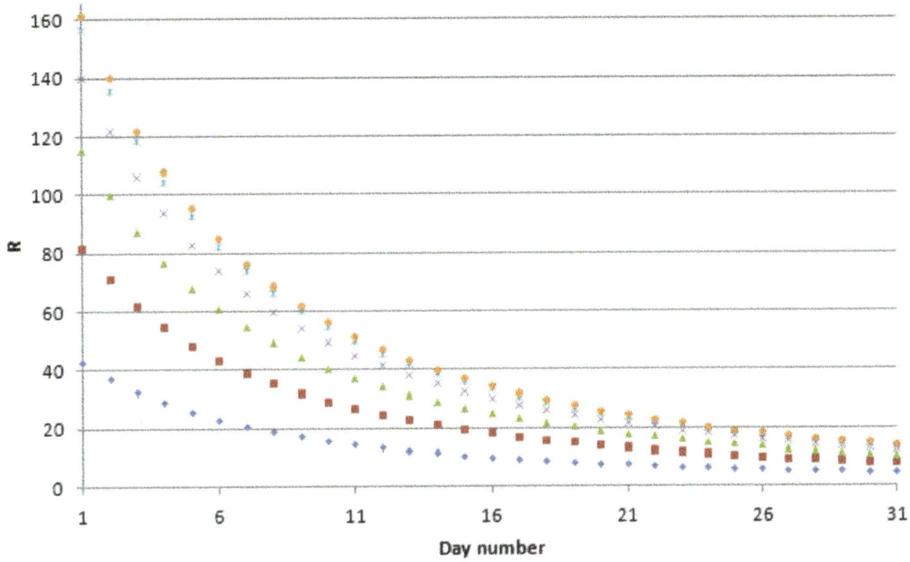

Figure 93. Daily variations of the geometric factor R for various slopes 15° (♦), 30° (■), 45° (▲), 60° (x), 75° (ж) and 90° (●) towards the Equator for the latitude 66.45°N during December

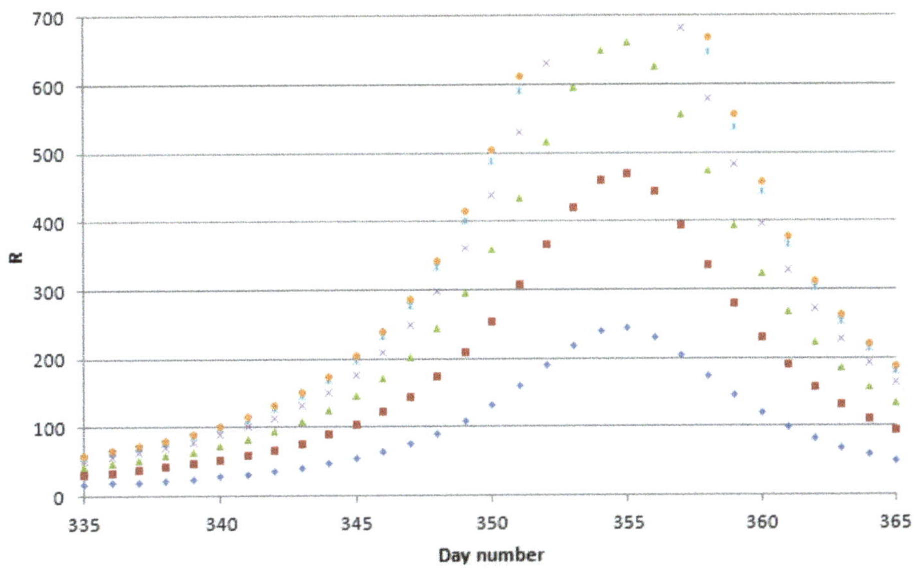

Figure 94. Daily variations of the geometric factor R for various slopes 15° (♦), 30° (■), 45° (▲), 60° (x), 75° (ж) and 90° (●) towards the Equator for the latitude 66.45°N during other months

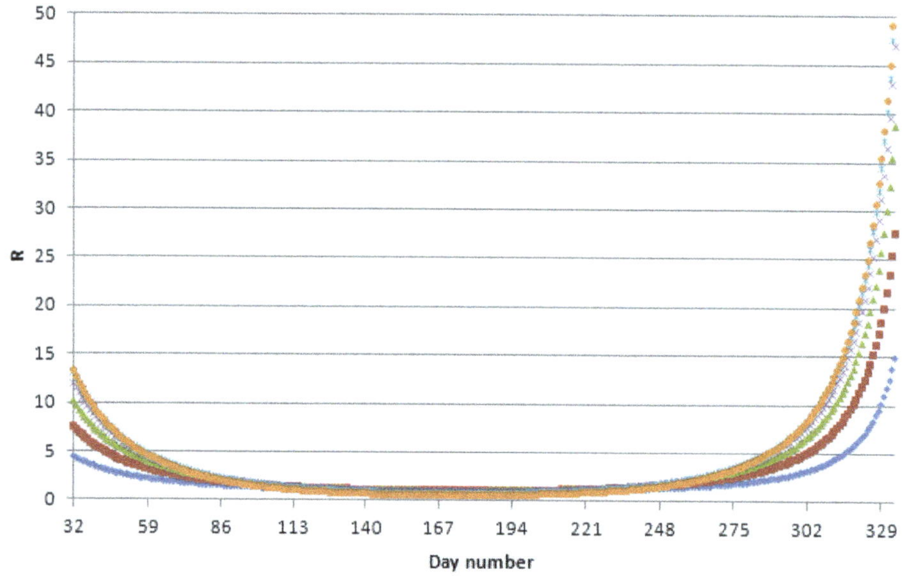

Figure 95. Calculated daily energy gain, on the base of daily (♦), monthly (■), seasonally (▲) and half-yearly (x) adjustments, for 43.45°N

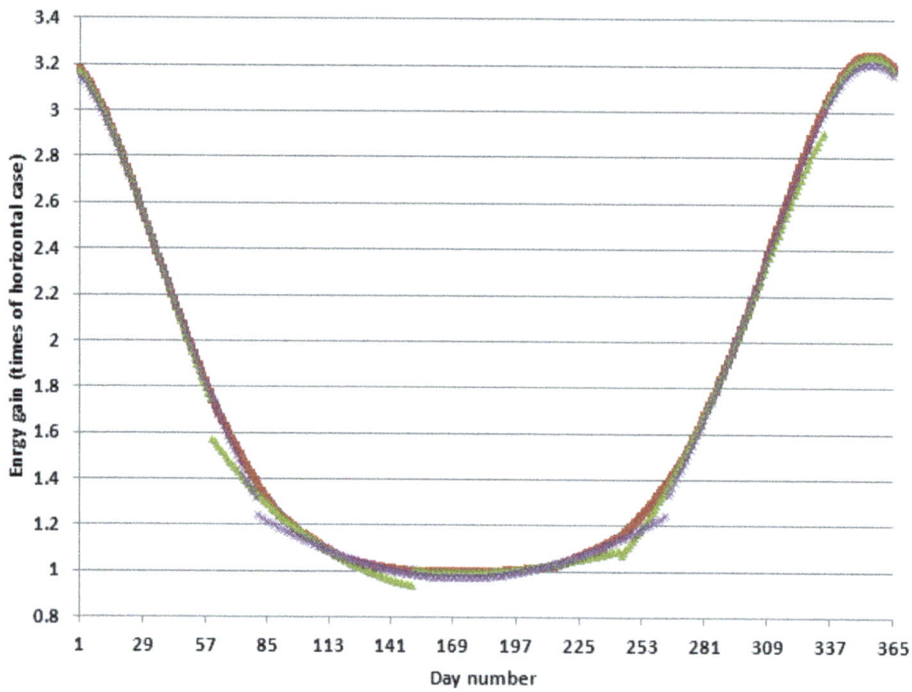

Figure 96. The calculated daily energy gain for $\beta = \beta_{opt,y}$ (x) and $\beta=\varphi$ (■) for 43.45°N

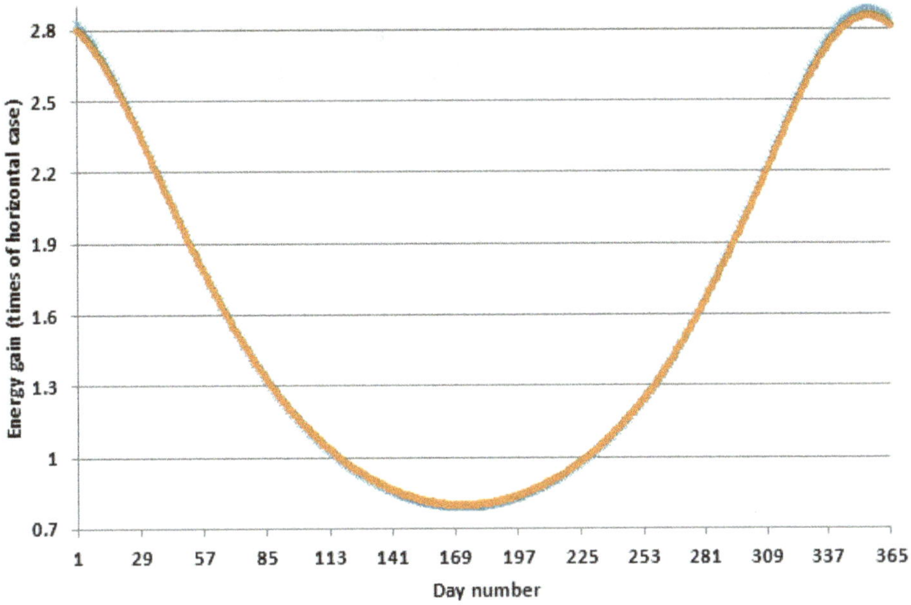

SOLAR ENERGY SHORT TERM COLLECTION

It is of interest to calculate the extraterrestrial solar radiation on a horizontal surface for an hour period. Integrating equation (11) for a period between hour angle ω_1 and ω_2 which define an hour (where ω_2 is the larger) leads to the following equation:

$$I_0 = \frac{12x3600}{\pi} G_s * \left\{ cos(\varphi) cos(\delta) \left[sin(\omega_2) - sin(\omega_1) \right] + \frac{\pi(\omega_2 - \omega_1)}{180^o} sin(\varphi) sin(\delta) \right\} \quad (47)$$

The hourly extraterrestrial solar radiation on a horizontal surface can be also approximated as follows:

$$I_0 = 3600 G_s * \left[cos(\varphi) cos(\delta) cos(\omega_{mid}) + sin(\varphi) sin(\delta) \right] \quad (48)$$

where

$$\omega_{mid} = \frac{\omega_2 + \omega_1}{2} \quad (49)$$

In order to demonstrate the validity of the equation (48) it is reasonable to compare its results with those of the precise equation (47). For this purpose, let's first consider some examples.

Example 12

What is the solar radiation on a horizontal surface in the absence of the atmosphere at latitude 33.51° on January 15 at solar noon and between the solar hours of 10 to 11?

Solution

The declination angle on January 15 is -21.28°. For January 15, the day number n in the year is 15 and the G_s value for this day is 1413.9 W/m^2 according to the equation (8). Using equation (47) with ω_1= -30° and ω_2= -15°,

$$I_0 = \frac{12x3600}{\pi}1413.9 * \left\{ \begin{matrix} cos\left(33.51^o\right)\cos\left(-21.28^o\right)\left[sin\left(-15^o\right) - sin\left(-30^o\right)\right] \\ + \frac{\pi\left(-15^o + 30^o\right)}{180^o} sin\left(33.51^o\right)\sin\left(-21.28^o\right) \end{matrix} \right\} = 2.61 MJ / m^2$$

On the other hand, using equation (48) with ω_{mid}= -22.5°,

$$I_0 = 3600 * 1413.9 * \left\{ \begin{matrix} cos\left(33.51^o\right)\cos\left(-21.28^o\right) cos\left(-22.5^o\right) \\ + sin\left(33.51^o\right)\sin\left(-21.28^o\right) \end{matrix} \right\} = 2.63 MJ / m^2$$

At solar noon, using equation (47) with ω_1= -7.5° and ω_2= +7.5°,

$$I_0 = \frac{12x3600}{\pi}1413.9 * \left\{ \begin{matrix} cos\left(33.51^o\right)\cos\left(-21.28^o\right)\left[sin\left(7.5^o\right) - sin\left(-7.5^o\right)\right] \\ + \frac{\pi\left(7.5^o + 7.5^o\right)}{180^o} sin\left(33.51^o\right)\sin\left(-21.28^o\right) \end{matrix} \right\} = 2.92 MJ / m^2$$

using equation (48) with ω_{mid}= 0°,

$$I_0 = 3600 * 1413.9 * \left\{ \begin{matrix} cos\left(33.51^o\right)\cos\left(-21.28^o\right) cos\left(0^o\right) \\ + sin\left(33.51^o\right)\sin\left(-21.28^o\right) \end{matrix} \right\} = 2.94 MJ / m^2$$

Example 13

What is the solar radiation on a horizontal surface in the absence of the atmosphere at latitude 33.51° on June 21 at solar noon and between the solar hours of 5 to 6?

Solution

The declination angle on June 21 is 23.46°. For June 21, the day number n in the year is 172 and the G_s value for this day is 1322.6 W/m² according to the equation (8). Using equation (47) with ω_1= -105° and ω_2= -90°,

$$I_0 = \frac{12x3600}{\pi}1322.6 * \begin{Bmatrix} cos\left(33.51^o\right)\cos\left(23.46^o\right)\left[sin\left(-90^o\right) - sin\left(-105^o\right)\right] \\ + \frac{\pi\left(-90^o + 105^o\right)}{180^o} sin\left(33.51^o\right)\sin\left(23.46^o\right) \end{Bmatrix} = 0.572 MJ / m^2$$

On the other hand, using equation (48) with ω_{mid}= -97.5°,

$$I_0 = 3600 * 1413.9 * \begin{Bmatrix} cos\left(33.51^o\right)\cos\left(23.46^o\right)cos\left(-97.5^o\right) \\ + sin\left(33.51^o\right)\sin\left(23.46^o\right) \end{Bmatrix} = 0.571 MJ / m^2$$

At solar noon, using equation (47) with ω_1= -7.5° and ω_2= +7.5°,

$$I_0 = \frac{12x3600}{\pi}1413.9 * \begin{Bmatrix} cos\left(33.51^o\right)\cos\left(23.46^o\right)\left[sin\left(7.5^o\right) - sin\left(-7.5^o\right)\right] \\ + \frac{\pi\left(7.5^o + 7.5^o\right)}{180^o} sin\left(33.51^o\right)\sin\left(23.46^o\right) \end{Bmatrix} = 4.678 MJ / m^2$$

using equation (48) with ω_{mid}= 0°,

$$I_0 = 3600 * 1413.9 * \begin{bmatrix} cos\left(33.51^o\right)\cos\left(23.46^o\right)cos\left(0^o\right) \\ + sin\left(33.51^o\right)\sin\left(23.46^o\right) \end{bmatrix} = 4.688 MJ / m^2$$

Example 14

What is the solar radiation on a horizontal surface in the absence of the atmosphere at latitude 50° on December 21 at solar noon and between the solar hours of 10 to 11?

Solution

The declination angle on December 21 is -23.46°. For December 21, the day number n in the year is 355 and the G_s value for this day is 1413.7 W/m² according to the equation (8). Using equation (47) with ω_1= -30° and ω_2= -15°,

$$I_0 = \frac{12x3600}{\pi}1413.7 * \left\{ \begin{matrix} cos\left(33.51^o\right)\cos\left(-23.46^o\right)\left[sin\left(-15^o\right) - sin\left(-30^o\right)\right] \\ + \frac{\pi\left(-15^o + 30^o\right)}{180^o} sin\left(33.51^o\right)\sin\left(-23.46^o\right) \end{matrix} \right\} = 1.221 MJ / m^2$$

On the other hand, using equation (48) with ω_{mid}= -22.5°,

$$I_0 = 3600 *1413.7 * \left\{ \begin{matrix} cos\left(33.51^o\right)\cos\left(-23.46^o\right)cos\left(-22.5^o\right) \\ +sin\left(33.51^o\right)\sin\left(-23.46^o\right) \end{matrix} \right\} = 1.213 MJ / m^2$$

At solar noon, using equation (47) with ω_1= -7.5° and ω_2= +7.5°,

$$I_0 = \frac{12x3600}{\pi}1413.9 * \left\{ \begin{matrix} cos\left(33.51^o\right)\cos\left(-23.46^o\right)\left[sin\left(7.5^o\right) - sin\left(-7.5^o\right)\right] \\ + \frac{\pi\left(7.5^o + 7.5^o\right)}{180^o} sin\left(33.51^o\right)\sin\left(-23.46^o\right) \end{matrix} \right\} = 1.440 MJ / m^2$$

using equation (48) with ω_{mid}= 0°,

$$I_0 = 3600 *1413.9 * \left\{ \begin{matrix} cos\left(33.51^o\right)\cos\left(-23.46^o\right)cos\left(0^o\right) \\ +sin\left(33.51^o\right)\sin\left(-23.46^o\right) \end{matrix} \right\} = 1.449 MJ / m^2$$

Example 15

What is the solar radiation on a horizontal surface in the absence of the atmosphere at latitude 50° on June 21 at solar noon and between the solar hours of 5 to 6?

Solution

The declination angle on June 21 is 23.46°. For June 21, the day number n in the year is 172 and the G_s value for this day is 1322.6 W/m² according to the equation (8). Using equation (47) with ω_1= -105° and ω_2= -90°,

$$I_0 = \frac{12x3600}{\pi}1322.6 * \left\{ \begin{matrix} cos\left(50^o\right)\cos\left(23.46^o\right)\left[sin\left(-90^o\right) - sin\left(-105^o\right)\right] \\ + \frac{\pi\left(-90^o + 105^o\right)}{180^o} sin\left(50^o\right)\sin\left(23.46^o\right) \end{matrix} \right\} = 4.038 MJ / m^2$$

On the other hand, using equation (48) with ω_{mid}= -97.5°,

$$I_0 = 3600 * 1413.9 * \begin{bmatrix} cos\left(33.51^o\right)\cos\left(23.46^o\right)cos\left(-97.5^o\right) \\ +sin\left(33.51^o\right)\sin\left(23.46^o\right) \end{bmatrix} = 4.045\, MJ / m^2$$

At solar noon, using equation (47) with $\omega_1 = -7.5^o$ and $\omega_2 = +7.5^o$,

$$I_0 = \frac{12x3600}{\pi} 1413.9 * \begin{Bmatrix} cos\left(50^o\right)\cos\left(23.46^o\right)\left[sin\left(7.5^o\right) - sin\left(-7.5^o\right)\right] \\ + \frac{\pi\left(7.5^o + 7.5^o\right)}{180^o} sin\left(50^o\right)\sin\left(23.46^o\right) \end{Bmatrix} = 4.251\, MJ / m^2$$

using equation (48) with $\omega_{mid} = 0^o$,

$$I_0 = 3600 * 1413.9 * \begin{Bmatrix} cos\left(50^o\right)\cos\left(23.46^o\right)cos\left(0^o\right) \\ +sin\left(50^o\right)\sin\left(23.46^o\right) \end{Bmatrix} = 4.259\, MJ / m^2$$

So, the differences between the hourly solar radiation, calculated by the equations (47) and (48), is very small and it does not exceed 1% whatever the latitude is and whatever the hour interval is located near the solar noon or near sunrise and sunset. The daily distribution of hourly solar extraterrestrial solar radiation at latitude of Damascus (33.51°N) and 50° is demonstrated in Figures 97 and 98 respectively.

Figure 97. The calculated daily hourly extraterrestrial solar radiation at latitude 33.51°. (●), (+), (-), (-), (♦), (■), (▲) and (x) stand for solar noon, 0.5h, 1.5h, 2.5h, 3.5h, 4.5h, 5.5h and 6.5h from solar noon respectively

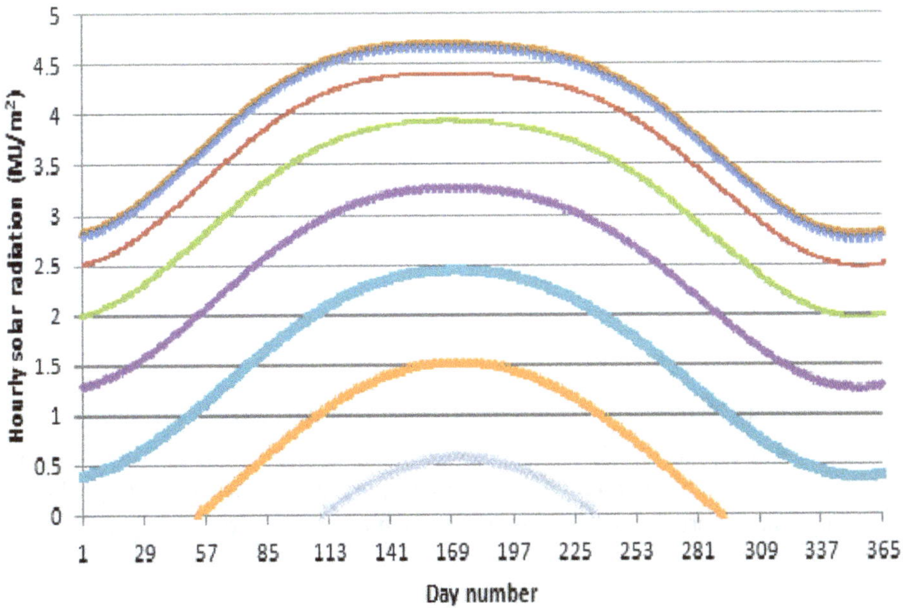

Figure 98. The calculated daily hourly extraterrestrial solar radiation at latitude 50°. (●), (+), (-), (-), (♦), (■), (▲) and (x) stand for solar noon, 0.5h, 1.5h, 2.5h, 3.5h, 4.5h, 5.5h and 6.5h from solar noon respectively

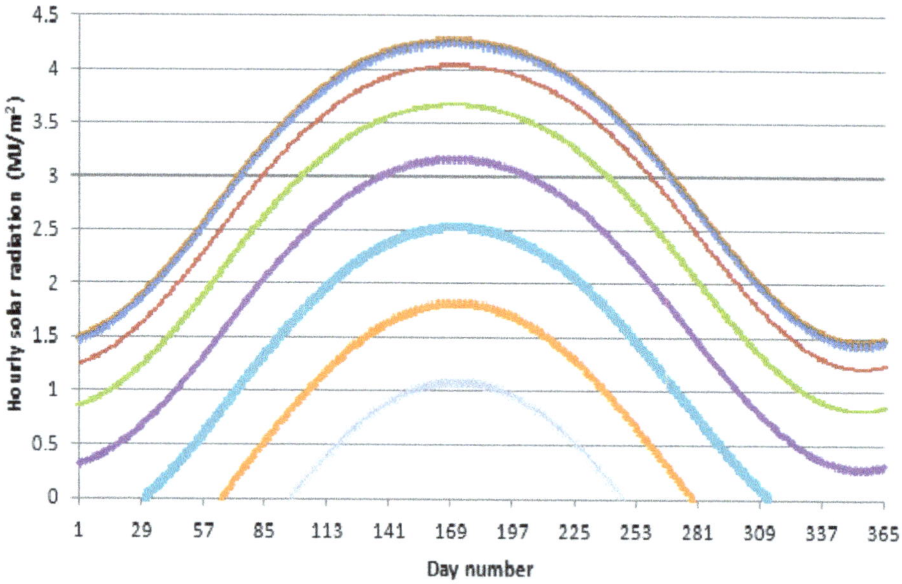

The single axis and dual axes solar tracking will be based on short term solar energy collection. Therefore, different modes of solar tracking in addition to short term solar energy collection will be used in details in chapter 6.

SOLAR ENERGY LONG TERM COLLECTION

The daily extraterrestrial solar radiation intercepted on a surface, tilted by an angle β to the horizontal plane and orientated by an angle γ from the Equator direction, can be expressed using equation (18). So, the extraterrestrial solar radiation intercepted on such a surface over a period of consecutive days could be expressed as follows:

$$H\left(n_1,n_2\right)=\sum_{n_1}^{n_2}H_0\left(n,\varphi,\beta,\gamma\right) \tag{50}$$

where N_1 and N_2 are, respectively, the day numbers of the first day and the last day of the period in the year. Thus, the equation (50) could be used for calculating the solar energy collected over a week, a fortnight, a month, a season, an half year, a year or any period of consecutive days. For a period of a month, the monthly solar energy, in the absence of atmosphere, could be calculated as:

$$H\left(N_1, N_2\right) = \frac{12x3600xG_s}{\pi} * \sum_{N_1}^{N_2} \left\{ \begin{array}{l} A_2\left(\sin\left(\omega_{ss}\right) - \sin\left(\omega_{sr}\right)\right) - \\ A_3\left[\cos\left(\omega_{ss}\right) - \cos\left(\omega_{sr}\right)\right] + \dfrac{\pi * \left(\omega_{ss} - \omega_{sr}\right)}{180^o} A_1 \end{array} \right\} \tag{51}$$

where N_1 and N_2 are given in the Table 19.

As an example of the long term solar energy collection let us consider the monthly and yearly solar energy collection using a solar energy flat plate collector tilted by an angle $\beta=\varphi$ and orientated towards the southwest by an angle $\gamma=15^{o.}$ For these circumstances, the monthly and yearly collections of the solar energy by the mentioned collector are presented in the Table 20.

When calculating the daily solar radiation on a surface maintained normal to the Sun's rays, the obtained results are plotted in Figure 99. On the other hand, Table 21 presents the daily average daily yearly solar radiation on a horizontal plane as well as on a surface maintained normal to the Sun's rays the results, for different latitudes.

The interesting result shown by Table 21 is that, the average daily yearly total solar radiation on a surface maintained normal to the Sun's rays is essentially the same regardless of the latitude (the average daily yearly over all latitudes is 59.06 MJ/m^2). This is so because anywhere on the earth there are 4,380 hours of daylight; or, taken over a year, the average length of daylight is 12 hours. Therefore, except for the slight difference caused by the winter extraterrestrial solar irradiance being about 6 percent greater than the summer solar irradiance, the total yearly extraterrestrial normal solar irradiance is essentially the same anywhere on the earth.

Note also that the yearly average of the daily normal solar radiation values is very close to the product of 12 (the average length of daylight) times the solar constant, which gives 59.1 MJ/m^2. Note also, in comparing the average daily yearly total solar radiation on a horizontal surface with its values on a normal surface, that over the year the cosine effect reduces solar radiation on a horizontal surface by 39

Table 19. The day numbers of the first and last days of the months

Month	N_1		N_2	
		Leap year		Leap year
Jan.	1	1	31	31
Feb.	32	32	59	60
March	60	61	90	91
April	91	92	120	121
May	121	122	151	152
June	152	153	181	182
July	182	183	212	213
Aug.	213	214	243	244
Sep.	244	245	273	274
Oct.	274	275	304	305
Nov.	305	306	334	335
Dec.	335	336	365	366

Table 20. The monthly and yearly collections (in MJ/m²) of the solar energy in the case of $\beta=\varphi$ and $\gamma=15^o$ in the Northern Hemisphere

Month	10°N	20°N	30°N	40°N	50°N	60°N	66.45°N
Jan.	1117.9	1101.0	1079.8	1041.9	973.7	814.8	500.9
Feb.	1044.7	1033.5	1023.6	1006.1	978.8	923.2	841.1
March	1171.3	1161.6	1150.0	1134.7	1118.8	1103.7	1092.1
April	1101.1	1095.0	1075.0	1052.8	1027.0	1000.3	984.3
May	1077.1	1073.5	1050.4	1028.0	1002.9	978.2	963.9
June	1004.9	1002.2	981.2	960.5	937.2	913.9	900.2
July	1051.0	1047.9	1025.7	1003.9	979.5	955.2	941.1
Aug.	1102.5	1097.5	1074.7	1051.2	1024.7	999.2	984.9
Sep.	1110.0	1101.8	1088.3	1071.4	1052.4	1032.1	1019.3
Oct.	1149.4	1138.2	1128.6	1113.3	1093.1	1058.2	1009.9
Nov.	1083.0	1068.0	1050.6	1019.3	964.3	840.7	622.4
Dec.	1098.8	1080.2	1054.7	1009.7	926.1	721.5	200.7
Year	13111.8	13000.5	12782.6	12493.1	12078.4	11341.1	10060.9

Figure 99. The calculated daily extraterrestrial solar radiation on a surface maintained normal to the Sun's rays at latitudes 10°N (-), 20°N (-), 30°N (+), 40°N (●), 50°N (ж), 60°N (x) and 66.45°N (▲)

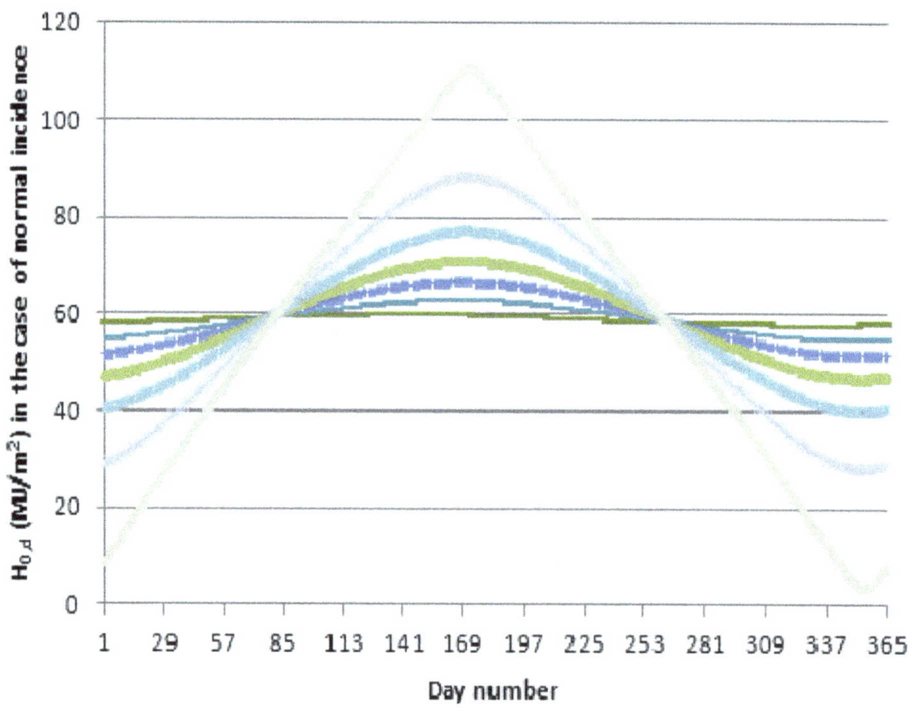

Table 21. Extraterrestrial solar radiation on a horizontal plane and on a surface maintained normal to the Sun's rays (dual axes tracking)

Latitude	H_0 on a Horizontal Plane (MJ/m^2)	H_0 on a Surface Maintained Normal to the Sun's Rays (MJ/m^2)
66.45	17.68	58.29
60	20.14	58.55
50	24.32	58.73
40	28.18	58.83
30	31.45	58.91
20	33.94	58.96
10	35.48	59.01
0	36.05	59.06
-10	35.64	59.11
-20	32.03	59.16
-30	31.9	59.22
-40	28.79	59.29
-50	25.02	59.39
-60	20.92	59.57
-66.45	18.50	59.83

percent at the equator, whereas the solar radiation is reduced by 52.1 percent at 40 degrees latitude and by 69.7 percent at 66.45 degrees latitude.

The interesting result shown by Figure 99 is that, the daily extraterrestrial solar radiation on a surface maintained normal to the Sun's rays for latitude 66.45°N is the possibility of approximating this radiation during the period 22/12 to 21/6 as a linear function of a day number in the year with a correlation factor of 0.9993 and positive tilt while this function is of a negative tilt and a correlation factor of 0.9996 during the period 22/6 to 21/12.

CONCLUSION

In this chapter we have outlined the basic characteristics of the solar radiation, noting that the solar constant, the mean radiation flux density outside of the Earth's atmosphere, is 1367 W/m^2. We have included in this chapter those topics that are based on extraterrestrial solar radiation. Different formulae for calculating the extraterrestrial solar radiation on surfaces with different orientations are provided. This is background information for chapter 4 which is concerned with effects of the atmosphere, radiation measurements and data manipulation. The main concepts of the direct geometric factor calculation using different modes of tracking are provided. The short term solar energy collection is also introduced. This is background information for chapter 6 which is concerned with the short term tracking. The long term solar energy collection is also introduced in order to be used lately in chapter 5 which is concerned with the long term tracking (in determining the optimum tilt angles and orientations of solar collectors).

The provided information in this chapter will used in chapter 8 which is concerned with the economic considerations of solar tracking. This understanding of extraterrestrial solar radiation on hypothetical surfaces above the earth's atmosphere has been discussed here to give the reader some idea of the solar energy resource and the effects of the mechanics of the earth-sun system. In the chapter 4 we will discuss the additional effects of the earth's atmosphere (water vapor, carbon dioxide, clouds, smog, particulates) on the solar radiation received below the atmosphere.

REFERENCES

Beringer, S., Schilke, H., Lohse, I., & Seckmeyer, G. (2011). Case study showing that the tilt angle of photovoltaic plants is nearly irrelevant. *Solar Energy*, *85*(3), 470–476. doi:10.1016/j.solener.2010.12.014

Delinger, W. G. (1976). The Definition of the Langley. *Solar Energy*, *18*(4), 369–370. doi:10.1016/0038-092X(76)90065-7

Duffie, J. A., & Beckman, W. A. (2013). *Solar engineering of thermal processes* (3rd ed.). New York, NY: Wiley & Sons; doi:10.1002/9781118671603

Duncan, C. H., Willson, R. C., Kendall, J. M., Harrison, R. G., & Hickey, J. R. (1982). Latest rocket measurements of the solar constant. *Solar Energy*, *28*(5), 385–387. doi:10.1016/0038-092X(82)90256-0

Eddy, J. A. (1979). *A New Sun, the Solar Results from Skylab*. NASA Report SP -402.

Fortunato, B., Torresi, M., & Deramo, A. (2014). Modeling, performance analysis and economic feasibility of a mirror-augmented photovoltaic system. *Energy Conversion and Management*, *80*(0), 276–286. doi:10.1016/j.enconman.2013.12.074

Fröhlich, C. (1991). History of solar radiometry and the world radiation reference. *Metrologia*, *28*(3), 111–115. doi:10.1088/0026-1394/28/3/001

Gueymard, C. A. (2004). The suns total and spectral irradiance for solar energy application and solar radiation models. *Solar Energy*, *76*(4), 423–453. doi:10.1016/j.solener.2003.08.039

Gueymard, C. A., & Myers, D. R. (2008). Solar radiation measurement: progress in radiometry for improved modeling. In V. Badescu (Ed.), *Modeling solar radiation at the earth surface*. Berlin: Springer. doi:10.1007/978-3-540-77455-6_1

Gunerhan, H., & Hepbasli, A. (2007). Determination of the optimum tilt angle of solar collectors for building applications. *Building and Environment*, *42*(2), 779–783. doi:10.1016/j.buildenv.2005.09.012

Hickey, J. R., Alton, B. M., Griffin, F. J., Jacobwitz, H., Pellegrino, P., Maschhoff, R. H., & Vonder Harr, T. H. et al. (1982). Extraterrestrial solar irradiance variability. Two and one-half years of measurements from Nimbus 7. *Solar Energy*, *29*(2), 125–127. doi:10.1016/0038-092X(82)90173-6

Hottel, H. C., & Woertz, B. B. (1942). Performance of flat-plate solar heat collectors. *Trans. ASME*, *64*, 91.

Iqbal, M. (1983). *An introduction to solar radiation*. Toronto: Academic Press.

Ismail, M., Moghavvemi, M., & Mahlia, T. (2013a). Design of an optimized photovoltaic and microturbine hybrid power system for a remote small community: Case study of Palestine. *Energy Conversion and Management*, *75*(0), 271–281. doi:10.1016/j.enconman.2013.06.019

Ismail, M., Moghavvemi, M., & Mahlia, T. (2013b). Analysis and evaluation of various aspects of solar radiation in the Palestinian territories. *Energy Conversion and Management*, *73*(0), 57–68. doi:10.1016/j.enconman.2013.04.026

Khorasanizadeh, H., Mohammadi, K. & Mostafaeipour, A. (2014). Establishing a diffuse solar radiation model for determining the optimum tilt angle of solar surfaces in Tabass, Iran. *Energy Conversion and Management*, *78*(0),805 - 814. doi.10. 1016/j.enconman.2013.11.048g

Moghadam, H., Tabrizi, F. F., & Sharak, A. Z. (2011). Optimization of solar flat collector inclination. *Desalination*, *265*(1-3), 107–111. doi:10.1016/j.desal.2010.07.039

NREL. (2012). National renewable energy laboratory. *Extraterrestrial Solar Spectrum*. Retrieved from http://rredc.nrel.gov/solar/spectra/am0

Sethi, V., Pal, D., & Sumathy, K. (2014). Performance evaluation and solar radiation capture of optimally inclined box type solar cooker with parallelepiped cooking vessel design. *Energy Conversion and Management*, *81*(0), 231–241. doi:10.1016/j.enconman.2014.02.041

Skeiker, K. (2009). Optimum tilt angle and orientation for solar collectorsin Syria. *Energy Conversion and Management*, *50*(9), 2439–2448. doi:10.1016/j.enconman.2009.05.031

Soulayman, S. (1991). On the optimum tilt of solar absorber plate. *Renewable Energy*, *1*(3/4), 551–554. doi:10.1016/0960-1481(91)90070-6

Soulayman, S., & Hammoud, M. (2016). Optimum tilt angle of solar collectors for building applications in mid-latitude zone. *Energy Conversion and Management*, *124*, 20–28. doi:10.1016/j.enconman.2016.06.066

Soulayman, S., & Sabbagh, W. (2015). Solar collector optimum tilt and orientation. *Open Journal of Renewable and Sustainable Energy*, *2*(1), 1–9. doi:10.12966/rse.03.01.2015g

Spencer, J. W. (1971). Fourier series representation of the position of the Sun. *Search*, *2*(5), 172.

Sultan, F., Ali, F. A., & Razaq, T. K. A. (2012). Tilt angle optimization of solar collectors for maximum radiation in three Iraqi cities. *International Journal of Engineering and Industries*, *3*(4), 99–107. doi:10.4156/ijei.vol3.issue4.11

Talebizadeh, P., Mehrabian, M., & Abdolzadeh, M. (2011). Prediction of the optimum slope and surface azimuth angles using the genetic algorithm. *Energy and Building*, *43*(11), 2998–3005. doi:10.1016/j.enbuild.2011.07.013

Tang, R., & Yang, Y. (2014). Nocturnal reverse flow in water-in-glass evacuated tube solar water heaters. *Energy Conversion and Management*, *80*, 173–177. doi:10.1016/j.enconman.2014.01.025

White, O. R. (Ed.). (1977). *The Solar Output and Its Variation*. Colorado Associated University Press Boulder.

Willson, R. C., Gulkis, S., Janssen, M., Hudson, H. S., & Chapman, G. A. (1981). Observations of Solar Irradiance Variability. *Science*, *2011*(4483), 700–702. doi:10.1126/science.211.4483.700 PMID:17776650

Yadav, P. C. S. S. (2013). Optimal slope angles for solar photovoltaic panels for maximum solar energy gain. *International Journal of Sustainable Development and Green Economics, 2*(2), 85-89.

Yan, R., Saha, T. K., Meredith, P., & Goodwin, S. (2013). Analysis of yearlong performance of differently tilted photovoltaic systems in Brisbane, Australia. *Energy Conversion and Management*, *74*(0), 102–108. doi:10.1016/j.enconman.2013.05.007

Zhang, X., You, S., Xu, W., Wang, M., He, T., & Zheng, X. (2014). Experimental investigation of the higher coeficient of thermal performance for water-in-glass evacuated tube solar water heaters in China. *Energy Conversion and Management*, *78*(0), 386–392. doi:10.1016/j.enconman.2013.10.070

ADDITIONAL READING

GAW. (2012). *Global atmospheric watch.* Retrieved from http://www.wmo.int/pages/prog/arep/gaw/gaw_home_en.html

GEWEX. (2012). *Global energy and water cycle experiment.* Retrieved from http://www.gewex.org/

Grell, G., Dudhia, J., & Stauffer, D. (1998). A Description of the fifth-generation penn state/NCAR mesoscale model (MM5). NCAR tech. note, NCAR/TN-398 STR, USA

APPENDIX A

The important characteristics of the Sun are summarized in the Table 22.

Table 22. Characteristics of the Sun

Characteristic	Value
Present age	4.5×10^9 years
Life expectancy	10×10^9 years
Mean distance to Earth	1.496×10^{11} m = 1.000AU
Variation of distance	1.016735 to 0.98329 AU
Diameter of photosphere	1.39×10^9 m
Angular diameter from the Earth	9.6×10^{-3} radians
Variation of the angular diameter	±1.7%
Volume of the photosphere	$1.41 \times 10^{27} m^3$
Mass	1.987×10^{30} kg
Hydrogen content	73.46%
Helium content	24.85%
Oxygen content	0.77%
Carbon content	0.29%
Iron content	0.16%
Neon content	0.12%
Nitrogen, silicon, magnesium, sulfur, etc. content	<0.1%
Mean density	14.1 kg/m^3
Density at center	1,600 kg/m^3
Solar radiation entire the Sun	3.83×10^{26} W
Solar radiation per unit area of its surface	6.33×10^7 W/m^2
Solar radiation at 1 AU (i.e. the solar constant)	1,367 W/m^2
Temperature at center	15000000 K
Temperature at surface (photosphere)	6050 K
Temperature at chromosphere	4300-50000 K
Temperature at corona	800000-3000000 K
Rotation at solar equator	26.8 days
Rotation at 30° latitude	28.3 days
Rotation at 60° latitude	30.8 days
Rotation at 75° latitude	31.8 days
Rate of mass loss	4.1×10^9 kg/s

Source: *(Eddy, 1979)*

APPENDIX B

Before providing Table 23 it is useful to mention that, the langley (Ly) was originally defined in terms of the mean gram calorie rather than the more common thermochemical gram calorie. However, current literature suggests that the thermochemical calorie be used to define the langley (Delinger, 1976). The langley unit has been dropped in the SI system of units.

Table 23. The solar constant in units commonly used in solar literature

Solar Constant	Unit
I_{sc} =	1367 W/m^2
I_{sc} =	136.7 mW/cm^2
I_{sc} =	0.1367 W/cm^2
I_{sc} =	1.367 x 10^6 erg/cm^2 s
I_{sc} =	127.0 W/ft^2
I_{sc} =	0.03267 cal/ cm^2 s
I_{sc} =	1.960 cal/ cm^2 min.
I_{sc} =	1.960 Ly/min. (thermochemical cal/cm^2 min.)
I_{sc} =	1.957 Ly/min. (mean cal/cm^2 min.)
I_{sc} =	433.4 Btu/ft^2 hr
I_{sc} =	0.1204 Btu/ft^2 s

Chapter 4
Terrestrial Solar Radiation

ABSTRACT

After shading a light on the extraterrestrial solar radiation in the chapter 3 it is important to evaluate the global terrestrial solar radiation and its components. The information on terrestrial solar radiation is required in several different forms depending on the kinds of calculations and kind of application that are to be done. Of course, terrestrial solar radiation on the horizontal plane depends on the different weather conditions such as cloud cover, relative humidity, and ambient temperature. Therefore, the impact of the atmosphere on solar radiation should be considered. One of the most important points of terrestrial solar radiation evaluation is its determination during clear sky conditions. Therefore, in this chapter, the equations that determine the air mass basing on available theories are given and the clear sky conditions are introduced with shading a light on the previous work in identifying clear sky conditions. Taking into consideration that, clear sky solar radiation estimation is of great importance for solar tracking, a detailed review of main available models is given in this chapter. As daily, monthly, seasonally, biannually and yearly mean daily solar radiations are required information for designing and installing long term tracking systems, different available methods are commented regarding their applicability for the estimation of solar radiation information in the desired format from the data that are available. An important accent is paid also on the assessment and comparison of monthly mean daily solar radiation estimation models.

INTRODUCTION

As solar radiation passes through the Earth's atmosphere, it is absorbed (the reason for some atmospheric heating), reflected (the reason astronauts can see the Earth from outer space), scattered (the reason one can read this book in the shade under a tree), and transmitted directly (the reason there are shadows). At the surface of the Earth, the sun has a lower intensity, a different color, and a different shape from that observed above the atmosphere. Therefore, in solar energy applications, the following parameters are commonly used in practice:

DOI: 10.4018/978-1-5225-2950-7.ch004

- Direct normal solar irradiation/irradiance is the component that is involved in thermal (concentrating solar power, CSP) and photovoltaic concentration technology (concentrated photovoltaic, CPV).
- Global horizontal solar irradiation/irradiance is the sum of direct and diffuse radiation received on a horizontal plane. Global horizontal irradiance is a reference radiation for the comparison of climatic zones; it is also essential parameter for calculation of radiation on a tilted plane.
- Global tilted solar irradiation/irradiance is the total solar radiation received on a surface with defined tilt and azimuth, fixed or sun-tracking. This is the sum of the scattered radiation, direct and reflected. It is a reference for photovoltaic (PV) applications, and can be occasionally affected by shadow.

On the surface of the Earth, we perceive a beam or direct solar irradiance, the component of solar radiation that is neither reflected nor scattered and which comes directly from the disc of the sun and directly reaches the surface; this is the component that produces shadows. The component of solar radiation that is scattered by the atmosphere, and which reaches the ground is called diffuse radiation. This component appears to come from all directions over the entire sky. In this text we will use the term direct to signify solar irradiance coming directly from the sun's disc, and the term diffuse to indicate solar irradiance coming from all other directions. We use the traditional subscript "*b*" to represent the direct component of solar irradiance and the subscript "*d*" to indicate the diffuse component. The sum of direct and diffuse solar irradiance is called the global or total solar irradiance and is identified by the traditional subscript "*t*". In the case of inclined surfaces, there is a small part of the radiation reflected by the surface and reaching an inclined plane is called the reflected radiation. These three components together create global radiation.

On a clear day, direct solar irradiance represents about 80 or 90 percent of the total amount of solar energy reaching the surface of the earth. Local blockage of the direct component of solar irradiance produces shadows. On a cloudy or foggy day when "you can't see the Sun," the direct component of solar irradiance is essentially zero and there are no shadows. The direct component of solar irradiance is of the greatest interest to designers of high-temperature solar energy systems because it can be concentrated on small areas using mirrors or lenses, whereas the diffuse component cannot.

The diffuse or scattered component of solar irradiance is what permits us to see in the shade. If there was no diffuse component of solar irradiance, the sky would appear black as at night and stars would be visible throughout the day. The first astronauts vividly described this phenomenon to us from the moon where there is no atmosphere to scatter the solar radiation.

The diffuse radiation is the result of downward scattering of solar irradiance by nitrogen, oxygen, and water molecules, water droplets, and dust particles in the atmosphere. The amount of this scattering depends on the amount of water and dust in the atmosphere and the altitude of the observer above sea level.

Since diffuse solar irradiance cannot be concentrated, only flat-plate (non-concentrating) solar collectors and some low-temperature types of concentrators (having wide acceptance angles) can collect diffuse solar irradiance. Few of the collectors used in industrial applications can utilize the diffuse component of solar radiation. In this book we will use the term global to indicate this sum.

The variation of some factors, especially that of water droplets (i.e. clouds) as they attenuate the direct component and change the diffuse component, is the major unknown parameter in the design of systems to collect solar energy. Consequently, a considerable amount of effort has been and is being spent in measuring, cataloging, and developing analytical models to predict these effects.

Estimation of global solar radiation is vital to solar energy system design everywhere where adequate observations are missing. Values of clear-sky solar radiation are useful for determination of the maximum performances of solar heating and photovoltaic plants as well as for sizing air conditioning equipment in buildings or for determining their thermal loading, for instance. In fact, it has been recently shown that inverters of photovoltaic plants with similar efficiency present different performances (in the sense of the quality of the electricity supply) under clear-sky and partially cloudy conditions, showing minimum total harmonic distortion for clear skies (Sidrach-de-Cardona and Carretero, 2005).

Other scientific fields such as agriculture or hydrology also demand global solar radiation estimations as they need knowledge of insolation levels for studying ecosystem fluxes of materials and energy (Lindsey and Farnsworth, 1997; Mahmood and Hubbard, 2002). One methodology to achieve this task is to calculate global solar radiation under cloudless skies and then amend this estimation taking into account the effect of the cloud cover using an appropriate transmission function. This methodology is, for instance, widely used to calculate global solar radiation from satellite images.

Another important point to have modeled clear-sky global solar radiation values is related to the need of splitting all weather solar radiation data bases into two categories: clear-sky and non clear-sky conditions. This task could be achieved using meteorological information about the cloud cover of the sky, or employing sunshine duration measurements or even by visual inspection of instantaneous measurements of either global or direct components of solar radiation every day. However, these methods are not usually easy to apply because cloud cover observations are often either unavailable or with a low frequency in a day, sunshine duration records are not measured along with actinometrical variables and the last case is time consuming.

Several alternatives based on threshold values of the clearness index (defined as the ratio between global and extraterrestrial global solar radiation) or clearness index based functions have been used to this end (Molineaux, Ineichen and Delaunay, 1995; Pedrós, Utrillas, Martínez-Lozano and Tena, 1999). However, these threshold values depend on the local climatology of the site as well as on the time of the year, and thus, they will not perform uniformly and accurately. An accurate clear-sky model would allow automatically comparing estimated against measured solar radiation values and selecting in this way the desired sky conditions.

Since the beam component (direct irradiance) is important in designing systems employing solar energy, such as high-temperature heat engines and high-intensity solar cells, emphasis is often put on modeling the beam component. There are two categories of solar radiation models, available in the literature, that predict the beam component or sky component based on other more readily measured quantities:

- Parametric models
- Decomposition models

Parametric models require detailed information of atmospheric conditions. Meteorological parameters frequently used as predictors include the type, amount, and distribution of clouds or other observations, such as the fractional sunshine, atmospheric turbidity and perceptible water content (ASHRAE, 1999; Iqbal, 1978; 1983; Davies and McKay, 1982; Sherry and Justus, 1984; Rao, Bradley and Lee, 1984; Carroll, 1985; Gueymard, 1993). A simpler method was adopted by the ASHRAE algorithm (ASHRAE, 1999) and very widely used by the engineering and architectural communities. At 1993 the (Iqbal, 1983) model offers extra-accuracy over more conventional models as reviewed by Gueymard (1993).

Development of correlation models that predict the beam or sky radiation using other solar radiation measurements is possible. Decomposition models usually use information only on global radiation to predict the beam and sky components. These relationships are usually expressed in terms of the irradiations which are the time integrals (usually over 1 h) of the radiant flux or irradiance. Decomposition models developed to estimate direct and diffuse irradiance from global irradiance data were found in the literature (see for example Liu and Jordan, 1960; Orgill and Hollands, 1977; Erbs, Klein and Duffie, 1982; Skartveit and Olseth, 1987; Reindl, Beckman and Duffie, 1990; Louche, Notton, Poggi and Simonnot, 1991; Spencer, 1982; Lam and Li, 1996; Batlles, Rubio, Tovar, Olmo and Alados-Arboledas, 2000; Ångström, 1924).

THE IMPACT OF THE ATMOSPHERE ON SOLAR RADIATION

The spectral distribution curve shows the relation between the radiant energy and the wavelength of the radiation. The spectral distribution of the solar radiation in the space is of the form presented in the Figure 1.

When solar radiation passes the Earth's atmosphere, its spectral distribution is modified by absorption and scattering. The scattering of light is either caused by molecules (Rayleigh-scattering) or by larger accumulations such as aerosols. The colouring of the sky is caused by the Rayleigh-scattering (blue or sunset) whereas aerosols cloud up the sky evenly. These effects lead to a modification of the solar spectrum and to a subdivision of the total radiation into direct and indirect (diffuse) parts. The contents of the atmosphere, such as steam, oxygen and carbon dioxide, cause the main part of the absorption which is wavelength-selective and therefore results in band gaps. Only ozone absorbs a wider spectrum of light (from 200 to 700 nm).

Figure 1. Spectra-Spectrum in space (ASTM E-490 AM0 Standard)

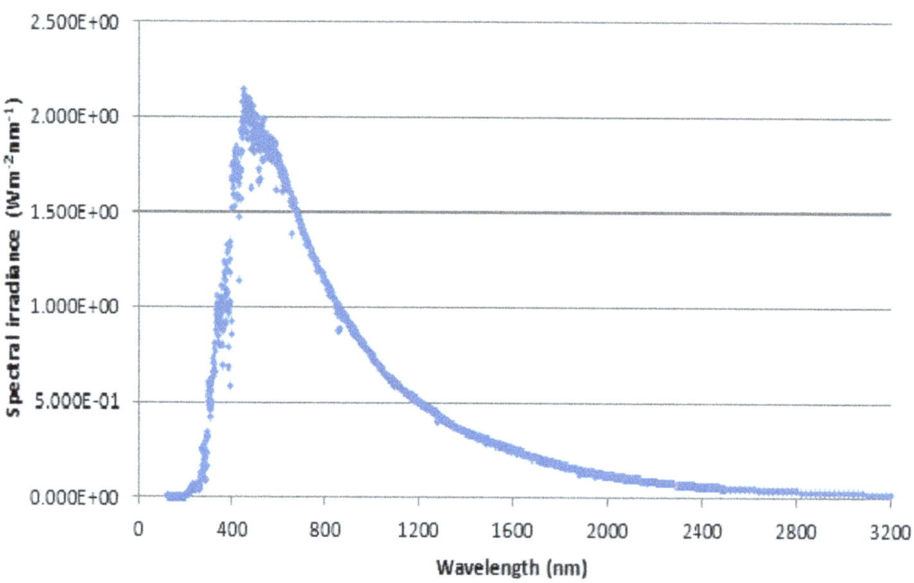

As the international norm for the measurement of solar cells, the value AM1.5 has generally been accepted, the spectral distribution of the solar radiation for AM1.5, after passing the Earth's atmosphere, is presented in the Figure 2. The AM1.5 global spectrum was defined by the Commision Internationale de l'Eclairage (CIE) and the American Society for Testing and Materials (ASTM) for testing the terrestrial solar cells. The standard assumes the following: the receiving surface is tilted toward the Equator, the solar zenith angle is 48°19', the total ozone column content is 0.34 cm atm, the Ångstrom turbidity coefficient at wavelength 0.5 μm is 0.084 and the water vapor column content is 1.42 g cm^{-2}.

It is seen from Figure 2 that, in addition to a reduction in intensity in comparison with AM0 standard, the spectrum of solar radiation reaching the surface of the Earth is also modified as it passes through the atmosphere. The processes taking place include Rayleigh and particulate (dust and water) scattering and absorption by ozone, water vapor, and carbon dioxide. All of these processes depend not only on the temporal condition of the atmosphere, but also on how much of the atmosphere the sunlight passes through.

Note in Figure 2 the effects of the strong water vapor and carbon dioxide absorption bands in the IR region (wavelength > 0.7 micrometers). Also note the reduction in blue and violet light (wavelength 0.3-0.4 micrometers) due to particulate and Rayleigh scattering and the reductions in the UV light (wavelength < 0.3 micrometers) due mostly to the ozone content of the upper atmosphere. This is why the sunrises and sunsets appear to be red, since the sunlight at these times must pass through more than 30 air masses. For small air mass values (in the mountains near noontime), there is an abundance of UV and short-wavelength visible light. This explains the need for strong eye and sunburn protection in the mountains and why photographs taken at high altitudes have a bluish tint.

The knowledge, of the relative amount of energy contained in sunlight of different wavelengths, permits the engineer to evaluate the impact of wavelength phenomena on total energy collection.

Figure 2. The spectrum of the Sun light after passing the atmosphere. ♦ is for AM0, ■ is for AM1.5

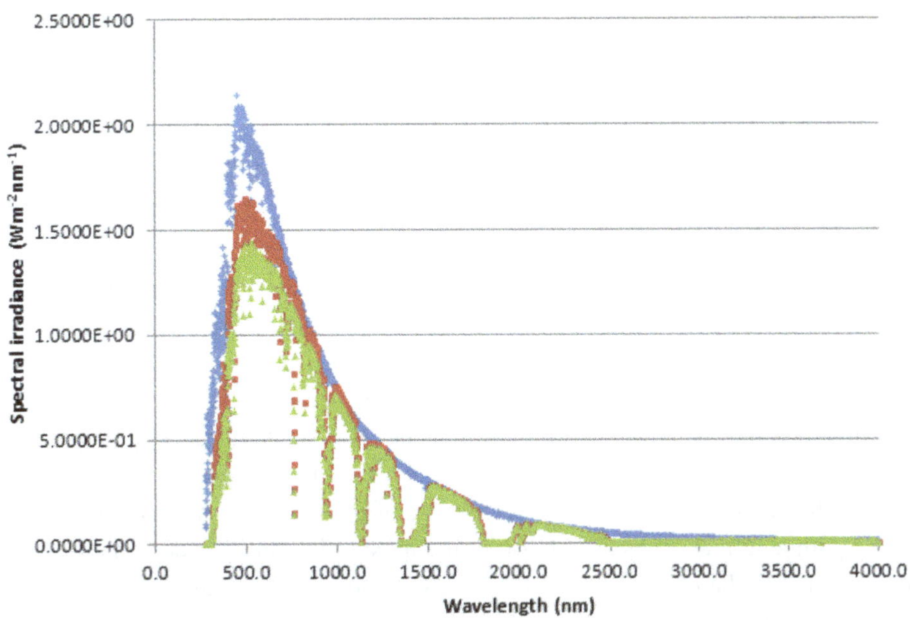

When integrating the results, presented in the Figure 2, over different intervals of the wavelength, it is interesting to note that about 45% of the Sun's energy comes to us at wavelengths in the visible spectrum, (nominally between 0.3 and 0.7 micrometers). Also, note that only a little more than 1% of the Sun's energy at shorter wavelengths (UV and X-solar radiation) and the rest (54%) is in the infrared (IR) region. The spectrum of the Sun's radiation changes slightly as it passes through the Earth's atmosphere.

The system designer must know how much solar irradiance is available in order to predict the rate of energy that will be incident on a solar collector aperture. To do this, the position of the sun relative to a collector that is not parallel to the surface of the Earth must be found. Combining the amount of solar irradiance falling on the collector, with the orientation of the collector relative to the Sun, the designer then knows the rate of solar energy being input into that collector.

Example 1

Consider a certain collector concept that uses a glass cover on its receiver aperture that transmits 90 percent of the Sun's energy at wavelengths below 1 micrometer and no energy at longer wavelengths. Determine the portion of the solar energy that can pass to the receiver.

Solution

By consulting Figure 2 one can determine that 69.5 percent of the energy coming from the Sun is coming at wavelengths shorter than 1 micrometer. A simple calculation will show that the glass receiver cover mentioned previously will pass 69.5 x0.9 = 62.6% of the Sun's energy into the receiver.

AIR MASS

The atmospheric attenuation is characterized by the term called air mass. It is defined as the ratio of the optical thickness, of the atmosphere through which the beam radiation passes, to the optical thickness if the Sun were at zenith. So, air mass refers to the optical path length through the atmosphere where light is scattered and absorbed. This is why objects closer to the horizon and with larger zenith angles appear less bright than when they are directly overhead. In solar energy, air mass actually refers to relative air mass that is measured relative to the path length at the zenith. When the sun is directly overhead, air mass has a value of one, AM=1. Irradiance at AM=0 is the extraterrestrial irradiance (without any atmosphere). Irradiance at AM=2 occurs when the solar energy is traveling through twice as much atmosphere with zenith angle approximately equal to 60°. Atmospheric attenuation of solar radiation is not the same for all wavelengths, so the solar spectrum also changes with air mass. The atmosphere also causes refraction of the sunlight which makes the sun appear higher above the horizon that it actually is and distorts the path length to be slightly longer (Young, 1994). Large value of air mass indicates that solar radiation travel greater distance in atmosphere. Hence is prone to attenuation. The correspondence between different air mass values and the maximum radiation intensity values is given in the Table 1.

Due to the Earth's rotation around the Sun the angle of incidence of the radiation changes and as a result of this the path length changes, too. In order to characterize this, the designation "air mass" AM has been introduced. This latter factor (the path length changes) is measured in terms of the air mass, which is simply the ratio of the distance that solar radiation travels through the Earth's atmosphere (path length), to the distance (path length) it would travel if the Sun were directly overhead. The radiation

Table 1. The maximum solar radiation intensity values for different air mass values

Air Mass	Maximum Solar Radiation Intensity (W/m²)
AM0	1367
AM1	935
AM1.5	853
AM2	698
AM4	601
AM7	418
AM10	306

coming from directly overhead, therefore, is said to pass through an air mass of 1.0 at sea level. The solar irradiance that coming from a zenith angle of 60°, would pass through approximately twice the perpendicular path length and hence an air mass of 2.0.

Air mass is often approximated for a constant density atmosphere and ignoring Earth's curvature using the geometry of a parallel plate. The following expression to approximate air mass at any zenith angle θ_z has been widely used (Duffie and Beckman, 2013):

$$AM = \frac{1}{cos\left(\theta_z\right)} \tag{1}$$

This simple approach is adequate for zenith angles as large as 80°, but at larger zenith angles and especially near the horizon, the accuracy degrades rapidly because AM goes to infinity at 90°. Many formulas have been developed by various authors to fit measured data as a function of zenith angle. Rapp-Arraras and Domingo-Santos (2011) present a good overview of 26 different functional forms of published air mass models. Some of the most commonly used air mass models for solar radiation modeling are highlighted below.

Young and Irvine (1967) proposed a function that is a better fit for zenith angles between 83° and 87° than the parallel plate equation, but at 90° the formula still diverges and goes to negative infinity:

$$AM = \frac{\left\{1 - 0.0012\left[sec^2\left(\theta_z\right) - 1\right]\right\}}{cos\left(\theta_z\right)} \tag{2}$$

This form is not commonly used due to this limitation, but it led to the development of several future models, such as two other models that Young was involved in developing. The first model was developed by Kasten and Young (1989):

$$AM = \frac{1}{cos\left(\theta_z\right) + 0.50572\left(96.07995 - \theta_z\right)^{-1.6364}} \tag{3}$$

while the other one was developed by Young himself (Young, 1994):

$$AM = \frac{1.002432cos^2\left(\theta_z\right) + 0.148386cos\left(\theta_z\right) + 0.0096467}{cos^3\left(\theta_z\right) + 0.149864cos^2\left(\theta_z\right) + 0.0102963cos\left(\theta_z\right) + 0.000303978} \tag{4}$$

Other published models that we consider here include that of Rodgers (1967):

$$AM = \frac{35}{\sqrt{1224cos^2\left(\theta_z\right) + 1}} \tag{5}$$

that of Badescu (1987) where the model verification was performed using measurement data from six Romanian actinometric stations:

$$AM = \frac{-cos\left(\theta_z\right) + \sqrt{cos^2\left(\theta_z\right) + f^2 - 1}}{f - 1} \tag{6}$$

where

$$f = 1 + \frac{h_{atm}}{R_E} \tag{7}$$

with R_E being the Earth's radius = 6371.2 km and h_{atm} is the atmospheric height at latitude 45°N = 11 km and that of Gueymard (1993):

$$AM = \frac{1}{cos\left(\theta_z\right) + 0.00176759\theta_z\left(94.37515 - \theta_z\right)^{-1.21563}} \tag{8}$$

The daily variation of the air mass for the latitude of Damascus (33.51°N) at solar noon is calculated basing on the equation (3) and plotted in Figure 3 as all proposed equations (1)-(6), (8) give similar results. On the other hand, the daily variation of the air mass for the latitude of Damascus (33.51°N) at an half hour before sunset time is calculated basing on the equations (1)-(6), (8) and plotted in Figure 4. The differences are important but the equations (3) and (8) give approximately the same results and equation (6) gives almost very near results to those of (3) and (8). Therefore, one of these equations could be used. During equinox, summer and winter solstices, the variations of air mass with time of the day for the latitude of Damascus are calculated using the equation (3) and plotted in Figure 5.

The Linke turbidity factor, proposed by Linke in 1922 (Linke, 1922), is a measure of the absorption by the water vapor and the absorption and scattering by the aerosol particles compared to a dry and clean atmosphere (Linke, 1922). Ineichen proposed an air mass independent formulation for the Linke turbid-

Figure 3. Daily values of air mass at solar noon in Damascus (33.51°N)

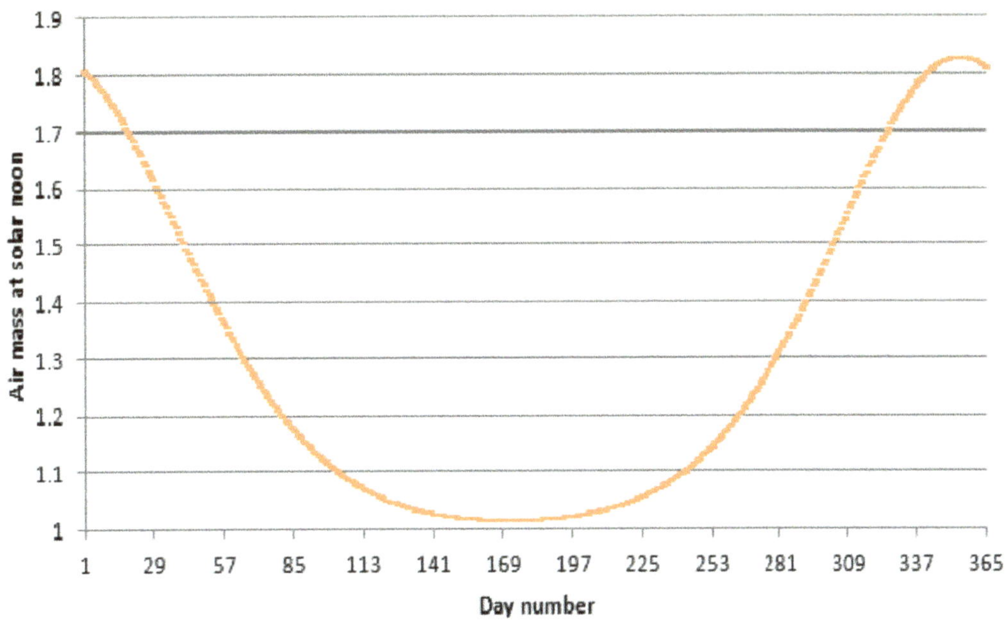

Figure 4. Daily values of air mass in Damascus (33.51°N) at half-hour before sunset using different methods. ♦ stands for equation (1); ■ stands for equation (2); ▲ stands for equation (3); x stands for equation (4); ж stands for equation (5); ● stands for equation (6) and + stands for equation (8)

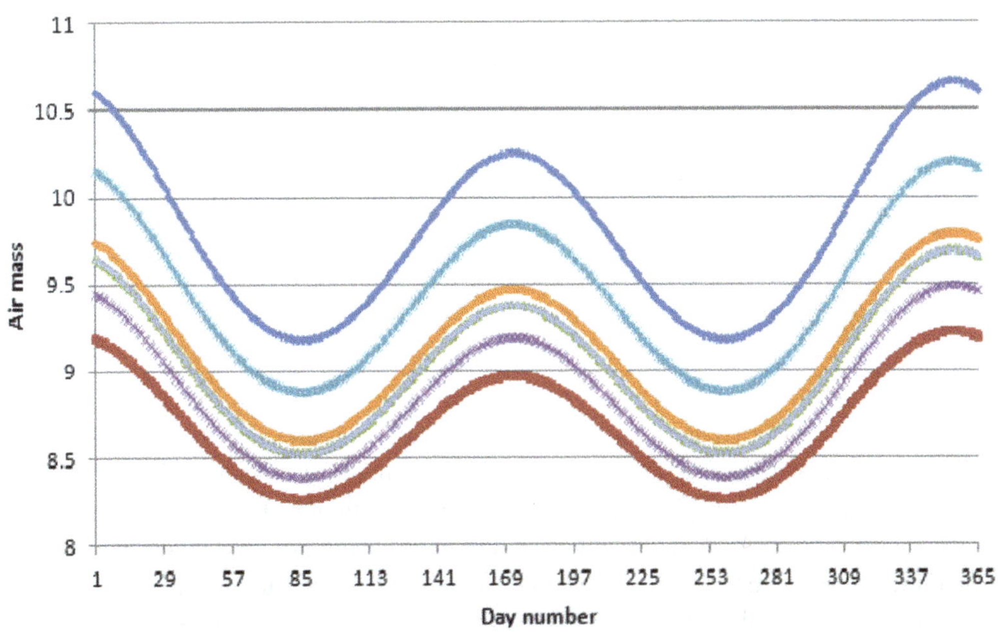

Figure 5. The use of equation (3) in calculating the variation of air mass with time in Damascus for equinox (21/3, ж), summer solstice (21/6, ▲) and winter solstice (21/12, x)

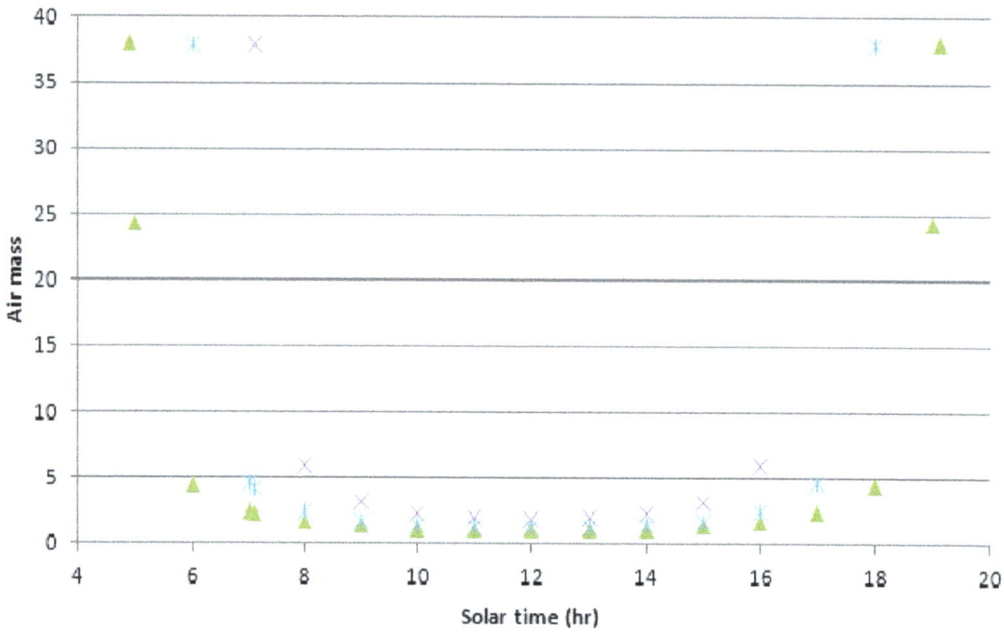

ity (Ineichen and Perez, 2002) that has the advantages of being solar altitude independent and matches the original Linke turbidity factor at air mass 2. Remund, Lefevre, Ranchin and Page (2003) calculated and produced Linke turbidity maps for the world for each month using a combination of ground measurement and satellite data. Monthly images of Linke turbidity for the world can be downloaded from either the HelioClim website (HelioClim, 2013) or Solar Radiation Data (SoDa) website (SoDa, 2011).

It is clear from Figure 3 that the air mass at solar noon reaches its maximum value on 21/12 while it reaches its minimal value on 21/6. Figure 5 shows clearly that in equinox the air mass is minimal when approaching to solar sunrise and sunset. Figure 5 shows that, the sunshine hour is shorter and air mass is higher for the winter solstice (December 21), in comparison with other days as expected.

CLEAR SKY CONDITION

A clear sky condition is defined generally as the absence of visible clouds across the entire sky dome, and clear sky irradiance is the irradiance occurring during these conditions. A more precise definition of clear sky irradiance relies to some extent on judgment. Even in the absence of visible clouds, aerosols, perceptible water, and other atmospheric conditions can vary in time and space and affect the transmittance of solar radiation through the atmosphere. Accordingly, there are several alternatives for precise definitions of the clear sky condition and clear sky irradiance (Younes and Muneer, 2007).

Methods are available which can identify clear sky condition from ground measurements. These methods are intended to identify only periods with clear sky irradiance (distinguishing periods with irradiance similar to clear sky irradiance) and often rely on measurements such as diffuse horizontal irradiance which are not typically available in conjunction with PV system monitoring data. Such method could be valuable to identify both clear sky periods and periods with irradiance similar to clear sky irradiance relying only on clear sky irradiance measurements.

The term 'clear sky equivalent condition' could be also used to refer to either clear sky irradiance during clear sky conditions or sky irradiance during all-sky conditions that is similar to clear sky irradiance. When clouds are present in the sky but are sufficiently separated from the direct path between the sun and a point of interest on the ground, the sky irradiance observed at that point may be indistinguishable from what the clear sky irradiance would be under clear skies. Reno and Hansen (2016) see that, both terms 'clear sky irradiance' and 'irradiance similar to clear sky irradiance' are meaningful because a PV system will produce similar levels of power under either condition. When monitoring a PV system, it is desirable to make assessments in as timely a manner as possible to avoid delays in identifying and correcting system faults.

Previous Work in Identifying Clear Sky Conditions

Many clear sky detection methods rely on measured irradiance quantities. Younes and Muneer (2007) presented an overview of several clear sky detection algorithms and evaluated the accuracy of nine methods. The simplest method for identifying clear sky conditions is to compare the clearness index K_T, defined as the ratio of global horizontal irradiance to the extraterrestrial irradiance on a horizontal plane, with a threshold value. Authors have used different thresholds for the value that defines clear sky: 0.6 (Muneer, Gul and Kubie, 2000), 0.65 (Cucumo, De Rosa, Ferraro, Kaliakatsos and Marinelli, 2008; Cucumo, De Rosa, Ferraro, Kaliakatsos and Marinelli, 2010), or 0.7 (Lam and Li, 1996; Li and Lam, 2001). The ratio of diffuse to global solar radiation can also be used to define clear skies (Rahim & Baharuddin, 2004), with a proposed value of less than 0.27 being clear (Lam and Li, 1996; Li and Lam, 2001). The ratio of zenith luminance to horizontal diffuse illuminance can also be used to determine clear skies (Li and Tang, 2008). Other authors have proposed using more refined measures of sky clearness; for example, a formulation of sky clearness ε have been proposed (Perez, Ineichen, Seals, Michalsky and Stewart, 1990; Perez, Ineichen, Seals and Zelenka, 1990) using the diffuse irradiance G_d and beam irradiance G_b that takes into account the zenith angle θ_z.

$$\varepsilon = \frac{\left[G_d + G_b / cos\left(\theta_z\right)\right] \Big/ G_d + 1.041 cos^3\left(\theta_z\right)}{1 + 1.041 cos^3\left(\theta_z\right)} \tag{9}$$

Perez and co-authors (Perez, Ineichen, Seals, Michalsky and Stewart, 1990; Perez, Ineichen, Seals and Zelenka, 1990) suggested a value ε = 6.2 as a definition of a clear sky.

In addition to distinguishing clear and non-clear periods, some authors have also classified cloud type using measured clear sky and diffuse irradiances. The clearness index, diffuse fraction, and the short-term variability of clear sky and diffuse irradiances were used to classify the cloudiness of the sky into categories using supervised classification techniques (Calbó, González and Pagès, 2001; Pages, Calbo and Gonzalez, 2003). Clear sky and diffuse irradiances were used to determine cloud amount and discriminate between strati form and convective cloud types (Harrison, Chalmers and Hogan, 2008).

DeFelice and Wylie (2001) use a four-band, ground-based, sun photometer to detect and classify clouds (DeFelice and Wylie, 2001). Other methods use the variability in a time-series of irradiance measurements to classify the period as clear using metrics such as variability index (Stein, Hansen and Reno, 2012) or probability of persistence (Kang and Tam, 2013). Duchon and O'Malley (1999) used the mean clearness index and standard deviation of irradiance over a 21 minute window to categorize seven types of clouds.

Long and Ackerman (2000) outline a detailed approach comprising four tests which use GHI and diffuse irradiance that together detect all cloud scenarios. The tests are: (1) the normalized total shortwave (i.e., global horizontal irradiance) limits test; (2) the maximum diffuse shortwave test; (3) the change in shortwave magnitude with time test; and (4) the normalized diffuse ratio variability test. Their method is designed for high frequency (sub-15 minute resolution) data and uses an iterative method for determining threshold values. Their method has become a de facto standard in the solar measurement community, for example, for data quality control.

Clear sky conditions can also be defined using instrumentation other than irradiance measurements. Markou, Bartzokas and Kambezidis (2007) classified the sky condition based on sky luminance scan data. Dupont, Haeffelin and Long (2008) compared the results of cloud detection using shortwave and longwave radiation and Lidar backscatter measurements. Hogan, Jakov and Illingworth (2001) estimated cloud cover using ground-based radar. Marty and Philipona (2000) use long wave down welling radiation along with air temperature and humidity to detect clear skies during the day and night. Orsini, Tomasi, Calzolari, Nardino, Cacciari and Georgiadis, (2002) used down welling shortwave radiation and down welling long wave radiation to detect cloud type, cloudiness, and cloud height. More recently, sky imagers have been used to detect clouds and classify their opacity (Ghonima, Urquhart, Chow, Shields, Cazorla and Kleissl, 2012).

Zenith Angle Dependent Clear Sky Models

The hourly and also instantaneous distribution of solar radiation is needed in various applications in the field of solar energy. The most reliable predictions of solar system performance are based on pyranometer data taken over a period of year at the place of interest. In many instances, however, available solar radiation data that have been collected are presented as a sum of integrated daily values. There are models to estimate hourly values using daily values and also information about the direct and the diffuse solar radiation transmittance. The simplest models are those which propose a solar zenith angle simple formula for calculating the clear sky beam radiation incident normally on a surface on the Earth's surface G_{cnb} (Wm^{-2}). Here, one can mention:

- **The Haurwitz (1945, 1946) Model:** In this model the clear sky beam radiation incident normally on a surface on the Earth's surface G_{cnb} (Wm^{-2}) could be calculated as follows:

$$G_{cnb} = 1098cos\left(\theta_z\right)exp\left[\frac{-0.057}{cos\left(\theta_z\right)}\right] \tag{10}$$

- **Meinel and Meinel (1976) Model:** According to this the clear sky beam radiation incident normally on a surface on the Earth's surface G_{cnb} (Wm^{-2}) could be calculated as follows:

$$G_{cnb} = G_s 0.7^{\left[\cos(\theta_z)\right]^{-0.678}} \tag{11}$$

- **The Paltridge and Proctor (1976) Model:** When analyzing the monthly mean solar radiation statics for Australia, Paltridge and Proctor (1976) proposed the following simple formula for calculating the clear sky beam radiation incident normally on a surface on the Earth's surface G_{cnb} (Wm^{-2}).

$$G_{cnb} = 950.2\left\{1 - exp\left[-0.075\left(90° - \theta_z\right)\right]\right\} \tag{12}$$

Daneshyar (1978) obtained the same constant values of (10) when treating the solar radiation statistics for Iran. According to (Paltridge and Proctor, 1976), the clear sky diffuse solar radiation incident on a horizontal surface on the Earth's surface G_{cd} (Wm^{-2}) is:

$$G_{cd} = 14.2921.04\left(\frac{\pi}{2} - \frac{\pi}{180^o}\theta_z\right) \tag{13}$$

and the clear sky global solar radiation incident on a horizontal surface on the Earth's surface G_c (Wm^{-2}) is given by:

$$G_c = G_{cb} cos\left(\theta_z\right) + G_{cd} \tag{14}$$

The calculation of the direct beam energy over a day, according to (Daneshyar, 1978), is simply a matter of calculating θ_z as a function of time and then integrating the equation (19) over a day, and reducing the integral value by the observed fractional cloud cover CF, since the presence of cloud in the solar beam normally reduces 1 to zero. Mathematically, the appropriate formula for the mean daily input of direct solar energy on a horizontal plane is:

$$H_b = \left(1 - CF\right)\sum_{t_{sr}}^{t_{ss}} G_{cb} cos\left(\theta_z\right) \tag{15}$$

- **Berger and Duffie (1979) Model:** In this model the clear sky beam radiation incident normally on a surface on the Earth's surface G_{cnb} (Wm^{-2}) could be calculated as follows:

$$G_{cnb} = 0.70 G_s cos\left(\theta_z\right) \tag{16}$$

- **Adnot, Bourges, Campana and Gicquel (1979) Model:** According to this model the clear sky beam radiation incident normally on a surface on the Earth's surface G_{cnb} (Wm^{-2}) could be calculated as follows:

$$G_{cnb} = 951.39\left[cos\left(\theta_z\right)\right]^{1.15} \quad (17)$$

- **Kasten and Czeplak (1980) Model:** This model proposed a simple formula for calculating the clear sky beam radiation incident normally on a surface on the Earth's surface G_{cnb} (Wm^{-2}).

$$G_{cnb} = 910cos\left(\theta_z\right) - 30 \quad (18)$$

- **Robledo and Soler (2000) Model:** In this model the clear sky beam radiation incident normally on a surface on the Earth's surface G_{cnb} (Wm^{-2}) could be calculated as follows:

$$G_{cnb} = 1159.24\left[cos\left(\theta_z\right)\right]^{1.179} exp\left[-0.0019\left(90° - \theta_z\right)\right] \quad (19)$$

Figure 6 presents the calculated results basing on these 7 models. It is seen from Figure 6 that the results of these models could be classified into two groups; the first group contains the equations (12) and (13) while the second group contains other equations. Moreover, the equation (18) leads to the negative results near to sunrise and sunset. Therefore, equation (18) could be eliminated from the second group.

Figure 6. The calculated solar beam irradiance with the azimuth angle variation using several models. ♦ stands for equation (10); ▲ stands for equation (11); x stands for equation (12); ж stands for equation (16); ● stands for equation (17); + stands for equation (18) and - stands for equation (19)

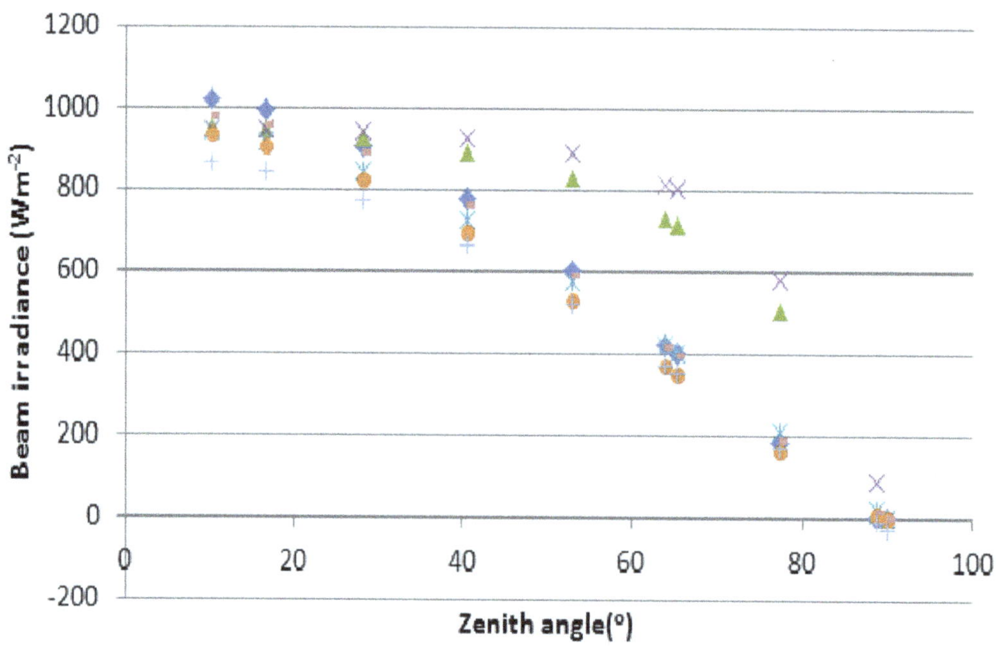

Other Simple Clear Sky Models

The next category of models are simple clear sky models that include, in addition to zenith angle, some basic parameters of the atmospheric state such as air pressure, temperature, relative humidity, aerosol content, and Rayleigh scattering. Some models propose formulae for calculating the clear sky beam radiation incident normally on a surface on the Earth's surface G_{cnb} basing on the observer elevation (h) and the zenith angle. Here, one can mention:

- **Laue (1970) Model:** According to this model the clear sky beam radiation incident normally on a surface on the Earth's surface G_{cnb} (Wm^{-2}) could be calculated as follows:

$$G_{cnb} = G_s \left[\left(1 - 0.14\ h\right) 0.7^{\left[cos\left(\theta_z\right)\right]^{-0.678}} + 0.14\ h \right] \tag{20}$$

where the observer elevation (h) is in meters. When comparing the results of Laue (1970) model with those of the first group models for an observer in Damascus ($h = 608m$), it was found (see Figure 7)) that, the results of Laue (1970) model are very close to those of The Paltridge and Proctor (1976). However, the model of Laue (1970) gives somehow higher results in the case of low zenith angle values and its results do not tend verse zero at sunrise and sunset.

- **Hottel (1976) Model:** Hottel (1976) presented a model, with a good accuracy and simple use, to estimate the clear-day transmittance of direct solar radiation through clear sky. In this general

Figure 7. The calculated solar beam irradiance with the azimuth angle variation using several models. ▲ *stands for equation (11); x stands for equation (12) and - stands for equation (20)*

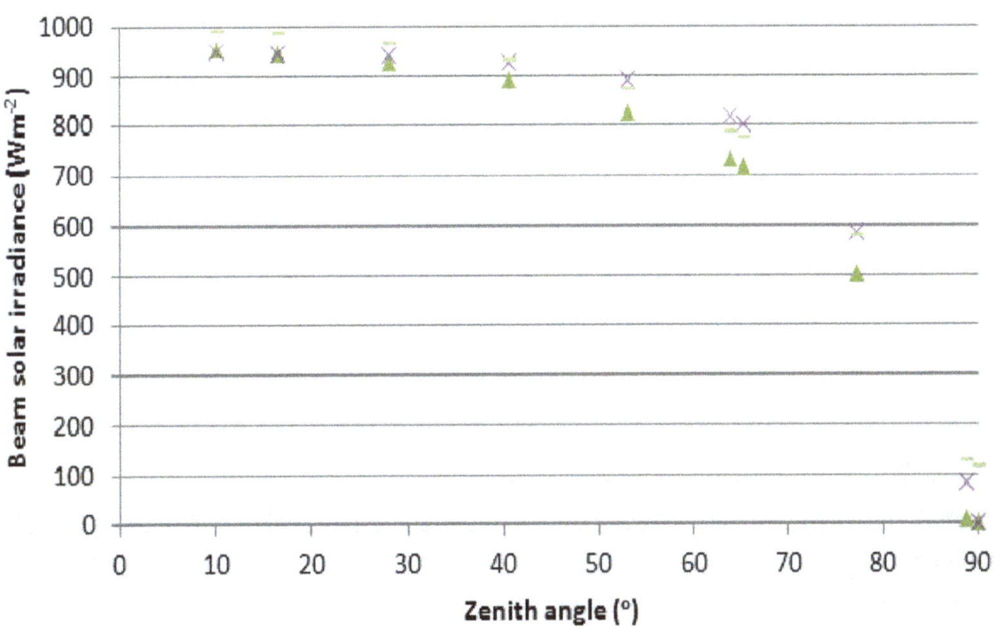

model, with taking into account the solar zenith θ_z angle and cite altitude A the transmittance to direct solar radiation is calculated using constants and corrections for four different climate zones in the globe. The transmittance to direct or beam solar radiation through the 1962 standard atmosphere with 23 km visibility to a surface at altitude A, as presented in the clear sky radiation model, can be written as:

$$\tau_b = a_0 + a_1 exp\left[-\frac{k}{cos\left(\theta_z\right)}\right] \tag{21}$$

where the coefficients a_0, a_1, and k are determined using the correction factors $r_0 = \frac{a_0}{a_0^*}$, $r_1 = \frac{a_1}{a_1^*}$ and

$r_k = \frac{k}{k^*}$, which are given for altitudes less than 2.5 km by:

$$a_0^* = 0.4237 - 0.0082\left(6 - A\right)^2 \tag{22}$$

$$a_1^* = 0.5055 + 0.00595\left(6.5 - A\right)^2 \tag{23}$$

$$k^* = 0.2711 - 0.01858\left(2.5 - A\right)^2 \tag{24}$$

where A is the latitude of the observer in kilometers (Hottel also gives equations for a_0^*, a_1^*, and k^* for a standard atmosphere with 5 km visibility.) and the correction factors are given in Table 2.

Thus, the transmittance of this standard atmosphere for beam radiation can be determined for any zenith angle and any altitude up to 2.5 km. The clear sky beam radiation incident normally on a surface on the Earth's surface G_{cnb} (Wm^{-2}) is then:

Table 2. Correction factors for climate types

Climate type	r_0	r_1	r_k
Tropical	0.95	0.98	1.02
Mid-latitude summer	0.97	0.99	1.02
Subarctic summer	0.99	0.99	1.01
Mid-latitude winter	1.03	1.01	1.00

Source: (Hottel, 1976)

$$G_{cnb} = \tau_b G_s \tag{25}$$

where G_s is obtained from equation (8) in the third chapter. The clear sky beam solar radiation incident on a horizontal surface on the Earth's surface G_{cb} (Wm^{-2}) is:

$$G_{cb} = \tau_b G_s cos\left(\theta_z\right) \tag{26}$$

It is also necessary to estimate the clear sky diffuse radiation on a horizontal surface to get the total radiation. Liu and Jordan (1960) developed an empirical relationship between the transmission coefficients for beam and diffuse radiation for clear days:

$$\tau_d = 0.271 - 0.294\tau_b \tag{27}$$

where τ_d is the ratio of diffuse radiation to the extraterrestrial radiation on the horizontal plane. The clear sky diffuse solar radiation incident on a horizontal surface on the Earth's surface G_{cd} (Wm^{-2}) is:

$$G_{cd} = \tau_d G_s cos\left(\theta_z\right) \tag{28}$$

The clear sky global solar radiation incident on a horizontal surface on the Earth's surface G_c (Wm^{-2}) is given by

$$G_c = G_{cb} + G_{cd} \tag{29}$$

Example 2

Calculate the transmittance for beam solar radiation of the standard clear sky atmosphere at Damascus (latitude 33.51°N and altitude 610m) on January 17 at 10:00 AM solar time. Estimate the intensity of beam solar radiation at that time and its component on a horizontal plane.

Solution

On January 17, the declination is -20.92°. The hour angle at 10:00 AM solar time is -30°. So, the cosine of the zenith angle is 0.473. Then, as the altitude of Damascus is 0.61km, from equations (22) – (24) the values for the standard atmosphere could be obtained:

$$a_0^* = 0.4237 - 0.0082\left(6 - 0.61\right)^2 = 0.185$$

$$a_1^* = 0.5055 + 0.00595\left(6.5 - 0.61\right)^2 = 0.299$$

$$k^{*} = 0.2711 - 0.01858\left(2.5 - 0.61\right)^{2} = 0.205$$

The climate-type correction factors could be obtained from Table 2 for mid-latitude winter. Equation (21) becomes:

$$\tau_{b} = 1.03 * 0.185 + 0.299 * 1.01 * exp\left(-\frac{0.205 * 1.00}{0.473}\right) = 0.386$$

On January 17, $G_{s} = 1413Wm^{-2}$. Then, from equation (25), we have:

$$G_{cb}' = 0.386 * 1413 = 545Wm^{-2}$$

The component on a horizontal plane is:

$$G_{cb} = 545 * 0.473 = 258Wm^{-2}$$

Example 3

Calculate the transmittance for diffuse solar radiation of the standard clear sky atmosphere at Damascus (latitude 33.51°N and altitude 610m) on January 17 at 10:00 AM solar time. Estimate the intensity of diffuse solar radiation at that time.

Solution

From example 4.2, $\tau_{b} = 0.386$. So:

$$\tau_{d} = 0.271 - 0.294 * 0.386 = 0.158$$

Thus,

$$G_{cd} = 0.158 * 1413 * 0.473 = 105Wm^{-2}$$

Example 4

Calculate the standard clear day radiation intensity at Damascus (latitude 33.51°N and altitude 610m) on January 17.

Solution

On January 17, the declination is -20.92° and the solar sunset hour angle is:

Table 3. Results from standard clear day radiation intensity at Damascus on January 17.

Hours From Solar Noon (hr)	θ_z $(degrees)$	τ_b	G_{cnb} Wm^{-2}	G_{cb} Wm^{-2}.	τ_d	G_{cd} Wm^{-2}	G_c Wm^{-2}
0	54.43	0.403	569.2	331.1	0.153	125.4	456.5
0.25	54.55	0.403	568.9	329.9	0.153	125.1	455.0
0.5	54.90	0.402	568.0	326.6	0.153	124.2	450.7
0.75	55.48	0.401	566.4	321.0	0.153	122.6	443.6
1	56.28	0.399	564.2	313.2	0.154	120.5	433.8
1.25	57.30	0.397	561.2	303.2	0.154	117.7	421.0
1.5	58.51	0.395	557.4	291.2	0.155	114.4	405.6
1.75	59.91	0.391	552.7	277.1	0.156	110.5	387.6
2	61.49	0.387	547.0	261.1	0.157	106.0	367.1
2.25	63.23	0.382	539.9	243.2	0.159	101.0	344.1
2.5	65.12	0.376	531.4	223.5	0.160	95.4	318.9
2.75	67.15	0.369	521.0	202.4	0.163	89.2	291.6
3	69.30	0.360	508.2	179.7	0.165	82.6	262.2
3.25	71.56	0.349	492.4	155.8	0.169	75.3	231.1
3.5	73.92	0.335	472.8	130.95	0.173	67.6	198.5
3.75	76.38	0.317	447.9	105.5	0.178	59.2	164.7
4	78.92	0.295	416.1	80.0	0.184	50.1	130.1
4.25	81.53	0.266	375.3	55.3	0.193	40.1	95.4
4.5	84.21	0.230	325.2	32.8	0.203	29.0	61.8
4.75	86.95	0.197	278.3	14.8	0.213	16.0	30.8
5	89.75	0.191	269.2	1.2	0.215	1.3	2.5

$$\omega_s = \arccos\left[-\tan(\delta)*\tan(\varphi)\right] = \arccos\left[-\tan\left(-20.92^o\right)*\tan\left(33.51^o\right)\right] = 75.34^o$$

The results are given in the following Table 3.

- **The Kasten (1980) Model:** This model accounts for atmospheric turbidity and elevation. The inputs to this model are air mass (*AM*), Linke turbidity (λ_L), and elevation (*h*):

$$G_{cnb} = 0.84G_s\, cos\left(\theta_z\right) exp\left[-0.027AM\left(f_{h1} + f_{h2}\lambda_L - f_{h2}\right)\right] \tag{30}$$

with f_{h1} = exp (-*h* /8000) and f_{h2} = exp (-*h* /1250). Ineichen and Perez (2002) added some additional correction terms to the Kasten model to improve the fit:

$$G_{cnb} = c_{g1} G_s cos\left(\theta_z\right) exp\left[- c_{g2} AM\left(f_{h1} + f_{h2}\lambda_L - f_{h2}\right) + 0.01AM^{1.8}\right] \quad (31)$$

where c_{g1} = 5.09e-5×h + 0.868 and c_{g2} = 3.92e-5×h + 0.0387. So, The inputs to this model are normal incidence extraterrestrial irradiance, solar zenith θ_z (radians), air mass *AM* (unitless), Linke turbidity factor λ_L (unitless), and cite elevation h (m).

Basing on the equation (31), Reno and Hansen (2016) proposed an algorithm that is dependent on the use of a clear sky model, which estimates the irradiance at the earth's surface at a given location and time under clear sky conditions. The proposed algorithm uses five criteria to classify a period of time as clear, i.e., with global horizontal irradiance equivalent to the same irradiance during clear sky conditions. These criteria are:

1. Mean value of global horizontal irradiance;
2. Maximum value of global horizontal irradiance;
3. Line length of irradiance vs. time curve;
4. Standard deviation of rate of change in global horizontal irradiance;
5. Maximum difference between changes in global horizontal irradiance and clear sky time series.

The time series of global horizontal irradiance and corresponding clear sky model output are divided into intervals, each of them contains n values, and each of the five quantities are calculated for each time period. Then, for each time period the calculated quantities are compared with threshold values, which depend on n, to classify the time period as clear or not. The clearness index k_t can be directly calculated from the extraterrestrial radiation, but clear sky irradiance is also dependent on atmospheric parameters.

Much research has gone into creating models for calculating surface irradiance under a clear sky. While none of these clear sky models are perfect, they provide a much more accurate representation of surface irradiance during clear skies than the extraterrestrial horizontal irradiance does. In this algorithm the clear sky model proposed by Ineichen and Perez (2002), which is based on the Kasten model (Kasten, 1980) and has been shown to be generally accurate and fairly easy to implement (Badescu, Gueymard, Cheval, Oprea, Baciu, Dumitrescu, Iacobescu, Milos and Rada, 2012; Gueymard, 2012; Ineichen, 2006; Reno, Hansen and Stein, 2012).

Complex Clear Sky Models

Complex models take into consideration various measurable atmospheric parameters like ozone, aerosols, and perceptible water. These are the most accurate clear sky models when they are properly calibrated, but they also require many inputs, which may not be readily available.

Many of the parameters can be estimated using a fixed constant value, but doing so will decrease the accuracy of the model and can be a tedious process to find the best constant to fit the model to the data for a location. Satellite data can also be used to help estimate many of the parameters.

A more accurate representation would be obtained by employing a full meteorological measurement station like that deployed by NREL to measure all needed parameters. Due to the complexity of these models, only an overview is provided here. See the references for the full description of the models.

Davies and McKay (1982, 1989) investigated the performance of models to estimate solar irradiance and its components using data for six Canadian stations for nine years (1968–1976). The model of Davies and McKay (1982, 1989) is known in literature as MAC model. A greatest emphasis is placed on the MAC model, which uses cloud information from different layers. The MAC model takes into account the absorption by the ozone layer, the Rayleigh scattering by molecules, the extinction by aerosols and the absorption by water vapor. The inputs to this model are air mass, humidity, temperature, and an aerosol transmittance factor. According to the MAC model, the clear sky beam radiation incident normally on a surface on the Earth's surface G_{cnb} (Wm^{-2}) could be calculated as follows:

$$G_{cnb} = G_s \left(T_0 T_r - a_w\right) T_a \tag{32}$$

where T_0 the transmittance of the ozone layer, T_a the transmittance for aerosols, T_r the transmittance of the atmosphere due to Rayleigh scattering and a_w the absorptivity of the water vapor. The transmittance T_0 of the ozone layer is computed by:

$$T_0 = 1 - a_0 \tag{33}$$

The absorption coefficient a_0 (mm) by ozone is given by:

$$a_0 = \frac{0.1082 x_0}{1 + 13.86 x_0^{0.805}} + \frac{0.00658 x_0}{1 + \left(10.36 x_0\right)^3} + \frac{0.00218}{1 + 0.0042 x_0 + 3.23\text{x}10^{-6} x_0^{\,2}} \tag{34}$$

where x_0 is the equivalent length of the radiation path through the depth of the ozone layer in the atmosphere μ_0 (mm):

$$x_0 = AM \mu_0 \tag{35}$$

The transmittance T_r of the atmosphere due to Rayleigh scattering by molecules is:

$$T_r = \begin{pmatrix} 0.9768 - 0.0874 AM + 0.010607552 AM^2 - 8.46205\text{x}10^{-4} AM^3 \\ +3.57246\text{x}10^{-5} AM^4 - 6.0176\text{x}10^{-7} AM^5 \end{pmatrix} \tag{36}$$

The absorptivity a_w of the water vapor is given by:

$$a_w = \frac{0.29 x_w}{\left(1 + 14.15 x_w\right)^{0.635} + 0.5925 x_w} \tag{37}$$

where x_w is the length of the radiation path through the equivalent thickness of perceptible water layer. x_w as a function of humidity R_H and ambient temperature T could be calculated as:

$$x_w = AM \frac{4.93 R_H}{T} exp\left(26.23 - \frac{5416}{T}\right) \tag{38}$$

The aerosol transmittance T_a depends on the airmass and is chosen between 0.84 and 0.91 to fit the data (Badescu, 2008). However, the aerosol transmittance expression is repeated below for the sake of subsequent discussions:

$$T_a = exp\left[t_a^{0.873}\left(1 + t_a - t_a^{0.7088}\right) AM^{0.9108}\right] \tag{39}$$

where t_a is the average aerosol optical depth. t_a . is specifically defined for MAC model from the spectral optical depths at 0.38 and 0.50 µm as:

$$t_a = 0.2758\ \tau_{a\lambda(0.38)} + 0.35\tau_{a\lambda(0.5)} \tag{40}$$

Because these spectral optical depths are generally not known, equation (40) has been modified here through the use of Ångström's law:

$$\tau_{a\lambda} = \beta_A \lambda^{\alpha} \tag{41}$$

where β_A is the Ångström's turbidity coefficient. For the single value of the wavelength exponent considered here, $\alpha = 1.3$, equation (40) becomes:

$$t_a = 1.832\ \beta_A \tag{42}$$

According to Davies and McKay (1982, 1989), the clear sky global solar radiation incident on a horizontal surface on the Earth's surface G_c (Wm^{-2}) in the MAC model is the sum of the direct component and diffuse components from Rayleigh scatter D_R and scattering by aerosol D_A. The MAC model comprises the following equations:

$$G_c = G_{cnb} cos\left(\theta_z\right) + D_R + D_A \tag{43}$$

$$D_R = 0.5 G_s cos\left(\theta_z\right) T_0 \left(1 - T_r\right) \tag{44}$$

$$D_A = G_s cos\left(\theta_z\right)\left(T_0 T_r - a_w\right)\left(1 - T_a\right)\omega_0 f_{ratio} \tag{45}$$

ω_0 is the spectrally-averaged single scattering albedo for aerosol and f_{ratio} is the ratio of forward to total scattering by aerosols.

Iqbal (1983) proposed a model in which the clear sky beam radiation incident normally on a surface on the Earth's surface G_{cnb} (Wm^{-2}) could be calculated as follows:

$$G_{cnb} = 0.9751\ G_s \tau_a \tau_g \tau_o \tau_r \tau_w \tag{46}$$

where the factor 0.9751 is included because the spectral interval considered is 0.3–3 μm; $\tau_{a,}$ τ_g, τ_0, τ_r, τ_w are the scattering transmittances for aerosols, gas, ozone, Rayleigh and water respectively. They are dimensionless. They are given by:

$$T_a = exp\left[t_a^{0.873}\left(1 + t_a - t_a^{0.7088}\right)AM^{0.9108}\right] \tag{47}$$

$$\tau_g = exp\left(t_a^{-0.0127}AM^{0.26}\right) \tag{48}$$

$$\tau_o = 1 - \left[0.1611U_3\left(1 + 139.48U_3\right)^{-0.3035}\right] - \frac{0.002715U_3}{1 + 0.044U_3 + 0.0003U_3^2} \tag{49}$$

$$\tau_r = exp\left[-0.0903\left(1 + AM - AM^{1.01}\right)AM^{0.84}\right] \tag{50}$$

$$\tau_w = 1 - \frac{2.4959U_1}{\left(1 + 79.034U_1\right)^{0.6828} + 6.385U_1} \tag{51}$$

where U_1 (cm) is the pressure-corrected relative optical-path length of perceptible water, as given by:

$$U_1 = wAM\frac{1013.25}{p} \tag{52}$$

p (mbar) is the local air-pressure, w (cm) is the perceptible water-vapour thickness reduced to the standard pressure (1013.25 mbar) and at the temperature T of 273 K. w is calculated with the perceptible water-vapour thickness under the actual condition w' (cm) by:

$$w = w'\left(\frac{p}{1013.25}\right)^{0.75}\left(\frac{273}{\mathrm{T}}\right)^{0.5} \tag{53}$$

U_3 (cm) is the ozone's relative optical-path length under the normal temperature and surface pressure (NTP) and is given by:

$$U_3 = x_{00}\,AM\,\frac{1013.25}{p} \tag{54}$$

where x_{00} (cm) is the vertical ozone-layer thickness, and the aerosol optical thickness t_a is given by the equation (40).

The clear sky beam solar radiation incident on a horizontal surface on the Earth's surface G_{cb} (Wm^{-2}) is:

$$G_{cb} = G_{cnb}cos\left(\theta_z\right) \tag{55}$$

The clear sky diffuse solar radiation incident on a horizontal surface on the Earth's surface G_{cd} (Wm^{-2}) is a combination of three individual components corresponding to the aerosols scattering after the first pass through the atmosphere D_A (Wm^{-2}); the multiple-reflection processes between the ground and sky D_M (Wm^{-2}); and the Rayleigh scattering after the first pass through the atmosphere D_R (Wm^{-2}):

$$G_{cd} = D_A + D_M + D_R \tag{56}$$

where:

$$D_A = \frac{0.79G_s cos\left(\theta_z\right)\tau_{aa}\tau_g\tau_o\tau_w F_c\left(1-\tau_{as}\right)}{1 - AM + AM^{1.02}} \tag{57}$$

$$D_M = \frac{\rho_g\rho_a\left[G_{cnb}cos\left(\theta_z\right) + D_A + D_R\right]}{1 - \rho_g\rho_a} \tag{58}$$

$$D_r = \frac{0.79G_s cos\left(\theta_z\right)\tau_{aa}\tau_g\tau_o\tau_w 0.5\left(1-\tau_r\right)}{1 - AM + AM^{1.02}} \tag{59}$$

where τ_{aa} (dimensionless) is the transmittance of direct radiation due to aerosol absorptance and is given by:

$$\tau_{aa} = 1 - \left(1 - \omega_0\right)\left(1 + AM - AM^{1.01}\right)\left(1 - \tau_a\right) \tag{60}$$

ω_0 (dimensionless) is the single-scattering albedo fraction of incident energy scattered to total attenuation by aerosols and is taken to be 0.9; F_c (dimensionless) is the fraction of forward scattering to total scattering and is taken to be 0.84; τ_{as} (dimensionless) is the fraction of the incident energy transmitted after the scattering effects of the aerosols and is given by:

$$\tau_{as} = \frac{\tau_a}{\tau_{aa}} \tag{61}$$

ρ_g (dimensionless) is the ground albedo; ρ_a (dimensionless) is the albedo of the cloudless sky and can be computed from:

$$\rho_a = 0.0685 + \left(1 - F_c\right)\left(1 - \tau_{as}\right) \tag{62}$$

where $\left(1 - F_c\right)$ is the back-scatterance. Consequently, the second term on the right hand side of the equation (62) represents the albedo of cloudless skies due to the presence of aerosols, whereas the first term represents the albedo of clean air.

The clear sky global (direct plus diffuse) solar radiation incident on a horizontal surface on the Earth's surface G_c (Wm^{-2}) is then the sum of the direct component and diffuse components from Rayleigh scatter D_R and scattering by aerosol D_A:

$$G_c = \frac{\left[G_{cnb} cos\left(\theta_z\right) + D_R + D_A\right]}{1 - \rho_g \rho_a} \tag{63}$$

ASHRAE (1999) model is a simpler procedure for solar-radiation evaluation is adopted and widely used by the engineering and architectural communities. According to ASHRAE (1999) model the clear sky beam solar radiation incident on a horizontal sea-level surface on the Earth's surface G_{cb} (Wm^{-2}) is:

$$G_{cnb} = A exp\left[-\frac{B}{cos\left(\theta_z\right)}\right] \tag{64}$$

where A(Wm^{-2}) is the apparent extraterrestrial irradiance, as given in Table 4, and which takes into account the variations in the Sun–Earth distance (ASHRAE, 1999). Therefore, $A = G_s$.

As equation (64) was developed for sea-level conditions, it can be adopted for other atmospheric pressures by:

Table 4. Values of A, B and C for the calculation of solar irradiance according to the ASHRAE handbook

Date	A(Wm^{-2})	B	C_n	Date	A(Wm^{-2})	B	C_n
21 January	1229.475	0.142	0.058	17 January	1229.882	0.142	0.058
21 February	1213.713	0.144	0.060	16 February	1216.255	0.144	0.060
21 March	1185.340	0.156	0.071	16 March	1190.407	0.153	0.068
21 April	1134.900	0.180	0.097	15 April	1144.663	0.175	0.092
21 May	1103.375	0.196	0.121	15 May	1109.680	0.192	0.116
21 June	1087.613	0.205	0.134	11 June	1092.697	0.202	0.130
21 July	1084.460	0.207	0.136	17 July	1084.880	0.207	0.136
21 August	1106.528	0.201	0.122	16 August	1102.986	0.202	0.124
21 September	1150.663	0.177	0.092	15 September	1142.120	0.182	0.098
21 October	1191.645	0.160	0.073	15 October	1183.449	0.164	0.077
21 November	1220.018	0.149	0.063	14 November	1213.611	0.151	0.065
21 December	1232.628	0.142	0.057	10 December	1228.004	0.145	0.059

Source: ASHRAE (1999)

$$G_{cnb} = Aexp\left[-\frac{B}{cos\left(\theta_z\right)}\left(\frac{p}{1013.25}\right)\right] = Aexp\left[-Bsec\left(\theta_z\right)\left(\frac{p}{1013.25}\right)\right] \quad (65)$$

where p (mbar) is the actual local-air pressure and 1013.25 mbar is the standard pressure.

The term $sec\left(\theta_z\right)\left(\frac{p}{1013.25}\right)$ approximates to the air mass, with the assumptions that the curvature of the Earth and the refraction of air are negligible.

The variable B (dimensionless) in Table 4 represents an overall broadband value of the atmospheric attenuation coefficient for the basic atmosphere of Threlkeld and Jordan (ASHRAE, 1999). C_n (dimensionless) is the clearness number and the map of C_n values for the USA is provided in the ASHRAE handbook: applications (ASHRAE, 1999). C_n is the ratio of the direct normal irradiance calculated with the local mean clear-day water-vapour to the direct normal irradiance calculated with water vapour according to the basic atmosphere.

The total solar irradiance G_{ct} (Wm^{-2}) of a terrestrial surface of any orientation and tilt with an incident angle θ_i (degrees) is the sum of the direct component $G_{cnb}\cos\left(\theta_i\right)$ plus the diffuse component G_{cd} coming from the sky, plus whatever amount of reflected short-wave radiation G_{cr} (Wm^{-2}) that may reach the surface from the Earth or from adjacent surfaces, i.e.,

$$G_{ct} = G_{cnb} cos(\theta_i) + G_{cd} + G_{cr} \tag{66}$$

The diffuse component is defined through a variable C_n (dimensionless) given in Table 4. The variable C_n is the ratio of the diffuse radiation falling on a horizontal surface under a cloudless sky G_{cd} to the direct normal irradiation at the Earth's surface on a clear day G_{cnb}. The diffuse radiation G_{cd} is given by:

$$G_{cd} = C_n G_{cnb} F_{ss} \tag{67}$$

where for a horizontal surface,

$$C_n = \frac{G_{cnb}}{G_{cd}} \tag{68}$$

F_{ss} (dimensionless), the angle factor between the surface and the sky, is given by:

$$F_{ss} = \frac{1 + cos\beta}{2} \tag{69}$$

where β (degrees) is the tilt angle of a surface measured from the horizontal. The reflected radiation G_{cr} (Wm^{-2}) from the foreground is given by:

$$G_{cr} = \rho_g G_{ct} (\theta = 0^o) F_{sg} \tag{70}$$

where ρ_g (dimensionless) is the reflectance of the foreground; and $G_c(\theta = 0^o)$ is the total horizontal irradiation at $(\theta = 0^o)$. F_{sg} - the angle factor between the surface and the ground, is given by:

$$F_{sg} = \frac{1 - cos(\beta)}{2} \tag{71}$$

Example 5

Basing on ASHRAE (1999) model, calculate the intensity of global solar radiation and its components, incident on a surface, orientated towards the Equator and tilted by an angle 45°, at Damascus (latitude 33.51°N and altitude 610m) on January 21 at 10:00 AM.

Solution

On January 21, A=1229.475 Wm^{-2}, B= 0.142 and . At altitude of Damascus the atmospheric pressure is 94208.38 Pa and at 10:00 AM. Therefore, the beam solar radiation intensity is:

$$G_{cnb} = A exp\left[-\frac{B}{cos\left(\theta_z\right)}\left(\frac{p}{1013.25}\right)\right] = 1229.475 * exp\left[-\left(\frac{0.142}{0.363}\right)\left(\frac{942.08}{1013.25}\right)\right] = 854.6 \text{ Wm}^{-2}$$

$$G_{cb} = G_{cnb} cos\left(\theta_z\right) = 854.6 * 0.363 = 310.2 \text{ Wm}^{-2}$$

$$G_{cd} = C_n G_{cnb} F_{ss} = C_n G_{cnb} * \left[\frac{1+cos\left(\beta\right)}{2}\right] = 0.058 * 854.6 * \left(\frac{1+0.71}{2}\right) = 42.4 \text{ Wm}^{-2}$$

$$G_{cr} = \rho_g G_{ct}(\theta = 0^o) F_{sg} = \rho_g (G_{cnb} + G_{cd}) \left[\frac{1-cos\left(\beta\right)}{2}\right]$$
$$= 0.2 * (854.6 + 42.4)((1-0.71)/2) = 26.0 Wm^{-2}$$

Example 6

Basing on (Hottel, 1976) model and ASHRAE (1999) model, calculate the standard clear day radiation intensity at Damascus (latitude 33.51°N and altitude 610m) on January 17.

Solution

On January 17, the declination is -20.92° and the solar sunset hour angle is:

$$\omega_s = \arccos\left[-\tan\left(' \right) * \tan\left(Æ\right)\right] = \arccos\left[-\tan\left(-20.92^o\right) * \tan\left(33.51^o\right)\right] = 75.34^o$$

The results are given in the following Table 5.

It is seen from these results that, the (Hottel, 1976) model underestimates beam solar irradiance.

Another more complicated and very well developed model is that of Bird and Hulstrom (1981). This model includes the Rayleigh transmittance, the ozone transmittance, the uniformly-mixed gas transmittance and the aerosol transmittance. This model formed the basis for Iqbal's Model (Iqbal, 1983) that was. It is also the basis for the more recent meteorological/statistical solar radiation model (METSTAT) model (Maxwell, 1998). The METSTAT model was used to estimate hourly values of direct normal, diffuse horizontal, and global horizontal solar radiation for those times and locations for which measured data were not available. The input parameters for METSTAT include total and opaque cloud cover, aerosol optical depth, perceptible water vapor, ozone, surface albedo, snow depth, days-since-last-snowfall,

Table 5. Results from calculation of the standard clear day radiation intensity at Damascus on January 17.

Hours From Solar Noon (hr)	θ_z $(degrees)$	G_{cnb} Wm^{-2}		G_{cd} Wm^{-2}		G_c Wm^{-2}	
		ASHRAE	Hottel	ASHRAE	Hottel	ASHRAE	Hottel
0	54.43	980.1	569.2	56.8	331.1	626.9	456.5
0.25	54.55	979.5	568.9	56.8	329.9	624.9	455.0
0.5	54.90	977.6	568.0	56.7	326.6	618.8	450.7
0.75	55.48	974.3	566.4	56.5	321.0	608.6	443.6
1	56.28	969.6	564.2	56.2	313.2	594.4	433.8
1.25	57.30	963.3	561.2	55.9	303.2	576.3	421.0
1.5	58.51	955.2	557.4	55.4	291.2	554.4	405.6
1.75	59.91	945.1	552.7	54.8	277.1	528.6	387.6
2	61.49	932.7	547.0	54.1	261.1	499.3	367.1
2.25	63.23	917.4	539.9	53.2	243.2	466.4	344.1
2.5	65.12	898.6	531.4	52.1	223.5	430.2	318.9
2.75	67.15	875.4	521.0	50.8	202.4	390.8	291.6
3	69.30	846.6	508.2	49.1	179.7	348.4	262.2
3.25	71.56	810.2	492.4	47.0	155.8	303.3	231.1
3.5	73.92	763.5	472.8	44.3	131.0	255.7	198.5
3.75	76.38	702.1	447.9	40.7	105.5	206.1	164.7
4	78.92	618.9	416.1	35.9	80.0	154.9	130.1
4.25	81.53	501.9	375.3	29.1	55.3	103.0	95.4
4.5	84.21	332.2	325.2	19.3	32.8	52.8	61.8
4.75	86.95	102.7	278.3	6.0	14.8	11.4	30.8
5	89.75	1.2E-10	269.2	6.8E-12	1.2	7.3E-12	2.5

atmospheric pressure and present weather. The model employs deterministic algorithms to generate accurate monthly means for each element for each hour and statistical algorithms to simulate the statistical and stochastic characteristics of multiyear solar radiation data sets.

Later on, the model of Bird and Hulstrom (1981) was modified by Bird (1984) and updated in 1986 by Bird and Riordan (1986). In its final form, this model includes transmittance of atmosphere for Rayleigh scattering, aerosol attenuation, water vapor absorption, ozone absorption, and uniformly mixed gas absorption. Each piece is a function of frequency to model the resulting spectrum of received light on the ground. This model requires a number of inputs and variables, including atmospheric turbidity, perceptible water vapor, surface pressure, and ground albedo.

The Atwater and Ball (1978, 1981a) clear sky model is another model considered in the analysis. This model uses perceptible water, pressure, ground albedo, sky albedo, air mass, and broadband aerosol optical depth to model the transmittance impacts of aerosols and water vapor. This model can also be used to estimate the irradiance with cloud cover (Atwater and Ball, 1981b).

Christian Gueymard's REST2 model (Gueymard, 2008) predicts cloudless-sky broadband irradiance, illuminance, and photosynthetically active radiation. This is the most recent model from Gueymard after his previous CPCR2 (Gueymard, 1989) and REST (Gueymard, 2003a; 2003b; 2004a; 2004b) models. REST2 is a two-band model with particular attention to the impact of aerosols and turbidity to the spectrum. For predictions of global solar radiation incident on a horizontal surface, RMSE is supposed to be within 2% (Gueymard, 2010). This detailed model requires a large number of atmospheric parameters, such as station pressure, ground albedo, aerosol optical depth, albedo reduced ozone vertical path length, Angstrom's turbidity coefficient, perceptible water, Angstrom's wavelength exponents, aerosol single-scattering albedo, and reduced NO2 vertical path length. An executable version of the model is available online (Gueymard). There are many other complex models that discussed in literature (see for example (Gueymard, 2003a) for more details on a main part of these models).

Clear Sky Models Fit to Measured Data

The existence is known of numerous radiative transfer based models to estimate solar radiation, which need detailed information about atmospheric constituents as water vapour, ozone, carbon dioxide, nitrogen dioxide etc. However, for most practical purposes and users, most of them show to be unusable due to the large amount of atmospheric information required or present a difficult software implementation. On the other hand, many simple clear-sky models have been proposed in the literature for calculating instantaneous values of global solar irradiance (Badescu, 1997).

Clear sky models can also be developed for a specific location with measured irradiance data. Because all simple clear sky models are regression fits between a formula and measured data, any formulation can be used to create a clear sky model for the location. Grigiante, Mottes, Zardi and de Franceschi (2011) proposed a method for fitting the Bird clear sky model to measured data. The proposed methodology (Grigiante, Mottes, Zardi and de Franceschi, 2011) improves the predictive Bird's real-sky model by introducing in it both atmospheric parameters, specifically defined for the analyzed site, and a local cloud cover factor, based on experimental data, to calculate the global real sky irradiance. The experimental data have been measured at the meteorological station of the University of Trento located in the city center. At first, a selection of the global irradiance measurements, representative of daily clear-sky conditions of each season, is presented and compared with the corresponding values obtained by the improved Bird's clear-sky model. Making use of the improved procedure, the monthly mean daily irradiation is then calculated and compared both with experimental measurements covering the years from 2003 to 2006 and available models as well as data banks. The results, presented in terms of statistical functions, demonstrate that the generalized calculation procedures usually adopted, also available from commercial software tools, reach a satisfactory accuracy if compared with an experimental methodology approach as the one proposed in (Grigiante, Mottes, Zardi and de Franceschi, 2011). In order to fit a clear sky model to measured data, only clear sky measured data can be used. Detecting clear days in the data can be done visually, or by other methods.

NREL's Sunny Days is an example of a program used to develop a clear sky model with measured irradiance data. It is largely based a method of clear sky detection and formulation done by Long and Ackerman (2000). The user must load in global horizontal irradiance, beam normal irradiance, and diffuse irradiance measurements to generate a clear sky model with four fitting coefficients (a, b, c, d) dependent on zenith angle:

$$G_c = a\left[cos\left(\theta_z\right)\right]^b \tag{72}$$

$$G_{cd} = c\left[cos\left(\theta_z\right)\right]^d \tag{73}$$

The coefficients (*a*, *b*, *c*, *d*) should be found for each clear day, and their values for other days are found by interpolating the coefficients between clear days. The local empirical determination of these coefficients constrains the accuracy and generality in the use of methods based on equations (72) and (73). This is the case of earlier versions of the Heliosat method (Dribssa, Cogliani, Lavagno and Tetrarca, 1999; Rigollier, Bauer and Wald, 2000). However, it was found that A and B can be expressed as functions depending on the perceptible water content, Ångström's turbidity coefficient, site elevation and ground albedo, avoiding in this sense the above noted locality (López, Batlles and Tovar-Pescador, 2007).

Review of Previous Analysis of Clear Sky Models

There are several papers previously published on different types of clear sky models, analysis, and validation for various measurement locations. Each paper has its own benefit for looking at differing complexities, models for each component of irradiance, and accuracy for a given location or geographic region. A review of some of those papers is discussed in different locations such as (Myers, 2005), (Gueymard, 2003a) and (Reno, Hansen and Stein, 2012).

Several broadband solar radiation model approaches, including some developed at the National Renewable Energy Laboratory, for estimating direct beam, total hemispherical and diffuse sky radiation are briefly reviewed. The latter include the Bird clear sky model for global, direct beam, and diffuse terrestrial solar radiation; the Direct Insolation Simulation Code (DISC) for estimating direct beam radiation from global measurements; and the METSTAT (Meteorological and Statistical) and Climatological Solar Radiation (CSR) models that estimate solar radiation from meteorological data

Bird and Hulstrom (1981) analyzed six atmospheric clear sky models for direct, diffuse sky, diffuse sky/ground, and global horizontal irradiance. The models were not compared to measured data, but instead were compared in a theoretical manner to a spectral baseline dependence on aerosol transmittance, transmittance after molecular (Rayleigh) scattering, the water vapor transmittance, and ozone transmittance. The analysis identified performance of various models (including the Atwater and Ball (1978, 1981a) model) and is notable in that it led to formulation of the Bird clear-sky model. Gueymard (2003a) analyzed eleven clear sky irradiance models for predicting beam, diffuse, and global radiation on a horizontal surface. Three types of analyses were performed to test the validity and performance of the models. First, the models were analyzed according to how atmospheric effects are modeled and the sensitivity of the models to these parameters, such as optical masses, Rayleigh scattering, ozone absorption, mixed gases absorption, water vapor absorption, and aerosol extinction. Second, the response to several benchmark spectral codes was investigated. Finally, the accuracy of each model was compared to data from seven sites in California, Canada, Belgium, Switzerland, France, and India.

Badescu (1997) looked at five very simple clear sky models for GHI for two cities in Romania. He also analyzed five simple cloudy sky models. Among the model's considered, he found the best model for the data in Romania to be the Adnot, Bourges, Campana and Gicquel (1979) model, which was calibrated to data from Western Europe, and that the performance of very simple models is comparable to that of more complicated ones.

Ineichen (2006) surveyed eight clear sky models and evaluated them compared to 16 independent data banks of measured irradiance covering 20 years/stations, altitudes from sea level to 1600 m and a large range of different climates. As all models he considered are complex, Ineichen also investigated the importance of the atmospheric parameters, finding that using climatic data banks instead of locally measured parameters resulted in regular underestimation. The author concluded that the accuracy is not highly dependent on the model, so model selection should be based on either implementation simplicity or input parameter availability.

Alam (2006) analyzed three parametric clear sky models for four Indian stations, namely New Delhi, Mumbai, Pune and Jaipur over the period of 1995–2002, under cloudless conditions. These stations have different climatic conditions. The beam radiation at normal incidence as well as global solar radiation at horizontal surface was computed for these locations during all seasons except monsoon (June to September). The computed values of beam and global irradiance was compared with reference values in case of beam and measured values in case of global solar radiation on the basis of percentage root mean square error (RMSE) and mean bias error (MBE).

He used average hourly clear sky irradiance data to calculate the error for each month not during monsoon season. The REST model (without considering transmittance due to nitrogen dioxide) was found to fit the data best with a RMSE of about 7% for both direct normal irradiance and global horizontal irradiance.

Younes and Muneer (2007) evaluated four clear sky models for six locations in UK, Spain, and India. The authors found that MRM (Kambezidis, 2008) performed the best, but it had to be locally calibrated. For non-calibrated models, REST2 was judged to be the most accurate.

Gueymard (2010) looked at five broadband radiative models that can predict the beam radiation at normal incidence under clear skies from atmospheric data. These models were compared to the results from 18 separation models used to predict the beam radiation at normal incidence from the global solar radiation at horizontal surface by decomposing beam solar irradiance and diffuse irradiance. The separation methods were found to be close but not highly accurate for clear conditions, and they were found to be highly inaccurate for non-clear conditions. Gueymard (2012) did a very detailed study on 18 broadband radiative clear sky models that predict direct, diffuse, and global irradiances under clear skies from atmospheric data.

According to the categories of models used in this report, four very simple, four simple, and ten complex clear sky models were compared to data from five sites, Oklahoma, Illinois, Colorado, Hawaii, and Saudi Arabia. The author conclude that whereas many models can predict the global horizontal irradiance within uncertainty limits similar to those of the radiation measurements, the models' predictions of direct irradiance are less accurate, and diffuse radiation is even harder to model. The author found that REST2 was most accurate, but the Ineichen model performed nearly as well.

LINKE TURBIDITY FACTOR

Some of the clear sky models account for atmospheric turbidity. One of the inputs to these models is the Linke turbidity coefficient (λ_L). Moreover, the Linke turbidity factor is a very convenient approximation to model the atmospheric absorption and scattering of the solar radiation under clear skies. It describes the optical thickness of the atmosphere due to water vapor and the aerosol particles relative to a dry and clean atmosphere. With larger Linke turbidity, there is more attenuation of the radiation by the clear sky atmosphere. Linke turbidity was proposed by Linke (1922) to express the total optical thickness of a cloudless atmosphere relative to the optical thickness t_a of a free atmosphere from water and aerosol. The observed transmission is achieved by multiplying the reference clear, dry atmosphere by the Linke turbidity coefficient λ_L, as:

$$G_{cnb} = G_s\, exp\left(-t_a \lambda_L \mathrm{AM}\right) \tag{74}$$

This definition of Linke turbidity depends on the theoretical value of t_a and air mass. Linke used the value of:

$$t_a = 0.128 - 0.054\, \log\left(AM\right) \tag{75}$$

which he computed from theoretical assumptions and a very pure, dry mountain atmosphere.

Kasten (1980) fitted the following equation to spectral data tables where both molecular scattering and absorption by the stratospheric ozone layer are taken into account:

$$t_a = \frac{1}{9.4 + 0.9AM} \tag{76}$$

Louche, Peri and Iqbal (1986) did a fourth order polynomial fit on computed spectral data of a clean and dry atmosphere:

$$t_a = \frac{1}{6.5567 + 1.7513AM - 0.1202AM^2 + 0.0065AM^3 - 0.00013AM^4} \tag{77}$$

Grenier, De La Casinière and Cabot (1994) had a similar method, but added some minor changes to the spectral absorption and scattering equations to obtain different constants in the equation (51):

$$t_a = \frac{1}{5.4729 + 3.0312AM - 0.6329AM^2 + 0.091AM^3 - 0.00512AM^4} \tag{78}$$

Ineichen (2008) also developed a conversion between Linke turbidity and the water vapor (*w*) and aerosol optical depth at 055 μm ($\tau_{a\lambda(0.55)}$), taking into account the altitude of the site, for a fixed air mass of 2:

$$\lambda_L = \begin{bmatrix} 3.91\exp\left(0.689\dfrac{p_0}{p}\right)\tau_{a\lambda(0.55)} + 0.376\ln\left(w\right) + 2 \\ +0.54\dfrac{p_0}{p} - 0.5\left(\dfrac{p_0}{p}\right)^2 + 0.16\left(\dfrac{p_0}{p}\right)^3 \end{bmatrix} \quad (79)$$

Some authors have tried to remove this dependence of Linke turbidity on air mass. Ineichen and Perez, (2002) proposed an air mass independent formulation for the Linke turbidity that has the advantages of being solar altitude independent and matches the original Linke turbidity factor at air mass 2. This formulation of air mass was used to generate a seasonal grid of Linke turbidity for the North American continent based on gridded climatological aerosol, ozone and water vapor data assembled in the 2005 National Solar Radiation Database (NSRDB) (Perez, Ineichen, Moore, Kmiecik, Chain, George and Vignola, 2002). The method could also be used to generate Linke turbidity from any ground monitoring stations.

Remund, Lefevre, Ranchin & Page (2003) calculated and produced Linke turbidity maps for the world for each month using a combination of ground measurement and satellite data. Figure 8 shows an example of one of the monthly images of Linke turbidity for the world that can be downloaded from either the (HelioClim, 2013) website or Solar Radiation Data (SoDa, 2011) website.

ESTIMATION OF AVERAGE GLOBAL SOLAR RADIATION

The growing demand in urban and rural areas for energy has necessitated the finding of alternative sources of energy. With the change in the rural scenario and agricultural practices, and the advent of gadgets like

Figure 8. An example of one of the monthly images of Linke turbidity for the world

televisions, mobile phones, and computers, the demand of energy has increased by a multitude. Many countries have already established or are in the process of establishing support programs to encourage the adoption of the renewable energy new technology.

Except for the monsoon months, solar radiation incidence is very encouraging, from the application point of view. Global solar radiation data are the best source of information for estimating average incident solar radiation. These data are required in different areas, such as solar water heating, wood drying, stoves, ovens, photovoltaic, atmospheric energy balance studies, thermal load analyses on buildings, agricultural studies, and meteorological forecasting which should be reliable and readily available for design, optimization and performance evaluation of solar technologies for any particular location (Bezir, Akkurt and Ozek, 2010a; 2010b; El-Sebaii, Al-Hazmi, Al-Ghamdi and Yaghmour, 2010).

Compared to measurements of other meteorological variables, the measurement of solar radiation is more prone to errors and often encounters more problems such as technical failure and operation related problems given by Tang, Yang, He and Qin (2010). These problems could be one of many: calibration problems, problems with dirt on the sensor, accumulated water, shading of the sensor by masts, etc. Even at stations where global solar radiation is observed, there could be many days when global solar radiation data are missing or lie outside the expected range due to these equipment failures and other problems as given by Rahimikhoob (2010). Nevertheless, for many developing countries, solar radiation measurements are not easily available because of the incapability to afford the measuring equipment and techniques involved as given by Bezir, Akkurt and Ozek (2010b). Therefore, it is necessary to develop methods to predict solar radiation For the efficient functioning and better performance of solar energy device, the information of solar radiation and its components at particular location is very essential for designing the solar energy devices. Therefore, over the years, several empirical correlations have been developed in order to estimate the more appropriate solar radiation all around the world.

Many models of solar radiation were presented in the literature. These methods can be mathematical such as linear and polynomial functions, heuristic methods, fuzzy logic techniques, or other individual methods such as Fourier series and Markov chain. However, recently, artificial intelligence techniques based models such as artificial neural networks (ANNs) were used for solar radiation prediction. In 1990s, ANNs were proposed for predicting monthly or daily solar radiation utilizing monthly or daily meteorological variables due to the availability of such data. According to (Mellit and Kalogirou, 2008, Khatib, Mohamed and Sopian, 2012) ANNs were used many times for solar radiation modeling, prediction, and forecasting. Different types of ANNs were utilized for this purpose. Examples for these models are feedback back forward ANN, cascade-forward back propagation ANN, generalized regression ANN, neurofuzzy ANN, and optimized ANN-genetic algorithm. In general, most of the conducted work was done for solar radiation prediction using ground measured meteorological variables such as ambient temperature, sunshine ratio, relative humidity, wind speed, and other solar geometry angles such as hour angle and angle of declination. The main purpose of the aforementioned models is to generate synthetic solar radiation data at a specific location where there are no measuring devices in order to be utilized in solar energy system design, to restore a solar radiation data set in case of having missing data due to monitoring system outages, or to predict the performance of a solar energy system.

Soulayman (1985) analyzed available formulas for estimating solar radiation on the horizontal surface on the Earth's surface and stated that the general form of such formula could be written as follows:

$$H_{t,m} = \eta H_{0,m} \tag{80}$$

where $H_{t,m}$ is the monthly average daily global solar radiation on a horizontal surface; $H_{o,m}$ is the extraterrestrial solar radiation and η is a dimensionless coefficient which is dependent on the site's geographic and climatic characteristics. Soulayman (1985) proposed an approximation of η to be in the form of:

$$\eta = \eta_1(\varphi)\eta_2\left(\frac{s}{S_o}\right)\eta_3(h) \tag{81}$$

where s is the monthly average daily hours of bright sunshine; S_o is the monthly average daily of the possible sunshine hours (i.e., the monthly daily average day length); h is the site's height in km; $\eta_1(\varphi)$; $\eta_2\left(\frac{s}{S_o}\right)$ and $\eta_3(h)$ are:

$$\eta_1(\varphi) = \left[cos(\varphi)\right]^{cos(\varphi)} \tag{82}$$

$$\eta_2\left(\frac{s}{S_o}\right) = \left(\frac{s_{max} - s_{min}}{6S_{o,max}} + 1\right)\left(\frac{s}{S_o}\right)^{0.5} \tag{83}$$

$$\eta_3(h) = 0.757 + \sum_{i=1}^{i=4} 0.002(5-i)h^i - 0.0025h^5 \tag{84}$$

where S_{max} is the maximum monthly average sunshine duration; s_{min} is the minimum monthly average sunshine duration and $S_{o,max}$ is the maximum possible daily sunshine duration. Later on, Soulayman Soulayman, S., Ananikyan, N. and Martoyan, G. (2010) modified the equation (81) by taking the influence of other parameters such as the monthly average ambient temperature T_a and the relative monthly average humidity *RH*. The new formula is of the form:

$$\eta = \left[\begin{array}{l} 0.5\eta_1(\varphi)\eta_2\left(\frac{s}{S_o}\right)\eta_3(h) + 0.207 + 0.2\left(\frac{s}{S_o}\right) \\ -0.00275RH\left(4.7923 + 0.3647T_a + 0.0055T_a^2 + 0.0003T_a^3\right) \end{array}\right] \tag{85}$$

Robaa (2009) reviewed and tested the applicability of 10 models available at that time for computing the monthly average daily global radiation on a horizontal surface in Egypt. The different meteorological data measured at nine stations during the period (1983–2006) were used.

Sunshine Hour Duration Based Methods

It is possible to use empirical relationships to estimate global solar radiation from hours of sunshine duration. Data on average sunshine durations or average percentage of possible sunshine durations are widely available from many hundreds of stations in many countries and are usually based on data taken with Campell-Stokes instruments.

The very first equation was given by Ångström (1924). The original Ångström-type regression equation related monthly average daily global solar radiation to clear day global solar radiation at the location in question and average fraction of possible sunshine hours:

$$\frac{H_{t,m}}{H_{c,m}} = a' + b' \frac{s}{S_o} \tag{86}$$

$H_{c,m}$ is the monthly average daily clear sky global solar radiation on a horizontal surface; *a'* and *b'* are empirical constants.

The estimated value of $H_{t,m}$ obtained for any station, therefore, depends on the accuracy with which these coefficients are determined which is neither a very accurate nor a very convenient way to calculate mean daily solar radiation and the exact evaluation of daily global radiation with a cloud free atmosphere is difficult. A basic difficulty with equation (86) lies in the ambiguity of the terms s/S_o and $H_{c,m}$. The former is an instrumental problem (records from sunshine recorders are open to interpretation). The letter seems from the uncertainty in the definition of a clear day. So, because there may be problems in calculating clear sky radiation accurately, this model was modified by Prescott (1940) by replacing clear sky radiation with extraterrestrial radiation $H_{o,m}$. This model popularly known as the Ångström-Prescott model is the most commonly used model and is given by:

$$\frac{H_{t,m}}{H_{0,m}} = a + b \frac{s}{S_o} \tag{87}$$

where *a* and *b* were the new regression parameters, established empirically for each location. Due to the wide spatial and temporal coverage of sunshine records and relatively less frequent global radiation measurements, such relationship has found immense use in the past. The constants *a* and *b* can assume a wide range of values depending on the local and seasonal variations, like cloudiness, effect of snow albedo, atmospheric constituents, latitude etc. If these constants cannot be estimated from measured data for the specific location, they can be inferred from regressions established at neighbouring locations (Driesse and Thevenard, 2002).

The Ångström-Prescott model's regression coefficients *a* and *b* have important physical interpretations with respect to the total insolation reaching the earth's surface at any place. The coefficient *a* is supposed to be related to the percentage of extraterrestrial insolation reaching the earth's surface on a completely cloud covered day that is the diffuse radiation. For a completely overcast day, s/S_o =0. Thus,

$$\frac{H_{t,m}}{H_{0,m}} = a \tag{88}$$

The sum of the regression coefficients a and b could be interpreted as the transmissivity of the atmosphere for global solar radiation under perfectly clear sky conditions (Revfeim, 1983). A clear day means that, $s/S_o = 1$, and then by:

$$\frac{H_{t,m}}{H_{0,m}} = a + b \tag{89}$$

So, b is related to the percentage of extraterrestrial insolation absorbed by the clouds on such a day.

Most of the investigations made so far have been based on monthly mean values of s and $H_{t,m}$. Even gave values, ranging from 0.06 to 0.44 for a and from 0.19 to 0.87 for b for 101 locations, the typical values of a published in literature range from 0.14 to 0.54 and those of b, from 0.18 to 0.73. The lower values of a are invariably associated with higher values of b and vice versa. The variability of $(a + b)$ was much less than that of either a or b. According to the regression coefficient b was more or less constant while the value of a showed marked variation. concluded in their model that for all practical purposes the coefficient b is considered constant. analyzed several published values of a and b and showed that a is related linearly, and b is related hyperbolically, to the appropriate mean value of s/S_o. He gave a new relation between daily global radiation with a cloud free atmosphere and daily total extraterrestrial solar radiation which is believed to be applicable anywhere in the world. In a model given by the method of least squares is used to find the value of regression coefficients. used the value of a and b as a function of latitude and altitude of the place.

Following a review of the literature on Ångström's equation and the properties of instantaneous solar radiation, Suehrcke (2000) derived the sunshine–radiation relationship from monthly average values of daily beam radiation which are used as a measure for the fraction of clear sky. The resulting (non-linear) relationship, unlike the Ångström–Prescott relation, does not contain empirical constants and only requires the monthly average clear sky atmospheric transmittance to account for the climate of a particular location which is typically between 0.65 and 0.75. Suehrcke (2000) verified the relationship for monthly intervals for Perth and Townsville, Australia and compared to existing Ångström–Prescott equations for a wide range of climates and locations and claimed the excellent agreement of the proposed relationship with the average value of data in the latter comparison suggests that it may be universally valid and that the Ångström–Prescott equation is a local (linear) approximation of the derived relationship.

There are several types of models (linear, quadratic, third degree polynomial and logarithmic) in literature for estimating the global radiation from extra-terrestrial irradiance and measured and theoretical daily sunshine duration. Almorox and Hontoria (2004) used these models, along with an exponential model for data from 16 meteorological stations in Spain to estimate monthly average daily global radiation. They found that the third degree models performed better than the others, however, suggested the linear model as optimum.

Iqbal (1979) developed linear and second order empirical equations which correlate monthly average daily values of diffuse and direct radiation with the fraction of maximum possible sunshine hours, based on three Canadian locations. The diffuse fraction is expressed as a first order function; whereas, both diffuse and direct transmittance indices are second order functions of fractional sunshine.

Abdalla and Feregh (1988) predicted diffuse radiation by two methods i.e. regression of diffuse fraction with clearness index and with fractional sunshine. They found a good agreement between the two. Soler (1990) investigated the dependence of monthly average hourly diffuse radiation on daily sunshine fraction for Uccle, Belgium. Following Iqbal's approach, similar regression equations have been developed to obtain monthly-average daily solar radiation from fraction of maximum possible sunshine hours for different locations by several authors, for example, estimation of diffuse radiation by Barbaro, Cannata, Coppolino-Leone and Sinagra (1981), diffuse and global by Jain (1990) and global, diffuse and direct solar radiation by Castro-Diez, Alados-Arboledas and Jiménez (1989), Ahmad, Burney and Husain (1991) and Tiris, Tiris and Türe (1996). El-Sebaii and Trabea (2003) proposed second order regression between diffuse transmittance index and sunshine fraction to estimate monthly average daily diffuse for four locations in Egypt and an overall regression for the country to predict annual averages of diffuse radiation.

Srivastava and Pandey (2013) estimated Angstrom-Prescott model parameters for seven different sites in India, and developed a correlation for India, which is found to be a good fit. Also they developed a correlation for predicting the solar radiation using only sunshine hour data. In this correlation, two new parabolic correlations were developed so that the coefficients of the Angstrom-Prescott models can be estimated even if only sunshine hour data is available:

$$a = -17.222\left(\frac{s}{S_o}\right)^2 + 27.18\frac{s}{S_o} - 10.533 \tag{90}$$

$$b = 18.676\left(\frac{s}{S_o}\right)^2 - 29.395\frac{s}{S_o} + 12.098 \tag{91}$$

As an alternative to conventional approaches, artificial neural networks (ANNs) have been successfully applied to solar radiation estimation and instead of being programmed in the traditional way, they are trained using past history data representing the behavior of a system. Mohandes, Rehman and Halawani (1998) estimated global solar radiation using artificial neural network and Hontoria, Aguilera and Zufiria (2002) used neural network multilayer perceptron model for hourly irradiation synthetic series. Tymvios, Jacovides, Michaelides and Scouteli (2005) conducted comparative study of Angstrom's and artificial neural network methodologies in estimating global solar radiation. Estimation of daily solar irradiation over a mountainous area using artificial neural networks was done by Bosch, Lópe, and Batlles (2008) and Lam, Wan and Yang (2008) done the modeling using ANNs for different climates in China. An integrated artificial neural networks approach for predicting global radiation was taken by Azadeh, Maghsoudi and Sohrabkhani (2009). Şenkal and Kuleli, (2009) introduced artificial neural networks (ANNs) for the estimation of solar radiation in Turkey (26–45 E and 36–42 N). Resilient propagation (RP), Scale conjugate gradient (SCG) learning algorithms and logistic sigmoid transfer function were used in the network. In order to train the neural network, meteorological data for the period from August

1997 to December 1997 for 12 cities (Antalya, Artvin, Edirne, Kayseri, Kütahya, Van, Adana, Ankara, İstanbul, Samsun, İzmir, Diyarbakır) spread over Turkey were used as training (nine stations) and testing (three stations) data. Meteorological and geographical data (latitude, longitude, altitude, month, mean diffuse radiation and mean beam radiation) are used in the input layer of the network. Solar radiation is the output. However, solar radiation has been estimated as monthly mean daily sum by using Meteosat-6 satellite C3 D data in the visible range over 12 cities in Turkey. Digital counts of satellite data were converted into radiances and these are used to calculate the albedos. Using the albedo, the cloud cover index of each pixel was constructed. Diffuse and direct component of horizontal irradiation were calculated as a function of optical air mass, turbidity factor and Rayleigh optical thickness for clear-sky. Using the relation between clear-sky index and cloud cover index, the solar irradiance for any pixel is calculated for Physical method.

Ambient Air Temperature Based Methods

In fact, the models estimating solar radiation from sunshine duration are generally more accurate than those involving other meteorological observations (Baigorria, Villegas, Trebejo, Carlos and Quiroz, 2004; Iziomon and Mayer, 2002; Chen, Ersi, Yang, Lu and Zhao, 2004; Rivington, Bellocchi, Matthews and Buchan, 2005). However, sunshine duration is not as readily available as air temperature data at standard meteorological stations (Rahimikhoob, 2010; Abraha and Savage, 2008). So, it is meaningful to elaborate models that estimate solar radiation based on air temperature as an alternative.

Two common approaches estimating solar radiation from air temperature use the methods of Hargreaves and Samani (1982) (HS) and Bristow and Campbell (1984) (BC). Since the establishment of the two models, many investigations concerning the HS and BC models have been carried out on the improvement in prediction accuracy and general validity, which were reviewed in detail by Liu, Mei, Li, Wang, Jensen, Zhang and Porter (2009). The HS model is primarily intended for application in monthly calculation (Allen, 1997). Although the BC model is superior to the HS model on daily global solar radiation calculation in some studies (Liu, Mei, Li, Wang, Jensen, Zhang and Porter, 2009; Baigorria, Villegas, Trebejo, Carlos and Quiroz, 2004; Almorox, 2011), however, it is not as good as the HS model in estimation of monthly average solar radiation (Bandyopadhyay, Bhadra, Raghuwanshi and Singh, 2008; Meza and Varas, 2000). The study of Bandyopadhyay, Bhadra, Raghuwanshi and Singh (2008) that estimates solar radiation for 29 stations across India showed that the HS model and its modifications (Annandale, Jovanovic, Benadé and Allen, 2002; Samani, 2000; Allen, 1997) models outperform the BC model in monthly calculation. Similarly, Meza and Varas (2000) demonstrated that the revised HS correlation, namely, (Allen, 1997) model, has a larger coefficient of determination than the BC model based on the monthly measured data from 21 stations in Chile. In addition, the HS model has been widely used because of its simplicity, and it is recommended in FAO-56 for solar radiation estimation (Liu, Mei, Li, Wang, Jensen, Zhang and Porter, 2009).

However, the performance of the HS and its modifications varies significantly in different locations (Liu, Mei, Li, Wang, Jensen, Zhang and Porter, 2009; Hargreaves and Samani, 1982). This limits the application of these models in a large country like China with diversities in climate and geography. The present work aims to propose a new simple and practical method that gives good estimates of monthly average daily global solar radiation from air temperature for all climatic regions. The performance of the proposed model is validated by comparing with the original HS model and its two modifications against the measured data at 65 meteorological stations in China using statistical error tests.

The HS model (Hargreaves and Samani, 1982) is the first procedure that calculates global solar radiation from T_{min} and T_{max} and defined as follows:

$$\frac{H_{t,m}}{H_{0,m}} = a_1 \Delta T^{0.5} \tag{92}$$

where $\Delta T = T_{\max} - T_{\min}$ is monthly average daily extraterrestrial radiation (MJ/m²), and a_1 is empirical coefficient.

Following Hargreaves and Samani (1982) pioneer work, Chen, Ersi, Yang, Lu and Zhao (2004) and Samani (2000) suggested the modifications in the form of (93) and (94), respectively,

$$\frac{H_{t,m}}{H_{0,m}} = a_2 \Delta T^{0.5} + b_2 \tag{93}$$

$$\frac{H_{t,m}}{H_{0,m}} = \left(a_3 + b_3 \Delta T + c_3 \Delta T^2\right) \Delta T^{0.5} \tag{94}$$

Where a_2, b_2, c_2, a_3, b_3 and c_3 are empirical coefficients.

The characteristic underlying equations (92)–(94) is that they explicitly account for solar radiation and air temperature and implicitly include the influence of relative humidity by means of ΔT, which is linearly related to relative humidity (Hargreaves and Samani, 1982). While these models succeed in some areas, the assumption in the HS model as well as its modifications could lead to a reduction in estimation accuracy in some conditions (Samani, 2000). The HS model assumes that ΔT is directly related to the fraction $H_{o,m}$ of received at the ground level. However, in fact, many other factors besides solar radiation, such as latitude, altitude, cloudiness, humidity, wind speed, precipitable water, aerosol, and proximity to a large body of water, can influence ΔT in a given location (Allen, 1997; Samani, 2000).

Among these factors, perceptible water has a considerable effect on solar radiation and then affects ΔT, especially in humid regions. The ways that precipitable water in the atmosphere affects solar radiation can be found in (Garrison, 1992). On the other hand, perceptible water is closely related to ambient temperature and relative humidity (Janjai, Praditwong and Moonin, 1996). In view of this, to improve estimation in simplicity, air temperature is added together with the relative humidity implicitly presented in the HS model to exert perceptible water's effects on calculating solar radiation, and the HS model is revised as follows:

$$\frac{H_{t,m}}{H_{0,m}} = \left(a_4 + b_4 T_a\right) \Delta T^{0.5} + c_4 \tag{95}$$

where Ta is monthly average air temperature (°C) and defined as Ta=0.5(Tmax+ Tmin) and a4, b4 and c4 are empirical coefficients.

Awachie and Okeke (1990) correlated the maximum ambient temperature, Tm with the global solar radiation for Nsukka as:

$$\frac{H_{t,m}}{H_{0,m}} = -8.7 + 0.8T_m \quad (96)$$

Bakirci (2009). Fadare (2009) done the modeling of solar energy potential in Nigeria using an artificial neural network model and assessment on diffuse solar energy under general sky condition using artificial neural network was done by Alam, Kaushik and Garg (2009). Martí and Gasque (2011) improved the temperature-based ANN models for solar radiation estimation through exogenous data assistance.

Cloud Cover Based Methods

Cloudiness is one of the most influential parameters for diffuse radiation since it accounts for scattering of solar radiation to a large extent. The fact that cloud observations are undertaken at a large number of stations as they usually require no measurements, led to the development of cloud radiation models. Available data on mean monthly cloud cover are expressed in tenth of the sky obscured by clouds. Empirical relationships have been derived to relate monthly average daily radiation $H_{t,m}/H_{o,m}$ to monthly average cloud cover C_m. These relationships are usually of the form:

$$\frac{H_{t,m}}{H_{0,m}} = a'' + b''C_m \quad (97)$$

where a'' and b'' are empirical coefficients.

Norris (1968) reviews several attempts to develop such a correlation. Bennet (1965) compared correlations of $H_{t,m}/H_{o,m}$ with C_m, with s/S_o and with the combination of the two variables and found the best correlation to be with s/S_o. Paltridge and Proctor (1976) have used cloud cover data to modify clear sky data for Australia and derived therefrom monthly averages of solar radiation which were in a good agreement with measured average data. One of the very initial attempts investigating the relative importance of cloud and other atmospheric constituents on scattering of direct radiation was made by Stanhill (1966) by comparing the diffuse fraction with local measurements of 'cloudiness index hours of bright sunshine' and observations of amount of cloud cover.

Rangarajan, Swaminathan and Mani (1984) fitted a cubic regression equation for monthly mean values of the fraction of the sky covered by clouds of all types and duration of bright sunshine. They computed the sunshine duration from cloud cover data and from the cloud derived sunshine data, monthly mean values of global and diffuse solar radiation were computed. Gul, Muneer and Kambezidis (1998) (who have also presented a brief review of such models in literature) extended the work of Kasten and Czeplak (1979) for UK, where the latter in their study of continuous hourly data (10 years) from Hamburg, showed that the ratio of global irradiance for any given cloud amount to global irradiance under a corresponding cloudless sky is independent of solar elevation. The equations for calculating and determining clear sky global irradiance and global irradiance for a given cloud amount are therefore, respectively represented as:

$$\frac{H_{t,m}}{H_{c,m}} = a_c + b_c cos\left(\theta_z\right) \tag{98}$$

$$\frac{H_{t,m}}{H_{c,m}} = a_c' + b_c' C_m \tag{99}$$

where a_c, b_c, a'_c and b'_c are empirical constants.

Satellite-Based Models

The amount of solar radiation reaching the earth's surface exhibits large spatial variations mainly due to geographical and climatological differences. It is not feasible to establish solar radiation monitoring networks at a density proportionate to this spatial complexity. In order to fill the gaps in surface measured data, approaches based on satellite observation turn into a suitable alternative. In the last decades, a number of methods for estimating solar radiation from satellite data have been developed. These methods are based on the data from metrological satellites. These satellites provide wide geographical coverage with high spatial resolution. Modeling techniques range from being totally subjective to those employing some of the most sophisticated irradiative transfer codes (Hay, 1993). Perez, Ineichen, Moore, Kmiecik, Chain, George and Vignola (2002) developed a model based on monitoring the dynamic range for satellite image pixels and assigning irradiance values corresponding to the relative brightness of the pixels. This cloud index acts as a quasi-linear modulation of a clear sky model. The modulating function was fitted to ground-based measurements grouped together with the data normalized by extraterrestrial irradiance (Vignola, Harlan, Perez and Kmiecik, 2005).

Most of the methods for deriving solar radiation from satellite observations employ meteorological geostationary satellite images. The geostationary satellites, orbiting at about 36,000 km, can offer a temporal resolution of up to 15 min and a spatial resolution of up to 1 km. The meteorological satellites collect images over a large area and with high time resolution allowing identification and forecasting the clouds evolution. This information is further processed, leading to the prediction of spatial variability of solar radiation at the ground level. Compared to ground measurements, satellite-derived hourly irradiation has been shown to be the most accurate option for locations that are further away with more than 25 km from a ground station (Zelenka, Perez, Seals and Renne, 1999). Thus, processing data collected by satellites can be a viable solution for forecasting solar radiation at the ground, aiming to proper operate the power grid.

However, there are a variety of problems faced by the users of satellite data to estimate solar radiation on earth's surface (Hay, 1993). One such issue was noted by Vignola, Harlan, Perez and Kmiecik (2005) when they compared the satellite-based model with ground based measurements; the former's inability to distinguish between the frost on the ground and low lying fogs and/or clouds on clear winter days. They concluded that ground based measurements are needed to augment the satellite data for comprehensive modeling. Also, the estimated diffuse radiation from their model was found to be sub-quality compared to the measured data. Besides, the technicalities, it is generally difficult to accommodate the satellite-based models within the existing earth's radiation and energy budget (Hay, 1993).

SOLAR RADIATION COMPONENTS AT THE GROUND LEVEL

The decomposition models decompose global into diffuse and direct/beam irradiance.

Distribution of Clear and Cloudy Days and Hours

The decomposition models, of global into diffuse and direct/beam irradiance, have extensively been developed based on inter-correlations between ratios such as clearness index K_t (global/extraterrestrial horizontal radiation). The monthly average clearness index $K_{t,m}$ is the ratio of the monthly average daily solar radiation on a horizontal surface $H_{t,m}$ to the monthly average daily solar extraterrestrial radiation $H_{o,m}$:

$$K_{t,m} = \frac{H_{t,m}}{H_{0,m}} \tag{100}$$

These ratios were originally referred to by Liu and Jordan (1960) as cloudiness indexes. As their values approach unity with increasing atmospheric clearness, they are also referred to as clearness indexes. The daily clearness index $K_{t,d}$ as the ratio of a particular day's total solar radiation on a horizontal surface $H_{t,d}$ to the solar extraterrestrial radiation $H_{o,d}$ for that day:

$$K_{t,d} = \frac{H_{t,d}}{H_{0,d}} \tag{101}$$

An hourly clearness index can be defined as (Duffie and Beckman, 2013):

$$k_t = \frac{I_t}{I_0} \tag{102}$$

Other transmittance indexes could be introduced such as diffuse transmittance index (diffuse/extra-terrestrial radiation) K_d. The monthly average daily ratio is:

$$K_{d,m} = \frac{H_{d,m}}{H_{0,m}} \tag{103}$$

and hourly beam transmittance index (direct/extraterrestrial radiation) K_b:

$$k_b = \frac{I_b}{I_0} \tag{104}$$

The data $H_{t,m}$, $H_{t,d}$ and I_b are from measurements of total solar radiation on a horizontal plane, that is, the commonly available pyranometer measurements. $H_{o,m}$, $H_{o,d}$ and I_o values can be calculated using the corresponding equations provided in chapter 3.

Note that small lettered $k_{(s)}$ specifically refer to hourly radiation values, whilst capital lettered K is used for values of long term periods and also for daily values.

The frequency of occurrence of periods of various radiation levels is of interest in two contexts. First, information on the frequency distribution is the link between two kinds of correlations, that of the daily fraction of diffuse solar radiation with daily solar radiation and that of the monthly average fraction of diffuse solar radiation with monthly average solar radiation. Second, the concept of utilizability depends on these frequency distributions.

Liu and Jordan (1960) found that the cumulative distribution K-K_t curves are very nearly identical for locations having the same values of K_t, even though the locations varied widely in latitude and elevation. On the basis of this information, they developed a set of generalized distribution curves versus f which are functions of K_t (see Table 6).

Bendt, Collares-Periera and Rabl (1981) have developed equations to represent Liu and Jordan distributions, based on 20 years of data from locations in the United States:

$$f\left(K_{t,d}\right) = \frac{exp\left(\gamma K_{t,d,min}\right) - exp\left(\gamma K_{t,d}\right)}{exp\left(\gamma K_{t,d,min}\right) - exp\left(\gamma K_{t,d,max}\right)} \tag{105}$$

where γ is determined from the following equation:

$$K_{t,m} = \frac{\left(K_{t,d,min} - \frac{1}{\gamma}\right) exp\left(\gamma K_{t,d,min}\right) - \left(K_{t,d,max} - \frac{1}{\gamma}\right) exp\left(\gamma K_{t,d,max}\right)}{exp\left(\gamma K_{t,d,min}\right) - exp\left(\gamma K_{t,d,max}\right)} \tag{106}$$

The correlation (105) represents the Liu and Jordan curves very well for $f\left(K_{t,d}\right) < 0.9$; above 0.9 correlation (105) over predict the frequency for given values of clearness index (Duffie and Beckman, 2013).

Solving the equation (106) for γ is not convenient. Therefore, an explicit equation for γ from the curve fit is given (Duffie and Beckman, 2013):

$$\gamma = -1.498 + \frac{1.184\left(\frac{K_{t,d,max} - K_{t,d,min}}{K_{t,d,max} - K_{t,m}}\right) - 27.182 exp\left(-1.5\frac{K_{t,d,max} - K_{t,d,min}}{K_{t,d,max} - K_{t,m}}\right)}{K_{t,d,max} - K_{t,d,min}} \tag{107}$$

Table 6. Coordinates of the Liu and Jordan generalized monthly cumulative distribution curves.

K_t	Values of $f(K_t)$ for $K_{t,m}$				
	0.3	**0.4**	**0.5**	**0.6**	**0.7**
0.04	0.073	0.015	0.001	0.000	0.000
0.08	0.162	0.070	0.023	0.008	0.000
0.12	0.245	0.129	0.045	0.021	0.007
0.16	0.299	0.190	0.082	0.039	0.007
0.20	0.395	0.249	0.121	0.053	0.007
0.24	0.460	0.298	0.160	0.076	0.007
0.28	0.513	0.346	0.194	0.101	0.013
0.32	0.579	0.379	0.234	0.126	0.013
0.36	0.628	0.438	0.277	0.152	0.027
0.40	0.687	0.493	0.323	0.191	0.034
0.44	0.748	0.545	0.358	0.153	0.047
0.48	0.793	0.601	0.400	0.269	0.054
0.52	0.824	0.654	0.460	0.310	0.081
0.56	0.861	0.719	0.509	0.360	0.128
0.60	0.904	0.760	0.614	0.410	0.161
0.64	0.936	0.827	0.703	0.467	0.228
0.68	0.953	0.888	0.792	0.538	0.295
0.72	0.967	0.931	0.873	0.648	0.517
0.76	0.979	0.967	0.945	0.758	0.678
0.80	0.986	0.981	0.980	0.884	0.859
0.84	0.993	0.997	0.993	0.945	0.940
0.88	0.995	0.999	1.000	0.985	0.980
0.92	0.998	0.999		0.996	1.000
0.96	0.998	1.000		0.999	
1.00	1.000			1.000	

$K_{t,d,min} = 0.05$ is recommended by Bendt, Collares-Periera and Rabl (1981). Hollands and Huget (1983) recommend:

$$K_{t,d,max} = 0.6313 + 0.267K_{t,m} - 11.9\left(K_{t,m} - 0.75\right)^8 \tag{108}$$

Figure 9 shows the generalized distribution of days with various values of K_t as a function of $K_{t,m}$.

Figure 9. The generalized distribution of days with various values of K_t as a function of $K_{t,m}$. ♦ stands for $K_{t,m} = 0.3$; ■ stands for $K_{t,m} = 0.4$; ▲ stands for $K_{t,m} = 0.5$; x stands for $K_{t,m} = 0.6$ and ж stands for $K_{t,m} = 0.7$

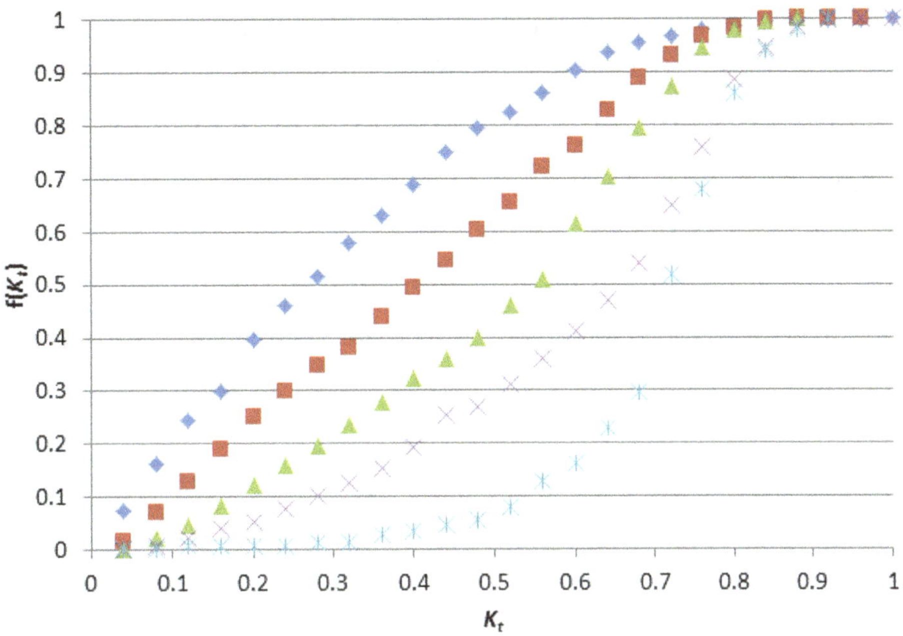

Diffuse Ratio- Clearness Index Regressions

One of the most common and popular estimation techniques in literature and for practical use in solar resource assessment has been the use of global radiation to estimate diffuse radiation. Studies of available daily solar radiation data have shown that the average ratio K (diffuse ratio (diffuse/global radiation H_d/H_t) is a function of K_t. In this regards, the first regression model based on K-K_t was proposed by Liu and Jordan (1960). Since then, this correlation relationship has evolved over the time and its coefficients have been modified for different locations and also different time scales, i.e. monthly-average, daily or hourly radiation models.

Page (1961) using data from 10 widely-spread sites in the 40°N to 40°S latitude belt derived a linear relationship between monthly-average K and K_t. Following Page's approach, many linear type monthly-average K- K_t models have been proposed in the literature. Erbs, Klein and Duffie (1982) proposed third degree polynomial regression for monthly diffuse fraction.

The pioneering work of Liu and Jordan (1960) involved development of a regression between daily clearness index and daily diffuse ratio using data from one location, Blue Hill, MA. In the same study, they also used a statistical distribution of global irradiation to develop a regression between monthly-average diffuse and global irradiation. Later, other researchers, like Choudhary (1963), Stanhill (1966), Tuller (1976) used the Liu-Jordan relationship for New Delhi, Gilat and four Canadian locations respectively. They came to the common conclusion that the diffuse ratio showed a significant departure (being higher than the modeled) from that suggested by Liu and Jordan. Both Choudhary and Stanhill attributed this deviation to the presence of higher dust content in the respective locations. In his study, Tuller found individual trend for each location, attributing it to the latitude effect. Collares-Periera and Rabl (1979) used pyroheliometer data from five US locations and found their results closer to those of Choudhary and Stanhill than that of Liu and Jordan (1960). They concluded that this discrepancy of Liu and Jordan is due to reliance on uncorrected measurements of diffuse radiation (shade ring misalignment). Collares-Periera and Rabl (1979) proposed the following correlation for the ratio of the monthly average daily diffuse solar radiation $H_{d,m}$ to the monthly average daily solar radiation on a horizontal surface $H_{t,m}$:

$$K_m = \frac{H_{d,m}}{H_{t,m}} = \left\{ \begin{matrix} 0.775 + 0.00606\left(\omega_s - 90\right) \\ -\left[0.505 + 0.00455\left(\omega_s - 90\right) cos\left(115K_{t,m} - 103\right)\right] \end{matrix} \right\} \tag{109}$$

where ω_s is the sunset hour angle in degrees. The correlation (109) could be expressed as follows (Collares-Periera and Rabl, 1979) for $\omega_s \leq 81.4^o$ and $0.3 \leq K_{t,m} \leq 0.8$:

$$K_m = \frac{H_{d,m}}{H_{t,m}} = 1.391 - 3.560K_{t,m} + 4.189K_{t,m}^2 - 2.137K_{t,m}^3 \tag{110}$$

and for $\omega_s > 81.4^o$ and $0.3 \leq K_{t,m} \leq 0.8$:

$$K_m = \frac{H_{d,m}}{H_{t,m}} = 1.311 - 3.022K_{t,m} + 3.427K_{t,m}^2 - 1.821K_{t,m}^3 \tag{111}$$

Example 7

The monthly daily average total solar radiation on a horizontal surface for Damascus (latitude 33.51°N and altitude 610m) on January is 10MJ/m^2. Using equation (109), estimate the fraction and amount that is diffuse.

Solution

On January the number of the characteristic day is 17, the sunset hour angle is 75.34° on this day and the extraterrestrial solar radiation is 19.21MJ/m^2. Thus $K_{t,m} = 10 / 19.21 = 0.52$. Then, according to equation (109), we have:

$$K_m = \left\{ \begin{matrix} 0.775 + 0.00606\left(\omega_s - 90^o\right) \\ -\left[0.505 + 0.00455\left(\omega_s - 90^o\right) cos\left(115^o K_{t,m} - 103^o\right)\right] \end{matrix} \right\} =$$
$$\left\{ \begin{matrix} 0.775 + 0.00606\left(75.34^o - 90^o\right) \\ -\left[0.505 + 0.00455\left(75.34^o - 90^o\right) cos\left(115^o * 0.52 - 103^o\right)\right] \end{matrix} \right\} = 0.23$$

Thus,

$$H_{d,m} = K_m H_{t,m} = 0.23 * 10 = 2.3 \text{ MJ/m}^2$$

Example 8

Redo example using equation (110) or (111).

Solution

On January 17, the sunset hour angle is 75.34° $\leq 81.4^o$ and $0.3 \leq K_{t,m} \leq 0.8$. So, the equation (110) should be used. According to this equation we have:

$$K_m = \frac{H_{d,m}}{H_{t,m}} = 1.391 - 3.560K_{t,m} + 4.189K_{t,m}^2 - 2.137K_{t,m}^3 = 0.372$$

Thus,

$$H_{d,m} = K_m H_{t,m} = 0.372 * 10 = 3.72 \text{ MJ/m}^2$$

On the other hand, Collares-Periera and Rabl (1979) proposed the following correlation for the ratio of the daily diffuse solar radiation $H_{d,d}$ to the daily solar radiation on a horizontal surface $H_{t,d}$:

$$K_d = \frac{H_{d,d}}{H_{t,d}} = \begin{cases} 0.99 \, for \, K_{t,d} \leq 0.17 \\ 1.188 - 2.272K_{t,d} + 9.473K_{t,d}^2 - 21.865K_{t,d}^3 + 14.648K_{t,d}^4 \\ for \, 0.17 < K_{t,d} < 0.75 \\ -0.54K_{t,d} + 0.632 \; for \, 0.75 < K_{t,d} < 0.80 \\ 0.2 \; for \, K_{t,d} \geq 0.80 \end{cases} \tag{112}$$

With taking into consideration the sunset hour angle ω_s which reflects the seasonal dependence the correlation for the ratio of the daily diffuse solar radiation $H_{d,d}$ to the daily solar radiation on a horizontal surface $H_{t,d}$ becomes (Collares-Periera and Rabl, 1979) for $\omega_s < 81.4^o$:

$$K_d = \frac{H_{d,d}}{H_{t,d}} = \begin{cases} 1.0 - 0.2727K_{t,d} + 2.4495K_{t,d}^2 - 11.9514K_{t,d}^3 + 9.3879K_{t,d}^4 \\ for \, K_{t,d} < 0.715 \\ 0.143 \, for \; K_{t,d} \geq 0.715 \end{cases} \tag{113}$$

and

$$K_d = \frac{H_{d,d}}{H_{t,d}} = \begin{Bmatrix} 1.0 - 0.2832K_{t,d} - 2.5557K_{t,d}^2 + 0.8448K_{t,d}^3 \\ for\, K_{t,d} < 0.722 \\ 0.175\, for\ K_{t,d} \geq 0.722 \end{Bmatrix} \tag{114}$$

for $\omega_s \geq 81.4^o$.

Muneer and Hawas (1984) developed a regression based on 3 years data from 13 Indian stations between diffuse fraction as a third order polynomial of clearness index. They also concluded in their study, that no single regression (from the above mentioned) is applicable to all regions as each region has its own characteristics. Rao, Bradley and Lee (1984) developed polynomial regressions based on daily data from Corvallis, Oregon. Saluja and Muneer (1985) developed a linear regression between daily diffuse fraction and daily clearness index, based on 3 years data for five diverse locations in the UK. Within the same study, they also proposed a single regression for the country, asserting that there is no latitude effect. Lalas, Petrakis and Papadopoulos (1987) using data from Greek islands, proposed a linear regression on a daily basis. Oliveira, Escobedo, Machado and Soares (2002) proposed fourth degree polynomial and linearly varying diffuse fraction as a function of clearness index on daily and monthly average basis, respectively for the city of Sao Paulo in Brazil. Paliatsos, Kambezidis and Antoniou (2003) found that a linear K-K_t model to be optimum for diffuse radiation estimation in Balkan Peninsula (Greece).

A common methodology underlying most of the daily diffuse estimation models is to propose a piecewise regression. This is achieved by developing piecewise fits to K-K_t data for overcast, partly-cloudy and clear sky. For overcast skies, the regression is generally given in a linear form (with some models assuming a constant value in the overcast regime). So, the mathematical representation would, thus, be:

$$K = a_d^{'} + b_d^{'} K_t \tag{115}$$

where a'_d and b'_d are empirical coefficients.

For partly cloudy skies, a third-order (sometimes, fourth-order) polynomial in K_t is generally used:

$$K = a_d^{'} + b_d^{'} K_t + c_d^{'} K_t^2 + d_d^{'} K_t^3 \tag{116}$$

where a'_d, b'_d, c'_d and d'_d are empirical coefficients. Finally, for 'clear sky' conditions, a constant value is the widely-accepted choice:

$$K = a_d^{'} \tag{117}$$

Some of the authors of similar approach have also extended their work to propose single regression-polynomial that best fits the data. One of the exceptions to this polynomial relationship between diffuse fraction and clearness index is model proposed by Bartoli, Cuomo, Amato, Barone and Mattarelli (1982) given by a single empirical equation with fixed constant values and exponential power of K_t.

$$K = 0.154 + 0.846 exp\left[\frac{-1.062K_t^{0.861}}{\left(1 - K_t\right)}\right] \tag{118}$$

Linear equations correlating monthly average diffuse transmittance index to daily average cloud cover (C_m, in eighths) have also been proposed. In his comparative analysis of various models, Bashahu (2003) used two such models of the following form:

$$K_{d,m} = a_d + b_d C_m \tag{119}$$

$$K_{d,m} = a_d + b_d \left(1 - C_m\right) \tag{120}$$

Some researchers have also investigated and reported seasonal variations for daily regressions, e.g., Tuller (1976), Collares-Periera and Rabl (1979), Erbs, Klein and Duffie (1982), Rao, Bradley and Lee (1984), Vignola and McDaniel (1984), Chandrasekaran and Kumar (1994) for a tropical site (Madras) and by Jacovides, Hadjioannou, Pashiardis and Stefanou (1996) for Cyprus data.

Orgill and Hollands (1977) followed the Liu and Jordan approach in correlating the diffuse ratio to clearness index except on an hourly basis. Their study was based upon four years data from Toronto (Canada). The Orgill and Hollands (1977) correlation has been widely used. It is represented by the following equation:

$$k_d = \frac{I_d}{I} = \begin{Bmatrix} 1.0 - 0.249k_t \quad for\, k_t < 0 \\ 1.557 - 1.84k_t \quad for\, 0.35 < k_t < 0.75 \\ 0.177\, for\;\; k_t \geq 0.75 \end{Bmatrix} \tag{121}$$

It produces results that are for practical purposes the same as those of Erbs, Klein and Duffie (1982). The Erbs, Klein and Duffie (1982) is:

$$k_d = \frac{I_d}{I} = \begin{Bmatrix} 1.0 - 0.09k_t \quad for\, k_t \leq 0.22 \\ 0.9511 - 0.1604k_t + 4.388k_t^2 - 16.638k_t^3 + 12.336k_t^4 \\ for\, 0.22 < k_t \leq 0.80 \\ 0.165\, for\;\; k_t > 0.80 \end{Bmatrix} \tag{122}$$

For values of k_t greater than 0.8, there are very few data. Bugler (1977) proposed a model that correlates the hourly diffuse ratio to the ratio of hourly global radiation and the estimated clear sky radiation. Erbs, Klein and Duffie (1982) used 65 months of data from 5 locations in USA to develop regression between diffuse ratio and clearness index, for hourly, daily as well as monthly average data. They proposed a fourth degree and a third degree polynomial for daily and monthly average diffuse ratio, respectively. Muneer and Saluja (1986) developed regression for hourly diffuse fraction, following their work for daily data from the same 5 UK locations. They found a third degree polynomial to be the optimum to relate diffuse ratio as a function of clearness index. They also investigated the effect of fractional possible sunshine and solar altitude and concluded that although the latter showed somewhat relevance, the former had no bearing on the regression. Newland (1989), using data from Macau, developed a fourth-order polynomial regression for daily diffuse fraction and a linear logarithmic one for monthly average.

De Miguel, Bilbao, Aguiar, Kambezidis and Negro (2001) reproduced CLIMED 1-1997 (daily regression) and CLIMED2-1997 (hourly regression) models for Mediterranean region. These models are essentially third degree polynomial fits for partly cloudy sky, and either constant or linear regression for overcast and clear sky regimes. They concluded that solar altitude plays a significant role on diffuse fraction and should be taken in account in future.

Iqbal (1980) used data from three Canadian and two French sites to develop a regression between hourly diffuse transmittance index (ratio of diffuse/Extraterrestrial radiation), instead of diffuse ratio, and hourly clearness index (ratio of global/extraterrestrial radiation). He found the models to be site-specific and concluded that they cannot be generalized. He recommended that the range of solar altitudes higher than 40 0 should be investigated.

Maxwell (1987) developed regressions between K_b (direct transmittance index) and K_t, including solar elevation dependence, where K_b is computed as a function of air mass along with clearness index. Perez, Ineichen, Seals and Zelenka (1990) enhanced the use of global radiation to improve estimation accuracy within the DISC model (Maxwell's quasi-physical model), by using a zenith-angle independent clearness index and also by utilizing time variability of global radiation. Such additional descriptors, as suggested by Perez and co-authors, could overcome the two limitations posed by solar elevation dependence of clearness index and its inability to account for abrupt changes in sky conditions from one hour to the next. Once the normal beam irradiance has been estimated from these models, diffuse irradiance is calculated as the difference between global irradiance and the product of normal beam irradiance and the solar altitude.

Parametric Diffuse Radiation Dependence

Several authors have demonstrated the dependence of diffuse radiation on one or more of the variables such as surface albedo, bright sunshine hours, perceptible water, the atmospheric turbidity, solar elevation, cloud cover, apart from the global irradiance (Bashahu, 2003). Some investigators also attempted at correlating the diffuse ratio with sunshine fraction along with clearness index. Gopinathan (1988) computed the monthly mean daily diffuse radiation from clearness index and percent possible sunshine. Al-Hamdani, AI-Riahi and Tahir (1989) developed a multi-linear regression equation for daily diffuse fraction as a function of clearness index and fractional sunshine duration to estimate for Baghdad, Iraq. Later Al-Riahi, AI-Hamdani and Tahir (1992) developed similar model to estimate hourly diffuse radiation.

Page (1986) developed a clear sky radiation model as part of the work undertaken for the development of European Solar Radiation Atlas, to estimate diffuse (or direct) irradiance as a function of solar altitude and air mass 2 Linke turbidity factor, after incorporating the standard corrections for mean solar distance and air mass adjustment for station height. Piece-wise regressions of diffuse radiation for overcast (k_t <0.3), intermediate (0.3< k_t <0.8) and clear skies (k_t >0.8) similar to the earlier work on Liu-Jordan type models, have been proposed based on clearness index, solar elevation, ambient temperature and relative humidity (Reindl, Beckman and Duffie, 1990; Chendo and Maduekwe, 1994). Gopinathan and Soler (1996) and Soler, Gopinathan and Robledo (1999) proposed a regression between monthly mean hourly diffuse to global radiation (k) with monthly mean hourly clearness index, mean monthly hourly sunshine fraction $\left(s / S_o\right)$, and monthly mean solar elevation at mid hour for four locations in Spain and UK, respectively. They concluded in their study that the diffuse radiation estimation can be improved by adding solar elevation or sunshine fraction along with kb yielding best results when all three variables are used. Gul and Muneer (1998) proposed a diffuse fraction model as a function of clearness index and product of precipitable water and air mass for several locations in UK. They claimed that their model outperformed the previous models, like those of Maxwell (1987), Reindl, Beckman and Duffie, (1990) and Muneer and Saluja (1986). Bashahu (2003) used for his comparative models analysis a model that correlates monthly average diffuse radiation as an exponential function of fractional sunshine and solar elevation. Monthly average diffuse fraction was correlated with monthly average sunshine fraction and monthly average value of water vapour content (Bashahu, 2003). Diffuse radiation has also been shown to depend on atmospheric turbidity by various researchers, which in turn varies with air mass origin as demonstrated by Rapti (2000). He investigated the effect of seasonal variations on air mass content (like, heavy water content owing to humidity, or continental dust content owing to winds) in a Mediterranean location and correlated it with diffuse radiation accordingly.

Estimation of Hourly Radiation From Daily Data

The knowledge of hourly values is very useful in many practical applications. In some cases the knowledge of averages is sufficient (e.g., for the calculation of long term performances of solar systems), but the knowledge of the actual distribution of hourly values is necessary to describe the behaviour of the system (e.g., overheating effects in solar passive buildings). The reliability of the solar power/thermal systems designed based on hourly solar radiation data is greater than systems designed based on daily or monthly solar radiation profiles (Erdinc and Uzunoglu, 2012). The need for hourly solar radiation data, for accurate system's design and control, led researchers to utilize hourly meteorological variables for predicting hourly solar radiation. In many stations only daily values are available. Therefore, there is a big debate regarding the availability of hourly meteorological data such as ambient temperature, relative humidity, and sunshine ratio for this purpose (Khatib, Mohamed and Sopian, 2012). So that it is very useful to have methods which allow calculate average values of hourly solar irradiation and their distribution starting from the values of daily solar irradiance.

Some of pioneer researchers have proposed empirical equations that can predict hourly solar radiation in terms of daily or monthly solar radiation, hour angle, and sunrise/sunset hour angle. Examples of these models are that of Liu and Jordan (1960):

$$r_t = \frac{I_t}{H_{t,d}} = r_0 = \frac{\pi}{24}\frac{cos\left(\omega\right) - cos\left(\omega_s\right)}{sin\left(\omega_s\right) - \omega_s cos\left(\omega_s\right)} \tag{123}$$

where ω_s is the solar sunset angle on a horizontal surface (in radians) and ω is the hour angle, that of Collares-Pereira and Rabel (1979):

$$r_t = \frac{I_t}{H_{t,d}} = \left[a + b cos\left(\omega\right)\right] r_0 \tag{124}$$

$$a = 0.409 + 0.5016 \sin\left(\omega_s - \frac{\pi}{3}\right) \tag{125}$$

$$b = 0.6609 - 0.4767 \sin\left(\omega_s - \frac{\pi}{3}\right) \tag{126}$$

that of Gueymard (2000):

$$r_t = \frac{I_t}{H_{t,d}} = r_0 \frac{\left\{1 + \frac{24 r_0 cos\left(\varphi\right) cos\left(\delta\right)\left[sin\left(\omega_s\right) - \omega_s cos\left(\omega_s\right)\right] a_1}{\pi b_1}\right\}}{\left\{1 + \frac{24 cos\left(\varphi\right) cos\left(\delta\right)\left[0.5\omega_s + \omega_s cos^2\left(\omega_s\right) - 0.75 sin\left(2\omega_s\right)\right] a_1}{\pi b_1\left[sin\left(\omega_s\right) - \omega_s cos\left(\omega_s\right)\right]}\right\}} \tag{127}$$

$$a_1 = 0.41341 K_t + 0.61197\ K_t^2 - 0.01886 \frac{24 K_t \omega_s}{\pi} + 0.00759 \frac{24\omega_s}{\pi} \tag{128}$$

$$b_1 = max\left\{\begin{array}{l} 0.54, 0.28116 + 2.2475 K_t - 0.61197\ K_t^2 \\ + 1.84535 \frac{cos\left(\varphi\right) cos\left(\delta\right)\left[sin\left(\omega_s\right) - \omega_s cos\left(\omega_s\right)\right]}{\omega_s} \\ + 1.6811\left\{\frac{cos\left(\varphi\right) cos\left(\delta\right)\left[sin\left(\omega_s\right) - \omega_s cos\left(\omega_s\right)\right]}{\omega_s}\right\}^3 \end{array}\right\} \tag{129}$$

Jain (1988) proposed a Gaussian function to fit the recorded data and established the following relation:

$$r_t = \frac{I_t}{H_{t,d}} = \frac{1}{\sigma\sqrt{2\pi}} exp\left[-\frac{\left(t - 12\right)^2}{2\sigma^2}\right] \tag{130}$$

t is the true solar time in hours, and σ is defined by:

$$\sigma = \frac{1}{r_t\left(t = 12\right)\sqrt{2\pi}} \tag{131}$$

where $r_t\left(t = 12\right)$ is the hourly ratio of the global irradiation at the midday true solar time. Baig, Akhter and Mufti (1991) modified Jain's model to fit the recorded data during the starting and ending periods of a given day better. In this model, r_t is estimated by:

$$r_t = \frac{I_t}{H_{t,d}} = \frac{1}{\sigma\sqrt{2\pi}}\left\{exp\left[-\frac{\left(t-12\right)^2}{2\sigma^2}\right]+cos\left[\frac{180\left(t-12\right)}{S_0}\right]\right\} \tag{132}$$

S_o is the daily length of a day, n, at a specific site. Kaplanis (2006) and Kaplanis and Kaplain (2007) proposed a model, in which the global solar irradiance is approximated as follows:

$$G_{ct} = a + bcos\left(\frac{2\pi t}{24}\right) \tag{133}$$

A boundary condition provides a relationship between a and b. That is, at $t=t_{ss}$, $G_{ct}=0$. Hence, from equation (133), one obtains:

$$a + bcos\left(\frac{2\pi t_{ss}}{24}\right) = 0 \tag{134}$$

Integrating equation (133) over t, from sunrise t_{sr}, to sunset t_{ss}, one obtains:

$$H_{t,d} = \int_{t_{sr}}^{t_{ss}} G_{ct}dt = aS_0 + \frac{24b}{\pi}sin\left(\frac{2\pi t_{ss}}{24}\right) \tag{135}$$

Then using the $H_{t,d}$ values that are taken from the recorded data and the boundary condition (134) the values of the equation (133) coefficients a and b could be determined from the equations (134) and (135). Then, the hourly values of global solar radiation on a horizontal surface, during any day in the year, could be determined by integrating the equation (133) over the hours of that day. So, proposing these equations Kaplanis (2006) and Kaplanis and Kaplain (2007) made a big advantage in predicting hourly solar radiation without the need for other meteorological variables.

Al-Sadah, Ragab and Arshad (1990) developed the following correlation to estimate the ratio of the monthly mean hourly (I_t) to monthly mean daily ($H_{t,d}$) global radiation on a horizontal surface as follows:

$$r_t = \frac{I_t}{H_{t,d}} = a_1 t^2 + b_1 t + c_1 \tag{136}$$

where t is the local time in hours $6 \le t \le 18$, a_1, b_1 and c_1 are coefficients that can be determined by any curve fitting tool. However, the drawback of this model is that it is a location dependent model whereas such a type of models is devoted to a specific region. This is because the coefficients a_1 , b_1 and c_1 are calculated based on a specific solar radiation profile. Later on Singh, Srivastava and Pandey (1997) applied the equation (136) for estimating the hourly global radiation on a horizontal surface Lucknow (latitude 26.75°N, longitude 80.50°E) and found that the equation (136) provides good estimates of the hourly global radiation except for the sunrise and sunset hours. The present results are comparable with the estimates of the Liu and Jordan (1960) and Collares-Pereira and Rabel (1979) models which also correlate hourly values and daily totals of the global radiation.

Here, it should be mentioned that, the above mentioned models give a good estimation of long term means of hourly irradiation as a function of the daily irradiation and of the day length, but do not give any information about the spread of data about the mean. This spread was found to be quite large and, moreover, the time series of hourly values of solar irradiation was found not stationary nor in mean or in variance.

More powerful approaches have been proposed using auto-correlated models of hourly data. Among these methods on can mention that of Amato, Bartoli, Coluzzi, Cuomo, Serio and Silvestrini (1986) which allows, as it was claimed, reproduce the distribution of hourly values of solar radiation in a satisfactory way starting from the daily values of solar radiation and that of Koussa, Malek and Haddadi (2009). Moreover, most of these models are either empirical or statistical models that implying complex calculations are required. Therefore, these empirical models can be further enhanced in terms of accuracy and simplicity by utilizing novel learning machine such as generalized artificial neural network (GRNN) which was proven to be able to generate hourly solar radiation data from daily solar radiation data at sites where only daily averages of solar radiation are available Khatib and Elmenreich (2015). These data can be used in optimal sizing of photovoltaic systems. The optimal sizing of such systems requires hourly prediction of system performance for at least one-year time in order to provide optimal sizes of photovoltaic array and storage units, for example. Moreover, such a model can be used to optimally manage photovoltaic based distributed generation (DG) units. The output of DG systems needs to be predicted in order to optimally operate the penetrated power system in terms of optimal power flow and system's stability, protection, and power quality.

The ratio of hourly diffuse solar radiation to daily diffuse solar radiation as a function of time and day length can be represented by the following (Liu and Jordan, 1960):

$$r_d = \frac{I_d}{H_{t,d}} = r_0 = \frac{\pi}{24} \frac{cos\left(\omega\right) - cos\left(\omega_s\right)}{sin\left(\omega_s\right) - \omega_s cos\left(\omega_s\right)} \tag{137}$$

Example 9

The monthly daily average total solar radiation on a horizontal surface for Damascus (latitude 33.51°N and altitude 610m) on January is 10MJ/m^2. Estimate the average diffuse, basing on (Liu and Jordan, 1960) method, the average beam, and the average total solar radiation, basing on the equation of Collares-Pereira and Rabel (1979), for the hours 10 to 11 and 13 to 14.

Solution

On January the number of the characteristic day is 17, the sunset hour angle is 75.34^o on this day and the extraterrestrial solar radiation is 19.21MJ/m^2. On the midpoint of hours from 10 to 11 the hour angle is -22.5° and 13:30 is the midpoint for the hours 13 to 14. So, the hour angle for 13:30 is 22.5°. Thus, from the equation (137) we have for the midpoint 10:30 hour:

$$r_d = \frac{I_d}{H_{d,d}} = r_0 = \frac{\pi}{24}\frac{cos\left(\omega\right) - cos\left(\omega_s\right)}{sin\left(\omega_s\right) - \omega_s cos\left(\omega_s\right)}$$
$$= \frac{\pi}{24}\frac{cos\left(-22.5^o\right) - cos\left(75.34^o\right)}{sin\left(75.34^o\right) - \frac{\left(75.34^o\right)\pi}{180^o}cos\left(75.34^o\right)} = 0.14$$

From the example 4.7 we obtained that:

$$H_{d,m} = K_m H_{t,m} = 2.3 \text{ MJ/m}^2$$

Thus,

$$I_d = 2.3 * 0.14 = 0.3 MJm^{-2}$$

For the midpoint 13:30 hour we have:

$$r_d = \frac{I_d}{H_{d,d}} = r_0 = \frac{\pi}{24}\frac{cos\left(\omega\right) - cos\left(\omega_s\right)}{sin\left(\omega_s\right) - \omega_s cos\left(\omega_s\right)}$$
$$= \frac{\pi}{24}\frac{cos\left(22.5^o\right) - cos\left(75.34^o\right)}{sin\left(75.34^o\right) - \frac{\left(75.34^o\right)\pi}{180^o}cos\left(75.34^o\right)} = 0.14$$

Thus,

$$I_d = 2.3 * 0.14 = 0.3 MJm^{-2}$$

On the other hand, from the equations (125) and (126) we have:

$$a = 0.409 + 0.5016 \sin\left(75.34^o - 60^o\right) = 0.54$$

$$b = 0.6609 - 0.4767 \sin\left(75.34^o - 60^o\right) = 0.53$$

Thus, from the equation (124) we have for the midpoint 10:30 hour:

$$r_t = \frac{I_t}{H_{t,d}} = \left[a + bcos\left(\omega\right)\right] r_0 = \left[0.54 + 0.53cos\left(-22.5^o\right)\right] * 0.14 = 0.14$$

So,

$$I_t = 10 * 0.14 = 1.4 MJm^{-2}$$

Accordingly,

$$I_b = I_t - I_d = 1.4 - 0.3 = 1.1 MJm^{-2}$$

TERRESTRIAL SOLAR RADIATION ON A TILTED SURFACE

Solar radiation data are the best source of information for estimating the average incident radiation, necessary for proper design and the assessment of solar energy conversion systems (Sabziparvar, 2008). The availability of more comprehensive solar radiation data is invaluable for the design and evaluation of solar-based conversion systems. Particularly, the basic solar radiation data for the surfaces of interests are not readily available in most developing countries (Li, Lam and Chu, 2008; El-Sebaii, Al-Hazmi, Al-Ghamdi and Yaghmour, 2010). Generally, the meteorological stations measure global and diffuse solar radiation intensities mostly on horizontal surfaces only (Evseev and Kudish, 2009). Whereas, the stationary solar conversion systems are tilted towards the sun in order to maximize the amount of solar radiation incident on the collector or module surface. But, the availability of required data on tilted surfaces is very rare (Pandey and Katiyar, 2009; Miguel, Bilbao, Aguiar, Kambezidis and Negro, 2001). The solar radiation incident on a tilted surface must be determined by converting the solar radiation intensities measured on a horizontal surface to that incident on the tilted surface of interest. There exist a large number of models designed to perform such a conversion. Therefore, the tilted surface irradiation in most cases is calculated from measured global horizontal irradiation by means of empirical models.

There are several forms of solar radiation data, which could be used for a variety of purposes in the design and development of solar energy systems. Daily data is often available and hourly radiation can be estimated from available daily data. Monthly total solar radiation on a horizontal surface can be used in some process design methods. However, the process performance is generally not linear with solar radiation.

The use of averages may lead to serious errors if non-linearities are not taken into account (Duffie and Beckman, 2013). So that, the measurements of solar radiation on tilted surfaces are important for determination of accurate input to solar photovoltaic (PV) systems or collectors (Loutzenhiser, Maz, Felsmann, Strachan, Frank and Maxwell, 2007). Basically, two types of solar radiation models are required to predict the tilted surface irradiation from global horizontal irradiation. One type of model predicts beam and diffuse components from global horizontal irradiation and the other estimates the incident radiation on tilted surfaces. The total radiation on a tilted surface consists of three components: beam, reflected radiation from the ground and diffuse from the all part of the sky. The direct and reflected components can be computed with good accuracy by using simple algorithms but the nature of diffuse part is more complicated. Calculations of diffuse radiation require information of both global and direct radiation incident on a horizontal surface at the same time period (Noorian, Moradi and Kamali, 2008; Notton, Cristofari, Muselli, Poggi and Heraud, 2006). Empirical models which were found in literature for the estimation of diffuse radiation is mostly based on the data collected from the meteorological stations of United States, Canada, Australia, and Northern European countries (Liu and Jordan, 1962).

Although, a large number of empirical models exist, and attempts were made to correlate the diffuse radiation on a tilted surface to that measured on horizontal surface according to local climatic conditions for a particular area. The abundant of such models indicated the complexity of the task for converting diffuse solar radiation measured on a horizontal surface to that on a tilted surface (Miguel, Bilbao, Aguiar, Kambezidis and Negro, 2001).

Calculation Methodology

In order to calculate global solar radiation on a tilted surface it is advised to follow step by step the following:

1. Determine the extraterrestrial solar radiation on the horizontal surface at the required site (for details see chapter 3).
2. Gather available daily data on terrestrial solar radiation on the horizontal surface at the required site.
3. Choose the tilt and the orientation of the tilted surface.
4. Determine the diffuse solar radiation on horizontal surfaces basing on one of methods, provided in (Diffuse ratio- clearness index regressions) of this chapter. For particularity, let us choose the method of Erbs, Klein and Duffie (1982) which is widely cited. When the sunset hour angle $\omega_s \leq 81.4^o$ for mean day of the month and $0.3 \leq K_t \leq 0.8$, then the diffuse solar radiation can be calculated from the following equation:

$$K_d = \frac{H_{d,m}}{H_{t,m}} = \left\{1.391 - 3.56K_t + 4.189K_t^2 - 2.137K_t^3\right\} \tag{138}$$

When the sunset hour angle $\omega_s > 81.4^o$ for mean day of the month and $0.3 \leq K_t \leq 0.8$, then the diffuse solar radiation can be calculated from the following equation:

$$K_d = \frac{H_{d,m}}{H_{t,m}} = \left\{1.311 - 3.022K_t + 3.427K_t^2 - 1.821K_t^3\right\} \tag{139}$$

5. Determine the incident solar radiation on tilted surface:

The incident solar irradiation on a tilted surface is the sum of a set of radiation streams including beam radiation, the three components of diffuse radiation from the sky, and the radiation reflected from the various surfaces seen by the tilted surface. The total incident radiation on tilted surface can be written as in the following form:

$$H_{t,m,\beta} = H_{b,m,\beta} + H_{d,m,\beta} + H_{r,m,\beta} \tag{140}$$

where $H_{t,m,\beta}$ is the monthly total incident radiation on a tilted surface, $H_{b,m,\beta}$ is the beam radiation, $H_{d,m,\beta}$ is the diffuse component and $H_{r,m,\beta}$ is the ground reflected component.

6. Calculate the beam radiation on the tilted surface:

The beam radiation on the tilted surface could be calculated from the beam radiation on the horizontal surface using the following equation:

$$H_{b,m,\beta} = H_{b,m} R_{b,m} \tag{141}$$

where $R_{b,m}$ is the ratio of mean daily beam radiation on the tilted surface to that on a horizontal surface.

$$R_{b,m} = \frac{\sum_{n_1}^{n_2} \int_{t_{sr'}}^{t_{ss'}} cos\left(\theta_i\right) dt}{\sum_{n_1}^{n_2} \int_{t_{sr}}^{t_{ss}} cos\left(\theta_z\right) dt} \tag{142}$$

n_1 being the number of the first day of the considered month in the year, n_2 is the number of the last day of the considered month in the year, t_{sr} is the true sunrise on the tilted surface and $t_{ss'}$ is the true sunset on the tilted surface. The numerator of the equation (142) denotes amount of the extraterrestrial radiation on tilted surface and the denominator is that on horizontal surface.

7. Calculate the ground reflected radiation on tilted surface:

The ground reflected radiation on tilted surface $H_{r,m,\beta}$ is composed of diffuse reflectance ρ_g from the ground (also called ground albedo) and a view factor F_{c-g}:

$$H_{r,m,\beta} = \rho_g F_{c-g} H_{t,m} = \rho_g \left(\frac{1 - cos\beta}{2} \right) H_{t,m} \tag{143}$$

The ground reflectance could be taken as 0.2 in a condition that the mean monthly temperature is greater than 0ºC and the measuring station is located on a roof top with a low reflectance. Its value could be taken as 0.7 if the temperature is less than -5ºC (Duffie and Beckman, 2013).

8. Determine the diffuse radiation component on a tilted surface.

Diffuse Radiation Component on a Tilted Surface

It is found from the literature, that there is an agreement among authors in terms of beam and reflected radiation (Noorian, Moradi and Kamali, 2008). However, the differences are largely in the defining and treating of diffuse radiation on tilted surface. Due to the complicated nature of diffuse fraction many researchers mostly use isotropic models to estimate the amount of diffuse radiation incident on tilted surfaces (Duffie and Beckman, 2013). Because, the isotropic sky models are easy to understand and make calculation of radiation on tilted surfaces simple. However, the anisotropic models have been developed which takes into account the circumsolar diffuse and horizon brightening components on a tilted surface. The models used to predict the diffuse radiation on a tilted surface are broadly classified as isotropic and anisotropic sky models. The isotropic models assume that the intensity of diffuse sky radiation is uniform over the sky dome. Hence, the diffuse radiation incident on a tilted surface depends on a fraction of the sky dome seen by it. The anisotropic models on the other hand, presume that the anisotropy of the diffuse sky radiation in the circumsolar region (sky near the solar disk) plus the isotropically distributed diffuse component from the rest of the sky dome (horizon brightening fraction) (Noorian, Moradi and Kamali, 2008). For this reason, six empirical models were chosen. Out of six, three isotropic models namely those of (Liu and Jordan, 1962), Koronakis (1986), and Badescu (2002), and three anisotropic models namely those of Hay and Davies (1980), Reindl, Beckman and Duffie (1990), and that known as HDKR model from the first letters of Hay-Davies-Klucher-Reindl (Klucher, 1979) were treated. Reindl, Beckman and Duffie (1990) add a horizon brightening term to the Hay and Davies (1980) model, as proposed by Klucher (1979).

A brief description of the selected isotropic and anisotropic sky models is given below:

- **The First Isotropic Model (Liu and Jordan, 1962):** In this model, the solar radiation on tilted surface is considered to be composed of three parts such as; beam, reflected from ground and diffuse fraction. It was assumed that the diffuse radiation is isotropic only; whereas, circumsolar and horizon brightening were taken as zero (Duffie and Beckman, 2013). So, the radiation view factor from the sky to the collector is:

$$F_{c-s} = \left(1 + cos\beta\right) / 2 \tag{144}$$

Hence,

$$I_{d,\beta} = I_d \left[\frac{1 + cos\left(\beta\right)}{2}\right] \tag{145}$$

and the overall formula for computing the total radiation on tilted surface is proposed as sum of beam, earth reflected and isotropic diffuse radiation. Thus, is given by:

$$I_{t,\beta} = I_b R_b + I_d \left[\frac{1 + cos\left(\beta\right)}{2}\right] + \rho_g \left[\frac{1 - cos\left(\beta\right)}{2}\right] I_t \tag{146}$$

where R_b is the ratio of mean hourly beam radiation on the tilted surface to that on a horizontal surface: $R_b = \int_{t_1}^{t_2} cos\left(\theta_i\right) dt \, / \int_{t_1}^{t_2} cos\left(\theta_z\right) dt$, t_1 and t_2 are the beginning and ending times of the considered hour respectively.

- **The Second Isotropic Model (Koronakis, 1986):** Koronakis (1986) analyzed the global radiation monthly average daily values by a simple correlation into diffuse and beam components. The diffuse and beam components are then used to calculate monthly average hourly and daily values on an inclined surface. The results of these calculations are tabulated and plotted against the angle of tilt for summer, winter and all-year-round intended use. Global radiation measurements used in this work come from (a) the Climatological Bulletin of the National Observatory of Athens, years 1957–1981, (b) the unpublished records of the National Weather Service of Greece (referred to as EMY), years 1977–1982, and (c) the Scientific Publications of the Public Power Co. (referred to as PPC) on measurements of solar potential of Greece, years 1982–1983; the latter was the source of solar diffuse data as well. In his correlation Koronakis (1986) modified the assumption of isotropic sky diffuse radiation and proposed that the slope $\beta=90^o$ provides 66.7% of diffuse solar radiation of the total sky dome, for example $F_{c-s} = \left(2 + cos\beta\right) / 3$. Thus, following correlation was suggested to measure incident radiation on tilted surface:

$$I_{t,\beta} = I_b R_b + I_d \left[\frac{2 + cos\left(\beta\right)}{3}\right] + \rho_g \left[\frac{1 - cos\left(\beta\right)}{2}\right] I_t \tag{147}$$

- **The Third Isotropic Model (Badescu, 2002):** Badescu (2002) characterized the approach of Liu and Jordan (1962) which allows computing isotropic solar diffuse irradiance on a tilted surface as a 2D approach as the position of a sky element, in this approach, is characterized by a single (zenith) angle. Badescu (2002) introduced a more realistic 3D model (that uses both zenith and azimuth angles to describe sky element's position) for both isotropic diffuse irradiance and ground reflected irradiance incident on an arbitrary oriented surface. Badescu (2002) claimed that the 3D formula predicts a lower diffuse irradiance than the 2D relationship while the ground reflected irradiance is higher in case of the 3D model than in case of the 2D approach. In case of a small tilt angle, the 2D and 3D approximations predict comparable values, higher than the mean of the results obtained with a (reference) non-isotropic model. However, the 3D model is slightly more precise. When a larger tilt angle is considered, the 3D model predicts a few percent larger value than the mean of the values estimated by the reference model while the 2D model gives a significantly higher value. So, Badescu demonstrated model for the solar diffuse radiation on a tilted surface, and considered the view factor $F_{c-s} = \left[3 + cos\left(2\beta\right)\right] / 4$. Therefore, the total radiation on a tilted surface was expressed as:

$$I_{t,\beta} = I_b R_b + I_d \left[\frac{3 + cos\left(2\beta\right)}{4}\right] + \rho_g \left[\frac{1 - cos\left(\beta\right)}{2}\right] I_t \tag{148}$$

- **The First Anisotropic Model (Hay and Davies, 1980):** Hay and Davies (1980) assumed that the diffuse radiation from the sky is composed of an isotropic and circumsolar component only, whereas, the horizon brightening part was not taken into account. It was assumed that the diffuse parts coming directly from the Sun's direction is circumsolar and the diffuse component reaching through the rest of the sky dome isotropically. These components were weighted according to an anisotropy index *A*. The anisotropy index was used to quantify a portion of diffuse radiation treated as circumsolar with remaining part of the diffuse radiation assumed to be isotropic. The reflected part is dealt with same as suggested by Liu and Jordan (1962). The total radiation on a tilted surface is proposed as follows:

$$I_{t,\beta} = \left(I_b + AI_d\right) R_b + I_d \left\{\left[\frac{3 + cos\left(2\beta\right)}{4}\right]\left(1 - A\right) + AR_b\right\} + \rho_g \left[\frac{1 - cos\left(\beta\right)}{2}\right] I_t \tag{149}$$

where *A* is anisotropy index, which is the function of transmittance of the atmosphere for beam radiation and defined as:

$$A = \frac{I_{bn}}{I_{0n}} = \frac{I_b}{I_0} \tag{150}$$

- **The Second Anisotropic Model is That of Reindl, Beckman and Duffie (1990):** In this model, horizon brightening factor was added to isotropic diffuse and circumsolar radiation component. Beam and reflected fraction of solar radiation was taken as same, which were proposed by Liu and Jordan and other authors. A definition of anisotropy index *A* was introduced as proposed by Hay and Davies. The modulating factor:

$$f = \sqrt{\frac{I_b}{I}} \tag{151}$$

was also added to multiply the term of $sin^3\left(\beta / 2\right)$ for horizon brightening factor. They considered all three components of diffuse fraction. So, the global solar radiation on a tilted surface is proposed as follows:

$$\begin{aligned} I_{t,\beta} &= \left(I_b + AI_d\right)R_b \\ &+I_d\left\{\left[\frac{1+cos\left(\beta\right)}{2}\right]\left(1-A\right)\left[1+sin^3\left(\frac{\beta}{2}\right)\sqrt{\frac{I_b}{I}}\right]+AR_b\right\} \\ &+\rho_g\left[\frac{1-cos\left(\beta\right)}{2}\right]I_t \end{aligned} \tag{152}$$

- **The Third Anisotropic Model is the HDKR Model:** If the beam, reflected and all terms of diffuse radiation such as isotropic, circumsolar and horizon brightening are added to the solar radiation equation, a new correlation develops called HDKR model (Duffie and Beckman, 2013). It is basically the combination of Hay and Davies, Klucher and Reindl models. The solar energy irradiation on tilted surface is then determined as:

$$\begin{aligned} I_{t,\beta} &= \left(I_b + AI_d\right)R_b \\ &+I_d\left\{\left[\frac{1+cos\left(\beta\right)}{2}\right]\left(1-A\right)\left[1+sin^3\left(\frac{\beta}{2}\right)\right]\right\}+\rho_g\left[\frac{1-cos\left(\beta\right)}{2}\right]I_t \end{aligned} \tag{153}$$

when applying the above mentioned models using data from different locations, it is found that all isotropic models estimated lower solar radiation availability in the worst months due to conservative results of these models in overcast skies, and executed higher results in the good weather conditions. Overall, both the (Hay and Davies, 1980) and HDKR models demonstrated same results and established slightly more values than Liu and Jordan (1962) model. This may be due to addition of the circumsolar component in diffuse radiation fraction in these models as compared to isotropic models. The Reindl, Beckman and Duffie (1990) model displayed highest estimated values among all models. This is because of the individual consideration of all diffuse components in their model and incorporation of modulating factor, which was multiplied by the term used for horizon brightening. The Badescu (2002) model demonstrated lowest results as compared to isotropic as well as anisotropic models. It is due to the factor used in the cosine of tilt angle which results the lower values of diffused radiation. Statistically, it is discovered that isotropic models executed the higher values than anisotropic models in good weather conditions and clear skies days. However, all examined models executed nearly 1% of mean difference in estimated results among each other.

Finally, one can mention that, the proposed a model by Perez, Stewart, Seals and Guertin (1988) for calculating the solar diffuse component of solar global radiation could be used also. The model of Perez,

Stewart, Seals and Guertin (1988) is based on a detailed analysis of the three diffuse components. The diffuse component on the tilted surface is given by:

$$I_{d,\beta} = I_d \left\{ \left[\left[\frac{1+\cos(\beta)}{2} \right] (1 - A_1) + A_1 \frac{a}{b} + A_2 \sin(\beta) \right] \right\} \tag{154}$$

where A_1 and A_2 are circumsolar and horizon brightness coefficients and a and b are terms that account for the angles of incidence of the cone of circumsolar radiation on the tilted and horizontal surfaces. The circumsolar radiation in considered to be from a point source at the Sun. The terms a and b are:

$$a = max\left[0, cos(\theta_i)\right] \tag{155}$$

$$b = max\left[cos85^o, cos(\theta_z)\right] \tag{156}$$

where θ_z is the zenith angle and θ_i is the incidence angle on the tilted surface. With this definition, $a / b = R_b$ for most hours when collectors will have useful outputs.

The brightness coefficients A_1 and A_2 are functions of three parameters that describe the sky conditions, the zenith angle θ_z, a clearness ε and a brightness Δ. The clearness ε is given by:

$$\varepsilon = \frac{\frac{I_d + I_n}{I_d} + 5.535x10^{-6}\theta_z^{\ 3}}{1 + 5.535x10^{-6}\theta_z^{\ 3}} \tag{157}$$

where θ_z in degrees. The brightness Δ is given by:

$$\Delta = AM \frac{I_d}{I_{0,n}} \tag{158}$$

where AM is the air mass and $I_{o,n}$ is the extraterrestrial normal incidence solar radiation. The brightness Coefficients A_1 and A_2 are functions of statistically derived coefficients for ranges of values of ε. A recommended set of these coefficients is shown in Table 7. The equation for calculating A_1 and A_2 are:

$$A_1 = max\left[0, (f_{11} + f_{12}\Delta + f_{13}\theta_z)\right] \tag{159}$$

$$A_2 = f_{21} + f_{22}\Delta + f_{23}\theta_z \tag{160}$$

Table 7. Brightness coefficients for anisotropic sky

Ranges of ε	f_{11}	f_{12}	f_{13}	f_{21}	f_{22}	f_{23}
0-1.065	-0.196	1.084	-0.006	-0.114	0.180	-0.019
1.065-1.230	0.236	0.519	-0.180	-0.011	0.020	-0.038
1.230-1.500	0.454	0.321	-0.255	0.072	-0.098	-0.046
1.500-1.950	0.866	-0.381	-0.375	0.203	-0.403	-0.049
1.950-2.800	1.026	-0.711	-0.426	0.273	-0.602	-0.061
2.800-4.500	0.978	-0.986	-0.350	0.280	-0.915	-0.024
4.500-6.200	0.748	-0.913	-0.236	0.173	-1.045	0.065
6.200- ↑	0.318	-0.757	0.103	0.062	-1.698	0.236

Source: (Perez, Stewart, Seals and Guertin, 1988)

If the beam and reflected terms are added to the equation (154) which calculate the three diffuse radiation components such as isotropic, circumsolar and horizon brightening the solar energy irradiation on tilted surface is then determined as:

$$I_{t,\beta} = I_b R_b + I_d \left\{ \left[\frac{1 + cos(\beta)}{2} \right] (1 - A_1) + A_1 \frac{a}{b} + A_2 \, sin(\beta) \right\} + \rho_g \left[\frac{1 - cos(\beta)}{2} \right] I_t \tag{161}$$

The total solar radiation on the tilted surface, according to equation (161), includes five terms: the beam, the isotropic diffuse, the circumsolar diffuse and the ground reflected term.

Average Global Radiation on a Tilted Surface

The calculation of total solar radiation on the tilted surfaces from measurements on a horizontal surface was discussed previously. For use in solar process design procedures, the monthly average daily solar radiation on the tilted surface is also required. The procedure for calculating $H_{t,m,\beta}$ is parallel to that for $I_{t,\beta}$.

In the isotropic model of Liu and Jordan (1962) the solar radiation on tilted surface could be calculated by assuming that the diffuse and ground-reflected solar radiations are each to be isotropic. Then in a manner analogous to equation (146), the monthly mean daily total solar radiation on unshaded tilted surface can be expressed as:

$$H_{t,m,\beta} = H_{b,m} R_{b,m} + H_{d,m} \left[\frac{1 + cos(\beta)}{2} \right] + \rho_g \left[\frac{1 - cos(\beta)}{2} \right] H_{t,m} \tag{162}$$

$R_{b,m}$ is given by equation (142). In a manner analogous to the equation (162), the monthly mean daily total solar radiation on unshaded tilted surface, according to isotropic model (Koronakis, 1986), can be expressed as:

$$H_{t,m,\beta} = H_{b,m} R_{b,m} + H_{d,m} \left[\frac{2 + cos\left(\beta\right)}{3} \right] + \rho_g \left[\frac{1 - cos\left(\beta\right)}{2} \right] H_{t,m} \tag{163}$$

Example 10

Basing on the isotropic model of Liu and Jordan (1962), calculate the monthly average daily global solar radiation, incident on a surface, orientated towards the Equator and tilted by an angle 45°, at Damascus (latitude 33.51°N and altitude 610m) on January. The ground reflectance $\rho_g = 0.2$ and the monthly average daily global solar radiation, incident on a horizontal surface is $10MJm^{-2}$.

Solution

On January the number of the characteristic day is 17, the sunset hour angle is 75.34^o on this day and the extraterrestrial solar radiation is $19.21MJ/m^2$. Thus $K_t = 10 / 19.21 = 0.52$. The first steps are to obtain $H_{d,m}$ and $R_{b,m}$. $H_{d,m}$ could be calculated using the equation (138):

$H_{d,m} = K_d H_{t,m} = \left\{1.391 - 3.56K_t + 4.189K_t^2 - 2.137K_t^3\right\} H_{t,m} = 0.372 * 10 = 3.72$ M J / m^2 $R_{b,m} = 1.4$

So, according to the equation (162) we have:

$$H_{t,m,45} = H_{b,m} R_{b,m} + H_{d,m} \left[\frac{1 + cos\left(\beta\right)}{2} \right] + \rho_g \left[\frac{1 - cos\left(\beta\right)}{2} \right] H_{t,m}$$

$$= \left(10 - 3.72\right) * 1.4 + 3.72 * \left(\frac{1 + 0.71}{2} \right) + 0.2 \left(\frac{1 - 0.71}{2} \right) * 10 = 12.26 \mathrm{MJ} / \mathrm{m}^2$$

In a manner analogous to the equation (162), the monthly mean daily total solar radiation on unshaded tilted surface, according to isotropic model (Badescu, 2002), can be expressed as:

$$H_{t,m,\beta} = H_{b,m} R_{b,m} + H_{d,m} \left[\frac{3 + cos\left(2\beta\right)}{4} \right] + \rho_g \left[\frac{1 - cos\left(\beta\right)}{2} \right] H_{t,m} \tag{164}$$

The monthly mean daily total solar radiation on unshaded tilted surface, according anisotropic model (Hay and Davies, 1980) can be expressed as:

$$H_{t,m,\beta} = \left(H_{b,m} + A_m H_{d,m}\right) R_{b,m} + H_{d,m}\left\{\left[\frac{3+\cos\left(2\beta\right)}{4}\right]\left(1-A_m\right) + A_m R_{b,m}\right\} + \rho_g \left[\frac{1-\cos\left(\beta\right)}{2}\right] H_{t,m} \tag{165}$$

where A_m is the monthly daily average anisotropy index, which is the function of transmittance of the atmosphere for beam radiation and defined as:

$$A_m = \frac{H_{b,m}}{H_{0,m}} \tag{166}$$

The monthly mean daily total solar radiation on unshaded tilted surface, according to anisotropic model (Reindl, Beckman and Duffie, 1990), can be expressed as:

$$H_{t,m,\beta} = \left(H_{b,m} + AH_{d,m}\right) R_{b,m} + H_{d,m}\left\{\left[\frac{1+\cos\left(\beta\right)}{2}\right]\left(1-A\right)\left[1+\sin^3\left(\frac{\beta}{2}\right)\sqrt{\frac{H_{b,m}}{H_{t,m}}}\right] + AR_{b,m}\right\} + \rho_g \left[\frac{1-\cos\left(\beta\right)}{2}\right] H_{t,m} \tag{167}$$

where $\sqrt{H_{b,m} / H_{t,m}}$ is the monthly daily average modulating factor.

The monthly mean daily total solar radiation on unshaded tilted surface, according to the HDKR anisotropic model, can be expressed as:

$$R_{b,m} = D + H_{d,m}\left\{\left[\frac{1+\cos\left(\beta\right)}{2}\right]\left(1-A_m\right)\left[1+\sin^3\left(\frac{\beta}{2}\right)\right]\right\} + \rho_g \left[\frac{1-\cos\left(\beta\right)}{2}\right] H_{t,m} \tag{168}$$

For extending the method of (Perez, Stewart, Seals and Guertin, 1988) for calculating the monthly mean daily total solar radiation on unshaded tilted surface, it is recommended to apply the equation (161) in characteristic days hour by hour and then summing the hourly contributions.

Finally, Klein and Theilacker (1981) developed a general equation that is valid for any surface azimuth angle:

$$H_{t,m,\beta} = H_{t,m} D + H_{d,m} \left[\frac{1+\cos\left(\beta\right)}{2}\right] + \rho_g \left[\frac{1-\cos\left(\beta\right)}{2}\right] H_{t,m} \tag{169}$$

$$D = \begin{cases} max\left[0, G\left(\omega_{ss}, \omega_{sr}\right)\right] if\, \omega_{ss} \geq \omega_{sr} \\ max\left[0, G\left(\omega_{ss}, -\omega_{s}\right) + G\left(\omega_{s}, \omega_{sr}\right)\right] if\, \omega_{ss} > \omega_{sr} \end{cases} \tag{170}$$

$$G\left(\omega_1, \omega_2\right) = \frac{1}{2d} \begin{cases} \left(\frac{bA}{2} - a'B\right)\left(\omega_1 - \omega_2\right) + \left(a'A - bB\right)\left[sin\left(\omega_1\right) - sin\left(\omega_2\right)\right] \\ -a'C\left[cos\left(\omega_1\right) - cos\left(\omega_2\right)\right] \\ +\frac{bA}{2}\left[sin\left(\omega_1\right)cos\left(\omega_1\right) - sin\left(\omega_2\right)cos\left(\omega_2\right)\right] \\ +\frac{bC}{2}\left[sin^2\left(\omega_1\right) - sin^2\left(\omega_2\right)\right] \end{cases} \tag{171}$$

where ω_s is sunset hour angle on a horizontal plane; ω_{ss} and ω_{sr} are the sunset and sunrise hour angles on a tilted surface:

$$\omega_{ss} = \begin{cases} \left|min\left[\omega_s,\ arccos\left(\frac{AB - C\sqrt{A^2 - B^2 + C^2}}{A^2 + C^2}\right)\right]\right| if\left(A > 0 and\, B > 0\right) or\left(A \geq B\right) \\ -\left|min\left[\omega_s,\ arccos\left(\frac{AB - C\sqrt{A^2 - B^2 + C^2}}{A^2 + C^2}\right)\right]\right| otherwise \end{cases} \tag{172}$$

$$a' = a - \frac{H_{d,m}}{H_{t,m}} \tag{173}$$

$$A = cos\left(\beta\right) + \tan\left(\varphi\right)cos\left(\gamma\right)sin\left(\beta\right) \tag{174}$$

$$B = \cos\left(\omega_s\right)cos\left(\beta\right) + \tan\left(\delta\right)cos\left(\gamma\right)sin\left(\beta\right) \tag{175}$$

$$C = \frac{sin\left(\beta\right)sin\left(\gamma\right)}{cos\left(\varphi\right)} \tag{176}$$

a and *b* are the coefficients of Collares-Pereira and Rabel (1979) equation (see equations (125) and (126)).

UTILIZABILITY

Original Method

An energy balance equation to represent the performance of a solar collector can be written. This states that the useful gain at any time is the difference between the solar energy absorbed and the losses from the collector. In the case of solar thermal collector, the losses depend on the difference in temperature between the collector plate and the ambient temperature and on a heat loss coefficient. Therefore, there is a value of incident radiation that is just enough so that the absorbed radiation equals the losses. This value of incident radiation is the critical radiation level, $I_{t,c}$, for that collector operating under those conditions.

The concept of the solar utilizability method is described in detail by Duffie and Beckman (2013) and is based on the premise that a critical solar radiation level can be defined below which no useful energy can be collected. Thus, utilizability is essentially a statistic that is a fraction of the total radiation that is received at an intensity value higher than a critical level. The monthly average hourly utilizability method was essentially introduced for flat-plate thermal collectors, and originally formulated by Whillier (1953) and later generalized through the characteristic distribution of insolation values by Liu and Jordan (1963). The hourly utilizability method yields the long term average performance of a solar collector with a performance characteristic that is operated at a constant critical radiation level. Using the steady state performance equation for a flat-plate thermal collector.

The average radiation for the period can then be multiplied by this fraction to find the total utilizable energy. Thus, the concept of utilizability is based on the following consideration. If only radiation above a critical intensity $I_{t,c}$ is useful, then it is possible to define a radiation statistic, called utilizability, Φ, as the fraction of the total radiation that is received at an intensity value higher than the critical level. The utilizability can be referred to different time basis (hour or day). The fraction of an hour's total energy that is above the critical level is the utilizability for that particular hour:

$$\Phi_h = \frac{\left(I_t - I_{t,c}\right)^+}{I_t} \tag{177}$$

where the superscript + indicates that the utilizable energy can be zero or positive but not negative. So, Φ_h can have values, which vary from zero to unity. The utilizability Φ for a particular hour for a month of N days in which the hour's average radiation is $I_{t,m}$ is useful.

$$\Phi = \frac{1}{N}\sum_1^N \frac{\left(I_t - I_{t,c}\right)^+}{I_{t,m}} \tag{178}$$

The monthly average utilizable energy for the any period of time can be found by multiplying the average radiation for the period by this fraction. For a period of one hour, the monthly average utilizable energy is the product $I_{t,m}$ Φ. The month's average utilizable energy for the hour is then:

$$q_{u,m,h} = NI_{t,m}\Phi \tag{179}$$

The calculation can be done for individual hours for the month and the result summed to get the month's utilizable energy.

In the case of a solar thermal collector, the critical irradiance level $G_{t,c}$ which must be exceeded in order for solar energy collection to occur, is given by:

$$G_{t,c} = \frac{U_L F_R \left(T_{i,m} - T_{a,m}\right)}{F_R \left(\tau\alpha\right)_{n,m}} \tag{180}$$

where $G_{t,c}$ is in units of W/m^2, F_R is the heat removal factor, U_L is the overall heat transfer coefficient for the collector (Wm^{-2}K^{-1}), $T_{i,m}$ is the monthly average inlet fluid temperature to the collector (^{o}C), $T_{a,m}$ is the monthly average ambient air temperature (^{o}C), and $\left(\tau\alpha\right)_{n,m}$ is the monthly transmittance-absorptance product. The parameters F_R, U_L and $\left(\tau\alpha\right)_{n,m}$ are readily determined from collector efficiency tests, where the subscript n denotes normal incidence.

Solar collectors are tested, rated, and certified by the National Solar Rating and Certification Corporation according to industry-accepted standards. The general procedure for testing solar collectors is to operate the collector under steady-state conditions of solar radiation at normal incidence, wind speed, ambient temperature, and inlet fluid temperature.

The result of a solar collector test is described in terms of collector efficiency, defined as useful energy gained by the circulating fluid divided by the measured solar radiation.

Collector efficiency is then plotted versus the ratio of $\left(T_i - T_a\right) / G_t$, where T_i (^{o}C) is the collector inlet fluid temperature, T_a (^{o}C) is the ambient air temperature, and G_t is the incident solar radiation (W/m^2). Such a plot results is a straight line having a slope equal to - $F_R U_L$ and a y-intercept equal to $F_R \left(\tau\alpha\right)_n$. The effect of the heat removal factor (F_R) is to reduce the calculated useful energy gain from what it would be if the whole collector were at the inlet fluid temperature to what it actually is with a fluid that increases in temperature as it flows through the collector.

The monthly useful energy collected ($q_{u,m}$) is given by:

$$q_{u,m} = A_c \; F_R \left(\tau\alpha\right)_m H_{t,m,\beta}\Phi \tag{181}$$

where $q_{u,m}$ is the monthly useful energy collected (J), A_c is the collector area (m^2), $H_{t,m,\beta}$ is the monthly average daily irradiance in the plane of the collector (Jm^{-2}day^{-1}), and Φ is the monthly average daily utilizability.

Generalized Method

The monthly distribution of daily total radiation is a unique function of K_t. Thus the effect of daily radiation distribution on Φ is related to a single variable, K_t.

Radiation data are available in several forms, with the most widely available being pyranometer measurements of total (beam-plus-diffuse) radiation on horizontal surfaces. These data are available on an hourly basis from a limited number of stations and on a daily basis for many stations. Nevertheless, solar radiation information is needed in several different forms, depending on the kinds of calculations that are to be done:

- Procedures based on detailed hour-by-hour basis for long time performance of a solar process system (hourly information of solar radiation and other meteorological measurements are needed).
- Procedures based on monthly average solar radiation. These are useful in estimating long-term performance of some kinds of solar processes.

There are methods for the estimation of solar radiation information in the desired format from the data that are available. These include estimation of beam and diffuse radiation from total radiation, time distribution of radiation in a day, and radiation on surfaces other than horizontal.

There are a number of intermediate steps necessary in the calculation of Φ . First, the monthly average daily radiation on horizontal plane $H_{t,d}$ must be known, which is available from weather data bases. Next, the monthly average clearness index $K_{t,m}$ should be calculated (see equation (100)). Next, the monthly average daily diffuse solar radiation $H_{d,m}$ is calculated from the global radiation and the clearness index $K_{t,m}$.

The monthly average daily beam radiation $H_{b,m}$ is computed simply from:

$$H_{b,m} = H_{t,m} - H_{d,m} \tag{182}$$

Then, the monthly average radiation in the plane of the collector $H_{t,m,\beta}$ can be computed (see related items for this purpose). The numerator of the geometric factor $\left(R / R_n\right)$ is defined as the ratio of the monthly average daily radiation on the tilted surface to that on a horizontal surface and is calculated as:

$$R = \frac{H_{t,m,\beta}}{H_{t,m}} \tag{183}$$

The denominator (R_n) of the geometric factor is the ratio of the hour centered at noon of radiation on the tilted surface to that on a horizontal surface for an average day of the month and is computed as:

$$R_n = \left(\frac{I_{t,m,\beta}}{I_{t,m}}\right)_n = \left(1 - \frac{r_{d,n} H_{d,m}}{r_{t,n} H_{t,m}}\right) R_{b,n} + \left(\frac{r_{d,n} H_{d,m}}{r_{t,n} H_{t,m}}\right)\left[\frac{1 + cos\left(\beta\right)}{2}\right] + \rho_g \left[\frac{1 - cos\left(\beta\right)}{2}\right] \tag{184}$$

where $r_{t,n}$ is the ratio of hourly total to daily total radiation for the hour centered around solar noon, and $r_{d,n}$ is the ratio of hourly diffuse to daily diffuse radiation for the hour centered around solar noon. $r_{t,n}$ and $r_{d,n}$ could be computed from the equation (124) which is the equation of Collares-Pereira and Rabel (1979) (see also (Duffie and Beckman, 2013)) and $r_{d,n}$ could be computed from the equation (137) (Liu and Jordan, 1960).

Now, the critical radiation level X_C can be computed from:

$$X_{C,m} = \frac{I_{tC}}{r_{t,n} H_{t,m} R_n} \tag{185}$$

Finally, the monthly average daily utilizability Φ can be computed from:

$$\Phi = exp\left[\left(a + b\left(\frac{R_{n,m}}{R_m}\right)\right)\left(X_{C,m} + cX_{C,m}^2\right)\right] \tag{186}$$

where

$$a = 2.943 - 9.271K_t + 4.031K_t^2 \tag{187}$$

$$b = -4.345 + 8.853K_t - 3.602K_t^2 \tag{188}$$

$$c = -0.170 - 0.306K_t + 2.936K_t^2 \tag{189}$$

Finally, it should be mentioned here that, it is convenient for computations to have an analytical representation of the utilizability function. A computationally simple algorithm is presented by Clark, Klein and Beckman (1983) for evaluating the hourly utilizability function Φ , defined as the fraction of the long-term, monthly-average, hourly solar radiation incident on a surface which exceeds a specified threshold intensity. The algorithm was developed by correlating values of φ obtained by numerical integration of hourly radiation for three locations. The algorithm was shown to compare well both with a more complex analytical expression for Φ and with results obtained numerically using many years of hourly horizontal radiation measurements in nine U.S. locations. In addition, the algorithm is shown to be applicable for surfaces of any orientation. According to Clark, Klein and Beckman (1983), the generalized utilizability functions are represented by:

$$\Phi = \left\{ \begin{array}{c} 0\, if\, X_{C,m} \geq X_m \\ \left(1 - \frac{X_{C,m}}{X_m}\right)^2 if\, X_m = 2 \\ \left\| g - \left[g^2 + \left(1 + 2g\right)\left(1 - \frac{X_{C,m}}{X_m}\right)^2 \right]^{0.5} \right| otherwise \end{array} \right\} \tag{190}$$

$$X_m = 1.85 + 0.169 \frac{R_{h,m}}{k_{t,m}^2} - 0.0696 \frac{cos\left(\beta\right)}{k_{t,m}^2} - 0.981 \frac{k_{t,m}}{cos^2\left(\delta\right)} \tag{191}$$

$$g = \frac{\left(X_m - 1\right)}{\left(2 - X_m\right)} \tag{192}$$

$$k_{t,m} = \frac{I_{t,m}}{I_{0,m}} = \frac{H_{t,m}}{H_{0,m}} \frac{r_t}{r_d} = \frac{r_t}{r_d} K_{t,m} = K_{t,m} \left[a + bcos\left(\omega\right)\right] \tag{193}$$

$$R_{h,m} = \frac{I_{t,m,\beta}}{I_{t,m}} = \frac{I_{t,m}}{r_t H_{t,m}} \tag{194}$$

COMPARISON TECHNIQUES

There are numerous works in literature which deal with the assessment and comparison of monthly mean daily solar radiation estimation models. The most popular statistical parameters are the mean bias error (*MBE*) and the root mean square error (*RMSE*). To evaluate the accuracy of the estimated data, from any model, the following statistical tests were used, *MBE, NMBE, RMSE, NRMSE,* mean percentage error (*MPE*) and coefficient of correlation (*r*), to test the linear relationship between predicted and measured values. For better data modeling, these statistics should be closer to zero, but coefficient of correlation, *r*, should approach to 1 as closely as possible. The Nash–Sutcliffe equation (*NSE*) is also selected as an evaluation criterion. A model is more efficient when *NSE* is closer to 1. However, these estimated errors provide reasonable criteria to compare models but do not objectively indicate whether a model's estimates are statistically significant. The t-statistic allows models to be compared and at the same time it indicates whether or not a model's estimate is statistically significant at a particular confidence level, so, t-test of the models should be carried out to determine statistical significance of the predicted values by the models.

The Mean Bias Error

The first most popular statistical parameter is the mean bias error (MBE):

$$MBE = \frac{1}{n}\sum_{1}^{n}\left(I_{t,calc} - I_{t,meas}\right) \tag{195}$$

This test provides information on long-term performance. A low *MBE* value is desired. A negative value gives the average amount of underestimation in the calculated value. So, one drawback of these two mentioned tests is that overestimation of an individual observation will cancel underestimation in a separate observation.

The Normalized Mean Bias Error

The normalized mean bias error (NMBE):

$$NMBE = \frac{\frac{1}{n}\sum_{1}^{n}\left(I_{t,calc} - I_{t,meas}\right)}{\frac{1}{n}\sum_{1}^{n}I_{t,meas}} \tag{196}$$

The Root Mean Square Error

The second most popular statistical parameter is the root mean square error (RMSE):

$$RMSE = \sqrt{\frac{1}{n}\sum_{1}^{n}\left(I_{t,calc} - I_{t,meas}\right)^{2}} \tag{197}$$

The value of *RMSE* is always positive, representing zero in the ideal case.

The Normalized Root Mean Square Error

The normalized root mean square error (NRMSE):

$$NRMSE = \frac{\sqrt{\frac{1}{n}\sum_{1}^{n}\left(I_{t,calc} - I_{t,meas}\right)^2}}{\frac{1}{n}\sum_{1}^{n} I_{t,meas}} \tag{198}$$

The NRMSE, given above, provides information on the short-term performance of the correlations by allowing a term-by-term comparison of the actual deviation between the predicted and measured values. The smaller the value is, the better the performance of the model is.

The Normalized Root Mean Square Error

The third most popular statistical parameter is the coefficient of correlation (*r*). The r can be used to determine the linear relationship between the measured and estimated values, which can be calculated from the following equation:

$$r = \sqrt{\frac{\sum_{1}^{n}\left(I_{t,calc} - I_{t,\langle calc\rangle}\right)\left(I_{t,meas} - I_{t,\langle meas\rangle}\right)}{\sqrt{\sum_{1}^{n}\left(I_{t,meas} - I_{t,\langle meas\rangle}\right)^2}\sqrt{\sum_{1}^{n}\left(I_{t,calc} - I_{t,\langle calc\rangle}\right)^2}}} \tag{199}$$

where $I_{t,\langle meas\rangle}$ is the average of the measured values and $I_{t,\langle calc\rangle}$ is the average of the calculated values, which are given by:

$$I_{t,\langle meas\rangle} = \frac{1}{n}\sum_{1}^{n} I_{t,meas} \; ; \; I_{t,\langle calc\rangle} = \frac{1}{n}\sum_{1}^{n} I_{t,calc} \tag{200}$$

The Mean Percentage Error

The fourth most popular statistical parameter is the Mean Percentage Error (MPE):

$$MBE(\%) = \frac{100}{n}\sum_{1}^{n}\left(\frac{I_{t,calc} - I_{t,meas}}{I_{t,meas}}\right) \tag{201}$$

A percentage error between -10% and $+10\%$ is considered acceptable.

The Nash–Sutcliffe Equation

The fifth most popular statistical parameter is the Nash–Sutcliffe Equation (NSE):

$$NSE = 1 - \frac{\sum_1^n \left(I_{t,calc} - I_{t,meas}\right)^2}{\sum_1^n \left(I_{t,meas} - I_{t,\langle meas\rangle}\right)^2} \tag{202}$$

A model is more efficient when *NSE* is closer to 1 (Chen, Ersi, Yang, Lu and Zhao, 2004).

The t-Statistic Test

The sixth most popular statistical parameter is the t-Statistic Test. The random variable *t* with $n-1$ degrees of freedom may be written here as follows (Bevington, 1969):

$$t = \left[\frac{(n-1)(MBE)^2}{(RMSE)^2 - (MBE)^2}\right]^{0.5} \tag{203}$$

The smaller the value of *t* the better is the performance. To determine whether a model's estimates are statistically significant, one simply has to determine, from standard statistical tables, the critical *t* value, i.e. $t\alpha/2$ at α level of significance and $(n-1)$ degrees of freedom. For the model's estimates to be judged statistically significant at the $(1-\alpha)$ confidence level, the calculated *t* value must be less than the critical value.

Finally, it should be mentioned here that, to evaluate the accuracy of the estimated data from the models described above, some statistical tests to verify the linear relationship between the predicted and measured values should be used. For better data modeling, these statistics should be close to zero, but *r* should approach one as closely as possible. In addition, the *t*-test for the models should be carried out to determine the statistical significance of the predicted values by the models.

TERRESTRIAL SOLAR RADIATION MEASUREMENTS

According to (Coulson, 1975), among the earliest inventions in radiation instrumentation is the electrical compensation pyrheliometer by Knut Ångström in 1899, which is still used as standard for absolute radiant energy determinations in many countries of the world. Later, Anders K. Ångström using the same principle as his predecessor constructed an instrument, what is now known as pyrgeometer, to measure nocturnal long-wave atmospheric radiation. Some examples of the pre-World War II solar radiation measuring instruments under the pyrheliometer and pyranometer categories, respectively, are:

1. Water-flow type, the silver-disk type (currently used for reference), Eppley, Linke-Feussner, Yanishevsky and Michelson pyrheliometers. Modified versions of the latter four are still operational, the basic designs of which were developed prior to 1940.
2. Kimball-Hobbs, Moll-Gorczynski and Robitzsch and Bellani pyranometers. Since 1945, many advances have taken place in instrumentation technology and new improved versions have been introduced. Incorporation of temperature circuits in Eppley type instruments, use of thermopiles

to replace the conventional resistance strips in Ångström-type pyrheliometers, redesign of collimator tube for pyrheliometers, sealing the enclosed air space and installation of black discs at the receiver of pyranometers are to name a few. Computerized data loggers/electronic data acquisition has revolutionized the traditional time-consuming methods of data handling. The general designs of a pyranometer and a pyranometer are shown in Figures 10 and 11 respectively.

Radiometry is the science of electromagnetic radiation measurement. The generic device is named radiometer. The detection of the optical electromagnetic radiation is primarily performed by conversion of the beam's energy in electric signals that subsequently can be measured by conventional techniques. Due to their nearly constant spectral sensitivity for the whole solar spectral range, radiometers equipped with thermal sensors are widely used to measure broadband solar irradiance. Temperature fluctuations (the instruments are placed outdoor and their temperature may vary between -20 and 70 °C), wind, rain, and snow are factors that affect the measurements. The minimization of these perturbations is a difficult task in the engineering of solar radiometers. The pyrheliometer is a radiometer that measures the direct beam irradiance. The pyranometer is a radiometer that uses for measuring the horizontal beam and diffuse irradiances.

Solar radiation measurements can broadly be classified as: ground-based measurements carried out at specific locations on earth and measurements derived from geostationary satellites which measure the energy reflected by the system (earth/atmosphere) in different wavelength bands. Since the input radiation of the models presented in this book is the ground-source, the foregoing literature review only focuses on ground-based measurements.

Figure 10. The schematic of a pyrheliometer

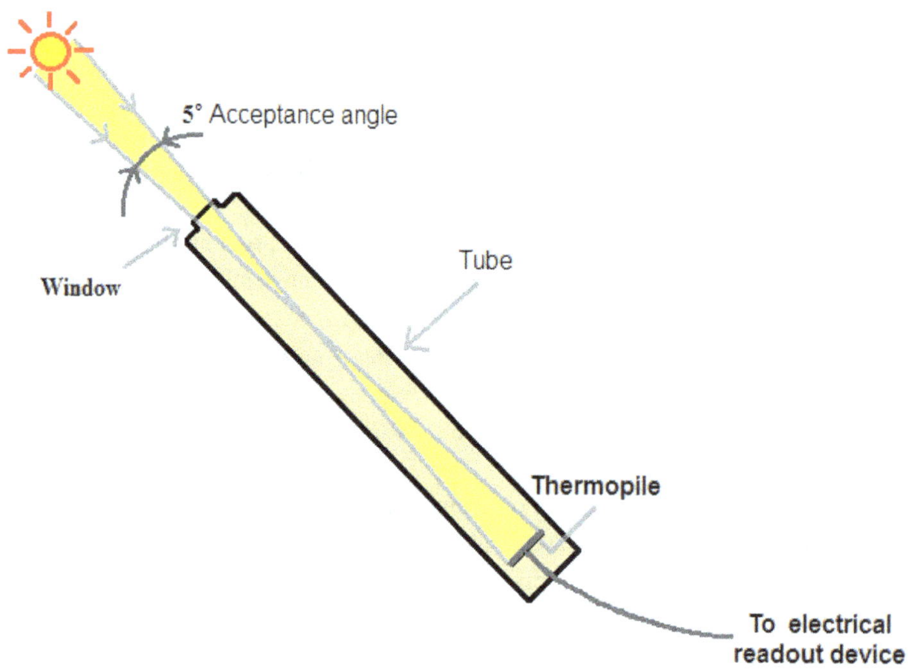

Figure 11. The schematic of a pyranometer

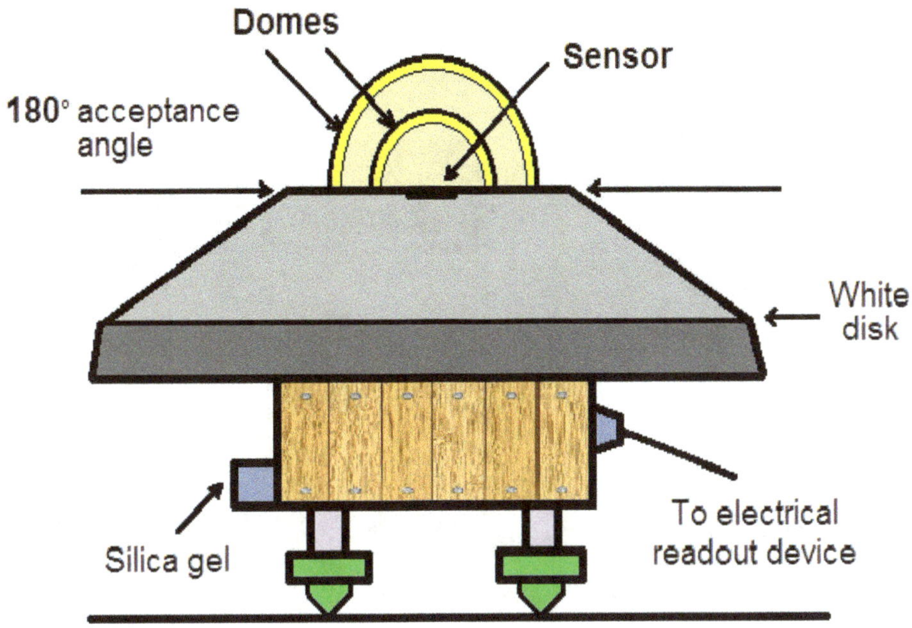

The Pyrheliometer

The pyrheliometer is a broadband instrument that measures the beam solar radiation component (W/m^2) at normal incidence G_{cnb} . Consequently, the instrument should be permanently pointed toward the Sun. A two-axis Sun tracking mechanism is most often used for this purpose. The pyrheliometer (see Figure 12) comprises of a narrow cavity tube known as collimator with an optimum aperture of about 5° to completely include sun's disc in the field of view excluding much of the scatter radiation at the same time. The detector is a fast-response multi-junction thermopile placed inside the collimator at the bottom of a collimating tube (Figure 10) provided with a quartz window to protect the instrument. The detector is coated with optical black paint (acting as a full absorber for solar energy in the wavelengths range 0.280–3 μm). Its temperature is compensated to minimize sensitivity of ambient temperature fluctuations. Consequently, radiation is received from the Sun and a limited circumsolar region, but all diffuse radiation from the rest of the sky is excluded. A readout device is used to give the instant value of the direct beam irradiance. Its scale is adapted to the sensitivity of the particular instrument in order to display the value in SI units, Wm^{-2}. Pyrheliometer is usually attached to a Sun tracking equatorial mount driven by electricity.

Measurement of beam normal irradiance is an expensive affair. The collection of pyrheliometric data can be very expensive not only in terms of equipment costs but also the high level maintenance costs that this type of instrument incurs. According to one estimate, the direct equipment cost of a pyreheliometer itself is almost six times the expense of alternate collection methods, e.g. shaded pyranometer measurements (Muneer, 2004). However, even though used world-wide, lesser expensive alternatives, compromise on the accuracy of radiation measurements compared to that obtained from pyrheliometers.

Figure 12. A photo of a first class pyrheliometer

The Pyranometer

Pyranometers are broadband instruments that measure global solar irradiance (W/m^2) received from the whole hemisphere (from a 2π solid angle) on a planar surface. This hemisphere is usually the complete sky dome. The instrument consists of a flat, blackened thermopile detector mounted on a base covered by two concentric hemispherical transparent covers made of glass. The two domes shield the sensor from thermal convection, protect it against weather threat (rain, wind, and dust) and limit the spectral sensitivity of the instrument in the wavelength range 0.29–2.8 µm. A cartridge of silica gel inside the dome absorbs water vapor and prevents moisture accumulation. Figure 13 is a picture of CM 11 type-pyranometer. A spherical pyranometer or a pyranometer in a tilted position (see Figure 14), to additionally measure the ground reflected radiation, can also be used. Here, the working principle of a pyranometer, in general, and of CM 11, in particular, is given as described by Muneer (2004). The detector responds to the total power, unselective to the spectral distribution of the radiation absorbed. The heat generated by the absorption of radiation by the black disc flows through a thermal resistance to the heat sink. The resultant temperature difference across the thermal resistance of the disc is converted into a voltage, which can be read by computer. Double glass construction of the CM11 minimizes temperature fluctuations from the natural elements and reduces thermal radiation losses to the atmosphere. The glass dome requires periodical cleaning to remove the debris that often gets collected over the time.

The technical Characteristics of pyranometers according to ISO 9060/1990 standard are provided in Table 8.

Figure 13. A photo of CM 11 type-pyranometer mounted on a horizontal surface

Figure 14. A photo of CM 11 type-pyranometer mounted on an inclined surface

Table 8. Characteristics of pyranometers, ISO 9060/1990 standard

ISO Specification	Secondary Standard	First Class
WMO characteristics	High quality	Good quality
Response time (to reach 95% of the final value)	<15 s	<30 s
Zero off-set response: —Response to 200 W/m^2 net radiation	7 Wm^{-2}	15 Wm^{-2}
—Response to 5 °C/h change in ambient temperature	±2 Wm^{-2}	±4 Wm^{-2}
Resolution	±1 Wm^{-2}	±5 Wm^{-2}
Stability (change in sensitivity per year)	±0.8%	±1.5%
Directional response for beam radiation (error due when assuming that the normal incidence response at1000 Wm^{-2} is valid for all directions)	±10 Wm^{-2}	±20 Wm^{-2}
Linearity (deviation from sensitivity at 500 Wm^{-2} over100–1000 Wm^{-2} irradiance range)	±0.5%	±1%
Spectral selectivity (deviation of the product of spectral absorptance and transmittance, respectively, from the mean) • ISO (0.35–1.5 μm)	±3%	±5%
• WMO (0.3–3 μm)	±2%	±5%
Temperature response (maximum relative error due to anychange of ambient temperature within a 50 °C interval)	±2%	±4%
Tilt response (percentage deviation from horizontal response when the tilt is changed from horizontal to vertical at 1000 Wm^{-2})	±0.5%	±2%
Achievable uncertainty, 95% confidence level • WMO hourly totals	3%	8%
• WMO daily totals	2%	5%

The Pyranometer With a Shading Device

A pyranometer can be also used to measure the diffuse solar irradiance G_{cd}, provided that the contribution of the direct beam component is eliminated. For this, a small shading disk can be mounted on an automated solar tracker to ensure that the pyranometer is continuously shaded. Alternatively, a shadow ring may prevent the direct component G_{cb} from reaching the sensor whole day long (see Figure 15). Because the daily maximum Sun elevation angle changes day by day, it is necessary to change periodically (days lag) the height of the shadow ring.

On the other hand, because the shadow ring also intercepts a part of the diffuse radiation, it is necessary to correct the measured values. The percentage of diffuse radiation intercepted by the shadow ring varies during the year with its position and atmospheric conditions (Siren, 1987).

Self-calibrating absolute radiometers (Reda, 1996) are used as primary standard, the other radiometers being calibrated against an absolute instrument. The uncertainty of the measured value depends on factors such as: resolution (the smallest change in the radiation quantity which can be detected by the instrument), nonlinearity of response (the change in sensitivity associated with incident irradiance level), deviation of the directional response (cosine response and azimuth response), time constant of the instrument (time to reach 95% of the final value), changes in sensitivity due to changes of weather variables (such as temperature, humidity, pressure, and wind), long-term drifts of sensitivity (defined as the ratio of electrical output signal to the irradiance applied). All the above uncertainties should be

Figure 15. A photo of CM 11 type-pyranometer with a shading device

known for a well-characterized instrument. Certain instruments perform better for particular climates, irradiances, and solar positions; therefore, the instruments should be selected according to their end use.

Occulting disk on the other hand is Sun-synchronous i.e. tracks along with the Sun and therefore, more expensive. It provides more accurate estimation of diffuse radiation because, it doesn't block off the portion of diffuse radiation as shade ring does.

The Albedometer

Albedometer measures both global solar radiation as well as reflected radiation, yielding albedo value as an output. This instrument is generally installed several meters above the surface to ensure precise measurement.

The Pyrgeometer

Pyrgeometer measures long-wave radiation. The dome is mirrored in a way to reflect maximum possible short-wave radiation from the sun and has filters which allow only infrared radiation (3-50 µm) to pass. It is often used in combination with a shading to ensure the exclusion of direct solar radiation.

The UV Meter

UV meter as the name suggests, measures the ultra-violet part (UV -A: 0.351-0.400 µm or UV-B: 0.280-0.315 µm) of the spectrum measurable at earth's surface.

All the above-listed instruments fall under the category of broadband instruments. Some examples of spectral instruments are: sky scanners, Sun photometers, multi-filter shadow band radiometers, rotating shadow band spectrometers, spectroradiometers, interfreometers and grating spectrometers (Gueymard and Kambezidis, 2004).

Shadow Band Correction

Shadow band for diffuse radiation measurement is an economical option but unfortunately also shades a portion of the sky. This shading induces underestimation of diffuse radiation thus significantly affecting measurement accuracy. It is noted by Muneer (2004) that shading can lead to a maximum error of up to 24% in the true diffuse irradiance value and hence a correction factor needs to be used.

For the case of solar total irradiance measurements, reliable values of its diffuse component on a horizontal surface, I_{dr}, can be obtained from the measurements recorded using a shadow band, I_{dm}, as follows (Batlles, Olmo and Alados-Arboledas, 1995):

$$I_{dr} = C I_{dm} \tag{204}$$

where C is a factor that corrects for the diffuse radiation blocked by the band. For this spectral interval, the correction factor has been widely studied and quantified (Drummond, 1956; Kasten, Dehne and Brettschneider, 1983; Stevens, 1984; LeBaron, Michalsky and Perez, 1990; Kudish and Ianetz; 1993; Batlles, Olmo and Alados-Arboledas, 1995; Muneer and Zhang, 2002).

Drummond (1956) developed a simple model for this correction factor based upon solar geometric calculations, which can be applied anywhere in the world. He calculated the proportion of the sky area that is subtended by the shadow band *f*, from which the correction factor C_D can be deduced by the means of following equation:

$$C_D = \frac{1}{1-f} \tag{205}$$

So, *f*, represents the diffuse irradiance fraction blocked by a shadow band with an axis parallel to the polar axis. This magnitude of C_D is calculated under the assumptions that (1) the band width is small compared to its radius, (2) the sky radiance distribution is isotropic, (3) the sensor is negligible in size compared to the other dimensions, and (4) there is no back-reflection of radiation from the inner surface of the band. Assuming these hypotheses, the expression obtained is:

$$f = \left(\frac{2b}{\pi r}\right)\left(\omega_s sin\varphi sin\delta + cos\varphi cos\delta sin\omega_s\right)cos^3\delta \tag{206}$$

where *r* is the radius of the band, *b* its width, ω_s the hour angle at sunset (in radians), δ the solar declination, and φ the latitude of the location. Drummond concluded that this geometric correction factor is sufficient for uniform or isotropic sky conditions but that an additional correction of up to 7% and 3% should be applied under cloudless and overcast/cloudy sky conditions, respectively. His correction factor, however, does not take into account the circumsolar radiation blocked by the shadow ring, which is by definition anisotropic with respect to the hemisphere as viewed by the horizontal surface. Stevens

(1984) developed a simple model of sky radiance as the sum of a uniform background and circumsolar component. The model proposed by Steven introduces an additional factor that takes into account the anisotropy in the diffuse irradiance distribution at the circumsolar region. The correction factor proposed by this model is:

$$C_S = \frac{I_{dr}}{I_{dm}} = \frac{1}{1 - fQ} \tag{207}$$

where *f* is the fraction of diffuse irradiance occluded by the band given by equation (148), and *Q* is the anisotropy correction factor. For the total solar region and under cloud-free conditions, *Q* is calculated as follows:

$$Q = 1 - C\xi' + \frac{C}{f'} \tag{208}$$

where *C* expresses the relative strength of the circumsolar component, and ξ' is the angular width of the circumsolar region. *C* and ξ' are obtained by linear fitting of equation (150). The function *f'* is defined as follows:

$$f' = \omega_s sin(\varphi) sin(\delta) + cos(\varphi) cos(\delta) sin(\omega_s) \tag{209}$$

Kudish and Ianetz (1993) presented the following review of various studies in literature that consider it necessary to apply an anisotropic correction factor, in addition to the geometric correction factor. Painter (1981) measured diffuse radiation by using an occulting disk and shadow ring pyranometer simultaneously. He observed that the magnitude of the anisotropic correction factor varied significantly with season and therefore correlated it with the declination angle and ratio of the diffuse radiation as measured by the shadow ring to the global radiation. Painter's results, as reported by him, were specific to conditions of his experiment; particular to the shadow ring pyranometer, specific latitude and for the general prevailing state of sky and atmospheric turbidity. Ineichen, Gremaud, Guisan and Mermoud (1984) determined the shadow band correction factor by means of two simple models; one was based on isotropy of diffuse radiation and the other on the diffuse radiation density as a function for solar altitude. The anisotropic correction factor of Kasten, Dehne and Brettschneider (1983) was a linear function of three parameters, namely, the diffuse ratio (obtained after applying isotropic correction to diffuse radiation), the solar declination, and the extinction coefficient for beam transmission. Kudish and Ianetz (1993) applied Steven's model to determine the shadow ring correction factors to diffuse sky radiation measurements at Beer Sheva, Israel. The former found the correction factor under anisotropic conditions to vary between 14 and 30%. The latter authors reported a variation range of 2.9 to 20.9% for the monthly average hourly anisotropic correction factor and that from 5.6 to 14% for geometric correction factor.

LeBaron, Michalsky and Perez (1990) developed a model using four parameters. One of the parameters is Drummond's correction factor and the other three are, zenith angle, and two dimensionless indices for sky's clearness and brightness- one as a function of cloud conditions and the other as a function of cloud thickness or aerosol loading, respectively. Thus, to determine the correction factor the LeBaron,

Michalsky and Perez (1990) model takes into account both isotropic and anisotropic conditions, characterized by the latter three parameter. Battles et al (1995) model is based upon multiple linear regression equations employing the four parameters used in LeBaron, Michalsky and Perez (1990)approach. In Battles model, the values of LeBaron, Michalsky and Perez (1990) sky clearness index determines which equation to be applied for calculating the correction factor i.e., each equation is applied only if this parameter falls within a corresponding given range.

Batlles, Olmo and Alados-Arboledas (1995) proposed two models to correct the diffuse irradiance measured with shadow-bands. The first one, called model Batlles A in this study, proposes the following expression for the correction factor:

$$C_{BA} = aC_D\, blog\Delta + clog\varepsilon + dexp\left(-\frac{1}{\theta_z}\right) \tag{210}$$

where C_D is the isotropy correction proposed by Drummond (1956) in equation (205), ε is related to cloud conditions, Δ is a brightness index which is a function of cloud thickness and aerosol loading, and θ_z is the solar zenith angle. These three parameters account for different sources of anisotropy in the diffuse irradiance. *a*, *b*, *c* and *d* are empirical coefficient that are obtained by regression analysis. The second model, called Batlles B model in this study, based on the dependence of the correction factor with respect to ε. This model consists of a multiple linear model similar to equation (210) for each interval of ε.

Muneer and Zhang (2002) relatively recently developed a model based upon an anisotropic sky-diffuse distribution theory. The sky distribution is two dimensional, a function of any given sky patch geometry (altitude and azimuth) and the position of the sun. The diffuse radiation (H_d) is obtained from the zenith radiance and two parameters corresponding to the radiance distribution indices for the two sky quadrants containing the sun and opposed to the sun. Their correction factor C_{M-Z} is given as:

$$C_{M-Z} = \frac{1}{1 - \frac{f}{I_d}} \tag{211}$$

where, *F* is defined by a complex empirical equation using declination, latitude, solar hour angle and view angle of the shadow ring subtended at the diffuse irradiance sensor. According to Muneer and Zhang (2002) their model represents a compromise between simplicity of an isotropic model and complexity of a two-dimensional model, besides the robust nature and general applicability.

CONCLUSION

Solar radiation information is required in several different forms depending on the kinds of calculations that are to be done. Therefore, after shading a light on the impact of the atmosphere on solar radiation and providing the equations that determine the air mass basing on available theories, the clear sky condition was introduced with shading a light on the previous work in identifying clear sky conditions.

Clear sky solar radiation estimation is of great importance for solar tracking. Therefore, a detailed review of main available models is given in this chapter. Daily, monthly, seasonally and yearly mean daily solar radiations are required information for designing and installing long term tracking systems. We have presented, in this chapter, methods (and commented on their applications) for the estimation of solar radiation information in the desired format from the data that are available. This includes estimation of average global solar radiation on the horizontal plane using sunshine hour duration based methods, ambient air temperature based methods, cloud cover based methods, satellite-based models and others. It includes also estimation of beam and diffuse solar radiation from total solar radiation on the horizontal plane as well as on the tilted surfaces.

Numerous works which deal with the assessment and comparison of monthly mean daily solar radiation estimation models are considered and the instruments that measure total solar radiation and its components are described in this chapter. The provided information in this chapter will be used in the following chapters, mainly in chapter 5.

REFERENCES

Abdalla, Y. A. G., & Feregh, G. M. (1988). Contribution to the study of solar radiation in Abu Dhabi. *Energy Conversion and Management*, *28*(1), 63–67. doi:10.1016/0196-8904(88)90013-1

Abraha, M. G., & Savage, M. J. (2008). Comparison of estimates of daily solar radiation from air temperature range for application in crop simulations. *Agricultural and Forest Meteorology*, *148*(3), 401–416. doi:10.1016/j.agrformet.2007.10.001

Ahmad, F., Burney, S. M. A., & Husain, S. A. (1991). Monthly average daily global beam and diffuse solar radiation and its correlation with hours of bright sunshine for Karachi, Pakistan. *Renewable Energy*, *1*(1), 115–118. doi:10.1016/0960-1481(91)90111-2

Al-Hamdani, N., Al-Riahi, M., & Tahir, K. (1989). Estimation of the diffuse fraction of daily and monthly average global radiation for Fudhaliyah, Baghdad (Iraq). *Solar Energy*, *42*(1), 81–85. doi:10.1016/0038-092X(89)90132-1

Al-Riahi, M., Al-Hamdani, N., & Tahir, K. (1992). An empirical method for estimation of hourly diffuse fraction of global radiation. *Renewable Energy*, 2(4-5), 451–456. doi:10.1016/0960-1481(92)90079-I

Al-Sadah, F. H., Ragab, F. M., & Arshad, M. K. (1990). Hourly solar radiation over Bahrain. *Energy*, *15*(5), 395–402. doi:10.1016/0360-5442(90)90036-2

Alam, S. (2006). Prediction of direct and global solar irradiance using broadband models: Validation of REST model. *Renewable Energy*, *31*(8), 1253–1263. doi:10.1016/j.renene.2005.06.009

Alam, S., Kaushik, S. C., & Garg, S. N. (2009). Assessment of diffuse solar energy under general sky condition using artificial neural network. *Applied Energy*, *86*(4), 554–564. doi:10.1016/j.apenergy.2008.09.004

Allen, R. G. (1997). Self-calibrating method for estimating solar radiation from air temperature. *Journal of Hydrologic Engineering*, 2(2), 56–67.

Almorox, J. (2011). Estimating global solar radiation from common meteorological data in Aranjuez, Spain. *Turkish Journal of Physics*, *35*(1), 53–64. doi:10.3906/fiz-0912-20

Almorox, J., & Hontoria, C. (2004). Global solar radiation estimation using sunshine duration in Spain. *Energy Conversion and Management*, *45*(9-10), 1529–1535. doi:10.1016/j.enconman.2003.08.022

Amato, U., Bartoli, B., Coluzzi, B., Cuomo, V., Serio, C., & Silvestrini, P. (1986). statistical correlation between hourly and daily values of solar radiation on horizontal surface at sea level in the Italian climate. *Revue de Physique Appliquée (Paris)*, *21*(3), 219–227. doi:10.1051/rphysap:01986002103021900

Annandale, J., Jovanovic, N., Benadé, N., & Allen, R. (2002). Software for missing data error analysis of Penman-Monteith reference evapotranspiration. *Irrigation Science*, *21*(2), 57–67. doi:10.1007/s002710100047

ASHRAE handbook: HVAC applications. (1999). Atlanta, GA: ASHRAE.

Atwater, M. A., & Ball, J. T. (1978). A numerical solar radiation model based on standard meteorological observations. *Solar Energy*, *21*(3), 163–170. doi:10.1016/0038-092X(78)90018-X

Atwater, M. A., & Ball, J. T. (1981a). A surface solar radiation model for cloudy atmospheres. *Monthly Weather Review*, *109*(4), 878–888. doi:10.1175/1520-0493(1981)109<0878:ASSRMF>2.0.CO;2

Atwater, M. A., & Ball, J. T. (1981b). Effects of clouds on insolation models. *Solar Energy*, *27*(1), 37–44. doi:10.1016/0038-092X(81)90018-9

Awachie, I. R. N., & Okeke, C. E. (1990). New empirical solar model and its use in predicting global solar irradiation. *Nigerian J. Solar Energy*, *9*, 143–156.

Azadeh, A., Maghsoudi, A., & Sohrabkhani, S. (2009). An integrated artificial neural networks approach for predicting global radiation. *Energy Conversion and Management*, *50*(6), 1497–1505. doi:10.1016/j.enconman.2009.02.019

Badescu, V. (1987). Can the model proposed by barbaro et al. be used to compute global solar radiation on the romanian territory? *Solar Energy*, *38*(4), 247–254. doi:10.1016/0038-092X(87)90046-6

Badescu, V. (1997). Verification of some very simple clear and cloudy sky models to evaluate global solar irradiance. *Solar Energy*, *61*(4), 251–264. doi:10.1016/S0038-092X(97)00057-1

Badescu, V. (2002). 3-D isotropic approximation for solar diffuse irradiance on tilted surfaces. *Renewable Energy*, *26*(2), 221–223. doi:10.1016/S0960-1481(01)00123-9

Badescu, V. (2008). Use of sunshine number for solar irradiance time series generation. In Modeling Solar Radiation at the Earth's Surface. Heidelberg, Germany: Springer-Verlag Berlin Heidelberg. doi:10.1007/978-3-540-77455-6_13

Badescu, V., Gueymard, C. A., Cheval, S., Oprea, C., Baciu, M., Dumitrescu, A., & Rada, C. et al. (2012). Computing global and diffuse solar hourly irradiation on clear sky. Review and testing of 54 models. *Renewable & Sustainable Energy Reviews*, *16*(3), 1636–1656. doi:10.1016/j.rser.2011.12.010

Baig, A., Akhter, P., & Mufti, A. (1991). A novel approach to estimate the clear day global radiation. *Renewable Energy*, *1*(1), 119–123. doi:10.1016/0960-1481(91)90112-3

Baigorria, G. A., Villegas, E. B., Trebejo, I., Carlos, J. F., & Quiroz, R. (2004). Atmospheric transmissivity: Distribution and empirical estimation around the central Andes. *International Journal of Climatology, 24*(9), 1121–1136. doi:10.1002/joc.1060

Bakirci, K. (2009). Models of solar radiation with hours of bright sunshine: A review. *Renewable & Sustainable Energy Reviews, 13*(9), 2580–2588. doi:10.1016/j.rser.2009.07.011

Bandyopadhyay, A., Bhadra, A., Raghuwanshi, N. S., & Singh, R. (2008). Estimation of monthly solar radiation from measured air temperature extremes. *Agricultural and Forest Meteorology, 148*(11), 1707–1718. doi:10.1016/j.agrformet.2008.06.002

Barbaro, S., Cannata, G., Coppolino-Leone, C., & Sinagra, E. (1981). Diffuse solar radiation statistics for Italy. *Solar Energy, 26*(5), 429–435. doi:10.1016/0038-092X(81)90222-X

Bartoli, B., Cuomo, V., Amato, U., Barone, G., & Mattarelli, P. (1982). Diffuse and beam components of daily global radiation in Genoa and Macerata. *Solar Energy, 28*(4), 307–311. doi:10.1016/0038-092X(82)90304-8

Bashahu, M. (2003). Statistical comparison of models for estimating the monthly average daily diffuse radiation at a subtropical African site. *Solar Energy, 75*(1), 43–51. doi:10.1016/S0038-092X(03)00213-5

Batlles, F. J., Olmo, F. J., & Alados-Arboledas, L. (1995). On shadow band correction methods for diffuse irradiance measurements. *Solar Energy, 54*(2), 105–114. doi:10.1016/0038-092X(94)00115-T

Batlles, F. J., Rubio, M. A., Tovar, J., Olmo, F. J., & Alados-Arboledas, L. (2000). Empirical modeling of hourly direct irradiance by means of hourly global irradiance. *Energy, 25*(7), 675–688. doi:10.1016/S0360-5442(00)00007-4

Bendt, P., Collares-Periera, M., & Rabl, A. (1981). The frequency distribution of daily radiation values. *Solar Energy, 27*(1), 1–5. doi:10.1016/0038-092X(81)90013-X

Bennett, I. (1965). Monthly maps of mean daily insolation for the United States. *Solar Energy, 9*(3), 145–158. doi:10.1016/0038-092X(65)90088-5

Bevington, P. R. (1969). *Data reduction and error analysis for the physical sciences* (1st ed.). New York: McGraw Hill Book Co.

Bezir, N. C., Akkurt, I., & Ozek, N. (2010a). Estimation of horizontal solar radiation in Isparta (Turkey). *Energy Sources, 32*(6), 512–517. doi:10.1080/15567030802624056

Bezir, N. C., Akkurt, I., & Ozek, N. (2010b). The development of a computer program for estimating solar radiation. *Energy Sources, 32*(11), 995–1003. doi:10.1080/15567030902937234

Bird, R. E. (1984). A simple, solar spectral model for direct-normal and diffuse horizontal Irradiance. *Solar Energy, 32*(4), 461–471. doi:10.1016/0038-092X(84)90260-3

Bird, R. E., & Hulstrom, R. L. (1981). Review, evaluation, and improvement of direct irradiance models. *Trans. ASME. Journal of Solar Energy Engineering, 103*(3), 182–192. doi:10.1115/1.3266239

Bird, R. E., & Riordan, C. (1986). Simple solar spectral model for direct and diffuse irradiance on horizontal and tilted planes at the Earths surface for cloudless atmospheres. *Journal of Climate and Applied Meteorology*, *25*(1), 87–97. doi:10.1175/1520-0450(1986)025<0087:SSSMFD>2.0.CO;2

Black, J. N., Bonython, G. W., & Prescott, J. A. (1954). Solar radiation and the duration of sunshine. *Quarterly Journal of the Royal Meteorological Society*, *80*(344), 231–235. doi:10.1002/qj.49708034411

Bosch, J. L., Lópe, G., & Batlles, F. J. (2008). Daily solar irradiation estimation over a mountainous area using artificial neural networks. *Renewable Energy*, *33*(7), 1622–1628. doi:10.1016/j.renene.2007.09.012

Bristow, K. L., & Campbell, G. S. (1984). On the relationship between incoming solar radiation and daily maximum and minimum temperature. *Agricultural and Forest Meteorology*, *31*(2), 159–166. doi:10.1016/0168-1923(84)90017-0

Bugler, J. W. (1977). The determination of hourly insolation on an inclined plane using a diffuse irradiance model based on hourly measured global horizontal insolation. *Solar Energy*, *19*(5), 477–491. doi:10.1016/0038-092X(77)90103-7

Calbó, J., González, J.-A., & Pagès, D. (2001). A Method for Sky-Condition Classification from Ground-Based Solar Radiation Measurements. *Journal of Applied Meteorology and Climatology*, *40*(12), 2193–2199. doi:10.1175/1520-0450(2001)040<2193:AMFSCC>2.0.CO;2

Carroll, J. J. (1985). Global transmissivity and diffuse fraction of solar radiation for clear and cloudy skies as measured and as predicted by bulk transmissivity models. *Solar Energy*, *35*(2), 105–118. doi:10.1016/0038-092X(85)90001-5

Castro-Diez, Y., Alados-Arboledas, L., & Jiménez, J. L. (1989). A model for climatological estimations of global, diffuse, and direct solar radiation on a horizontal surface. *Solar Energy*, *42*(5), 417–424. doi:10.1016/0038-092X(89)90060-1

Chandel, S. S., Aggarwl, R. K., & Pandey, A. N. (2002). A new approach to estimate global solar radiation on horizontal surfaces using temperature data. *SESI Journal*, *12*(2), 109–114.

Chandrasekaran, J., & Kumar, S. (1994). Hourly diffuse fraction correlation at a tropical location. *Solar Energy*, *53*(6), 505–510. doi:10.1016/0038-092X(94)90130-T

Chen, R., Ersi, K., Yang, J., Lu, S., & Zhao, W. (2004). Validation of five global radiation models with measured daily data in China. *Energy Conversion and Management*, *45*(11-12), 1759–1769. doi:10.1016/j.enconman.2003.09.019

Chendo, M. A. C., & Maduekwe, A. A. L. (1994). Hourly global and diffuse radiation of Lagos, Nigeria-correlation with some atmospheric parameters. *Solar Energy*, *52*(3), 247–251. doi:10.1016/0038-092X(94)90491-X

Choudhary, N. K. (1963). Solar radiation at New Delhi. *Solar Energy*, *7*(2), 44–52. doi:10.1016/0038-092X(63)90004-5

Clark, D. R., Klein, S. A., & Beckman, W. A. (1983). Algorithm for evaluating the hourly radiation utilizability function. Trans. ASME. *Journal of Solar Energy Engineering*, *105*(3), 281–287. doi:10.1115/1.3266379

Collares-Pereira, M., & Rabl, A. (1979). The average distribution of solar radiation-correlations between diffuse and hemispherical and between daily and hourly insolation values. *Solar Energy*, *22*(2), 155–164. doi:10.1016/0038-092X(79)90100-2

Coulson, K. L. (1975). *Solar and Terrestrial Radiation*. New York: Academic Press.

Cucumo, M., De Rosa, A., Ferraro, V., Kaliakatsos, D., & Marinelli, V. (2008). Correlations of global and diffuse solar luminous efficacy for all sky conditions and comparisons with experimental data of five localities. *Renewable Energy*, *33*(9), 2036–2047. doi:10.1016/j.renene.2007.11.015

Cucumo, M., De Rosa, A., Ferraro, V., Kaliakatsos, D., & Marinelli, V. (2010). Correlations of direct solar luminous efficacy for all sky, clear sky and intermediate sky conditions and comparisons with experimental data of five localities. *Renewable Energy*, *35*(10), 2143–2156. doi:10.1016/j.renene.2010.04.004

Daneshyar, M. (1978). Solar radiation statistics for Iran. *Solar Energy*, *21*(4), 345–349. doi:10.1016/0038-092X(78)90013-0

Davies, J. A., & McKay, D. C. (1982). Estimating solar irradiance and components. *Solar Energy*, *29*(1), 55–64. doi:10.1016/0038-092X(82)90280-8

Davies, J. A., & McKay, D. C. (1989). Evaluation of selected models for estimating solar radiation on horizontal surfaces. *Solar Energy*, *43*(3), 153–168. doi:10.1016/0038-092X(89)90027-3

De Miguel, A., Bilbao, J., Aguiar, R., Kambezidis, H., & Negro, E. (2001). Diffuse solar radiation model evaluation in the North Mediterranean belt area. *Solar Energy*, *70*(2), 143–153. doi:10.1016/S0038-092X(00)00135-3

DeFelice, T. P., & Wylie, B. K. (2001). Sky type discrimination using a ground-based sun photometer. *Atmospheric Research*, *59-60*, 313–329. doi:10.1016/S0169-8095(01)00122-3

Dribssa, E., Cogliani, E., Lavagno, E., & Tetrarca, S. (1999). A modification of the Heliosat method to improve its performance. *Solar Energy*, *65*(6), 369–377. doi:10.1016/S0038-092X(99)00014-6

Driesse, A., & Thevenard, D. (2002). A test of Suehrckes sunshine-radiation relationship using a global data set. *Solar Energy*, *72*(2), 167–175. doi:10.1016/S0038-092X(01)00082-2

Drummond, A. I. (1956). On the measurement of sky radiation. *Arch. Meteor. Geophys. Bioklim.*, *B7*(3-4), 413–436. doi:10.1007/BF02242969

Duchon, C. E., & OMalley, M. S. (1999). Estimating cloud type from pyranometer observations. *Journal of Applied Meteorology and Climatology*, *38*(1), 132–141. doi:10.1175/1520-0450(1999)038<0132:ECTFPO>2.0.CO;2

Duffie, J. A., & Beckman, W. A. (2013). *Solar engineering of thermal processes* (3rd ed.). New York, NY: Wiley & Sons; doi:10.1002/9781118671603

Dupont, J. C., Haeffelin, M., & Long, C. N. (2008). Evaluation of cloudless-sky periods detected by shortwave and longwave algorithms using lidar measurements. *Geophysical Research Letters*, *35*(10), L10803. doi:10.1029/2008GL033658

El-Sebaii, A. A., Al-Hazmi, F. S., Al-Ghamdi, A. A., & Yaghmour, S. J. (2010). Global, direct and diffuse solar radiation on horizontal and tilted surfaces in Jeddah, Saudi Arabia. *Applied Energy*, *87*(2), 568–576. doi:10.1016/j.apenergy.2009.06.032

El-Sebaii, A. A., & Trabea, A. A. (2003). Estimation of horizontal diffuse solar radiation in Egypt. *Energy Conversion and Management*, *44*(15), 2471–2482. doi:10.1016/S0196-8904(03)00004-9

Erbs, D. G., Klein, S. A., & Duffie, J. A. (1982). Estimation of the diffuse radiation fraction for hourly, daily and monthly-average global radiation. *Solar Energy*, *28*(4), 293–302. doi:10.1016/0038-092X(82)90302-4

Erdinc, O., & Uzunoglu, M. (2012). Optimum design of hybrid renewable energy systems: Overview of different approaches. *Renewable & Sustainable Energy Reviews*, *16*(3), 1412–1425. doi:10.1016/j.rser.2011.11.011

Evseev, E. G., & Kudish, A. I. (2009). The assessment of different models to predict the global solar radiation on a surface tilted to the south. *Solar Energy*, *83*(3), 377–388. doi:10.1016/j.solener.2008.08.010

Fadare, D. A. (2009). Modelling of solar energy potential in Nigeria using an artificial neural network model. *Applied Energy*, *86*(9), 1410–1422. doi:10.1016/j.apenergy.2008.12.005

Ghonima, M. S., Urquhart, B., Chow, C. W., Shields, J. E., Cazorla, A., & Kleissl, J. (2012). A method for cloud detection and opacity classification based on ground based sky imagery. *Atmos. Meas. Tech.*, *5*(11), 2881–2892. doi:10.5194/amt-5-2881-2012

Glover, J., & McCulloch, J. S. G. (1958). The empirical relation between solar radiation and hours of sunshine. *Q J Roy. Met. Soc.*, *84*(360), 172–175. doi:10.1002/qj.49708436011

Gopinathan, K. K. (1988). Computing the monthly mean daily diffuse radiation from clearness index and percent possible sunshine. *Solar Energy*, *41*(4), 379–385. doi:10.1016/0038-092X(88)90034-5

Gopinathan, K. K., & Soler, A. (1996). The determination of monthly mean hourly diffuse radiation on horizontal surfaces using equations based on hourly clearness index, sunshine fraction and solar elevation. *Int. J. Solar Energy*, *18*(2), 115–124. doi:10.1080/01425919608914310

Grenier, J. C., De La Casinière, A., & Cabot, T. (1994). A spectral model of Linkes turbidity factor and its experimental implications. *Solar Energy*, *52*(4), 303–313. doi:10.1016/0038-092X(94)90137-6

Grigiante, M., Mottes, F., Zardi, D., & de Franceschi, M. (2011). Experimental solar radiation measurements and their effectiveness in setting up a real-sky irradiance model. *Renewable Energy*, *36*(1), 1–8. doi:10.1016/j.renene.2010.04.039

Gueymard, C. A. (1989). A two-band model for the calculation of clear sky solar irradiance, illuminance, and photosynthetically active radiation at the earths surface. *Solar Energy*, *43*(5), 253–265. doi:10.1016/0038-092X(89)90113-8

Gueymard, C. A. (1993). Critical analysis and performance assessment of clear sky solar irradiance models using theoretical and measured data. *Solar Energy*, *51*(2), 121–138. doi:10.1016/0038-092X(93)90074-X

Gueymard, C. A. (2000). Prediction and Performance Assessment of Mean Hourly Global Radiation. *Solar Energy*, *68*(3), 285–303. doi:10.1016/S0038-092X(99)00070-5

Gueymard, C. A. (2003a). Direct solar transmittance and irradiance predictions with broadband models. Part I: Detailed theoretical performance assessment. *Solar Energy*, *74*(5), 355–379. doi:10.1016/S0038-092X(03)00195-6

Gueymard, C. A. (2003b). Direct solar transmittance and irradiance predictions with broadband models. Part II: Validation with high-quality measurements. *Solar Energy*, *74*(5), 381–395. doi:10.1016/S0038-092X(03)00196-8

Gueymard, C. A. (2004a). Corrigendum to Direct solar transmittance and irradiance predictions with broadband models. Part I: Detailed theoretical performance assessment. *Solar Energy*, *76*(4), 513. doi:10.1016/j.solener.2003.11.002

Gueymard, C. A. (2004b). Corrigendum to Direct solar transmittance and irradiance predictions with broadband models. Part II: Validation with high-quality measurements. *Solar Energy*, *76*(4), 515. doi:10.1016/j.solener.2003.11.003

Gueymard, C. A. (2008). REST2: High-performance solar radiation model for cloudless-sky irradiance, illuminance, and photosynthetically active radiation – Validation with a benchmark dataset. *Solar Energy*, *82*(3), 272–285. doi:10.1016/j.solener.2007.04.008

Gueymard, C. A. (2010). *Progress in direct irradiance modeling and validation.* The Solar 2010 Conference, Phoenix, AZ.

Gueymard, C. A. (2012). Clear-sky irradiance predictions for solar resource mapping and large-scale applications: Improved validation methodology and detailed performance analysis of 18 broadband radiative models. *Solar Energy*, *86*(8), 2145–2169. doi:10.1016/j.solener.2011.11.011

Gueymard, C. A. (n.d.). *REST2 Model.* Retrieved from http://www.solarconsultingservices.com/REST2_model.php

Gueymard, C. A., & Kambezidis, H. D. (2004). Chapter 5: Solar Spectral Radiation. In Solar Radiation and Daylight Models. Oxford, UK: Elsevier.

Gul, M. S., & Muneer, T. (1998). Solar diffuse irradiance: Estimation using air mass and precipitable water data. *Building Servo Eng.*, *19*(2), 79–85. doi:10.1177/014362449801900204

Gul, M. S., Muneer, T., & Kambezidis, H. D. (1998). Models for obtaining solar radiation from other meteorological data. *Solar Energy*, *64*(1-3), 99–108. doi:10.1016/S0038-092X(98)00048-6

Hargreaves, G. H., & Samani, Z. A. (1982). Estimating potential evapotranspiration. *Journal of the Irrigation and Drainage Division*, *108*(3), 225–230.

Harrison, R. G., Chalmers, N., & Hogan, R. J. (2008). Retrospective cloud determinations from surface solar radiation measurements. *Atmospheric Research*, *90*(1), 54–62. doi:10.1016/j.atmosres.2008.04.001

Haurwitz, B. (1945). Insolation in relation to cloudiness and cloud density. *Journal of Meteorology*, *2*(3), 154–166. doi:10.1175/1520-0469(1945)002<0154:IIRTCA>2.0.CO;2

Haurwitz, B. (1946). Insolation in Relation to Cloud Type. *Journal of Meteorology*, *3*(4), 123–124. doi:10.1175/1520-0469(1946)003<0123:IIRTCT>2.0.CO;2

Hay, J. E. (1993). Satellite based estimates of solar irradiance at the earths surface-I. Modelling approaches. *Renewable Energy*, *3*(4/5), 381–393. doi:10.1016/0960-1481(93)90105-P

Hay, J. E., & Davies, J. A. (1980). Calculation of the solar radiation incident on an inclined surface. In *Proc. First Canadian Solar Radiation Data Workshop*. Ministry of Supply and Services Canada.

HelioClim. (2013). *HelioClim Solar Radiation*. Retrieved from <http://www.helioclim.org/linke/linke_helioserve.html>

Hogan, R. J., Jakov, C., & Illingworth, A. J. (2001). Comparison of ECMWF winter-season cloud fraction with radar-derived values. *Journal of Applied Meteorology and Climatology*, *40*(3), 513–525. doi:10.1175/1520-0450(2001)040<0513:COEWSC>2.0.CO;2

Hollands, K. G. T., & Huget, R. G. (1983). A probability density function for the clearness index, with applications. *Solar Energy*, *30*(3), 195–209. doi:10.1016/0038-092X(83)90149-4

Hontoria, L., Aguilera, J., & Zufiria, P. (2002). Generation of hourly irradiation synthetic series using the neural network multilayer perceptron. *Solar Energy*, *72*(5), 441–446. doi:10.1016/S0038-092X(02)00010-5

Hottel, H. C. (1976). A simple model for estimating the transmittance of direct solar radiation through clear atmospheres. *Solar Energy*, *18*(2), 129–134. doi:10.1016/0038-092X(76)90045-1

Hukseflux. (2012). *Radiation measurement sensors*. Available online http://www.hukseflux usa.com/radiation-measurement.php

Ineichen, P. (2006). Comparison of eight clear sky broadband models against 16 independent data banks. *Solar Energy*, *80*(4), 468–478. doi:10.1016/j.solener.2005.04.018

Ineichen, P. (2008). Conversion function between the Linke turbidity and the atmospheric water vapor and aerosol content. *Solar Energy*, *82*(11), 1095–1097. doi:10.1016/j.solener.2008.04.010

Ineichen, P., Gremaud, J. M., Guisan, O., & Mermoud, A. (1984). Study of the corrective factor involved when measuring the diffuse solar radiation by use of the ring method. *Solar Energy*, *32*(5), 585–590. doi:10.1016/0038-092X(84)90133-6

Ineichen, P., & Perez, R. (2002). A new airmass independent formulation for the Linke turbidity coefficient. *Solar Energy*, *73*(3), 151–157. doi:10.1016/S0038-092X(02)00045-2

Iqbal, M. (1978). Estimation of the monthly average of the diffuse component of total insolation on a horizontal surface. *Solar Energy*, *20*(1), 101–105. doi:10.1016/0038-092X(78)90149-4

Iqbal, M. (1979). Correlation of average diffuse and beam radiation with hours of bright sunshine. *Solar Energy*, *23*(2), 169–173. doi:10.1016/0038-092X(79)90118-X

Iqbal, M. (1980). Prediction of hourly diffuse solar radiation from measured hourly global radiation on a horizontal surface. *Solar Energy*, *24*(5), 491–503. doi:10.1016/0038-092X(80)90317-5

Iqbal, M. (1983). An Introduction to Solar Radiation. Academic Press.

Iziomon, M. G., & Mayer, H. (2002). Assessment of some global solar radiation parameterizations. *Journal of Atmospheric and Solar-Terrestrial Physics*, *64*(15), 1631–1643. doi:10.1016/S1364-6826(02)00131-1

Jacovides, C. P., Hadjioannou, L., Pashiardis, S., & Stefanou, L. (1996). On the diffuse fraction of daily and monthly global radiation for the Island of Cyprus. *Solar Energy*, *56*(6), 565–572. doi:10.1016/0038-092X(96)81162-5

Jain, P. C. (1988). Estimation of monthly average hourly global and diffuse irradiation. *Solar & Wind Technology*, *5*(1), 7–14. doi:10.1016/0741-983X(88)90085-9

Janjai, S., Praditwong, P., & Moonin, C. (1996). A new model for computing monthly average daily diffuse radiation for Bangkok. *Renewable Energy*, *9*(1–4), 1283–1286. doi:10.1016/0960-1481(96)88511-9

Kambezidis, H. D. (2008). The meteorological radiation model (MRM): Advancements and Applications. In V. Badescu (Ed.), *Modeling Solar Radiation at the Earth's Surface*. Heidelberg, Germany: Springer-Verlag Berlin Heidelberg. doi:10.1007/978-3-540-77455-6_14

Kang, B. O., & Tam, K. S. (2013). A new characterization and classification method for daily sky conditions based on ground-based solar irradiance measurement data. *Solar Energy*, *94*, 102–118. doi:10.1016/j.solener.2013.04.007

Kaplanis, S., & Kaplain, E. (2007). A model to predict expected mean and stochastic hourly global solar radiation I (h;nj) values. *Renewable Energy*, *32*(8), 1414–1425. doi:10.1016/j.renene.2006.06.014

Kaplanis, S. N. (2006). New methodologies to estimate the hourly global solar radiation: Comparisons with existing models. *Renewable Energy*, *31*(6), 781–790. doi:10.1016/j.renene.2005.04.011

Kasten, F. (1980). A simple parameterization of the pyrheliometric formula for determining the Linke turbidity factor. *Meteorologische Rundschau*, *33*, 124–127.

Kasten, F., & Czeplak, G. (1980). Solar and terrestrial radiation dependent on the amount and type of cloud. *Solar Energy*, *24*(2), 177–189. doi:10.1016/0038-092X(80)90391-6

Kasten, F., Dehne, K., & Brettschneider, W. (1983). Improvement of measurement of diffuse solar radiation. Solar Radiation Data, Ser. F, 2, 221–225.

Kasten, F., & Young, A. T. (1989). Revised optical air mass tables and approximation formula. *Applied Optics*, *28*(22), 4735–4738. doi:10.1364/AO.28.004735 PMID:20555942

Khatib, T. & Elmenreich, W. (2015). A Model for Hourly Solar Radiation Data Generation from Daily Solar Radiation Data Using a Generalized Regression Artificial Neural Network. *International Journal of Photoenergy, 2015*, Article ID 968024. 10.1155/2015/968024

Khatib, T., Mohamed, A., & Sopian, K. (2012). A review of solar energy modeling techniques. *Renewable & Sustainable Energy Reviews*, *16*(5), 2864–2869. doi:10.1016/j.rser.2012.01.064

Klein, S. A., & Theilacker, J. C. (1981). An algorithm for calculating monthly-average radiation on inclined surfaces. *Trans. ASME Journal of Solar Energy Engineering*, *103*(1), 29–33. doi:10.1115/1.3266201

Klucher, T. M. (1979). Evaluation of models to predict insolation on tilted surfaces. *Solar Energy*, *23*(2), 11–114. doi:10.1016/0038-092X(79)90110-5

Koronakis, P. S. (1986). On the choice of angle of tilt for south facing solar collectors in Athens Basin Area. *Solar Energy*, *36*(3), 217–225. doi:10.1016/0038-092X(86)90137-4

Koussa, M., Malek, A., & Haddadi, M. (2009). Statistical comparison of monthly mean hourly and daily diffuse and global solar irradiation models and a Simulink program development for various Algerian climates. *Energy Conversion and Management*, *50*(5), 1227–1235. doi:10.1016/j.enconman.2009.01.035

Kudish, A. I., & Ianetz, A. (1993). Analysis of diffuse radiation data for Beer Sheva: Measured (shadow band) versus calculated (global- horizontal beam) values. *Solar Energy*, *51*(6), 495–503. doi:10.1016/0038-092X(93)90134-A

Lalas, D. P., Petrakis, M., & Papadopoulos, C. (1987). Correlations for the estimation of the diffuse radiation component in Greece. *Solar Energy*, *39*(5), 455–458. doi:10.1016/S0038-092X(87)80065-8

Lam, J. C., & Li, D. H. W. (1996). Correlation between global solar radiation and its direct and diffuse components. *Building and Environment*, *31*(6), 527–535. doi:10.1016/0360-1323(96)00026-1

Lam, J. C., Wan, K. K. W., & Yang, L. (2008). Solar radiation modeling using ANNs for different climates in China. *Energy Conversion and Management*, *49*(5), 1080–1090. doi:10.1016/j.enconman.2007.09.021

Laue, E. G. (1970). The measurement of solar spectral irradiance at different terrestrial elevations. *Solar Energy*, *13*(1), 43–50. doi:10.1016/0038-092X(70)90006-X

LeBaron, B. A., Michalsky, J. J., & Perez, R. (1990). A new simplified procedure for correcting shadow band data for all sky conditions. *Solar Energy*, *44*(5), 249–256. doi:10.1016/0038-092X(90)90053-F

Li, D. H. W., & Lam, J. C. (2001). An analysis of climatic parameters and sky condition classification. *Building and Environment*, *36*(4), 435–445. doi:10.1016/S0360-1323(00)00027-5

Li, D. H. W., Lam, T. N. T., & Chu, V. W. C. (2008). Relationship between the total solar radiation on tilted surfaces and the sunshine hours in Hong Kong. *Solar Energy*, *82*(12), 1220–1228. doi:10.1016/j.solener.2008.06.002

Li, D. H. W. & Tang, H. L. (2008). Standard skies classification in Hong Kong. *Journal of Atmospheric and Solar-Terrestrial Physics, 70*(8-9), 1222-1230. 10.1016/j.jastp.2008.03.004

Lindsey, S. D., & Farnsworth, R. K. (1997). Sources of solar radiation estimates and their effect on daily potential evaporation for use in stream flow modeling. *Journal of Hydrology (Amsterdam)*, *201*(1-4), 348–366. doi:10.1016/S0022-1694(97)00046-2

Linke, F. (1922). Transmissions-Koeffizient und Trubungs faktor. *Beitr. Phys. fr. Atmos*, *10*, 91–103.

Liu, B. Y. H., & Jordan, R. C. (1960). The interrelationship and characteristic distribution of direct, diffuse and total solar radiation. *Solar Energy*, *4*(3), 1–19. doi:10.1016/0038-092X(60)90062-1

Liu, B. Y. H., & Jordan, R. C. (1962). Daily insolation on surfaces tilted toward the Equator. *ASHRAE Journal*, *3*(10), 53–59.

Liu, B. Y. H., & Jordan, R. C. (1963). A rational procedure for predicting the long-term average performance of flat plate solar energy collectors. *Solar Energy*, *7*(2), 53–70. doi:10.1016/0038-092X(63)90006-9

Liu, X., Mei, X., Li, Y., Wang, Q., Jensen, J. R., Zhang, Y., & Porter, J. R. (2009). Evaluation of temperature-based global solar radiation models in China. *Agricultural and Forest Meteorology*, *149*(9), 1433–1446. doi:10.1016/j.agrformet.2009.03.012

Long, C. N., & Ackerman, T. P. (2000). Identification of clear skies from broadband pyranometer measurements and calculation of downwelling shortwave cloud effects. *Journal of Geophysical Research, D, Atmospheres*, *105*(D12), 15609–15626. doi:10.1029/2000JD900077

López, G., Batlles, F. J., & Tovar-Pescador, J. (2007). A new simple parameterization of daily clear-sky global solar radiation including horizon effects. *Energy Conversion and Management*, *48*(1), 226–233. doi:10.1016/j.enconman.2006.04.019

Louche, A., Notton, G., Poggi, P., & Simonnot, G. (1991). Correlations for direct normal and global horizontal irradiations on a French Mediterranean site. *Solar Energy*, *46*(4), 261–266. doi:10.1016/0038-092X(91)90072-5

Louche, A., Peri, G., & Iqbal, M. (1986). An analysis of Linke turbidity factor. *Solar Energy*, *37*(6), 393–396. doi:10.1016/0038-092X(86)90028-9

Loutzenhiser, P. G., Maz, H., Felsmann, C., Strachan, P. A., Frank, T., & Maxwell, G. M. (2007). Empirical validation of models to compute solar irradiance on inclined surfaces for building energy simulation. *Solar Energy*, *81*(2), 254–267. doi:10.1016/j.solener.2006.03.009

Mahmood, R., & Hubbard, K. G. (2002). Effect of time of temperature observation and estimation of daily solar radiation for the Northern Great Plains. *USA Agron. J.*, *94*(4), 723–733. doi:10.2134/agronj2002.7230

Markou, M. T., Bartzokas, A., & Kambezidis, H. D. (2007). A new statistical methodology for classification of sky luminance distributions based on scan data. *Atmospheric Research*, *86*(3-4), 261–277. doi:10.1016/j.atmosres.2007.06.001

Martí, P., & Gasque, M. (2011). Improvement of temperature-based ANN models for solar radiation estimation through exogenous data assistance. *Energy Conversion and Management*, *52*(2), 990–1003. doi:10.1016/j.enconman.2010.08.027

Martínez-Lozano, J. A., Tena, F., Onrubia, J. E., & De La Rubia, J. (1984). The historical evolution of the Ångström formula and its modifications: Review and bibliography. *Agricultural and Forest Meteorology*, *33*(2-3), 109–128. doi:10.1016/0168-1923(84)90064-9

Marty, C., & Philipona, R. (2000). The Clear-Sky Index to separate clear-sky from cloudy-sky situations in climate research. *Geophysical Research Letters*, *27*(17), 2649–2652. doi:10.1029/2000GL011743

Maxwell, E. L. (1987). *A quasi-physical model for converting hourly global horizontal to direct normal insolation.* Report SERI/TR- 215- 3087. Solar Energy Research Institute, Golden, CO.

Maxwell, E. L. (1998). METSTAT — The solar radiation model used in the production of the National Solar Radiation Data Base (NSRDB). *Solar Energy*, *62*(4), 263–279. doi:10.1016/S0038-092X(98)00003-6

Mellit, A., & Kalogirou, S. A. (2008). Artificial intelligence techniques for photovoltaic applications: A review. *Progress in Energy and Combustion Science*, *34*(5), 574–632. doi:10.1016/j.pecs.2008.01.001

Meza, F., & Varas, E. (2000). Estimation of mean monthly solar global radiation as a function of temperature. *Agricultural and Forest Meteorology*, *100*(2-3), 231–241. doi:10.1016/S0168-1923(99)00090-8

Miguel, D., Bilbao, J., Aguiar, R., Kambezidis, H., & Negro, E. (2001). Diffuse solar irradiation model evaluation in the north mediterranean belt area. *Solar Energy*, *70*(2), 143–153. doi:10.1016/S0038-092X(00)00135-3

Mohandes, M., Rehman, S., & Halawani, T. O. (1998). Estimation of global solar radiation using artificial neural networks. *Renewable Energy*, *14*(1–4), 179–184. doi:10.1016/S0960-1481(98)00065-2

Molineaux, B., Ineichen, P., & Delaunay, J. J. (1995). Direct luminous efficacy and atmospheric turbidity: Improving model performance. *Solar Energy*, *55*(2), 125–137. doi:10.1016/0038-092X(95)00035-P

Muneer, T. (2004). *Solar Radiation and Daylight Models* (1st ed.). Oxford, UK: Elsevier.

Muneer, T., Gul, M. S., & Kubie, J. (2000). Models for estimating solar radiation and illuminance from meteorological parameters. *Journal of Solar Energy Engineering*, *122*(3), 146–153. doi:10.1115/1.1313529

Muneer, T., & Hawas, M. (1984). Re-examination of ASHRAE insolation models with reference to Indian locations. *Energy Conversion and Management*, *24*(3), 243–246. doi:10.1016/0196-8904(84)90042-6

Muneer, T., & Saluja, G. S. (1986). Correlation between hourly diffuse and global irradiance for the UK. *Building Serv. Eng.*, *7*(1), 37–43. doi:10.1177/014362448600700106

Muneer, T., & Zhang, X. (2002). A new method for correcting shadow band diffuse irradiance data. *Journal of Solar Energy Engineering*, *124*(1), 34–43. doi:10.1115/1.1435647

Myers, D. R. (2005). Solar radiation modeling and measurements for renewable energy applications: Data and model quality. *Energy*, *30*(9), 1517–1531. doi:10.1016/j.energy.2004.04.034

Newland, F. J. (1989). A study of solar radiation modes for the coastal region of South China. *Solar Energy*, *43*(4), 227–235. doi:10.1016/0038-092X(89)90022-4

Ångström, A. (1924). Solar and terrestrial radiation. *Quarterly Journal of Royal Meteorological Society*, *50*, 121–126.

Nguyen, B. T., & Pryor, T. L. (1997). The relationship between global solar radiation and sunshine duration in Vietnam. *Renewable Energy*, *11*(1), 47–60. doi:10.1016/S0960-1481(96)00122-X

Noorian, M., Moradi, I., & Kamali, G. A. (2008). Evaluation of 12 models to estimate hourly diffuse irradiation on inclined surfaces. *Renewable Energy*, *33*(6), 1406–1412. doi:10.1016/j.renene.2007.06.027

Norris, D. J. (1968). Correlations of solar radiation with clouds. *Solar Energy*, *12*(1), 107–112. doi:10.1016/0038-092X(68)90029-7

Notton, G., Cristofari, C., Muselli, M., Poggi, P., & Heraud, N. (2006). Hourly solar irradiations estimation from horizontal measurements to inclined data. *First International Symposium on Environment Identities and Mediterranean Area, ISEIMA, IEEE Transactions*, 234-239. doi:10.1109/ISEIMA.2006.344958

Oliveira, A. P., Escobedo, J. F., Machado, A. J., & Soares, J. (2002). Correlation models of diffuse solar-radiation applied to the city of Sao Paulo, Brazil. *Applied Energy*, *71*(1), 59–73. doi:10.1016/S0306-2619(01)00040-X

Orgill, J. F., & Hollands, K. G. T. (1977). Correlation equation for hourly diffuse radiation on a horizontal surface. *Solar Energy*, *19*(4), 357–359. doi:10.1016/0038-092X(77)90006-8

Orsini, A., Tomasi, C., Calzolari, F., Nardino, M., Cacciari, A., & Georgiadis, T. (2002). Cloud cover classification through simultaneous ground-based measurements of solar and infrared radiation. *Atmospheric Research*, *61*(4), 251–275. doi:10.1016/S0169-8095(02)00003-0

Page, J. K. (1961). The estimation of monthly mean values of daily total shortwave radiation on vertical and inclined surfaces from sunshine records for latitudes 40°N-40°S. *Proc. UN Conf. On New Sources of Energy*.

Page, J. K. (1986). Prediction of solar radiation on inclined surfaces. In *Solar Energy R&D in the European Community. Solar Radiation Data, Series F-3*. Dordrecht: D. Reidel.

Pages, D., Calbo, J., & Gonzalez, J. A. (2003). Using routine meteorological data to derive sky conditions. *Annales Geophysicae*, *21*(3), 649–654. doi:10.5194/angeo-21-649-2003

Painter, H. E. (1981). The shade ring correction for diffuse irradiance measurements. *Solar Energy*, *26*(4), 361–363. doi:10.1016/0038-092X(81)90182-1

Paliatsos, A. G., Kambezidis, H. D., & Antoniou, A. (2003). Diffuse solar irradiation at a location in the Balkan Peninsula. *Renewable Energy*, *28*(13), 2147–2156. doi:10.1016/S0960-1481(03)00077-6

Paltridge, G. W., & Proctor, D. (1976). Monthly mean solar radiation statistics for Australia. *Solar Energy*, *18*(3), 235–243. doi:10.1016/0038-092X(76)90022-0

Pandey, C. K., & Katiyar, A. K. (2009). A note on diffuse solar radiation on a tilted surface. *Energy*, *34*(11), 1764–1769. doi:10.1016/j.energy.2009.07.006

Pedrós, R., Utrillas, M. P., Martínez-Lozano, J. A., & Tena, F. (1999). Values of broad band turbidity coefficients in a Mediterranean site. *Solar Energy*, *66*(1), 11–20. doi:10.1016/S0038-092X(99)00015-8

Perez, R., Ineichen, P., Moore, K., Kmiecik, M., Chain, C., George, R., & Vignola, F. (2002). A new operational model for satellite-derived irradiances: Description and validation. *Solar Energy*, *73*(5), 307–317. doi:10.1016/S0038-092X(02)00122-6

Perez, R., Ineichen, P., Seals, R., Michalsky, J., & Stewart, R. (1990). Modeling daylight availability and irradiance components from direct and global irradiance. *Solar Energy*, *44*(5), 271–289. doi:10.1016/0038-092X(90)90055-H

Perez, R., Ineichen, P., Seals, R., & Zelenka, A. (1990). Making full use of the clearness index for parameterizing hourly insolation conditions. *Solar Energy*, *45*(2), 111–114. doi:10.1016/0038-092X(90)90036-C

Perez, R., Seals, R., Zelenka, A., & Ineichen, P. (1990). Climatic evaluation of models that predict hourly direct irradiance from hourly global irradiance: Prospects- for performance improvements. *Solar Energy*, *44*(2), 99–108. doi:10.1016/0038-092X(90)90071-J

Perez, R., Stewart, R., Seals, R., & Guertin, T. (1988). *The development and verification of the Perez diffuse radiation model*. Sandia National Contractor Report SAN88-7030.

Prescott, J. A. (1940). Evaporation from water surface in relation to solar Radiation. *Transactions of the Royal Society of South Australia*, *46*, 114–118.

Rahim, R., Baharuddin, M. R., & Mulyadi, R. (2004). Classification of daylight and radiation data into three sky conditions by cloud ratio and sunshine duration. *Energy and Building*, *36*(7), 660–666. doi:10.1016/j.enbuild.2004.01.012

Rahimikhoob, A. (2010). Estimating global solar radiation using artificial neural network and air temperature data in a semi-arid environment. *Renewable Energy*, *35*(9), 2131–2135. doi:10.1016/j.renene.2010.01.029

Rangarajan, S., Swaminathan, M. S., & Mani, A. (1984). Computation of solar radiation from observations of cloud cover. *Solar Energy*, *32*(4), 553–556. doi:10.1016/0038-092X(84)90270-6

Rao, C. R. N., Bradley, W. A., & Lee, T. Y. (1984). The diffuse component of the daily global solar-irradiation at Corvallis, Oregon (USA). *Solar Energy*, *32*(5), 637–641. doi:10.1016/0038-092X(84)90140-3

Rapp-Arraras, I., & Domingo-Santos, J. M. (2011). Functional forms for approximating the relative optical air mass. *Journal of Geophysical Research, D, Atmospheres*, *116*(D24), D24308. doi:10.1029/2011JD016706

Rapti, A. S. (2000). Atmospheric transparency, atmospheric turbidity and climatic parameters. *Solar Energy*, *69*(2), 99–111. doi:10.1016/S0038-092X(00)00053-0

Reda, I. (1996). *Calibration of a solar absolute cavity radiometer with traceability to the world radiometric reference. Tech. Rep. NREL TP-463-20619*. Golden, CO: National Renewable Energy Laboratory. doi:10.2172/15000940

Reindl, D. T., Beckman, W. A., & Duffie, J. A. (1990). Diffuse fraction corrections. *Solar Energy*, *45*(1), 1–7. doi:10.1016/0038-092X(90)90060-P

Remund, J. W. L., Lefevre, M., Ranchin, T., & Page, J. (2003). Worldwide Linke turbidity information. *Proceedings of ISES Solar World Congress*.

Reno, M. J., & Hansen, C. W. (2016). Identification of periods of clear sky irradiance in time series of GHI Measurements. *Renewable Energy*, *90*, 520–531. doi:10.1016/j.renene.2015.12.031

Reno, M. J., Hansen, C. W. & Stein, J. S. (2012). *Global Horizontal Irradiance Clear Sky Models: Implementation and Analysis*. Sandia National Laboratories, Sandia Report, SAND2012-2389.

Revfeim, K. J. A. (1983). An interpretation of the coefficients of the Ångström equation. *Solar Energy*, *31*(4), 415–416. doi:10.1016/0038-092X(83)90141-X

Rietveld, M. R. (1978). A new method for estimating the regression coefficients in the formula relating solar radiation to sunshine. *Agricultural Meteorology*, *19*(2-3), 243–252. doi:10.1016/0002-1571(78)90014-6

Rigollier, C., Bauer, O., & Wald, L. (2000). On the clear sky model of the ESRA –European Solar Radiation Atlas- with respect to the Heliosat method. *Solar Energy*, *68*(1), 33–48. doi:10.1016/S0038-092X(99)00055-9

Rivington, M., Bellocchi, G., Matthews, K. B., & Buchan, K. (2005). Evaluation of three model estimations of solar radiation at 24 UK stations. *Agricultural and Forest Meteorology, 132*(3-4), 228–243. doi:10.1016/j.agrformet.2005.07.013

Robaa, S. M. (2009). Validation of the existing models for estimating global solar radiation over Egypt. *Energy Conversion and Management, 50*(1), 184–193. doi:10.1016/j.enconman.2008.07.005

Robledo, L., & Soler, A. (2000). Luminous efficacy of global solar radiation for clear skies. *Energy Conversion and Management, 41*(16), 1769–1779. doi:10.1016/S0196-8904(00)00019-4

Rodgers, C. D. (1967). *The radiative heat budget of the troposphere and lower stratosphere. Massachusetts Institute of Technology (M.I.T.).*

Sabziparvar, A. A. (2008). A simple formula for estimating global solar radiation in central arid deserts of Iran. *Renewable Energy, 33*(5), 1002–1010. doi:10.1016/j.renene.2007.06.015

Saluja, G. S., & Muneer, T. (1985). Correlation between daily diffuse and global irradiation for the UK. *Building Service Engineering, 6*(3), 103–108. doi:10.1177/014362448500600302

Samani, Z. (2000). Estimating solar radiation and evapotranspiration using minimum climatological data. *Journal of Irrigation and Drainage Engineering, 126*(4), 265–267. doi:10.1061/(ASCE)0733-9437(2000)126:4(265)

Şenkal, O., & Kuleli, T. (2009). Estimation of solar radiation over Turkey using artificial neural network and satellite data. *Applied Energy, 86*(7–8), 1222–1228. doi:10.1016/j.apenergy.2008.06.003

Sherry, J. E., & Justus, C. G. (1984). A simple hourly all-sky solar-radiation model based on meteorological parameters. *Solar Energy, 32*(2), 195–204. doi:10.1016/S0038-092X(84)80036-5

Sidrach-de-Cardona, M., & Carretero, J. (2005). Analysis of the current total harmonic distortion for different single-phase inverters for gridconnected pv-systems. *Solar Energy Materials and Solar Cells, 87*(1-4), 529–540. doi:10.1016/j.solmat.2004.08.016

Singh, O. P., Srivastava, S. K., & Pandey, G. N. (1997). Estimation of hourly global solar radiation in the plane areas of Uttar Pradesh, India. *Energy Conversion and Management, 38*(8), 779–785. doi:10.1016/S0196-8904(96)00088-X

Siren, K. E. (1987). The shadow band correction for diffuse irradiation based on a two component sky radiance model. *Solar Energy, 39*(5), 433–438. doi:10.1016/S0038-092X(87)80062-2

Skartveit, A., & Olseth, J. A. (1987). A model for the diffuse fraction of hourly global radiation. *Solar Energy, 38*(4), 271–274. doi:10.1016/0038-092X(87)90049-1

SoDa. (2011). *Solar Radiation Data Service.* Retrieved from http://www.soda-is.com/eng/index.html

Soler, A. (1990). The dependence of the distribution of the monthly average hourly diffuse radiation on the values of the daily sunshine fraction. *Solar & Wind Technology, 7*(5), 545–547. doi:10.1016/0741-983X(90)90061-6

Soler, A., Gopinathan, K. K., & Robledo, L. (1999). Comparison of monthly mean hourly sunshine fraction estimation techniques from calculated diffuse radiation values. *Renewable Energy*, *17*(2), 227–234. doi:10.1016/S0960-1481(98)00116-5

Soulayman, S. (1985). On the solar radiation dependence on the site's geographic parameters. [English translation of Geliotekhnika]. *Applied Solar Energy*, *5*, 68–71.

Soulayman, S., Ananikyan, N., & Martoyan, G. (2010). *The solar budget of Armenia*. ICRE2010, Damascus, Syria.

Spencer, J. W. (1982). A comparison of methods for estimating hourly diffuse solar-radiation from global solar-radiation. *Solar Energy*, *29*(1), 19–32. doi:10.1016/0038-092X(82)90277-8

Srivastava, R. C. & Pandey, H. (2013). Estimating Angstrom-Prescott coefficients for India and developing a correlation between sunshine hours and global solar radiation for India. *ISRN Renewable Energy*, *2013*, Article ID 403742, 7 pages. 10.1155/2013/403742

Stanhill, G. (1966). Diffuse sky and cloud radiation in Israel. *Solar Energy*, *10*(2), 96–101. doi:10.1016/0038-092X(66)90044-2

Stein, J. S., Hansen, C. W., & Reno, M. J. (2012). *The variability index: a new and novel metric for quantifying irradiance and PV output variability*. World Renewable Energy Forum (WREF), Denver, CO.

Steven, M. D. (1984). The anisotropy of diffuse solar radiation determined from shade-ring measurements. *Quart. J. R. Met. Soc.*, *110*(463), 261–270. doi:10.1002/qj.49711046317

Suehrcke, H. (2000). On the relationship between duration of sunshine and solar radiation on the earths surface: Ångströms equation revisited. *Solar Energy*, *68*(5), 417–425. doi:10.1016/S0038-092X(00)00004-9

Tang, W., Yang, K., He, J., & Qin, J. (2010). Quality control and estimation of global solar radiation in China. *Solar Energy*, *84*(3), 466–475. doi:10.1016/j.solener.2010.01.006

Tiris, M., Tiris, Ç., & Türe, İ. E. (1996). Correlations of monthly-average daily global, diffuse and beam radiations with hours of bright sunshine in Gebze, Turkey. *Energy Conversion and Management*, *37*(9), 1417–1421. doi:10.1016/0196-8904(95)00227-8

Tuller, S. E. (1976). The relationship between diffuse, total and extraterrestrial solar radiation. *Solar Energy*, *18*(3), 259–263. doi:10.1016/0038-092X(76)90025-6

Tymvios, F. S., Jacovides, C. P., Michaelides, S. C., & Scouteli, C. (2005). Comparative study of Ångströms and artificial neural networks methodologies in estimating global solar radiation. *Solar Energy*, *78*(6), 752–762. doi:10.1016/j.solener.2004.09.007

Vignola, F., Harlan, P., Perez, R., & Kmiecik, M. (2005). Analysis of Satellite derived beam and global solar radiation data. *Proceedings of Solar World Congress (ISES and ASES) Conf. Proc.*

Vignola, F., & McDaniels, D. K. (1984). Diffuse-global correlation: Seasonal variations. *Solar Energy*, *33*(5), 397–402. doi:10.1016/0038-092X(84)90191-9

Whillier, A. (1953). *Solar energy collection and its utilization for house heating* (Ph.D. thesis). Mechanical Engineering, M.I.T., Cambridge, MA.

Younes, S., & Muneer, T. (2007). Clear-sky classification procedures and models using a world-wide data-base. *Applied Energy*, *84*(6), 623–645. doi:10.1016/j.apenergy.2007.01.016

Young, A. T. (1994). Air-mass and refraction. *Applied Optics*, *33*(6), 1108–1110. doi:10.1364/AO.33.001108 PMID:20862124

Young, A. T., & Irvine, W. M. (1967). Multicolor photoelectric photometry of the brighter Planets. *The Astronomical Journal*, *72*(8), 945–950. doi:10.1086/110366

Zelenka, A., Perez, R., Seals, R., & Renne, D. (1999). Effective accuracy of the satellite-derived hourly irradiance. *Theoretical and Applied Climatology*, *62*(3), 199–207. doi:10.1007/s007040050084

Chapter 5
Optimum Tilt Angle Determine

ABSTRACT

After treating extraterrestrial and terrestrial solar radiations in the previous chapters, the use of this information in treating an important question regarding the installation of fixed solar systems, namely the tilt and orientation of the solar receivers, becomes possible. There are several rules that guide designers in this field. These rules are called the rules of thumb. There are two rules that are directly related to the subject of this chapter. One of these two rules says that a solar collector should be orientated towards Equator. The other one says that solar collector should have a latitude tilt value. Are these two rules valid all over the world? The present chapter focuses on presenting an algorithm for determining the optimum tilt angle all over the world and for any collector azimuth angle. The Earth surface, located between latitudes 66.45°S and 66.45°N, is divided into 3 characteristic zones. The first zone is the tropical between latitudes 23.45°S and 23.45°N. The second zone is the mid-latitude zone between 23.45°N and 43.45°N and between 23.45°S and 43.45°S. The third zone is the high-latitude zone between 43.45°N and 66.45°N and between 43.45°S and 66.45°S. For each of these zones an adequate method is proposed for calculating the solar collector optimum tilt. Moreover, four simple equations are proposed for predicting daily optimum tilt angle and optimum tilt angle for any number of consecutive days. It is found that the above mentioned rules of thumb are not applicable in the tropical zone while they could be applied with a sufficient accuracy when dealing with fixed installations all over the year in the mid- and high latitude zones.

INTRODUCTION

Due to pollution problems related to fossil fuels, such as being non-renewable, their impact on environment and increasing their price due to uncertainties in future, the global investment to utilize renewable energy sources is rapidly growing. The total global investment in renewable energy from $220 billion in 2010 reached to $257 billion in 2011. Also by the end of 2011 the total renewable energy capacity reached to 1,360 GW with the most contribution belong to hydro power and wind power summing up to 970 GW and 238 GW, respectively.

DOI: 10.4018/978-1-5225-2950-7.ch005

Most countries in the world have realized the need for reduction of gases emission to contrast the adverse global climatic change, encouraging the use of renewable and sustainable sources of energy. Indeed, large quantities of carbon dioxide, nitrogen, and sulfur oxides are emitted in the world by conventional energy sources, which are released to the earth's atmosphere contributing to climate change. Furthermore, the world will soon run out of its conventional energy resources because of the rapid depletion of fossil fuel reserves. This future scenario and the risks associated with CO2 emissions and global warming have increased the interest in renewable energy.

The major renewable energy systems include solar photovoltaic (PVs), solar thermal, wind, biomass, hydroelectric, and geothermal. However, among various renewable energy sources, the photovoltaic technology for power generation is considered well-suited technology, particularly for distributed power generation. Solar panel is the energy conversion fundamental component of PV systems or solar collectors. Solar panels use light energy from the sun to generate electricity through the photovoltaic effect, whereas solar thermal systems generate heat. The amount of electrical power produced from PV systems is related to the amount of solar irradiation projecting on the modules. Hence, the global solar irradiation on tilted surfaces facing in different directions should be considered to estimate thermal and electrical power obtained in architectural planning.

Renewable energy obtained from the sun is very important because of the fact that it is free and environment-friendly. The importance of detailed knowledge of solar radiation received from the sun at a site in the design and selection of solar devices cannot be overstated. Solar energy is the most important clean, free and unending renewable energy source which can be utilized in many parts of the world. The limitation and shortage of fossil fuels and the issues resulted from changes of world environmental and weather conditions have created a good opportunity for solar energy to compete with fossil based fuels. This is more important, in countries with high potential of solar radiation to be benefited from green and clean energy. Utilization of different kinds of solar energy technologies such as solar photovoltaic, concentrating solar thermal power, solar hot water/space heating systems, solar dryers, solar stills and solar ovens are becoming rapidly widespread. Aligned with recent augmented deployment of these technologies, many studies have been undertaken to enhance the performance of such technologies (Khorasanizadeh, Aghaei, Ehteram, Dehghani Yazdeli and Hataminasar, 2014). In order to optimize solar isolation on solar collectors, appropriate method to determine solar tilt angles at any given time is essential to increase the efficiencies of the collectors and that of the devices connected to them (Gunerhan and Hepbasli, 2007). The position of the earth relative to the sun changes with time; the change must be monitored adequately in order to increase the amount of energy being received by solar devices (Gavin, 2007). The magnitude of solar radiation received by a collector is a function of many factors such as location latitude, the declination angle (the angular position of the sun at solar noon with respect to the plane of the equator), tilt angle, the sunrise hour angle and the azimuth angle (Kumarasamy, Tulika, Guddy and Manicam, 2013).

Flat-plate solar surfaces, such as thermal collectors and PV panels, have wide applications in domestic and industrial solar systems. A solar system, like any other system, must perform with the highest possible efficiency. This needs the correct design, manufacturing and installation of different components of the system. As for flat solar collectors, the best possible angle of the radiation incidence depends on their orientation and tilt angle; thus determining the proper installation is very vital in order to improve their performance. In an urban application, the optimum tilt angle can be affected by the surrounding obstacles. Consequently, new concerns such as shading or sky blocking effects have to be taken into consideration. The optimum tilt angle is influenced by different factors such as the latitude, the clearness index, the air pollution and the distribution of the sunny days throughout the year (Skeiker, 2009; Keyanpour-Rad, Haghgou, Bahar and Afshari, 2000; Siraki and Pillay, 2012). According to Benghanem (2011), both the orientation and tilt angles have significant effects on the magnitude of the solar radiation reaching the surface of a collector. Therefore, in order to design a solar energy system utilized by flat collectors in a region, performing solar radiation evaluation is essential for attaining suitable information about these factors. A study of this type has been undertaken by Khorasanizadeh, Mohammadi and Aghaei (2014) for Yazd province of Iran.

The literature provides that solar power supplied by the modules depends on many extrinsic factors, such as insolation levels, temperature, load conditions, and orientation of the panel. The solar radiation is also a function of the nature and extent of cloud cover and of the atmosphere's water vapour content, because solar radiation entering Earth's atmosphere is scattered by atmospheric gases, aerosols, and clouds. Indeed, meteorological parameters used as predictors include the amount and distribution of clouds or other observations such as the fractional sunshine and water content (Iqbal, 1978). Aerosols can either absorb or scatter the radiation and alter the energy balance of Earth, especially under clear skies (Gueymard, 2005).

These parameters, obviously, cannot be modified, whereas other variables can be changed to maximize the solar energy acquired by the panel. In fact, the design of a solar energy module involves complex tradeoffs due to the interaction of several factors such as the characteristics of the solar cells, power supply requirements and power management features of the embedded system, application behavior, inclination, and orientation of the panel. Hence, it is essential to understand and exploit these factors in order to maximize the energy efficiency of a solar module. In particular, for fixed absorber surfaces, solar energy gain is strictly related to the slope and azimuth angles of a solar panel.

The global solar radiation for inclined surfaces can be calculated by the values of direct and diffuse solar radiation on the corresponding horizontal surface. Meteorological data from all parts of the world are needed to know the horizontal global solar radiation, and, for some regions, measured data may only be applied within a radius of about 50 km from weather stations. This circumstance leads to interpolate parameters between stations. Furthermore, spectral irradiance is usually not measured routinely, so that the energy instantaneous production of a solar plant is to rely on appropriate models. Otherwise, the users need accurate computations of the slope and orientation of solar panels in order to maximize the solar energy that is collected by fixed solar panels.

Advantages of fixed solar panels are mainly related to their tolerance to misalignment, as approximately 20% of the incident solar radiation is diffuse light, available at any angle of misalignment with the direct sun. In contrast, the main advantage of tracking systems is to collect solar energy for the longest period of the day with the most accurate alignment as sun's position shifts with the seasons. Indeed, daily solar energy collected was calculated to be 19%–24% higher by a solar PV panel with one axis east-west tracking system than by a fixed system (Vilela, Fraidenraich and Tiba, 2003).

Nevertheless, since solar tracking systems have high operation and maintenance costs and are not always applicable, it is often convenient to set the solar collector at a fixed value of an optimum tilt angle (Elminir, Ghitas, El-Hussainy, Hamid, Beheary and Abdel-Moneim,2006). However, some algorithms for the minimization of the energy loss generated by the driving actuator were proposed (Seme and Štumberger, 2011; Ioniță and Alexandru, 2012).

Thus, the tilt angle of a solar energy system is one of the important parameters for capturing maximum solar radiation falling on the solar panels. This angle is site specific as it depends on the daily, monthly and yearly path of the sun. The accurate determination of the optimum tilt angle for the location of interest is essential for maximum energy production by the system. A number of methods have been used for determining the tilt angle at different locations worldwide.

PREVIOUS WORK IN OPTIMUM TILT ANGLE DETERMINATION

There have been many studies on the optimum tilt and azimuth angles of solar surfaces around the globe, from which references are made here only to some of them. In the previous studies, it was concluded that in the Northern Hemisphere, the optimum orientation is Equator facing and the optimum tilt angle depends only on the latitude. No definite value is accepted by all researchers for the optimum tilt angle. Therefore, several attempts were made to determine, or at least to estimate, optimum tilt angle β_{opt} theoretically and experimentally. For example, Hottel (1954) suggested $\beta_{opt} = \varphi + 20°$, Taybout and Löf (1970) proposed $\beta_{opt} = \varphi + (10 \rightarrow 20°)$, Heywood (1971) concluded that $\beta_{opt} = \varphi - 10°$, Löf and Taybout (1973) proposed $\beta_{opt} = \varphi + (10 \rightarrow 30°)$, Yellott (1973) reported $\beta_{opt} = \varphi \pm 20°$, Kern and Harris (1975) suggested $\varphi + 10°$, Lunde (1980) and Garge (1982) suggested $\beta_{opt} = \varphi \pm 15°$, Lewis (1987) reported $\beta_{opt} = \varphi \pm 8°$, where φ is latitude of the location and where plus and minus signs are used in winter and summer, respectively. Theoretical models for β_{opt} were suggested by Lewis (1987), who considered two different models for β_{opt}

Some other research works suggested two values for optimum tilt angle, one for summer (rainy season in tropical region) and the other for winter (dry season in tropical region): $\varphi \pm 8°$ (Pavlović, Pavlović, Pantić and Kostić, 2010; Lewis, 1987) and $\varphi \pm 5°$ (Pavlović, Pavlović, Pantić and Kostić, 2010; Garp and Gupta, 1978) ('+' is for winter or dry season, while '–' is for summer or rainy season).

The disagreement among these values may be due to two main reasons:

1. Firstly, the different methods of calculation that were used for the determination of the optimum slope value of a solar panel;
2. Secondly, the different empirical models that were considered for the determination of diffuse solar radiance and its link with the amount of global solar radiation.

Kern and Harris (1975) explained the calculations related to the optimum slope angle based on the beam radiation. El-Naggar and Chiou (1986) carried out an investigation for many regions of the world using different techniques to attain the relation between the optimum slope angle and the latitude. Lewis (1987) suggested two theoretical models for determining the optimum angle. In one model, he considered the cloudy and clearness indexes as variables and in another one he used the model of Kern and Harris (1975). El-Kassaby (1988) and El-Kassaby and Hassab (1994) introduced an analytical equation for finding the optimum slope angle in any latitude and showed that it can be integrated to calculate the optimum slope angle for any period of time. Soulayman (1991) by correcting the proposed method of El-Kassaby (1988) presented an analytical method for predicting solar collector optimum tilt angle for south facing solar collectors. Gopinathan (1991) presented the monthly average daily global radiation on surfaces tilted towards the equator and also inclined at various azimuth angles for three locations in the South African region. They found that Maximum energy occurs at an azimuth of 180° (facing equator) at any slope, because these African cities are located at the Southern Hemisphere.

Morcos (1994) developed a mathematical model for calculating the total radiation on a sloped surface. The model was then used to determine the optimum tilt angles for a flat plate collector and the optimum tilt and surface azimuth angles (orientation) for concentrating solar collectors in Assiut, Egypt on a daily basis, as well as for a specified period. These optimum angles were found by searching for the values for which the total radiation on the collector is a maximum for a particular day or a specified period. The results show that changing the tilt angle eight times in a year is necessary to maintain the total radiation on the collector in Assiut near its maximum value. This achieves a yearly gain in total radiation of 6.85% more than the case of a flat plate collector with a fixed slope of 27°, which is equal to the latitude angle of Assiut. Also, it is observed that changing the hourly surface azimuth angle twelve times and the tilt angle six times in a year for a concentrating solar collector maintains the total radiation near its maximum value (found by changing the surface azimuth and tilt angles daily to their optimum values). This achieves a yearly gain in total radiation of 29.18% more than the case of a concentrating solar collector facing south at fixed tilt equal to the latitude angle. Abdulaziz (1994) computed the optimum slope angle for latitudes of 10° to 50° north and concluded that if the collector adjusted by the seasonally optimum angles, a gain of 10% in energy is received compared with the zero slope angle. Oladiran (1995) determined the average global radiation on flat surfaces for three zones in Nigeria. The total radiation was obtained while the surface azimuth angle was varied between 0° and 75° at 15° intervals and he presented the results for three slope angles of the collector surface and found that the mean annual radiation increased for a surface with the slope angle of 10° less than the latitude angle.

Nijegorodov, Devan, Jain and Carlsson (1997) presented 12 equations (one for each month), for determining optimum tilt angle for any location that lies between latitude 60°S to 60°N. Hartley, Martinez-Lozano, Utrillas, Tena and Pedro (1999) calculated the optimum slope angle for Valencia, Spain basing on Hay and Davies (1980) model which is mentioned by the authors that it most accurately reproduces the variation in irradiation on all vertical surfaces. They used Hay and Davies (1980) model to find the hourly variation in the optimum tilt angle for a south-facing solar collector in Valencia, Spain, and also to calculate the yearly average of this angle. This method has been compared with the results provided by another model that uses average monthly values of daily irradiation derived from the same experimental data, to calculate average monthly values of the optimum tilt angle. They showed that the amount of irradiation loss received using the yearly average optimum tilt angle is only 6% when compared with the monthly average tilt angle, and thus using the yearly optimum angle may be preferred because it would involve cheaper equipment and less work to keep tilt angle the same all year round. Information on the Hay and Davies (1980) could be found in Hay and McKay (1985) where a variety of numerical models for calculating the solar irradiance for an inclined surface are described and evaluated using data for Vancouver, B.C., Canada. While all the hourly models have a common approach for calculating the direct component of the solar irradiance there is a variety of methods for calculating the diffuse irradiance based on the portion of the sky hemisphere within the field of view of the surface. A less significant distinction between the models is in the methods used to calculate the amount of radiation received as a result of reflection from adjacent surfaces.

Shariah, Al-Akhras and Al-Omari (2002), by employing the computer program TRNSYS (Transient System Simulation) found the optimum slope angle for a thermosyphon solar water heater installed in northern and southern parts of Jordan, showed that the optimum inclination angle for the maximum solar fraction is about $\varphi + (0 \rightarrow 10°)$ for the northern region (represented by Amman) and about $\varphi + (0 \rightarrow 20°)$ for the southern region (represented by the town of Aqaba). These values are greater than those for maximum solar radiation (which is commonly used as an indicator) at the top of the collector by about 5° to 8°.

Hj Mohd Yakup and Malik (2001) computed the monthly optimum slope angle for Brunei, Darussalam. Their results had significant difference with Nijegorodov equations in some months because the ambient condition of Hartley, Martinez-Lozano, Utrillas, Tena and Pedro (1999) is not the same as assumptions of Nijegorodov, Devan, Jain and Carlsson (1997). Skeiker (2009) studied the issue of optimum slope angle in Syria. Hj Mohd Yakup and Malik (2001) and Skeiker (2009) mentioned that in Northern Hemisphere the best orientation for the collectors is tilting them toward the south and the optimum tilt is only dependent on the latitude. Hussein, Ahmad and El-Ghetany (2004) used the software TRNSYS and showed that, for Cairo, Egypt, the optimum value of yearly maximum output energy (i.e. the yearly output energy at the maximum power point) can be obtained from PV modules oriented facing south with a tilt angle in the range of 20–30°. Besides, the yearly maximum output energy of the PV modules mounted at different tilt angles and orientations is obtained as a fraction of its optimum value at the optimum tilt angle and orientation. Tang and Wu (2004) presented an estimation of the optimal tilt angle for maximizing its energy based on the monthly global and diffuse radiation on a horizontal surface. They employed a mathematical model for the estimation of the optimal tilt angle of a collector and presented a contour map of the optimal tilt angles of the south-facing collectors used for the whole year in China, based on monthly horizontal radiation of 152 places around the country. By comparing the optimal tilt angles of collectors obtained from expected monthly diffuse radiation and that from the actual monthly diffuse radiation they showed that their method gives a good estimation of the optimal tilt angle, except for places with a considerably lower clearness index.

Ulgen (2006) used a mathematical model and determined the monthly, seasonally and yearly optimum slope angles for collectors in Izmir, Turkey. He mentioned that the optimum tilt angle changes between 0° (June) and 61° (December) throughout the year. In winter (December, January, and February) the tilt should be 55.7°, in spring (March, April, and May) 18.3°, in summer (June, July, and August) 4.3°, and in autumn (September, October, and November) 43°. The yearly average of this value was found to be 30.3° and this would be the optimum fixed tilt throughout the year. Elminir, Ghitas, El-Hussainy, Hamid, Beheary and Abdel-Moneim (2006) studied the optimum slope angle theoretically in Helwan, Egypt and compared the results of different mathematical models with experimental results. The results have a little deviation with the experimental results. Gunerhan and Hepbasli (2007) calculated the daily optimum slope angle for Izmir, Turkey and compared the results with the results achieved from equations of Nijegorodov, Devan, Jain and Carlsson (1997), even though, the ambient condition in Izmir is different from Nijegorodov, Devan, Jain and Carlsson (1997) assumptions. They offered for increasing the efficiency of solar collectors, they should be mounted at the monthly average tilt angle. They recommended that if a collector in Izmir is adjusted fixed for the whole year, its optimum slope angle is equal to the latitude. Furthermore, for seasonal fixed adjustment, they reported that the optimal slope angle is +15° for winter and -15° for summer. Calabro (2009) proposed an approach of employing sky radiance models for determining optimal tilt angle values of solar collectors with respect to a set of geographic latitudes was conducted by using a data archive of the daily global solar radiation collected by the Italian Institute of ENEA in the area of southern-Italy (Sicily). The results provided a set of tilt angles, for winter months, that enables a solar panel to absorb the maximum amount of global solar radiation and another set of smaller tilt angle values for summer months as well. The great difference between these two sets suggested to us to plan semi-fixed solar panels whose tilt angle can be changed twice a year. The relative gain in solar radiation collectable should not be negligible.

Skeiker (2009) presented his study aiming to develop an analytical procedure for calculating optimum tilt angle for any period of time. Moreover, he aimed to obtain formulae which require the least number of parameters to determine β_{opt} for any chosen day, latitude in either hemisphere and for any value of surface azimuth angle. This goal was not achieved by him because: a) His procedure is essentially the same procedure of El-Kassaby (1988), and b) His effort to compute the optimum tilt angle for the main Syrian zones is based on a fatal error in his main formulae. He claimed that he developed an equation for calculating the optimum daily slope angle and computed the optimum daily slope angle for some cities in Syria. In fact, he repeated the same equations of El-Kassaby (1988). For detailed comments on Skeiker (2009) work it is advised to read (Soulayman and Sabbagh, 2014) and (Soulayman and Sabbagh, 2015a).

Chenga, Jimenez and M. Lee (2009) analyzed the correlation between the optimal angle for a fixed Building Integrated Photovoltaic (BIPV) system and the latitude of the system's site as measured in degrees according to estimates made with PVSYST 3.41 software. Calculations were made for a BIPV south orientated tilted roof at 20 different locations in 14 countries, ranging from 0° to 85° latitude in the Northern Hemisphere. In order to prove the reliability of using the latitude angle as the angle for the tilted panel, the correlation was made between the performance obtained with the system using the optimal angle and the system with the site's location angle. Results indicate that an average of 98.6% a system's performance with the optimal angle can be obtained using the latitude angle for the tilted panel.

Talebizadeh, Mehrabian and Abdolzadeh (2011) proposed the genetic algorithm technique for determining the surface azimuth and optimum tilt angles of collectors for installation in Iran. An area of Iran (Kerman) is selected to verify the results of this algorithm. The optimum angles and the collector input solar energies for these angles are calculated in hourly, daily, monthly, seasonally and yearly bases respectively. Then, the influence of different combinations of solar radiation components on the optimum slope angle and the energy gain is investigated. The results show that the daily, monthly and yearly optimum surface azimuth angles for receiving the maximum solar energy are zero. Adjusting the collector at the daily optimum slope angle slightly increases the collector input energy compared with the case of monthly optimum slope angle so that the gain of solar energy is almost the same. The results also show that the hourly optimum surface azimuth angle is not zero and mounting the solar collector at the hourly optimum slope and azimuth angles increases the input energy significantly compared with the case of daily optimum angles. It is shown that the optimum slope angles are mostly dependent on the beam solar radiation. Furthermore, the results indicate that the optimum slope angles of solar collector and Photovoltaic panels are almost the same. Talebizadeh, Abdolzadeh and Mehrabian (2011) determined the daily, monthly, seasonally, and yearly optimum slope angles of solar collectors for areas of Iran and presented new correlations to determine the daily, monthly, seasonally, and yearly optimum slope angles of solar collectors for latitudes between 20° to 40° North.

By assuming that the collector surface is always facing toward the equator, Benghanem (2011) determined the optimum slope of flat solar collectors in Madinah, Saudi Arabia. In order to calculate the optimum tilt angle, he used the measured values of daily total and diffuse solar radiations on a horizontal surface. He reported that annual optimum tilt angle is approximately equal to latitude, i.e. 23.5°. Also, he found that the optimum seasonal tilt angle in winter and summer are 37° and 12°, respectively. He reported that the loss of receiving energy is around 8% when adjusting the collector fixed with the annual optimum tilt angle compared with that of a collector adjusted monthly with the optimum tilt angle.

Kaldellis, Kavadias and Zafirakis (2012) mentioned that several remote consumers that could cover their needs on the basis of PV-based stand-alone applications exploiting the high quality local solar potential. Therefore, optimum sizing of such installations involves investigation of the optimum panels' tilt angle, which opposite to grid-connected applications is required to provide year-round energy autonomy rather than maximization of the annual energy yield. Kaldellis, Kavadias and Zafirakis (2012) provided theoretical investigation that based on the validation of the assumption that the optimum angle for such applications coincides with the angle that provides maximum exploitation of solar potential during winter months. The common approach used by researchers has been to calculate the tilt angle which maximizes the amount of solar radiation received by the collector.

The optimization of tilt angles was performed using solar radiation data measured for eight big provinces in Turkey Bakirci (2012). The optimum angle for tilted surfaces varying from 0° to 90° in steps of 1° was calculated by searching for the values of which the daily total solar radiation was at a maximum for a specific period. It was found that the optimum tilt angle changed between 0° and 65° throughout the year in Turkey. It was seen that the optimum tilt angle reached a minimum of 0° in June and July and, the monthly average daily total radiation at this angle was generally at a maximum. In addition, the optimum tilt angle increased during the winter months and reached a mamum in December in all provinces. Likewise, general correlations were developed to estimate the optimum tilt angle of solar collectors used in Turkey and their accuracies were compared on the basis of statistical error tests of Mean Bias Error (MBE), Root Mean Square Error (RMSE), t-statistic (t-stat) and correlation coefficient (r).

Diaz, Ngo, Pascual, Planes and Chua (2014) used a mathematical model to estimate the daily extraterrestrial radiation received by a unit area in order to generate the optimal PV tilt angles in the Philippines. In the past years, computer programs have been used and the results have shown that the optimum yearly tilt angle is almost equal to the latitude. Their result is in a good agreement with those of Gopinathan (1991) and Morcos (1994).

Khorasanizadeh, Mohammadi and Mostafaeipour (2014) developed a horizontal diffuse radiation model for city of Tabass in Iran and determined the optimum tilt angle for south-facing solar surfaces for the fixed monthly, seasonal, semi-yearly and yearly adjustments. Khorasanizadeh, Mohammadi and Mostafaeipour (2014) found that the monthly optimum tilt varies from 0° in June and July up to 64° in December and the yearly optimum tilt is around 32°, which is very close to latitude of Tabass (33.36°). For different adjustments, particularly for a vertically mounted surface, the received monthly mean daily solar radiation components and the annual solar energy gains were calculated and compared. Total yearly extra solar gain for the monthly, seasonal, semi-yearly and yearly optimally adjusted surfaces compared to that of horizontal surface are 23.15%, 21.55%, 21.23% and 13.76%, respectively. The semi-yearly tilt adjustment of 10° for warm period (April–September) and 55° for cold period (October–March) is highly recommended, since it provides almost the same level of annual solar energy gain as those of monthly and seasonal adjustments.

Therefore, it is of the great importance to be able to determine the optimum slope of the collector at any latitude, for any surface azimuth angle, and on any day or any period of the year in this zone. In this context, Soulayman (1991) proposed a general algorithm for calculating β_{opt} for south facing collector. Furthermore, Soulayman and Sabbagh (2015a) proposed an algorithm which allowed the determination of β_{opt} at any latitude, φ, and for any direction (surface azimuth angle, γ). Stanciu and Stanciu (2014) proposed a simple formula for determining the optimum tilt of south facing collector at latitudes from 0° to 80°. Nijegorodov, Devan, Jain and Carlsson (1997) presented 12 equations (one for each month), for determining optimum tilt angle for any location that lies between latitude 60°S to 60°N. Calabrò (2013) proposed an algorithm to calculate the optimum tilt angle of solar panels by means of global horizontal solar radiation data, provided from Earth-based meteorological stations. This mathematical modeling is based on the maximization of the theoretical expression of the global solar irradiation impinging on an inclined surface, with respect to the slope and orientation of the panel and to the solar hour angle. A set of transcendent equations resulted, whose solutions give the optimum tilt and orientation of a solar panel. A simulation was carried out using global horizontal solar radiation data from the European Solar Radiation Atlas and some empirical models of diffuse solar radiation. The optimum tilt angle resulted was related to latitude by a linear regression with significant correlation coefficients. The standard error of the mean values resulted increased significantly with latitude, suggesting that unreliable values can be provided at high latitudes.

The objective of the present chapter is to present a general algorithm for treating B_{opt} over all latitudes and to shed a light on different suggested methods and provided results.

AVAILABLE THEORIES

El-Kassaby Method

El-Kassaby (1988) presented an analytical method for predicting solar collector optimum tilt angle for south facing solar collectors. According to El-Kassaby (1988) the daily optimum tilt angle could be determined as follows:

$$\beta_{opt,d} = \varphi - arctan\left[\omega_s \; sin\left(\delta\right) / \; cos\left(\delta\right) sin\left(\omega_s\right)\right] \tag{1}$$

El-Kassaby (1988) verified the validity of his formula experimentally during the period laid between 22/9 and 21/3. Some comments regarding El-Kassaby (1988) method are given by Soulayman (1991). Later on, Skeiker (2009) used the same formulae of El-Kassaby (1988) for determining the optimum tilt angles in Syria. (Soulayman and Sabbagh, 2014) and (Soulayman and Sabbagh, 2015a) gave detailed comments on the works of Skeiker (2009) and El-Kassaby (1988). Basing on results of equation (1) found that, the monthly $\beta_{opt,m}$ in terms of latitude φ and Julian month number M is given by:

1. From January to March (1≤M≤3):

$$\begin{aligned} \beta_{opt,m} &= 60.00012^o + 1.49986\mathrm{M} + 3.49996M^2 \\ &+\left(\varphi - 30^o\right)\left(0.7901 + 0.01794\ \mathrm{M} + 0.0165\ M^2\right) \end{aligned} \tag{2}$$

2. From April to June (4≤M≤6):

$$\begin{aligned} \beta_{opt,m} &= 216.0786^o - 72.03219^o\mathrm{M} + 6.00312^o M^2 \\ &+\left(\varphi - 40^o\right)\left(1.07515 + 0.11244\ \mathrm{M} - 0.03749\ M^2\right) \end{aligned} \tag{3}$$

3. From July to September (7≤M≤9):

$$\begin{aligned} \beta_{opt,m} &= 29.118311^o - 20.52981^o\mathrm{M} + 2.50186^o M^2 \\ &+\left(\varphi - 50^o\right)\left(-11.17256 + 2.70569\ \mathrm{M} - 0.15035\ M^2\right) \end{aligned} \tag{4}$$

4. From October to December (10≤M≤12):

$$\beta_{opt,m} = -441.2385^o - 84.54332^o \mathrm{M} - 3.50196^o M^2 + \left(\varphi - 40^o\right)\left(4.2137 - 0.54834\mathrm{M} + 0.0223M^2\right) \tag{5}$$

Providing that if $\beta_{opt,m} \leq 0^o$ then $\beta_{opt,m} = 0^o$. The design of solar collector at negative tilt angle is significant between latitude 0° and $\mp 25^o$.

Skeiker (2009) rewrote the equation (1) in the form:

$$\beta_{opt,d} = \varphi - arctan\left[\omega_s \; tan\left(\delta\right) / \; sin\left(\omega_s\right)\right] \tag{1a}$$

and proposed to calculate the monthly $\beta_{opt,m}$ using the following equation:

$$\beta_{opt,m} = \varphi - arctan\left[\frac{\sum_{n_1}^{n_2}\left(\frac{24}{\pi}\right) G_{sc} sin\left(\delta\right)\omega_s}{\sum_{n_1}^{n_2}\left(\frac{24}{\pi}\right) G_{sc} cos\left(\delta\right) sin\left(\omega_s\right)}\right] \tag{6}$$

where n_1 and n_2 are the first and last days of the mth month as counted from January. The yearly optimum tilt angle is 30.56° and energy gains are 28% and 26% at monthly and seasonally tilt angle respectively.

Elsayed Method

Elsayed (1989) presented long-term analyses to predict the optimum tilt angle of an absorber plate at any surface azimuth angle γ. The clearness index, K_t, defined as the ratio of earth's surface global irradiance over the extraterrestrial global irradiance, was introduced as a norm to characterize the optimum tilt of south facing collectors at a given point in time when only the global irradiance is known. The analyses include the effects of number of glass covers, latitude angle, monthly average clearness index K_t, month, and ground reflectivity ρ. The effects of each of these parameters on the optimum tilt of a south-facing surface are studied. Two numerical correlations, for ground reflectivity equal 0.2 and 0.7, are developed to predict the monthly optimum tilt of a surface. The two correlations are used to predict the optimum tilt of a surface over any specified period of time that extends from one month up to several months or a year. The analyses are also extended to predict the optimum tilt angle and azimuth angles of surfaces exposed to shading by surrounding objects. Elsayed (1989) proposed a formula that correlates the optimum tilt angle $\beta_{opt,d}$ in terms of clearness index K_t, latitude φ and day number n as follows:

$$\beta_{opt,m} = f\left(\varphi,\, n, K_t,^3, \rho\right) = \left(6 - 4.8K_t + 0.86K_t^{0.27}\varphi + 0.0021\varphi^2\right) + \left(31K_t^{0.37} + 0.094K_t^{0.46}\varphi + 0.00063K_t^{-0.17}\varphi^2\right) cos\left[\frac{360^o\left(n_{cd} + 11.5\right)}{365}\right] \tag{7}$$

n_{cd} is the characteristic day which is the Julian day of the mean day of each month.

Nijegorodov, Devan, Jain and Carlsson (NDJC) Method

Nijegorodov, Devan, Jain and Carlsson (1997) proposed two atmospheric transmittance models to calculate the diurnal profile of solar radiation intensity. These two, together with six other models proposed by earlier researchers, are used to obtain analytical formulae for the optimum tilt angle for a plane absorber plate at any latitude in either hemisphere. In all cases two sky models, namely the isotropic model and the anisotropic model proposed by Hay and Davies (1980), are used. This gives a total of 16 different analytical formulae for the optimum tilt angle. The optimum slopes calculated from these formulae for Kabul, Afghanistan (φ = 34.5°) for certain days of the year, when the surface's azimuth angle $\gamma = 0^o$ and $\gamma = \pm 45°$, are compared with the experimentally measured optimum slopes for those days. A simple empirical formula which is a function of latitude φ and the day of a year n is also proposed to calculate the optimum slope when $\gamma = 0$. For any given day of the year, the empirical formula reduces to a linear correlation between the optimum tilt and the latitude. The linear correlations for the monthly optimum slopes (corresponding to the Julian days) and latitude are given. The mean year optimum slope and the mean heating season slope for Gaborone, Botswana (φ = −24.5°) are calculated. Nijegorodov, Devan, Jain and Carlsson (1997) presented 12 linear equations (one for each month), for determining optimum tilt angle for any location that lies between latitude 60°S to 60°N. These equations are of the form:

$$\beta_{opt,m} = \mathrm{a}\ \varphi + b \tag{8}$$

where φ is positive in the Northern Hemisphere and negative in the Southern Hemisphere. a and b are constants. They depend on the considered month. The values of a and b are given in the Table 1. $\beta_{opt,m}$ is positive for south facing collector and negative for north facing collector. Therefore, $\beta_{opt,m} \geq 0^o$ for Equator facing collector in the Northern Hemisphere and $\beta_{opt,m} \leq 0^o$ for Equator facing collector in the Southern Hemisphere.

Here it should be mentioned that, the presented 12 equations of Nijegorodov, Devan, Jain and Carlsson (1997) for calculating the monthly optimum slope angle were widely used in subsequent studies for validation of other researchers' results. Nijegorodov, Devan, Jain and Carlsson (1997) used the atmospheric transmittance models to obtain analytical formulae for the optimum angle. The atmospheric transmittance models may not be accurate for all climates. He also used some simplifying assumptions for employing the equations. The equations are therefore not too accurate and have a big deviation from the exact values in some latitudes and especially in some months. In fact, the optimal slope angle is related to the local climatic condition, geographic latitude and the period of its use. Hence, different places will have different optimal tilt angles for a yearly-used solar collector.

Particle Swarm Optimization (PSO) Method

Particle swarm optimization is one of the famous meta- heuristic algorithms proposed firstly by Kennedy and Eberhart (1995), inspired mainly by the social behavior patterns of animals that live and interact within large groups. PSO begins with random population of particles in search space and incor-

Table 1. The coefficients of Nijegorodov, Devan, Jain and Carlsson (1997) equations

Month	a	*b* (°)
January	0.89	29
February	0.97	17
March	1.00	4
April	1.00	-10
May	0.93	-24
June	0.87	-34
July	0.89	-30
August	0.97	-17
September	1.00	-2
October	1.00	12
November	0.93	25
December	0.87	34

porates swarming behaviors such as fish schooling, bird flocking, etc. It stochastically assigns direction and velocity vectors to each particle, each particle then moves or flies through the search space following its velocity vector, which is adjusted by the directions and velocities of other particles in its neighborhood. How much influence a particular particle has on other particles is estimated by its objective function. These localized interactions with neighboring particles propagates through the entire swarm of potential solutions, each particle keeps track of its own coordinates. This process is repeated until some condition is met. On the other hand, the PSO finds the global best solution by simply adjusting the trajectory of each individual particle toward its own best location and toward the best particle of the entire swarm at each iteration (generation). The position and the velocity vector of the ith particle in N dimensional search space can be expressed as $X_i = \left[x_{i1}, x_{i2}, x_{i3}, \ldots, x_{iN}\right]$ and $V_i = \left[v_{i1}, v_{i2}, v_{i3}, \ldots, v_{iN}\right]$ respectively. The amount of solar energy captured by solar collector is set to be the objective function (ε):

$$\varepsilon = max\left(I_t\right) \tag{9}$$

The position of particle represents the candidate solution of the tilt angle (β) of solar collector. Note that the tilt angle (β) is the only searching parameter while doing PSO, other variables such as the study period, geographic latitude, atmospheric parameter and so on are available in this paper. According to the defined objective function, the best position for the ith particle (local best) is $P_i = \left[p_{i1}, p_{i2}, p_{i3}, \ldots, p_{iN}\right]$, and the fittest particle found so far (global best) is $P_g = \left[p_{g1}, p_{g2}, p_{g3}, \ldots, p_{gN}\right]$. The new velocities and positions of the particles for the next fitness evaluation are updated as follows:

$$v_{i,N+1} = v_{iN} + c_1 Rand_1(\ )\left(p_{iN} - x_{iN}\right) + c_2 Rand_2(\ )\left(p_{gN} - x_{iN}\right) \tag{10}$$

$$x_{i,N+1} = x_{iN} + v_{iN} \tag{11}$$

where c_1 and c_2 are the acceleration coefficients representing the cognitive and social parameter respectively, $Rand_1(\ )$ and $Rand_2(\ )$ are random numbers uniformly distributed within the range of [0,1].

Liu Feng-Jiao and Chang Tian-Pau (2015) applied an empirical model suitable for clear sky is applied to calculate the solar radiation for different locations given geographic latitudes. A heuristic algorithm, particle swarm optimization, is adopted to determine the best installation angle of solar collector in Taiwan under clear sky taking into consideration of various time periods. The results show that the particle swarm optimization is a powerful method in optimizing the design of solar collector; the optimal tilt angles are positive for most of the months of year while negative for summer months from May to July. The annual best tilt angles for station Taipei, Taichung, Tainan, Kaohsiung, Hualien and Taitung were found to be 22.4°, 21.5°, 20.5°, 20.2°, 21.3° and 20.5° respectively. Furthermore, it was found that the annual received energy is greater than the one incident upon the ground surface by 6.5%, 6.0%, 5.9%, 5.6%, 6.0% and 5.5% respectively.

Soulayman Method

Soulayman and Sabbagh (2015b), basing on the algorithm, provided by Soulayman (1991), gave a nonlinear algebraic equation for daily optimum tilt $\beta_{opt,d}$ determination:

$$C(n)\left\{\begin{array}{l}\left(\dfrac{\partial A_2}{\partial\beta}\right)\left[\sin\left(\omega_{ss}\right)-\sin\left(\omega_{sr}\right)\right]+A_2\left[\cos\left(\omega_{ss}\right)\left(\dfrac{\partial\omega_{ss}}{\partial\beta}\right)-\cos\left(\omega_{sr}\right)\left(\dfrac{\partial\omega_{sr}}{\partial\beta}\right)\right]+\\ \left(\dfrac{\partial A_1}{\partial\beta}\right)\left(\omega_{ss}-\omega_{sr}\right)+A_1\left(\dfrac{\partial\omega_{ss}}{\partial\beta}-\dfrac{\partial\omega_{sr}}{\partial\beta}\right)-\left(\dfrac{\partial A_3}{\partial\beta}\right)\left[\cos\left(\omega_{ss}\right)-\cos\left(\omega_{sr}\right)\right]\\ +A_3\left[\sin\left(\omega_{ss}\right)\left(\dfrac{\partial\omega_{ss}}{\partial\beta}\right)-\sin\left(\omega_{sr}\right)\left(\dfrac{\partial\omega_{sr}}{\partial\beta}\right)\right]\end{array}\right\}=0 \tag{12}$$

The solution of (23) in relation to the surface tilt β determines $\beta_{opt,d}$. In (23) *C(n)* is the nth day correction factor for Sun-Earth average distance:

$$\mathrm{C}(\mathrm{n}) = 1 + 0.034\cos\left(\frac{2\pi\mathrm{n}}{365}\right) \tag{13}$$

δ is the solar declination angle which could be calculated using the equation of Cooper (1969):

$$\delta = 23.45^{o} \, sin\left[\frac{2\pi\left(n+284\right)}{365}\right] \tag{14}$$

ω_{ss} (rad) is the sunset hour angle on tilted surface:

$$\omega_{ss} = min\left\{arccos\left[-tan\left(\delta\right)tan\left(\varphi\right)\right], arccos\left(-\frac{A_1}{A_4}\right) + arcsin\left(\frac{A_3}{A_4}\right)\right\} \tag{15}$$

ω_{sr} (rad) is the sunrise hour angle on tilted surface:

$$\omega_{sr} = \max\left\{-\arccos\left[-\tan\left(\delta\right)\tan\left(\varphi\right)\right], -\arccos\left(-\frac{A_1}{A_4}\right) + \arcsin\left(\frac{A_3}{A_4}\right)\right\} \tag{16}$$

A_1, A_2, A_3 and A_4 are functions of solar and collector angles:

$$A_1 = \sin\left(\delta\right)\left[\sin\left(\varphi\right)\cos\left(\beta\right) - \sin\left(\beta\right)cos\left(\varphi\right)\cos\left(\gamma\right)\right] \tag{17}$$

$$A_2 = \cos\left(\delta\right)\left[\cos\left(\varphi\right)\cos\left(\beta\right) + \sin\left(\beta\right)sin\left(\varphi\right)\cos\left(\gamma\right)\right] \tag{18}$$

$$A_3 = \cos\left(\delta\right)\sin\left(\beta\right)\sin\left(\gamma\right);\; A_4 = \sqrt{A_2^2 + A_3^2} \tag{19}$$

The analytical solution of Equation (2), in the case of ω_{ss} tilt angle independence, is:

$$\beta_{opt,d} = \varphi - arctan\left[\omega_{ss} tan\left(\delta\right) / \, sin\left(\omega_{ss}\right)\right] \tag{20}$$

As for Equator facing (*EF*) and Pole facing (*PF*) surfaces in both Northern Hemisphere (*NH*) and Southern Hemisphere (*SH*) $\omega_{ss} = -\omega_{sr}$ and sin(γ) =0. So, $A_3 = 0$. ω_{ss} is independent of tilt angle on *PF* surfaces at equator and for *EF* and *PF* surfaces for other latitudes in *NH* and *SH* for the period starting on 22/9 and ending on 21/3 in *NH* and for the period starting on 22/3 and ending on 21/9 in *SH*. For other periods Newton's iteration scheme could be applied for searching $\beta_{opt,d}$.

By analogy, the optimum tilt over any period of consecutive days, $\beta_{opt,p}$ could be derived directly from an equation similar to equation (12):

$$\sum_{n_1}^{n_2} C(n) \left\{ \begin{array}{l} \left(\frac{\partial A_2}{\partial\beta}\right)\left[\sin(\omega_{ss}) - \sin(\omega_{sr})\right] + A_2\left[\cos(\omega_{ss})\left(\frac{\partial\omega_{ss}}{\partial\beta}\right) - \cos(\omega_{sr})\left(\frac{\partial\omega_{sr}}{\partial\beta}\right)\right] + \\ \left(\frac{\partial A_1}{\partial\beta}\right)(\omega_{ss}-\omega_{sr}) + A_1\left(\frac{\partial\omega_{ss}}{\partial\beta} - \frac{\partial\omega_{sr}}{\partial\beta}\right) - \left(\frac{\partial A_3}{\partial\beta}\right)\left[\cos(\omega_{ss}) - \cos(\omega_{sr})\right] \\ +A_3\left[\sin(\omega_{ss})\left(\frac{\partial \omega_{ss}}{\partial\beta}\right) - \sin(\omega_{sr})\left(\frac{\partial\omega_{sr}}{\partial\beta}\right)\right] \end{array} \right\} = 0 \tag{21}$$

where the summation covers the period in consideration, n_1 is the first day of the period while n_2 is its last day. In the case of ω_{ss} tilt angle independence, the analytical solution of Equation (21) is:

$$\beta_{opt,p} = \varphi - arctan\left[\frac{\sum_{n_1}^{n_2} C(n)\,\omega_{ss}\; sin(\delta)}{\sum_{n_1}^{n_2} C(n) cos(\delta) sin(\omega_{ss})}\right] \tag{22}$$

This case takes place on Equator all over the year and in *NH*, for the period starting on 22/9 and ending on 21/3, and in *SH*, for the period starting on 22/3 and ending on 21/9. For other periods Newton's iteration scheme could be applied for searching $\beta_{opt,p}$. When studying the optimum tilt angle for Equator facing collectors Soulayman and Sabbagh (2015c) found that, the obtained monthly, seasonally, biannually and yearly results could be approximated using a set of linear equations on latitude. These equations are given in the Tables 2 and 3.

Later on, Soulayman, Alhelou and Sabbagh (2016) divided the Earth surface, located between latitudes 66.45°S and 66.45°N, into 3 characteristic zones. The first zone is the tropical between latitudes 23.45°S and 23.45°N. The second zone is the mid-latitude zone between 23.45°N and 43.45°N and between 23.45°S and 43.45°S. The third zone is the high-latitude zone between 43.45°N and 66.45°N and between 43.45°S and 66.45°S. For each of these zones an adequate method is proposed for calculating the solar collector optimum tilt. Moreover, four simple equations are proposed for predicting daily optimum tilt angle and optimum tilt angle for any number of consecutive days. The yearly possible energy gain in relation that received by a horizontal surface and latitude tilted surface is also calculated on the basis of daily optimum tilts.

When applying the above mentioned algorithm (Soulayman method) on the mid-latitudes zone (23.45°S<φ<43.45°S and 23.45°N<φ<43.45°N) and analyzing the obtained results as a function of δ and L it was found that, the equation (20) could be applied with a reasonable accuracy for determining $B_{opt,d}$ all over the year. The absolute difference between the results of the equation (20) and precise results, obtained from the equation (12), increases with φ but it does not exceed 2.6°. Therefore, the precise results calculated for mid- and high-latitude zones in NH and SH were fitted in a polynomial form:

$$\beta_{opt,d} = a\ \delta^2 + \mathrm{b}\delta + c \tag{23}$$

where "a", "b" and "c" are functions of φ. These functions are of the following form:

Table 2. Different sets of equations for monthly optimum tilt angle $\beta_{opt,m}$ $(^o)$

Month	Northern Hemisphere	Southern Hemisphere
January	$\beta_{opt,m} = 0.878\,\varphi\ +31.09^o$	$\beta_{opt,m} = \left[\varphi + 36.3^o\right]^-$
February	$\beta_{opt,m} = 0.936\,\varphi\ +20.49^o$	$\beta_{opt,m} = \left[\varphi + 21.50^o\right]^-$
March	$\beta_{opt,m} = 0.992\,\varphi\ +3.656^o$	$\beta_{opt,m} = \left[0.998\varphi + 3.63^o\right]^-$
April	$\beta_{opt,m} = \left[0.996\varphi - 15.14^o\right]^+$	$\beta_{opt,m} = 0.959\,\varphi\ -15.02^o$
May	$\beta_{opt,m} = \left[0.960\varphi - 29.87^o\right]^+$	$\beta_{opt,m} = 0.891\,\varphi\ -28.466^o$
June	$\beta_{opt,m} = \left[1.048\varphi - 44.52^o\right]^+$	$\beta_{opt,m} = 0.858\,\varphi\ -34.10^o$
July	$\beta_{opt,m} = \left[0.945\varphi - 34.00^o\right]^+$	$\beta_{opt,m} = 0.873\,\varphi\ -31.48^o$
August	$\beta_{opt,m} = \left[0.980\varphi - 20.53^o\right]^+$	$\beta_{opt,m} = 0.935\,\varphi\ -20.456^o$
September	$\beta_{opt,m} = \left[0.979\varphi - 2.17^o\right]^+$	$\beta_{opt,m} = 0.993\,\varphi\ -2.996^o$
October	$\beta_{opt,m} = 0.958\,\varphi\ +15.55^o$	$\beta_{opt,m} = \left[\varphi + 16.04^o\right]^-$
November	$\beta_{opt,m} = 0.890\,\varphi\ +28.86^o$	$\beta_{opt,m} = \left[\varphi + 32.60^o\right]^-$
December	$\beta_{opt,m} = 0.860\,\varphi\ +34.10^o$	$\beta_{opt,m} = \left[0.91\varphi + 36.90^o\right]^-$
Latitude range	$\left[0^o, 60^o\right]$	$\left[-60^o, 0^o\right]$

Note: Positive sign over brackets means the value should be taken as zero when the value is negative while negative sign over brackets means the value should be taken as zero when the value is positive.

Table 3. Different sets of equations for monthly optimum tilt angle $\beta_{opt,s}$ (°), $\beta_{opt,b}$ (°), $\beta_{opt,y}$ (°)

Month	Northern Hemisphere	Southern Hemisphere
Winter (22/12-21/3)	$\beta_{opt,s} = 0.873\varphi + 24.77^o$	$\beta_{opt,s} = \left[\varphi + 25.40^o\right]^-$
Spring (22/3-21/6)	$\beta_{opt,s} = \left[\varphi - 24.30^o\right]^+$	$\beta_{opt,s} = 0.879\varphi - 23.75^o$
Summer (22/6-21/9)	$\beta_{opt,s} = \left[0.999\varphi - 24.87^o\right]^+$	$\beta_{opt,s} = 0.875\varphi - 24.27^o$
Autumn (22/9-21/12)	$\beta_{opt,s} = 0.875\varphi + 24.54^o$	$\beta_{opt,s} = \left[\varphi + 25.20^o\right]^-$
First biannual (21/3-20/9)	$\beta_{opt,b} = \left[\varphi - 24.60^o\right]^+$	$\beta_{opt,b} = 0.877\varphi - 24.01^o$
Second biannual (21/9-20/3)	$\beta_{opt,b} = 0.873\varphi + 24.65^o$	$\beta_{opt,b} = \left[\varphi + 25.30^o\right]^-$
year	$\beta_{opt,y} = 0.916\varphi + 1.17^o$	$\beta_{opt,y} = 0.875\varphi - 1.85^o$
Latitude range	$\left[0^o, 60^o\right]$	$\left[-60^o, 0^o\right]$

Note: Positive sign over brackets means the value should be taken as zero when the value is negative while negative sign over brackets means the value should be taken as zero when the value is positive.

$$\mathrm{a} = 0.0000027\left(-1\right)^{\mathrm{i}}\varphi^2 + 0.0001\varphi + 0.01247\left(-1\right)^{\mathrm{i}} \tag{24}$$

$$\mathrm{b} = 0.00007328\varphi^2 + 0.0036486\left(-1\right)^{\mathrm{i}}\varphi - 1.48819 \tag{25}$$

$$\mathrm{c} = 0.0000026783\left(-1\right)^{\mathrm{i}}\varphi^2 + 1.01971\varphi + 0.0104399\left(-1\right)^{\mathrm{i}} \tag{26}$$

where i=0, 1 for SH and NH, respectively. The absolute difference between the results of the equation (23) and precise results, obtained from the equation (12) does not exceed 1.5°.

When integrating the equation (23) for obtaining optimum tilt angle $\beta_{opt,p}$ over any period of time one obtains the following formula:

$$\beta_{opt,p}\left(^{o}\right)=\left[\frac{n_2-n_1}{n_2-n_1+1}\right]$$
$$\left\{c+549.9a[0.5-\left(\frac{365}{4\pi}\right)\cos\left[\left(\frac{2\pi}{365}\right)\left(n_2+n_1+568\right)\right]*\sin\left[\left(\frac{2\pi}{365}\right)\left(n_2-n_1\right)\right]\right\} \quad (27)$$
$$+\left[\frac{23.45\left(\frac{365}{\pi}\right)b}{n_2-n_1+1}\right]\sin\left[\left(\frac{\pi}{365}\right)\left(n_2-n_1\right)\right]*sin\left[\left(\frac{\pi}{365}\right)\left(n_2+n_1+568\right)\right]$$

where n_1 and n_2 are the day numbers of the period beginning and ending respectively.

When applying the equations (21), (22) and (27) in determining $\beta_{opt,p}$ for the latitudes of the mid-latitude zone it was found that, the agreement between their results is of the same order as that for optimum daily tilt angle $\beta_{opt,d}$ in NH and SH. The absolute difference between the results of the equation (27) and precise results, obtained from the equation (21), does not exceed 2.6°.

Direct Calculation Method

Several methods were proposed in this field, where the main contribution of the authors is in modeling the solar radiation on the tilted surface in the studied region as precise as possible and to search the tilt angle that makes solar radiation maximal. It is known that the received solar radiation changes with the time of the year, and orientation and inclination of the receiving surface. So, in each of these models, different factors affecting the amount of received radiation will be investigated. Moreover, some basic assumptions should be given. In this context, the works of Altarawneh, Rawadieh, Tarawneh, Alrowwad and Rimawi (2016) and Jafarkazemi and Saadabadi (2013) are good examples.

In the work of Altarawneh, Rawadieh, Tarawneh, Alrowwad and Rimawi (2016), the assumption of clear sky (cloudless environment conditions) was used. This assumption allows the assessment of the potential of solar energy in Ma'an area. In its turn, the optimal tilt angles for maximizing the collected radiation could be determined. Altarawneh, Rawadieh, Tarawneh, Alrowwad and Rimawi (2016) determined the trajectories of monthly, seasonal, bi-annual, and annual optimal tilt angles using their model for calculating solar irradiation on south facing tilted surfaces. Their results showed that an annual optimal fixed tilt angle of 28.7° would result in 0.20% and 16.1% extra collected radiation when compared with that collected by a surface with the latitude fixed tilt of 30.2° and horizontal surface, respectively. The increases in the yearly average collected radiation were 22.0% for monthly, 20.9% for seasonally, 17.0% for semi-annual, 40.5% for one axis tracking, and 45.9% for two axes tracking compared with that on a horizontal surface.

In the work of Jafarkazemi and Saadabadi (2013), monthly average daily diffuse radiations on horizontal surfaces in Abu Dhabi were calculated based on monthly average daily total radiations on a horizontal surface from the 22-year average data provided by NASA Surface Meteorology and Solar-Energy model (Islam, Alili, Kubo and Ohadi, 2010). Then, total solar radiation on the inclined surface was estimated for different orientations ($-90^o \leq \gamma \leq 90^o$) at different tilt angles ($0^o \leq \beta \leq 90^o$). For this purpose, a Matlab-code based software was developed. Optimum tilt angle for each orientation was characterized by comparing solar radiation at different orientations. Jafarkazemi and Saadabadi (2013) found that, the yearly optimum tilt angle ($\beta_{opt,y} = 22^o$) was almost close to Abu Dhabi latitude ($\varphi = 24.4^o$) and the optimum orientation angle was in the south direction. The results showed that monthly and yearly optimum tilt angles reduced by the diversion of surface orientation from the south. Monthly optimum tilt angle for south-facing surface changes from -9° in June to 52° in December. Based on the calculation results, it is advised to change the tilt angle, at least twice a year.

Proposed Equations for Calculating Optimum Tilt

Ertekin, Evrendilek and Kulcu (2008) quantified optimal tilt angles for solar collectors based on the monthly global and diffuse solar radiation on a horizontal surface across Turkey. The dataset of monthly average daily global solar radiation was obtained from 158 places, and monthly diffuse radiation data were estimated using an empirical model in the related literature. Our results showed that high tilt angles during the autumn (September to November) and winter (December to February) and low tilt angles during the summer (March to August) enabled the solar collector surface to absorb the maximum amount of solar radiation. Monthly optimum tilt angles were estimated devising a sinusoidal function of latitude and day of the year, and their validation resulted in a high R^2 value of 98.8%, with root mean square error (RMSE) of 2.06°. To model monthly optimum tilt angles of the south facing collectors as a function of latitude, and day of the year over Turkey Ertekin, Evrendilek and Kulcu (2008) proposed the following equation:

$$\beta_{opt,m} = 25.521438^o + 26.838291^o \cos\left(-0.017844\varphi + 1.013901 n_{cd} + 7.527742\right) \quad (28)$$

where n_{cd} is the number of the characteristic day of each month in the years. It is evident that the daily optimum tilt in this case should be given using the following equation:

$$\beta_{opt,d} = 25.521438^o + 26.83829^o \cos\left(-0.017844\varphi + 1.01390 n + 7.52774\right) \quad (29)$$

where n is the day number in the year.

Talebizadeh, Mehrabian and Abdolzadeh (2011) determined the daily, monthly, seasonally, and yearly optimum slope angles of solar collectors for areas of Iran, and new correlations are developed to calculate the monthly, seasonally, and yearly optimum slope angles for latitudes of 20° to 40° North. To achieve this purpose, the values of monthly optimum slope angles of different locations in the above range of latitudes are needed. Thus, the slope and surface azimuth angles of solar collectors for receiving maximum solar radiation are determined in some Iranian cities in different days, months, seasons, and the whole year, employing different models. According to the optimum slope angles predicted in this article and using the optimum slope angles achieved by other researchers at locations out of Iran but in the same range of latitudes, the correlations are obtained. The results showed that the optimum azimuth angle is zero for receiving maximum solar energy, so the maximum daily or monthly solar energy is independent of the azimuth angle. In addition, the results of the monthly optimum slope angles achieved using these correlations show more accurate results when compared with reference data. The above mentioned equations are given in the Tables 4 for monthly optimum tilt angles $\beta_{opt,m}$ and 5 for

Table 4. The Talebizadeh, Mehrabian and Abdolzadeh (2011) equations for $\beta_{opt,m}$

Month	The Equation
January	$\beta_{opt,m} = 0.9901\varphi + 24.631^o$
February	$\beta_{opt,m} = 0.6613\varphi + 26.283^o$
March	$\beta_{opt,m} = 1.2657\varphi - 8.6368^o$
April	$\beta_{opt,m} = 0.89\varphi - 11.878^o$
May	$\beta_{opt,m} = 0.381\varphi - 9.3689^o$
June	$\beta_{opt,m} = 0.0235\varphi - 2.9196^o$
July	$\beta_{opt,m} = 0.138\varphi - 4.2233^o$
August	$\beta_{opt,m} = 0.3931\varphi - 0.4064^o$
September	$\beta_{opt,m} = 0.1767\varphi + 23.08^o$
October	$\beta_{opt,m} = 0.6592\varphi + 23.08^o$
November	$\beta_{opt,m} = 0.9975\varphi + 23.192^o$
December	$\beta_{opt,m} = 0.9236\varphi + 29.184^o$

seasonally $\beta_{opt,s}$ and yearly optimum tilt angles $\beta_{opt,y}$. Note that, according to Talebizadeh, Mehrabian and Abdolzadeh (2011), winter consists of January, February, and March; spring consists of April, May, and June; summer consists of July, August, and September; and autumn consists of October, November, and December.

Bakirci (2012) searched the optimum angle for tilted surfaces in the interval expanding from 0° to 90° in steps of 1° by searching for the values of which the daily total solar radiation was at a maximum for a specific period, and approximated his results using three equations:

$$\beta_{opt,d} = 32.619^o - 1.3629\delta \tag{30}$$

$$\beta_{opt,d} = 34.784^o - 1.3621\delta - 0.0081\delta^2 \tag{31}$$

Table 5. The Talebizadeh, Mehrabian and Abdolzadeh (2011) equations for $\beta_{opt,s}$ and $\beta_{opt,y}$

	The Equation
Winter	$\beta_{opt,s} = 1.073\varphi + 10.3^o$
Spring	$\beta_{opt,s} = 0.4885\varphi - 10.27^o$
Summer	$\beta_{opt,s} = 0.2631\varphi + 4.961^o$
Autumn	$\beta_{opt,s} = 0.8966\varphi + 23.81^o$
Year	$\beta_{opt,y} = 0.6804\varphi + 7.203^o$

$$\beta_{opt,d} = 34.783^o - 1.4317\delta - 0.0081\delta^2 + 0.0002\delta^3 \quad (32)$$

Bakirci (2012) used equations (30) to (32) to calculate the monthly optimum tilt angles $\beta_{opt,m}$ by taking the values of the declination angle on characteristic days δ_{cd} as follows:

$$\beta_{opt,m} = 32.619^o - 1.3629\delta_{cd} \quad (33)$$

$$\beta_{opt,m} = 34.784^o - 1.3621\delta_{cd} - 0.0081\delta_{cd}^2 \quad (34)$$

$$\beta_{opt,m} = 34.783^o - 1.4317\delta_{cd} - 0.0081\delta_{cd}^2 + 0.0002\delta_{cd}^3 \quad (35)$$

It is easily to see that the equations (30) to (32) lead to negative results for several days during the months of June and July for some latitudes over Turkish territory. This means that the proposed values are out of the range of tilt angle changing. Moreover, the equations (33) to (35) lead to negative results for June. Bakirci (2012) confirmed that by noting that, it is noticeable that the optimum tilt angle for the month of June is negative; the negative sign determines the orientation of the solar collector, which means that the solar collector is faced towards the north. A positive sign indicates that the solar collector is directed towards the south.

Alsadi, Nassar and Amer (2016) applied ASHRAE clear sky model to calculate the solar radiation; and the fundamental solar energy equations were programmed to determine optimum tilt angles in any location on the earth. The optimum tilt angle was presented in a polynomial form to enable the most convenient use of the function of latitude angle on a Julian day in a monthly, seasonal and an annual manner. According to the comparison between the obtained results with those of the local measured authoritative data and those of NASA published data, it was mentioned that, it can be safely recommended that these polynomials be used especially in mid and high latitudes (>20°) in the two hemispheres. The general form of the polynomial that prescribed the optimum tilt angle is expressed in fifth order three dimensional as:

$$\beta_{opt} = P_{00} + P_{10}x + P_{01}y + P_{20}x^2 + P_{11}xy + P_{02}y^2 + P_{30}x^3 + P_{21}x^2y + P_{12}y^2x + P_{03}y^3 + P_{40}x^4 + P_{31}x^3y + P_{13}y^3x + P_{22}x^2y^2 + P_{04}y^4 + P_{50}x^5 + P_{05}y^5 + P_{41}x^4y + P_{32}x^3y^2 + P_{23}x^2y^3 + P_{14}xy^4 \quad (36)$$

The equation (36) is proposed to calculate daily, seasonally and yearly optimum tilts. In each case, the coefficients $P_{00}, P_{10}, \ldots$, and P_{05} and the variables x and y should be taken to correspond the taken case. The coefficients $P_{00}, P_{10}, \ldots$, and P_{05} and the variables x and y are defined below in the Table 6. Here it should be noted that, for the Southern Hemisphere another arrangement must be taken to estimate the monthly optimum angles; one can use the same polynomial with the same latitude but the months will be reversed. That means December in the Northern Hemisphere will be June in the Southern Hemisphere. The months must be shifted every 6 months after the ordinary calendar. Table 7 described the Julian day n for both Hemispheres for monthly tilt angle optimization, and the optimum tilt angle for Southern Hemisphere is negative sing of the optimum tilt angle for Northern Hemisphere for the same latitude.

Moreover, in literatures, there are two classifications for seasons: Heating and cooling group interesting with two seasons heating season and cooling season. Another classification may be adopted for solar energy group dependent on the position of the sun in the sky, they classified seasons to four seasons in 12 months. Alsadi, Nassar and Amer (2016) adopted the second classification. Each season consists of 91 days, therefore, winter (6th November to 4th February), spring (5th February to 5th May), summer (6th May to 5th August), and autumn (6th August to 5th November).

SOLAR COLLECTOR OPTIMUM TILT DETERMINATION IN THE TROPICAL ZONE

Is the rule of thumb, which says that solar collector should be orientated towards Equator, is valid for tropical region? The present item focuses on determining the optimum tilt in the tropical region as for Equator facing and for Pole oriented collectors. Moreover, two simple equations are proposed: one is for predicting daily optimum tilt angle while the other is for predicting optimum tilt angle for any number of consecutive days which could be used for calculating weekly, fortnightly and monthly optimum tilt angles. The yearly possible energy gain in relation to horizontal surface was also calculated on the basis of daily and monthly optimum tilts. It was found that the rule of thumb, which says that solar collector should be orientated towards the Equator, is not applicable for a number of consecutive days in the year. The value of this number depends on the latitude value but it reaches its maximum value (6 months) for Equator (latitude = 0°). A comparison with available data is provided.

The main objective of the present item is to propose a simple and easy procedure for finding $B_{opt,d}$ and $B_{opt,p}$ at tropical region and to review the most relevant available literature results related to this region.

Table 6. The definition of the coefficients and the independent variables x, y for the offered polynomials

Coefficient	$\beta_{opt,d}$	$\beta_{opt,m}$	$\beta_{opt,s}$	$\beta_{opt,y}$
P_{00}	24.58	29.0	33.89	1.50
P_{10}	0.6836	0.155	0.1616	0
P_{01}	1.238	1.011	1.035	1.35
P_{20}	-2.04×10^{-2}	-1.162×10^{-2}	-1.148×10^{-2}	0
P_{11}	-1.3×10^{-3}	3.4×10^{-3}	3.059×10^{-3}	0
P_{02}	-1.252×10^{-2}	-4.87×10^{-3}	-5.567×10^{-3}	1.069×10^{-2}
P_{30}	1.241×10^{-4}	6.45×10^{-5}	6.406×10^{-5}	0
P_{21}	9.717×10^{-6}	-1.3×10^{-5}	-1.281×10^{-5}	0
P_{12}	1.733×10^{-4}	3.8×10^{-5}	4.633×10^{-5}	0
P_{03}	8.573×10^{-5}	0	0	0
P_{40}	-2.688×10^{-7}	-9.31×10^{-8}	-9.326×10^{-8}	0
P_{31}	-2.558×10^{-8}	1.15×10^{-8}	1.162×10^{-8}	0
P_{22}	-6.578×10^{-7}	-1.11×10^{-7}	-1.252×10^{-7}	0
P_{13}	0	0	0	0
P_{04}	-5.034×10^{-7}	0	0	0
P_{50}	1.879×10^{-10}	0	0	0
P_{41}	1.691×10^{-11}	0	0	0
P_{32}	5.226×10^{-10}	0	0	0
P_{23}	$2.756 \times 10-9$	0	0	0
P_{14}	-8.043×10^{-10}	0	0	0
P_{05}	2.759×10^{-9}	0	0	0
x	n	n	n	0
y	φ	φ	φ	φ

Table 7. Description of Julian day for Northern and Southern Hemispheres (NH and SH) that used in polynomial Alsadi, Nassar and Amer (2016)

Month	Jan.	Feb.	March	April	May	June	July	Aug.	Sep.	Oct.	Nov.	Dec.
n for NH	16	45	74	105	135	166	196	227	258	288	319	350
n for SH	196	227	258	288	319	350	16	45	74	105	135	166

Available Theories for Solar Collector Optimum Tilt Determination

As the goal of this item is to treat this question regarding the tropical region, it is reasonable to restrict ourselves to main available literature concerning directly or indirectly this zone.

In this context, Bari (2000) proposed an algorithm for determining the optimum tilt of solar collector at latitudes 1°, 3°, 5° and 7° in the Malaysian territory. Later on, Bari (2001a) applied this algorithm for determining the optimum tilt of solar collector at latitudes of odd values of interval [5° to 19°] in the Philippines territory. At the same year Bari, Lim and Yu (2001) applied this algorithm for determining the optimum tilt of solar collector at latitudes of even values of interval [4° to 20°] in the Thailand territory. After that, Bari (2001b) applied his algorithm to the low latitude countries.

$$\beta_{opt,d} = a_0 + a_1 n + a_2 n^2 + a_3 n^3 + a_4 n^4 + a_5 n^5 + a_6 n^6 \tag{37}$$

where n is day number in the year and $a_o \ldots a_6$ are constants for each latitude. The values of these coefficients are given in the Table 8. Then, the average optimum tilt angle over several days from n_1 to n_2 may be obtained by integrating the equation (21) according to Bari (2000). By integrating the equation (21) over a period from n_1 to n_2 one obtains:

$$\beta_{opt,n_1-n_2} = \frac{a_0\left(n_2 - n_1\right) + \frac{a_1}{2}\left(n_2^2 - n_1^2\right) + \frac{a_2}{3}\left(n_2^3 - n_1^3\right) + \frac{a_3}{4}\left(n_2^4 - n_1^4\right)}{\left(n_2 - n_1\right)} + \frac{\frac{a_4}{5}\left(n_2^5 - n_1^5\right) + \frac{a_5}{6}\left(n_2^6 - n_1^6\right) + \frac{a_6}{7}\left(n_2^7 - n_1^7\right)}{\left(n_2 - n_1\right)} \tag{38}$$

According to Bari (2000) the accuracy is $\pm 1^o$. Bari (2000) mentioned that, interpolation for other latitudes is also allowed. He provided several examples for demonstrating the results of such interpolation.

Stanciu and Stanciu (2014) proposed a simple formula:

$$\beta_{opt,m} = \varphi - \delta \tag{39}$$

for determining the optimum tilt of solar collector at latitudes from 0° to 80°. Soulayman and Sabbagh (2015a) proposed an algorithm for determining $B_{opt,d}$ and $B_{opt,p}$ at any latitude, φ, and for any direction (surface azimuth angle, γ). They applied this algorithm for determining daily $\beta_{opt,d}$, monthly, $\beta_{opt,m}$ and for any number of days $\beta_{opt,p}$, in the tropical region (Soulayman and Sabbagh, 2015b) and introduced new idea regarding Equator facing, North Pole facing and South Pole facing conditions. Nijegorodov, Devan, Jain and Carlsson (1997) presented 12 equations (one for each month, see Table 1) for determining optimum tilt angle for any location that lies between latitude 60° south to 60° north. Oko and Nnamchi (2012) studied theoretically the optimum tilt angles for the territory of Nigeria (φ=4.86- 13.02°N) and provided also expressions for different optimum tilt angles with respect to the low latitudes. The monthly,

Table 8. The coefficients of Bari polynomial

Latitude $\varphi(^o)$	a_0	$a_1(10^{-3})$	$a_2(10^{-3})$	$a_3(10^{-5})$	$a_4(10^{-8})$	$a_5(10^{-10})$	$a_6(10^{-13})$
1	30.146092	-7.4892138	-6.6664353	2.564799	6.377595	-3.693354	3.825501
3	31.606477	4.5949306	-6.7246877	2.509423	6.909207	-3.838659	3.953538
4	32.594476	8.3103295	-6.8461363	2.595986	6.680662	-3.818319	3.953185
5	33.600463	-0.9629276	-6.5729081	2.231348	8.825556	-4.374669	4.476573
6	34.505679	-4.3112815	-6.4848525	2.165456	9.019478	-4.393123	4.472939
7	35.575855	-16.503088	6.0870657	1.697486	11.5497	-5.034189	5.087468
8	36.628871	-35.179625	-5.614046	1.196697	14.0407	-5.611624	5.591772
9	37.190437	-23.188657	-5.8419331	1.4426260	12.7502	-5.304180	5.323549
10	38.131919	-14.046041	-5.9173448	1.355318	13.75302	-5.623887	5.655031
11	39.034538	-33.459046	-5.3995866	0.8424641	16.1829	-6.178317	6.142052
12	40.110005	-46.261753	-4.9732294	0.3284376	18.84232	-6.794297	6.669735
13	40.681009	-21.617636	-5.4681545	0.7503501	17.09614	-6.443707	6.392753
14	41.71824	-27.192194	-5.3024919	0.5689059	17.97585	-6.637929	6.552741
15	42.752797	-28.693874	-5.3648293	0.7072915	17.00667	-6.357901	6.265473
16	43.629829	-42.489781	-4.7904305	-0.0884811	21.66767	-7.570405	7.424208
17	44.728512	-54.827119	-4.483977	-0.4223136	23.41355	-7.999192	7.820645
18	45.131112	-33.569592	-4.6785222	-0.4465124	24.26167	-8.309209	8.16273
19	46.036917	-18.618458	-5.1373747	0.03745259	21.93921	-7.78666	7.712748
20	47.0618	-18.060321	-5.2963093	0.2792282	20.56008	-7.443239	7.39828

Source: *(see* Bari (2000), Bari, Lim and Yu (2001) *and* Bari (2001a))

seasonal and yearly average daily values of insolation were calculated for tilt angles ranging from 0-40°. Microsoft Excel was used to fit equations on monthly, seasonal and yearly average daily insolation with respect to tilt angles. The calculus method of optimization was employed to establish the optimum tilt angle for low latitudes, 4.86-13.02°N, spanning the territory of Nigeria. Expressions for the optimum tilt angles with respect to the low latitudes were also obtained (see Tables 9 and 10).

Idowu, Olarenwaju and Ifedayo (2013) optimized the collection of solar energy within the period of its availability in order to increase its utilisation and also to enhance performance of heating systems that depend on it through appropriate determination of optimum solar collector tilt angles. Fundamental solar radiation equations were programmed to determine optimum tilt angles in locations within latitudes 1° and 14°. A set of data recorded from a pryanometer located on latitude 6.45° north of the equator was used to generate average monthly radiation over the latitudes. Graphs obtained from latitude 6° and 13°

Table 9. The Oko and Nnamchi (2012) equations for $\beta_{opt,m}$

Month	The Equation
January	$\beta_{opt,m} = 3.1731^o + 1.8629\varphi - 0.04360\varphi^2$
February	$\beta_{opt,m} = 0.0376^o + 1.186\varphi - 0.0069\varphi^2$
March	$\beta_{opt,m} = 2.8649^o + 0.8606\varphi + 0.0087\varphi^2$
April	$\beta_{opt,m} = 5.1027^o + 1.5296\varphi - 0.0271\varphi^2$
May	$\beta_{opt,m} = 3.2078^o + 1.8402\varphi - 0.04148\varphi^2$
June	$\beta_{opt,m} = 1.3594^o + 1.3692\varphi - 0.0131\varphi^2$
July	$\beta_{opt,m} = 0.7960^o + 0.5101\varphi + 0.0305\varphi^2$
August	$\beta_{opt,m} = 2.4705^o + 0.6925\varphi + 0.0198\varphi^2$
September	$\beta_{opt,m} = 8.1841^o + 2.6203\varphi - 0.0881\varphi^2$
October	$\beta_{opt,m} = 2.2886^o + 1.6459\varphi - 0.0304\varphi^2$
November	$\beta_{opt,m} = 0.8236^o + 0.6640\varphi + 0.0227\varphi^2$
December	$\beta_{opt,m} = 2.2578^o + 0.7214\varphi + 0.0164\varphi^2$

Table 10. The Oko and Nnamchi (2012) equations for $\beta_{opt,s}$ and $\beta_{opt,y}$

	The Equation
Harmattan	$\beta_{opt,s} = 3.9743^o + 1.6770\varphi - 0.0374\varphi^2$
Rain	$\beta_{opt,s} = 2.6136^o + 1.3634\varphi - 0.0156\varphi^2$
Year	$\beta_{opt,y} = 2.9489^o + 1.4050\varphi + 0.0190\varphi^2$

data were analyzed to investigate solar radiation on some tilt angles. The optimum tilt angles for solar heating for periodic tracking of the sun in the region within latitudes 1° and 14° were predicted as φ + 25° for November, December and January; φ + 15° for February, September and October; φ – 15° for August; φ – 25° for May, June and July; and φ for March and April.

Soulayman and Sabbagh (2016) used their results (Soulayman and Sabbagh, 2015b) in calculating the optimum tilt angle in the tropical zone. When applying the above mentioned algorithm on the latitudes of the tropical region and analyzing the obtained results as a function of δ and φ it was found that $\beta_{opt,d}$ (°) can be calculated using the following equation:

$$\beta_{opt,d} = -1.457\delta + 0.936\varphi + 0.055 \tag{40}$$

Moreover, it was found that the equation (40) could be applied with a high accuracy for determining $B_{opt,d}$ all over the year. The absolute difference between the results of the equation (40) and precise results does not exceed 0.5°.

By integrating the equation (40) over several consecutive days or any number of days, starting from day number n_1 and ending on day number n_2 inclusively, the average optimum tilt angle over this period multiplied by the number of days in the studied period may be obtained. Then, dividing the obtained result by the number of days included in the studied period, the optimum tilt over this period, β_{opt,n_1-n_2} is determined. So, β_{opt,n_1-n_2} could be determined using the following equation:

$$\beta_{opt,n_1-n_2} = 0.936\varphi + 0.055 - \frac{3969.6}{n_2 - n_1 + 1}\sin\left[\frac{\pi\left(N1 + n_2 + 568\right)}{365}\right]\sin\left[\frac{\pi\left(n_2 - n_1\right)}{365}\right] \tag{41}$$

Moreover, it was found that the equation (41) could be applied with a high accuracy for determining β_{opt,n_1-n_2} all over the year. The absolute difference between the results of the equation (40) and precise results does not exceed 0.5°.

On the other hand, When applying the above mentioned Soulayman algorithm on the latitudes of the tropical region (23.5°S<φ<23.5°N) and analyzing the obtained results as a function of δ and φ it is found that the equation (20) could be applied with a high accuracy for determining $\beta_{opt,d}$ all over the year. The absolute difference between the results of the equation (20) and precise results, obtained from the equation (12), does not exceed 0.2°. Figures 1 to 3 show the results of applying the equation (20) and the equation (12) in determining $\beta_{opt,d}$ for three different latitudes ($\varphi = 0^o$, $\varphi = 15^o N$ and $\varphi = 15^o S$) respectively. It is seen from Figure 1 that the results are identic for $\varphi = 0^o$ while Figures 2 and 3 show that, the differences between the results of these two equations are negligible. Moreover, when calculating the optimum tilt angle over a period of several consecutive days, it was found that the equation (22) could be applied with a high accuracy for determining $\beta_{opt,p}$ all over the year. The absolute difference between the results of the equation (22) and precise results, obtained using the equation (21), does not exceed 0.2°. As the daily optimum tilt angle in the tropical zone could be positive (south facing orientation) and negative (north facing orientation), it is necessary to divide the calendar year into periods in such a manner that $\beta_{opt,d}$ is of the same sign over each period. This division does not coincide with the

Figure 1. A comparison between the results of the equations (12) and (20) for Equator ($\varphi = 0^o$). ■ is for the results of the equation (20) and ♦ is for the results of the equation (12)

habitual division of the calendar year into months. However, the calendar year could be divided into 12 periods. In order to use the usual months it is possible to divide each of two months into periods as during these two months $\beta_{opt,p}$ changes its sign. Table 11 shows the mentioned periods for the studied latitudes ($\varphi = 0^o$, $\varphi = 15^o N$ and $\varphi = 15^o S$). Tables 12 and 13 show the results of applying the equations (21) and (22) in determining $\beta_{opt,p}$ for three different latitudes ($\varphi = 0^o$, $\varphi = 15^o N$ and $\varphi = 15^o S$) respectively.

It is seen from Tables 11 and 12 that the results are identic for $\varphi = 0^o$ while the differences between the results of these two equations are negligible (less than 0.3°) for $\varphi = 15^o N$ and $\varphi = 15^o S$. Therefore, it is reasonable to compare all available calculated results with of the equations (20) and (22).

When comparing the results of Bari with those of the equation (20) for $\varphi = 7^o$ and $\varphi = 20^o N$, the results presented in Figures 4 and 5 are obtained. It is seen from Figure 4 that, for $\varphi = 7^o N$ Bari's method underestimates the daily optimum tilt during the heating season while it overestimates this angle during summer. The same situation is observed in the case of $\varphi = 20^o N$. Moreover, it is observed that, the differences between the calculated results using the equation (20) and Bari's method are not practically sensitive to the value of the latitude (see Figure 6).

Available Country Cases

When trying to divide the calendar year with regard to calendar months, it was found that parts of the calendar months at the beginning and the end of each characteristic period (where the daily optimum tilt angle changes its sign) are included (see some cases in the Tables 11 to 13). So, when calculating the monthly optimum tilt, $\beta_{opt,m}$, for Equator facing, NP facing and SP facing solar collectors, the results

Figure 2. A comparison between the results of the equations (12) and (20) for Equator ($\varphi = 15^{o} N$). x is for the results of the equation (20) and ж is for the results of the equation (12)

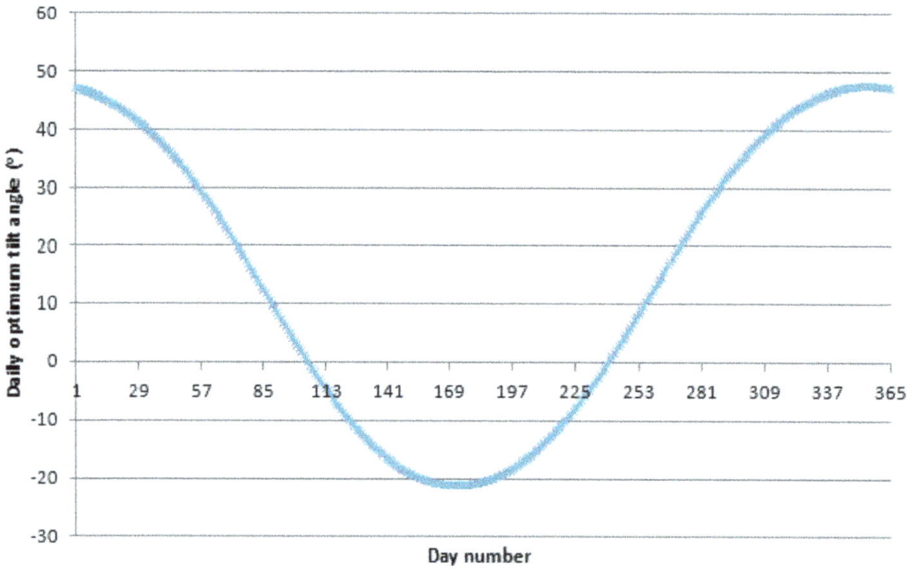

Figure 3. A comparison between the results of the equations (12) and (20) for Equator ($\varphi = 15^{o} S$). + is for the results of the equation (20) and - is for the results of the equation (12)

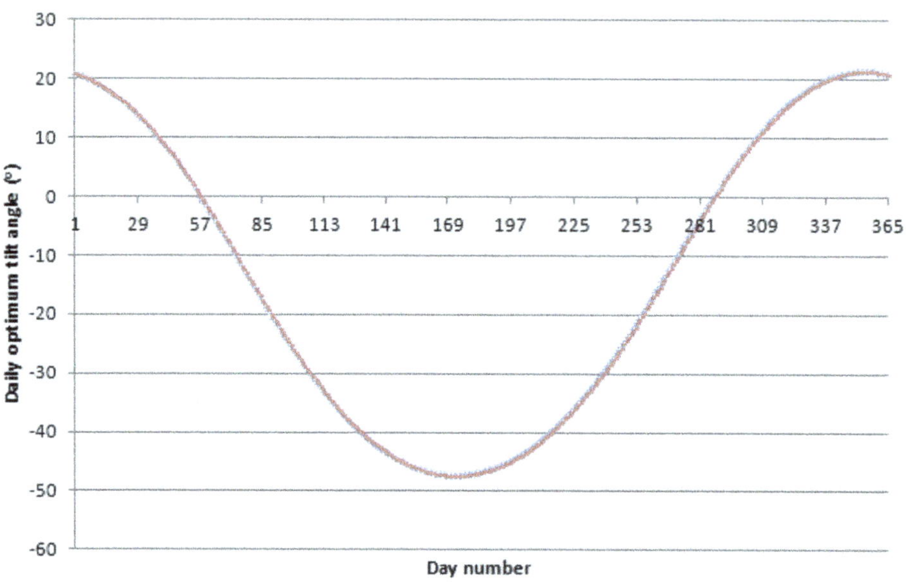

Table 11. The proposed periods of the calendar year

$\varphi = 0^o$	$\varphi = 15^o N$	$\varphi = 15^o S$
January	January	January
February	February	February 1 to February 25
		February 26 to February 28
March 1 to March 21	March	March
March 22 to March 31		
April	April 1 to April 15	April
	April 16 to April 30	
May	May	May
June	June	June
July	July	July
August	August 1 to August 27	August
	August 28 to August 31	
September 1 to September 20	September	September
September 21 to September 30		
October	October	October 1 to October 14
		October 15 to October 31
November	November	November
December	December	December

for these months should reflect this fact and indicate the number of days corresponding to each of the suitable orientations. Here it should be mentioned that this fact is ignored by all other authors, except Soulayman and Sabbagh (2015b), when treating this subject.

Nijegorodov, Devan, Jain and Carlsson (1997), Idowu, Olarenwaju and Ifedayo (2013) and Oko and Nnamchi (2012) proposed sets of equations for calculating $\beta_{opt,m}$. When comparing their results with the results of Soulayman and Sabbagh (2015b) at different latitudes it was found that the results based on the set of equations of Oko and Nnamchi (2012) are not compatible with the results of set of equations of Nijegorodov, Devan, Jain and Carlsson (1997) and Idowu, Olarenwaju and Ifedayo (2013) and those of the present work and they could not be justified. Therefore, the results of set of equations in Oko and Nnamchi (2012) are not included in Table 14 which contains the results of set of equations in Nijegorodov, Devan, Jain and Carlsson (1997) and Idowu, Olarenwaju and Ifedayo (2013) and those of Soulayman and Sabbagh (2015b). From Table 14 it is seen that the results of set of equations in Idowu, Olarenwaju and Ifedayo (2013) are too approximate in relation to those of Nijegorodov, Devan, Jain and Carlsson (1997) and those of Soulayman and Sabbagh (2015b). A relatively very good agreement is observed between the results of Soulayman and Sabbagh (2015b) and those of Nijegorodov, Devan, Jain

Table 12. Precise values of $\beta_{opt,p}$ (^{o}) over the proposed periods of the calendar year

Periods	$\varphi = 0^o$	$\varphi = 15^o N$	$\varphi = 15^o S$
January	30.9	44.3	17.6
February	19.7	34.5	6.9 (1/2-25/2); -0.8(26/2-28/2)
March	6.9 (1/3-21/3); -3.2(22/3-31/3)	18.5	-11.4
April	-14.7	4.2(1/4-15/4); -4.5(16/4-30/4)	-29.4
May	-28.1	-14.7	-41.9
June	-33.8	-20.9	-47.0
July	-31.2	-17.9	-44.6
August	-20.3	-7.2(1/8-27/8); 1.4(28/8-31/8)	-34.4
September	-6.3 (1/9-20/9); 3.1(21/9-30/9)	12.0	-17.9
October	15.2	29.9	-4.2(1/10-14/10); 4.8(14/10-31/10)
November	28.4	42.2	15.0
December	33.8	47.0	20.9

and Carlsson (1997). Here it should be mentioned that, for each of the studied latitudes, Soulayman and Sabbagh (2015b) gives two values of different signs for optimum tilt angle for two months (see Table 14). For $\varphi = 10^o N$ these months are April and September while for $\varphi = 20^o N$ these months are April and August. In addition, each tilt angle value is accompanied by a number in brackets. This should be read as follows: first value relates to the number of days in brackets for which the orientation of solar collector coincides with the anterior period while the second value relates to the number of days in brackets for which the orientation of solar collector coincides with the posterior period.

Iran is located at latitudes from 20° to 40° North. So, a part of Iran territory is located in the tropical zone. Therefore it is possible to apply set of equations Talebizadeh, Mehrabian and Abdolzadeh (2011) in determining the monthly optimum tilt in this part and to compare its results with those of theories which are announced to be applicable in the tropical zone such as Soulayman and Sabbagh (2015b), Nijegorodov, Devan, Jain and Carlsson (1997), Bari, Lim and Yu (2001) and Stanciu and Stanciu (2014). The application of all these theories on the latitude $\varphi = 20^o N$ leads to the results presented in the Table 15.

It is clearly demonstrated that, the formula of Stanciu and Stanciu (2014) underestimates $\beta_{opt,p}$ during winter (December, January and February) and overestimates it during summer (May, June and July). The set of equations of Talebizadeh, Mehrabian and Abdolzadeh (2011)_overestimates $\beta_{opt,p}$ during summer (May, June and July).

Table 13. The calculated values of $\beta_{opt,p}$ (°) using the equation (22)

Periods	$\varphi = 0^o$	$\varphi = 15^o N$	$\varphi = 15^o S$
January	30.9	44.3	17.6
February	19.7	34.6	6.98 (1/2-25/2); -0.88(26/2-28/2)
March	6.9 (1/3-21/3); -3.2(22/3-31/3)	18.7	-11.2
April	-14.7	4.3(1/4-15/4); -4.5(16/4-30/4)	-29.2
May	-28.1	-14.6	-41.8
June	-33.8	-20.9	-47.0
July	-31.2	-18.0	-44.6
August	-20.3	-7.2(1/8-27/8); 1.2(28/8-31/8)	-34.6
September	-6.3 (1/9-20/9); 3.1(21/9-30/9)	11.8	-18.1
October	15.2	29.7	-4.3(1/10-14/10); 5.0(14/10-31/10)
November	28.4	42.1	14.9
December	33.8	47.0	20.9

Figure 4. the calculated daily optimum tilt angle using Bari method (see Table 8) (ж) and the equation (20) (●) for $\varphi = 7^o N$

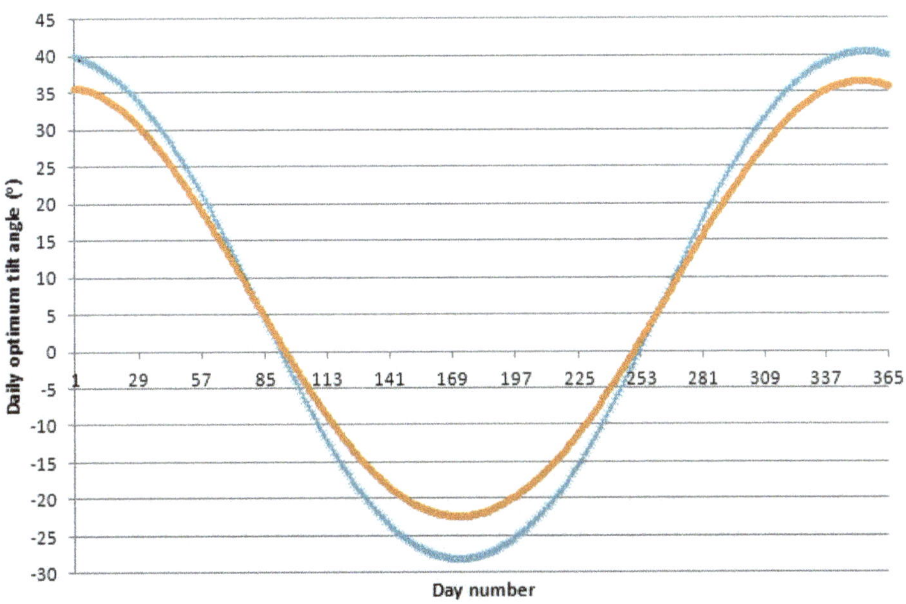

Figure 5. the calculated daily optimum tilt angle using Bari method (see Table 8) (■) and the equation (20) (▲) for $\varphi = 20^o N$

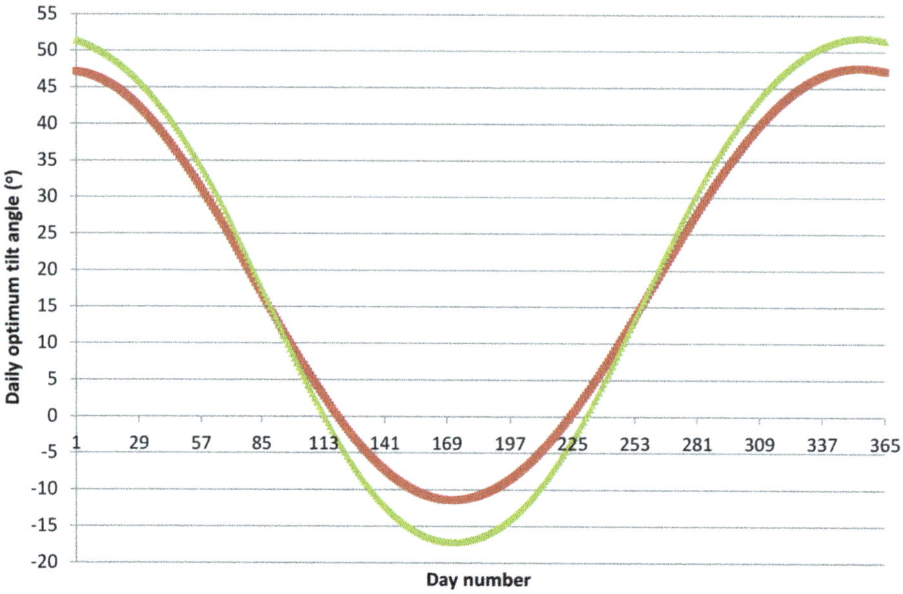

Figure 6. The differences between the calculated $\beta_{opt,d}$ (^{o}) using Bari method (see Table 8) and the equation (20) for $\varphi = 20^o N$ (-) and $\varphi = 7^o N$ (+)

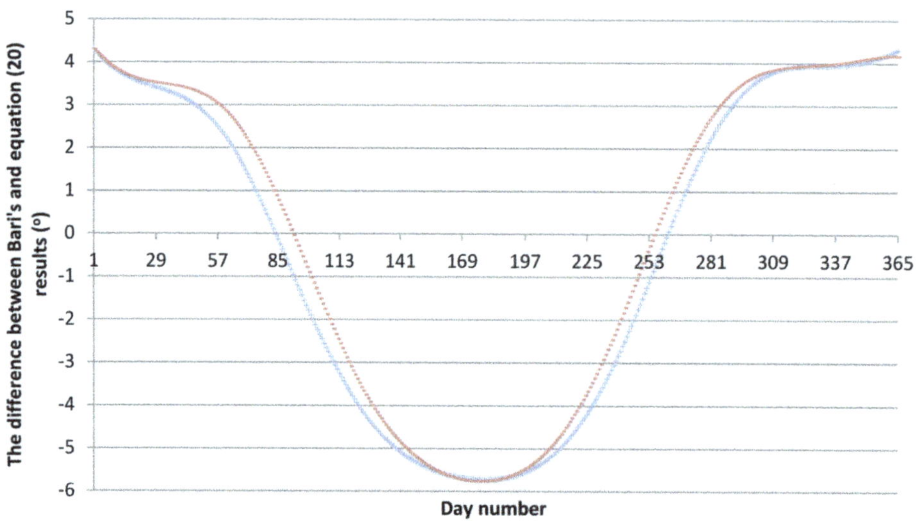

Table 14. A comparison between different approaches in calculating $\beta_{opt,m}$ (^{o})

Month	$\varphi = 10^o N$			$\varphi = 20^o N$		
	NDJC	IOI	SS	NDJC	IOI	SS
January	37.9	35	39.2	46.8	45	48.811
February	26.7	25	28.4	36.4	35	39.369
March	14	10	12.8	24	20	23.624
April	0	10	1.7(7); -3.7(23)	10	20	6.757(1/4-23/4); -1.956(24/4-30/4)
May	-14.7	-15	-17.4	-5.4	-5	-10.152
June	-25.3	-15	-23.5	-16.6	-5	-16.669
July	-21.1	-15	-20.7	-12.2	-5	-13.695
August	-7.3	-5	-9.6	2.4	5	-4.655(1/8-19/8); +3.666(20/8-31/8)
September	8	25	-1.1(4); 7.6(26)	18	35	16.7748
October	22	25	23.5	32	35	34.579
November	34.3	35	36.6	43.6	45	46.652
December	42.7	35	39.2	51.4	45	51.415

Note: NDJC stands for Nijegorodov, Devan, Jain and Carlsson (1997); SS stands for Soulayman and Sabbagh (2015b) IOI stands for Idowu, Olarenwaju and Ifedayo (2013).

Table 15. The calculated values of $\beta_{opt,p}$ (^{o}) using different methods at $\varphi = 15^o N$

Month	TMA	NDJC	SS	Bari	Stanciu
January	44.433	46.8	48.811	45.053	40.849
February	39.509	36.4	39.369	36.085	33.3289
March	16.677	24	23.624	21.923	22.3939
April	5.922	10	6.757(1/4-23/4); -1.956(24/4-30/4)	6.394	10.511
May	-1.7489	-5.4	-10.152	-5.646	1.1972
June	-2.450	-16.6	-16.669	-11.021	-3.077
July	-1.463	-12.2	-13.695	-8.311	-1.104
August	3.798	2.4	-4.655(1/8-19/8); +3.666(20/8-31/8)	1.846	6.699
September	26.614	18	16.7748	16.392	18.000
October	36.264	32	34.579	31.431	30.837
November	43.142	43.6	46.652	42.747	39.046
December	47.644	51.4	51.415	47.326	43.095

Note: TMA stands for Talebizadeh, Mehrabian and Abdolzadeh (2011); NDJC. stands for Nijegorodov, Devan, Jain and Carlsson (1997); SS stands for Soulayman and Sabbagh (2015b); Bari stands for Bari, Lim and Yu (2001) and Stanciu stands for Stanciu and Stanciu (2014).

Yakup and Malik (2001) used a mathematical model for estimating the total (global) solar radiation on a tilted surface, and to determine the optimum tilt angle and orientation (surface azimuth angle) for the solar collector in Brunei Darussalam on a daily basis, as well as for a specific period. The optimum angle was computed by searching for the values for which the total radiation on the collector surface is a maximum for a particular day or a specific period. The results reveal that changing the tilt angle 12 times in a year (i.e. using the monthly-averaged optimum tilt angle) maintains approximately the total amount of solar radiation near the maximum value that is found by changing the tilt angle daily to its optimum value. This achieves a yearly gain in solar radiation of 5% more than the case of a solar collector fixed on a horizontal surface.

When calculating $\beta_{opt,d}$ using the equation (20) for the solar collector in Brunei Darussalam ($\varphi = 4.5352^o N$) on characteristic days which reflects according to Yakup and Malik (2001) the monthly values and $\beta_{opt,m}$ using the equation (22), the results are presented in the Table 16 which contains also the results of Yakup and Malik (2001). It is seen from Table 16 that, the use of characteristic days in the tropical zone is possible except the months where $\beta_{opt,d}$ changes its sign. When excluding these two months, the agreement between the results of the equation (20) and those of Yakup and Malik (2001) is good.

Waziri, Usman, Enaburekhan and Babakano (2014) calculated the $\beta_{opt,m}$ (°) for Kano, Nigeria, $\varphi = 12.1^o N$ for months of November to April. When calculating $\beta_{opt,m}$ (°) for the same months for Kano, Nigeria, $\varphi = 12.1^o N$ using the equation (22) and comparing the obtained results with those of

Table 16. The calculated values of $\beta_{opt,m}$ (°) using different methods at $\varphi = 4.5352^o N$

Month	Characteristic Day Number	$\beta_{opt,cd}$ (°)	$\beta_{opt,m}$ (°)	Yakup and Malik (2001) (°)
January	17	35.04	34.92	28.3
February	47	24.20	24.67	20.5
March	75	8.33	8.87(1/3-29/3) -0.84(30/3-31/3)	5
April	105	-10.18	-10.25	-6.5
May	135	-24.0	-23.98	-19.3
June	162	-29.85	-29.84	-24.3
July	198	-27.31	-27.17	-24.3
August	228	-16.3	-16.00	-12
September	258	1.04	-3.94(1/9-13/9) 5.47(14/9-30/9)	1.6
October	288	19.28	19.58	10.7
November	318	32.41	32.56	28.7
December	344	37.73	37.79	32.3

Soulayman and Sabbagh (2015b) and those of Waziri, Usman, Enaburekhan and Babakano (2014), the results presented in the Table 17 are obtained.

It is clear from Table 17 that, the results of Waziri, Usman, Enaburekhan and Babakano (2014), except April, coincide well with the results of Soulayman and Sabbagh (2015b) and those of the equation (22). Moreover, the results of the equation (22) are in excellent agreement with those of Soulayman and Sabbagh (2015b).

Eke (2011) calculated the $\beta_{opt,m}$ (°) for Zaria, Kaduna State, Nigeria $\varphi = 11.13°$ N. $\beta_{opt,m}$ (°) for Zaria, Kaduna State, Nigeria $\varphi = 11.13°$ N were calculated using the equation (22) and the obtained results with those of Soulayman and Sabbagh (2015b) and those of Eke (2011) are given in the Table 18.

When comparing Eke (2011) results with those of $\beta_{opt,m}$ (°) obtained by Soulayman and Sabbagh (2015b)_and those of the equation (22) (see Table 18) with taking into consideration those of Idowu, Olarenwaju and Ifedayo (2013) three comments could be given: a) The results of Eke (2011) are too approximate regarding the period starting from 22/9 to 21/3; b) The sign from May to August, in results of Eke (2011), should be negative indicating that solar collector is orientated towards the North Pole during this period; c) The results of Eke (2011) for April and September could not be understood.

Uba and Sarsah (2013) found, for WA, Ghana ($\varphi = 10.01° N$), that collector tilt should be changed three times a year: January – March (26.4°), April – August (29.7°) and September – December (25.9°). When calculating $\beta_{opt,m}$ (°) for these periods approximately using the proposed methodology of Soulayman and Sabbagh (2015b) one obtains: January 1st – April 7th (23.8°), April 8th – September 4th (-16.8°) and September 5th – December 31 (28.8°). So, the results of Soulayman and Sabbagh (2015b) are in a good agreement with those of Uba and Sarsah (2013) for the first and third periods while the result of Uba and Sarsah (2013) should be corrected for the second period because 1) The solar collector should be orientated towards the North Pole during April – August period and 2) The proposed value by Uba and Sarsah (2013) is too high.

Ng, Adam, Inayatullah and Ab Kadir (2014) calculated the $\beta_{opt,m}$ (°) for Bangi, Malaysia (φ= 3°). When calculating $\beta_{opt,m}$ (°) for Bangi, Malaysia (φ = 3°) using the equation (22) and comparing the obtained results with those of Ng, Adam, Inayatullah and Ab Kadir (2014), those of Bari (2000) and

Table 17. The calculated values of $\beta_{opt,m}$ (°) using different methods at $\varphi = 12.1° N$

Month	$\beta_{opt,m}$ (°)	Waziri, Usman, Enaburekhan and Babakano (2014) (°)	Soulayman and Sabbagh (2015b) (°)
January	41.71	35	41.2
February	31.86	25	30.4
March	15.77	15	14.8
April	2.94(1/4-10/4) -5.84(11/4-30/4)	15	2.8(1/4-10/4) -4.8(11/4-30/4)
November	39.45	35	38.6
December	44.45	45	44.4

Table 18. The calculated values of $\beta_{opt,m}$ (^{o}) using different methods at $\varphi = 11.13^o N$.

Month	$\beta_{opt,m}$ (^{o})	Eke (2011) (^{o})	Soulayman and Sabbagh (2015b) (^{o})
January	40.84	26.5	40.4
February	30.94	24.5	29.6
March	14.80	10.0	14.1
April	2.60(1/4-8/4) -6.20(9/4-30/4)	19.5	4.5(1/4-15/4) -4.3(16/4-30/4)
May	-18.04	26.0	-16.2
June	-24.15	30.0	-22.3
July	-21.36	24.0	-19.5
August	-9.77	21.0	-8.4
September	-0.81(1/9-2/9) 8.58(3/9-30/9)	11.5	-0.5(1/9-2/9) 8.3(3/9-30/9)
October	25.98	19.5	24.7
November	38.57	27.0	37.8
December	43.60	30.0	43.6

those of Soulayman and Sabbagh (2015b) (see Table 19) it was found that in comparison with the results of (22), those of Bari (2000) and those of Soulayman and Sabbagh (2015b), even the orientation is determined correctly by Ng, Adam, Inayatullah and Ab Kadir (2014), $\beta_{opt,m}$ (^{o}) is underestimated for October to March and overestimated for April to September. Moreover, the agreement between the results of the equation (22) and those of Soulayman and Sabbagh (2015b) is very good. The comment with regard to months where $\beta_{opt,m}$ (^{o}) changes its sign could be repeated in relation to results of Ng, Adam, Inayatullah and Ab Kadir (2014) and Bari (2000).

Kamanga, Mlatho, Mikeka and Kamunda (2014) proposed a study to determine the optimum tilt angle for installing photovoltaic solar panels in Zomba district, Malawi. The study was conducted at Chancellor College Meteorological Station in Zomba district, Malawi, located at latitude of -15.387°. The goal of Kamanga, Mlatho, Mikeka and Kamunda (2014) was to determine the optimum monthly tilt angles of PV solar panels and the seasonal adjustments needed for the panels in order to collect maximum solar radiation throughout the year. Kamanga, Mlatho, Mikeka and Kamunda (2014) measured global solar radiation on four tilted Equator-facing surfaces of 0°, 15°, 20°, and 25°. Using obtained data, it was concluded that the optimum tilt angle is 0° from October to February and 25° from March to September and that two seasonal tilt adjustments were suggested: one is at the end of February and the other is at the end of September. The yearly optimum tilt angle for north facing surfaces was proposed to be 25° by Kamanga, Mlatho, Mikeka and Kamunda (2014).

Table 19. The calculated values of $\beta_{opt,m}$ (^{o}) using different methods at $\varphi = 3^o N$

Month	$\beta_{opt,m}$ (^{o})	Ng, Adam, Inayatullah and Ab Kadir (2014) (^{o})	Bari (2000) (^{o})	Soulayman and Sabbagh (2015b) (^{o})
January	33.57	22	29.66	32.6
February	23.27	16	20.31	21.8
March	8.32(1/3-26/3) -1.42(27/3-31/3)	5	6.04	7.7(1/3-26/3) -1.3(27/3-31/3)
April	-11.7756	-8	-9.06	-10.7
May	-25.4062	-18	-20.48	-24.0
June	-31.1731	-22	-25.53	-30.1
July	-28.5565	-21	-23.05	-27.3
August	-17.5129	-11	-13.57	-16.1
September	-4.83(1/9-15/9) 4.57(16/9-30/9)	0	0.34	-4.5(1/9-15/9) 4.3(16/9-30/9)
October	18.18	11	15.31	16.9
November	31.20	19	27.29	30.3
December	36.44	24	32.41	35.9

When providing $\beta_{opt,m}$ (^{o}) and $\beta_{opt,p}$ (^{o}) calculation for the same region of Kamanga, Mlatho, Mikeka and Kamunda (2014) using the procedure proposed by Soulayman and Sabbagh (2015b), the following results were obtained:

1. $\beta_{opt,m}$ (^{o}) = 15.4°, 6.1° (25) -1.0° (3), -10.9°, -27.9°, -41.2°, -47.3°, -44.5°, -33.3°, -17.2°, -4.1° (15) +4.2° (16) , 12.8°, 18.6° for January to December respectively;
2. $\beta_{opt,57-288}$ (^{o}) = -30.23° and $\beta_{opt,289-56}$ = 7.7°.

So, as Kamanga, Mlatho, Mikeka and Kamunda (2014) deals with Equator facing case only, the optimum tilt during the period starting from 16/10 to 25/2 should be zero. This result coincides well with the finding of Kamanga, Mlatho, Mikeka and Kamunda (2014) and it gives more precisely the starting and ending dates of period. On the other hand, as -25° is the maximum absolute tilt angle value used by Kamanga, Mlatho, Mikeka and Kamunda (2014) in measuring global solar radiation, the obtained result in this work coincides with the finding of Kamanga, Mlatho, Mikeka and Kamunda (2014).

Finally, Diaz, Ngo, Pascual, Planes and Chua (2014) determined $\beta_{opt,d}$ theoretically in Philippine (φ from 4° to 21°) using the equation (20) and verified their calculations at φ = 14.56° experimentally during days of February. The results of Diaz, Ngo, Pascual, Planes and Chua (2014) confirm the validity of equation (20) in the tropical zone.

Comments on Optimum Tilt Angle in The Tropical Zone

As mentioned before, the daily optimum tilt angle in the tropical zone could be positive (south facing orientation) and negative (north facing orientation), and it is necessary to divide the calendar year into periods in such a manner that $\beta_{opt,d}$ is of the same sign over each period.

Let us imagine now an axis of solar noon incident angle on a horizontal surface, $\theta_{noon} = \varphi\text{-}\delta$. Supposing that, θ_{noon} is positive when solar rays are incident from south direction and θ_{noon} is negative when solar rays are incident from north direction. Then the sign of θ_{noon} could determine the period where the solar collector should be oriented, for receiving solar rays at noon, to Equator or to opposite direction. This could be done by studying the dependence of θ_{noon} as a function of n. The period of positive θ_{noon} does not match the period of positive $\beta_{opt,d}$. This means that, there are periods where the product of θ_{noon} and $\beta_{opt,d}$ is negative. During these periods the solar rays can incident on the both sides of the solar collectors during the day from sunrise to sunset. Tables 20 and 21 show the mentioned periods for the tropical zone while Table 22 gives the periods of different signs of the product θ_{noon} x $\beta_{opt,d}$.

More details on the daily, monthly, seasonally and yearly optimum tilt angles in the tropical zone are given in the appendixes. As during the mentioned months and seasons the daily optimum tilt angle $\beta_{opt,d}$ (^{o}) changes its sign, the corresponded tables show the optimum tilt angle for these two periods and the number of days of each period.

OPTIMUM TILT ANGLE DETERMINATION IN THE MID-LATITUDE ZONE

The majority of installations are with fixed mountings. Therefore, it is often practicable to orient the solar collector at an optimum tilt angle, β_{opt} and to correct the tilt from time to time. So, optimally orienting the collector maximizes the solar energy collected. For this purpose, one should be able to determine the optimum slope of the collector at any latitude, for any surface azimuth angle, and on any day or any period of the year. Various schemes have been proposed for optimizing the tilt angle and orientation of solar collectors designed for different geographical latitudes or possible utilization periods. However, the tilt angle optimization has been extensively addressed in many articles and several attempts were made to determine, or at least to estimate, the optimum tilt angle β_{opt} theoretically and experimentally.

Available Theories

There are number of studies that were carried out in order to find the optimum tilt angle of solar collectors around the mid-latitude zone (Carbondale, Illinois, Gong and Kulkarni (2005), Izmir, Turkey, Gunerhan and Hepbasli (2007), Sanliurfa, Turkey, Kacira, Simsek, Babur and Demirkol (2004), Dhaka, Ghosh, Bhowmik and Hussain (2010), 30 cities in China, Tang and Wu (2004), Madinah, Saudi Arabia, Benghanem (2011), Jordan, Shariah, Al-Akhras and Al-Omari (2002), Helwan, Egypt, Elminir, Ghitas, El-Hussainy, Hamid, Beheary and Abdel-Moneim (2006), Syria, Skeiker (2009), Cyprus, Ibrahim (1995), Burgos, Spain, De Miguel, Bilbao and Diez (1995), Brisbane, Australia, Yan, Saha, Meredith and Goodwin (2013), Athens basin area, Koronakis (1986), Mediterranean region Darhmaoui and Lahjouji

Table 20. The periods of positive and negative θ_{noon}

	$\theta_n \geq 0$					$\theta_n < 0$		
φ	Days	From	To	From	To	Days	From	To
-23.45	0	0	0	0	0	365	1	365
-23	23	1	1	344	365	342	2	343
-22	41	1	10	335	365	324	11	334
-21	53	1	16	329	365	312	17	328
-20	63	1	21	324	365	302	22	323
-19	73	1	26	319	365	292	27	318
-18	81	1	30	315	365	284	31	314
-17	88	1	34	312	365	277	35	311
-16	95	1	37	308	365	270	38	307
-15	101	1	40	305	365	264	41	304
-14	108	1	44	302	365	257	45	301
-13	114	1	47	299	365	251	48	298
-12	120	1	50	296	365	245	51	295
-11	125	1	52	293	365	240	53	292
-10	131	1	55	290	365	234	56	289
-9	137	1	58	287	365	228	59	286
-8	142	1	61	285	365	223	62	284
-7	147	1	63	282	365	218	64	281
-6	153	1	66	279	365	212	67	278
-5	157	1	68	277	365	208	69	276
-4	163	1	71	274	365	202	72	273
-3	167	1	73	272	365	198	74	271
-2	173	1	76	269	365	192	77	268
-1	177	1	78	267	365	188	79	266
0	183	1	81	264	365	182	82	263
1	187	1	83	262	365	178	84	261
2	193	1	86	259	365	172	87	258
3	197	1	88	257	365	168	89	256
4	203	1	91	254	365	162	92	253

continued on following page

Table 20. Continued

	$\theta_n \geq 0$					$\theta_n < 0$		
φ	**Days**	**From**	**To**	**From**	**To**	**Days**	**From**	**To**
5	207	1	93	252	365	158	94	251
6	213	1	96	249	365	152	97	248
7	217	1	98	247	365	148	99	246
8	223	1	101	244	365	142	102	243
9	229	1	104	241	365	136	105	240
10	233	1	106	239	365	132	107	238
11	239	1	109	236	365	126	110	235
12	245	1	112	233	365	120	113	232
13	251	1	115	230	365	114	116	229
14	257	1	118	227	365	108	119	226
15	263	1	121	224	365	102	122	223
16	269	1	124	221	365	96	125	220
17	277	1	128	217	365	88	129	216
18	284	1	132	214	365	81	133	213
19	293	1	136	209	365	72	137	208
20	301	1	140	205	365	64	141	204
21	311	1	145	200	365	54	146	199
22	324	1	152	194	365	41	153	193
23	343	1	161	184	365	22	162	183
23.45	365	1	182	183	365	0	182	182

Table 21. The periods of positive and negative $\beta_{opt,d}$

	$\beta_{opt} \geq 0$					$\beta_{opt} < 0$		
φ	**Days**	**From**	**To**	**From**	**To**	**Days**	**From**	**To**
-23.45	105	1	42	303	365	260	43	302
-23	107	1	43	302	365	258	44	301
-22	111	1	45	300	365	254	46	299
-21	115	1	47	298	365	250	48	297
-20	117	1	48	297	365	248	49	296
-19	121	1	50	295	365	244	51	294
-18	125	1	52	293	365	240	53	292
-17	127	1	53	292	365	238	54	291

continued on following page

Table 21. Continued

	$\beta_{opt} \geq 0$					$\beta_{opt} < 0$		
φ	Days	From	To	From	To	Days	From	To
-16	131	1	55	290	365	234	56	289
-15	135	1	57	288	365	230	58	287
-14	137	1	58	287	365	228	59	286
-13	141	1	60	285	365	224	61	284
-12	145	1	62	283	365	220	63	282
-11	147	1	63	282	365	218	64	281
-10	151	1	65	280	365	214	66	279
-9	153	1	66	279	365	212	67	278
-8	157	1	68	277	365	208	69	276
-7	161	1	70	275	365	204	71	274
-6	163	1	71	274	365	202	72	273
-5	167	1	73	272	365	198	74	271
-4	169	1	74	271	365	196	75	270
-3	173	1	76	269	365	192	77	268
-2	176	1	78	268	365	189	79	267
-1	179	1	79	266	365	186	80	265
0	183	1	81	264	365	182	82	263
1	185	1	82	263	365	180	83	262
2	189	1	84	261	365	176	85	260
3	191	1	85	260	365	174	86	259
4	195	1	87	258	365	170	88	257
5	199	1	89	256	365	166	90	255
6	201	1	90	255	365	164	91	254
7	205	1	92	253	365	160	93	252
8	207	1	93	252	365	158	94	251
9	211	1	95	250	365	154	96	249
10	215	1	97	248	365	150	98	247
11	217	1	98	247	365	148	99	246
12	221	1	100	245	365	144	101	244
13	223	1	101	244	365	142	102	243
14	227	1	103	242	365	138	104	241

continued on following page

Table 21. Continued

	$\beta_{opt} \geq 0$					$\beta_{opt} < 0$		
φ	Days	From	To	From	To	Days	From	To
15	231	1	105	240	365	134	106	239
16	233	1	106	239	365	132	107	238
17	237	1	108	237	365	128	109	236
18	241	1	110	235	365	124	111	234
19	243	1	111	234	365	122	112	233
20	247	1	113	232	365	118	114	231
21	251	1	115	230	365	114	116	229
22	254	1	117	229	365	111	118	228
23	257	1	118	227	365	108	119	226
23.45	259	1	119	226	365	106	120	225

Table 22. The periods of different signs of the product θ_{noon} x $\beta_{opt,d}$

	$\theta_n, \beta_{opt,d} \geq 0$					$\theta_n, \beta_{opt,d} < 0$		
Φ	Days	From	To	From	To	Days	From	To
-23.45	0	0	0	0	0	260	43	302
-23	23	1	1	344	365	258	44	301
-22	41	1	10	335	365	254	46	299
-21	53	1	16	329	365	250	48	297
-20	63	1	21	324	365	248	49	296
-19	73	1	26	319	365	244	51	294
-18	81	1	30	315	365	240	53	292
-17	88	1	34	312	365	238	54	291
-16	95	1	37	308	365	234	56	289
-15	101	1	40	305	365	230	58	287
-14	108	1	44	302	365	228	59	286
-13	114	1	47	299	365	224	61	284
-12	120	1	50	296	365	220	63	282
-11	125	1	52	293	365	218	64	281
-10	131	1	55	290	365	214	66	279
-9	137	1	58	287	365	212	67	278

continued on following page

Table 22. Continued

	$\theta_n, \beta_{opt,d} \geq 0$					$\theta_n, \beta_{opt,d} < 0$		
Φ	**Days**	**From**	**To**	**From**	**To**	**Days**	**From**	**To**
-8	142	1	61	285	365	208	69	276
-7	147	1	63	282	365	204	71	274
-6	153	1	66	279	365	202	72	273
-5	157	1	68	277	365	198	74	271
-4	163	1	71	274	365	196	75	270
-3	167	1	73	272	365	192	77	268
-2	173	1	76	269	365	189	79	267
-1	177	1	78	267	365	186	80	265
0	183	1	81	264	365	182	82	263
1	185	1	82	263	365	178	84	261
2	189	1	84	261	365	172	87	258
3	191	1	85	260	365	168	89	256
4	195	1	87	258	365	162	92	253
5	199	1	89	256	365	158	94	251
6	201	1	90	255	365	152	97	248
7	205	1	92	253	365	148	99	246
8	207	1	93	252	365	142	102	243
9	211	1	95	250	365	136	105	240
10	215	1	97	248	365	132	107	238
11	217	1	98	247	365	126	110	235
12	221	1	100	245	365	120	113	232
13	223	1	101	244	365	114	116	229
14	227	1	103	242	365	108	119	226
15	231	1	105	240	365	102	122	223
16	233	1	106	239	365	96	125	220
17	237	1	108	237	365	88	129	216
18	241	1	110	235	365	81	133	213
19	243	1	111	234	365	72	137	208
20	247	1	113	232	365	64	141	204
21	251	1	115	230	365	54	146	199
22	254	1	117	229	365	41	153	193
23	257	1	118	227	365	22	162	183
23.45	259	1	119	226	365	1	182	182

(2013), Ma'an, Jordan, Altarawneh, Rawadieh, Tarawneh, Alrowwad and Rimawi (2016), Tabass, Iran, Khorasanizadeh, Mohammadi and Mostafaeipour (2014), Hamirpur, India, Sinha and Chandel (2016), South Africa, Le Roux (2016), and many more).

Soulayman (1991) proposed a general algorithm for calculating daily $\beta_{opt,d}$, monthly $\beta_{opt,m}$, seasonally $\beta_{opt,s}$, and yearly $\beta_{opt,y}$, for a south facing collector at any latitude from 0°N to 60°N. Soulayman and Sabbagh (2014) proposed an algorithm for determining $\beta_{opt,d}$, $\beta_{opt,m}$, $\beta_{opt,s}$, and $\beta_{opt,y}$ at any latitude, φ, and for any direction (surface azimuth angle, γ). Stanciu, and Stanciu (2014) proposed a simple formula for determining the optimum tilt of a south facing collector at latitudes from 0°N to 80°N. Nijegorodov, Devan, Jain and Carlsson (1997) presented 12 equations (one for each month), for determining optimum tilt angle for any location that lies between latitude 60°S to 60°N. Mujahid (1994) computed the optimum slope angle for latitude of 10°N to 50°N and concluded that if the collector is adjusted by the seasonally optimum angles, 10% more energy is received compared with the zero slope angle. Calabrò (2013) proposed an algorithm to calculate the optimum tilt angle of solar panels by means of global horizontal solar radiation data, provided from Earth-based meteorological stations. Some comments, in regard to the applicability of the formula $\beta_{opt,d} = \varphi - \delta$ proposed by Stanciu, and Stanciu (2014), are given by Soulayman (2015). The effects of latitude, solar reflectivity, and clearness index were considered by Elsayed (1989) in determining the optimal tilt angle analytically. Tang and Wu (2004) used the monthly horizontal radiation to develop a simple mathematical procedure allowing the determination of the optimal tilt angle. For the cold seasons, Chiou and El-Naggar (1986) developed a method for calculating the optimal tilt angle for south orientating collectors. Darhmaoui and Lahjouji (2013) proposed to determine the yearly optimum tilt $\beta_{opt,y}$ as a function of latitude φ as follows $\beta_{opt,y} = 1.25351\,\varphi - 0.00728944\varphi^2$. So, in the previous studies, no definite value or relation has been found by the researchers for the optimum tilt angle. Moreover, it can be noticed from the literature, that the optimal tilt angle is considered in many studies as location-dependent.

Lubitz (2011) used hourly typical meteorological year (TMY3) data was utilized with the Perez, Ineichen, Seals, Michalsky and Stewart (1990) radiation model to simulate solar radiation on fixed at 217 geographically diverse temperate latitude sites across the contiguous United States of America. The optimum tilt angle for maximizing annual irradiation on a fixed south-facing panel varied from being equal to the latitude at low latitude, high clearness sites, to up to 14° less than the latitude at a northwestern coastal site with very low clearness index. Across the United States, the optimum tilt angle for an azimuth tracking panel was found to be on average 19° closer to vertical than the optimum tilt angle for a fixed, south-facing panel at the same site. Introduction of manual surface tilt changes during the year produced a greater impact for non-tracking surfaces than it did for azimuth tracking surfaces.

Lubitz (2011) calculated firstly the mean annual irradiance for the base case of a fixed south-facing panel oriented at optimum tilt angle, as predicted using the Perez, Ineichen, Seals, Michalsky and Stewart (1990) model (see Figure 7). The points in Figure 7 indicate the location of the simulated sites, while the contours are based on interpolation between these points. Because of this, it is not recommended that conclusions be drawn for regions beyond the coverage of TMY3 sites, such as in Canada, Mexico or at sea. As expected, annual irradiance within the contiguous United States shows a strong dependence on geographic location. Higher irradiance is associated with lower latitude and clearer skies in the desert southwest and central plains, while the lowest irradiance (approximately half as much) occurs at locations with the lowest clearness indexes, such as the Pacific Northwest and Great Lakes regions.

Figure 7. Mean annual irradiance on a fixed south-facing surface at optimum tilt angle predicted using Perez model. Filled circles indicate TMY3 Class 1 stations

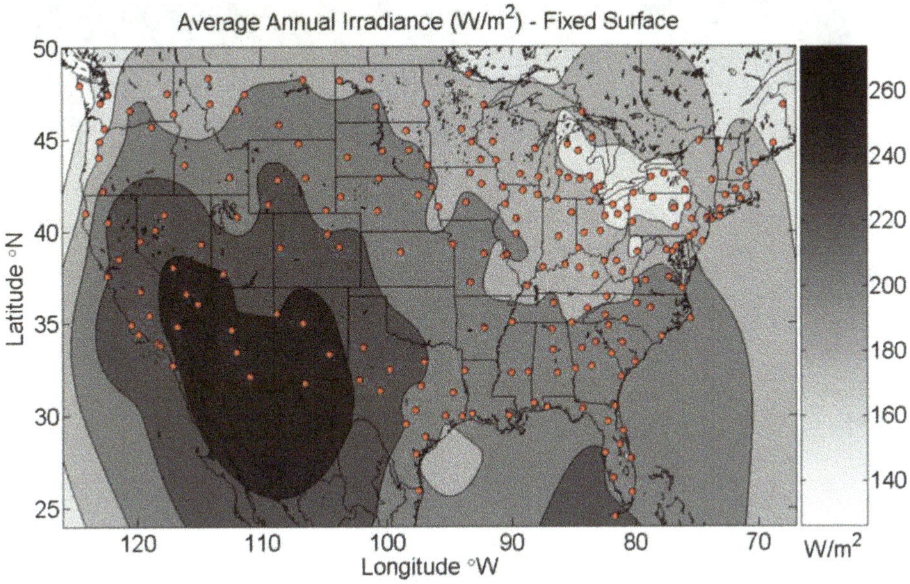

The optimum tilt angle $\beta_{opt,y}$ for a fixed, south-facing panel is an important practical quantity when installing any solar panel. One common recommendation for temperate latitudes is that $\beta_{opt,y} \approx \varphi$, however in practice there is usually a difference between the latitude and optimum tilt angle (Nann, 1990). Christensen and Barker (2001) defined this difference as:

$$w = \varphi - \beta_{opt,y} \tag{42}$$

where $\beta_{opt,y}$ is the yearly optimum tilt angle for the panel. In a simulation of 229 TMY2 sites in the contiguous United States, Christensen and Barker (2001) found that values of w for fixed south-facing surfaces ranged between 0° and 16°, with higher w values associated with higher latitudes and lower annual average clearness index. Very similar results were observed in these simulations: for fixed south-facing panels, w ranged between 0° and 14°, with similar patterns of geographic variation (Figure 8).

When changing panel tilt twice per year, the optimum times to tilt the panel may not coincide with the equinoxes. To investigate this, a range of possible tilt dates were considered for each site. Since the TMY3 data is a composite of real weather data, it is not possible to optimize these dates precisely. Pairs of tilt change dates were searched that were symmetric about the equinoxes, with each pair differing by 1 week, and the pair of dates that resulted in the greatest annual irradiation were considered the optimum. The most common optimum tilt change times were to change tilt at each equinox, resulting in equal time at each of the two angles over the course of the year, however, this was not the case for all sites. Considering fixed surfaces at all sites, the duration of the ‘‘summer tilt’’ time period averaged 168 days (46% of the year); at individual sites it ranged from 140 days (20 weeks) to 196 days (28 weeks).

Figure 8. Geographic distribution of w for fixed, south-facing surfaces

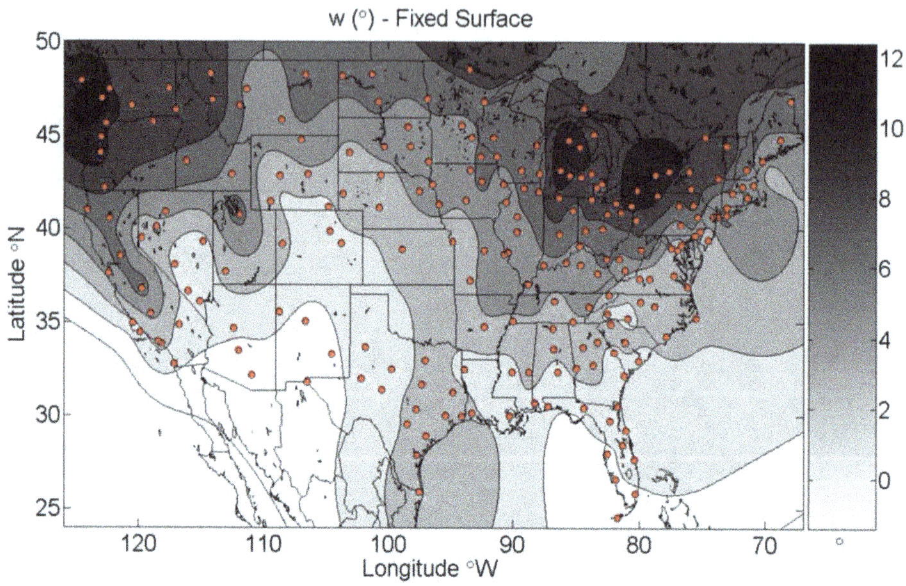

The impact of increasing the number of tilt changes per year to six or 12 was investigated for both fixed and azimuth tracking panels. Tilt angles that resulted in maximum annual incident irradiation were determined assuming monthly tilt changes on the 6th day of each month, or every second month, and annual irradiation incident on the surface was computed.

Figure 9 shows the percent increase in annual irradiation achieved by implementing monthly tilt changes, relative to using a single annual-optimum tilt angle. For all sites, changing tilt angle during the year increases annual irradiation, and the effect becomes more pronounced as the frequency of tilt changes increases. Greater increases are typically seen for the central portion of the country, and lower increases in areas with lower clearness like the areas around the Great Lakes and the northwest Pacific coast. The increase in annual irradiation gained by tilting is important. The average irradiation increase for fixed surfaces when monthly tilting was implemented was 4.8%. Even monthly tilt changes only resulted in an average annual irradiation increase of 5% for fixed panels, relative to using a single optimized tilt angle in each case. In practice, the decision whether to manually tilt panels requires balancing the added cost in labor and the panel support versus the extra energy generation and the cost value of that energy.

It should be noted while regional patterns are evident in Figure 6 the increase in irradiation gains due to tilt changes was not found to be strongly correlated to φ or other single variables. It is likely that local effects not reflected in the TMY3 database, such as sources of fog or cloud near a station, account for some of the variability between otherwise similar sites. The TMY3 data itself may also introduce some artificial variation between sites, especially in seasonal data. The TMY3 file for each site is a composite of actual data drawn from many years of observations to produce an annual dataset that is "typical" across a range of meteorological parameters. Files for nearby sites may be composed of data from different time periods, and individual parameters may not be exactly "typical" for the site in order to ensure that the overall set of parameters is as representative as possible, resulting in differences in specific variables between sites. However, manual tilting of panels between two and 12 times per year produced several times greater gains in irradiation for an otherwise fixed south-facing panel.

Figure 9. Percent annual irradiation increase due to implementing monthly tilt adjustments for a fixed surface

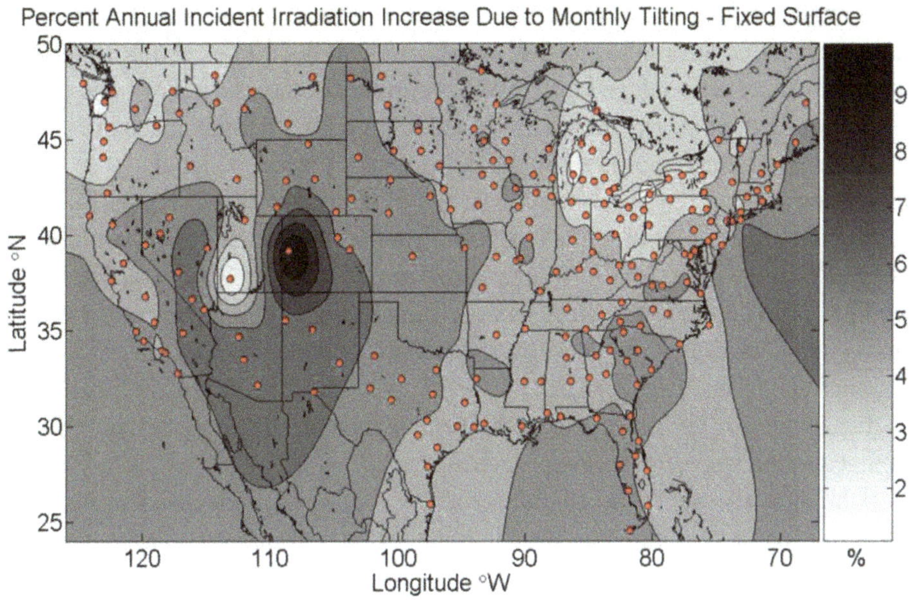

Here it should be noted that, there may be cases where it is worth considering. For example, if panels must be tilted to vertical for regular cleaning, tilting them back to an angle optimized until the next cleaning interval would be feasible and incur virtually no additional cost. It should also be noted that if the tariff paid for solar electricity is high enough, it would likely be worthwhile changing tilt angles.

Finally, it is possible to regroup all available theories, which treat the optimum tilt angle and orientation of the solar collectors, into three major branches:

- Clearness index based branch.
- Extraterrestrial solar radiation based branch.
- Beam radiation based branch

Clearness Index Based Theories

The clearness index based branch is widely used. Moreover, various simulation tools, based on the clearness index, with interactive maps related to renewable energy are available. NREL-PVWatts Viewer is one of the popular one. It has an interactive map-based interface for providing access to multiple data sets which link directly to their corresponding PVWatts calculators. It can instantly determine the energy production and cost savings of grid-connected PV energy systems throughout the world. NREL-IMBY is another tool for estimating the amount of electricity that can be produced for household applications. It also provides generation analysis given the average system loads for a given location. Solar Calculator, powered by National Oceanic and Atmospheric Administration Earth System Research Laboratory (NOAA ESRL), allows user to pick the location of interest through an interactive map and helps finding the solar noon, sunrise and sunset times for any date. SunCalc is another similar but much more interactive application for determining the Sun positions at sunrise, sunset and any specified time.

Optimum tilt angle for solar collectors at any location is a function of latitude (Æ) and the quantity of solar radiations available at that location. If the above mentioned information is authentically provided, a simple method can be deployed for finding optimum tilt angles for any given span of the year. In this context, determination of geocode (latitude and longitude) of location, acquiring solar radiation data and the method for finding optimum tilt angle is discussed below in details.

The map interface provides an easy and user-friendly interaction for picking up location of interest instead of entering the values of latitude and longitude. For the development of map interface in website, Google Maps JavaScript API V3 (About Google Maps) along with The Google Geocoding API (About Google Geocoding API) was used. It is a free service, available for any web site that is free to consumers.

By default, the map object is set to display „Street Map" at zoom level of „3", so that country level information (names etc.) could be displayed. The map object is allowed to switch between "Satellite" and "Terrain" display modes. User may also zoom the map up to the desired level, provided that Google map supports that zoom level in the region of interest.

A marker (bubble) is also set to the default location geocode (latitude and longitude). The marker is able to be dragged so that user can drag-drop it to the location of interest. On dropping the marker at any location, the geocode which is the input to the solar data acquiring system, is updated. Location's real address (or name) will be obtained through Google reverse geocoding (address lookup).

Another feature added to the interface is "Find location". User will be required to enter the location's real address (or name). Resulting locations matching the query will be listed. On selecting any result, the marker on the map will dynamically move to the location along with updating the geocode. User may also enter the values of latitude and longitude.

NASA's Surface meteorology and Solar Energy (SSE) data is used as the backend database for acquiring solar radiation data of the location of interest. NASA SSE data has been used in various research activities around the globe for such studies. Out of many other parameters, these data include long-term estimates of global solar radiations on horizontal surface in kWh/m^2/day averaged over each month and trustworthy set of polynomials for estimating diffused horizontal radiations in terms of clearness index (K_t), sunset hour angle and noon solar altitude angle from the horizon in degrees. These satellite and model-based products have also been shown to be accurate enough to provide reliable solar resource data over regions where surface measurements are sparse or nonexistent, and it also offers two unique features i.e. the data is global and, in general, contiguous in time (Surface meteorology and Solar Energy (SSE) Release 6.0 Methodology, 2012). This database has been used as benchmark by different researchers for comparing their derived correlations and empirical models for estimating the climatic conditions especially solar radiations. In this context we can mention the following examples:

- Zawilska and Brooks (2011) presents a year-long data record of the solar flux intensity for the city of Durban (29°58′N 30°55′E). Global horizontal irradiance (G_t), direct normal irradiance (G_{DN}), diffuse horizontal irradiance (G_d) and daily average clearness index (K_T) are used. The data were recorded at the Solar Thermal Applications Research Laboratory (STARlab) at Mangosuthu University of Technology. Ground-based measurements obtained from STARlab are compared with data from a variety of sources including NASA's SSE database and the literature.

- El-Sebaii, Al-Hazmi, Al-Ghamdi, and Yaghmour (2010) analyzed the measured data of global and diffuse solar radiation on a horizontal surface, the number of bright sunshine hours, mean daily ambient temperature, maximum and minimum ambient temperatures, relative humidity and amount of cloud cover for Jeddah (lat. 21°42′37′′N, long. 39°11′12′′E), Saudi Arabia, during the period (1996–2007). The monthly averages of daily values for these meteorological variables have been calculated. The data are then divided into two sets. The sub-data set I (1996–2004) are employed to develop empirical correlations between the monthly average of daily global solar radiation fraction (H/H_0) and the various weather parameters. The sub-data set II (2005–2007) are then used to evaluate the derived correlations. Furthermore, the total solar radiation on horizontal surfaces is separated into the beam and diffuses components. Empirical correlations for estimating the diffuse solar radiation incident on horizontal surfaces have been proposed. The total solar radiation incident on a tilted surface facing south $H_{t,\beta}$ with different tilt angles is then calculated using both Liu and Jordan isotropic model and Klucher's anisotropic model. It is inferred that the isotropic model is able to estimate $H_{t,\beta}$ more accurate than the anisotropic one. At the optimum tilt angle, the maximum value of $H_{t,\beta}$ is obtained as ~36 (MJ/m^2 day) during January. Comparisons with 22 years average data of NASA SSE Model showed that the proposed correlations are able to predict the total annual energy on horizontal and tilted surfaces in Jeddah with a reasonable accuracy. It is also found that at Jeddah, the solar energy devices have to be tilted to face south with a tilt angle equals the latitude of the place in order to achieve the best performance all year round.
- Islam, Alili, Kubo, and Ohadi (2010) presented actual measurements of direct solar radiation in Abu Dhabi (24.43°N, 54.45°E) with the existing meteorological conditions encountered during the measurement throughout the year. High resolution, real-time solar radiation and other meteorological data were collected and processed. Daily and monthly statistics of direct solar radiation were calculated from the one-minute average recorded by a Middleton Solar DN5-E Pyroheliometer. The highest daily and monthly mean solar radiation values were recorded as 730 and 493.5 W/m^2, respectively. The highest one-minute average daily solar radiation was recorded as 937 W/m^2. In addition to direct beam radiation, the daily average clearness indexes, surface temperature variations, wind speeds and relative humidity variations are discussed. When possible, direct beam radiation and some meteorological data are compared with corresponding data of the 22-year average of NASA's surface meteorology and solar-energy model. The measured data (direct beam radiation and meteorological) are in close agreement with the NASA SSE model with some discrepancy.

The optimum tilt angle β_{opt} for a particular span of time, at any location, will be the one at which, if collector is tilted, will collect maximum solar energy compared with any other angle. After determining the latitude φ and the global solar radiation on horizontal surface using any of methods mentioned in chapter 4 of this book or using (Surface meteorology and Solar Energy (SSE) Release 6.0 Methodology, 2012), the following correlations provided in NASA SSE Methodology could be used to find out the monthly diffused (H_d) component on horizontal surface (Surface meteorology and Solar Energy (SSE) Release 6.0 Methodology, 2012):

For latitudes between 0° and 45° North and South:

$$\frac{H_d}{H} = 0.96268 - 1.452K_{t,m} + 0.27365K_{t,m}^2 + 0.04279K_{t,m}^3 + 0.000246\omega_{ss} + 0.001189\alpha_{noon} \tag{42}$$

where ω_{ss} is the sunset hour angle on the tilted surface and α_{noon} is solar altitude angle at solar noon:

$$\alpha_{noon} = 90^o - \left|\varphi - \delta\right| \text{ For Northern Hemisphere} \tag{43}$$

$$\alpha_{noon} = 90^o - \left|\delta - \varphi\right| \text{For Southern Hemisphere} \tag{44}$$

For latitudes between 45° and 90° North and South:

$$\frac{H_d}{H} = \begin{Bmatrix} 1.441 - 3.6839K_{t,m} + 6.4927K_{t,m}^2 - 4.147K_{t,m}^3 + 0.0008\omega_{ss} \\ +0.008275\alpha_{noon}\ if\ \omega_{ss} = 0^o - 81.4^o \\ 1.6821 - 2.5866K_{t,m} + 2.373K_{t,m}^2 - 0.5294K_{t,m}^3 - 0.00277\omega_{ss} \\ -0.004233\alpha_{noon}\ if\ \omega_{ss} = 81.4^o - 100.0^o \\ 0.3498 + 3.8035K_{t,m} - 11.765K_{t,m}^2 + 9.1748K_{t,m}^3 + 0.001575\omega_{ss} \\ -0.002837\alpha_{noon}\ if\ \omega_{ss} = 100.0^o - 125.0^o \\ 1.6586 - 4.412K_{t,m} + 5.8K_{t,m}^2 - 3.1223K_{t,m}^3 + 0.000144\omega_{ss} \\ -0.000829\alpha_{noon}\ if\ \omega_{ss} = 125.0^o - 150.0^o \\ 0.6563 - 2.893K_{t,m} + 4.594K_{t,m}^2 - 3.23K_{t,m}^3 + 0.004\omega_{ss} \\ -0.23\alpha_{noon}\ if\ \omega_{ss} = 150.0^o - 180.0^o \end{Bmatrix} \tag{45}$$

Using Liu and Jordan (1962), the beam component on tilted surface is determined using $R_{b,m}$ which is the ratio of the average beam radiation on tilted surface to that on horizontal surface for a month (see the equation (142) of the 4th chapter). Assuming isotropic sky with ground reflectance (ρ_g), total radiations on tilted surface could be determined using the equation (162) of the fourth chapter:

$$H_{t,m,\beta} = H_{b,m}R_{b,m} + H_{d,m}\left[\frac{1 + cos\left(\beta\right)}{2}\right] + \rho_g\left[\frac{1 - cos\left(\beta\right)}{2}\right]H_{t,m}$$

For finding for a single month, the values of the above equation must be looped from -90° to +90°.

The input to the function was span of months, latitude and data for solar radiation on horizontal, of the location of interest. The function was returning optimum tilt angle and estimate of energy to be collected, for that span of given months.

Extraterrestrial Solar Radiation Based Theories

The first method, based on the extraterrestrial solar radiation, and proposed to calculate the daily and monthly optimum tilt angle, is that proposed by El-Kassaby (1988). As seen before in the fourth chapter, the extraterrestrial solar radiation on the surface of tilt β and orientated with an azimuth angle γ is:

The first method, based on the extraterrestrial solar radiation, and proposed to calculate the daily and monthly optimum tilt angle, is that proposed by El-Kassaby (1988). As seen before in the third chapter, the daily extraterrestrial solar radiation on the surface of tilt β and orientated towards the Equator is $H_{0,d}$(Jm^{-2}):

$$H_{0,\beta d} = \frac{24x3600xG_s}{\pi} * \left[cos(\varphi-\beta)cos(\delta)sin(\omega_{ss}) + \frac{\pi * \omega_{ss}}{180} sin(\varphi-\beta)sin(\delta)\right] \tag{46}$$

where ω_{ss} is the sunset hour angle, in degrees, on the Equator facing tilted surface. El-Kassaby (1988) assumed that, the sunset hour angle on the Equator facing tilted surface ω_{ss} is equal to that on the horizontal surface $\omega_s, \omega_{ss} = \omega_s$. Thus, as ω_s does not depend on β, the differentiation of the equation (46) with respect to β and by equating the result to zero, the daily optimum tilt angle could be determined:

$$\frac{\partial H_{0,\beta d}}{\partial \beta} = 0; ==> \beta_{opt,d} = \varphi - arctan\left[\omega_s \; sin(\delta) / cos(\delta) sin(\omega_s)\right] \tag{47}$$

Soulayman (1991) corrected the El-Kassaby (1988) method and made it applicable all over the year. According to Soulayman (1991), the extraterrestrial solar radiation on the tilted surface with different orientations could be written as follows (Soulayman, 1991):

$$H_{0,\beta d} = \frac{12x3600xG_s}{\pi} * \left\{\begin{matrix} A_2\left[sin(\omega_{ss}) - sin(\omega_{sr})\right] - \\ A_3\left[cos(\omega_{ss}) - cos(\omega_{sr})\right] + \dfrac{\pi * (\omega_{ss} - \omega_{sr})}{180^o} A_1 \end{matrix}\right\} \tag{48}$$

where

$$\omega_{ss} = min\left\{arccos\left[-tan(\delta)tan(\varphi)\right], arccos\left(-\frac{A_1}{A_4}\right) + arcsin\left(\frac{A_3}{A_4}\right)\right\} \tag{49}$$

$$\omega_{sr} = \max\left\{-\arccos\left[-\tan(\delta)\tan(\varphi)\right], -\arccos\left(-\frac{A_1}{A_4}\right) + \arcsin\left(\frac{A_3}{A_4}\right)\right\} \tag{50}$$

$$A_1 = \sin(\delta)\left[\sin(\varphi)\cos(\beta) - \sin(\beta)cos(\varphi)\cos(\gamma)\right] \tag{51}$$

$$A_2 = \cos(\delta)\left[\cos(\varphi)\cos(\beta) + \sin(\beta)sin(\varphi)\cos(\gamma)\right] \tag{52}$$

$$A_3 = \cos(\delta)\sin(\beta)\sin(\gamma) \tag{53}$$

$$A_4 = \sqrt{A_2^2 + A_3^2} \tag{54}$$

Thus, the daily optimum tilt angle $\beta_{opt,d}$ is the solution of the nonlinear algebraic equation:

$$\frac{\partial H_{0,\beta d}}{\partial \beta} = 0; ==> \left(\frac{\partial A_2}{\partial \beta}\right)\left[\sin(\omega_{ss}) - \sin(\omega_{sr})\right]$$
$$+A_2\left[\frac{\partial \omega_{ss}(n)}{\partial \beta}\cos(\omega_{ss}) - \frac{\partial \omega_{sr}(n)}{\partial \beta}\cos(\omega_{sr})\right]+$$
$$\left(\frac{\partial A_1}{\partial \beta}\right)\left[\omega_{ss}(n) - \omega_{sr}(n)\right] + A_1\left[\frac{\partial \omega_{ss}(n)}{\partial \beta} - \frac{\partial \omega_{sr}(n)}{\partial \beta}\right] - \left(\frac{\partial A_3}{\partial \beta}\right)\left[\cos(\omega_{ss}) - \cos(\omega_{sr})\right]+ \tag{55}$$
$$A_3\left[\frac{\partial \omega_{ss}(n)}{\partial \beta}\sin(\omega_{ss}) - \frac{\partial \omega_{sr}(n)}{\partial \beta}\sin(\omega_{sr})\right] = 0$$

where n is the day number of the considered day in the year, $\omega_{ss}(n)$ and $\omega_{sr}(n)$ are the sunset and sunrise in considered day in the year. The solution of the equation (55), for south facing solar collector during the period from 22/9 to 21/3, is:

$$\beta_{opt,d} = \varphi - arctan\left\{\frac{\omega_s(n)\ sin[\delta(n)]}{cos[\delta(n)]sin[\omega_s(n)]}\right\} \tag{56}$$

where $\delta(n)$ is declination angle in considered day in the year. According to Soulayman (1991), the equation (47) is correct for south facing solar collector during the period from 22/9 to 21/3 only as during this period, $\omega_{ss} = \omega_s$ and $\omega_{ss}(n)$ does not depend on β. Later on, the algorithm of Soulayman (1991) was applied in the different cases (see for example, (Soulayman and Sabbagh, 2014) and (Soulayman and Sabbagh, 2015a)). For a period of N days, started on n_1 and ended on n_2, the extraterrestrial solar radiation during this period is:

For a period of N days, started on n_1 and ended on n_2, the extraterrestrial solar radiation during this period is:

$$H_{0,\beta p} = \sum_{n_1}^{n_2} H_{0,\beta d}\left(n\right) = \frac{24x3600}{\pi} * \sum_{n_1}^{n_2} \left\{ \begin{matrix} A_2\left[sin\left(\omega_{ss}\right) - sin\left(\omega_{sr}\right)\right] - A_3\left[cos\left(\omega_{ss}\right) - cos\left(\omega_{sr}\right)\right] \\ + \frac{\pi * \left[\omega_{ss}\left(n\right) - \omega_{sr}\left(n\right)\right]}{180} A_1 \end{matrix} \right\} G_s\left(n\right) \tag{57}$$

The differentiation of the equation (57) with respect to β and by equating the result to zero, the optimum tilt angle, over the mentioned period, could be determined:

$$\frac{\partial H_{0,\beta p}}{\partial \beta} = 0; ==> \sum_{n_1}^{n_2} \left\{ \begin{bmatrix} \left(\frac{\partial A_2}{\partial \beta}\right)\left[\sin \omega_{ss}\left(n\right) - \sin \omega_{sr}\left(n\right)\right] \\ +A_2\left[\frac{\partial \omega_{ss}\left(n\right)}{\partial \beta}\cos\left(\omega_{ss}\right) - \frac{\partial \omega_{sr}\left(n\right)}{\partial \beta}\cos\left(\omega_{sr}\right)\right] + \\ \left(\frac{\partial A_1}{\partial \beta}\right)\left[\omega_{ss}\left(n\right) - \omega_{sr}\left(n\right)\right] \\ +A_1\left[\frac{\partial \omega_{ss}\left(n\right)}{\partial \beta} - \frac{\partial \omega_{sr}\left(n\right)}{\partial \beta}\right] - \\ \left(\frac{\partial A_3}{\partial \beta}\right)\left[\cos\left(\omega_{ss}\right) - \cos\left(\omega_{sr}\right)\right] \\ +A_3\left[\frac{\partial \omega_{ss}\left(n\right)}{\partial \beta}\sin\left(\omega_{ss}\right) - \frac{\partial \omega_{sr}\left(n\right)}{\partial \beta}\sin\left(\omega_{sr}\right)\right] \end{bmatrix} \right\} G_s\left(n\right) = 0 \tag{58}$$

For south facing solar collector, ω_{ss} does not depend on β and $\omega_{ss} = \omega_s$ during the period from 22/9 to 21/3 only. Then, the analytical solution of the equation (58) is:

$$\beta_{opt,p} = \varphi - arctan\left[\frac{\sum_{n_1}^{n_2} G_s\left(n\right)\omega_{ss}\left(n\right)\ sin\left(\delta\right)}{\sum_{n_1}^{n_2} G_s\left(n\right)cos\left(\delta\right)sin\left(\omega_{ss}\right)}\right] \tag{59}$$

The equation (59) is valid on Equator all over the year and in *NH*, for the period starting on 22/9 and ending on 21/3, and in *SH*, for the period starting on 22/3 and ending on 21/9. For other periods Newton's iteration scheme could be applied for searching the solution of the equation (58), $\beta_{opt,p}$. When n_1 and n_2 are the day numbers of the month beginning and month end, the equation (59) gives the monthly optimum tilt angle, $\beta_{opt,m}$. When n_1 and n_2 are the day numbers of the season beginning and season end, the equation (59) gives the seasonally optimum tilt angle, $\beta_{opt,s}$. Finally, the equation (59) gives the biannually optimum tilt angle, $\beta_{opt,b}$ and yearly optimum tilt angle, $\beta_{opt,y}$ when the studied period covers half-year and whole year respectively.

Beam Radiation Based Theories

Radiation data are the best source of information for estimating average incident radiation. However, sometimes in the lack of such data, it is possible to apply empirical relationships by using data providing the ratio (s/S_0), where s is the monthly average daily hours of bright sunshine and S_0 is the day length of the average day of the month (the most popular and most commonly used model is known as the Angstrom-Prescott model (Prescott, 1940)). Solar radiation is an important parameter for the proper design of solar energy conversion devices and building energy systems. As air temperature is readily available at most meteorological stations, the models based on air temperature for estimating solar radiation are more attractive and practical. Therefore, Samani (2000) introduced a procedure to estimate solar radiation and subsequently reference crop evapotranspiration using minimum climatological data - the ambient temperature (T). Samani (2000) describes a modification to an original equation that uses maximum and minimum temperature to estimate solar radiation and reference crop evapotranspiration. The proposed modification allows for the correction of errors associated with indirect climatological parameters affecting the local temperature range. The proposed modification also improves the accuracy of estimates of solar radiation from temperature.

Bristow and Campbell (1984) proposed a relationship between atmospheric transmittance and the daily range of air temperature. The relationship is:

$$\tau_t = A\left[1 - exp\left(-B\Delta T^c\right)\right] \tag{60}$$

where τ_t is the daily total atmospheric transmittance, ΔT is the daily range of air temperature, and A, B, and C are empirical coefficients, determined for a particular location from measured solar radiation data. Tests on three data sets indicate that 70–90% of the variation in daily solar radiation can be accounted for by this simple model.

To improve the air temperature-based models', such as that of Bristow and Campbell (1984) (BC), performance in humid regions, the effects of seasonality and relative humidity are introduced to these correlations, and thus six candidate models are proposed by Li, Cao, Bu, and Zhao (2015). These models are tested at Guangzhou station in China with humid subtropical climate influenced by the Asian monsoon as a case study. Results show that relative humidity (RH) has a greater effect than that of seasonality on the air temperature-based models for daily global solar radiation estimation, and the models of BC family including the effects of seasonality and/or relative humidity give better estimates. Accordingly, a strategy for selecting an optimum model for estimating daily global solar radiation in the humid regions is proposed. However, even the relative humidity and air temperature are widely available from many hundreds of stations in most countries, the main part of empirical relationships is restricted to sunshine duration s. Thus, these relationships could be written as:

$$H_m = f\left(\frac{s}{S_0}, R_H, T\right) H_{0,m} \tag{61}$$

where H_m = monthly average daily radiation on a horizontal surface and $H_{0,m}$ = monthly average daily extraterrestrial solar radiation on a horizontal surface. Taking into consideration that on Northern Hemisphere (*NH*) φ is positive, tilt angle β is positive when surface is oriented toward Equator and negative for the opposite direction and on Southern Hemisphere (*SH*) φ is negative, β is negative when surface is oriented toward Equator and positive for the opposite direction.

The monthly average daily total solar radiation H on a horizontal plane, provided from meteorological station, includes direct H_b and diffuse H_d components while the monthly average daily total solar radiation H_t on a tilted surface can be divided into direct, diffuse and ground reflected components. The H_d and H_b components of H can be estimated using empirical relationships H_d=f(H) by means of clearness index K_T (Duffie and Beckman, 2013). If the diffuse and ground reflected H_g are each assumed to be isotropic, the total solar radiation on a tilted surface H_t, can be obtained using the equation of Liu and Jordan (1962) as follows:

$$H_{t,d,\beta} = H_{b,d} R_{b,d} + H_{d,d}\left[\frac{1+cos(\beta)}{2}\right] + \rho_g\left[\frac{1-cos(\beta)}{2}\right] H_{t,d} \tag{62}$$

where the index "d" denotes the daily values. When assuming the beam radiation, received by a surface tilted by an angle β and orientated by an angle γ with respect to the south, is a linear function of the extraterrestrial solar radiation, received by the same surface outside the Earth's atmosphere:

$$H_{b,d} = f'\left(\frac{s}{S_0}, R_H, T\right) H_{0,d} \tag{63}$$

Thus, by neglecting the influence of the diffuse and ground reflected solar radiation on the optimum tilt angle, the equations (62) and (63) could be regrouped in one equation As follows:

$$H_{t,d,\beta} = H_{b,d,\beta} = f'\left(\frac{s}{S_0}, R_H, T\right) H_{0,\beta,d} \tag{64}$$

where $H_{0,\beta d}$ is given by the equation (48). Then, for any given azimuth angle, the daily optimum tilt angle could be found by solving the following equation:

$$\frac{\partial H_{t,d,\beta}}{\partial \beta} = 0; ==> \beta_{opt,d} \tag{65}$$

The solution of the equation (65) for the Equator facing collectors, during the period 22/9-21/3, is the same as the equation (56), while the Newton iteration method could be applied for searching the nonlinear algebraic equation (65).

In the general case, the optimum daily tilt angle and azimuth angle could be found by solving a system of two nonlinear algebraic equations:

$$\frac{\partial H_{t,d,\beta,\gamma}}{\partial \beta} = 0 \tag{66}$$

$$\frac{\partial H_{t,d,\beta,\gamma}}{\partial \gamma} = 0 \tag{67}$$

It is to mention here that, beam solar radiation based theory and extraterrestrial solar radiation based theories lead, approximately, to the same results with regard to the optimum tilt angle and optimum azimuth angle.

On the Beam Radiation Based Theories Assumption

When the daily values of total solar radiation on horizontal plane are available, it is important to determine the solar radiation intensity distribution during these days and to estimate the daily solar radiation components on the tilted plane. If these data are not available, it is useful to define a standard clear sky and calculate the daily solar radiation which could be received on a tilted surface under these standard conditions. Accordingly, the daily total solar radiation on a tilted Equator facing plane could be calculated as:

$$H_{t,d,\beta} = \frac{12 * 3.6}{\pi} * 1.367 * C(n) \left\{ \begin{array}{l} \int_{\omega_{sr}}^{\omega_{ss}} \tau_b \cos(\theta)\, d\omega \\ + \int_{0}^{\omega_s} \left[\begin{array}{l} \tau_b Á[1 - \cos(\beta)] \\ + \tau_d [1 - \cos(\beta)] \end{array} \right] \cos(\theta_z)\, d\omega \end{array} \right\} \tag{68}$$

where τ_b is the atmospheric transmittance for beam radiation, τ_d is the atmospheric transmittance for diffuse radiation and θ_z is the solar incidence angle on the horizontal plane. The dependence of the atmospheric transmittance on the solar hour angle ω makes the intensity of solar radiation variable from sunrise to sunset.

Then, differentiating the equation (68) with the respect to the β, a nonlinear algebraic equation is obtained. By equating the left part of the derived equation to zero, the daily optimum tilt angle $\beta_{opt,d}$ is its solution:

$$\int_{\omega_{sr}}^{\omega_{ss}} \tau_b \frac{\partial \cos\left(\theta\right)}{\partial \beta} d\omega$$
$$+\int_{0}^{\omega_s} (\tau_b Á - \tau_d) \sin\left(\beta\right) cos\left(\theta_z\right) d\omega + \tau_b\left(\omega_{ss}\right) \cos\left[\theta\left(\omega_{ss}\right)\right] \frac{\partial \omega_{ss}}{\partial \beta} \tag{69}$$
$$-\tau_b\left(\omega_{sr}\right) \cos\left[\theta\left(\omega_{sr}\right)\right] \frac{\partial \omega_{sr}}{\partial \beta} = 0$$

In the case of ω_{ss} tilt angle independent, the solution of the equation (69) could be approximated with a high accuracy by the equation (56). The precise solution of the equation (69), for different stations in the mid-latitude zone, clearly demonstrates that, it could be negative $\beta_{opt,d} \leq 0^o$ during a relatively small period and positive $\beta_{opt,d} \geq 0^o$ during a relatively long period in the NH and it could be positive $\beta_{opt,d} \geq 0^o$ during a relatively small period and negative $\beta_{opt,d} \leq 0^o$ during a relatively long period in the SH. The length of the small period decreases with increasing the latitude value. Tables 23 and 24 show the above mentioned periods in the mid-latitude zone.

When the daily monthly average total solar radiation on a horizontal plane H is available, the daily monthly average total solar radiation on a tilted surface $H_{t,m}$, can be obtained as follows:

$$H_{t,m} = H\left(1 - \frac{H_{d,m}}{H}\right) R_{b,m} + H_{d,m}\left[\frac{1 + \cos\left(\beta\right)}{2}\right] + \rho H\left[\frac{1 - \cos\left(\beta\right)}{2}\right] \tag{70}$$

if the daily monthly average diffuse $H_{d,m}$ and ground reflected $H_{g,m}$ are each assumed to be isotropic, where ρ is the diffuse reflectance of the surroundings and $R_{b,m}$ is geometric factor:

$$R_b = 0.5 \frac{\sum\left\{A_1\left(\omega_{ss} - \omega_{sr}\right) + A_2\left[\sin\left(\omega_{ss}\right) - \sin\left(\omega_{sr}\right)\right] + A_3\left[\cos\left(\omega_{ss}\right) - \cos\left(\omega_{sr}\right)\right]\right\}}{\sum\left[\omega_s \sin\left(\delta\right)\sin\left(\varphi\right) + \cos\left(\delta\right)\sin\left(\varphi\right) \sin\left(\omega_s\right)\right]} \tag{71}$$

Then, differentiating the equation (70) with the respect to the surface tilt angle β, a nonlinear algebraic equation is obtained. By equating the left part of the derived equation to zero, the optimum tilt angle over a period of several consecutive days $\beta_{opt,p}$ is its solution:

Table 23. The periods of positive and negative $\beta_{opt,d}$ in the NH

	$\beta_{opt,d} \geq 0^o$			$\beta_{opt,d} < 0^o$			$\beta_{opt,d} \geq 0^o$		
aφ(°)	**Days**	**From**	**To**	**Days**	**From**	**To**	**Days**	**From**	**To**
23.45	119	1	119	106	120	225	140	226	365
24	120	1	120	104	121	224	141	225	365
25	122	1	122	100	123	222	143	223	365
26	124	1	124	96	125	220	145	221	365
27	126	1	126	93	127	219	146	220	365
28	127	1	127	90	128	217	148	218	365
29	129	1	129	86	130	215	150	216	365
30	131	1	131	82	132	213	152	214	365
31	133	1	133	78	134	211	154	212	365
32	135	1	135	74	136	209	156	210	365
33	137	1	137	70	138	207	158	208	365
34	139	1	139	66	140	205	160	206	365
35	142	1	142	61	143	203	162	204	365
36	144	1	144	56	145	200	165	201	365
37	146	1	146	52	147	198	167	199	365
38	149	1	149	47	150	196	169	197	365
39	151	1	151	42	152	193	172	194	365
40	154	1	154	36	155	190	175	191	365
41	157	1	157	30	158	187	178	188	365
42	161	1	161	22	162	183	182	184	365
43	169	1	169	12	170	181	184	182	365
43.45	181	1	181	0			184	182	365

Table 24. The periods of positive and negative $\beta_{opt,d}$ in the SH

	$\beta_{opt,d} \geq 0^o$			$\beta_{opt,d} < 0^o$			$\beta_{opt,d} \geq 0^o$		
φ(°)	Days	From	To	Days	From	To	Days	From	To
-23.45	42	1	42	260	43	302	63	303	365
-24	42	1	42	261	43	303	62	304	365
-25	40	1	40	264	41	304	61	305	365
-26	38	1	38	268	39	306	59	307	365
-27	36	1	36	272	37	308	57	309	365
-28	34	1	34	276	35	310	55	311	365
-29	32	1	32	280	33	312	53	313	365
-30	30	1	30	284	31	314	51	315	365
-31	28	1	28	288	29	316	49	317	365
-32	26	1	26	292	27	318	47	319	365
-33	24	1	24	296	25	320	45	321	365
-34	22	1	22	300	23	322	43	323	365
-35	20	1	20	304	21	324	41	325	365
-36	18	1	18	308	19	326	39	327	365
-37	16	1	16	312	17	328	37	329	365
-38	13	1	13	318	14	331	34	332	365
-39	11	1	11	322	12	333	32	334	365
-40	8	1	8	328	9	336	29	337	365
-41	5	1	5	334	6	339	26	340	365
-42	1	1	1	342	2	343	22	344	365
-43	0			351	1	351	14	352	365
-43.45	0			358	1	358	7	359	365

$$\sum\left\{\begin{array}{l}\left(\frac{\partial A_2}{\partial\beta}\right)\left[\sin\left(\omega_{ss}\right)-\sin\left(\omega_{sr}\right)\right]+A_2\left[\cos\left(\omega_{ss}\right)\left(\frac{\partial\omega_{ss}}{\partial\beta}\right)-\cos\left(\omega_{sr}\right)\left(\frac{\partial\omega_{sr}}{\partial\beta}\right)\right]\\+\left(\frac{\partial A_1}{\partial\beta}\right)\left(\omega_{ss}-\omega_{sr}\right)+A_1\left(\frac{\partial\omega_{ss}}{\partial\beta}-\frac{\partial\omega_{sr}}{\partial\beta}\right)-\left(\frac{\partial A_3}{\partial\beta}\right)\left[\cos\left(\omega_{ss}\right)-\cos\left(\omega_{sr}\right)\right]+\\A_3\left[\sin\left(\omega_{ss}\right)\left(\frac{\partial\omega_{ss}}{\partial\beta}\right)-\sin\left(\omega_{sr}\right)\left(\frac{\partial\omega_{sr}}{\partial\beta}\right)\right]\end{array}\right\}G_s(n)$$
$$-0.5\sin\left(\beta\right)\left(K_T-\rho\right)/\left(1-K_T\right)=0 \tag{72}$$

In the case of Equator facing (*EF*) and Pole facing (*PF*) surfaces, the sunset hour angle ω_{ss} is tilt angle independent over several months (and this is always the case when $\varphi=0^o$ while for $\varphi\neq0^o$ it takes place for a half year) and the solution of the equation (72) could be approximated with a high accuracy by:

$$\beta_{opt,p} = \varphi - arctan\left[\frac{\Sigma G_s\left(n\right)\omega_s\ sin\left(\delta\right)}{\Sigma G_s\left(n\right)\ cos\left(\delta\right)sin\left(\omega_s\right)}\right] \tag{73}$$

where the summation in the equation (73) covers the period under consideration.

When ω_{ss} is a function of tilt angle, the equation (72) could be solved using Newton iteration method.

When the daily values of total solar radiation on horizontal plane are available, it is important to determine the solar radiation intensity distribution during these days and to estimate the daily solar radiation components on the tilted plane.

On the Applicability of Different Formulas

Generally, for surfaces of different orientations, $\sin(\gamma)\neq0$ and $\omega_{ss}\left(n\right)\neq-\omega_{sr}\left(n\right)$. Furthermore, for these surfaces the equations (56) and (59) are not applicable for Abs(γ)> 20° even as approximate solutions. On the other hand, for *EF* surfaces in both *NH* and *SH*, $\omega_{ss}\left(n\right)=-\omega_{sr}\left(n\right)$ and $\sin(\gamma)=0$. So, $A_3=0$ and $\omega_{ss}\left(n\right)$ is independent of tilt angle on *EF* surfaces for the period starting on 22/9 and ending on 21/3 in *NH* and for the period starting on 22/3 and ending on 21/9 in *SH*. Therefore, the equations (56) and (59) are excellent approximate solutions of the equations (55) and (58) during these two periods while they are acceptable approximate solutions for other periods.

For Equator facing surface, when calculating $\beta_{opt,d}$ for different latitudes of the mid-latitude zone (23.45°N<φ<43.45°N and 23.45°S<φ<43.45°S) using the equation (55), it was found that (see Tables 23 and 24): a) There are several days in the year where $\beta_{opt,d}\leq0^o$ in the NH and $\beta_{opt,d}\geq0^o$ in the SH and b) The number of these days decreases from 106 to 0 days with φ increase from 23.45°N to 43.45°N or from 23.45°S to 43.45°S. Moreover, when calculating $\beta_{opt,d}$ for the same latitudes using the equation (55) it was found that, the difference between the precise values of $\beta_{opt,d}$ and those deriving from the equation (56) is zero for the cases where φ=23.45° N, 23.45°S and during half year for other φ. The maximum difference value increases from 0° to 2.5° when φ increases from 23.45°N to 43.45 °N or from

23.45°S to 43.45 °S. This value is found to be 0.64° for φ=30°N. Figure 10 demonstrates the above mentioned difference for latitudes 43.45 °N and 43.45 °S, where this difference is maximal. Therefore, this suggests that the equation (56) could be applied to calculate $\beta_{opt,d}$ with a good accuracy in mid-latitudes zone.

The practical design and installation of a solar system where the solar collectors could be orientated towards south and north directions requires additional costs which are not evident to be covered by the solar system during its exploitation in comparison with the other design and installation where the solar collectors could be orientated towards south or north directions as the energy losses are not considerable. Therefore, as the substitution of the negative values of $\beta_{opt,d}$ by 0° in the NH and the positive values of $\beta_{opt,d}$ by 0° in the SH does not lead to a considerable losses in the solar gain, it is reasonable from practical point of view to substitute the negative values of $\beta_{opt,d}$ by 0° in the NH and the positive values of $\beta_{opt,d}$ by 0° in the SH. This solution could be programmed using the following condition:

$$sign\left(\beta_{opt,d}\right) = sign\left(\varphi - \delta\right) elsewhere\, \beta_{opt,d} = 0^o \tag{74}$$

When calculating $\beta_{opt,d}$ for different latitudes of the mid-latitude zone in NH (23.45°N<φ<43.45°N) using the equation (55) and the condition (70) for Equator facing surface, the results presented in Figure 11 are obtained. Figure 12 gives the results of calculations for the boundary latitudes of mid-latitude zone in both NH and SH (Soulayman and Hammoud, 2016).

Figure 10. The differences between the calculated $\beta_{opt,d}$ (°) using the equations (55) and (56) for $\varphi = 43.45^o N$ (♦) and $\varphi = 43.45^o S$ (■).

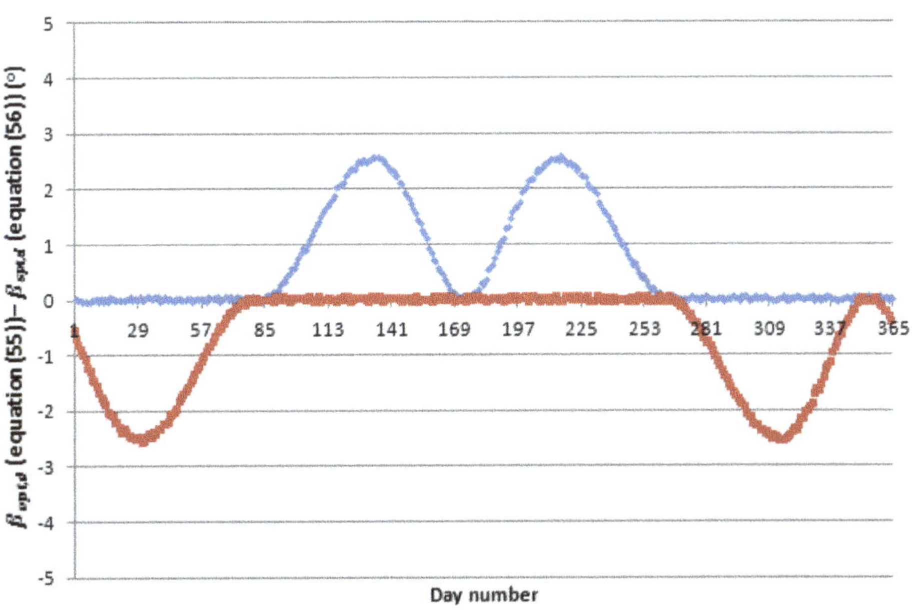

Figure 11. Daily optimum tilts of solar collectors at mid-latitude zone in NH

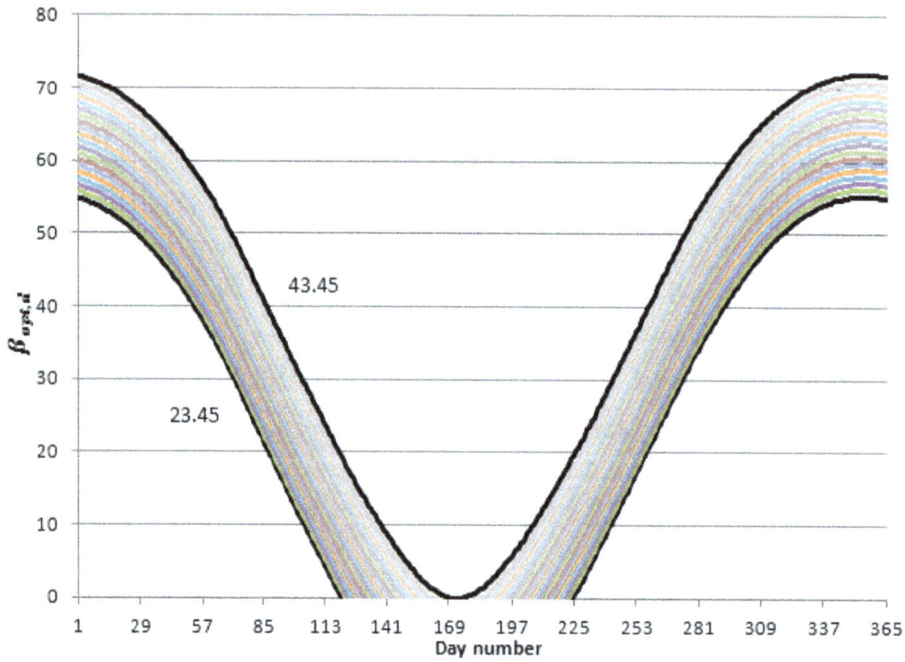

Figure 12. $\beta_{opt,d}$ for φ= 23.45°N (-), 23.45°S (♦), 43.45°N (●) and 43.45°S (■)

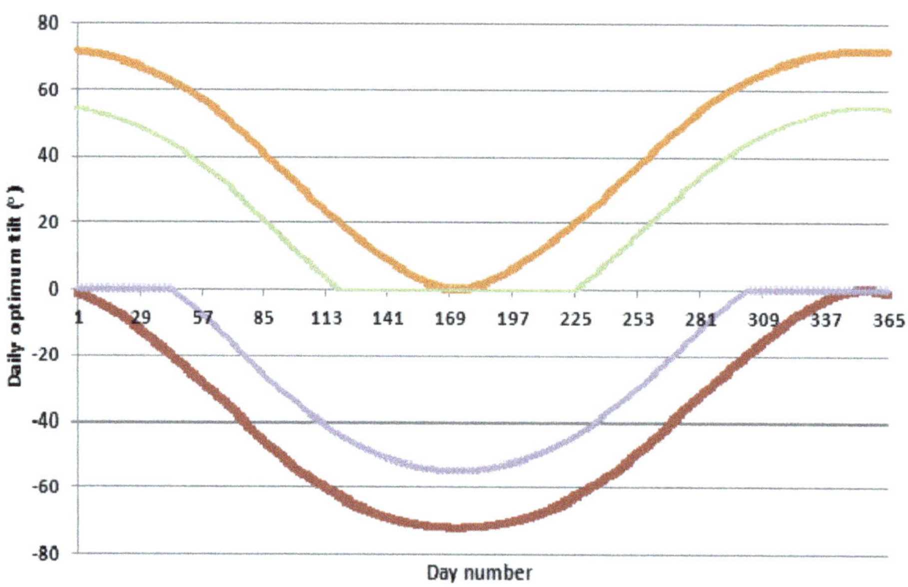

On the other hand, when calculating $\beta_{opt,m}$ for different latitudes of the mid-latitude zone in NH (23.45°N<L<43.45°N) using the equations (58) and (59) conjointly with the condition (74) it was found that: a) There is at least one month in the year where $\beta_{opt,m} \approx 0°$ (see Figure 13); b) The number of months, where $\beta_{opt,m} \approx 0°$, decreases from 3 to 1 when φ increases from 23.45°N to 43.45 °N or from 23.45°S to 43.45 °S (see Figure 13). Klein (1977) introduced 12 representative days (one for each month in the year) which could be called the characteristic days. The daily optimum tilt angle for characteristic day $\beta_{opt,cd}$ of each month is very near to monthly optimum tilt angle $\beta_{opt,m}$ and the difference between them does not exceed 1° (see Figure 14). Dividing the calendar year into 4 quarters and renaming these quarters as the known seasons where winter period starts from January 1 until March 31, spring period starts from April 1 until June 30, summer period starts from July 1 until September 30 and autumn period starts from October 1 until December 31 the calculated $\beta_{opt,s}$ for the mid-latitude zone is given on Figures 15 to 18.

For surface with different orientations, when calculating $\beta_{opt,m}$ for different latitudes of the mid-latitude zone using the equation (58) conjointly with condition (74), it was found that: a) $\beta_{opt,m}$ decreases when γ increases; b) For $\gamma=\pm 90°$, $\beta_{opt,m}=0°$ for a number of months; c) The number of these months decreases from 12 (see Figure 19) to 7 months (see Figure 20) with φ increase from 23.45°N to 43.45 °N or from 23.45°S to 43.45 °S; and d) For Abs(γ)$\leq$ 20° the $\beta_{opt,m}$ changes are not considerable (see Figures 19 and 20).

Finally, it should be mentioned here that, the proposed polynomial equation (see equation (3)) could be applied for calculating the daily optimum tilt in the mid-latitude zone with an accuracy of $\mp 1^o$. Figure 21 demonstrates this applicability for two latitude values: φ =30°S and 43.45°S.

Figure 13. $\beta_{opt,m}$ for φ= 23.45°N (▲), 23.45°S (x), 43.45°N (■) and 43.45°S (♦)

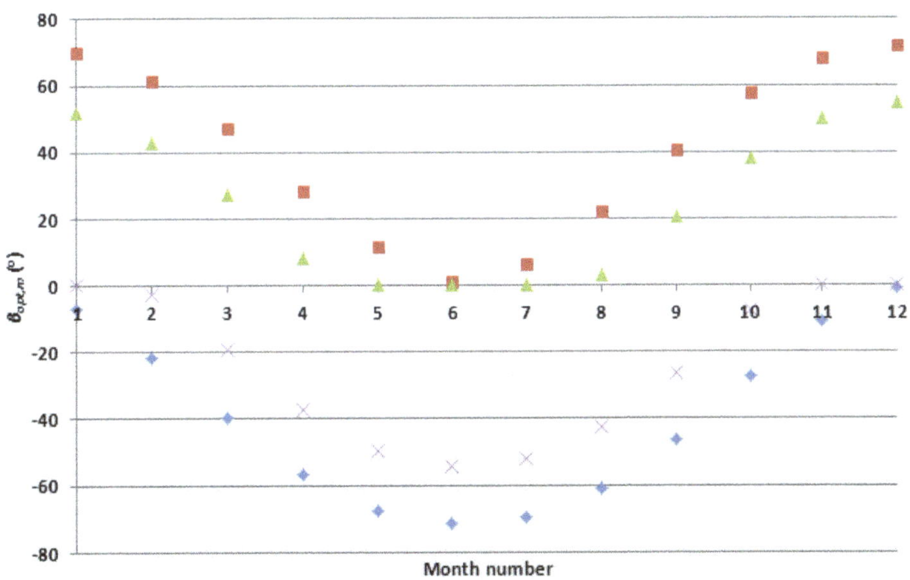

Figure 14. $\beta_{opt,m} - \beta_{opt,cd}$ *for 43.45°S (ж) and 43.45°N (●)*

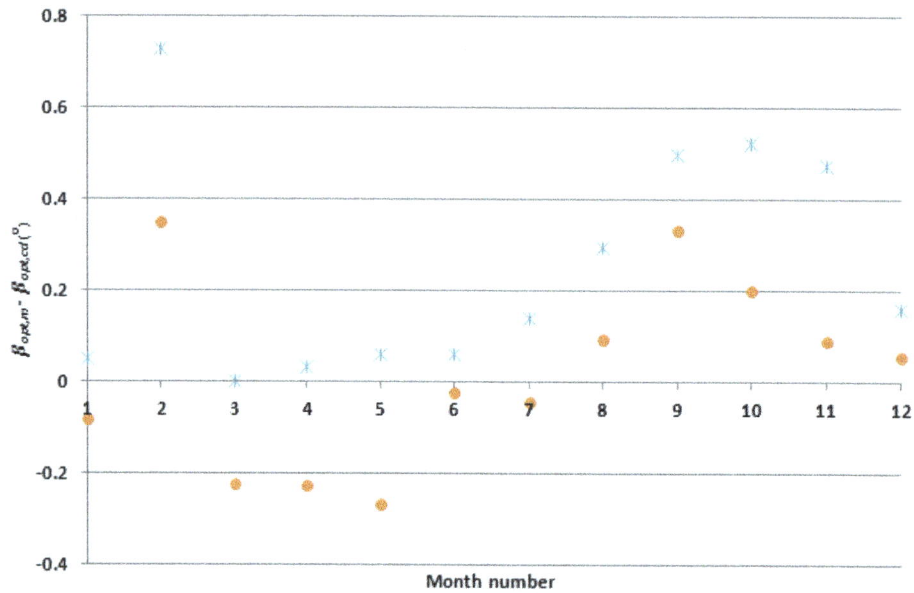

Figure 15. $\beta_{opt,d}$, $\beta_{opt,m}$, $\beta_{opt,s}$, $\beta_{opt,b}$ *and* $\beta_{opt,y}$ *for* φ*= 23.45°N*

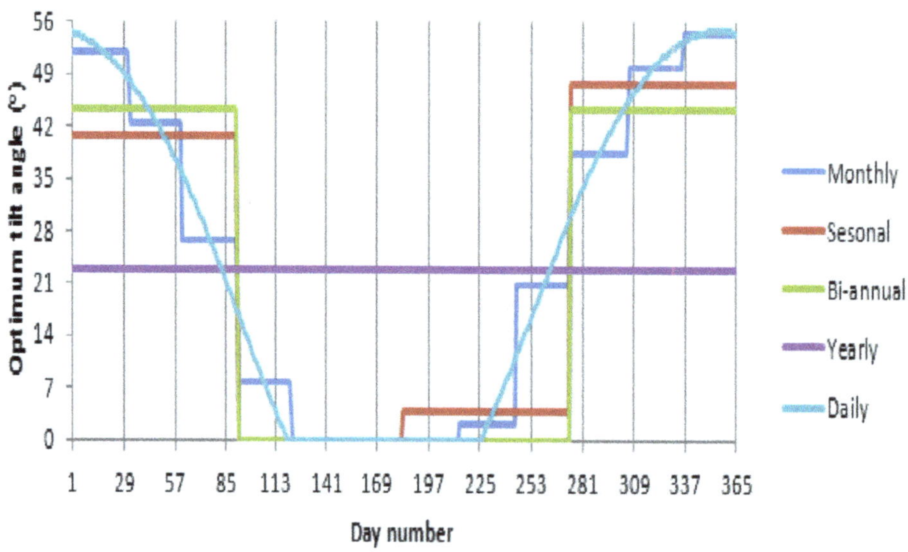

Figure 16. $\beta_{opt,d}$, $\beta_{opt,m}$, $\beta_{opt,s}$, $\beta_{opt,b}$ and $\beta_{opt,y}$ for φ = 23.45°S

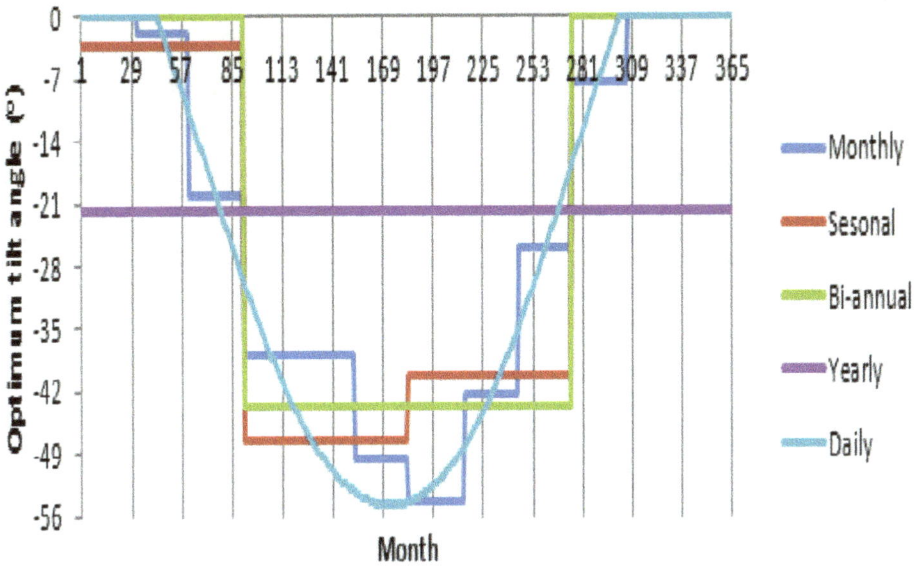

Figure 17. $\beta_{opt,d}$, $\beta_{opt,m}$, $\beta_{opt,s}$, $\beta_{opt,b}$ and $\beta_{opt,y}$ for φ = 43.45°N

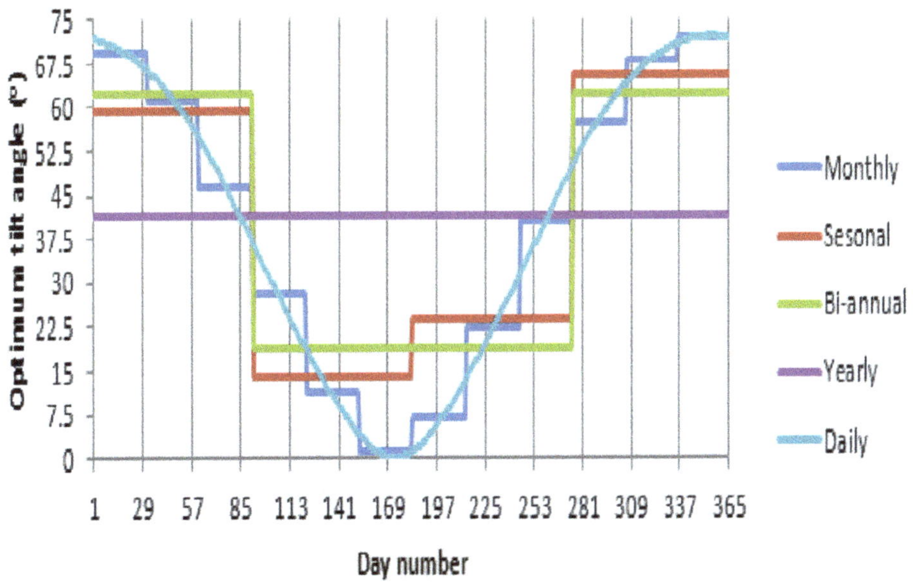

Figure 18. $\beta_{opt,d}$, $\beta_{opt,m}$, $\beta_{opt,s}$, $\beta_{opt,b}$ *and* $\beta_{opt,y}$ *for* $\varphi = 43.45°S$

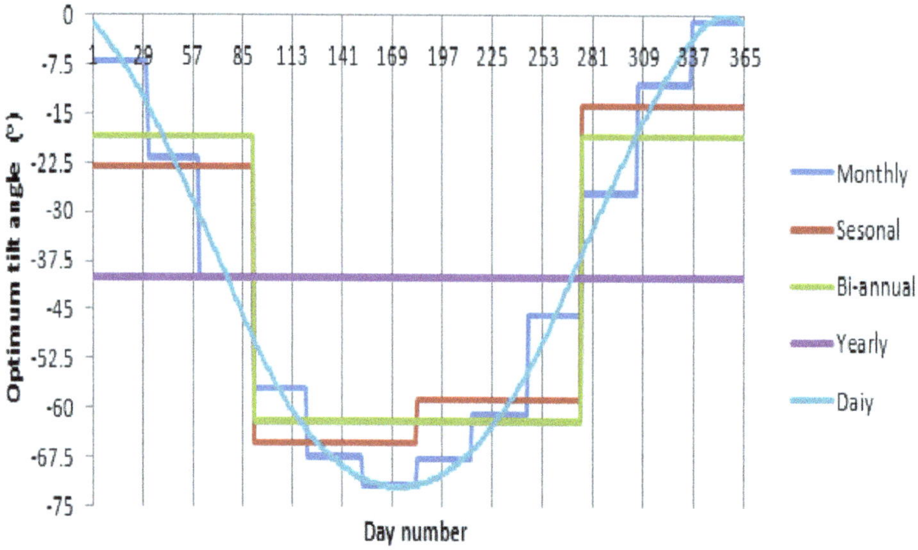

Figure 19. $\beta_{opt,m}$ *for 23.45°N at different orientations*

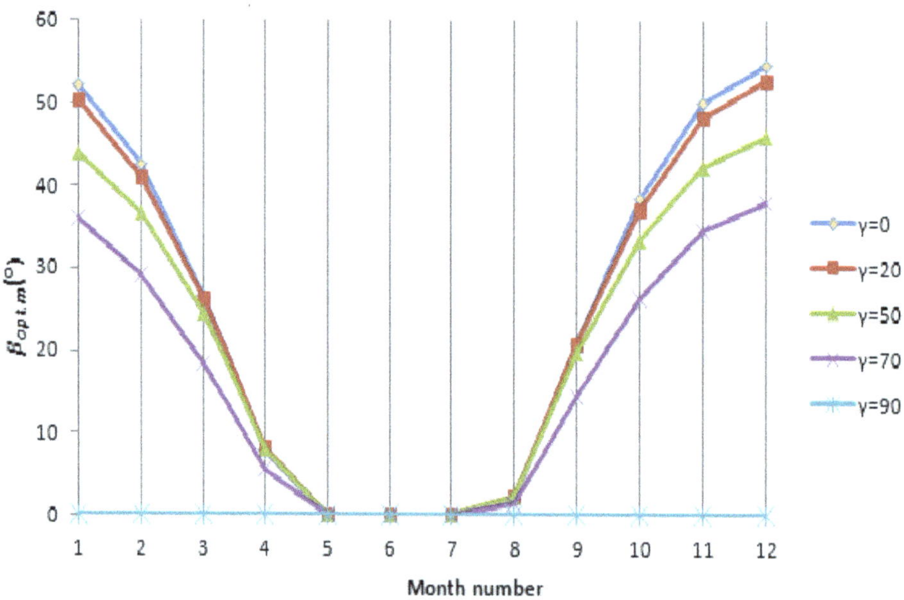

Figure 20. $\beta_{opt,m}$ for 43.45°N at different orientations

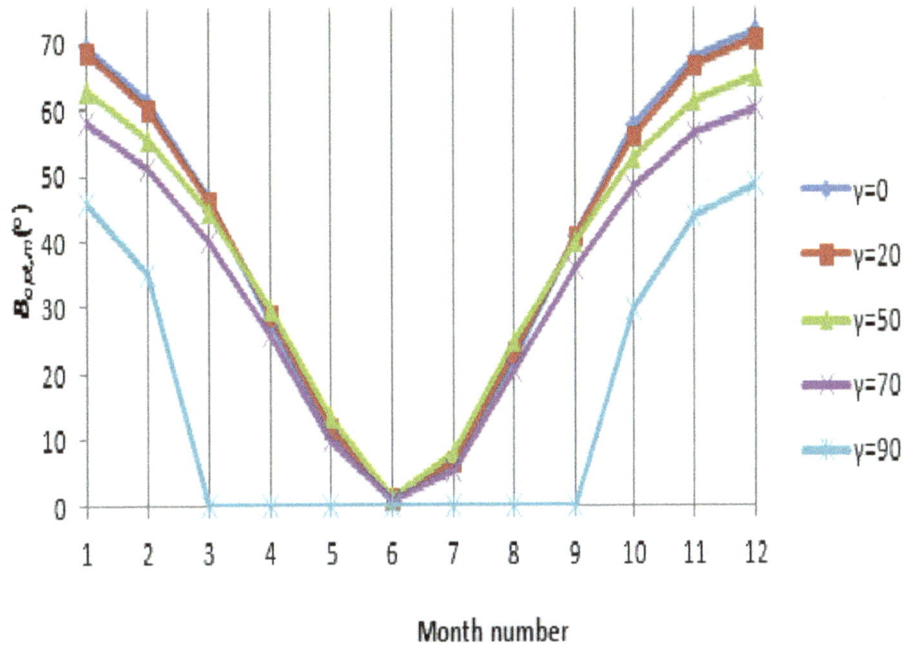

Figure 21. The differences between the calculated results of $\beta_{opt,d}$ using the (Soulayman, 1991) approach (precise) and the polynomial fit (see equation (23)) for φ =30°S and 43.45°S

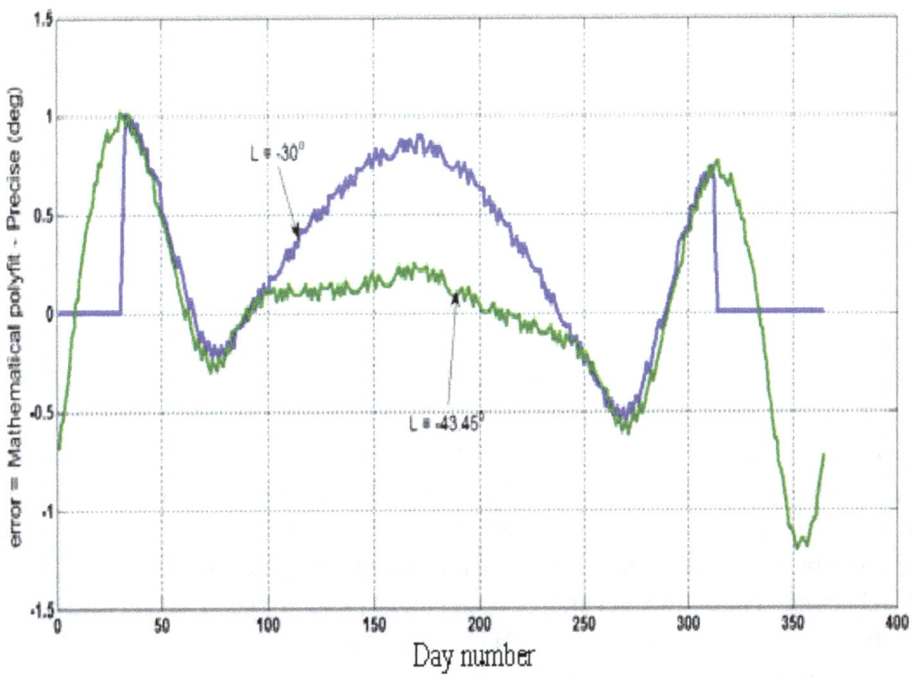

Comparison With Available Previous Results

Daily Optimum Tilt

Stanciu and Stanciu (2014) proposed $\beta_{opt,d} = \varphi\text{-}\delta$ for determining the optimum tilt of south facing collector at latitudes from 0° to 80°. As $\theta_{noon} = \varphi\text{-}\delta$ it is expected from their proposal that $\beta_{opt,d} = \theta_{noon}$. Comments on (Stanciu and Stanciu, 2014) are given in (Soulayman, 2015). So, no need to compare daily optimum tilt results with those mentioned in (Stanciu and Stanciu, 2014). El-Kassaby (1988) was the first who applied the equation (56) for calculating $\beta_{opt,d}$ for latitude of 0°N to 60°N without the condition (74). As consequences, this has led to get unacceptable $\beta_{opt,d}$ values over a period of time for high latitudes. Such results are meaningless for these periods and approximate for half year duration. Gunerhan and Hepbasli (2007) gave graphically the $\beta_{opt,d}$ dependence on year day number for Izmir (φ=38.46°) and provided numerical values of $\beta_{opt,d}$ for characteristic days. A remarkable agreement is achieved between the results of (Soulayman and Hammoud, 2016) and those of (Gunerhan and Hepbasli, 2007) with regard to $\beta_{opt,d}$. Since it is mentioned in (Gunerhan and Hepbasli, 2007) that, the experimental data measured in the Solar-Meteorological Station of Solar Institute in Ege University were used to determine the optimum tilt angle, the excellent agreement with the results based on such data in (Gunerhan and Hepbasli, 2007) could be considered as an experimental validation of the proposed algorithm of (Soulayman and Hammoud, 2016).

Monthly Optimum Tilt

The values of $\beta_{opt,m}$ obtained from (Soulayman and Hammoud, 2016) for Izmir compared with those found in (Gunerhan and Hepbasli, 2007) and derived from (Nijegorodov, Devan, Jain and Carlsson, 1997). The results are illustrated in Table 25. It is clear that the results obtained from (Gunerhan and Hepbasli, 2007) and (Soulayman and Hammoud, 2016) are in excellent agreement (the differences are related to the assumption taken in (Gunerhan and Hepbasli, 2007) that $\beta_{opt,m} = \beta_{opt,cd}$ while they are in acceptable agreement with those of (Nijegorodov, Devan, Jain and Carlsson, 1997). It's worthwhile to give two comments on (Nijegorodov, Devan, Jain and Carlsson, 1997) results. The first one is related to the negative $\beta_{opt,m}$ obtained for June which is not acceptable when considering the condition (74) and the second one is related to the relatively higher values of $\beta_{opt,m}$ for the months of April, May, and August as compared with those obtained in the other works, although a very good agreement is found for other months. Here it should be mentioned that, the equations used in (Nijegorodov, Devan, Jain and Carlsson, 1997), were optimized using mathematical techniques without taking into account the localized patterns of solar radiation falling over a particular location. As consequences, one or more assumptions of (Nijegorodov, Devan, Jain and Carlsson, 1997) are responsible for this disagreement. Furthermore, the assumption of constant clearness index (k_T=0.7) in calculating the hourly total radiation before integrating them to obtain the total daily solar radiation may be responsible for these slight differences given that the clearness index is not generally constant and varies for each area between two values.

Talebizadeh, Mehrabian and Abdolzadeh (2011) applied Genetic Algorithm to calculate the optimum slope and surface azimuth angles for solar collectors to receive maximum solar radiation. An Iranian

Table 25. A comparison of $\beta_{opt,m}$ (°) of the (Soulayman and Hammoud, 2016) results with those of (Gunerhan and Hepbasli, 2007) and (Nijegorodov, Devan, Jain and Carlsson, 1997) in Izmir

Month	(Gunerhan and Hepbasli, 2007)	(Nijegorodov, Devan, Jain and Carlsson, 1997)	(Soulayman and Hammoud, 2016)
January	65.3	63.2	65.3
February	56.4	54.3	56.7
March	42.2	42.5	41.6
April	23.4	28.5	22.9
May	6.6	11. 8	6.2
June	0.0	-0.5	0
July	1.3	4.2	1.7
August	16.6	20.3	17.2
September	35.0	36.5	35.7
October	52.2	50.5	52.7
November	63.2	60. 8	63.5
December	67.4	67.5	67.5
$B_{opt,y}$(°)	35.8	36.6	35.9

station (φ=32.5°), for which the data of the solar radiation components on a horizontal surface are available, is chosen to verify the results of this algorithm. The values of $\beta_{opt,m}$ derived from (Soulayman and Hammoud, 2016) for the Iranian station and obtained in (Talebizadeh, Mehrabian and Abdolzadeh, 2011) and derived from (Nijegorodov, Devan, Jain and Carlsson, 1997) are in a good agreement as shown in Table 26. The comments mentioned above on the results of (Nijegorodov, Devan, Jain and Carlsson, 1997) remain justifiable in spite of the agreement achieved between the three results. Moreover, in the mid-latitudes zone solar rays fall always on the solar collector surface from Equator direction. So, negative $\beta_{opt,m}$ obtained in (Talebizadeh, Mehrabian and Abdolzadeh, 2011) for June and July are not acceptable when considering the condition (74). A similar approach was applied to calculate the $\beta_{opt,m}$ for Abu Dhabi (φ=24.4°) and the resulting values are compared with those of (Jafarkazemi and Saadabadi, 2013) and (Nijegorodov, Devan, Jain and Carlsson, 1997) in Table 27.

Although the data in Table 27 shows a satisfactory agreement between the three results, the same reproaches mentioned above remain still valid on the data derived from (Nijegorodov, Devan, Jain and Carlsson, 1997) and (Jafarkazemi and Saadabadi, 2013) works. The negative $\beta_{opt,m}$ values obtained in (Jafarkazemi and Saadabadi, 2013) for May, June and July are not acceptable when considering the condition (74). The negative $\beta_{opt,m}$ values are a consequence of negative $\beta_{opt,d}$ which in turn leads to a decrease of $\beta_{opt,m}$ for some summer months, $\beta_{opt,s}$ for spring and summer and $\beta_{opt,b}$ for cooling half year values.

When neglecting the distribution of solar intensity from sunrise to sunset, one can find that a small northward tilt would capture the extra direct radiation in the morning and afternoon that a southward tilt would not. During the times when the Sun is in the southern half of the sky, the penalty for the non-optimal northward tilt could be more than offset by the extra hours contributed to the irradiation seen

Table 26. A comparison of $\beta_{opt,m}$ (°) of (Soulayman and Hammoud, 2016) results with those of (Talebizadeh, Mehrabian and Abdolzadeh, 2011) and (Nijegorodov, Devan, Jain and Carlsson, 1997) for Iranian Station

Month	(Talebizadeh, Mehrabian and Abdolzadeh, 2011)	(Nijegorodov, Devan, Jain and Carlsson, 1997)	(Soulayman and Hammoud, 2016)
January	56.72	57.54	60.95
February	47.59	48.53	49.91
March	32.50	36.5	36.09
April	16.65	22.5	17.10
May	2.98	6.23	2.51
June	−3.91	-5.73	0
July	−0.97	-1.08	0.40
August	11.32	14.53	10.83
September	28.21	30.5	29.25
October	44.04	44.5	46.97
November	54.72	55.23	56.94
December	58.8	62.23	63.50
$B_{opt,y}$(°)	29.46	30.96	29.89

Table 27. A comparison of $\beta_{opt,m}$ (°) of (Soulayman and Hammoud, 2016) results with those of (Jafarkazemi and Saadabadi, 2013) and (Nijegorodov, Devan, Jain and Carlsson, 1997) for Abu Dhabi

Month	(Jafarkazemi and Saadabadi, 2013)	(Nijegorodov, Devan, Jain and Carlsson, 1997)	(Soulayman and Hammoud, 2016)
January	50	50.7	52.8
February	39	40.7	43.3
March	25	28.4	27.4
April	10	14.4	8.6
May	-3	-1.3	0
June	-9	-12.8	0
July	-6	-8.3	0
August	5	6.7	3.4
September	20	22.4	21.9
October	36	36.4	39.4
November	48	47.7	50.9
December	52	55.2	55.3
$B_{opt,y}$(°)	22	23.4	25.3

by the solar PV panel in the morning and evening. Therefore, it is expected that the ignorance of solar intensity distribution leads to negative $\beta_{opt,d}$ which in turn leads to observed disagreement between the results of (Soulayman and Hammoud, 2016) with those of (Talebizadeh, Mehrabian and Abdolzadeh, 2011) and (Jafarkazemi and Saadabadi, 2013).

A comparative power generation analysis of different configurations of hybrid systems with fixed tilt, monthly optimum tilt, yearly optimum tilt and 6 different sun tracking photovoltaic systems is carried out using hybrid optimization model for electric renewables at CEEE (Lat. 31.59°N, Long. 76.52°E; altitude 875 m), Hamirpur, India, where measured solar radiation, wind speed and temperature data from the weather monitoring station of four year period (July 2011–June 2015) with 1 min duration are used (Sinha and Chandel, 2016). The proposed approach in the (Soulayman and Hammoud, 2016) work was applied to calculate the $\beta_{opt,m}$ for CEEE (φ=31.59°N) and the resulting values are compared with those of (Sinha and Chandel, 2016) and (Nijegorodov, Devan, Jain and Carlsson, 1997) in Table 28 where a very good agreement is achieved between all results.

In (Khorasanizadeh, Mohammadi and Mostafaeipour, 2014), long term daily global solar radiation on a horizontal surface as well as sunshine hours for the period of 1988–2000, taken from Iranian Meteorological Organization (IMO), have been utilized for establishing a proper model to estimate diffuse solar radiation in Tabass, Iran (φ= 33.6°N and elevation of 711 m above the sea level) and determining the optimum tilt angles. The calculated, in the present work monthly $\beta_{opt,m}$ and yearly $\beta_{opt,y}$ are compared with those of (Khorasanizadeh, Mohammadi and Mostafaeipour, 2014) and (Nijegorodov, Devan, Jain and Carlsson, 1997) in Table 29. The absence of negative β_{opt} in the results of (Khorasanizadeh, Mohammadi and Mostafaeipour, 2014), contrary to (Talebizadeh, Mehrabian and Abdolzadeh, 2011) and (Jafarkazemi and Saadabadi, 2013), should be mentioned.

Table 28. A comparison of $\beta_{opt,m}$ (°) of (Soulayman and Hammoud, 2016) results with those of (Sinha and Chandel, 2016) and (Nijegorodov, Devan, Jain and Carlsson, 1997) for Hamirpur

Month	(Sinha and Chandel, 2016)	(Nijegorodov, Devan, Jain and Carlsson, 1997)	(Soulayman and Hammoud, 2016)
January	57	57.1	59.2
February	47	47.6	50.3
March	34	35.6	35.1
April	16	21.6	15.8
May	4	5.4	1.4
June	0	-6.5	0
July	0	-1.9	0. 1
August	8	13.6	9.5
September	27	29.6	28.5
October	43	43.6	46.0
November	55	54.4	57.3
December	60	61.5	61.6
$B_{opt,y}$(°)	29.25	30.1	30.27

Table 29. A comparison of $\beta_{opt,m}$ (^{o}) of (Soulayman and Hammoud, 2016) results with those of (Khorasanizadeh, Mohammadi and Mostafaeipour, 2014) and (Nijegorodov, Devan, Jain and Carlsson, 1997) for Tabass

Month	(Khorasanizadeh, Mohammadi and Mostafaeipour, 2014)	(Nijegorodov, Devan, Jain and Carlsson, 1997)	(Soulayman and Hammoud, 2016)
January	62	58.0	60.7
February	53	48.6	51.8
March	38	36.6	36.2
April	19	22.6	16.8
May	2	6.3	2.5
June	0	-5.6	0
July	0	-1.0	0.3
August	12	14.6	11.6
September	32	30.6	30.9
October	49	44.6	48.1
November	60	55.3	59.1
December	64	62.4	63.2
$B_{opt,y}$ (°)	32	31.1	31.6

Elminir, Ghitas, El-Hussainy, Hamid, Beheary and Abdel-Moneim (2006) examined the theoretical aspects of choosing a tilt angle for the solar flat-plate collectors used in Helwan site (29.8483° N, 31.3369° E) and made recommendations on how the collected energy can be increased by varying the tilt angle. The obtained result has been compared with the results provided by other models that use declination (Tiris and Tiris, 1998), daily clearness index and ground reflectivity (Elsayed, 1989) and provided experimental results based on weekly measurements obtained at a multi-position test facility. The tests were started at noon time when the PV module is vertical and facing north (i.e., the tilt angle is -90°). The tests continue to the horizontal mode and are completed to south facing with the tilt angle of +90°. At every five degrees variation in the tilt angle, the PV module parameters (i.e., current and voltage) were recorded. Using a computer program for fitting the data smoothly for each day of the measurements, the optimum tilt angles are determined. Here, the optimum tilt angle is the angle at which the maximum average output is obtained. The results of (Elminir, Ghitas, El-Hussainy, Hamid, Beheary and Abdel-Moneim, 2006) as well as those derived using (Soulayman and Hammoud, 2016) method are given in the Table 30. Here, it should be mentioned that, the following formulas that use declination (Tiris and Tiris, 1998) are proposed.

$$\beta_{opt,m} = 33.24 - 1.31\ \delta \tag{75}$$

$$\beta_{opt,m} = 33.15 - 1.37\ \delta - 0.007\delta^2 \tag{76}$$

Table 30. A comparison of $\beta_{opt,m}$ (°) of (Soulayman and Hammoud, 2016) results with those of (Elminir, Ghitas, El-Hussainy, Hamid, Beheary and Abdel-Moneim, 2006), (Elsayed, 1989), (Tiris and Tiris, 1998), (Nijegorodov, Devan, Jain and Carlsson, 1997) and experiment for Helwan, Egypt

Month	Elsayed	Tiris (75)	Elminir	Nijegorodov	Soulayman	Experiment
January	50.9	60.5	55	55.6	57.8	51
February	41.1	50.3	45	46.0	48.9	48
March	29.5	35.9	30	33.9	33.7	33
April	15.1	20.7	15	19.9	14.9	21
May	3.8	8.7	5	3.8	3.0	4
June	-1.4	3.0	5	-8.08	0.0	4
July	1.3	5.4	5	-3.48	0.0	7
August	11.2	15.1	15	12.0	8.4	20
September	25.2	30.7	25	27.9	26.6	32
October	39.9	44.4	40	41.9	44.1	48
November	50.0	57.1	50	52.8	55.6	53
December	53.8	63.3	55	60.0	60.2	55

Note: Elsayed stands for (Elsayed, 1989), Tiris (71) stands for the equation (71) of (Tiris and Tiris, 1998), Elminir stands for (Elminir, Ghitas, El-Hussainy, Hamid, Beheary and Abdel-Moneim, 2006), Nijegorodov stands for (Nijegorodov, Devan, Jain and Carlsson, 1997), Soulayman stands for (Soulayman and Hammoud, 2016) and experiment stands for the experimental results.

$$\beta_{opt,m} = 33.15 - 1.39\ \ \delta - 0.007\delta^2 + 4.26x10^{-5}\delta^3 \tag{77}$$

$$\beta_{opt,m} = 22.09 + 25.79K_t - 1.49\ \ \delta \tag{78}$$

Regarding the results obtained from Table 30, the experimental optimum tilt angles seem to agree quite well with the predicted values obtained from the different methods except in the winter where the variation of diffuse fraction is a maximum.

When applying the clearness index method for calculating the monthly optimum tilt angle $\beta_{opt,m}$ (°) and comparing the resulting tilt angles for different regions with the results which were published by other researchers in literature, the effectiveness of this method could be demonstrated. Table 31 contains the results of $\beta_{opt,m}$ (°) for solar collectors, facing south, calculated using the clearness index method and reported for Izmir, Turkey (38.4188°N, 27.1287°E) (Ulgen, 2006) with those of (Soulayman and Hammoud, 2016), (Nijegorodov, Devan, Jain and Carlsson, 1997). When taking the experimental results of (Gunerhan and Hepbasli, 2007) for the same city (see Table 25) into consideration, it is possible to conclude that the experimental values of $\beta_{opt,m}$ seem to agree quite well with the predicted values obtained from the different methods all over the year.

Table 32 contains the results of $\beta_{opt,m}$ (°) for solar collectors, facing south, calculated using the clearness index method reported for Beijing, China (39.9042°N, 116.4074°E) (Tang and Wu, 2004) with those of (Soulayman and Hammoud, 2016), (Nijegorodov, Devan, Jain and Carlsson, 1997). It is seen

Table 31. A comparison of $\beta_{opt,m}$ (^{o}) reported for Izmir, Turkey (Ulgen, 2006) with those of (Soulayman and Hammoud, 2016), (Nijegorodov, Devan, Jain and Carlsson, 1997) and clearness index method

Month	(Ulgen, 2006)	(Nijegorodov, Devan, Jain and Carlsson, 1997)	(Soulayman and Hammoud, 2016)	Clearness index method
January	58	63.2	65.3	62
February	48	54.3	56.7	52
March	34	42.5	41.9	39
April	17	28.5	22.9	21
May	4	11.8	6.2	6
June	0	-0.5	0	-2
July	1	4.2	1.7	1
August	12	20.3	17.2	16
September	28	36.5	35.7	34
October	45	50.5	52.7	50
November	56	60.8	63.5	60
December	61	67.5	67.5	63

Table 32. A comparison of $\beta_{opt,m}$ (^{o}) reported for Beijing, China (Tang and Wu, 2004) with those of (Soulayman and Hammoud, 2016), (Nijegorodov, Devan, Jain and Carlsson, 1997) and clearness index method

Month	(Tang and Wu, 2004)	(Nijegorodov, Devan, Jain and Carlsson, 1997)	(Soulayman and Hammoud, 2016)	Clearness Index Method
January	66	64.5	66.2	67
February	57	55.7	57.7	58
March	42	43.9	42.6	42
April	24	29.9	23.9	23
May	8	13.1	7.3	7
June	0	0.7	0	-1
July	3	5.5	2.8	2
August	17	21.7	18.3	15
September	36	37.9	36.8	33
October	52	51.9	53.8	52
November	63	62.11	64.4	64
December	68	68.7	68.5	69

from Table 32 that, the predicted values of $\beta_{opt,m}$ using different methods agree very well with each other all over the year. Table 33 gives the results of comparison of monthly tilt angles (in degrees) for solar collectors, facing south, calculated using the clearness index method and reported for Khatkar Kalan (Punjab), India (31.1647°N, 76.0239°E) (Kumar, Thakur, Makade, and Shivhare, 2011). The results presented in Table 33 show that the agreement between the results of different methods is very high.

From the results presented in the Tables 31 to 33, one can conclude that, the results of (Nijegorodov, Devan, Jain and Carlsson, 1997) are relatively high for the months of April, May, and August as compared with those obtained in the other works, although a very good agreement is found for other months. Therefore, this method could be considered as a good method except the above mentioned months.

Finally, it should be mentioned here that, in the mid-latitude zone and for Equator facing solar collectors, the Elkassaby method (Elkassaby, 1988) could be applied with a good accuracy as an approximate approach whatever the latitude is while the method of Soulayman (1991) could be applied with a high accuracy for any location in this zone. The developed version of this method (Soulayman and Hammoud, 2016) could be applied with a high accuracy for any location in this zone whatever the orientation of the receiver is. The different versions of the clearness index methods, mainly with regard to the equations (42) and (45) and sources of solar radiation data, are responsible for the observed difference between the results of different authors. This method could be applied with a good accuracy whatever the latitude is and whatever the orientation is.

Seasonally Optimum Tilt

When considering the seasons to be consisted of November, December and January for winter, of February, March and April for spring, of May, June and July for summer and of August, September and

Table 33. Comparison of monthly tilt angles (in degrees) for solar collectors, facing south, reported for Khatkar Kalan (Punjab), India (31.1647°N, 76.0239°E) (Kumar, Thakur, Makade, and Shivhare, 2011)

Month	(Tang and Wu, 2004)	(Nijegorodov, Devan, Jain and Carlsson, 1997)	(Soulayman and Hammoud, 2016)	Clearness Index Method
January	60.5	56.7	59.2	58
February	50.5	47.2	50.3	48
March	36.5	35.2	35.0	34
April	18	21.2	15.8	16
May	0	5.0	1.4	-1
June	0	-6.9	0	-7
July	0	-2.3	0.1	-4
August	8.5	13.2	9.5	8
September	26	29.27	28.5	26
October	56.5	43.2	46.0	46
November	58.5	54.0	57.3	57
December	62.5	68.7	61.6	61

October for autumn, $\beta_{opt,s}$ were reported for Hamirpur, India in (Sinha and Chandel, 2016) to be 57.3°, 32.33 °, 1.53° and 26° respectively. The corresponding values calculated in (Soulayman and Hammoud, 2016) are 59.3°, 33.7°, 0.5° and 27.97° respectively. The $\beta_{opt,s}$ for Abu Dhabi was calculated for *EF* surface in (Jafarkazemi and Saadabadi, 2013) and the values 39°, -1°, 6° and 45° were reported for winter, spring, summer and autumn respectively. In the present work these values were found to be 41.7°, 2.9 °, 8.4° and 48.5° for winter, spring, summer and autumn respectively. So, the good agreement is achieved. Mujahid (1994) concluded that, for non-tracking solar collection systems, the tilt angle has a predominant effect on the quantity of energy that the system can intercept and developed a computational algorithm for the calculation of the optimum tilt angle that would orient a non-tracking solar collection system (concentrating or non-concentrating) in its best position for the maximum average daily, monthly seasonal or yearly intercepted radiation. Mujahid (1994) computed the optimum slope angle for latitude of 10°N to 50°N on monthly, seasonal and yearly bases and concluded that if the collector is adjusted by the seasonally optimum angles, 10% more energy is received compared with the horizontal case. A case study is applied on Riyadh City (latitude 24.9°N) to investigate the sensitivity of intercepted radiation when the tilt angle varies from that of the optimum value. The results show that, on a monthly basis when the collector is mounted at the yearly optimum tilt angle, the loss of radiation intercepted is less than 10% as compared to the monthly optimum tilt angle. The optimum seasonal tilt angle reduces the incident radiation by less than 2% from that of the monthly optimum tilt angle. According to (Soulayman and Hammoud, 2016), the yearly energy gain daily average resulting from adjusting collector tilt angle seasonally reaches 1.705 for 43.45°N which means that 70.5% more energy is received compared with the horizontal case while the daily energy gain reaches 3.24.

Biannually Optimum Tilt

The $\beta_{opt,b}$ for Abu Dhabi was calculated previously for *EF* surface in (Jafarkazemi and Saadabadi, 2013) and the heating and cooling half year values were 42° and 2° respectively. The ones calculated in (Soulayman and Hammoud, 2016) are 47.9° and 5.7° respectively and are in satisfactory agreement. For the cooling period (from October to March) and warm period (from April to September) the reported values of $\beta_{opt,b}$ for Tabass, Iran in (Khorasanizadeh, Mohammadi and Mostafaeipour, 2014) are 55° and 10° respectively. The corresponding values calculated in (Soulayman and Hammoud, 2016) are 53.2° and 10.35° respectively. So, a good agreement is achieved. In the NH it is suggested that the optimum tilt angle for summer (from April to September) and for winter (from October to March) are respectively $\beta_{opt,b} = \varphi - 15°$ and $\beta_{opt,b} = \varphi + 15°$ (Gunerhan and Hepbasli, 2007). When analyzing results of the optimum tilt angle in the mid-latitude zone of NH, it was found (Soulayman and Hammoud, 2016) that, $\beta_{opt,b}$ could be approximated as follows: $\beta_{opt,b} = [\varphi\text{-}24.6°]^+$ for the first half year (22/3-21/9) and $\beta_{opt,b} = 0.873\varphi+24.65°$ for the second half year (22/9-21/3) where positive sign over the brackets means the value should be taken as zero when the value is negative. Therefore, it is reasonable to demonstrate the influence of $\varphi - 15°$, $\varphi + 15°$ and $\beta_{opt,b}$ solar collector tilting on their corresponding energy gain.

Yearly Optimum Tilt

It is suggested that for systems, which utilize solar energy throughout the year, $\beta_{opt,y} = \varphi$. Le Roux (2016) used measured data from nine measuring stations in the Southern African Universities Radiometric Network (SAURAN) equipped with pyranometers and pyrheliometers to calculate the annual solar insolation on fixed collectors at all possible installation angles. Roux (Le Roux, 2016) reported that the $\beta_{opt,y}$ found do compare well with the latitude angle for all studied locations ($\beta_{opt,y} = \varphi \pm 2.6^o$). When calculating $\beta_{opt,y}$ for Durban (29.816942 °S, 30.944917 °E, 200m) it was found (Soulayman and Hammoud, 2016) that, $\beta_{opt,y}$ = -29.01° instead of -29° reported in (Le Roux, 2016). This means that the agreement is excellent.

The analysis of the optimum tilt angle results in the mid-latitudes zone of NH seems suggesting that $\beta_{opt,y} = 0.916\varphi + 1.171^o$ (Soulayman and Hammoud, 2016). Darhmaoui and Lahjouji (2013) proposed to define the yearly optimum tilt $\beta_{opt,y}$ as a function of latitude φ as follows $\beta_{opt,y} = 1.25351\varphi - 0.00728944\varphi^2$. In this case, the values of $\beta_{opt,y}$ obtained, for Izmir, the Iranian station, Abu Dhabi, Tabass and hamirpur are 37.43°, 33.04°, 26.2°, 33.9° and 32.32° respectively. When comparing these values with those listed in the previous Tables, the maximum difference between them was found to be within the range $\varphi \pm 2.4^o$. However, the yearly daily energy gain is not highly sensitive to $\beta_{opt,y}$ changes and a change of $\beta_{opt,y}$ by 3° leads to a change in energy gain by 0.02 only. The obtained, in the present work, result coincides will with that of (Altarawneh, Rawadieh, Tarawneh, Alrowwad and Rimawi, 2016) and (Soulayman and Hammoud, 2016).

Finally, it should be mentioned here that the trajectories of monthly, seasonal, bi-annual, and annual optimal tilt angles specified in the present work (see Figures 22 to 25) were also specified in (Khorasanizadeh, Mohammadi and Mostafaeipour, 2014), (Talebizadeh, Mehrabian and Abdolzadeh, 2011), (Jafarkazemi and Saadabadi, 2013) and (Altarawneh, Rawadieh, Tarawneh, Alrowwad and Rimawi, 2016). In (Altarawneh, Rawadieh, Tarawneh, Alrowwad and Rimawi, 2016), it is showed that $\beta_{opt,y}$ = 28.7°. This obtained result, in the present work, $\beta_{opt,y}$ = 29.26° agrees well with that of (Altarawneh, Rawadieh, Tarawneh, Alrowwad and Rimawi, 2016) and (Soulayman and Hammoud, 2016).

More details on the daily optimum tilt angles in the mid-latitude zone are given in the appendix C. As during the some months and seasons the daily optimum tilt angle $\beta_{opt,d}$ (°) changes its sign, the corresponded tables show the optimum tilt angle for these two periods and the number of days of each period.

OPTIMUM TILT ANGLE DETERMINATION IN THE HIGH-LATITUDE ZONE

On the Applicability of Some Methods

When applying the above mentioned algorithms to the high-latitude zone (43.5°S $\leq \varphi \leq$ 66.45°S and 43.5°N $\leq \varphi \leq$ 66.45°N) and analyzing the obtained results as a function of δ and φ it is found that the equation (56) could not be applied with an acceptable accuracy for determining $\beta_{opt,d}$ all over the year.

Figure 22. The differences between the calculated results of $\beta_{opt,d}$ using the precise approach (equation (55)) and the approximate equation (56) for latitudes: $\varphi = 44^o N$ (♦), $60^o N$ (-) and $66.45^o N$ (●)

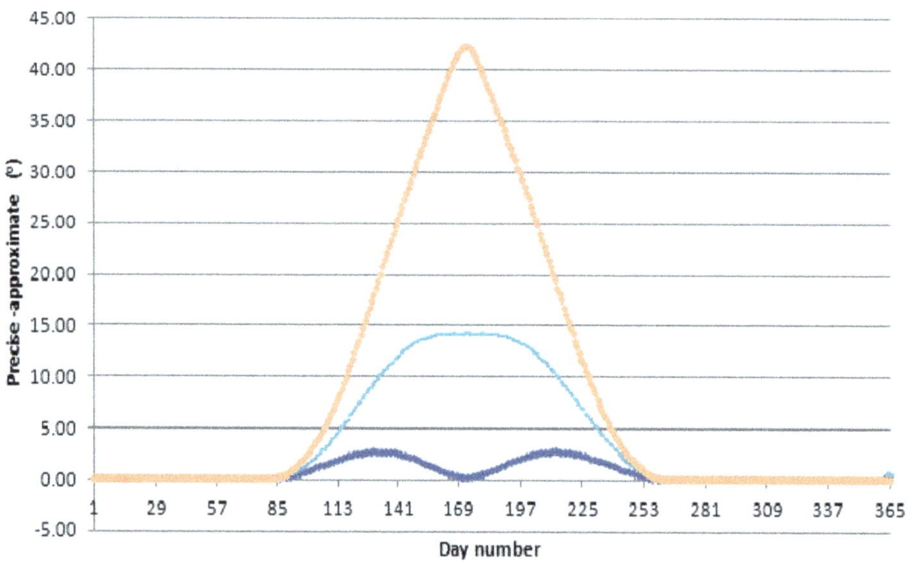

Figure 23. The differences between the calculated results of $\beta_{opt,d}$ using the precise approach (equation (55)) and the approximate equation (56) for latitudes: $\varphi = 44^o S$ (●), $60^o S$ (-) and $66.45^o S$ (♦)

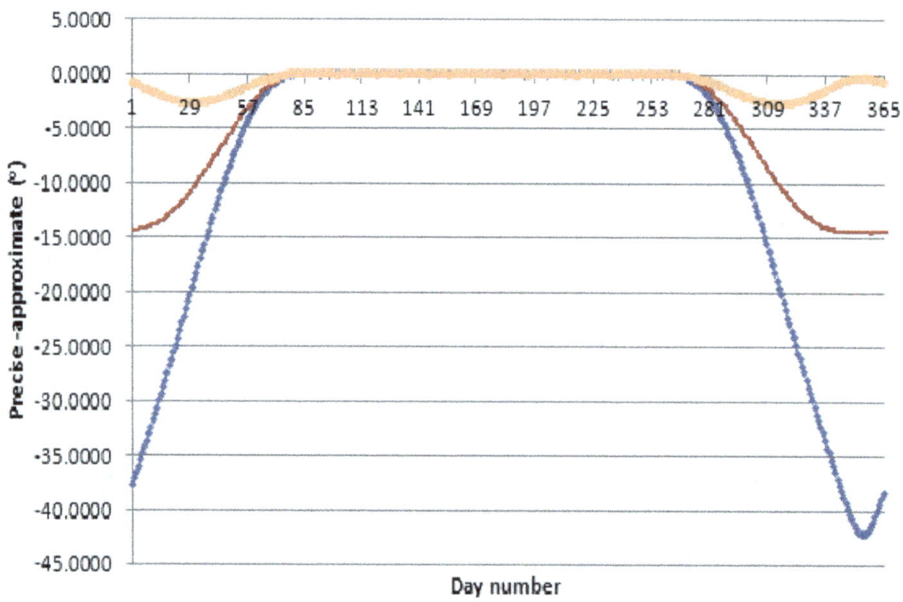

Figure 24. The differences between the calculated results of $\beta_{opt,d}$ using the precise approach and the mathematical polynomial fitting for latitudes 50°N and 66°N

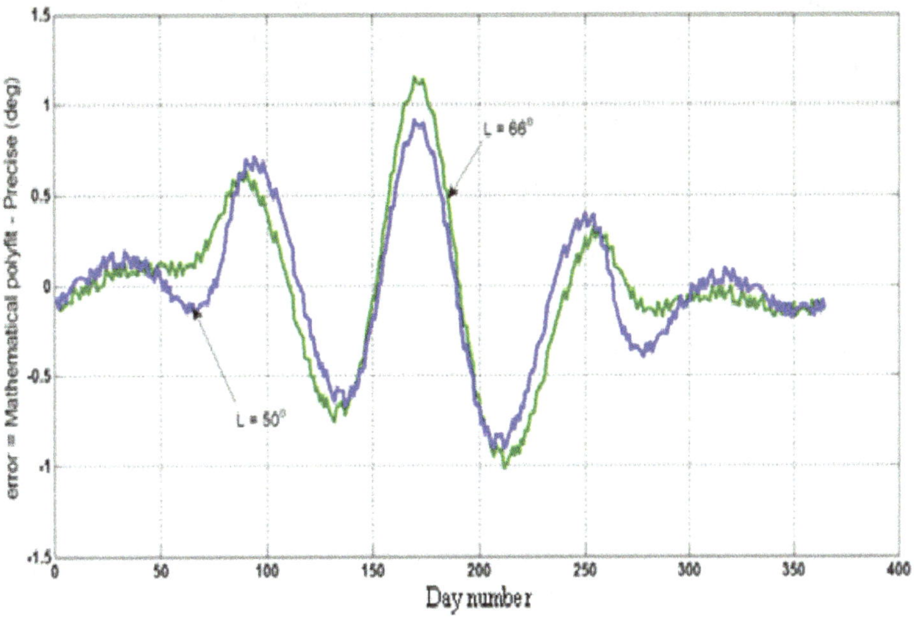

Figure 25. Daily optimum tilt angle for latitudes 60° S in SH (lower curves: (x) for Soulayman and SAbbagh (2014) and (ж) for polynomial fitting) and 60°N in NH (upper curves: (●) for (Soulayman, 1991), (▲) for polynomial fitting and (■) for Soulayman and SAbbagh (2014))

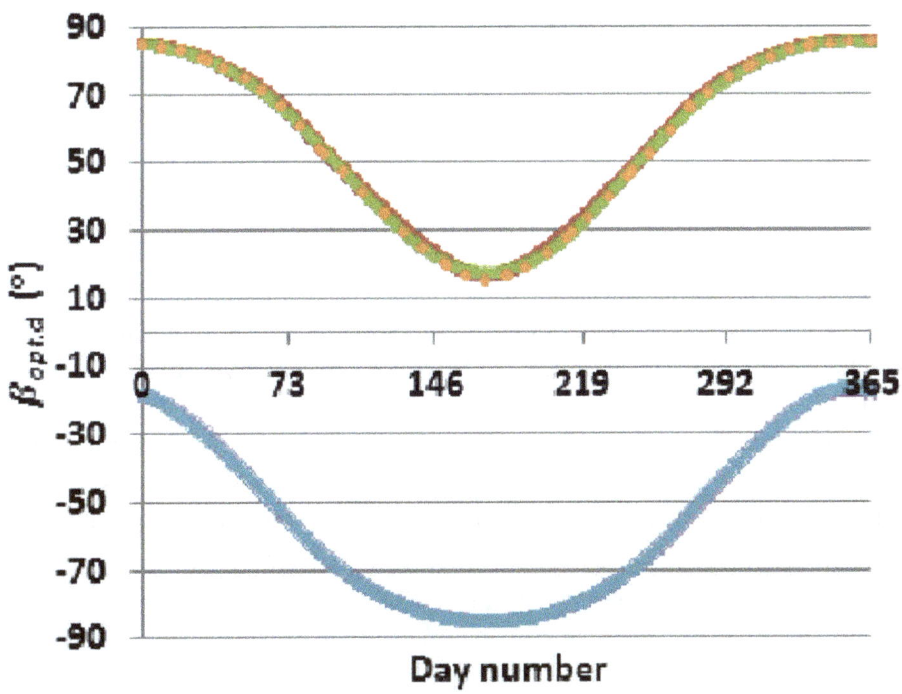

The equation (56) is a precise solution of equation (55) over a half-year (from 22/9 to 21/3 in NH and from 22/3 to 21/9 in SH) while the polynomial fit, equation (23), describes the $\beta_{opt,d}$ dependence on latitude and day number very well. Figures 22 and 23 show the results of applying these three approaches in NH and SH, respectively. It is seen from these Figures that, the applicability of the equation (56) is restricted to a period the length of which decreases with increasing φ, while the polynomial fit (equation (23)) is a very good approximation.

It is seen from Figures 22 and 23 that, the absolute difference, between the results of the equation (56) and the results of the equation (55), increases with φ increase and it reaches 42.2°, while this difference does exceed 1.2° in the case of the polynomial fit (see Figure 24). Therefore, the polynomial fit could be applied for predicting $\beta_{opt,d}$ all over the year in the high-latitude one. This means that the optimum tilt angle over any period of consecutives days could be calculated using the equation (27) with a very good accuracy. So, the equation (27) could be applied for calculating the weekly, fortnightly, monthly, seasonally, half-yearly and yearly optimum tilt angles.

A Comparison With Available Results

El-Kassaby (1988) proposed a formula for determining $\beta_{opt,d}$ at latitudes up to 60° in NH by verifying the applicability of his formula during the period from 22/9 to 21/3 but his formula suffered from uncertainty during the other period of the year (22/3 to 21/9) (see (Soulayman, 1991) for more details). Skeiker (2009) presented a study aiming to develop an analytical procedure to obtain a formula for determining $\beta_{opt,d}$ for any chosen day at any latitude in NH but he repeated the same formula of El-Kassaby (1988) (see (Soulayman, 1991) for more details). So, no need to compare the results of the present work with those of (El-Kassaby, 1988) and (Skeiker, 2009). Figure 25 shows the results of applying the equations (23) and (55) for determining $\beta_{opt,d}$ for 60°S. Figure 25 shows also the results of applying the equation (55), the algorithm of Soulayman and SAbbagh (2014) and the results of (Soulayman, 1991) for 60°N. It is seen from Figure 25 that the results of the present work agree very well with those of (Soulayman, 1991) and the algorithm of Soulayman and SAbbagh (2014). Moreover, the equation (23) gives an excellent fit of the calculated results of the algorithm in this work.

When calculating the monthly optimum tilt angle at 60°S using equation (55) and comparing the obtained results with those of set of equations in (Nijegorodov, Devan, Jain and Carlsson, 1997) and set of equations in (Soulayman and Sabbagh, 2015c) with taking into consideration the latitudes interval of the applicability of the mentioned references it is observed that in the SH the results of the present work are in a very good agreement with those of (Nijegorodov, Devan, Jain and Carlsson, 1997), and (Soulayman and Sabbagh, 2015c) (see Table 34). However, the equations of (Nijegorodov, Devan, Jain and Carlsson, 1997) give a little bit higher values of optimum angle for the months of April, May, and August as compared with those of (Soulayman and Sabbagh, 2015c) while the agreement for other months is very good.

More details on the optimum tilt angles in the high-latitude zone for the days given in the appendix D.

Table 34. Monthly optimum tilt angles at 60° S

Month	φ=60°S		
	Equation (27)	(Nijegorodov, Devan, Jain and Carlsson, 1997)	(Soulayman and Sabbagh, 2015c)
January	-22.3	-24.4	-23.7
February	-36.6	-41.2	-38.5
March	-54.6	-56	-56.3
April	-69.9	-70	-72.6
May	-78.9	-79.8	-81.9
June	-82.1	-86.2	-85.6
July	-80.7	-83.4	-83.9
August	-74.0	-75.2	-76.6
September	-60.8	-62	-62.6
October	-42.8	-48	-44.0
November	-25.9	-30.8	-27.4
December	-17.6	-18.2	-17.7
MBE		2.41766	1.425014
RMSE		2.98281	2.234652
t		4.58982	2.745675
R^2		0.994	0.999

CONCLUSION

In this chapter, a general algorithm is proposed for treating β_{opt} all over the world. The theoretical aspects that determine the optimal tilt angle, regarding to maximum solar energy collection, are examined. A computer program is used in determining the optimal tilt angle of any site at the Earth's surface between 66.45°S and 66.45°N. A regression analysis using site's latitude, solar declination angle and its corresponding optimal tilt angle is conducted to develop a mathematical model that allows the determination of the optimal tilt angle at which maximum solar radiation is collected using only the site's latitude and the day number of the year. A comparison with available experimental and theoretical results from other researchers is provided.

A rule of thumb says that a solar collector should be orientated towards Equator; is it valid all over the world? A rule of thumb says that solar collector should have a latitude tilt value; is it valid all over the world and all over the year? The present chapter focuses on presenting an algorithm for determining the optimum tilt angle all over the world and for any collector azimuth angle. The Earth surface, located between latitudes 66.45°S and 66.45°N, is divided into 3 characteristic zones. The first zone is the tropical between latitudes 23.45°S and 23.45°N. The second zone is the mid-latitude zone between 23.45°N and 43.45°N and between 23.45°S and 43.45°S. The third zone is the high-latitude zone between 43.45°N and 66.45°N and between 43.45°S and 66.45°S. For each of these zones an adequate method is proposed for calculating the solar collector optimum tilt. Moreover, four simple equations are proposed for predicting daily optimum tilt angle and optimum tilt angle for any number of consecutive days. The yearly possible energy gain in relation that received by a horizontal surface and latitude tilted surface is also calculated on the basis of daily optimum tilts. It is found that the above mentioned rules of thumb are not applicable for a number of consecutive days in the year.

The results of this chapter allow calculating the possible energy gain of any orientation as well as determine the maximum energy gain for long term tracking of different periods. It is shown that, the optimum tilt angle for solar systems must be determined accurately for each location. The information presented in this chapter will be useful for designers and researchers to select suitable methodology for determining optimal tilt angle for solar systems at any site.

REFERENCES

Abdulaziz, M. (1994). Optimum tilt angle for solar collection systems. *International Journal of Solar Energy*, *14*(4), 191–202. doi:10.1080/01425919408909810

Alsadi, S. Y., Nassar, Y. F., & Amer, Kh. A. (2016). General polynomial for optimizing the tilt angle of flat solar energy harvesters based on ASHRAE clear sky model in mid and high latitudes. *Energy and Power*, *6*(2), 29–38. doi:10.5923/j.ep.20160602.01

Altarawneh, I. S., Rawadieh, S. I., Tarawneh, M. S., Alrowwad, S. M., & Rimawi, F. (2016). Optimal tilt angle trajectory for maximizing solar energy potential in Maan area in Jordan. *Journal of Renewable and Sustainable Energy*, *8*(3), 033701. doi:10.1063/1.4948389

Bakirci, K. (2012). General models for optimum tilt angles of solar panels: Turkey case study. *Renewable & Sustainable Energy Reviews*, *16*(8), 6149–6159. doi:10.1016/j.rser.2012.07.009

Bari, S. (2000). Optimum slope angle and orientation of solar collectors for different periods of possible utilization. *Energy Conversion and Management*, *41*(8), 855–860. doi:10.1016/S0196-8904(99)00154-5

Bari, S. (2001a). Seasonal orientation of solar collectors for the Philippines. *ASEAN Journal on Science and Technology for Development*, *18*(1), 45–54.

Bari, S. (2001b). Optimum orientation of domestic solar water heaters for the low latitude countries. *Energy Conversion and Management*, *42*(10), 1205–1214. doi:10.1016/S0196-8904(00)00135-7

Bari, S., Lim, T. H., & Yu, C. W. (2001). Slope angle for seasonal applications of solar collectors in Thailand. *Int. Energy Journal*, *2*(1), 43–51.

Benghanem, M. (2011). Optimization of tilt angle for solar panel: Case study for Madinah, Saudi Arabia. *Applied Energy*, *88*(4), 1427–1433. doi:10.1016/j.apenergy.2010.10.001

Bristow, K. L., & Campbell, G. S. (1984). On the relationship between incoming solar radiation and daily maximum and minimum temperature. *Agricultural and Forest Meteorology*, *31*(2), 159–166. doi:10.1016/0168-1923(84)90017-0

Calabro, E. (2009). Determining optimum tilt angles of photovoltaic panels at typical north-tropical latitudes. *Journal of Renewable and Sustainable Energy*, *1*(3), 033104. doi:10.1063/1.3148272

Calabrò, E. (2013). An algorithm to determine the optimum tilt angle of a solar panel from global horizontal solar radiation. *Journal of Renewable Energy, 2013*, Article ID 307547, 12 pages. 10.1155/2013/307547

Chenga, C. L., Jimenez, C. S. S., & Lee, M. (2009). Research of BIPV optimal tilted angle, use of latitude concept for south orientated plans. *Renewable Energy, 34*(6), 1644–1650. doi:10.1016/j.renene.2008.10.025

Chiou, J. P., & El-Naggar, M. M. (1986). Optimum slope for solar insolation on a flat surface titled toward the equator in heating season. *Solar Energy, 36*(5), 471–478. doi:10.1016/0038-092X(86)90096-4

Christensen, C. B., & Barker, G. M. (2001). Effects of tilt and azimuth on annual incident solar radiation for United States locations. *Proc solar Forum 2001: Solar Energy: The Power to Choose.*

Cooper, P. L. (1969). The absorption of radiation in solar stills. *Solar Energy, 12*(3), 333–346. doi:10.1016/0038-092X(69)90047-4

Darhmaoui, H., & Lahjouji, D. (2013). Latitude based model for tilt angle optimization for solar collectors in the Mediterranean region. *Energy Procedia, 42*, 426–435. doi:10.1016/j.egypro.2013.11.043

De Miguel, A., Bilbao, J., & Diez, M. (1995). Solar radiation incident on tilted surfaces in Burgos, Spain: Isotropic models. *Energy Conversion and Management, 36*(10), 945–951. doi:10.1016/0196-8904(94)00067-A

Diaz, F., Ngo, J., Pascual, M., Planes, A., & Chua, A. (2014). Enhanced energy conversion performance of Philippine photovoltaic panels through tilt angle adjustments: A mechatronics approach. *Philippine Science Letters, 7*(1), 7–12.

Duffie, J. A., & Beckman, W. A. (2013). Solar engineering of thermal processes (3rd ed.). New York: John Wiley & Sons. doi:10.1002/9781118671603

Eke, A. B. (2011). Prediction of optimum angle of inclination for flat plate solar collector in Zaria, Nigeria. *Agricultural Engineering International: CIGR Journal, 13*, Manuscript No. 1840.

El-Kassaby, M. M. (1988). Monthly and daily optimum tilt angle for south facing solar collectors; theoretical model, experimental and empirical correlations. *Solar & Wind Technology, 5*(6), 589–596. doi:10.1016/0741-983X(88)90054-9

El-Kassaby, M. M., & Hassab, M. H. (1994). Investigation of a variable tilt angle Australian type solar collector. *Renewable Energy, 4*(3), 327–332. doi:10.1016/0960-1481(94)90036-1

Elminir, H. K., Ghitas, A. E., El-Hussainy, F., Hamid, R., Beheary, M. M., & Abdel-Moneim, K. M. (2006). Optimum solar flat-plate collector slope: Case study for Helwan, Egypt. *Energy Conversion and Management, 47*(5), 624–637. doi:10.1016/j.enconman.2005.05.015

El-Naggar, M. M., & Chiou, J. P. (1986). Optimum slope for solar insolation on a flat surface tilted towards the equator in heating season. *Solar Energy, 36*(5), 471–478. doi:10.1016/0038-092X(86)90096-4

Elsayed, M. M. (1989). Optimum orientation of absorber plates. *Solar Energy, 42*(2), 89–102. doi:10.1016/0038-092X(89)90136-9

El-Sebaii, A., Al-Hazmi, F., Al-Ghamdi, A., & Yaghmour, S. (2010). Global, direct and diffuse solar radiation on horizontal and tilted surfaces in Jeddah, Saudi Arabia. *Applied Energy*, *87*(2), 568–576. doi:10.1016/j.apenergy.2009.06.032

Ertekin, C., Evrendilek, F., & Kulcu, R. (2008). Modeling spatio-temporal dynamics of optimum tilt angles for solar collectors in Turkey. *Sensors (Basel, Switzerland)*, *8*(5), 2913–2931. doi:10.3390/s8052913 PMID:27879857

Garg, H. (1980). *Treatise on solar energy. In Fundamentals of solar energy* (Vol. 1). New York: Wiley.

Garp, H. P., & Gupta, G. L. (1978). *Flat plate collectors: Experimental studies and design data for India.* International Solar Energy Congress, New Delhi, India.

Gavin, D. J. N. (2007). *Solar Energy Projects for the evil Genius* (1st ed.). Chicago: McGraw-Hill.

Ghosh, H., Bhowmik, N., & Hussain, M. (2010). Determining seasonal optimum tilt angles, solar radiations on variously oriented, single and double axis tracking surfaces at Dhaka. *Renewable Energy*, *35*(6), 1292–1297. doi:10.1016/j.renene.2009.11.041

Gong, X., & Kulkarni, M. (2005). Design optimization of a large-scale rooftop photovoltaic system. *Solar Energy*, *78*(3), 362–374. doi:10.1016/j.solener.2004.08.008

Google Geocoding API. (n.d.). Retrieved 07 24, 2012, from https://developers.google.com/maps/documentation/geocoding/

Gopinathan, K. K. (1991). Solar radiation on variously oriented sloping surfaces. *Solar Energy*, *47*(3), 173–179. doi:10.1016/0038-092X(91)90076-9

Gueymard, C. A. (2005). Importance of atmospheric turbidity and associated uncertainties in solar radiation and luminous efficacy modeling. *Energy*, *30*(9), 1603–1621. doi:10.1016/j.energy.2004.04.040

Gunerhan, H., & Hepbasli, A. (2007). Determination of the optimum tilt angle of solar collectors for building applications. *Building and Environment*, *42*(2), 779–783. doi:10.1016/j.buildenv.2005.09.012

Hartley, L. E., Martinez-Lozano, J. A., Utrillas, M. P., Tena, F., & Pedro, R. (1999). The optimization of the angle of inclination of a solar collector to maximize the incident solar radiation. *Renewable Energy*, *6*(3), 180–298. doi:10.1016/S0960-1481(98)00763-0

Hay, J. E., & Davies, J. A. (1980). Calculation of the solar radiation on an inclined surface. In *Pros. First Canadian Solar Radiation Data Workshop*. Ministry of supply and Services Canada.

Hay, J. E., & McKay, D. C. (1985). Estimating Solar Irradiance on Inclined Surfaces: A Review and Assessment of Methodologies. *International Journal of Solar Energy*, *3*(4-5), 203–240. doi:10.1080/01425918508914395

Yakup, M. A. H. M., & Malik, A. Q. (2001). Optimum tilt angle and orientation for solar collector in Brunei Darussalam. *Renewable Energy*, *24*(2), 223–234. doi:10.1016/S0960-1481(00)00168-3

Hottel, H. C. (1954). Performance of flat-plate energy collectors. In *Space Heating with Solar Energy, Proc. Course Symp*. Cambridge, MA: MIT Press.

Hussein, H. M. S., Ahmad, G. E., & El-Ghetany, H. H. (2004). Performance evaluation of photovoltaic modules at different tilt angles and orientations. *Energy Conversion and Management*, *45*(15-16), 2441–2452. doi:10.1016/j.enconman.2003.11.013

Hyewood, H. (1971). Operating experience with solar water heating. *Journal of the Institution of Heating and Ventilation Engineers*, *39*(6), 63–69.

Ibrahim, D. (1995). Optimum tilt angle for solar collectors used in Cyprus. *Renewable Energy*, *6*(7), 813–819. doi:10.1016/0960-1481(95)00070-Z

Idowu, O. S., Olarenwaju, O. M., & Ifedayo, O. I. (2013). Determination of optimum tilt angles for solar collectors in low-latitude tropical region. *International Journal of Energy and Environmental Engineering*, *4*(1), 29. doi:10.1186/2251-6832-4-29

Ionită, M. A. & Alexandru, C. (2012). Dynamic optimization of the tracking system for a pseudo-azimuthal photovoltaic platform. *Journal of Renewable and Sustainable Energy, 4*, Article ID 053117, 15 pages, 2012. 10.1063/1.4757630

Iqbal, M. (1978). Estimation of the monthly average of the diffuse component of total insolation on a horizontal surface. *Solar Energy*, *20*(1), 101–105. doi:10.1016/0038-092X(78)90149-4

Islam, M., Alili, A., Kubo, I., & Ohadi, M. (2010). Measurement of solar-energy (direct beam radiation) in Abu Dhabi, UAE. *Renewable Energy*, *35*(2), 515–519. doi:10.1016/j.renene.2009.07.019

Jafarkazemi, F., & Saadabadi, S. A. (2013). Optimum tilt angle and orientation of solar surfaces in Abu Dhabi, UAE. *Renewable Energy*, *56*, 44–49. doi:10.1016/j.renene.2012.10.036

Kacira, M., Simsek, M., Babur, Y., & Demirkol, S. (2004). Determining optimum tilt angles and orientations of photovoltaic panels in Sanliurfa, Turkey. *Renewable Energy*, *29*(8), 1265–1275. doi:10.1016/j.renene.2003.12.014

Kaldellis, J., Kavadias, K., & Zafirakis, D. (2012). Experimental validation of the optimum photovoltaic panels tilt angle for remote consumers. *Renewable Energy*, *46*, 179–191. doi:10.1016/j.renene.2012.03.020

Kamanga, B., Mlatho, J. S. P., Mikeka, C. & Kamunda, C. (2014). Optimum tilt angle for photovoltaic solar panels in Zomba district, Malawi. *Journal of Solar Energy, 2014*, Article ID 132950, 9 pages. 10.1155/2014/132950

Kennedy, J., & Eberhart, R. (1995). Particle Swarm Optimization. *Proc. IEEE Int. Conf. Neural Networks*, 1942-1948. doi:10.1109/ICNN.1995.488968

Kern, J., & Harris, I. (1975). On the optimum tilt of a solar collector. *Solar Energy*, *17*(2), 97–102. doi:10.1016/0038-092X(75)90064-X

Keyanpour-Rad, M., Haghgou, H. R., Bahar, F., & Afshari, E. (2000). Feasibility study of the application of solar heating systems in Iran. *Renewable Energy*, *20*(3), 333–345. doi:10.1016/S0960-1481(99)00088-9

Khorasanizadeh, H., Aghaei, A., Ehteram, H., Dehghani Yazdeli, R., & Hataminasar, N. (2014). Attaining optimum tilts of flat solar surfaces utilizing measured solar data: Case study for Ilam, Iran. *Iranica Journal of Energy & Environment*, *5*(3), 224–232. doi:10.5829/idosi.ijee.2014.05.03.01

Khorasanizadeh, H., Mohammadi, K., & Aghaei, A. (2014). The potential and characteristics of solar energy in Yazd province, Iran. *Iranica Journal of Energy & Environment*, *5*(2), 173–183. doi:10.5829/idosi.ijee.2014.05.02.09

Khorasanizadeh, H., Mohammadi, K., & Mostafaeipour, A. (2014). Establishing a diffuse solar radiation model for determining the optimum tilt angle of solar surfaces in Tabass, Iran. *Energy Conversion and Management*, *78*, 805–814. doi:10.1016/j.enconman.2013.11.048

Klein, S. A. (1977). Calculation of monthly average insolation on tilted surfaces. *Solar Energy*, *19*(4), 325–329. doi:10.1016/0038-092X(77)90001-9

Koronakis, P. S. (1986). On the choice of the angle of tilt for south facing solar collectors in the Athens basin area. *Solar Energy*, *36*(3), 217–225. doi:10.1016/0038-092X(86)90137-4

Kumar, A., Thakur, N., Makade, R., & Shivhare, M. K. (2011). Optimization of tilt angle for photovoltaic array. *International Journal of Engineering Science and Technology, 3*(4), 3153-3161.

Kumarasamy, S., Tulika, S., Guddy, S., & Manicam, P. (2013). Modelling and estimation of photosynthetically active incident radiation based on global irradiance in Indian latitudes. *IJEEE*, *4*, 21.

Le Roux, W. G. (2016). Optimum tilt and azimuth angles for fixed solar collectors in South Africa using measured data. *Renewable Energy, 96*(A), 603-612. doi:10.1016/j.renene.2016.05.003

Lewis, G. (1987). Optimum tilt of solar collectors. *Solar & Wind Technology*, *4*(3), 407–410. doi:10.1016/0741-983X(87)90073-7

Li, H., Cao, F., Bu, X., & Zhao, L. (2015). Models for calculating daily global solar radiation from air temperature in humid regions—A case study. *Environmental Progress & Sustainable Energy*, *34*(2), 595–599. doi:10.1002/ep.12018

Feng-Jiao, L., & Tian-Pau, C. (2015). Optimizing the tilt angle of solar collector under clear sky by particle swarm optimization method. *Journal of Applied Science and Engineering*, *18*(2), 167–172. doi:10.6180/jase.2015.18.2.09

Liu, B. Y. H., & Jordan, R. C. (1962). Daily insolation on surfaces tilted toward the Equator. *ASHRAE Journal*, *3*(10), 53–59.

Löf, G. O. G., & Taybout, R. A. (1973). Cost of house heating with solar energy. *Solar Energy*, *14*(3), 253–278. doi:10.1016/0038-092X(73)90094-7

Lubitz, W. D. (2011). Effect of manual tilt adjustments on incident irradiance on fixed and tracking solar panels. *Applied Energy*, *88*(5), 1710–1719. doi:10.1016/j.apenergy.2010.11.008

Lunde, P. (1980). *Solar thermal engineering*. New York: Wiley.

Morcos, V. H. (1994). Optimum tilt angle and orientation for solar collectors in Assiut, Egypt. *Renewable Energy*, *4*(3), 291–298. doi:10.1016/0960-1481(94)90032-9

Mujahid, A. M. (1994). Optimum tilt angle for solar collection systems. *International Journal of Solar Energy*, *14*(4), 191–202. doi:10.1080/01425919408909810

Nann, S. (1990). Potentials for tracking photovoltaic systems and V-troughs in moderate climates. *Solar Energy*, *45*(6), 385–393. doi:10.1016/0038-092X(90)90160-E

NASA Surface meteorology and Solar Energy (SSE) Release 6.0 Methodology. (2012, March 1). Retrieved July 7, 2012, from http://eosweb.larc.nasa.gov/cgi-bin/sse/sse.cgi?rets@nrcan.gc.ca+s06#s06

Ng, Kh. M., Adam, N. M., & Inayatullah, O. & Ab Kadir, M. Z. A. (2014). Assessment of solar radiation on diversely oriented surfaces and optimum tilts for solar absorbers in Malaysian tropical latitude. *International Journal of Energy and Environmental Engineering*, *5*(5), 1–13.

Nijegorodov, N., Devan, K. R. S., Jain, P. K., & Carlsson, S. (1997). Atmospheric transmittance models and an analytical method to predict the optimum slope on an absorber plate. *Renewable Energy*, *4*(5), 529–543. doi:10.1016/0960-1481(94)90215-1

NREL PVWatts. (n.d.). Retrieved 07 23, 2012, from NREL: http://www.nrel.gov/rredc/pvwatts/about.html

Oko, C. O. C., & Nnamchi, S. N. (2012). Optimum collector tilt angles for low latitudes. *The Open Renewable Energy Journal*, *5*, 7–14. doi:10.2174/1876387101205010007

Oladiran, M. T. (1995). Mean global radiation captured by inclined collectors at various surface azimuth angles in Nigeria. *Applied Energy*, *52*(4), 317–330. doi:10.1016/0306-2619(95)00016-X

Pavlović, T., Pavlović, Z. L., Pantić, L., & Kostić, J. (2010). Determining optimum tilt angles and orientations of photovoltaic panels in Nis Serbia. *Contemporary Materials.*, *2*(2), 151–156. doi:10.5767/anurs.cmat.100102.en.151P

Perez, R., Ineichen, P., Seals, R., Michalsky, J., & Stewart, R. (1990). Modeling daylight availability and irradiance components from direct and global irradiance. *Solar Energy*, *44*(5), 271–289. doi:10.1016/0038-092X(90)90055-H

Prescott, J. A. (1940). Evaporation from Water Surface in Relation to Solar Radiation. *Transactions of the Royal Society of South Australia*, *40*, 114–118.

Samani, Z. (2000). Estimating solar radiation and evapotranspiration using minimum climatological data. *Journal of Irrigation and Drainage Engineering*, *126*(4), 265–267. doi:10.1061/(ASCE)0733-9437(2000)126:4(265)

Seme, S., & Štumberger, G. (2011). A novel prediction algorithm for solar angles using solar radiation and differential evolution for dual-axis sun tracking purposes. *Solar Energy*, *85*(11), 2757–2770. doi:10.1016/j.solener.2011.08.031

Shariah, A. M., Al-Akhras, A., & Al-Omari, I. A. (2002). Optimizing the tilt angle of solar collectors. *Renewable Energy*, *26*(4), 587–598. doi:10.1016/S0960-1481(01)00106-9

Sinha, S., & Chandel, S. S. (2016). Analysis of fixed tilt and sun tracking photovoltaic–micro wind based hybrid power systems. *Energy Conversion and Management*, *115*, 265–275. doi:10.1016/j.enconman.2016.02.056

Siraki, A. (2012). Study of optimum tilt angles for solar panels in different latitudes for urban applications. *Solar Energy*, *86*(6), 1920–1928. doi:10.1016/j.solener.2012.02.030

Skeiker, K. (2009). Optimum tilt angle and orientation for solar collectors in Syria. *Energy Conversion and Management*, *50*(9), 2439–2448. doi:10.1016/j.enconman.2009.05.031

Soulayman, S. Sh. (1991). On the optimum tilt of solar absorber plates. *Renewable Energy*, *1*(3-4), 551–554. doi:10.1016/0960-1481(91)90070-6

Soulayman, S. (2015). Comments on Optimum tilt angle for flat plate collectors all over the World – A declination dependence formula and comparisons of three solar radiation models by Stanciu, C., Stanciu, D. *Energy Conversion and Management*, *93*, 448–449. doi:10.1016/j.enconman.2015.01.005

Soulayman, S., Alhelou, M., & Sabbagh, W. (2016). All-world solar collector optimum slope determination basing on site latitude and day number. *Open Access Library Journal*, *3*, e2385. doi:10.4236/oalib.1102385

Soulayman, S., & Hammoud, M. (2016). Optimum tilt angle of solar collectors for building applications in mid-latitude zone. *Energy Conversion and Management*, *124*, 20–28. doi:10.1016/j.enconman.2016.06.066

Soulayman, S., & Sabbagh, W. (2014). An algorithm for determining optimum tilt angle and orientation for solar collectors. *Journal of Solar Energy Research Updates*, *1*(1), 19–30. doi:10.15377/2410-2199.2014.01.01.3

Soulayman, S., & Sabbagh, W. (2015a). Comment on Optimum tilt angle and orientation for solar collectors in Syria by Skeiker, K. *Energy Conversion and Management*, *89*, 1001–1002. doi:10.1016/j.enconman.2014.10.023

Soulayman, S., & Sabbagh, W. (2015b). Optimum tilt angle at tropical region. *Int. Journal of Renewable Energy Development*, *4*(1), 48–54. doi:10.14710/ijred.4.1.48-54

Soulayman, S., & Sabbagh, W. (2015c). Solar collector optimum tilt and orientation. *Open Journal of Renewable and Sustainable Energy*, *2*(1), 1–9. doi:10.12966/rse.03.01.2015

Soulayman, S., & Sabbagh, W. (2016). General algorithm for optimum tilt angles determination at tropical region. *British Journal of Renewable Energy*, *1*(1), 32–37. Retrieved from www.measpublishing.co.uk/BJRE

Stanciu, C., & Stanciu, D. (2014). Optimum tilt angle for flat plate collectors all over the World – A declination dependence formula and comparisons of three solar radiation models. *Energy Conversion and Management*, *81*, 133–143. doi:10.1016/j.enconman.2014.02.016

Talebizadeh, P., Abdolzadeh, M., & Mehrabian, M. A. (2011). Determination of optimum slope angles of solar collectors based on new correlations. *Energy Sources. Part A*, *33*(17), 1567–1580. doi:10.1080/15567036.2010.551253

Talebizadeh, P., Mehrabian, M. A., & Abdolzadeh, M. (2011). Prediction of the optimum slope and surface azimuth angles using the Genetic Algorithm. *Energy and Building*, *43*(11), 2998–3005. doi:10.1016/j.enbuild.2011.07.013

Tang, R., & Wu, T. (2004). Optimal tilt-angles for solar collectors used in China. *Applied Energy*, *79*(3), 239–248. doi:10.1016/j.apenergy.2004.01.003

Taybout, R. A., & Löf, G. O. G. (1970). Solar house heating. *Natural Resources Journal, 10*(2), 268–326.

Tiris, M., & Tiris, C. (1998). Optimum collector slope and model evaluation: Case study for Gebze, Turkey. *Energy Conversion and Management, 39*(3-4), 167–172. doi:10.1016/S0196-8904(96)00229-4

Uba, F. A., & Sarsah, E. A. (2013). Optimization of tilt angle for solar collectors in WA, Ghana. Pelagia Research Library, Advances in Applied Science Research, 4(4), 108-114.

Ulgen, K. (2006). Optimum tilt angle for solar collectors. *Energy Sources, Part A: Recovery. Utilization and Environmental Effects, 28*(13), 1171–1180. doi:10.1080/00908310600584524

Vilela, O. C., Fraidenraich, N., & Tiba, C. (2003). Photovoltaic pumping systems driven by tracking collectors. Experiments and simulation. *Solar Energy, 74*(1), 45–52. doi:10.1016/S0038-092X(03)00109-9

Waziri, N. H., Usman, A. M., Enaburekhan, J. S., & Babakano, A. (2014). Determination of Optimum Tilt Angle and Orientation of a Flat Plate Solar Collector for Different Periods in Kano. *Scirj, 2*(2), 34-40.

Yakup, M. A., & Malik, A. Q. (2001). Optimum tilt angle and orientation for solar collector in Brunei Darussalam. *Renewable Energy, 24*(2), 223–234. doi:10.1016/S0960-1481(00)00168-3

Yan, R., Saha, T. K., Meredith, P., & Goodwin, S. (2013). Analysis of yearlong performance of differently tilted photovoltaic systems in Brisbane, Australia. *Energy Conversion and Management, 74*, 102–108. doi:10.1016/j.enconman.2013.05.007

Yellott, H. (1973). Utilization of sun and sky radiation for heating cooling of buildings. *ASHRAE Journal, 15*(12), 31–42.

Zawilska, E., & Brooks, M. (2011). An assessment of the solar resource for Durban, South Africa. *Renewable Energy, 36*(12), 3433–3438. doi:10.1016/j.renene.2011.05.023

APPENDIX A

The day numbers given in all tables of these appendixes correspond the dates given in Table 35.

Table 35. The chosen dates.

Day	Day/Month	Remark
1	1/1	-
10	10/1	-
17	17/1	Characteristic day
20	20/1	-
30	30/1	-
32	1/2	-
41	10/2	-
47	16/2	Characteristic day
51	20/2	-
59	28/2	-
60	1/3	-
69	10/3	-
75	16/3	Characteristic day
79	20/3	-
89	30/3	-
91	1/4	-
100	10/4	-
105	15/4	Characteristic day
110	20/4	-
120	30/4	-
121	1/5	-
130	10/5	-
135	15/5	Characteristic day
140	20/5	-
150	30/5	-
152	1/6	-
161	10/6	-
162	11/6	Characteristic day
171	20/6	-
181	30/6	-
182	1/7	-
191	10/7	-
198	17/7	Characteristic day

continued on following page

Table 35. Continued

Day	Day/Month	Remark
201	20/7	-
211	30/7	-
213	1/8	-
222	10/8	-
228	16/8	Characteristic day
232	20/8	-
242	30/8	-
244	1/9	-
253	10/9	-
258	15/9	Characteristic day
263	20/9	-
273	30/9	-
274	1/10	-
283	10/10	-
288	15/10	Characteristic day
293	20/10	-
303	30/10	-
305	1/11	-
314	10/11	-
318	14/11	Characteristic day
324	20/11	-
334	30/11	-
335	1/12	-
344	10/12	Characteristic day
355	20/12	-
365	31/12	-

APPENDIX B

Optimum Tilt Angle β_{opt} (°) at Tropical Zone

Tables 36 to 41 give the daily optimum tilt angle $\beta_{opt,d}$ (°) at tropical zone. Tables 42 to 47 give the monthly optimum tilt angle $\beta_{opt,m}$ (°) at tropical zone. As during some calendar months the daily optimum tilt angle $\beta_{opt,d}$ (°) changes its sign, Tables 42 to 47 show the optimum tilt angle for these two periods

Table 36. Daily optimum tilt angle $\beta_{opt,d}$ (°) for latitudes [23.45°S-16 °S] for characteristic and other days

	Latitudes [23.45°S-16 °S]								
n	**-23.45**	**-23**	**-22**	**-21**	**-20**	**-19**	**-18**	**-17**	**-16**
1	13.77	14.13	14.95	15.77	16.60	17.43	18.26	19.10	19.94
10	12.29	12.66	13.49	14.33	15.17	16.01	16.85	17.70	18.55
17	10.56	10.94	11.79	12.63	13.49	14.34	15.20	16.06	16.93
20	9.67	10.05	10.90	11.76	12.62	13.48	14.34	15.21	16.08
30	6.04	6.44	7.32	8.20	9.09	9.97	10.87	11.76	12.65
32	5.20	5.60	6.49	7.37	8.27	9.16	10.06	10.96	11.86
41	0.98	1.39	2.30	3.22	4.14	5.06	5.99	6.91	7.84
47	-2.21	-1.78	-0.85	0.09	1.03	1.97	2.91	3.85	4.79
51	-4.47	-4.04	-3.09	-2.14	-1.19	-0.24	0.71	1.67	2.62
59	-9.26	-8.82	-7.85	-6.88	-5.91	-4.94	-3.96	-2.99	-2.02
60	-9.88	-9.44	-8.47	-7.49	-6.52	-5.54	-4.57	-3.59	-2.62
69	-15.59	-15.14	-14.15	-13.16	-12.17	-11.18	-10.18	-9.19	-8.20
75	-19.45	-19.00	-18.01	-17.01	-16.01	-15.01	-14.01	-13.02	-12.02
79	-22.02	-21.57	-20.57	-19.57	-18.57	-17.57	-16.57	-15.57	-14.57
89	-28.27	-27.82	-26.82	-25.83	-24.83	-23.83	-22.83	-21.84	-20.84
91	-29.47	-29.02	-28.03	-27.04	-26.04	-25.05	-24.05	-23.06	-22.06
100	-34.64	-34.20	-33.22	-32.24	-31.25	-30.27	-29.29	-28.30	-27.32
105	-37.30	-36.87	-35.89	-34.92	-33.95	-32.97	-32.00	-31.02	-30.05
110	-39.80	-39.36	-38.40	-37.44	-36.47	-35.51	-34.55	-33.58	-32.62
120	-44.23	-43.81	-42.87	-41.93	-40.98	-40.04	-39.10	-38.15	-37.21
121	-44.64	-44.21	-43.27	-42.33	-41.39	-40.45	-39.51	-38.57	-37.63
130	-47.90	-47.49	-46.57	-45.65	-44.73	-43.80	-42.88	-41.96	-41.03
135	-49.45	-49.04	-48.12	-47.21	-46.30	-45.39	-44.47	-43.56	-42.65
140	-50.79	-50.39	-49.48	-48.58	-47.68	-46.77	-45.87	-44.96	-44.05
150	-52.90	-52.50	-51.61	-50.72	-49.83	-48.94	-48.05	-47.16	-46.27

continued on following page

Table 36. Continued

	Latitudes [23.45°S-16 °S]								
n	**-23.45**	**-23**	**-22**	**-21**	**-20**	**-19**	**-18**	**-17**	**-16**
152	-53.23	-52.83	-51.95	-51.06	-50.17	-49.28	-48.39	-47.50	-46.61
161	-54.33	-53.94	-53.06	-52.18	-51.29	-50.41	-49.53	-48.65	-47.77
162	-54.42	-54.02	-53.14	-52.26	-51.38	-50.50	-49.62	-48.74	-47.86
171	-54.83	-54.44	-53.56	-52.68	-51.81	-50.93	-50.05	-49.17	-48.29
181	-54.57	-54.17	-53.30	-52.42	-51.54	-50.66	-49.78	-48.90	-48.02
182	-54.50	-54.11	-53.23	-52.35	-51.47	-50.59	-49.71	-48.83	-47.95
191	-53.55	-53.15	-52.26	-51.38	-50.49	-49.61	-48.72	-47.83	-46.94
198	-52.38	-51.97	-51.08	-50.19	-49.29	-48.40	-47.50	-46.61	-45.71
201	-51.76	-51.35	-50.46	-49.56	-48.66	-47.76	-46.86	-45.96	-45.07
211	-49.19	-48.78	-47.87	-46.95	-46.04	-45.12	-44.21	-43.29	-42.38
213	-48.58	-48.17	-47.25	-46.34	-45.42	-44.50	-43.58	-42.66	-41.74
222	-45.46	-45.04	-44.11	-43.17	-42.24	-41.30	-40.36	-39.43	-38.49
228	-43.04	-42.61	-41.66	-40.71	-39.76	-38.82	-37.87	-36.92	-35.97
232	-41.26	-40.83	-39.88	-38.92	-37.96	-37.01	-36.05	-35.09	-34.13
242	-36.33	-35.89	-34.91	-33.93	-32.96	-31.98	-31.00	-30.02	-29.04
244	-35.26	-34.81	-33.83	-32.85	-31.87	-30.89	-29.91	-28.93	-27.95
253	-30.14	-29.70	-28.70	-27.71	-26.71	-25.72	-24.73	-23.73	-22.74
258	-27.12	-26.67	-25.67	-24.68	-23.68	-22.68	-21.68	-20.68	-19.69
263	-24.00	-23.55	-22.55	-21.55	-20.55	-19.55	-18.55	-17.55	-16.55
273	-17.60	-17.15	-16.16	-15.16	-14.17	-13.17	-12.18	-11.18	-10.19
274	-16.96	-16.51	-15.51	-14.52	-13.53	-12.53	-11.54	-10.54	-9.55
283	-11.21	-10.77	-9.79	-8.81	-7.83	-6.85	-5.87	-4.89	-3.91
288	-8.11	-7.67	-6.71	-5.74	-4.77	-3.81	-2.84	-1.87	-0.90
293	-5.12	-4.69	-3.74	-2.79	-1.83	-0.88	0.08	1.03	1.99
303	0.40	0.81	1.73	2.66	3.58	4.50	5.43	6.36	7.29
305	1.42	1.83	2.74	3.66	4.57	5.49	6.41	7.33	8.26
314	5.57	5.97	6.85	7.74	8.63	9.52	10.42	11.31	12.21
318	7.18	7.57	8.45	9.32	10.20	11.08	11.96	12.85	13.74
324	9.31	9.69	10.55	11.41	12.27	13.13	14.00	14.87	15.74
334	12.05	12.42	13.25	14.09	14.93	15.77	16.62	17.47	18.33
335	12.26	12.64	13.47	14.30	15.14	15.98	16.83	17.68	18.53
344	13.75	14.12	14.93	15.76	16.58	17.41	18.25	19.09	19.93
354	14.40	14.77	15.58	16.39	17.22	18.04	18.87	19.70	20.54
365	13.90	14.27	15.08	15.91	16.73	17.56	18.39	19.23	20.07

Table 37. Daily optimum tilt angle $\beta_{opt,d}$ (°) for latitudes [15°S-8°S]

		Latitudes [15°S-8°S]							
	n	-15	-14	-13	-12	-11	-10	-9	-8
1/1	1	20.79	21.63	22.48	23.34	24.19	25.05	25.91	26.77
10/1	10	19.41	20.27	21.13	21.99	22.86	23.72	24.59	25.46
17/1	17	17.79	18.66	19.54	20.41	21.29	22.16	23.04	23.93
20/1	20	16.96	17.83	18.71	19.59	20.47	21.36	22.24	23.13
30/1	30	13.55	14.45	15.35	16.25	17.16	18.06	18.97	19.88
1/2	32	12.76	13.66	14.57	15.48	16.38	17.29	18.21	19.12
10/2	41	8.77	9.69	10.62	11.55	12.49	13.42	14.35	15.29
16/2	47	5.74	6.68	7.63	8.58	9.52	10.47	11.42	12.37
20/2	51	3.58	4.53	5.49	6.44	7.40	8.36	9.32	10.28
29/2	59	-1.04	-0.07	0.91	1.88	2.86	3.84	4.81	5.79
1/3	60	-1.64	-0.66	0.31	1.29	2.27	3.25	4.22	5.20
10/3	69	-7.21	-6.22	-5.23	-4.23	-3.24	-2.25	-1.26	-0.26
16/3	75	-11.02	-10.02	-9.03	-8.03	-7.03	-6.03	-5.03	-4.04
20/3	79	-13.57	-12.57	-11.57	-10.57	-9.57	-8.57	-7.57	-6.57
30/3	89	-19.84	-18.85	-17.85	-16.85	-15.86	-14.86	-13.86	-12.87
1/4	91	-21.06	-20.07	-19.07	-18.08	-17.08	-16.09	-15.09	-14.10
10/4	100	-26.34	-25.35	-24.37	-23.39	-22.40	-21.42	-20.44	-19.45
15/4	105	-29.07	-28.10	-27.13	-26.15	-25.18	-24.20	-23.23	-22.25
20/4	110	-31.65	-30.69	-29.72	-28.76	-27.79	-26.83	-25.86	-24.89
30/4	120	-36.27	-35.32	-34.38	-33.43	-32.49	-31.55	-30.60	-29.66
1/5	121	-36.69	-35.74	-34.80	-33.86	-32.92	-31.98	-31.03	-30.09
10/5	130	-40.11	-39.19	-38.26	-37.34	-36.42	-35.49	-34.57	-33.64
15/5	135	-41.73	-40.82	-39.90	-38.99	-38.07	-37.16	-36.24	-35.33
20/5	140	-43.15	-42.24	-41.34	-40.43	-39.52	-38.62	-37.71	-36.81
30/5	150	-45.37	-44.48	-43.59	-42.70	-41.80	-40.91	-40.02	-39.13
1/6	152	-45.72	-44.83	-43.94	-43.05	-42.16	-41.27	-40.38	-39.49
10/6	161	-46.88	-46.00	-45.12	-44.24	-43.35	-42.47	-41.59	-40.71
11/6	162	-46.97	-46.09	-45.21	-44.33	-43.45	-42.56	-41.68	-40.80
20/6	171	-47.41	-46.53	-45.65	-44.77	-43.90	-43.02	-42.14	-41.26
30/6	181	-47.14	-46.25	-45.37	-44.49	-43.61	-42.73	-41.85	-40.97
1/7	182	-47.06	-46.18	-45.30	-44.42	-43.54	-42.66	-41.78	-40.90
10/7	191	-46.06	-45.17	-44.28	-43.39	-42.50	-41.62	-40.73	-39.84
17/7	198	-44.82	-43.92	-43.03	-42.13	-41.23	-40.34	-39.44	-38.55
20/7	201	-44.17	-43.27	-42.37	-41.47	-40.57	-39.67	-38.77	-37.87
30/7	211	-41.46	-40.55	-39.63	-38.71	-37.80	-36.88	-35.97	-35.05
1/8	213	-40.82	-39.90	-38.98	-38.06	-37.15	-36.23	-35.31	-34.39

continued on following page

Table 37. Continued

		Latitudes [15°S-8°S]							
	n	-15	-14	-13	-12	-11	-10	-9	-8
10/8	222	-37.55	-36.62	-35.68	-34.74	-33.80	-32.86	-31.93	-30.99
16/8	228	-35.02	-34.07	-33.12	-32.17	-31.22	-30.27	-29.31	-28.36
20/8	232	-33.17	-32.21	-31.26	-30.30	-29.34	-28.38	-27.42	-26.46
30/8	242	-28.07	-27.09	-26.11	-25.13	-24.15	-23.18	-22.20	-21.22
1/9	244	-26.97	-25.99	-25.00	-24.02	-23.04	-22.06	-21.08	-20.09
10/9	253	-21.75	-20.75	-19.76	-18.76	-17.77	-16.77	-15.78	-14.79
15/9	258	-18.69	-17.69	-16.69	-15.69	-14.69	-13.70	-12.70	-11.70
20/9	263	-15.55	-14.55	-13.55	-12.55	-11.55	-10.55	-9.55	-8.55
30/9	273	-9.19	-8.19	-7.20	-6.20	-5.21	-4.21	-3.22	-2.22
1/10	274	-8.55	-7.56	-6.57	-5.57	-4.58	-3.58	-2.59	-1.59
10/10	283	-2.93	-1.95	-0.97	0.01	0.99	1.98	2.96	3.94
15/10	288	0.07	1.04	2.01	2.98	3.96	4.93	5.90	6.87
20/10	293	2.95	3.91	4.87	5.83	6.79	7.75	8.71	9.67
30/10	303	8.22	9.15	10.08	11.02	11.95	12.89	13.82	14.76
1/11	305	9.18	10.11	11.04	11.96	12.89	13.82	14.76	15.69
10/11	314	13.11	14.01	14.91	15.82	16.73	17.63	18.54	19.45
14/11	318	14.63	15.52	16.41	17.31	18.20	19.10	20.00	20.90
20/11	324	16.62	17.50	18.38	19.26	20.14	21.03	21.92	22.81
30/11	334	19.18	20.04	20.90	21.77	22.63	23.50	24.37	25.25
1/12	335	19.38	20.24	21.10	21.97	22.83	23.70	24.57	25.44
10/12	344	20.77	21.62	22.47	23.32	24.18	25.04	25.90	26.76
20/12	354	21.38	22.22	23.07	23.92	24.77	25.62	26.48	27.34
31/12	365	20.91	21.76	22.61	23.46	24.32	25.17	26.03	26.89

Table 38. Daily optimum tilt angle $\beta_{opt,d}$ (°) for latitudes [7°S-1°S]

		Latitudes [7°S-1°S]						
Date	n	-7	-6	-5	-4	-3	-2	-1
1/1	1	27.64	28.50	29.37	30.24	31.11	31.98	32.85
10/1	10	26.34	27.21	28.09	28.97	29.84	30.72	31.61
17/1	17	24.81	25.69	26.58	27.47	28.36	29.25	30.14
20/1	20	24.02	24.91	25.80	26.69	27.58	28.48	29.37
30/1	30	20.78	21.69	22.61	23.52	24.43	25.34	26.26
1/2	32	20.03	20.95	21.86	22.78	23.70	24.61	25.53
10/2	41	16.22	17.16	18.09	19.03	19.97	20.91	21.84
16/2	47	13.32	14.27	15.22	16.17	17.12	18.07	19.03

continued on following page

Table 38. Continued

		Latitudes [7°S-1°S]						
Date	**n**	**-7**	**-6**	**-5**	**-4**	**-3**	**-2**	**-1**
20/2	51	11.24	12.20	13.16	14.12	15.08	16.04	17.00
29/2	59	6.76	7.74	8.72	9.69	10.67	11.65	12.63
1/3	60	6.18	7.16	8.14	9.12	10.09	11.07	12.05
10/3	69	0.73	1.72	2.71	3.71	4.70	5.69	6.68
16/3	75	-3.04	-2.04	-1.04	-0.04	0.95	1.95	2.95
20/3	79	-5.57	-4.57	-3.57	-2.57	-1.57	-0.57	0.43
30/3	89	-11.87	-10.87	-9.87	-8.88	-7.88	-6.88	-5.89
1/4	91	-13.10	-12.11	-11.11	-10.12	-9.12	-8.13	-7.13
10/4	100	-18.47	-17.48	-16.50	-15.52	-14.53	-13.55	-12.57
15/4	105	-21.27	-20.30	-19.32	-18.35	-17.37	-16.40	-15.42
20/4	110	-23.93	-22.96	-22.00	-21.03	-20.07	-19.10	-18.14
30/4	120	-28.71	-27.77	-26.82	-25.88	-24.94	-23.99	-23.05
1/5	121	-29.15	-28.21	-27.27	-26.32	-25.38	-24.44	-23.50
10/5	130	-32.72	-31.80	-30.87	-29.95	-29.03	-28.11	-27.19
15/5	135	-34.42	-33.50	-32.59	-31.68	-30.76	-29.85	-28.94
20/5	140	-35.90	-35.00	-34.09	-33.19	-32.28	-31.38	-30.48
30/5	150	-38.24	-37.35	-36.46	-35.57	-34.68	-33.79	-32.90
1/6	152	-38.60	-37.72	-36.83	-35.94	-35.05	-34.17	-33.28
10/6	161	-39.83	-38.95	-38.07	-37.19	-36.31	-35.43	-34.55
11/6	162	-39.92	-39.04	-38.16	-37.28	-36.40	-35.53	-34.65
20/6	171	-40.38	-39.51	-38.63	-37.75	-36.88	-36.00	-35.13
30/6	181	-40.09	-39.21	-38.33	-37.46	-36.58	-35.70	-34.83
1/7	182	-40.02	-39.14	-38.26	-37.38	-36.50	-35.63	-34.75
10/7	191	-38.95	-38.07	-37.18	-36.30	-35.41	-34.53	-33.65
17/7	198	-37.65	-36.76	-35.87	-34.97	-34.08	-33.19	-32.30
20/7	201	-36.97	-36.07	-35.17	-34.27	-33.38	-32.48	-31.59
30/7	211	-34.13	-33.22	-32.30	-31.39	-30.48	-29.56	-28.65
1/8	213	-33.47	-32.55	-31.63	-30.71	-29.79	-28.87	-27.96
10/8	222	-30.05	-29.11	-28.18	-27.24	-26.30	-25.37	-24.43
16/8	228	-27.41	-26.46	-25.51	-24.56	-23.61	-22.66	-21.71
20/8	232	-25.50	-24.54	-23.59	-22.63	-21.67	-20.71	-19.75
30/8	242	-20.24	-19.26	-18.28	-17.30	-16.33	-15.35	-14.37
1/9	244	-19.11	-18.13	-17.15	-16.17	-15.18	-14.20	-13.22
10/9	253	-13.79	-12.80	-11.80	-10.81	-9.81	-8.82	-7.83
15/9	258	-10.70	-9.70	-8.71	-7.71	-6.71	-5.71	-4.71
20/9	263	-7.55	-6.55	-5.55	-4.55	-3.55	-2.55	-1.55
30/9	273	-1.22	-0.23	0.77	1.76	2.76	3.75	4.75

continued on following page

Table 38. Continued

		Latitudes [7°S-1°S]						
Date	**n**	**-7**	**-6**	**-5**	**-4**	**-3**	**-2**	**-1**
1/10	274	-0.60	0.40	1.39	2.39	3.38	4.38	5.37
10/10	283	4.92	5.90	6.89	7.87	8.85	9.84	10.82
15/10	288	7.84	8.82	9.79	10.76	11.74	12.71	13.69
20/10	293	10.63	11.59	12.55	13.52	14.48	15.44	16.41
30/10	303	15.70	16.64	17.57	18.51	19.45	20.40	21.34
1/11	305	16.62	17.55	18.49	19.42	20.36	21.29	22.23
10/11	314	20.37	21.28	22.19	23.11	24.02	24.94	25.85
14/11	318	21.81	22.71	23.61	24.52	25.43	26.33	27.24
20/11	324	23.70	24.59	25.48	26.38	27.27	28.17	29.07
30/11	334	26.12	27.00	27.88	28.75	29.63	30.52	31.40
1/12	335	26.31	27.19	28.06	28.94	29.82	30.70	31.58
10/12	344	27.62	28.49	29.36	30.22	31.10	31.97	32.84
20/12	354	28.20	29.06	29.92	30.79	31.65	32.52	33.39
31/12	365	27.76	28.62	29.49	30.36	31.23	32.10	32.97

Table 39. Daily optimum tilt angle $\beta_{opt,d}$ (°) for latitudes [0°-7°N]

		Latitudes [0o-7oN]							
Date	**n**	**0**	**1**	**2**	**3**	**4**	**5**	**6**	**7**
1/1	1	33.73	34.60	35.48	36.36	37.24	38.12	38.99	39.88
10/1	10	32.49	33.37	34.26	35.14	36.03	36.91	37.80	38.69
17/1	17	31.03	31.92	32.81	33.71	34.60	35.50	36.40	37.29
20/1	20	30.27	31.17	32.07	32.96	33.86	34.76	35.67	36.57
30/1	30	27.17	28.09	29.01	29.92	30.84	31.76	32.68	33.59
1/2	32	26.45	27.37	28.29	29.21	30.13	31.06	31.98	32.90
10/2	41	22.78	23.72	24.66	25.60	26.54	27.48	28.42	29.37
16/2	47	19.98	20.93	21.88	22.84	23.79	24.74	25.70	26.65
20/2	51	17.96	18.92	19.88	20.84	21.81	22.77	23.73	24.69
29/2	59	13.60	14.58	15.56	16.54	17.51	18.49	19.47	20.45
1/3	60	13.03	14.01	14.99	15.97	16.95	17.93	18.91	19.89
10/3	69	7.68	8.67	9.66	10.66	11.65	12.64	13.63	14.63
16/3	75	3.95	4.95	5.95	6.94	7.94	8.94	9.94	10.94
20/3	79	1.43	2.43	3.42	4.42	5.42	6.42	7.42	8.42
30/3	89	-4.89	-3.89	-2.89	-1.90	-0.90	0.10	1.09	2.09
1/4	91	-6.14	-5.14	-4.14	-3.15	-2.15	-1.16	-0.16	0.83
10/4	100	-11.58	-10.60	-9.61	-8.63	-7.65	-6.66	-5.68	-4.70

continued on following page

Table 39. Continued

		Latitudes [0o-7oN]							
Date	**n**	**0**	**1**	**2**	**3**	**4**	**5**	**6**	**7**
15/4	105	-14.45	-13.48	-12.50	-11.53	-10.55	-9.58	-8.60	-7.63
20/4	110	-17.17	-16.21	-15.24	-14.28	-13.32	-12.35	-11.39	-10.43
30/4	120	-22.11	-21.17	-20.22	-19.28	-18.34	-17.40	-16.46	-15.52
1/5	121	-22.56	-21.62	-20.68	-19.74	-18.80	-17.86	-16.93	-15.99
10/5	130	-26.26	-25.34	-24.42	-23.51	-22.59	-21.67	-20.75	-19.84
15/5	135	-28.03	-27.12	-26.21	-25.30	-24.39	-23.49	-22.58	-21.68
20/5	140	-29.58	-28.68	-27.78	-26.88	-25.98	-25.08	-24.19	-23.29
30/5	150	-32.02	-31.13	-30.25	-29.36	-28.48	-27.60	-26.72	-25.84
1/6	152	-32.40	-31.52	-30.63	-29.75	-28.87	-27.99	-27.12	-26.24
10/6	161	-33.68	-32.80	-31.93	-31.06	-30.19	-29.32	-28.45	-27.58
11/6	162	-33.78	-32.90	-32.03	-31.16	-30.29	-29.42	-28.55	-27.69
20/6	171	-34.26	-33.39	-32.52	-31.65	-30.78	-29.92	-29.05	-28.19
30/6	181	-33.95	-33.08	-32.21	-31.34	-30.47	-29.60	-28.74	-27.87
1/7	182	-33.88	-33.00	-32.13	-31.26	-30.39	-29.52	-28.65	-27.79
10/7	191	-32.77	-31.88	-31.01	-30.13	-29.25	-28.37	-27.50	-26.63
17/7	198	-31.41	-30.52	-29.63	-28.74	-27.85	-26.97	-26.09	-25.20
20/7	201	-30.69	-29.80	-28.90	-28.01	-27.12	-26.23	-25.34	-24.46
30/7	211	-27.74	-26.82	-25.91	-25.00	-24.09	-23.18	-22.28	-21.37
1/8	213	-27.04	-26.13	-25.21	-24.30	-23.38	-22.47	-21.56	-20.65
10/8	222	-23.49	-22.56	-21.62	-20.69	-19.76	-18.82	-17.89	-16.96
16/8	228	-20.77	-19.82	-18.87	-17.92	-16.97	-16.02	-15.08	-14.13
20/8	232	-18.79	-17.84	-16.88	-15.92	-14.96	-14.01	-13.05	-12.10
30/8	242	-13.39	-12.41	-11.43	-10.46	-9.48	-8.50	-7.52	-6.55
1/9	244	-12.24	-11.26	-10.28	-9.29	-8.31	-7.33	-6.35	-5.37
10/9	253	-6.83	-5.84	-4.84	-3.85	-2.85	-1.86	-0.87	0.13
15/9	258	-3.71	-2.72	-1.72	-0.72	0.28	1.28	2.28	3.27
20/9	263	-0.55	0.45	1.45	2.45	3.45	4.45	5.45	6.45
30/9	273	5.75	6.74	7.74	8.73	9.73	10.73	11.72	12.72
1/10	274	6.37	7.36	8.36	9.35	10.35	11.34	12.34	13.33
10/10	283	11.80	12.78	13.77	14.75	15.73	16.72	17.70	18.68
15/10	288	14.66	15.63	16.61	17.58	18.56	19.53	20.51	21.48
20/10	293	17.37	18.34	19.30	20.26	21.23	22.19	23.16	24.12
30/10	303	22.28	23.22	24.16	25.10	26.05	26.99	27.93	28.88
1/11	305	23.17	24.11	25.04	25.98	26.92	27.86	28.80	29.74
10/11	314	26.77	27.69	28.61	29.53	30.45	31.37	32.29	33.21
14/11	318	28.15	29.06	29.97	30.89	31.80	32.71	33.62	34.53

continued on following page

Table 39. Continued

		Latitudes [0o-7oN]							
Date	**n**	**0**	**1**	**2**	**3**	**4**	**5**	**6**	**7**
20/11	324	29.97	30.86	31.76	32.67	33.57	34.47	35.37	36.27
30/11	334	32.28	33.17	34.05	34.94	35.83	36.72	37.60	38.49
1/12	335	32.47	33.35	34.23	35.12	36.01	36.89	37.78	38.67
10/12	344	33.72	34.59	35.47	36.35	37.22	38.10	38.98	39.86
20/12	354	34.26	35.14	36.01	36.88	37.76	38.63	39.51	40.39
31/12	365	33.84	34.72	35.59	36.47	37.35	38.23	39.11	39.99

Table 40. Daily optimum tilt angle $\beta_{opt,d}$ (^{o}) for latitudes [8°-15^oN]

		Latitudes [8o-15oN]							
Date	**n**	**8**	**9**	**10**	**11**	**12**	**13**	**14**	**15**
1/1	1	40.76	41.64	42.52	43.40	44.28	45.17	46.05	46.93
10/1	10	39.58	40.47	41.36	42.25	43.13	44.02	44.91	45.80
17/1	17	38.19	39.09	39.98	40.88	41.78	42.68	43.58	44.47
20/1	20	37.47	38.37	39.27	40.17	41.07	41.98	42.88	43.78
30/1	30	34.51	35.43	36.35	37.27	38.19	39.11	40.03	40.95
1/2	32	33.82	34.74	35.67	36.59	37.51	38.44	39.36	40.28
10/2	41	30.31	31.25	32.19	33.13	34.07	35.01	35.95	36.89
16/2	47	27.61	28.56	29.51	30.47	31.42	32.38	33.33	34.28
20/2	51	25.66	26.62	27.58	28.54	29.50	30.47	31.43	32.39
29/2	59	21.43	22.40	23.38	24.36	25.34	26.31	27.29	28.27
1/3	60	20.87	21.85	22.83	23.81	24.79	25.77	26.75	27.73
10/3	69	15.62	16.61	17.61	18.60	19.59	20.58	21.58	22.57
16/3	75	11.93	12.93	13.93	14.93	15.93	16.92	17.92	18.92
20/3	79	9.42	10.42	11.42	12.42	13.42	14.42	15.42	16.42
30/3	89	3.09	4.08	5.08	6.08	7.07	8.07	9.07	10.07
1/4	91	1.83	2.82	3.82	4.81	5.81	6.80	7.80	8.79
10/4	100	-3.71	-2.73	-1.75	-0.77	0.22	1.20	2.18	3.16
15/4	105	-6.66	-5.68	-4.71	-3.74	-2.77	-1.79	-0.82	0.15
20/4	110	-9.46	-8.50	-7.54	-6.58	-5.62	-4.66	-3.70	-2.74
30/4	120	-14.58	-13.64	-12.71	-11.77	-10.83	-9.90	-8.97	-8.03
1/5	121	-15.05	-14.12	-13.18	-12.25	-11.32	-10.38	-9.45	-8.52
10/5	130	-18.92	-18.01	-17.10	-16.19	-15.28	-14.37	-13.46	-12.56
15/5	135	-20.77	-19.87	-18.97	-18.07	-17.17	-16.28	-15.38	-14.49
20/5	140	-22.40	-21.51	-20.62	-19.73	-18.84	-17.96	-17.07	-16.19

continued on following page

Table 40. Continued

		Latitudes [8o-15oN]							
Date	**n**	**8**	**9**	**10**	**11**	**12**	**13**	**14**	**15**
30/5	150	-24.97	-24.09	-23.22	-22.35	-21.48	-20.61	-19.75	-18.89
1/6	152	-25.37	-24.50	-23.63	-22.76	-21.89	-21.03	-20.17	-19.31
10/6	161	-26.72	-25.86	-25.00	-24.14	-23.28	-22.43	-21.58	-20.73
11/6	162	-26.82	-25.96	-25.10	-24.24	-23.39	-22.54	-21.69	-20.84
20/6	171	-27.33	-26.47	-25.62	-24.76	-23.91	-23.06	-22.22	-21.37
30/6	181	-27.01	-26.15	-25.29	-24.44	-23.58	-22.73	-21.88	-21.04
1/7	182	-26.93	-26.07	-25.21	-24.35	-23.50	-22.64	-21.79	-20.95
10/7	191	-25.76	-24.89	-24.02	-23.15	-22.29	-21.43	-20.57	-19.72
17/7	198	-24.32	-23.44	-22.57	-21.69	-20.82	-19.95	-19.08	-18.21
20/7	201	-23.57	-22.69	-21.80	-20.92	-20.04	-19.17	-18.29	-17.42
30/7	211	-20.47	-19.56	-18.66	-17.76	-16.86	-15.96	-15.06	-14.17
1/8	213	-19.74	-18.83	-17.92	-17.01	-16.11	-15.21	-14.31	-13.41
10/8	222	-16.03	-15.10	-14.17	-13.24	-12.31	-11.39	-10.46	-9.54
16/8	228	-13.19	-12.24	-11.30	-10.35	-9.41	-8.47	-7.53	-6.58
20/8	232	-11.14	-10.18	-9.23	-8.28	-7.32	-6.37	-5.42	-4.47
30/8	242	-5.57	-4.59	-3.61	-2.64	-1.66	-0.69	0.29	1.27
1/9	244	-4.39	-3.41	-2.43	-1.45	-0.47	0.51	1.49	2.47
10/9	253	1.12	2.12	3.11	4.10	5.10	6.09	7.08	8.08
15/9	258	4.27	5.27	6.27	7.27	8.27	9.26	10.26	11.26
20/9	263	7.44	8.44	9.44	10.44	11.44	12.44	13.44	14.44
30/9	273	13.71	14.71	15.71	16.70	17.70	18.69	19.69	20.69
1/10	274	14.33	15.32	16.32	17.31	18.31	19.30	20.30	21.29
10/10	283	19.67	20.65	21.63	22.62	23.60	24.58	25.56	26.55
15/10	288	22.45	23.43	24.40	25.38	26.35	27.33	28.30	29.27
20/10	293	25.09	26.05	27.02	27.98	28.94	29.91	30.87	31.84
30/10	303	29.82	30.76	31.71	32.65	33.60	34.54	35.48	36.43
1/11	305	30.68	31.62	32.56	33.49	34.43	35.37	36.31	37.25
10/11	314	34.13	35.05	35.97	36.89	37.81	38.73	39.66	40.58
14/11	318	35.45	36.36	37.28	38.19	39.10	40.02	40.93	41.84
20/11	324	37.18	38.08	38.98	39.89	40.79	41.70	42.60	43.50
30/11	334	39.38	40.27	41.16	42.05	42.94	43.83	44.73	45.62
1/12	335	39.56	40.45	41.33	42.22	43.11	44.00	44.89	45.78
10/12	344	40.74	41.63	42.51	43.39	44.27	45.15	46.04	46.92
20/12	354	41.26	42.14	43.02	43.90	44.78	45.66	46.54	47.42
31/12	365	40.87	41.75	42.63	43.51	44.39	45.27	46.15	47.03

Table 41. Daily optimum tilt angle $\beta_{opt,d}$ (°) for latitudes [16°-23.45°N]

	Latitudes [16°-23.45°N]								
n	**16**	**17**	**18**	**19**	**20**	**21**	**22**	**23**	**23.45**
1	47.81	48.69	49.58	50.46	51.34	52.22	53.10	53.98	54.38
10	46.69	47.58	48.47	49.36	50.25	51.14	52.02	52.91	53.31
17	45.37	46.27	47.17	48.06	48.96	49.86	50.75	51.65	52.05
20	44.68	45.59	46.49	47.39	48.29	49.19	50.09	50.99	51.39
30	41.87	42.78	43.70	44.62	45.54	46.45	47.37	48.29	48.70
32	41.20	42.13	43.05	43.97	44.89	45.81	46.73	47.65	48.07
41	37.83	38.78	39.72	40.66	41.60	42.53	43.47	44.41	44.83
47	35.24	36.19	37.14	38.09	39.05	40.00	40.95	41.90	42.33
51	33.35	34.31	35.28	36.24	37.20	38.16	39.12	40.08	40.51
59	29.25	30.22	31.20	32.18	33.16	34.13	35.11	36.08	36.52
60	28.70	29.68	30.66	31.64	32.62	33.60	34.58	35.55	35.99
69	23.56	24.55	25.55	26.54	27.53	28.52	29.51	30.51	30.95
75	19.92	20.92	21.91	22.91	23.91	24.91	25.91	26.90	27.35
79	17.42	18.42	19.42	20.42	21.42	22.42	23.42	24.42	24.87
89	11.06	12.06	13.06	14.05	15.05	16.05	17.04	18.04	18.49
91	9.79	10.78	11.78	12.77	13.77	14.76	15.76	16.75	17.20
100	4.14	5.12	6.10	7.08	8.07	9.05	10.02	11.00	11.44
105	1.12	2.09	3.06	4.03	5.00	5.97	6.94	7.90	8.34
110	-1.78	-0.82	0.14	1.09	2.05	3.00	3.96	4.91	5.34
120	-7.10	-6.17	-5.24	-4.32	-3.39	-2.46	-1.54	-0.62	-0.21
130	-11.65	-10.75	-9.85	-8.95	-8.06	-7.16	-6.27	-5.38	-4.98
135	-13.60	-12.71	-11.82	-10.94	-10.06	-9.18	-8.30	-7.43	-7.04
140	-15.31	-14.44	-13.56	-12.69	-11.82	-10.96	-10.10	-9.24	-8.85
150	-18.03	-17.17	-16.32	-15.47	-14.62	-13.78	-12.94	-12.11	-11.73
152	-18.45	-17.60	-16.75	-15.91	-15.06	-14.22	-13.39	-12.56	-12.18
161	-19.88	-19.04	-18.20	-17.37	-16.54	-15.71	-14.89	-14.07	-13.71
171	-20.53	-19.70	-18.87	-18.04	-17.21	-16.39	-15.57	-14.76	-14.40
181	-20.19	-19.35	-18.52	-17.69	-16.86	-16.03	-15.21	-14.40	-14.03
182	-20.11	-19.27	-18.43	-17.60	-16.77	-15.94	-15.12	-14.31	-13.94
191	-18.86	-18.01	-17.17	-16.32	-15.49	-14.65	-13.82	-12.99	-12.62
198	-17.35	-16.49	-15.63	-14.77	-13.92	-13.07	-12.23	-11.39	-11.01
201	-16.55	-15.68	-14.82	-13.96	-13.10	-12.24	-11.39	-10.54	-10.16
211	-13.27	-12.38	-11.50	-10.61	-9.72	-8.84	-7.96	-7.09	-6.70
213	-12.51	-11.61	-10.72	-9.82	-8.94	-8.05	-7.16	-6.28	-5.89
222	-8.61	-7.69	-6.77	-5.85	-4.94	-4.02	-3.11	-2.20	-1.79
228	-5.65	-4.71	-3.77	-2.83	-1.90	-0.96	-0.03	0.90	1.32
232	-3.52	-2.57	-1.62	-0.67	0.28	1.22	2.17	3.11	3.54

continued on following page

Table 41. Continued

n	Latitudes [16°-23.45°N]								
	16	17	18	19	20	21	22	23	23.45
242	2.24	3.22	4.19	5.16	6.14	7.11	8.08	9.05	9.49
244	3.45	4.43	5.41	6.39	7.37	8.34	9.32	10.30	10.74
253	9.07	10.07	11.06	12.05	13.05	14.04	15.03	16.03	16.47
258	12.26	13.26	14.25	15.25	16.25	17.25	18.25	19.24	19.69
263	15.44	16.44	17.44	18.44	19.44	20.44	21.44	22.44	22.89
273	21.68	22.68	23.67	24.67	25.66	26.66	27.65	28.65	29.10
274	22.29	23.28	24.28	25.27	26.27	27.26	28.25	29.25	29.70
283	27.53	28.51	29.50	30.48	31.46	32.44	33.42	34.41	34.85
288	30.25	31.22	32.19	33.17	34.14	35.11	36.09	37.06	37.50
293	32.80	33.77	34.73	35.69	36.66	37.62	38.58	39.54	39.98
303	37.37	38.31	39.25	40.20	41.14	42.08	43.02	43.96	44.39
305	38.19	39.13	40.07	41.01	41.94	42.88	43.82	44.75	45.18
314	41.50	42.42	43.34	44.26	45.18	46.10	47.02	47.93	48.35
318	42.76	43.67	44.58	45.50	46.41	47.32	48.23	49.14	49.55
324	44.41	45.31	46.21	47.12	48.02	48.92	49.82	50.72	51.13
334	46.51	47.40	48.29	49.18	50.07	50.96	51.84	52.73	53.13
335	46.67	47.56	48.45	49.34	50.23	51.12	52.00	52.89	53.29
344	47.80	48.68	49.57	50.45	51.33	52.21	53.09	53.97	54.36
354	48.29	49.17	50.05	50.93	51.81	52.69	53.56	54.44	54.83
365	47.92	48.80	49.68	50.56	51.44	52.32	53.20	54.08	54.47

Table 42. Monthly optimum tilt $\beta_{opt,m}$ (°) for January and February

Month	January				February				February			
φ	$\beta_{opt,m}$	Days	Start	End	$\beta_{opt,m}$	Days	Start	End	$\beta_{opt,m}$	Days	Start	End
-23.45	10.43	31	1	31	2.90	11	32	42	-4.54	17	43	59
-23	10.81	31	1	31	3.06	12	32	43	-4.39	16	44	59
-22	11.66	31	1	31	3.47	14	32	45	-4.00	14	46	59
-21	12.51	31	1	31	3.88	16	32	47	-3.62	12	48	59
-20	13.36	31	1	31	4.54	17	32	48	-2.95	11	49	59
-19	14.21	31	1	31	4.94	19	32	50	-2.56	9	51	59
-18	15.07	31	1	31	5.35	21	32	52	-2.18	7	53	59
-17	15.94	31	1	31	6.02	22	32	53	-1.50	6	54	59

continued on following page

Table 42. Continued

Month	January				February				February			
φ	$\beta_{opt,m}$	Days	Start	End	$\beta_{opt,m}$	Days	Start	End	$\beta_{opt,m}$	Days	Start	End
-16	16.80	31	1	31	6.42	24	32	55	-1.12	4	56	59
-15	17.67	31	1	31	6.82	26	32	57	-0.74	2	58	59
-14	18.54	31	1	31	7.49	27	32	58	-0.07	1	59	59
-13	19.41	31	1	31	8.16	28	32	59				
-12	20.29	31	1	31	9.11	28	32	59				
-11	21.16	31	1	31	10.05	28	32	59				
-10	22.04	31	1	31	10.99	28	32	59				
-9	22.92	31	1	31	11.94	28	32	59				
-8	23.80	31	1	31	12.88	28	32	59				
-7	24.69	31	1	31	13.83	28	32	59				
-6	25.57	31	1	31	14.77	28	32	59				
-5	26.46	31	1	31	15.72	28	32	59				
-4	27.35	31	1	31	16.67	28	32	59				
-3	28.24	31	1	31	17.62	28	32	59				
-2	29.13	31	1	31	18.56	28	32	59				
-1	30.02	31	1	31	19.51	28	32	59				
0	30.91	31	1	31	20.46	28	32	59				
1	31.80	31	1	31	21.41	28	32	59				
2	32.70	31	1	31	22.36	28	32	59				
3	33.59	31	1	31	23.31	28	32	59				
4	34.49	31	1	31	24.26	28	32	59				
5	35.38	31	1	31	25.21	28	32	59				
6	36.28	31	1	31	26.16	28	32	59				
7	37.18	31	1	31	27.11	28	32	59				
8	38.08	31	1	31	28.06	28	32	59				
9	38.97	31	1	31	29.01	28	32	59				
10	39.87	31	1	31	29.96	28	32	59				
11	40.77	31	1	31	30.91	28	32	59				
12	41.67	31	1	31	31.86	28	32	59				
13	42.57	31	1	31	32.81	28	32	59				
14	43.46	31	1	31	33.76	28	32	59				

continued on following page

Table 42. Continued

Month	January				February				February			
φ	$\beta_{opt,m}$	Days	Start	End	$\beta_{opt,m}$	Days	Start	End	$\beta_{opt,m}$	Days	Start	End
15	44.36	31	1	31	34.71	28	32	59				
16	45.26	31	1	31	35.67	28	32	59				
17	46.16	31	1	31	36.62	28	32	59				
18	47.06	31	1	31	37.56	28	32	59				
19	47.95	31	1	31	38.51	28	32	59				
20	48.85	31	1	31	39.46	28	32	59				
21	49.75	31	1	31	40.41	28	32	59				
22	50.64	31	1	31	41.36	28	32	59				
23	51.54	31	1	31	42.31	28	32	59				
23.45	51.94	31	1	31	42.74	28	32	59				

Table 43. Monthly optimum tilt $\beta_{opt,m}$ (°) for March and April

Month	March				March				April			
φ	$\beta_{opt,m}$	Days	Start	End	$\beta_{opt,m}$	Days	Start	End	$\beta_{opt,m}$	Days	Start	End
-23.45	-19.42	31	60	90					-38.80	30	91	120
-23	-18.98	31	60	90					-38.34	30	91	120
-22	-17.98	31	60	90					-37.34	30	91	120
-21	-16.99	31	60	90					-36.34	30	91	120
-20	-16.00	31	60	90					-35.34	30	91	120
-19	-15.00	31	60	90					-34.33	30	91	120
-18	-14.01	31	60	90					-33.33	30	91	120
-17	-13.02	31	60	90					-32.33	30	91	120
-16	-12.02	31	60	90					-31.32	30	91	120
-15	-11.03	31	60	90					-30.32	30	91	120
-14	-10.04	31	60	90					-29.32	30	91	120
-13	-9.36	30	60	89	0.31	1	90	90	-28.31	30	91	120
-12	-8.99	28	60	87	0.69	3	88	90	-27.31	30	91	120
-11	-8.30	27	60	86	1.37	4	87	90	-26.31	30	91	120
-10	-7.93	25	60	84	1.74	6	85	90	-25.30	30	91	120
-9	-7.25	24	60	83	2.42	7	84	90	-24.30	30	91	120

continued on following page

Table 43. Continued

Month	March				March				April			
φ	$\beta_{opt,m}$	Days	Start	End	$\beta_{opt,m}$	Days	Start	End	$\beta_{opt,m}$	Days	Start	End
-8	-6.88	22	60	81	2.80	9	82	90	-23.29	30	91	120
-7	-6.52	20	60	79	3.17	11	80	90	-22.29	30	91	120
-6	-5.83	19	60	78	3.85	12	79	90	-21.29	30	91	120
-5	-5.47	17	60	76	4.22	14	77	90	-20.28	30	91	120
-4	-4.78	16	60	75	4.90	15	76	90	-19.28	30	91	120
-3	-4.42	14	60	73	5.28	17	74	90	-18.27	30	91	120
-2	-4.05	12	60	71	5.65	19	72	90	-17.27	30	91	120
-1	-3.37	11	60	70	6.33	20	71	90	-16.27	30	91	120
0	-3.00	9	60	68	6.70	22	69	90	-15.26	30	91	120
1	-2.32	8	60	67	7.38	23	68	90	-14.26	30	91	120
2	-1.95	6	60	65	7.75	25	66	90	-13.26	30	91	120
3	-1.27	5	60	64	8.43	26	65	90	-12.25	30	91	120
4	-0.90	3	60	62	8.80	28	63	90	-11.25	30	91	120
5	-0.53	1	60	60	9.17	30	61	90	-10.25	30	91	120
6	9.85	31	60	90	-9.25	30	91	120	No	0		
7	10.84	31	60	90	-8.87	28	91	118	0.52	2	119	120
8	11.84	31	60	90	-8.18	27	91	117	1.20	3	118	120
9	12.83	31	60	90	-7.80	25	91	115	1.57	5	116	120
10	13.83	31	60	90	-7.43	23	91	113	1.94	7	114	120
11	14.82	31	60	90	-6.73	22	91	112	2.62	8	113	120
12	15.81	31	60	90	-6.36	20	91	110	2.99	10	111	120
13	16.81	31	60	90	-5.66	19	91	109	3.68	11	110	120
14	17.80	31	60	90	-5.28	17	91	107	4.05	13	108	120
15	18.80	31	60	90	-4.91	15	91	105	4.42	15	106	120
16	19.79	31	60	90	-4.21	14	91	104	5.10	16	105	120
17	20.78	31	60	90	-3.83	12	91	102	5.48	18	103	120
18	21.78	31	60	90	-3.45	10	91	100	5.86	20	101	120
19	22.77	31	60	90	-2.75	9	91	99	6.53	21	100	120
20	23.77	31	60	90	-2.37	7	91	97	6.91	23	98	120
21	24.76	31	60	90	-2.01	5	91	95	7.29	25	96	120
22	25.75	31	60	90	-1.70	3	91	93	7.67	27	94	120
23	26.75	31	60	90	-0.93	2	91	92	8.35	28	93	120
23.45	27.20	31	60	90	-0.93	1	91	91	8.49	29	92	120

Table 44. Monthly optimum tilt $\beta_{opt,m}$ (°) May and June

Month	May				June			
φ	$\beta_{opt,m}$	Days	Start	End	$\beta_{opt,m}$	Days	Start	End
-23.45	-49.42	31	121	151	-54.42	30	152	181
-23	-49.01	31	121	151	-54.02	30	152	181
-22	-48.10	31	121	151	-53.14	30	152	181
-21	-47.19	31	121	151	-52.26	30	152	181
-20	-46.27	31	121	151	-51.38	30	152	181
-19	-45.36	31	121	151	-50.50	30	152	181
-18	-44.45	31	121	151	-49.62	30	152	181
-17	-43.53	31	121	151	-48.74	30	152	181
-16	-42.62	31	121	151	-47.85	30	152	181
-15	-41.71	31	121	151	-46.97	30	152	181
-14	-40.79	31	121	151	-46.09	30	152	181
-13	-39.88	31	121	151	-45.21	30	152	181
-12	-38.96	31	121	151	-44.33	30	152	181
-11	-38.05	31	121	151	-43.44	30	152	181
-10	-37.14	31	121	151	-42.56	30	152	181
-9	-36.22	31	121	151	-41.68	30	152	181
-8	-35.31	31	121	151	-40.80	30	152	181
-7	-34.39	31	121	151	-39.92	30	152	181
-6	-33.48	31	121	151	-39.04	30	152	181
-5	-32.57	31	121	151	-38.16	30	152	181
-4	-31.66	31	121	151	-37.28	30	152	181
-3	-30.74	31	121	151	-36.40	30	152	181
-2	-29.83	31	121	151	-35.53	30	152	181
-1	-28.92	31	121	151	-34.65	30	152	181
0	-28.01	31	121	151	-33.78	30	152	181
1	-27.10	31	121	151	-32.90	30	152	181
2	-26.19	31	121	151	-32.03	30	152	181
3	-25.28	31	121	151	-31.16	30	152	181
4	-24.38	31	121	151	-30.29	30	152	181

continued on following page

Table 44. Continued

Month	May				June			
φ	$\beta_{opt,m}$	Days	Start	End	$\beta_{opt,m}$	Days	Start	End
5	-23.47	31	121	151	-29.42	30	152	181
6	-22.57	31	121	151	-28.55	30	152	181
7	-21.66	31	121	151	-27.68	30	152	181
8	-20.76	31	121	151	-26.82	30	152	181
9	-19.86	31	121	151	-25.96	30	152	181
10	-18.96	31	121	151	-25.10	30	152	181
11	-18.06	31	121	151	-24.24	30	152	181
12	-17.16	31	121	151	-23.39	30	152	181
13	-16.27	31	121	151	-22.53	30	152	181
14	-15.37	31	121	151	-21.68	30	152	181
15	-14.48	31	121	151	-20.84	30	152	181
16	-13.59	31	121	151	-19.99	30	152	181
17	-12.70	31	121	151	-19.15	30	152	181
18	-11.82	31	121	151	-18.32	30	152	181
19	-10.93	31	121	151	-17.48	30	152	181
20	-10.05	31	121	151	-16.65	30	152	181
21	-9.17	31	121	151	-15.83	30	152	181
22	-8.30	31	121	151	-15.00	30	152	181
23	-7.43	31	121	151	-14.19	30	152	181
23.45	-7.04	31	121	151	-13.82	30	152	181

Table 45. Monthly optimum tilt $\beta_{opt,m}$ (°) for July and August

Month	July				August			
φ	$\beta_{opt,m}$	Days	Start	End	$\beta_{opt,m}$	Days	Start	End
-23.45	-52.26	31	182	212	-42.73	31	213	243
-23	-51.86	31	182	212	-42.31	31	213	243
-22	-50.96	31	182	212	-41.36	31	213	243
-21	-50.07	31	182	212	-40.41	31	213	243
-20	-49.17	31	182	212	-39.46	31	213	243
-19	-48.28	31	182	212	-38.51	31	213	243
-18	-47.38	31	182	212	-37.56	31	213	243

continued on following page

Table 45. Continued

Month	July				August			
φ	$\beta_{opt,m}$	Days	Start	End	$\beta_{opt,m}$	Days	Start	End
-17	-46.49	31	182	212	-36.61	31	213	243
-16	-45.59	31	182	212	-35.67	31	213	243
-15	-44.69	31	182	212	-34.72	31	213	243
-14	-43.80	31	182	212	-33.77	31	213	243
-13	-42.90	31	182	212	-32.82	31	213	243
-12	-42.00	31	182	212	-31.86	31	213	243
-11	-41.11	31	182	212	-30.91	31	213	243
-10	-40.21	31	182	212	-29.96	31	213	243
-9	-39.32	31	182	212	-29.01	31	213	243
-8	-38.42	31	182	212	-28.06	31	213	243
-7	-37.53	31	182	212	-27.11	31	213	243
-6	-36.63	31	182	212	-26.16	31	213	243
-5	-35.74	31	182	212	-25.21	31	213	243
-4	-34.84	31	182	212	-24.26	31	213	243
-3	-33.95	31	182	212	-23.31	31	213	243
-2	-33.06	31	182	212	-22.36	31	213	243
-1	-32.16	31	182	212	-21.42	31	213	243
0	-31.27	31	182	212	-20.47	31	213	243
1	-30.38	31	182	212	-19.52	31	213	243
2	-29.49	31	182	212	-18.57	31	213	243
3	-28.61	31	182	212	-17.62	31	213	243
4	-27.72	31	182	212	-16.67	31	213	243
5	-26.83	31	182	212	-15.73	31	213	243
6	-25.95	31	182	212	-14.78	31	213	243
7	-25.07	31	182	212	-13.83	31	213	243
8	-24.19	31	182	212	-12.89	31	213	243
9	-23.31	31	182	212	-11.94	31	213	243
10	-22.43	31	182	212	-11.00	31	213	243
11	-21.55	31	182	212	-10.06	31	213	243
12	-20.68	31	182	212	-9.12	31	213	243
13	-19.81	31	182	212	-8.17	31	213	243

Table 46. Monthly optimum tilt $\beta_{opt,m}$ (°) for July and August

Month	July				August				August			
φ	$\beta_{opt,m}$	Days	Start	End	$\beta_{opt,m}$	Days	Start	End	$\beta_{opt,m}$	Days	Start	End
14	-18.94	31	182	212	-7.77	29	213	241	0.59	2	242	243
15	-18.07	31	182	212	-7.37	27	213	239	0.97	4	240	243
16	-17.20	31	182	212	-6.70	26	213	238	1.65	5	239	243
17	-16.34	31	182	212	-6.30	24	213	236	2.03	7	237	243
18	-15.48	31	182	212	-5.89	22	213	234	2.41	9	235	243
19	-14.63	31	182	212	-5.22	21	213	233	3.08	10	234	243
20	-13.77	31	182	212	-4.81	19	213	231	3.46	12	232	243
21	-12.92	31	182	212	-4.40	17	213	229	3.85	14	230	243
22	-12.08	31	182	212	-3.74	16	213	228	4.52	15	229	243
23	-11.24	31	182	212	-3.33	14	213	226	4.91	17	227	243
23.45	-10.86	31	182	212	-3.17	13	213	225	5.05	18	226	243

and the number of days of each period. Then, by dividing the calendar year into 4 quarters and denoting the seasons by quarters as follows:

- Winter is January, February and March,
- Spring is April, May and June,
- Summer is July, August and September,
- Autumn is October, November and December.

The seasonally optimum tilt angle $\beta_{opt,s}$ (°), for different latitudes at the tropical zone, is given in the Tables 48 and 49. As during the mentioned seasons the daily optimum tilt angle $\beta_{opt,d}$ (°) changes its sign, Tables 50 and 51 show the optimum tilt angle for these two periods and the number of days of each period.

Table 47. Monthly optimum tilt $\beta_{opt,m}$ (°) for September and October

Month	September				October				October			
φ	$\beta_{opt,m}$	Days	Start	End	$\beta_{opt,m}$	Days	Start	End	$\beta_{opt,m}$	Days	Start	End
-23.45	-26.67	30	244	273	-8.26	29	274	302	0.66	2	303	304
-23	-26.23	30	244	273	-8.12	28	274	301	0.81	3	302	304
-22	-25.23	30	244	273	-7.74	26	274	299	1.21	5	300	304
-21	-24.24	30	244	273	-7.36	24	274	297	1.61	7	298	304
-20	-23.24	30	244	273	-6.68	23	274	296	2.27	8	297	304
-19	-22.25	30	244	273	-6.30	21	274	294	2.67	10	295	304
-18	-21.25	30	244	273	-5.93	19	274	292	3.07	12	293	304
-17	-20.26	30	244	273	-5.25	18	274	291	3.73	13	292	304
-16	-19.26	30	244	273	-4.87	16	274	289	4.13	15	290	304
-15	-18.27	30	244	273	-4.50	14	274	287	4.52	17	288	304
-14	-17.27	30	244	273	-3.82	13	274	286	5.19	18	287	304
-13	-16.28	30	244	273	-3.44	11	274	284	5.58	20	285	304
-12	-15.28	30	244	273	-3.07	9	274	282	5.97	22	283	304
-11	-14.29	30	244	273	-2.39	8	274	281	6.65	23	282	304
-10	-13.29	30	244	273	-2.02	6	274	279	7.04	25	280	304
-9	-12.30	30	244	273	-1.34	5	274	278	7.71	26	279	304
-8	-11.30	30	244	273	-0.97	3	274	276	8.10	28	277	304
-7	-10.31	30	244	273	-0.60	1	274	274	8.48	30	275	304

Table 48. Monthly optimum tilt $\beta_{opt,m}$ (°) for September and October

Month	September				September				October			
φ	$\beta_{opt,m}$	Days	Start	End	$\beta_{opt,m}$	Days	Start	End	$\beta_{opt,m}$	Days	Start	End
-6	-9.31	30	244	273					9.16	31	274	304
-5	-8.95	28	244	271	0.45	2	272	273	10.13	31	274	304
-4	-8.26	27	244	270	1.14	3	271	273	11.10	31	274	304
-3	-7.90	25	244	268	1.50	5	269	273	12.07	31	274	304
-2	-7.21	24	244	267	2.19	6	268	273	13.03	31	274	304
-1	-6.85	22	244	265	2.55	8	266	273	14.01	31	274	304

continued on following page

Table 48. Continued

Month	September				September				October			
φ	$\beta_{opt,m}$	Days	Start	End	$\beta_{opt,m}$	Days	Start	End	$\beta_{opt,m}$	Days	Start	End
0	-6.48	20	244	263	2.92	10	264	273	14.98	31	274	304
1	-5.79	19	244	262	3.60	11	263	273	15.95	31	274	304
2	-5.43	17	244	260	3.97	13	261	273	16.92	31	274	304
3	-4.74	16	244	259	4.65	14	260	273	17.89	31	274	304
4	-4.37	14	244	257	5.02	16	258	273	18.86	31	274	304
5	-4.00	12	244	255	5.39	18	256	273	19.83	31	274	304
6	-3.32	11	244	254	6.07	19	255	273	20.80	31	274	304
7	-2.95	9	244	252	6.44	21	253	273	21.77	31	274	304
8	-2.27	8	244	251	7.12	22	252	273	22.74	31	274	304
9	-1.90	6	244	249	7.49	24	250	273	23.71	31	274	304
10	-1.52	4	244	247	7.86	26	248	273	24.68	31	274	304
11	-0.84	3	244	246	8.54	27	247	273	25.65	31	274	304
12	-0.47	1	244	244	8.91	29	245	273	26.63	31	274	304
13					9.59	30	244	273	27.60	31	274	304
14					10.58	30	244	273	28.57	31	274	304
15					11.58	30	244	273	29.54	31	274	304
16					12.57	30	244	273	30.51	31	274	304
17					13.57	30	244	273	31.48	31	274	304
18					14.56	30	244	273	32.45	31	274	304
19					15.56	30	244	273	33.42	31	274	304
20					16.55	30	244	273	34.39	31	274	304
21					17.54	30	244	273	35.36	31	274	304
22					18.54	30	244	273	36.33	31	274	304
23					19.53	30	244	273	37.30	31	274	304
23.45					19.98	30	244	273	37.73	31	274	304

Table 49. Monthly optimum tilt $\beta_{opt,m}$ (°) for November and December

Month	November				December			
φ	$\beta_{opt,m}$	Days	Start	End	$\beta_{opt,m}$	Days	Start	End
-23.45	7.38	30	305	334	13.85	31	335	365
-23	7.77	30	305	334	14.21	31	335	365
-22	8.64	30	305	334	15.03	31	335	365
-21	9.52	30	305	334	15.85	31	335	365
-20	10.39	30	305	334	16.68	31	335	365
-19	11.27	30	305	334	17.51	31	335	365
-18	12.15	30	305	334	18.34	31	335	365
-17	13.03	30	305	334	19.18	31	335	365
-16	13.92	30	305	334	20.02	31	335	365
-15	14.81	30	305	334	20.86	31	335	365
-14	15.70	30	305	334	21.71	31	335	365
-13	16.59	30	305	334	22.56	31	335	365
-12	17.48	30	305	334	23.41	31	335	365
-11	18.38	30	305	334	24.26	31	335	365
-10	19.27	30	305	334	25.12	31	335	365
-9	20.17	30	305	334	25.98	31	335	365
-8	21.07	30	305	334	26.84	31	335	365
-7	21.97	30	305	334	27.71	31	335	365
-6	22.88	30	305	334	28.57	31	335	365
-5	23.78	30	305	334	29.44	31	335	365
-4	24.68	30	305	334	30.31	31	335	365
-3	25.59	30	305	334	31.18	31	335	365
-2	26.50	30	305	334	32.05	31	335	365
-1	27.40	30	305	334	32.92	31	335	365
0	28.31	30	305	334	33.80	31	335	365
1	29.22	30	305	334	34.67	31	335	365
2	30.13	30	305	334	35.55	31	335	365
3	31.04	30	305	334	36.42	31	335	365
4	31.95	30	305	334	37.30	31	335	365

continued on following page

Table 49. Continued

Month	November				December			
φ	$\beta_{opt,m}$	Days	Start	End	$\beta_{opt,m}$	Days	Start	End
5	32.86	30	305	334	38.18	31	335	365
6	33.77	30	305	334	39.06	31	335	365
7	34.68	30	305	334	39.94	31	335	365
8	35.59	30	305	334	40.82	31	335	365
9	36.51	30	305	334	41.70	31	335	365
10	37.42	30	305	334	42.58	31	335	365
11	38.33	30	305	334	43.46	31	335	365
12	39.24	30	305	334	44.35	31	335	365
13	40.16	30	305	334	45.23	31	335	365
14	41.07	30	305	334	46.11	31	335	365
15	41.98	30	305	334	46.99	31	335	365
16	42.89	30	305	334	47.87	31	335	365
17	43.81	30	305	334	48.76	31	335	365
18	44.72	30	305	334	49.64	31	335	365
19	45.63	30	305	334	50.52	31	335	365
20	46.54	30	305	334	51.40	31	335	365
21	47.45	30	305	334	52.28	31	335	365
22	48.36	30	305	334	53.16	31	335	365
23	49.27	30	305	334	54.04	31	335	365
23.45	49.68	30	305	334	54.43	31	335	365

Table 50. Seasonally optimum tilt angle $\beta_{opt,s}$ (°) for winter and spring

Season	Winter				Winter				Spring			
φ	$\beta_{opt,s}$	Days	Start	End	$\beta_{opt,s}$	Days	Start	End	$\beta_{opt,s}$	Days	Start	End
-23.45	8.46	42	1	42	-15.10	48	43	90	-47.07	91	91	181
-23	8.65	43	1	43	-14.97	47	44	90	-46.66	91	91	181
-22	9.11	45	1	45	-14.61	45	46	90	-45.74	91	91	181
-21	9.57	47	1	47	-14.25	43	48	90	-44.82	91	91	181
-20	10.23	48	1	48	-13.55	42	49	90	-43.90	91	91	181
-19	10.69	50	1	50	-13.19	40	51	90	-42.98	91	91	181
-18	11.15	52	1	52	-12.85	38	53	90	-42.05	91	91	181

continued on following page

Table 50. Continued

Season	Winter				Winter				Spring			
φ	$\beta_{opt,s}$	Days	Start	End	$\beta_{opt,s}$	Days	Start	End	$\beta_{opt,s}$	Days	Start	End
-17	11.82	53	1	53	-12.13	37	54	90	-41.13	91	91	181
-16	12.27	55	1	55	-11.79	35	56	90	-40.21	91	91	181
-15	12.72	57	1	57	-11.45	33	58	90	-39.29	91	91	181
-14	13.40	58	1	58	-10.73	32	59	90	-38.36	91	91	181
-13	13.84	60	1	60	-10.39	30	61	90	-37.44	91	91	181
-12	14.29	62	1	62	-10.06	28	63	90	-36.52	91	91	181
-11	14.97	63	1	63	-9.34	27	64	90	-35.60	91	91	181
-10	15.41	65	1	65	-9.02	25	66	90	-34.67	91	91	181
-9	16.09	66	1	66	-8.29	24	67	90	-33.75	91	91	181
-8	16.53	68	1	68	-7.97	22	69	90	-32.83	91	91	181
-7	16.96	70	1	70	-7.67	20	71	90	-31.91	91	91	181
-6	17.64	71	1	71	-6.93	19	72	90	-30.98	91	91	181
-5	18.08	73	1	73	-6.64	17	74	90	-30.06	91	91	181
-4	18.76	74	1	74	-5.89	16	75	90	-29.14	91	91	181
-3	19.19	76	1	76	-5.62	14	77	90	-28.22	91	91	181
-2	19.62	78	1	78	-5.40	12	79	90	-27.30	91	91	181
-1	20.30	79	1	79	-4.63	11	80	90	-26.38	91	91	181
0	20.72	81	1	81	-4.50	9	82	90	-25.46	91	91	181
1	21.40	82	1	82	-3.71	8	83	90	-24.54	91	91	181
2	21.83	84	1	84	-3.90	6	85	90	-23.62	91	91	181
3	22.51	85	1	85	-3.17	5	86	90	-22.71	91	91	181
4	22.93	87	1	87	0	3			-21.79	91	91	181
5	23.35	89	1	89	0	1			-20.88	91	91	181
6	24.03	90	1	90	-19.96	91	91	181	0	0		
7	24.97	90	1	90	-19.49	89	91	179	0.52	2	180	181
8	25.92	90	1	90	-18.80	88	91	178	1.20	3	179	181
9	26.87	90	1	90	-18.32	86	91	176	1.57	5	177	181
10	27.82	90	1	90	-17.84	84	91	174	1.94	7	175	181
11	28.76	90	1	90	-17.14	83	91	173	2.62	8	174	181
12	29.71	90	1	90	-16.66	81	91	171	2.99	10	172	181

continued on following page

Table 50. Continued

Season	Winter				Winter				Spring			
φ	$\beta_{opt,s}$	Days	Start	End	$\beta_{opt,s}$	Days	Start	End	$\beta_{opt,s}$	Days	Start	End
13	30.66	90	1	90	-15.97	80	91	170	3.68	11	171	181
14	31.61	90	1	90	-15.48	78	91	168	4.05	13	169	181
15	32.56	90	1	90	-14.99	76	91	166	4.42	15	167	181
16	33.50	90	1	90	-14.30	75	91	165	5.10	16	166	181
17	34.45	90	1	90	-13.80	73	91	163	5.48	18	164	181
18	35.40	90	1	90	-13.30	71	91	161	5.86	20	162	181
19	36.34	90	1	90	-12.62	70	91	160	6.53	21	161	181
20	37.29	90	1	90	-12.12	68	91	158	6.91	23	159	181
21	38.24	90	1	90	-11.61	66	91	156	7.29	25	157	181
22	39.18	90	1	90	-11.10	64	91	154	7.67	27	155	181
23	40.13	90	1	90	-10.42	63	91	153	8.35	28	154	181
23.45	40.55	90	1	90	-10.21	62	91	152	8.49	29	153	181

Table 51. Seasonally optimum tilt angle $\beta_{opt,s}$ (°) for summer and autumn

Season	Summer				Autumn				Autumn			
φ	$\beta_{opt,s}$	Days	Start	End	$\beta_{opt,s}$	Days	Start	End	$\beta_{opt,s}$	Days	Start	End
-23.45	-40.71	92	182	273	-8.26	29	274	302	10.35	63	303	365
-23	-40.28	92	182	273	-8.12	28	274	301	10.57	64	302	365
-22	-39.34	92	182	273	-7.74	26	274	299	11.08	66	300	365
-21	-38.39	92	182	273	-7.36	24	274	297	11.59	68	298	365
-20	-37.45	92	182	273	-6.68	23	274	296	12.27	69	297	365
-19	-36.50	92	182	273	-6.30	21	274	294	12.78	71	295	365
-18	-35.55	92	182	273	-5.93	19	274	292	13.29	73	293	365
-17	-34.61	92	182	273	-5.25	18	274	291	13.97	74	292	365
-16	-33.66	92	182	273	-4.87	16	274	289	14.47	76	290	365
-15	-32.71	92	182	273	-4.50	14	274	287	14.97	78	288	365
-14	-31.77	92	182	273	-3.82	13	274	286	15.66	79	287	365
-13	-30.82	92	182	273	-3.44	11	274	284	16.16	81	285	365
-12	-29.88	92	182	273	-3.07	9	274	282	16.65	83	283	365
-11	-28.93	92	182	273	-2.39	8	274	281	17.34	84	282	365
-10	-27.98	92	182	273	-2.02	6	274	279	17.82	86	280	365
-9	-27.04	92	182	273	-1.34	5	274	278	18.52	87	279	365
-8	-26.09	92	182	273	-0.97	3	274	276	19.00	89	277	365

continued on following page

Table 51. Continued

Season	Summer				Autumn				Autumn			
φ	$\beta_{opt,s}$	Days	Start	End	$\beta_{opt,s}$	Days	Start	End	$\beta_{opt,s}$	Days	Start	End
-7	-25.14	92	182	273	-0.60	1	274	274	19.48	91	275	365
-6	-24.20	92	182	273	0	0			20.17	92	274	365
-5	-23.78	90	182	271	0.45	2	272	273	21.09	92	274	365
-4	-23.09	89	182	270	1.14	3	271	273	22.00	92	274	365
-3	-22.67	87	182	268	1.50	5	269	273	22.92	92	274	365
-2	-21.99	86	182	267	2.19	6	268	273	23.83	92	274	365
-1	-21.57	84	182	265	2.55	8	266	273	24.75	92	274	365
0	-21.14	82	182	263	2.92	10	264	273	25.67	92	274	365
1	-20.46	81	182	262	3.60	11	263	273	26.58	92	274	365
2	-20.03	79	182	260	3.97	13	261	273	27.50	92	274	365
3	-19.35	78	182	259	4.65	14	260	273	28.42	92	274	365
4	-18.91	76	182	257	5.02	16	258	273	29.34	92	274	365
5	-18.48	74	182	255	5.39	18	256	273	30.26	92	274	365
6	-17.80	73	182	254	6.07	19	255	273	31.18	92	274	365
7	-17.36	71	182	252	6.44	21	253	273	32.10	92	274	365
8	-16.68	70	182	251	7.12	22	252	273	33.02	92	274	365
9	-16.24	68	182	249	7.49	24	250	273	33.95	92	274	365
10	-15.79	66	182	247	7.86	26	248	273	34.87	92	274	365
11	-15.11	65	182	246	8.54	27	247	273	35.79	92	274	365
12	-14.67	63	182	244	8.91	29	245	273	36.71	92	274	365
13	-13.99	62	182	243	9.59	30	244	273	37.63	92	274	365
14	-13.54	60	182	241	9.96	32	242	273	38.56	92	274	365
15	-13.09	58	182	239	10.33	34	240	273	39.48	92	274	365
16	-12.41	57	182	238	11.01	35	239	273	40.40	92	274	365
17	-11.96	55	182	236	11.38	37	237	273	41.32	92	274	365
18	-11.50	53	182	234	11.76	39	235	273	42.24	92	274	365
19	-10.83	52	182	233	12.44	40	234	273	43.16	92	274	365
20	-10.37	50	182	231	12.81	42	232	273	44.08	92	274	365
21	-9.91	48	182	229	13.19	44	230	273	45.00	92	274	365
22	-9.24	47	182	228	13.87	45	229	273	45.92	92	274	365
23	-8.78	45	182	226	14.24	47	227	273	46.84	92	274	365
23.45	-8.59	44	182	225	14.38	48	226	273	47.26	92	274	365

APPENDIX C

Optimum Tilt Angle β_{opt} (°) at Mid-Latitude Zone

Tables 52 to 58 give the daily optimum tilt angle $\beta_{opt,d}$ (°) at mid-latitude zone. Tables 59 to 70 give the monthly optimum tilt angle $\beta_{opt,m}$ (°) at mid-latitude zone. As during some calendar months the daily optimum tilt angle $\beta_{opt,d}$ (°) changes its sign, Tables 59 to 70 show the optimum tilt angle for these two periods and the number of days of each period. Then, by dividing the calendar year into 4 quarters and denoting the seasons by quarters as follows:

- Winter is January, February and March,
- Spring is April, May and June,
- Summer is July, August and September,
- Autumn is October, November and December.

The seasonally optimum tilt angle $\beta_{opt,s}$ (°), for different latitudes at the mid-latitude zone, is given in the Tables 71 and 72. As during the mentioned seasons the daily optimum tilt angle $\beta_{opt,d}$ (°) changes its sign, Tables 73 to 76 show the optimum tilt angle for these two periods and the number of days of each period.

Table 52. The daily optimum tilt angle $\beta_{opt,d}$ (°) for latitudes [23.45°N to 30°N]

Day number	φ(°)							
	23.45	24	25	26	27	28	29	30
1	54.4	54.9	55.7	56.6	57.5	58.4	59.2	60.1
10	53.3	53.8	54.7	55.6	56.4	57.3	58.2	59.1
17	52	52.5	53.4	54.3	55.2	56.1	57	57.9
20	51.4	51.9	52.8	53.7	54.6	55.5	56.4	57.3
30	48.7	49.2	50.1	51	51.9	52.9	53.8	54.7
32	48.1	48.6	49.5	50.4	51.3	52.2	53.2	54.1
41	44.8	45.3	46.3	47.2	48.2	49.1	50	51
47	42.3	42.9	43.8	44.8	45.7	46.7	47.6	48.5
51	40.5	41	42	43	43.9	44.9	45.8	46.8
59	36.5	37.1	38	39	40	41	41.9	42.9
60	36	36.5	37.5	38.5	39.5	40.4	41.4	42.4
69	31	31.5	32.5	33.5	34.5	35.5	36.5	37.4
75	27.4	27.9	28.9	29.9	30.9	31.9	32.9	33.9
79	24.9	25.4	26.4	27.4	28.4	29.4	30.4	31.4
89	18.5	19.1	20.1	21.1	22.1	23.1	24.1	25.1

continued on following page

Table 52. Continued

Day number	φ(°)							
	23.45	**24**	**25**	**26**	**27**	**28**	**29**	**30**
91	17.3	17.8	18.8	19.8	20.8	21.8	22.8	23.8
100	11.7	12.2	13.2	14.2	15.2	16.2	17.2	18.2
105	8.6	9.2	10.2	11.2	12.2	13.2	14.2	15.2
110	5.6	6.1	7.1	8.1	9.1	10.1	11.1	12.1
120	-0.2	0.3	1.3	2.3	3.3	4.3	5.3	6.3
121	-0.7	-0.2	0.8	1.8	2.8	3.8	4.8	5.8
130	-5	-4.5	-3.6	-2.7	-1.9	-1	-0.1	0.9
135	-7	-6.6	-5.7	-4.8	-4	-3.1	-2.3	-1.4
140	-8.9	-8.4	-7.5	-6.7	-5.8	-5	-4.2	-3.3
150	-11.7	-11.3	-10.4	-9.6	-8.8	-8	-7.2	-6.4
152	-12.2	-11.7	-10.9	-10.1	-9.3	-8.5	-7.7	-6.9
161	-13.7	-13.3	-12.5	-11.6	-10.9	-10.1	-9.3	-8.5
162	-13.8	-13.4	-12.6	-11.8	-11	-10.2	-9.4	-8.6
171	-14.4	-14	-13.2	-12.4	-11.6	-10.8	-10	-9.2
181	-14	-13.6	-12.8	-12	-11.2	-10.4	-9.6	-8.9
182	-13.9	-13.5	-12.7	-11.9	-11.1	-10.3	-9.5	-8.8
191	-12.6	-12.2	-11.4	-10.5	-9.7	-8.9	-8.1	-7.3
198	-11	-10.5	-9.7	-8.9	-8.1	-7.2	-6.4	-5.6
201	-10.2	-9.7	-8.9	-8	-7.2	-6.4	-5.5	-4.7
211	-6.7	-6.2	-5.3	-4.5	-3.6	-2.8	-1.9	-1.1
213	-5.9	-5.4	-4.5	-3.7	-2.8	-1.9	-1.1	-0.2
222	-1.8	-1.3	-0.4	0.6	1.6	2.6	3.6	4.6
228	1.4	2	3	4	5	6	7	8
232	3.7	4.3	5.3	6.3	7.3	8.3	9.3	10.3
242	9.7	10.3	11.3	12.3	13.3	14.3	15.3	16.3
244	11	11.5	12.5	13.5	14.5	15.5	16.5	17.5
253	16.6	17.1	18.1	19.1	20.1	21.1	22.1	23.1
258	19.7	20.3	21.3	22.3	23.3	24.3	25.3	26.3
263	22.9	23.4	24.4	25.4	26.4	27.4	28.4	29.4
273	29.1	29.6	30.6	31.6	32.6	33.6	34.6	35.6
274	29.7	30.2	31.2	32.2	33.2	34.2	35.2	36.2
283	34.8	35.4	36.4	37.3	38.3	39.3	40.3	41.3
288	37.5	38	39	40	40.9	41.9	42.9	43.9
293	40	40.5	41.5	42.4	43.4	44.3	45.3	46.3
303	44.4	44.9	45.8	46.8	47.7	48.7	49.6	50.5
305	45.2	45.7	46.6	47.6	48.5	49.4	50.4	51.3
314	48.3	48.9	49.8	50.7	51.6	52.5	53.4	54.3

continued on following page

Table 52. Continued

Day number	φ(°)							
	23.45	24	25	26	27	28	29	30
318	49.6	50.1	51	51.9	52.8	53.7	54.6	55.5
324	51.1	51.6	52.5	53.4	54.3	55.2	56.1	57
334	53.1	53.6	54.5	55.4	56.3	57.2	58	58.9
335	53.3	53.8	54.7	55.5	56.4	57.3	58.2	59.1
344	54.4	54.8	55.7	56.6	57.5	58.4	59.2	60.1
354	54.8	55.3	56.2	57.1	57.9	58.8	59.7	60.6
365	54.5	55	55.8	56.7	57.6	58.5	59.3	60.2

Table 53. The daily optimum tilt angle $\beta_{opt,d}$ (°) for latitudes [31°N to 38°N]

	φ(°)							
Day number	31	32	33	34	35	36	37	38
1	61	61.9	62.7	63.6	64.5	65.3	66.2	67
10	60	60.9	61.7	62.6	63.5	64.3	65.2	66.1
17	58.8	59.7	60.6	61.4	62.3	63.2	64.1	65
20	58.2	59	59.9	60.8	61.7	62.6	63.5	64.4
30	55.6	56.5	57.4	58.3	59.2	60.1	61	61.9
32	55	55.9	56.8	57.7	58.6	59.5	60.4	61.4
41	51.9	52.8	53.8	54.7	55.6	56.5	57.5	58.4
47	49.5	50.4	51.4	52.3	53.3	54.2	55.2	56.1
51	47.7	48.7	49.7	50.6	51.6	52.5	53.5	54.4
59	43.9	44.9	45.8	46.8	47.8	48.7	49.7	50.7
60	43.4	44.3	45.3	46.3	47.3	48.2	49.2	50.2
69	38.4	39.4	40.4	41.4	42.4	43.4	44.4	45.4
75	34.9	35.9	36.9	37.9	38.9	39.9	40.9	41.9
79	32.4	33.4	34.4	35.4	36.4	37.4	38.4	39.4
89	26.1	27.1	28.1	29.1	30.1	31.1	32.1	33.1
91	24.8	25.8	26.8	27.8	28.8	29.8	30.8	31.8
100	19.2	20.2	21.2	22.2	23.2	24.2	25.2	26.2
105	16.2	17.2	18.2	19.2	20.2	21.2	22.2	23.2
110	13.1	14.1	15.1	16.1	17.1	18.1	19.1	20.1
120	7.3	8.3	9.3	10.3	11.3	12.3	13.3	14.3
121	6.8	7.8	8.8	9.8	10.8	11.8	12.8	13.8
130	1.9	2.9	3.9	4.9	5.9	6.9	7.9	8.9
135	-0.6	0.3	1.3	2.3	3.3	4.3	5.3	6.3
140	-2.5	-1.7	-0.9	-0.1	0.9	1.9	2.9	3.9
150	-5.6	-4.8	-4.1	-3.3	-2.5	-1.8	-1	-0.3

continued on following page

Table 53. Continued

	φ(°)							
Day number	**31**	**32**	**33**	**34**	**35**	**36**	**37**	**38**
152	-6.1	-5.3	-4.6	-3.8	-3	-2.3	-1.6	-0.9
161	-7.7	-7	-6.2	-5.5	-4.8	-4	-3.3	-2.6
162	-7.9	-7.1	-6.4	-5.6	-4.9	-4.2	-3.5	-2.8
171	-8.5	-7.7	-7	-6.3	-5.5	-4.8	-4.1	-3.5
181	-8.1	-7.3	-6.6	-5.9	-5.1	-4.4	-3.7	-3
182	-8	-7.2	-6.5	-5.7	-5	-4.3	-3.6	-2.9
191	-6.6	-5.8	-5	-4.3	-3.5	-2.8	-2.1	-1.4
198	-4.8	-4	-3.3	-2.5	-1.7	-1	-0.2	0.7
201	-3.9	-3.1	-2.3	-1.5	-0.8	0	1	2
211	-0.2	0.7	1.7	2.7	3.7	4.7	5.7	6.7
213	0.8	1.8	2.8	3.8	4.8	5.8	6.8	7.8
222	5.6	6.6	7.6	8.6	9.6	10.6	11.6	12.6
232	11.3	12.3	13.3	14.3	15.3	16.3	17.3	18.3
242	17.3	18.3	19.3	20.3	21.3	22.3	23.3	24.3
244	18.5	19.5	20.5	21.5	22.5	23.5	24.5	25.5
253	24.1	25.1	26.1	27.1	28.1	29.1	30.1	31.1
258	27.3	28.3	29.3	30.3	31.3	32.3	33.3	34.3
263	30.4	31.4	32.4	33.4	34.4	35.4	36.4	37.4
273	36.6	37.6	38.6	39.6	40.6	41.6	42.6	43.6
274	37.2	38.2	39.2	40.2	41.2	42.2	43.2	44.2
283	42.3	43.2	44.2	45.2	46.2	47.1	48.1	49.1
288	44.8	45.8	46.8	47.7	48.7	49.7	50.6	51.6
293	47.2	48.2	49.1	50.1	51.1	52	53	53.9
303	51.5	52.4	53.3	54.3	55.2	56.1	57.1	58
305	52.2	53.2	54.1	55	55.9	56.9	57.8	58.7
306	52.6	53.5	54.4	55.4	56.3	57.2	58.1	59.1
307	53	53.9	54.8	55.7	56.7	57.6	58.5	59.4
308	53.3	54.2	55.2	56.1	57	57.9	58.8	59.7
314	55.3	56.2	57.1	58	58.9	59.8	60.7	61.6
318	56.4	57.3	58.2	59.1	60	60.9	61.8	62.7
324	57.9	58.8	59.7	60.6	61.5	62.4	63.2	64.1
334	59.8	60.7	61.6	62.4	63.3	64.2	65.1	65.9
335	60	60.8	61.7	62.6	63.5	64.3	65.2	66.1
344	61	61.8	62.7	63.6	64.4	65.3	66.2	67
354	61.4	62.3	63.2	64	64.9	65.7	66.6	67.5
365	61.1	61.9	62.8	63.7	64.5	65.4	66.3	67.1

Table 54. The daily optimum tilt angle $\beta_{opt,d}$ (°) for latitudes [39°N to 43.45°N]

	φ(°)					
Day number	**39**	**40**	**41**	**42**	**43**	**43.45**
1	67.9	68.8	69.6	70.5	71.3	71.7
10	67	67.8	68.7	69.5	70.4	70.8
17	65.8	66.7	67.6	68.4	69.3	69.7
20	65.2	66.1	67	67.9	68.7	69.1
30	62.8	63.7	64.6	65.5	66.4	66.8
32	62.3	63.2	64.1	65	65.8	66.3
41	59.3	60.2	61.2	62.1	63	63.4
47	57	58	58.9	59.8	60.8	61.2
51	55.4	56.3	57.3	58.2	59.2	59.6
59	51.7	52.6	53.6	54.6	55.5	56
60	51.2	52.1	53.1	54.1	55	55.5
69	46.4	47.3	48.3	49.3	50.3	50.8
75	42.9	43.9	44.9	45.9	46.8	47.3
79	40.4	41.4	42.4	43.4	44.4	44.9
89	34.1	35.1	36.1	37.1	38.1	38.5
91	32.8	33.8	34.8	35.8	36.8	37.3
100	27.2	28.2	29.2	30.2	31.2	31.7
105	24.2	25.2	26.2	27.2	28.2	28.6
110	21.1	22.1	23.1	24.1	25.1	25.6
120	15.3	16.3	17.3	18.3	19.3	19.8
121	14.8	15.8	16.8	17.8	18.8	19.2
130	9.9	10.9	11.9	12.9	13.9	14.3
135	7.3	8.3	9.3	10.3	11.3	11.8
140	4.9	5.9	6.9	7.9	8.9	9.3
150	0.6	1.6	2.6	3.6	4.6	5
152	-0.1	0.8	1.8	2.8	3.8	4.2
161	-2	-1.3	-0.6	0	1	1.4
162	-2.1	-1.4	-0.8	-0.2	0.7	1.2
171	-2.8	-2.1	-1.5	-0.9	-0.3	0
181	-2.3	-1.7	-1.1	-0.4	0.3	0.7
182	-2.2	-1.6	-0.9	-0.3	0.5	0.9
191	-0.7	0	1	2	3	3.5
198	1.7	2.7	3.7	4.7	5.7	6.2
201	3	4	5	6	7	7.5
211	7.7	8.7	9.7	10.7	11.7	12.2
213	8.8	9.8	10.8	11.8	12.8	13.2

continued on following page

Table 54. Continued

	φ(°)					
Day number	**39**	**40**	**41**	**42**	**43**	**43.45**
222	13.6	14.6	15.6	16.6	17.6	18
228	17	18	19	20	21	21.4
232	19.3	20.3	21.3	22.3	23.3	23.7
242	25.3	26.3	27.3	28.3	29.3	29.7
244	26.5	27.5	28.5	29.5	30.5	31
253	32.1	33.1	34.1	35.1	36.1	36.6
258	35.3	36.3	37.3	38.3	39.3	39.7
263	38.4	39.4	40.4	41.4	42.4	42.9
273	44.6	45.6	46.6	47.5	48.5	49
274	45.1	46.1	47.1	48.1	49.1	49.6
283	50.1	51	52	53	54	54.4
288	52.6	53.5	54.5	55.5	56.4	56.8
293	54.9	55.8	56.8	57.7	58.7	59.1
303	58.9	59.8	60.8	61.7	62.6	63
305	59.6	60.6	61.5	62.4	63.3	63.7
314	62.5	63.4	64.3	65.2	66.1	66.5
318	63.6	64.5	65.4	66.3	67.1	67.5
324	65	65.9	66.8	67.6	68.5	68.9
334	66.8	67.7	68.5	69.4	70.2	70.6
335	66.9	67.8	68.7	69.5	70.4	70.8
344	67.9	68.7	69.6	70.4	71.3	71.7
354	68.3	69.2	70	70.9	71.7	72.1
365	68	68.8	69.7	70.5	71.4	71.8

Table 55. The daily optimum tilt angle $\beta_{opt,d}$ (°) for latitudes [24°S to 31°S]

	φ(°)							
Day number	**-24**	**-25**	**-26**	**-27**	**-28**	**-29**	**-30**	**-31**
1	13.3	12.5	11.7	10.9	10.1	9.3	8.6	7.8
10	11.8	11	10.2	9.4	8.6	7.8	7	6.2
17	10.1	9.3	8.4	7.6	6.8	6	5.2	4.4
20	9.2	8.4	7.5	6.7	5.9	5	4.2	3.4
30	5.6	4.7	3.8	2.9	2.1	1.2	0.4	-0.6
32	4.7	3.8	3	2.1	1.2	0.3	-0.6	-1.6
41	0.5	-0.5	-1.5	-2.5	-3.5	-4.5	-5.5	-6.5

continued on following page

Table 55. Continued

	φ(°)							
Day number	**-24**	**-25**	**-26**	**-27**	**-28**	**-29**	**-30**	**-31**
47	-2.9	-3.9	-4.9	-5.9	-6.9	-7.9	-8.9	-9.9
51	-5.3	-6.3	-7.3	-8.3	-9.3	-10.3	-11.3	-12.3
59	-10.1	-11.1	-12.1	-13.1	-14.1	-15.1	-16.1	-17.1
60	-10.7	-11.7	-12.7	-13.7	-14.7	-15.7	-16.7	-17.7
69	-16.3	-17.3	-18.3	-19.3	-20.3	-21.3	-22.3	-23.3
75	-20	-21	-22	-23	-24	-25	-26	-27
79	-22.6	-23.6	-24.6	-25.6	-26.6	-27.6	-28.6	-29.6
89	-28.8	-29.8	-30.8	-31.8	-32.8	-33.8	-34.8	-35.8
91	-30	-31	-32	-33	-34	-35	-36	-37
100	-35.2	-36.2	-37.1	-38.1	-39.1	-40.1	-41.1	-42.1
105	-37.8	-38.8	-39.8	-40.8	-41.7	-42.7	-43.7	-44.6
110	-40.3	-41.3	-42.3	-43.2	-44.2	-45.1	-46.1	-47.1
120	-44.8	-45.7	-46.6	-47.6	-48.5	-49.4	-50.4	-51.3
121	-45.2	-46.1	-47	-48	-48.9	-49.8	-50.8	-51.7
130	-48.4	-49.3	-50.2	-51.2	-52.1	-53	-53.9	-54.8
135	-49.9	-50.9	-51.8	-52.7	-53.6	-54.5	-55.4	-56.3
140	-51.3	-52.2	-53.1	-54	-54.9	-55.8	-56.7	-57.6
150	-53.4	-54.3	-55.2	-56.1	-56.9	-57.8	-58.7	-59.6
152	-53.7	-54.6	-55.5	-56.4	-57.3	-58.1	-59	-59.9
161	-54.8	-55.7	-56.6	-57.4	-58.3	-59.2	-60.1	-60.9
162	-54.9	-55.8	-56.7	-57.5	-58.4	-59.3	-60.2	-61
171	-55.3	-56.2	-57.1	-57.9	-58.8	-59.7	-60.5	-61.4
181	-55.1	-55.9	-56.8	-57.7	-58.6	-59.4	-60.3	-61.2
182	-55	-55.9	-56.7	-57.6	-58.5	-59.4	-60.2	-61.1
191	-54	-54.9	-55.8	-56.7	-57.6	-58.4	-59.3	-60.2
198	-52.9	-53.8	-54.6	-55.5	-56.4	-57.3	-58.2	-59.1
201	-52.3	-53.1	-54	-54.9	-55.8	-56.7	-57.6	-58.5
211	-49.7	-50.6	-51.5	-52.4	-53.3	-54.2	-55.2	-56.1
213	-49.1	-50	-50.9	-51.8	-52.7	-53.7	-54.6	-55.5
222	-46	-46.9	-47.8	-48.8	-49.7	-50.6	-51.6	-52.5
228	-43.6	-44.5	-45.4	-46.4	-47.3	-48.3	-49.2	-50.2
232	-41.8	-42.7	-43.7	-44.7	-45.6	-46.6	-47.5	-48.5
242	-36.9	-37.8	-38.8	-39.8	-40.8	-41.7	-42.7	-43.7
244	-35.8	-36.8	-37.8	-38.7	-39.7	-40.7	-41.7	-42.6
253	-30.7	-31.7	-32.7	-33.7	-34.7	-35.7	-36.6	-37.6
258	-27.7	-28.7	-29.7	-30.7	-31.7	-32.7	-33.7	-34.7

continued on following page

Table 55. Continued

	φ(°)							
Day number	**-24**	**-25**	**-26**	**-27**	**-28**	**-29**	**-30**	**-31**
263	-24.6	-25.6	-26.6	-27.6	-28.6	-29.6	-30.6	-31.6
273	-18.2	-19.2	-20.2	-21.2	-22.2	-23.2	-24.2	-25.2
274	-17.6	-18.6	-19.6	-20.6	-21.6	-22.6	-23.6	-24.6
283	-12	-13	-14	-15	-16	-17	-18	-19
288	-8.9	-9.9	-10.9	-11.9	-12.9	-13.9	-14.9	-15.9
293	-5.9	-6.9	-7.9	-8.9	-9.9	-10.9	-11.9	-12.9
303	-0.1	-1.1	-2.1	-3.1	-4.1	-5.1	-6.1	-7.1
305	0.9	0	-1	-2	-3	-4	-5	-6
314	5.1	4.2	3.3	2.5	1.6	0.7	-0.1	-1.1
318	6.7	5.8	5	4.1	3.3	2.4	1.6	0.7
324	8.8	8	7.2	6.3	5.5	4.7	3.8	3
334	11.6	10.8	10	9.1	8.3	7.5	6.7	6
335	11.8	11	10.2	9.4	8.6	7.8	7	6.2
344	13.3	12.5	11.7	10.9	10.1	9.3	8.6	7.8
354	14	13.2	12.4	11.6	10.8	10	9.2	8.5
365	13.5	12.7	11.9	11.1	10.3	9.5	8.7	7.9

Table 56. The daily optimum tilt angle $\beta_{opt,d}$ (°) for latitudes [32°S to 39°S]

	φ(°)							
Day number	**-32**	**-33**	**-34**	**-35**	**-36**	**-37**	**-38**	**-39**
1	7	6.3	5.6	4.8	4.1	3.4	2.7	2
10	5.4	4.7	3.9	3.2	2.4	1.7	1	0.3
17	3.6	2.8	2	1.2	0.5	-0.4	-1.4	-2.4
20	2.6	1.8	1	0.2	-0.7	-1.7	-2.7	-3.7
30	-1.6	-2.6	-3.6	-4.6	-5.6	-6.6	-7.6	-8.6
32	-2.6	-3.6	-4.6	-5.6	-6.6	-7.6	-8.6	-9.6
41	-7.5	-8.5	-9.5	-10.5	-11.5	-12.5	-13.5	-14.5
47	-10.9	-11.9	-12.9	-13.9	-14.9	-15.9	-16.9	-17.9
51	-13.3	-14.3	-15.3	-16.3	-17.3	-18.3	-19.3	-20.3
59	-18.1	-19.1	-20.1	-21.1	-22.1	-23.1	-24.1	-25.1
60	-18.7	-19.7	-20.7	-21.7	-22.7	-23.7	-24.7	-25.7
69	-24.3	-25.3	-26.3	-27.3	-28.3	-29.3	-30.3	-31.3
75	-28	-29	-30	-31	-32	-33	-34	-35
79	-30.6	-31.6	-32.6	-33.6	-34.6	-35.6	-36.6	-37.6

continued on following page

Table 56. Continued

	φ(°)							
Day number	**-32**	**-33**	**-34**	**-35**	**-36**	**-37**	**-38**	**-39**
89	-36.8	-37.8	-38.8	-39.8	-40.8	-41.8	-42.8	-43.8
91	-38	-39	-40	-41	-41.9	-42.9	-43.9	-44.9
100	-43	-44	-45	-46	-46.9	-47.9	-48.9	-49.9
105	-45.6	-46.6	-47.5	-48.5	-49.5	-50.5	-51.4	-52.4
110	-48	-49	-49.9	-50.9	-51.8	-52.8	-53.8	-54.7
120	-52.3	-53.2	-54.1	-55.1	-56	-56.9	-57.9	-58.8
121	-52.6	-53.6	-54.5	-55.4	-56.4	-57.3	-58.2	-59.1
130	-55.7	-56.7	-57.6	-58.5	-59.4	-60.3	-61.2	-62.1
135	-57.2	-58.1	-59	-59.9	-60.8	-61.7	-62.6	-63.5
140	-58.5	-59.4	-60.3	-61.2	-62	-62.9	-63.8	-64.7
150	-60.5	-61.3	-62.2	-63.1	-64	-64.8	-65.7	-66.6
152	-60.8	-61.7	-62.5	-63.4	-64.3	-65.1	-66	-66.9
161	-61.8	-62.7	-63.6	-64.4	-65.3	-66.1	-67	-67.9
162	-61.9	-62.8	-63.6	-64.5	-65.4	-66.2	-67.1	-67.9
171	-62.3	-63.1	-64	-64.9	-65.7	-66.6	-67.5	-68.3
181	-62	-62.9	-63.8	-64.6	-65.5	-66.4	-67.2	-68.1
182	-62	-62.8	-63.7	-64.6	-65.4	-66.3	-67.2	-68
191	-61.1	-62	-62.8	-63.7	-64.6	-65.4	-66.3	-67.2
198	-60	-60.9	-61.7	-62.6	-63.5	-64.4	-65.2	-66.1
201	-59.4	-60.3	-61.2	-62	-62.9	-63.8	-64.7	-65.6
211	-57	-57.9	-58.8	-59.7	-60.6	-61.5	-62.4	-63.3
213	-56.4	-57.3	-58.2	-59.1	-60	-60.9	-61.8	-62.7
222	-53.4	-54.4	-55.3	-56.2	-57.1	-58.1	-59	-59.9
228	-51.1	-52.1	-53	-53.9	-54.9	-55.8	-56.8	-57.7
232	-49.4	-50.4	-51.3	-52.3	-53.2	-54.2	-55.1	-56.1
242	-44.7	-45.6	-46.6	-47.6	-48.6	-49.5	-50.5	-51.5
244	-43.6	-44.6	-45.6	-46.6	-47.5	-48.5	-49.5	-50.5
253	-38.6	-39.6	-40.6	-41.6	-42.6	-43.6	-44.6	-45.6
258	-35.7	-36.7	-37.6	-38.6	-39.6	-40.6	-41.6	-42.6
263	-32.6	-33.6	-34.6	-35.6	-36.6	-37.6	-38.6	-39.6
273	-26.2	-27.2	-28.2	-29.2	-30.2	-31.2	-32.2	-33.2
274	-25.6	-26.6	-27.6	-28.6	-29.6	-30.6	-31.6	-32.6
283	-20	-21	-22	-23	-24	-25	-26	-27
288	-16.9	-17.9	-18.9	-19.9	-20.9	-21.9	-22.9	-23.9
293	-13.9	-14.9	-15.9	-16.9	-17.9	-18.9	-19.9	-20.9
303	-8.1	-9.1	-10.1	-11.1	-12.1	-13.1	-14.1	-15.1

continued on following page

Table 56. Continued

	φ(°)							
Day number	**-32**	**-33**	**-34**	**-35**	**-36**	**-37**	**-38**	**-39**
305	-7	-8	-9	-10	-11	-12	-13	-14
314	-2.1	-3.1	-4.1	-5.1	-6.1	-7.1	-8.1	-9.1
318	-0.1	-1.1	-2.1	-3.1	-4.1	-5.1	-6.1	-7.1
324	2.2	1.4	0.6	-0.2	-1.2	-2.2	-3.2	-4.2
334	5.2	4.4	3.6	2.9	2.1	1.4	0.7	0
335	5.4	4.6	3.9	3.1	2.4	1.7	0.9	0.2
344	7	6.3	5.5	4.8	4.1	3.4	2.7	2
354	7.7	7	6.3	5.5	4.8	4.1	3.5	2.8
365	7.2	6.4	5.7	5	4.3	3.6	2.9	2.2

Table 57. The daily optimum tilt angle $\beta_{opt,d}$ (°) for latitudes [40°S to 43.45°S]

	φ(°)				
Day number	**-40**	**-41**	**-42**	**-43**	**-43.45**
1	1.4	0.7	0.1	-0.9	-1.3
10	-0.6	-1.6	-2.6	-3.6	-4.1
17	-3.4	-4.4	-5.4	-6.4	-6.9
20	-4.7	-5.7	-6.7	-7.7	-8.2
30	-9.6	-10.6	-11.6	-12.6	-13
32	-10.6	-11.6	-12.6	-13.6	-14.1
41	-15.5	-16.5	-17.5	-18.5	-18.9
47	-18.9	-19.9	-20.9	-21.9	-22.4
51	-21.3	-22.3	-23.3	-24.3	-24.7
59	-26.1	-27.1	-28.1	-29.1	-29.5
60	-26.7	-27.7	-28.7	-29.7	-30.1
69	-32.3	-33.3	-34.3	-35.3	-35.7
75	-36	-37	-38	-39	-39.5
79	-38.6	-39.6	-40.6	-41.6	-42
89	-44.8	-45.7	-46.7	-47.7	-48.2
91	-45.9	-46.9	-47.9	-48.9	-49.3
100	-50.9	-51.8	-52.8	-53.8	-54.2
105	-53.3	-54.3	-55.3	-56.2	-56.7
110	-55.7	-56.6	-57.6	-58.5	-58.9
120	-59.7	-60.6	-61.6	-62.5	-62.9
121	-60.1	-61	-61.9	-62.8	-63.2

continued on following page

Table 57. Continued

	φ(°)				
Day number	**-40**	**-41**	**-42**	**-43**	**-43.45**
130	-63	-63.9	-64.8	-65.7	-66.1
135	-64.4	-65.3	-66.2	-67	-67.4
140	-65.6	-66.5	-67.3	-68.2	-68.6
150	-67.5	-68.3	-69.2	-70	-70.4
152	-67.7	-68.6	-69.5	-70.3	-70.7
161	-68.7	-69.6	-70.4	-71.3	-71.6
162	-68.8	-69.6	-70.5	-71.3	-71.7
171	-69.2	-70	-70.9	-71.7	-72.1
181	-68.9	-69.8	-70.6	-71.5	-71.8
182	-68.9	-69.7	-70.6	-71.4	-71.8
191	-68	-68.9	-69.7	-70.6	-71
198	-67	-67.9	-68.7	-69.6	-70
201	-66.4	-67.3	-68.2	-69.1	-69.4
211	-64.2	-65.1	-65.9	-66.8	-67.2
213	-63.6	-64.5	-65.4	-66.3	-66.7
222	-60.8	-61.7	-62.7	-63.6	-64
228	-58.6	-59.6	-60.5	-61.4	-61.8
232	-57	-57.9	-58.9	-59.8	-60.3
242	-52.4	-53.4	-54.4	-55.3	-55.8
244	-51.4	-52.4	-53.4	-54.4	-54.8
253	-46.6	-47.6	-48.6	-49.5	-50
258	-43.6	-44.6	-45.6	-46.6	-47.1
263	-40.6	-41.6	-42.6	-43.6	-44
273	-34.2	-35.2	-36.2	-37.2	-37.7
283	-28	-29	-30	-31	-31.4
288	-24.9	-25.9	-26.9	-27.9	-28.4
293	-21.9	-22.9	-23.9	-24.9	-25.4
303	-16.1	-17.1	-18.1	-19.1	-19.6
305	-15	-16	-17	-18	-18.4
314	-10.1	-11.1	-12.1	-13.1	-13.6
318	-8.1	-9.1	-10.1	-11.1	-11.6
324	-5.2	-6.2	-7.2	-8.2	-8.7
334	-1	-2	-3	-4	-4.5
335	-0.7	-1.7	-2.7	-3.7	-4.1
344	1.3	0.7	0.1	-0.9	-1.3
354	2.1	1.5	0.9	0.3	0
365	1.5	0.9	0.3	-0.6	-1

Table 58. The monthly optimum tilt angle $\beta_{opt,m}$ (°) for January and February in NH

φ(°)	January	Days	From	To	February	Days	From	To
23.45	51.95	31	1	31	42.73	28	32	59
24	52.43	31	1	31	43.26	28	32	59
25	53.32	31	1	31	44.21	28	32	59
26	54.22	31	1	31	45.15	28	32	59
27	55.11	31	1	31	46.10	28	32	59
28	56.01	31	1	31	47.05	28	32	59
29	56.91	31	1	31	47.98	28	32	59
30	57.79	31	1	31	48.93	28	32	59
31	58.68	31	1	31	49.87	28	32	59
32	59.57	31	1	31	50.82	28	32	59
33	60.45	31	1	31	51.76	28	32	59
34	61.32	31	1	31	52.70	28	32	59
35	62.22	31	1	31	53.64	28	32	59
36	63.10	31	1	31	54.57	28	32	59
37	63.97	31	1	31	55.51	28	32	59
38	64.85	31	1	31	56.46	28	32	59
39	65.72	31	1	31	57.39	28	32	59
40	66.60	31	1	31	58.33	28	32	59
41	67.47	31	1	31	59.26	28	32	59
42					60.19	28	32	59
43					61.13	28	32	59
43.45					61.55	28	32	59

Table 59. The monthly optimum tilt angle $\beta_{opt,m}$ (^{o}) for March and April in NH]

φ(°)	March	Days	From	To	April	Days	From	To	April	Days	From	To
23.45	27.21	31	60	90	8.68	29	91	119	-0.20	1	120	120
24	27.75	31	60	90	8.93	30	91	120				
25	28.75	31	60	90	9.93	30	91	120				
26	29.75	31	60	90	10.93	30	91	120				
27	30.74	31	60	90	11.93	30	91	120				
28	31.73	31	60	90	12.93	30	91	120				
29	32.73	31	60	90	13.93	30	91	120				
30	33.72	31	60	90	14.93	30	91	120				
31	34.71	31	60	90	15.93	30	91	120				
32	35.71	31	60	90	16.93	30	91	120				
33	36.70	31	60	90	17.93	30	91	120				
34	37.70	31	60	90	18.93	30	91	120				
35	38.69	31	60	90	19.93	30	91	120				
36	39.68	31	60	90	20.93	30	91	120				
37	40.68	31	60	90	21.93	30	91	120				
38	41.67	31	60	90	22.93	30	91	120				
39	42.66	31	60	90	23.93	30	91	120				
40	43.65	31	60	90	24.93	30	91	120				
41	44.64	31	60	90	25.93	30	91	120				
42	45.64	31	60	90	26.93	30	91	120				
43	46.62	31	60	90	27.93	30	91	120				
43.45	47.07	31	60	90	28.38	30	91	120				

Table 60. The monthly optimum tilt angle $\beta_{opt,m}$ (°) for May and June in NH

φ(°)	May	Days	From	To	May	Days	From	To	June	Days	From	To
23.45	-7.03	31	121	151	-13.82	30	152	181				
24	-6.56	31	121	151	-13.38	30	152	181				
25	0.50	2	121	122	-6.11	29	123	151	-12.58	30	152	181
26	0.93	4	121	124	-5.67	27	125	151	-11.77	30	152	181
27	1.37	6	121	126	-5.24	25	127	151	-10.98	30	152	181
28	2.10	7	121	127	-4.58	24	128	151	-10.19	30	152	181
29	2.56	9	121	129	-4.14	22	130	151	-9.40	30	152	181
30	3.02	11	121	131	-3.69	20	132	151	-8.62	30	152	181
31	3.48	13	121	133	-3.27	18	134	151	-7.87	30	152	181
32	3.96	15	121	135	-2.82	16	136	151	-7.10	30	152	181
33	4.44	17	121	137	-2.36	14	138	151	-6.36	30	152	181
34	4.93	19	121	139	-1.91	12	140	151	-5.61	30	152	181
35	5.18	22	121	142	-1.66	9	143	151	-4.89	30	152	181
36	5.69	24	121	144	-1.19	7	145	151	-4.17	30	152	181
37	6.21	26	121	146	-0.76	5	147	151	-3.47	30	152	181
38	6.50	29	121	149	-0.45	2	150	151	-2.78	30	152	181
39	7.04	31	121	151					-2.09	30	152	181
40	8.04	31	121	151	0.43	3	152	154	-1.63	27	155	181
41	9.04	31	121	151	0.92	6	152	157	-1.13	24	158	181
42	10.04	31	121	151	1.30	10	152	161	-0.68	20	162	181
43	11.04	31	121	151	1.39	18	152	169	-0.23	12	170	181
43.45	11.50	31	121	151	1.14	30	152	181	No	0		

Table 61. The monthly optimum tilt angle $\beta_{opt,m}$ (°) for July and August in NH

φ(°)	July	Days	From	To	August	Days	From	To	August	Days	From	To
23.45	-10.86	31	182	212	5.24	18	213	230	-3.18	13	231	243
24	-10.40	31	182	212	5.53	19	213	231	-2.93	12	232	243
25	-9.56	31	182	212	5.94	21	213	233	-2.51	10	234	243
26	-8.74	31	182	212	6.37	23	213	235	-2.10	8	236	243
27	-7.91	31	182	212	7.08	24	213	236	-1.46	7	237	243
28	-7.10	31	182	212	7.51	26	213	238	-1.02	5	239	243
29	-6.28	31	182	212	7.94	28	213	240	-0.63	3	241	243
30	-5.47	31	182	212	8.38	30	213	242	-0.20	1	243	243
31	0.30	1	182	182	-4.84	30	183	212	9.11	31	213	243
32	0.73	3	182	184	-4.37	28	185	212	10.11	31	213	243
33	1.24	5	182	186	-3.90	26	187	212	11.11	31	213	243
34	1.76	7	182	188	-3.43	24	189	212	12.11	31	213	243
35	2.28	9	182	190	-2.95	22	191	212	13.11	31	213	243
36	2.57	12	182	193	-2.63	19	194	212	14.11	31	213	243
37	3.11	14	182	195	-2.14	17	196	212	15.11	31	213	243
38	3.66	16	182	197	-1.69	15	198	212	16.11	31	213	243
39	4.00	19	182	200	-1.33	12	201	212	17.11	31	213	243
40	4.37	22	182	203	-0.96	9	204	212	18.11	31	213	243
41	4.77	25	182	206	-0.57	6	207	212	19.11	31	213	243
42	5.02	29	182	210	-0.25	2	211	212	20.11	31	213	243
43	5.67	31	182	212					21.11	31	213	243
43.45	6.12	31	182	212					21.54	31	213	243

Table 62. The monthly optimum tilt angle $\beta_{opt,m}$ (°) for September and October in NH

φ(°)	September	Days	From	To	October	Days	From	To
23.45	20.06	30	244	273	37.74	31	274	304
24	20.59	30	244	273	38.26	31	274	304
25	21.59	30	244	273	39.23	31	274	304
26	22.59	30	244	273	40.21	31	274	304
27	23.59	30	244	273	41.16	31	274	304
28	24.59	30	244	273	42.14	31	274	304
29	25.59	30	244	273	43.11	31	274	304
30	26.59	30	244	273	44.06	31	274	304
31	27.59	30	244	273	45.04	31	274	304
32	28.59	30	244	273	46.01	31	274	304
33	29.59	30	244	273	46.97	31	274	304
34	30.59	30	244	273	47.93	31	274	304
35	31.59	30	244	273	48.90	31	274	304
36	32.59	30	244	273	49.86	31	274	304
37	33.59	30	244	273	50.82	31	274	304
38	34.59	30	244	273	51.78	31	274	304
39	35.59	30	244	273	52.75	31	274	304
40	36.59	30	244	273	53.70	31	274	304
41	37.59	30	244	273	54.66	31	274	304
42	38.59	30	244	273	55.63	31	274	304
43	39.59	30	244	273	56.58	31	274	304
43.45	40.04	30	244	273	57.01	31	274	304

Table 63. The monthly optimum tilt angle $\beta_{opt,m}$ (°) for November and December in NH

φ(°)	November	Days	From	To	December	Days	From	To
23.45	49.68	30	305	334	54.43	31	335	365
24	50.19	30	305	334	54.92	31	335	365
25	51.09	30	305	334	55.79	31	335	365
26	52.00	30	305	334	56.67	31	335	365
27	52.91	30	305	334	57.55	31	335	365
28	53.81	30	305	334	58.43	31	335	365
29	54.71	30	305	334	59.29	31	335	365
30	55.62	30	305	334	60.17	31	335	365
31	56.52	30	305	334	61.04	31	335	365
32	57.43	30	305	334	61.92	31	335	365
33	58.33	30	305	334	62.77	31	335	365
34	59.23	30	305	334	63.65	31	335	365
35	60.13	30	305	334	64.52	31	335	365
36	61.03	30	305	334	65.36	31	335	365
37	61.92	30	305	334	66.24	31	335	365
38	62.80	30	305	334	67.10	31	335	365
39	63.70	30	305	334	67.95	31	335	365
40	64.59	30	305	334	68.81	31	335	365
41	65.48	30	305	334	69.65	31	335	365
42	66.35	30	305	334	70.50	31	335	365
43	67.24	30	305	334	71.35	31	335	365
43.45	67.64	30	305	334	71.74	31	335	365

Table 64. The monthly optimum tilt angle $\beta_{opt,m}$ (°) for January and February in SH

φ(°)	January	Days	From	To	February	Days	From	To	February	Days	From	To
-23.45	10.43	31	1	31	2.91	11	32	42	-4.74	17	43	59
-24	9.97	31	1	31	2.41	11	32	42	-5.29	17	43	59
-25	9.14	31	1	31	1.99	9	32	40	-5.71	19	41	59
-26	8.29	31	1	31	1.60	7	32	38	-6.13	21	39	59
-27	7.48	31	1	31	1.16	5	32	36	-6.56	23	37	59
-28	6.66	31	1	31	0.77	3	32	34	-6.99	25	35	59
-29	5.84	31	1	31	0.30	1	32	32	-7.43	27	33	59
-30	5.19	30	1	30	-0.10	1	31	31	-8.15	28	32	59
-31	4.72	28	1	28	-0.60	3	29	31	-9.15	28	32	59
-32	4.26	26	1	26	-1.10	5	27	31	-10.15	28	32	59
-33	3.79	24	1	24	-1.60	7	25	31	-11.15	28	32	59
-34	3.32	22	1	22	-2.10	9	23	31	-12.15	28	32	59
-35	2.85	20	1	20	-2.61	11	21	31	-13.15	28	32	59
-36	2.39	18	1	18	-3.13	13	19	31	-14.15	28	32	59
-37	1.92	16	1	16	-3.66	15	17	31	-15.15	28	32	59
-38	1.58	13	1	13	-3.98	18	14	31	-16.15	28	32	59
-39	1.12	11	1	11	-4.54	20	12	31	-17.15	28	32	59
-40	0.78	8	1	8	-4.89	23	9	31	-18.15	28	32	59
-41	0.38	5	1	5	-5.28	26	6	31	-19.15	28	32	59
-42	0.10	1	1	1	-5.51	30	2	31	-20.15	28	32	59
-43					-6.33	31	1	31	-21.15	28	32	59
-43.45					-6.77	31	1	31	-21.60	28	32	59

Table 65. The monthly optimum tilt angle $\beta_{opt,m}$ (°) for March and April in SH

φ(°)	March	Days	From	To	April	Days	From	To
-23.45	-19.51	31	60	90	-37.30	30	91	120
-24	-20.05	31	60	90	-37.84	30	91	120
-25	-21.05	31	60	90	-38.81	30	91	120
-26	-22.05	31	60	90	-39.78	30	91	120
-27	-23.05	31	60	90	-40.75	30	91	120
-28	-24.05	31	60	90	-41.72	30	91	120
-29	-25.05	31	60	90	-42.69	30	91	120
-30	-26.05	31	60	90	-43.67	30	91	120
-31	-27.05	31	60	90	-44.62	30	91	120
-32	-28.05	31	60	90	-45.60	30	91	120
-33	-29.05	31	60	90	-46.57	30	91	120
-34	-30.05	31	60	90	-47.52	30	91	120
-35	-31.05	31	60	90	-48.49	30	91	120
-36	-32.05	31	60	90	-49.45	30	91	120
-37	-33.05	31	60	90	-50.42	30	91	120
-38	-34.05	31	60	90	-51.39	30	91	120
-39	-35.05	31	60	90	-52.34	30	91	120
-40	-36.05	31	60	90	-53.32	30	91	120
-41	-37.04	31	60	90	-54.27	30	91	120
-42	-38.04	31	60	90	-55.23	30	91	120
-43	-39.04	31	60	90	-56.20	30	91	120
-43.45	-39.50	31	60	90	-56.62	30	91	120

Table 66. The monthly optimum tilt angle $\beta_{opt,m}$ (°) for May, June and July in SH

φ(°)	May	Days	From	To	June	Days	From	To	July	Days	From	To
-23.45	-49.41	31	121	151	-54.41	30	152	181	-52.26	31	182	212
-24	-49.91	31	121	151	-54.90	30	152	181	-52.75	31	182	212
-25	-50.83	31	121	151	-55.78	30	152	181	-53.64	31	182	212
-26	-51.73	31	121	151	-56.66	30	152	181	-54.54	31	182	212
-27	-52.65	31	121	151	-57.52	30	152	181	-55.42	31	182	212
-28	-53.55	31	121	151	-58.40	30	152	181	-56.30	31	182	212
-29	-54.46	31	121	151	-59.28	30	152	181	-57.20	31	182	212
-30	-55.37	31	121	151	-60.15	30	152	181	-58.09	31	182	212
-31	-56.27	31	121	151	-61.03	30	152	181	-58.98	31	182	212
-32	-57.17	31	121	151	-61.89	30	152	181	-59.85	31	182	212
-33	-58.08	31	121	151	-62.77	30	152	181	-60.75	31	182	212
-34	-58.98	31	121	151	-63.63	30	152	181	-61.63	31	182	212
-35	-59.88	31	121	151	-64.49	30	152	181	-62.50	31	182	212
-36	-60.78	31	121	151	-65.35	30	152	181	-63.38	31	182	212
-37	-61.68	31	121	151	-66.23	30	152	181	-64.27	31	182	212
-38	-62.58	31	121	151	-67.08	30	152	181	-65.13	31	182	212
-39	-63.46	31	121	151	-67.94	30	152	181	-66.02	31	182	212
-40	-64.36	31	121	151	-68.79	30	152	181	-66.88	31	182	212
-41	-65.25	31	121	151	-69.64	30	152	181	-67.75	31	182	212
-42	-66.13	31	121	151	-70.50	30	152	181	-68.62	31	182	212
-43	-67.01	31	121	151	-71.34	30	152	181	-69.48	31	182	212
-43.45	-67.41	31	121	151	-71.72	30	152	181	-69.86	31	182	212

Table 67. The monthly optimum tilt angle $\beta_{opt,m}$ (°) for August and September in SH

φ(°)	August	Days	From	To	September	Days	From	To
-23.45	-42.74	31	213	243	-26.68	30	244	273
-24	-43.26	31	213	243	-27.24	30	244	273
-25	-44.20	31	213	243	-28.24	30	244	273
-26	-45.14	31	213	243	-29.23	30	244	273
-27	-46.09	31	213	243	-30.22	30	244	273
-28	-47.04	31	213	243	-31.22	30	244	273
-29	-47.99	31	213	243	-32.22	30	244	273
-30	-48.92	31	213	243	-33.21	30	244	273
-31	-49.88	31	213	243	-34.20	30	244	273
-32	-50.82	31	213	243	-35.20	30	244	273
-33	-51.75	31	213	243	-36.19	30	244	273
-34	-52.70	31	213	243	-37.18	30	244	273
-35	-53.63	31	213	243	-38.17	30	244	273
-36	-54.57	31	213	243	-39.16	30	244	273
-37	-55.51	31	213	243	-40.16	30	244	273
-38	-56.45	31	213	243	-41.15	30	244	273
-39	-57.38	31	213	243	-42.15	30	244	273
-40	-58.31	31	213	243	-43.14	30	244	273
-41	-59.25	31	213	243	-44.14	30	244	273
-42	-60.19	31	213	243	-45.13	30	244	273
-43	-61.11	31	213	243	-46.12	30	244	273
-43.45	-61.54	31	213	243	-46.57	30	244	273

Table 68. The monthly optimum tilt angle $\beta_{opt,m}$ (°) for October and November in SH

φ(°)	Oct.	Days	From	To	Nov.	Days	From	To	Nov.	Days	From	To
-23.45	-8.46	29	274	302	0.65	2	303	304	7.38	30	305	334
-24	-8.70	30	274	303	0.40	1	304	304	6.91	30	305	334
-25	-9.40	31	274	304					6.04	30	305	334
-26	-10.40	31	274	304	-0.70	2	305	306	5.61	28	307	334
-27	-11.40	31	274	304	-1.15	4	305	308	5.17	26	309	334
-28	-12.40	31	274	304	-1.62	6	305	310	4.71	24	311	334
-29	-13.40	31	274	304	-2.08	8	305	312	4.26	22	313	334
-30	-14.40	31	274	304	-2.54	10	305	314	3.82	20	315	334
-31	-15.40	31	274	304	-3.01	12	305	316	3.38	18	317	334
-32	-16.40	31	274	304	-3.49	14	305	318	2.93	16	319	334
-33	-17.40	31	274	304	-3.97	16	305	320	2.48	14	321	334
-34	-18.40	31	274	304	-4.47	18	305	322	2.03	12	323	334
-35	-19.40	31	274	304	-4.97	20	305	324	1.61	10	325	334
-36	-20.40	31	274	304	-5.47	22	305	326	1.14	8	327	334
-37	-21.40	31	274	304	-5.99	24	305	328	0.72	6	329	334
-38	-22.40	31	274	304	-6.28	27	305	331	0.40	3	332	334
-39	-23.40	31	274	304	-6.82	29	305	333	0.00	1	334	334
-40	-24.40	31	274	304	-7.59	30	305	334				
-41	-25.40	31	274	304	-8.59	30	305	334				
-42	-26.40	31	274	304	-9.59	30	305	334				
-43	-27.40	31	274	304	-10.59	30	305	334				
-43.45	-27.87	31	274	304	-11.05	30	305	334				

Table 69. The monthly optimum tilt angle $\beta_{opt,m}$ (°) for December in SH

φ(°)	December	Days	From	To	December	Days	From	To
-23.45					13.85	31	335	365
-24					13.40	31	335	365
-25					12.60	31	335	365
-26					11.80	31	335	365
-27					11.01	31	335	365
-28					10.22	31	335	365
-29					9.43	31	335	365
-30					8.66	31	335	365
-31					7.89	31	335	365
-32					7.12	31	335	365
-33					6.38	31	335	365
-34					5.64	31	335	365
-35					4.92	31	335	365
-36					4.20	31	335	365
-37					3.48	31	335	365
-38					2.80	31	335	365
-39					2.13	31	335	365
-40	-0.50	2			1.58	29	335	365
-41	-0.96	5	335	339	1.10	26	340	365
-42	-1.33	9	335	343	0.63	22	344	365
-43	-1.44	17	335	351	0.20	14	352	365
-43.45	-1.40	24	335	358	0.00	7	359	365

Table 70. The Seasonally optimum tilt angle $\beta_{opt,s}$ (°) for winter and spring in NH

Season	Winter	Days	From	To	Spring	Days	From	To	Spring	Days	From	To
23.45	40.56	90	1	90	8.68	29	91	119	-10.20	62	120	181
24	41.08	90	1	90	8.93	30	91	120	-9.91	61	121	181
25	42.02	90	1	90	9.34	32	91	122	-9.40	59	123	181
26	42.97	90	1	90	9.75	34	91	124	-8.88	57	125	181
27	43.91	90	1	90	10.17	36	91	126	-8.37	55	127	181
28	44.86	90	1	90	10.88	37	91	127	-7.70	54	128	181
29	45.80	90	1	90	11.31	39	91	129	-7.18	52	130	181
30	46.74	90	1	90	11.73	41	91	131	-6.65	50	132	181
31	47.68	90	1	90	12.17	43	91	133	-6.14	48	134	181
32	48.63	90	1	90	12.61	45	91	135	-5.61	46	136	181
33	49.57	90	1	90	13.05	47	91	137	-5.09	44	138	181
34	50.50	90	1	90	13.50	49	91	139	-4.55	42	140	181
35	51.45	90	1	90	13.69	52	91	142	-4.14	39	143	181
36	52.38	90	1	90	14.16	54	91	144	-3.61	37	145	181
37	53.32	90	1	90	14.63	56	91	146	-3.08	35	147	181
38	54.26	90	1	90	14.86	59	91	149	-2.63	32	150	181
39	55.18	90	1	90	15.35	61	91	151	-2.09	30	152	181
40	56.12	90	1	90	15.60	64	91	154	-1.63	27	155	181
41	57.05	90	1	90	15.88	67	91	157	-1.13	24	158	181
42	57.98	90	1	90	15.95	71	91	161	-0.68	20	162	181
43	58.91	90	1	90	15.26	79	91	169	-0.23	12	170	181
43.45	59.33	90	1	90	13.65	91	91	181				

Table 71. The Seasonally optimum tilt angle $\beta_{opt,s}$ (°) for summer and autumn in NH

Season	Summer	Days	From	To	Summer	Days	From	To	Autumn	Days	From	To
23.45	-8.59	44	182	225	14.50	48	226	273	47.26	92	274	365
24	-8.32	43	182	224	14.75	49	225	273	47.76	92	274	365
25	-7.84	41	182	222	15.15	51	223	273	48.68	92	274	365
26	-7.37	39	182	220	15.55	53	221	273	49.60	92	274	365
27	-6.72	38	182	219	16.25	54	220	273	50.51	92	274	365
28	-6.25	36	182	217	16.66	56	218	273	51.43	92	274	365
29	-5.78	34	182	215	17.07	58	216	273	52.35	92	274	365
30	-5.30	32	182	213	17.49	60	214	273	53.26	92	274	365
31	-4.84	30	182	211	17.91	62	212	273	54.18	92	274	365
32	-4.37	28	182	209	18.33	64	210	273	55.09	92	274	365
33	-3.90	26	182	207	18.76	66	208	273	56.00	92	274	365
34	-3.43	24	182	205	19.20	68	206	273	56.91	92	274	365
35	-2.95	22	182	203	19.64	70	204	273	57.82	92	274	365
36	-2.63	19	182	200	19.81	73	201	273	58.73	92	274	365
37	-2.14	17	182	198	20.26	75	199	273	59.63	92	274	365
38	-1.69	15	182	196	20.72	77	197	273	60.54	92	274	365
39	-1.33	12	182	193	20.93	80	194	273	61.44	92	274	365
40	-0.96	9	182	190	21.15	83	191	273	62.34	92	274	365
41	-0.57	6	182	187	21.39	86	188	273	63.24	92	274	365
42	-0.25	2	182	183	21.41	90	184	273	64.14	92	274	365
43					21.93	92	182	273	65.03	92	274	365
43.45					22.38	92	182	273	65.44	92	274	365

Table 72. The Seasonally optimum tilt angle $\beta_{opt,s}$ (^{o}) for winter and spring in SH

φ(°)	Winter	Days	From	To	Winter	Days	From	To	Spring	Days	From	To
-23.45	8.46	42	1	42	-14.28	48	43	90	-47.07	91	91	181
-24	7.99	42	1	42	-14.82	48	43	90	-47.57	91	91	181
-25	7.53	40	1	40	-15.22	50	41	90	-48.50	91	91	181
-26	7.06	38	1	38	-15.62	52	39	90	-49.42	91	91	181
-27	6.60	36	1	36	-16.03	54	37	90	-50.34	91	91	181
-28	6.14	34	1	34	-16.43	56	35	90	-51.25	91	91	181
-29	5.67	32	1	32	-16.85	58	33	90	-52.17	91	91	181
-30	5.19	30	1	30	-17.26	60	31	90	-53.09	91	91	181
-31	4.72	28	1	28	-17.69	62	29	90	-54.00	91	91	181
-32	4.26	26	1	26	-18.11	64	27	90	-54.91	91	91	181
-33	3.79	24	1	24	-18.54	66	25	90	-55.83	91	91	181
-34	3.32	22	1	22	-18.98	68	23	90	-56.73	91	91	181
-35	2.85	20	1	20	-19.42	70	21	90	-57.65	91	91	181
-36	2.39	18	1	18	-19.87	72	19	90	-58.55	91	91	181
-37	1.92	16	1	16	-20.32	74	17	90	-59.47	91	91	181
-38	1.58	13	1	13	-20.51	77	14	90	-60.37	91	91	181
-39	1.12	11	1	11	-20.98	79	12	90	-61.27	91	91	181
-40	0.78	8	1	8	-21.20	82	9	90	-62.18	91	91	181
-41	0.38	5	1	5	-21.43	85	6	90	-63.08	91	91	181
-42	0.10	1	1	1	-21.44	89	2	90	-63.98	91	91	181
-43	No	0			-22.20	90	1	90	-64.87	91	91	181
-43.45	No	0			-22.65	90	1	90	-65.27	91	91	181

Table 73. The Seasonally optimum tilt angle $\beta_{opt,s}$ (°) for summer and autumn in SH

φ(°)	Summer	Days	From	To	Autumn	Days	From	To	Autumn	Days	From	To
-23.45	-40.71	92	182	273	10.35	63	274	336	-8.46	29	337	365
-24	-41.23	92	182	273	10.05	62	274	335	-8.70	30	336	365
-25	-42.18	92	182	273	9.37	61	274	334	-9.40	31	335	365
-26	-43.12	92	182	273	8.86	59	274	332	-9.82	33	333	365
-27	-44.06	92	182	273	8.34	57	274	330	-10.23	35	331	365
-28	-45.00	92	182	273	7.81	55	274	328	-10.65	37	329	365
-29	-45.95	92	182	273	7.29	53	274	326	-11.08	39	327	365
-30	-46.89	92	182	273	6.76	51	274	324	-11.51	41	325	365
-31	-47.83	92	182	273	6.23	49	274	322	-11.94	43	323	365
-32	-48.77	92	182	273	5.69	47	274	320	-12.38	45	321	365
-33	-49.71	92	182	273	5.17	45	274	318	-12.83	47	319	365
-34	-50.65	92	182	273	4.63	43	274	316	-13.28	49	317	365
-35	-51.58	92	182	273	4.11	41	274	314	-13.74	51	315	365
-36	-52.51	92	182	273	3.57	39	274	312	-14.21	53	313	365
-37	-53.45	92	182	273	3.04	37	274	310	-14.68	55	311	365
-38	-54.38	92	182	273	2.59	34	274	307	-14.90	58	308	365
-39	-55.32	92	182	273	2.06	32	274	305	-15.39	60	306	365
-40	-56.25	92	182	273	1.58	29	274	302	-15.64	63	303	365
-41	-57.19	92	182	273	1.10	26	274	299	-15.91	66	300	365
-42	-58.12	92	182	273	0.63	22	274	295	-15.98	70	296	365
-43	-59.04	92	182	273	0.20	14	274	287	-15.28	78	288	365
-43.45	-59.46	92	182	273	0.00	7	274	280	-14.46	85	281	365

APPENDIX D

Optimum Tilt Angle β_{opt} (°) at High-Latitude Zone

Tables 74 to 79 give the daily optimum tilt angle $\beta_{opt,d}$ (°) at high-latitude zone. Tables 80 and 81 give the monthly optimum tilt angle $\beta_{opt,m}$ (°) at high-latitude zone. Then, by dividing the calendar year into 4 quarters and denoting the seasons by quarters as follows:

- Winter is January, February and March,
- Spring is April, May and June,
- Summer is July, August and September,
- Autumn is October, November and December.

The seasonally optimum tilt angle $\beta_{opt,s}$ (°), for different latitudes at the high-latitude zone, is given in the Tables 82 and 83.

Table 74. The daily optimum tilt angle $\beta_{opt,d}$ (°) for latitudes [43.45°N to 50°N]

	φ(°)							
Day number	**43.45**	**44**	**45**	**46**	**47**	**48**	**49**	**50**
1	71.7	72.1	73	73.8	74.7	75.5	76.3	77.1
10	70.8	71.2	72.1	72.9	73.8	74.6	75.5	76.3
17	69.7	70.2	71	71.9	72.7	73.6	74.4	75.3
20	69.1	69.6	70.5	71.3	72.2	73	73.9	74.7
30	66.8	67.3	68.2	69.1	69.9	70.8	71.7	72.6
32	66.3	66.7	67.6	68.5	69.4	70.3	71.2	72
41	63.4	63.9	64.8	65.7	66.7	67.6	68.5	69.4
47	61.2	61.7	62.6	63.6	64.5	65.4	66.4	67.3
51	59.6	60.1	61	62	62.9	63.9	64.8	65.7
59	56	56.5	57.5	58.4	59.4	60.3	61.3	62.3
60	55.5	56	57	57.9	58.9	59.9	60.8	61.8
69	50.8	51.3	52.3	53.3	54.3	55.2	56.2	57.2
75	47.3	47.8	48.8	49.8	50.8	51.8	52.8	53.8
79	44.9	45.4	46.4	47.4	48.4	49.4	50.4	51.4
89	38.5	39.1	40.1	41.1	42.1	43.1	44.1	45.1
91	37.3	37.8	38.8	39.8	40.8	41.8	42.8	43.8
100	31.7	32.2	33.2	34.2	35.2	36.2	37.2	38.2
105	28.6	29.2	30.2	31.2	32.2	33.2	34.2	35.2

continued on following page

Table 74. Continued

	φ(°)							
Day number	**43.45**	**44**	**45**	**46**	**47**	**48**	**49**	**50**
110	25.6	26.1	27.1	28.1	29.1	30.1	31.1	32.1
120	19.8	20.3	21.3	22.3	23.3	24.3	25.3	26.3
121	19.2	19.8	20.8	21.8	22.8	23.8	24.8	25.8
130	14.3	14.9	15.9	16.9	17.9	18.9	19.9	20.9
135	11.8	12.3	13.3	14.3	15.3	16.3	17.3	18.3
140	9.3	9.9	10.9	11.9	12.9	13.9	14.9	15.9
141	8.9	9.4	10.4	11.4	12.4	13.4	14.4	15.4
142	8.4	9	10	11	12	13	14	15
143	8	8.5	9.5	10.5	11.5	12.5	13.5	14.5
150	5	5.6	6.6	7.6	8.6	9.6	10.6	11.6
152	4.2	4.8	5.8	6.8	7.8	8.8	9.8	10.8
161	1.4	2	3	4	5	6	7	8
162	1.2	1.7	2.7	3.7	4.7	5.7	6.7	7.7
171	0	0.5	1.5	2.5	3.5	4.5	5.5	6.5
181	0.7	1.3	2.3	3.3	4.3	5.3	6.3	7.3
182	0.9	1.5	2.5	3.5	4.5	5.5	6.5	7.5
191	3.5	4	5	6	7	8	9	10
198	6.2	6.7	7.7	8.7	9.7	10.7	11.7	12.7
201	7.5	8	9	10	11	12	13	14
211	12.2	12.7	13.7	14.7	15.7	16.7	17.7	18.7
213	13.2	13.8	14.8	15.8	16.8	17.8	18.8	19.8
222	18	18.6	19.6	20.6	21.6	22.6	23.6	24.6
228	21.4	22	23	24	25	26	27	28
232	23.7	24.3	25.3	26.3	27.3	28.3	29.3	30.3
242	29.7	30.3	31.3	32.3	33.3	34.3	35.3	36.3
244	31	31.5	32.5	33.5	34.5	35.5	36.5	37.5
253	36.6	37.1	38.1	39.1	40.1	41.1	42.1	43.1
258	39.7	40.3	41.3	42.3	43.3	44.3	45.3	46.3
263	42.9	43.4	44.4	45.4	46.4	47.4	48.4	49.4
273	49	49.5	50.5	51.5	52.5	53.5	54.5	55.5
274	49.6	50.1	51.1	52.1	53.1	54.1	55.1	56
283	54.4	54.9	55.9	56.9	57.9	58.8	59.8	60.8
288	56.8	57.4	58.3	59.3	60.3	61.2	62.2	63.1
293	59.1	59.6	60.6	61.5	62.4	63.4	64.3	65.3
303	63	63.5	64.4	65.4	66.3	67.2	68.1	69
305	63.7	64.2	65.1	66	67	67.9	68.8	69.7

continued on following page

Table 74. Continued

	φ(°)							
Day number	**43.45**	**44**	**45**	**46**	**47**	**48**	**49**	**50**
314	66.5	67	67.9	68.8	69.6	70.5	71.4	72.3
318	67.5	68	68.9	69.8	70.7	71.5	72.4	73.3
324	68.9	69.4	70.2	71.1	72	72.8	73.7	74.5
334	70.6	71.1	71.9	72.8	73.6	74.5	75.3	76.1
335	70.8	71.2	72.1	72.9	73.8	74.6	75.4	76.3
344	71.7	72.1	73	73.8	74.6	75.5	76.3	77.1
354	72.1	72.5	73.4	74.2	75	75.9	76.7	77.5
365	71.8	72.6	73.9	74.7	75.6	76.4	77.2	78

Table 75. The daily optimum tilt angle $\beta_{opt,d}$ (°) for latitudes [51°N to 58°N]

	φ(°)							
Day number	**51**	**52**	**53**	**54**	**55**	**56**	**57**	**58**
1	78	78.8	79.6	80.4	81.2	81.9	82.7	83.5
10	77.1	77.9	78.7	79.6	80.4	81.2	82	82.7
17	76.1	76.9	77.8	78.6	79.4	80.2	81	81.8
20	75.6	76.4	77.3	78.1	78.9	79.7	80.5	81.3
30	73.4	74.3	75.1	76	76.9	77.7	78.5	79.4
32	72.9	73.8	74.6	75.5	76.4	77.2	78.1	78.9
41	70.3	71.2	72.1	72.9	73.8	74.7	75.6	76.5
47	68.2	69.1	70	70.9	71.8	72.7	73.6	74.5
51	66.7	67.6	68.5	69.4	70.4	71.3	72.2	73.1
59	63.2	64.2	65.1	66.1	67	68	68.9	69.9
60	62.7	63.7	64.7	65.6	66.6	67.5	68.5	69.4
69	58.2	59.2	60.2	61.2	62.1	63.1	64.1	65.1
70	57.7	58.6	59.6	60.6	61.6	62.6	63.6	64.6
75	54.8	55.8	56.8	57.8	58.8	59.8	60.8	61.8
79	52.4	53.4	54.4	55.4	56.4	57.4	58.4	59.4
89	46.1	47.1	48.1	49.1	50.1	51.1	52.1	53.1
91	44.8	45.8	46.8	47.8	48.8	49.8	50.8	51.8
100	39.2	40.2	41.2	42.2	43.2	44.2	45.2	46.2
105	36.2	37.2	38.2	39.2	40.2	41.2	42.2	43.2
110	33.1	34.1	35.1	36.1	37.1	38.1	39.1	40.1
120	27.3	28.3	29.3	30.3	31.3	32.3	33.3	34.3
121	26.8	27.8	28.8	29.8	30.8	31.8	32.8	33.8
130	21.9	22.9	23.9	24.9	25.9	26.9	27.9	28.9
135	19.3	20.3	21.3	22.3	23.3	24.3	25.3	26.3

continued on following page

Table 75. Continued

	φ(°)							
Day number	**51**	**52**	**53**	**54**	**55**	**56**	**57**	**58**
140	16.9	17.9	18.9	19.9	20.9	21.9	22.9	23.9
150	12.6	13.6	14.6	15.6	16.6	17.6	18.6	19.6
152	11.8	12.8	13.8	14.8	15.8	16.8	17.8	18.8
161	9	10	11	12	13	14	15	16
162	8.7	9.7	10.7	11.7	12.7	13.7	14.7	15.7
171	7.5	8.5	9.5	10.5	11.5	12.5	13.5	14.5
181	8.3	9.3	10.3	11.3	12.3	13.3	14.3	15.3
182	8.5	9.5	10.5	11.5	12.5	13.5	14.5	15.5
191	11	12	13	14	15	16	17	18
198	13.7	14.7	15.7	16.7	17.7	18.7	19.7	20.7
201	15	16	17	18	19	20	21	22
211	19.7	20.7	21.7	22.7	23.7	24.7	25.7	26.7
213	20.8	21.8	22.8	23.8	24.8	25.8	26.8	27.8
222	25.6	26.6	27.6	28.6	29.6	30.6	31.6	32.6
223	26.1	27.1	28.1	29.1	30.1	31.1	32.1	33.1
228	29	30	31	32	33	34	35	36
232	31.3	32.3	33.3	34.3	35.3	36.3	37.3	38.3
242	37.3	38.3	39.3	40.3	41.3	42.3	43.3	44.3
244	38.5	39.5	40.5	41.5	42.5	43.5	44.5	45.5
253	44.1	45.1	46.1	47.1	48.1	49.1	50.1	51.1
258	47.3	48.3	49.3	50.3	51.3	52.3	53.3	54.3
263	50.4	51.4	52.4	53.4	54.4	55.4	56.4	57.4
273	56.5	57.5	58.5	59.4	60.4	61.4	62.4	63.4
274	57	58	59	60	61	62	63	63.9
283	61.7	62.7	63.7	64.6	65.6	66.5	67.5	68.5
288	64.1	65	66	66.9	67.9	68.8	69.7	70.7
293	66.2	67.1	68.1	69	69.9	70.8	71.8	72.7
303	69.9	70.8	71.7	72.6	73.5	74.4	75.2	76.1
305	70.6	71.4	72.3	73.2	74.1	75	75.9	76.7
314	73.1	74	74.9	75.7	76.6	77.4	78.3	79.1
318	74.1	75	75.8	76.7	77.5	78.3	79.2	80
324	75.4	76.2	77.1	77.9	78.7	79.5	80.3	81.2
334	77	77.8	78.6	79.4	80.2	81	81.8	82.6
344	77.9	78.8	79.6	80.4	81.2	81.9	82.7	83.5
354	78.3	79.1	79.9	80.7	81.5	82.3	83.1	83.8
365	78.8	79.6	80.4	81.2	82	82.8	83.6	84.3

Table 76. The daily optimum tilt angle $\beta_{opt,d}$ (°) for latitudes [59°N to 66.45°N]

	φ(°)								
Day number	**59**	**60**	**61**	**62**	**63**	**64**	**65**	**66**	**66.45**
1	84.3	85	85.8	86.5	87.2	88	88.7	89.3	89.6
10	83.5	84.3	85	85.8	86.5	87.3	88	88.7	89
17	82.6	83.4	84.2	84.9	85.7	86.4	87.2	87.9	88.2
20	82.1	82.9	83.7	84.5	85.3	86	86.8	87.5	87.8
30	80.2	81	81.8	82.6	83.4	84.2	85	85.8	86.1
32	79.7	80.6	81.4	82.2	83	83.8	84.6	85.4	85.7
41	77.3	78.2	79	79.9	80.7	81.6	82.4	83.2	83.6
47	75.4	76.3	77.2	78.1	78.9	79.8	80.7	81.5	81.9
51	74	74.9	75.8	76.7	77.6	78.5	79.4	80.2	80.6
59	70.8	71.7	72.7	73.6	74.5	75.5	76.4	77.3	77.7
60	70.4	71.3	72.3	73.2	74.1	75.1	76	76.9	77.3
69	66.1	67	68	69	70	70.9	71.9	72.9	73.3
75	62.8	63.8	64.8	65.8	66.7	67.7	68.7	69.7	70.2
79	60.4	61.4	62.4	63.4	64.4	65.4	66.4	67.4	67.8
89	54.1	55.1	56.1	57.1	58.1	59.1	60.1	61.1	61.5
91	52.8	53.8	54.8	55.8	56.8	57.8	58.8	59.8	60.3
100	47.2	48.2	49.2	50.2	51.2	52.2	53.2	54.2	54.7
105	44.2	45.2	46.2	47.2	48.2	49.2	50.2	51.2	51.6
110	41.1	42.1	43.1	44.1	45.1	46.1	47.1	48.1	48.6
120	35.3	36.3	37.3	38.3	39.3	40.3	41.3	42.3	42.8
121	34.8	35.8	36.8	37.8	38.8	39.8	40.8	41.8	42.2
130	29.9	30.9	31.9	32.9	33.9	34.9	35.9	36.9	37.3
135	27.3	28.3	29.3	30.3	31.3	32.3	33.3	34.3	34.8
140	24.9	25.9	26.9	27.9	28.9	29.9	30.9	31.9	32.3
150	20.6	21.6	22.6	23.6	24.6	25.6	26.6	27.6	28
152	19.8	20.8	21.8	22.8	23.8	24.8	25.8	26.8	27.2
161	17	18	19	20	21	22	23	24	24.4
162	16.7	17.7	18.7	19.7	20.7	21.7	22.7	23.7	24.2
171	15.5	16.5	17.5	18.5	19.5	20.5	21.5	22.5	22.9
181	16.3	17.3	18.3	19.3	20.3	21.3	22.3	23.3	23.7
182	16.5	17.5	18.5	19.5	20.5	21.5	22.5	23.5	23.9
191	19	20	21	22	23	24	25	26	26.5
198	21.7	22.7	23.7	24.7	25.7	26.7	27.7	28.7	29.2
201	23	24	25	26	27	28	29	30	30.5
211	27.7	28.7	29.7	30.7	31.7	32.7	33.7	34.7	35.2
213	28.8	29.8	30.8	31.8	32.8	33.8	34.8	35.8	36.2

continued on following page

Table 76. Continued

	φ(°)								
Day number	**59**	**60**	**61**	**62**	**63**	**64**	**65**	**66**	**66.45**
222	33.6	34.6	35.6	36.6	37.6	38.6	39.6	40.6	41
228	37	38	39	40	41	42	43	44	44.4
232	39.3	40.3	41.3	42.3	43.3	44.3	45.3	46.3	46.7
242	45.3	46.3	47.3	48.3	49.3	50.3	51.3	52.3	52.7
244	46.5	47.5	48.5	49.5	50.5	51.5	52.5	53.5	54
253	52.1	53.1	54.1	55.1	56.1	57.1	58.1	59.1	59.6
258	55.3	56.3	57.3	58.3	59.3	60.3	61.3	62.3	62.7
263	58.4	59.4	60.4	61.4	62.4	63.4	64.4	65.4	65.9
273	64.4	65.4	66.4	67.3	68.3	69.3	70.3	71.3	71.7
274	64.9	65.9	66.9	67.9	68.9	69.8	70.8	71.8	72.2
283	69.4	70.4	71.3	72.3	73.2	74.1	75.1	76	76.4
288	71.6	72.5	73.5	74.4	75.3	76.2	77.1	78	78.4
293	73.6	74.5	75.4	76.3	77.2	78.1	79	79.8	80.2
303	77	77.9	78.7	79.6	80.4	81.3	82.1	82.9	83.3
305	77.6	78.4	79.3	80.1	81	81.8	82.6	83.4	83.8
314	79.9	80.8	81.6	82.4	83.2	84	84.8	85.5	85.9
318	80.8	81.6	82.4	83.2	84	84.8	85.6	86.3	86.6
324	82	82.7	83.5	84.3	85.1	85.8	86.6	87.3	87.6
334	83.4	84.2	84.9	85.7	86.4	87.1	87.9	88.6	88.9
335	83.5	84.3	85	85.8	86.5	87.2	88	88.7	89
344	84.3	85	85.8	86.5	87.2	87.9	88.6	89.3	89.6
354	84.6	85.3	86.1	86.8	87.5	88.2	88.9	89.6	89.9
365	85.1	85.8	86.6	87.3	88	88.7	89.4	89.7	89.7

Table 77. The daily optimum tilt angle $\beta_{opt,d}$ (°) for latitudes [43.45°S to 50°S]

	φ(°)							
Day number	**-43.45**	**-44**	**-45**	**-46**	**-47**	**-48**	**-49**	**-50**
1	-1.3	-1.9	-2.9	-3.9	-4.9	-5.9	-6.9	-7.9
10	-4.1	-4.6	-5.6	-6.6	-7.6	-8.6	-9.6	-10.6
17	-6.9	-7.4	-8.4	-9.4	-10.4	-11.4	-12.4	-13.4
20	-8.2	-8.7	-9.7	-10.7	-11.7	-12.7	-13.7	-14.7
30	-13	-13.6	-14.6	-15.6	-16.6	-17.6	-18.6	-19.6
32	-14.1	-14.6	-15.6	-16.6	-17.6	-18.6	-19.6	-20.6

continued on following page

Table 77. Continued

	φ(°)							
Day number	**-43.45**	**-44**	**-45**	**-46**	**-47**	**-48**	**-49**	**-50**
41	-18.9	-19.5	-20.5	-21.5	-22.5	-23.5	-24.5	-25.5
47	-22.4	-22.9	-23.9	-24.9	-25.9	-26.9	-27.9	-28.9
51	-24.7	-25.3	-26.3	-27.3	-28.3	-29.3	-30.3	-31.3
59	-29.5	-30.1	-31.1	-32.1	-33.1	-34.1	-35.1	-36.1
60	-30.1	-30.7	-31.7	-32.7	-33.7	-34.7	-35.7	-36.7
69	-35.7	-36.3	-37.3	-38.3	-39.3	-40.3	-41.3	-42.3
75	-39.5	-40	-41	-42	-43	-44	-45	-46
79	-42	-42.6	-43.6	-44.6	-45.6	-46.6	-47.6	-48.6
89	-48.2	-48.7	-49.7	-50.7	-51.7	-52.7	-53.7	-54.7
91	-49.3	-49.9	-50.9	-51.9	-52.9	-53.9	-54.8	-55.8
100	-54.2	-54.8	-55.7	-56.7	-57.7	-58.6	-59.6	-60.6
105	-56.7	-57.2	-58.2	-59.1	-60.1	-61	-62	-62.9
110	-58.9	-59.5	-60.4	-61.3	-62.3	-63.2	-64.2	-65.1
120	-62.9	-63.4	-64.3	-65.2	-66.1	-67.1	-68	-68.9
121	-63.2	-63.8	-64.7	-65.6	-66.5	-67.4	-68.3	-69.2
130	-66.1	-66.6	-67.5	-68.4	-69.3	-70.2	-71	-71.9
135	-67.4	-67.9	-68.8	-69.7	-70.6	-71.4	-72.3	-73.2
140	-68.6	-69.1	-70	-70.8	-71.7	-72.5	-73.4	-74.3
150	-70.4	-70.9	-71.7	-72.6	-73.4	-74.3	-75.1	-76
152	-70.7	-71.2	-72	-72.9	-73.7	-74.6	-75.4	-76.2
161	-71.6	-72.1	-73	-73.8	-74.6	-75.5	-76.3	-77.1
162	-71.7	-72.2	-73	-73.9	-74.7	-75.5	-76.3	-77.2
171	-72.1	-72.5	-73.4	-74.2	-75	-75.9	-76.7	-77.5
181	-71.8	-72.3	-73.2	-74	-74.8	-75.6	-76.5	-77.3
182	-71.8	-72.3	-73.1	-73.9	-74.8	-75.6	-76.4	-77.2
191	-71	-71.4	-72.3	-73.1	-74	-74.8	-75.6	-76.5
198	-70	-70.4	-71.3	-72.2	-73	-73.9	-74.7	-75.5
201	-69.4	-69.9	-70.8	-71.6	-72.5	-73.3	-74.2	-75
211	-67.2	-67.7	-68.6	-69.5	-70.3	-71.2	-72.1	-73
213	-66.7	-67.2	-68.1	-69	-69.8	-70.7	-71.6	-72.5
222	-64	-64.5	-65.4	-66.3	-67.2	-68.1	-69	-69.9
228	-61.8	-62.3	-63.3	-64.2	-65.1	-66	-67	-67.9
232	-60.3	-60.8	-61.7	-62.6	-63.6	-64.5	-65.4	-66.4
242	-55.8	-56.3	-57.3	-58.2	-59.2	-60.2	-61.1	-62.1
244	-54.8	-55.3	-56.3	-57.3	-58.2	-59.2	-60.2	-61.1
253	-50	-50.5	-51.5	-52.5	-53.5	-54.5	-55.5	-56.5

continued on following page

Table 77. Continued

	φ(°)							
Day number	**-43.45**	**-44**	**-45**	**-46**	**-47**	**-48**	**-49**	**-50**
258	-47.1	-47.6	-48.6	-49.6	-50.6	-51.6	-52.6	-53.6
263	-44	-44.6	-45.6	-46.6	-47.6	-48.6	-49.6	-50.6
273	-37.7	-38.2	-39.2	-40.2	-41.2	-42.2	-43.2	-44.2
274	-37.1	-37.6	-38.6	-39.6	-40.6	-41.6	-42.6	-43.6
283	-31.4	-32	-33	-34	-35	-36	-37	-38
288	-28.4	-28.9	-29.9	-30.9	-31.9	-32.9	-33.9	-34.9
293	-25.4	-25.9	-26.9	-27.9	-28.9	-29.9	-30.9	-31.9
303	-19.6	-20.1	-21.1	-22.1	-23.1	-24.1	-25.1	-26.1
305	-18.4	-19	-20	-21	-22	-23	-24	-25
314	-13.6	-14.1	-15.1	-16.1	-17.1	-18.1	-19.1	-20.1
318	-11.6	-12.1	-13.1	-14.1	-15.1	-16.1	-17.1	-18.1
324	-8.7	-9.2	-10.2	-11.2	-12.2	-13.2	-14.2	-15.2
334	-4.5	-5	-6	-7	-8	-9	-10	-11
335	-4.1	-4.7	-5.7	-6.7	-7.7	-8.7	-9.7	-10.7
344	-1.3	-1.9	-2.9	-3.9	-4.9	-5.9	-6.9	-7.9
354	0	-0.5	-1.5	-2.5	-3.5	-4.5	-5.5	-6.5
365	-1	-1.6	-2.6	-3.6	-4.6	-5.6	-6.6	-7.6

Table 78. The daily optimum tilt angle $\beta_{opt,d}$ (°) for latitudes [51°S to 58°S]

	φ(°)							
Day number	**-51**	**-52**	**-53**	**-54**	**-55**	**-56**	**-57**	**-58**
1	-8.9	-9.9	-10.9	-11.9	-12.9	-13.9	-14.9	-15.9
10	-11.6	-12.6	-13.6	-14.6	-15.6	-16.6	-17.6	-18.6
17	-14.4	-15.4	-16.4	-17.4	-18.4	-19.4	-20.4	-21.4
20	-15.7	-16.7	-17.7	-18.7	-19.7	-20.7	-21.7	-22.7
30	-20.6	-21.6	-22.6	-23.6	-24.6	-25.6	-26.6	-27.6
32	-21.6	-22.6	-23.6	-24.6	-25.6	-26.6	-27.6	-28.6
41	-26.5	-27.5	-28.5	-29.5	-30.5	-31.5	-32.5	-33.5
47	-29.9	-30.9	-31.9	-32.9	-33.9	-34.9	-35.9	-36.9
51	-32.3	-33.3	-34.3	-35.3	-36.3	-37.3	-38.3	-39.3
59	-37.1	-38.1	-39.1	-40.1	-41.1	-42.1	-43.1	-44.1
60	-37.7	-38.7	-39.7	-40.7	-41.7	-42.7	-43.7	-44.7
61	-38.3	-39.3	-40.3	-41.3	-42.3	-43.3	-44.3	-45.3
69	-43.3	-44.3	-45.3	-46.3	-47.3	-48.3	-49.3	-50.3

continued on following page

Table 78. Continued

	φ(°)							
Day number	**-51**	**-52**	**-53**	**-54**	**-55**	**-56**	**-57**	**-58**
75	-47	-48	-49	-50	-51	-52	-53	-54
79	-49.6	-50.6	-51.6	-52.6	-53.6	-54.6	-55.6	-56.6
89	-55.7	-56.7	-57.7	-58.7	-59.7	-60.7	-61.6	-62.6
91	-56.8	-57.8	-58.8	-59.8	-60.8	-61.8	-62.8	-63.7
100	-61.6	-62.5	-63.5	-64.4	-65.4	-66.4	-67.3	-68.3
105	-63.9	-64.8	-65.8	-66.7	-67.7	-68.6	-69.6	-70.5
110	-66	-67	-67.9	-68.8	-69.8	-70.7	-71.6	-72.5
120	-69.8	-70.7	-71.6	-72.5	-73.4	-74.2	-75.1	-76
121	-70.1	-71	-71.9	-72.8	-73.7	-74.6	-75.4	-76.3
130	-72.8	-73.7	-74.5	-75.4	-76.2	-77.1	-77.9	-78.8
135	-74	-74.9	-75.7	-76.6	-77.4	-78.3	-79.1	-79.9
140	-75.1	-76	-76.8	-77.6	-78.5	-79.3	-80.1	-80.9
150	-76.8	-77.6	-78.4	-79.2	-80.1	-80.9	-81.7	-82.4
161	-77.9	-78.7	-79.5	-80.3	-81.1	-81.9	-82.7	-83.5
171	-78.3	-79.1	-79.9	-80.7	-81.5	-82.3	-83.1	-83.8
181	-78.1	-78.9	-79.7	-80.5	-81.3	-82.1	-82.9	-83.6
182	-78.1	-78.9	-79.7	-80.5	-81.3	-82	-82.8	-83.6
191	-77.3	-78.1	-78.9	-79.7	-80.5	-81.3	-82.1	-82.9
198	-76.4	-77.2	-78	-78.8	-79.7	-80.5	-81.3	-82.1
201	-75.9	-76.7	-77.5	-78.4	-79.2	-80	-80.8	-81.6
211	-73.8	-74.7	-75.5	-76.4	-77.2	-78.1	-78.9	-79.7
213	-73.3	-74.2	-75.1	-75.9	-76.8	-77.6	-78.4	-79.3
222	-70.8	-71.7	-72.6	-73.5	-74.3	-75.2	-76.1	-76.9
228	-68.8	-69.7	-70.6	-71.5	-72.4	-73.3	-74.2	-75.1
232	-67.3	-68.2	-69.1	-70.1	-71	-71.9	-72.8	-73.7
242	-63	-64	-65	-65.9	-66.9	-67.8	-68.8	-69.7
244	-62.1	-63.1	-64	-65	-65.9	-66.9	-67.9	-68.8
253	-57.5	-58.4	-59.4	-60.4	-61.4	-62.4	-63.4	-64.4
258	-54.6	-55.6	-56.6	-57.6	-58.6	-59.6	-60.6	-61.6
263	-51.6	-52.6	-53.6	-54.6	-55.6	-56.6	-57.6	-58.6
273	-45.2	-46.2	-47.2	-48.2	-49.2	-50.2	-51.2	-52.2
274	-44.6	-45.6	-46.6	-47.6	-48.6	-49.6	-50.6	-51.6
283	-39	-40	-41	-42	-43	-44	-45	-46
288	-35.9	-36.9	-37.9	-38.9	-39.9	-40.9	-41.9	-42.9
293	-32.9	-33.9	-34.9	-35.9	-36.9	-37.9	-38.9	-39.9
303	-27.1	-28.1	-29.1	-30.1	-31.1	-32.1	-33.1	-34.1
305	-26	-27	-28	-29	-30	-31	-32	-33

continued on following page

Table 78. Continued

	φ(°)							
Day number	**-51**	**-52**	**-53**	**-54**	**-55**	**-56**	**-57**	**-58**
314	-21.1	-22.1	-23.1	-24.1	-25.1	-26.1	-27.1	-28.1
318	-19.1	-20.1	-21.1	-22.1	-23.1	-24.1	-25.1	-26.1
324	-16.2	-17.2	-18.2	-19.2	-20.2	-21.2	-22.2	-23.2
334	-12	-13	-14	-15	-16	-17	-18	-19
335	-11.7	-12.7	-13.7	-14.7	-15.7	-16.7	-17.7	-18.7
344	-8.9	-9.9	-10.9	-11.9	-12.9	-13.9	-14.9	-15.9
354	-7.5	-8.5	-9.5	-10.5	-11.5	-12.5	-13.5	-14.5
365	-8.6	-9.6	-10.6	-11.6	-12.6	-13.6	-14.6	-15.6

Table 79. The daily optimum tilt angle $\beta_{opt,d}$ (o) for latitudes [59oS to 66.45oS]

	φ(o)								
Day number	**-59**	**-60**	**-61**	**-62**	**-63**	**-64**	**-65**	**-66**	**-66.45**
1	-16.9	-17.9	-18.9	-19.9	-20.9	-21.9	-22.9	-23.9	-24.3
10	-19.6	-20.6	-21.6	-22.6	-23.6	-24.6	-25.6	-26.6	-27.1
17	-22.4	-23.4	-24.4	-25.4	-26.4	-27.4	-28.4	-29.4	-29.9
20	-23.7	-24.7	-25.7	-26.7	-27.7	-28.7	-29.7	-30.7	-31.2
30	-28.6	-29.6	-30.6	-31.6	-32.6	-33.6	-34.6	-35.6	-36
32	-29.6	-30.6	-31.6	-32.6	-33.6	-34.6	-35.6	-36.6	-37.1
41	-34.5	-35.5	-36.5	-37.5	-38.5	-39.5	-40.5	-41.5	-41.9
47	-37.9	-38.9	-39.9	-40.9	-41.9	-42.9	-43.9	-44.9	-45.4
51	-40.3	-41.3	-42.3	-43.3	-44.3	-45.3	-46.3	-47.3	-47.7
59	-45.1	-46.1	-47.1	-48.1	-49.1	-50.1	-51.1	-52.1	-52.5
60	-45.7	-46.7	-47.7	-48.7	-49.7	-50.7	-51.7	-52.7	-53.1
69	-51.3	-52.3	-53.3	-54.3	-55.3	-56.3	-57.3	-58.3	-58.7
75	-55	-56	-57	-58	-59	-60	-61	-62	-62.5
79	-57.6	-58.6	-59.6	-60.6	-61.6	-62.6	-63.6	-64.6	-65
89	-63.6	-64.6	-65.6	-66.6	-67.6	-68.6	-69.6	-70.5	-71
91	-64.7	-65.7	-66.7	-67.7	-68.7	-69.6	-70.6	-71.6	-72
100	-69.2	-70.2	-71.1	-72.1	-73	-74	-74.9	-75.9	-76.3
105	-71.4	-72.4	-73.3	-74.2	-75.2	-76.1	-77	-77.9	-78.3
110	-73.4	-74.4	-75.3	-76.2	-77.1	-78	-78.8	-79.7	-80.1
120	-76.9	-77.7	-78.6	-79.5	-80.3	-81.2	-82	-82.8	-83.2
121	-77.2	-78	-78.9	-79.8	-80.6	-81.4	-82.3	-83.1	-83.4
130	-79.6	-80.4	-81.3	-82.1	-82.9	-83.7	-84.5	-85.2	-85.6

continued on following page

Table 79. Continued

	φ(o)								
Day number	**-59**	**-60**	**-61**	**-62**	**-63**	**-64**	**-65**	**-66**	**-66.45**
135	-80.7	-81.6	-82.4	-83.2	-83.9	-84.7	-85.5	-86.2	-86.6
140	-81.7	-82.5	-83.3	-84.1	-84.9	-85.6	-86.4	-87.1	-87.4
150	-83.2	-84	-84.8	-85.5	-86.3	-87	-87.7	-88.4	-88.7
152	-83.5	-84.2	-85	-85.7	-86.5	-87.2	-87.9	-88.6	-88.9
161	-84.2	-85	-85.7	-86.5	-87.2	-87.9	-88.6	-89.3	-89.6
162	-84.3	-85.1	-85.8	-86.5	-87.3	-88	-88.7	-89.4	-89.7
171	-84.6	-85.3	-86.1	-86.8	-87.5	-88.2	-88.9	-89.6	-89.9
181	-84.4	-85.2	-85.9	-86.6	-87.4	-88.1	-88.8	-89.5	-89.8
182	-84.4	-85.1	-85.9	-86.6	-87.3	-88	-88.7	-89.4	-89.7
191	-83.7	-84.4	-85.2	-86	-86.7	-87.4	-88.1	-88.8	-89.1
198	-82.8	-83.6	-84.4	-85.2	-85.9	-86.7	-87.4	-88.1	-88.4
201	-82.4	-83.2	-84	-84.7	-85.5	-86.2	-87	-87.7	-88
211	-80.6	-81.4	-82.2	-83	-83.8	-84.5	-85.3	-86.1	-86.4
213	-80.1	-80.9	-81.7	-82.6	-83.4	-84.1	-84.9	-85.7	-86
222	-77.8	-78.7	-79.5	-80.3	-81.2	-82	-82.8	-83.6	-84
228	-76	-76.8	-77.7	-78.6	-79.5	-80.3	-81.2	-82	-82.4
232	-74.6	-75.5	-76.4	-77.3	-78.2	-79	-79.9	-80.8	-81.1
242	-70.6	-71.6	-72.5	-73.5	-74.4	-75.3	-76.2	-77.2	-77.6
244	-69.8	-70.7	-71.7	-72.6	-73.5	-74.5	-75.4	-76.3	-76.8
253	-65.3	-66.3	-67.3	-68.3	-69.3	-70.2	-71.2	-72.2	-72.6
258	-62.6	-63.6	-64.5	-65.5	-66.5	-67.5	-68.5	-69.5	-70
263	-59.6	-60.6	-61.6	-62.6	-63.6	-64.6	-65.5	-66.5	-67
273	-53.2	-54.2	-55.2	-56.2	-57.2	-58.2	-59.2	-60.2	-60.7
274	-52.6	-53.6	-54.6	-55.6	-56.6	-57.6	-58.6	-59.6	-60.1
283	-47	-48	-49	-50	-51	-52	-53	-54	-54.4
288	-43.9	-44.9	-45.9	-46.9	-47.9	-48.9	-49.9	-50.9	-51.4
293	-40.9	-41.9	-42.9	-43.9	-44.9	-45.9	-46.9	-47.9	-48.4
303	-35.1	-36.1	-37.1	-38.1	-39.1	-40.1	-41.1	-42.1	-42.6
305	-34	-35	-36	-37	-38	-39	-40	-41	-41.4
314	-29.1	-30.1	-31.1	-32.1	-33.1	-34.1	-35.1	-36.1	-36.6
318	-27.1	-28.1	-29.1	-30.1	-31.1	-32.1	-33.1	-34.1	-34.6
324	-24.2	-25.2	-26.2	-27.2	-28.2	-29.2	-30.2	-31.2	-31.7
334	-20	-21	-22	-23	-24	-25	-26	-27	-27.5
335	-19.7	-20.7	-21.7	-22.7	-23.7	-24.7	-25.7	-26.7	-27.1
344	-16.9	-17.9	-18.9	-19.9	-20.9	-21.9	-22.9	-23.9	-24.3
354	-15.5	-16.5	-17.5	-18.5	-19.5	-20.5	-21.5	-22.5	-22.9
365	-16.6	-17.6	-18.6	-19.6	-20.6	-21.6	-22.6	-23.6	-24

Table 80. The monthly optimum tilt angle $\beta_{opt,m}$ (°) in NH

φ(°)	Jan.	Feb.	March	April	May	June	July	Aug.	Sep.	Oct.	Nov.	Dec.
43.45	69.59	61.55	47.07	28.38	11.50	1.14	6.12	21.54	40.04	57.01	67.64	71.74
44	70.07	62.05	47.61	28.93	12.04	1.67	6.67	22.11	40.59	57.53	68.13	72.23
45	70.93	62.98	48.60	29.93	13.04	2.67	7.67	23.11	41.58	58.49	68.99	73.12
46	71.79	63.91	49.59	30.93	14.04	3.67	8.67	24.11	42.58	59.45	69.88	73.95
47	72.64	64.83	50.58	31.93	15.04	4.67	9.67	25.11	43.58	60.41	70.76	74.78
48	73.50	65.74	51.58	32.93	16.04	5.67	10.67	26.11	44.58	61.35	71.62	75.62
49	74.35	66.66	52.57	33.93	17.04	6.67	11.67	27.11	45.58	62.30	72.49	76.45
50	75.18	67.58	53.55	34.93	18.04	7.67	12.67	28.11	46.57	63.25	73.35	77.26
51	76.02	68.49	54.55	35.93	19.04	8.67	13.67	29.11	47.57	64.19	74.21	78.07
52	76.85	69.41	55.54	36.93	20.04	9.67	14.67	30.11	48.57	65.14	75.06	78.87
53	77.68	70.32	56.52	37.93	21.04	10.67	15.67	31.11	49.57	66.09	75.91	79.69
54	78.51	71.22	57.52	38.93	22.04	11.67	16.67	32.11	50.56	67.03	76.75	80.49
55	79.33	72.12	58.50	39.93	23.04	12.67	17.67	33.11	51.56	67.96	77.59	81.29
56	80.15	73.02	59.49	40.93	24.04	13.67	18.67	34.11	52.56	68.90	78.43	82.07
57	80.95	73.92	60.47	41.93	25.04	14.67	19.67	35.11	53.56	69.84	79.26	82.85
58	81.73	74.80	61.47	42.93	26.04	15.67	20.67	36.11	54.56	70.76	80.08	83.61
59	82.53	75.69	62.45	43.93	27.04	16.67	21.67	37.11	55.56	71.70	80.90	84.39
60	83.31	76.58	63.43	44.93	28.04	17.67	22.67	38.11	56.56	72.62	81.70	85.13
61	84.08	77.45	64.42	45.93	29.04	18.67	23.67	39.11	57.56	73.54	82.51	85.89
62	84.86	78.31	65.40	46.93	30.04	19.67	24.67	40.11	58.55	74.47	83.30	86.62
63	85.61	79.18	66.38	47.93	31.04	20.67	25.67	41.11	59.55	75.37	84.08	87.34
64	86.36	80.04	67.36	48.93	32.04	21.67	26.67	42.11	60.55	76.29	84.86	88.05
65	87.10	80.90	68.35	49.93	33.04	22.67	27.67	43.11	61.55	77.18	85.63	88.75
66	87.82	81.73	69.33	50.93	34.04	23.67	28.67	44.11	62.55	78.09	86.37	89.41
66.45	88.14	82.11	69.75	51.38	34.50	24.13	29.12	44.54	63.01	78.48	86.72	89.68

Table 81. The monthly optimum tilt angle $\beta_{opt,m}$ (^{o}) in SH

φ(°)	Jan.	Feb.	March	April	May	June	July	Aug.	Sep.	Oct.	Nov.	Dec.
-43.45	-6.8	-22	-40	-57	-67	-72	-70	-62	-47	-28	-11	-1.4
-44	-7.3	-22	-40	-57	-68	-72	-70	-62	-47	-28	-12	-1.6
-45	-8.3	-23	-41	-58	-69	-73	-71	-63	-48	-29	-13	-2.6
-46	-9.3	-24	-42	-59	-70	-74	-72	-64	-49	-30	-14	-3.6
-47	-10	-25	-43	-60	-71	-75	-73	-65	-50	-31	-15	-4.6
-48	-11	-26	-44	-61	-71	-76	-74	-66	-51	-32	-16	-5.6
-49	-12	-27	-45	-62	-72	-76	-75	-67	-52	-33	-17	-6.6
-50	-13	-28	-46	-63	-73	-77	-75.	-68	-53	-34	-18	-7.6
-51	-14	-29	-47	-64	-74	-78	-76	-68	-54	-35	-19	-8.6
-52	-15	-30	-48	-65	-75	-79	-77	-69	-55	-36	-20	-9.6
-53	-16	-31	-49	-66	-76	-80	-78	-70	-56	-37	-21	-11
-54	-17	-32	-50	-67	-77	-80	-79	-71	-57	-38	-22	-12
-55	-18	-33	-51	-68	-77	-81	-80	-72	-58	-39	-23	-13
-56	-19	-34	-52	-69	-78	-82	-80	-73	-59	-40	-24	-14
-57	-20	-35	-53	-69	-79	-83	-81	-74	-60	-41	-25	-15
-58	-21	-36	-54	-70	-80	-84	-82	-75	-61	-42	-26	-16
-59	-22	-37	-55	-71	-81	-84	-83	-76	-62	-43	-27	-17
-60	-23	-38	-56	-72	-82	-85	-84	-77	-63	-44	-28	-18
-61	-24	-39	-57	-73	-82	-86	-84	-77	-64	-45	-29	-19
-62	-25	-40	-58	-74	-83	-87	-85	-78	-65	-46	-30	-20
-63	-26	-41	-59	-75	-84	-87	-86	-79	-66	-47	-31	-21
-64	-27	-42	-60	-76	-85	-88	-87	-80	-67	-48	-32	-22
-65	-28	-43	-61	-77	-85	-89	-87	-81	-68	-49	-33	-23
-66	-29	-44	-62	-78	-86	-89	-88	-82	-69	-50	-34	-24
-66.45	-30	-45	-62	-78	-87	-90	-88	-82	-69	-51	-34	-24

Table 82. The Seasonally optimum tilt angle $\beta_{opt,s}$ (°) in NH

φ(°)	Winter	Spring	Summer	Autumn
43.45	59.33	13.65	22.38	65.44
44	59.84	14.19	22.93	65.94
45	60.77	15.19	23.93	66.84
46	61.69	16.19	24.93	67.74
47	62.61	17.19	25.93	68.63
48	63.53	18.19	26.93	69.51
49	64.45	19.19	27.93	70.39
50	65.37	20.19	28.93	71.27
51	66.28	21.19	29.93	72.13
52	67.19	22.19	30.93	73.00
53	68.10	23.19	31.92	73.87
54	69.01	24.19	32.92	74.73
55	69.91	25.19	33.92	75.59
56	70.81	26.19	34.92	76.45
57	71.71	27.19	35.92	77.30
58	72.59	28.19	36.92	78.13
59	73.48	29.19	37.92	78.97
60	74.37	30.19	38.92	79.80
61	75.25	31.19	39.92	80.63
62	76.12	32.19	40.92	81.44
63	76.99	33.19	41.92	82.24
64	77.85	34.19	42.92	83.04
65	78.71	35.19	43.92	83.83
66	79.56	36.19	44.92	84.60
66.45	79.93	36.65	45.37	84.94

Table 83. The Seasonally optimum tilt angle $\beta_{opt,s}$ (°) in SH

φ(°)	Winter	Spring	Summer	Autumn
-43.45	-22.65	-65.27	-59.46	-14.46
-44	-23.20	-65.77	-59.97	-13.90
-45	-24.20	-66.66	-60.90	-14.90
-46	-25.20	-67.55	-61.82	-15.90
-47	-26.20	-68.44	-62.74	-16.90
-48	-27.20	-69.32	-63.66	-17.90
-49	-28.20	-70.21	-64.57	-18.90
-50	-29.20	-71.08	-65.49	-19.90
-51	-30.20	-71.96	-66.41	-20.90
-52	-31.20	-72.83	-67.31	-21.90
-53	-32.20	-73.70	-68.22	-22.90
-54	-33.20	-74.55	-69.12	-23.90
-55	-34.20	-75.42	-70.03	-24.90
-56	-35.20	-76.28	-70.92	-25.90
-57	-36.20	-77.13	-71.82	-26.90
-58	-37.20	-77.96	-72.71	-27.90
-59	-38.20	-78.80	-73.60	-28.90
-60	-39.20	-79.63	-74.47	-29.90
-61	-40.20	-80.46	-75.35	-30.90
-62	-41.20	-81.27	-76.23	-31.90
-63	-42.20	-82.09	-77.09	-32.90
-64	-43.20	-82.89	-77.94	-33.90
-65	-44.20	-83.68	-78.80	-34.90
-66	-45.20	-84.47	-79.65	-35.90
-66.45	-45.64	-84.81	-80.03	-36.36

Chapter 6
Solar Tracking

ABSTRACT

Basing on the provided information in the first to four chapters where it was clearly demonstrated that, the maximum utilization of solar energy depends upon determining the exact location of the sun position. By proper calculation and computer programs, the solar path for any geographical location can be tracked. Thus, a clear idea about the prospects of solar energy for any location can be obtained and accordingly decisions and, measures can be adopted for harnessing solar energy in that area. Basing on this information a general formula for on-axis sun-tracking system has been derived using coordinate transformation method. The derived sun-tracking formula is the most general form of mathematical solution for various kinds of arbitrarily oriented on-axis sun tracker, where azimuth-elevation and tilt-roll tracking formulas are specific cases. The application of the general formula is to improve the sun tracking accuracy because the misalignment of solar collector from an ideal azimuth-elevation or tilt-roll tracking during the installation can be corrected by a straightforward application of the general formula. Moreover, the rotation angle in a single axis tracker is calculated. The calculated rotation angle can be used to determine the number of motor revolutions to move the tracker to its optimum position.

INTRODUCTION

Photovoltaic energy involves the conversion of Sunlight into electricity. The efficiency of converting radiant solar energy into electrical energy is the critical point that influences the choice of solar energy as a form of alternative energy. The energy generated from PV panels is related with temperature, irradiance and incident angle of the solar radiation and so on. There are two ways to improve the PV technology performance. One is to use different materials or add other dopants to manufacture the PV modules. The other one is to use a tracker as the device for orienting a solar PV module toward the Sun. In order to achieve the highest conversion efficiency, the Sun light has to impinge the module surface perpendicularly. The earth not only has one year rotational motion around the Sun, but also a daily motion around its own axes. Therefore, different kinds of solar tracking could be used. Here, one can mention two types of tracking:

DOI: 10.4018/978-1-5225-2950-7.ch006

- Long term tracking, where the most common ones are those related to daily, fortnightly, seasonally, half-yearly and yearly tracking by determining the optimum solar collector orientation according to these periods. This kind of tracking is treated in details in Chapter 5 and this kind of tracking is widely in use all over the world. Moreover, it is common practice for traditional photovoltaic (PV) modules to be installed at a fixed angle in a wide and flat area so as to generate the maximum amount of power. In this case the optimum tilt angle and azimuth angle should be determined. There are many techniques developed to determine the optimum tilt angles for different latitudes and surface azimuth angles. Soulayman and Hammoud (2016) presented a modified general algorithm to optimize the tilt angle for mid-latitude zone. Stanci and Stanci (2014) proposed an equation to optimize the tilt angle at latitudes from 0° to 80°. Benghanem (2011) proposed a method to calculate the optimum tilt angle in Madinah, Saudi Arabia. Yan, Saha, Meredith and Goodwin (2013) determined the optimum tilt angle and orientation in Brisbane, Australia. Furthermore, Optimum tilt angles for other areas such as Carbondale, Illinois, USA (Gong and Kulkarni, 2005), Sanliurfa, Turkey (Kacira, Simsek, Babur and Demirkol, 2004), Taipei, Taiwan (Chang, 2009a), Burgos, Spain (De Miguel, Bilbao and Diez, 1995), etc. have been reported. However, the fixed tilt PV system using yearly optimum tilt angle has relatively lower efficiency, but daily, weekly, fortnightly, monthly and even seasonally adjustment of the tilt angle can increase the efficiency of a solar system remarkably.
- Short term tracking, where a device for orienting a solar PV module toward the Sun during the day is used. The efficiency of a PV panel can easily be increased by sun tracking systems which are investigated by many researches. The generated power is directly proportional with the collected solar radiation in a solar system. Maximum sun power collection is possible by adjusting solar system position with respect to the Sun's location. This adjustment can be realized more easily with two axes Sun tracking systems than single ones which is cheaper and simpler to design. The tracking systems include closed and open loop control mechanisms. The PV panels are positioned by the help of photo sensors and feedback controllers. The disadvantage of the closed loop system is that the system spends more energy than the generated one in the case of quick weather changes. The open loop one is based on calculations of the seasonal weather and the sun position. The hybrid control is made up of both closed and open loop tracking systems (Seme and Stumberger, 2011).

The different aspects of short term tracking will be treated in this chapter.

TRACKING PHILOSOPHY

A solar tracking system orients solar PV modules, reflectors, lenses or other optical devices toward the Sun at all times that the Sun is visible in the sky. It shows better effectiveness compared to modules in fixed positions. In standard photovoltaic applications, it was predicted in 2008-2009 that trackers could be used in at least 85% of commercial installations greater than 1MW from 2009 to 2012. Sun-tracking system plays an important role in the development of solar energy applications, especially for the high solar concentration systems that directly convert the solar energy into thermal or electrical energy.

In the 1978, Neville. (1978) examined the maximum solar energy available to an Earth-surface collector as a function of latitude, the north-south tilt of the collector from the Earth's surface (θ), and whether the collector is an ideal tracker (follows the Sun both north-south and east-west), an east-west tracker (follows the Sun east-west but is fixed in the north-south direction) or a fixed type. It is shown that the ideal tracker gives maximum potentially available energy, the use of an east-west tracking device results in 5–10 per cent degradation in potential performance, while performance of fixed collectors is degraded by close to 50 per cent.

There are number of studies showing that tracking systems enable significant amount of solar energy compared to fixed systems. Nann (1990) derived the irradiance received and the energy costs for tracking photovoltaic systems and V-trough concentrators relative to the costs of a fixed system. Radiation data from 28 sites, especially in moderate climates, are investigated with regard to the surplus of radiation achieved with one- and two-axes tracking surfaces. By use of the Perez model the fraction of the supplementary irradiance is computed. A comprehensive computer code describing the optical performance of a low concentrating V-trough was developed. It is used to calculate the annual average concentration for diffuse and direct irradiance under different climatic conditions. Applying a life-cycle cost analysis relative to the costs for a fixed array, the results are freed of political-economic assumptions (inflation or discount rate). The resulting relative energy costs are dependent mainly on module costs. A sensitivity analysis shows that the results are stable against variations in meteorological data, balance-of-system or module efficiencies, and power related balance-of-system costs. The most economical system turns out to be the one-axis tracking PV system for moderate climates. It promises cost advantages for module costs even down to 250 $\$/m^2$ if the tracker costs do not exceed the area-related balance-of-system costs of fixed systems by more than 100%.

Tomson (2008) reported increasing of seasonal energy yield by 10-20% if using the two-positional tracking system that positions collectors in the morning and in the afternoon. A very detailed review of energy gain of different trackers is done in Mousazadeh, Keyhani, Javadi, Mobli, Abrinia and Sharifi (2009). In that paper the authors report a boost of collected solar energy by means of a tracking system in the range of 10-100% depending on different time periods and geographical conditions. However, Sun-tracking systems are quite expensive and energy intensive. Therefore, Sun tracking is quite cumbersome and inconvenient practically and Sun trackers are not recommended for use with small solar panels (see Mousazadeh, Keyhani, Javadi, Mobli, Abrinia and Sharifi (2009) and Tomson (2008)).

Kim (2007) proposed a robust maximum power point tracker (MPPT) using sliding mode controller for the three-phase grid-connected photovoltaic system has been. The proposed system consists of MPPT controller and current controller for tight regulation of the current. The proposed MPPT controller generates current reference directly from the solar array power information and the current controller uses the integral sliding mode for the tight control of current. The proposed system can prevent the current overshoot and provide optimal design for the system components. The structure of the proposed system is simple, and it shows robust tracking property against modeling uncertainties and parameter variations. One way of Sun tracking is flat PV system. Two axis tracking PV theoretically propose 41% energy entrance improvement in a mid-latitude region with respect to the fixed PV panel. Besides, the improvement for a one-axis tracking system is 36%. Captured solar energy by Sun tracking system is related with region and meteorological conditions (Clifford and Eastwood, 2004). Many times, the improvement of a solar system is achieved using tracking controller structures by increasing the collected solar energy ratio. The tracking systems include microprocessors or other control mechanisms to determine the best position. The widespread samples of tracking systems are dual and single axis tracking (Clifford and Eastwood, 2004) which cannot exactly determine the best position for efficient collection of the solar energy. Two-axis tracking systems need two motors to follow the Sun in the azimuth and tilt angles.

Although they are more expensive, they propose better performance with respect to the single-axis systems even in the morning and evening (see for example, Mavromatakis and Franghiadakis (2008) and Seme and Stumberger (2011)).

A Sun tracking system with microprocessors does not need sensors since it estimates the best location via the written algorithm. It moves to the best location by step motors or optical encoders. The systems with microcontrollers can be used to direct arrays of sun tracking systems. The self-calibration of the microprocessor controlled trackers is possible by electro-optic sensors (Edwards, 1978) or a search routine without a sensor (Maish, 1988). Although the microprocessor based trackers provide good Sun light collection, it is hard to realize the detailed installation conditions.

Beside this, using two axes Sun tracking system is inevitable in some solar system and it can be adapted to all solar systems to improve collection of solar radiation (Abdallah and Nijmeh, 2004). It was found experimentally that, the measured collected solar energy on the moving surface was significantly larger than that on a fixed surface. The two axes tracking surface showed a better performance with an increase in the collected energy of up to 41.34% compared with the fixed surface tilted at 32° towards the south (Abdallah and Nijmeh, 2004).

A sun tracking system with microprocessors does not need sensors since it estimates the best location via the written algorithm. It moves to the best location by step motors or optical encoders. The systems with microcontrollers can be used to direct arrays of sun tracking systems. The self-calibration of the microprocessor controlled trackers is possible by electro-optic sensors (Edwards, 1978) or a search routine without a sensor (Maish, 1988). Although the microprocessor based trackers provide good Sun light collection, it is hard to realize the detailed installation conditions.

The performance of the electro-optical tracking systems such as Sun sensor composed of four photo resistors with cylindrical shades is better depending on good weather conditions (Rumala, 1986). This system is controlled by differential amplifiers, comparators and output components. Another electro-optical system with fixed and moving sensors is suggested better accuracy and reliability (Lynch and Salameh, 1990).

There are many types of solar trackers. Photovoltaic trackers can be classified into two types: standard photovoltaic (PV) trackers and concentrated photovoltaic (CPV) trackers. Each of these tracker types can be further categorized by the number and orientation of their axes, their actuation architecture and drive type, their intended applications, their vertical supports and foundation. The most commonly used Sun-tracking systems for short time tracking are the passive and active solar trackers. The passive tracking system uses, usually, the open-loop approach while the active tracking system uses, usually, the closed-loop approach. For the passive tracking system, the tracker will perform calculation to identify the Sun's position and to determine the rotational angles of the two tracking axes using a specific Sun-tracking formula in order to drive the solar collector towards the Sun. On the other hand, for the active tracking system, the Sun tracker normally will sense the direct solar radiation falling on a photo-sensor as a feedback signal to ensure that the solar collector is tracking the sun all the time. Figure 1 shows the scheme of a single-axis passive solar tracker while Figure 2 shows the scheme of a single-axis active tracker. The active and passive dual-axis trackers are also widely available. Passive tracker only meets the requirement of the change in day and night but not the seasons while active dual-axis tracker meets the requirement of the change in day and night and yearly revolution of the Earth around the Sun. Moreover, passive trackers use the Sun's heat to move liquid from side to side inside the tracker, allowing gravity to turn it and follow the Sun, using no motors, no gears and no controls to fail. Dual axis trackers have two degrees of freedom that act as axes of rotation. These axes are typically normal to one

Figure 1. The scheme of a passive single-axis solar tracker

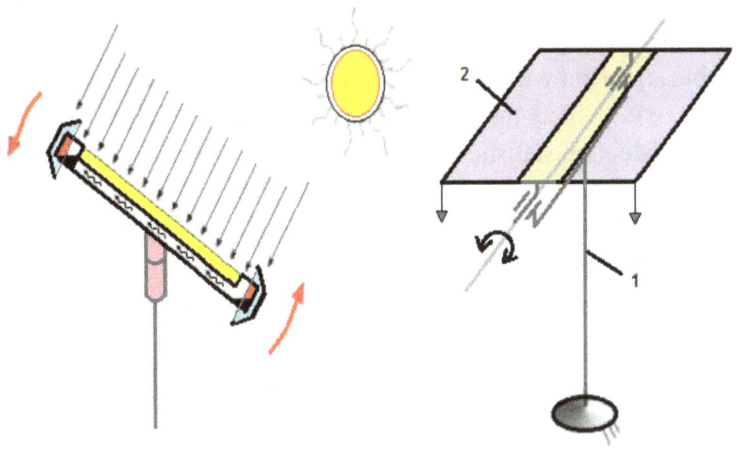

Figure 2. The scheme of an active single-axis solar tracker

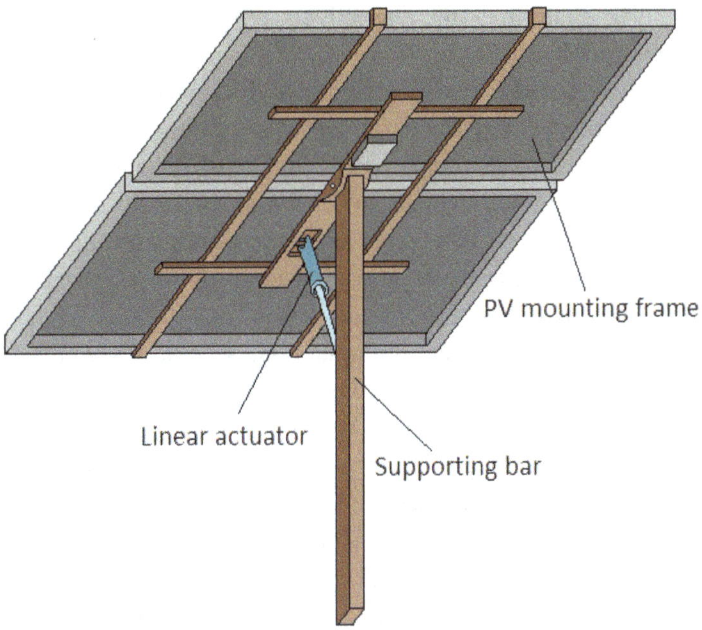

another. The axis that is fixed with respect to the ground can be considered a primary axis. The axis that is referenced to the primary axis can be considered a secondary axis. Moreover, dual axis trackers allow for optimum solar energy levels due to their ability to follow the Sun vertically and horizontally. No matter where the Sun is in the sky, dual axis trackers are able to angle themselves to be in direct contact with the Sun. Dual axis trackers use motors and gear trains to direct the tracker as commanded by a controller responding to the solar direction. In order to control and manage the movement of these massive structures special slewing drives are designed and rigorously tested. The technologies used to direct the tracker are constantly evolving and recent developments. Instead of the above options, some authors have also designed a hybrid system that contains both the active and passive tracking system to achieve a good tracking accuracy (Poulek and Libra, 1998).

There is considerable argument within the industry whether the small difference in yearly collection between single and dual-axis trackers makes the added complexity of a two-axis tracker worthwhile. A recent review of actual production statistics from southern Ontario suggested the difference was about 4% in total, which was far less than the added costs of the dual-axis systems. This compares unfavorably with the 24-32% improvement between a fixed-array and single-axis tracker.

An experimental study was performed by Abdallah (2004) to investigate the effect of using different types of Sun tracking systems on the voltage–current characteristics and electrical power generation at the output of flat plate photovoltaics (FPPV). Four electromechanical Sun tracking systems, two axes, one axis vertical, one axis east–west and one axis north–south, were designed and constructed for the purpose of investigating the effect of tracking on the electrical values, current, voltage and power, according to the different loads (variable resistance). The above mentioned variables were measured at the output of the FPPV and compared with those on a fixed surface. The results indicated that the volt–ampere characteristics on the tracking surfaces were significantly greater than that on a fixed surface. There were increases of electrical power gain up to 43.87%, 37.53%, 34.43% and 15.69% for the two axes, east–west, vertical and north–south tracking, respectively, as compared with the fixed surface inclined 32° to the south in Amman, Jordan.

The properly installed fixed panel is also operating less efficient at some point compared to tracking panel due to Sun's motion on daily and yearly basis. Therefore, employing a tracking panel will generate higher electricity and more efficient as compared to fixed panel (Chang, 2009a). In order to increase the power output, a Sun-tracking system which can orient the PV modules facing the Sun is developed to receive the maximum irradiation and then generate the maximum power output. Thus, solar trackers are devices used to orient photovoltaic panels, reflectors, lenses or other optical devices toward the Sun. Since the Sun's position in the sky changes with the seasons and the time of day, trackers are used to align the collection system to maximize energy production. Several factors must be considered when determining the use of trackers. Some of these include: the solar technology being used, the amount of direct solar irradiation, feed-in tariffs in the region where the system is deployed, and the cost to install and maintain the trackers. The Sun-tracking system can be classified into a single-axis Sun tracker and a dual-axis Sun tracker. The single-axis Sun tracker only tracks the Sun in one direction. The most of the single-axis Sun tracker only tracks the Sun from east to west, which is the azimuthal direction. Consequently, the tilt angle with respect to the horizontal is constant and optimized for the local latitude. The dual-axis Sun tracker can track the Sun in azimuthal and south- north (altitude) directions, which ensures the PV modules precisely follow the motions of the Sun. Concentrated applications like concentrated photovoltaic panels (CPV) or concentrated solar power (CSP) require a high degree of accuracy to ensure the sunlight is directed precisely at the focal point of the reflector or lens. Non-concentrating applications don't require tracking but using a tracker can improve the total power produced by the system. Photovoltaic systems using high efficiency panels with trackers can be very effective. There are many types of solar trackers, of varying costs, sophistication, and performance. The two basic categories of trackers are single axis and dual axis. Single axis solar trackers can either have a horizontal or a vertical axis. The horizontal type is used in tropical regions where the Sun gets very high at noon, but the days are short. The vertical type is used in high latitudes where the Sun does not get very high, but summer days can be very long. In concentrated solar power applications, single axis trackers are used with parabolic and linear Fresnel mirror designs.

The application of high concentration solar cells technology allows a significant increase in the amount of energy collected by solar arrays per unit area. However, to make it possible, more severe specifications on the Sun pointing error are required. In fact, the performance of solar cells with concentrators decreases drastically if this error is greater than a small value. These specifications are not fulfilled by simple tracking systems due to different sources of errors (e.g., small misalignments of the structure with respect to geographical north) that appear in practice in low cost, domestic applications. Therefore, Rubio, Ortega, Gordillo and Lopez-Martinez (2007)_presented a control application of a Sun tracker that is able to follow the Sun with high accuracy without the necessity of either a precise procedure of installation or recalibration. A hybrid tracking system that consists of a combination of open loop tracking strategies based on solar movement models and closed loop strategies using a dynamic feedback controller is presented. Energy saving factors are taken into account, which implies that, among other factors, the Sun is not constantly tracked with the same accuracy, to prevent energy overconsumption by the motors. Simulation and experimental results with a low cost two axes solar tracker are exposed, including a comparison between a classical open loop tracking strategy and the proposed hybrid one.

Lubitz (2011) used hourly typical meteorological year (TMY3) data with the Perez radiation model to simulate solar radiation on fixed, azimuth tracking and two axis tracking surfaces at 217 geographically diverse temperate latitude sites across the contiguous United States of America. The optimum tilt angle for maximizing annual irradiation on a fixed south-facing panel varied from being equal to the latitude at low-latitude, high clearness sites, to up to 14° less than the latitude at a north-western coastal site with very low clearness index. Across the United States, the optimum tilt angle for an azimuth tracking panel was found to be on average 19° closer to vertical than the optimum tilt angle for a fixed, south-facing panel at the same site. Azimuth tracking increased annual solar irradiation incident on a surface by an average of 29% relative to a fixed south-facing surface at optimum tilt angle. Two axis tracking resulted in an average irradiation increase of 34% relative to the fixed surface. Introduction of manual surface tilt changes during the year produced a greater impact for non-tracking surfaces than it did for azimuth tracking surfaces. Even monthly tilt changes only resulted in an average annual irradiation increase of 5% for fixed panels and 1% for azimuth tracked surfaces, relative to using a single optimized tilt angle in each case. In practice, the decision whether to manually tilt panels requires balancing the added cost in labor and the panel support versus the extra energy generation and the cost value of that energy.

Over the past three decades, various types of Sun-tracking mechanisms have been proposed to enhance the solar energy harnessing performance of solar collectors. Although the degree of accuracy required depends on the specific characteristics of the solar concentrating system being analyzed, generally the higher the system concentration the higher the tracking accuracy that will be needed (Blanco-Muriel, Alarcon-Padilla, Lopez-Moratalla and Lara-Coira, 2001). To achieve good tracking accuracy, sun-tracking systems normally employ sensors to feedback error signals to the control system in order to continuously receive maximum solar irradiation on the receiver. The two common types of sensors used for this purpose are closed-loop sensors and open-loop sensors.

Two most commonly used configurations in two-axis sun-tracking system are azimuth-elevation and tilt-roll (or polar) tracking system. Inspired by an ordinary optical mirror mount, azimuth-elevation system is among the most popular sun-tracking system employed in various solar energy applications (Roth, Georgiev and Boudinov, 2004). In the azimuth-elevation tracking, the collector must be free to rotate about the zenith-axis and the axis parallel to the surface of the earth. The tracking angle about the zenith-axis is the solar azimuth angle and the tracking angle about the horizontal axis is the solar elevation angle. Therefore, the accuracy of the azimuth-elevation tracking system highly relied on how well the azimuth-axis is aligned to be parallel with the zenith-axis.

Alternatively, tilt-roll (or polar) tracking system adopts an idea of driving the collector to follow the sun-rising in the east and sun-setting in the west from morning to evening as well as changing the tilting angle of the collector due to the yearly change of sun path (Nuwayhid, Mrad, and Abu-Said, 2001; Sharan and Prateek, 2006). Hence, for the tilt-roll tracking system, one axis of rotation is aligned parallel with the earth's polar-axis that is aimed towards the star Polaris. This gives it a tilt from the horizon equal to the local latitude angle. The other axis of rotation is perpendicular to this polar-axis (Poulek and Libra, 1998, 2000). The tracking angle about the polar-axis is equal to the sun's hour angle and the tracking angle about the perpendicular axis is dependent on the declination angle. The advantage of tilt-roll tracking is that the tracking velocity is almost constant at 15 degrees per hour and therefore the control system is easy to be designed. The accuracy of the tilt-roll tracking system relies strongly upon how well the roll-axis can be aligned in parallel with the polar-axis, which is also latitude dependent.

SOME POPULAR SOLAR TRACKERS

Spectroscopic analyses of direct incident sunlight are commonly used in atmospheric research. Such experiments make use of the Sun as a light source to quantify molecular absorptions in the atmosphere and then retrieve trace gas abundances. Stratospheric ozone (Barret, De Mazière and Demoulin, 2002) and greenhouse gases (De Mazière, Vigouroux, Gardiner, Coleman, Woods, Ellingsen, Gauss, Isaksen, Blumenstock, Hase, Kramer, Camy-Peyret, Chelin, Mahieu, Demoulin, Duchatelet, Mellqvist, Strandberg, Velazco, Notholt, Sussmann, Stremme and Rockmann, 2005) are routinely measured with this technique from ground-based Fourier transform infrared (FTIR) spectrometers, e.g., within the Network for the Detection of Atmospheric Composition Change. In the UV-visible range, light scattering is more important and enables spectroscopic studies of the atmosphere in other geometries such as zenith measurements (Van Roozendael, Peeters, Roscoe, Backer, Jones, Bartlett, Vaughan, Goutail, Pommereau, Kyro, Wahlstrom, Braathen and Simon, 1998). However, direct sunlight is also used (Wang, Pongetti, Sander, Spinei, Mount, Cede and Herman, 2010), its unique and unambiguous light path making it advantageous for some applications. Beside the spectrometer, the main part of the involved apparatus in direct sunlight spectrometry is the solar tracker, required to compensate for the Sun's diurnal motion.

Several kinds of solar trackers, sometimes referred to as heliostats, are used for atmospheric spectrometry, based on setups of one or several rotating mirrors. Some of them are equatorially mounted, like in Table Mountain Facility (Cageao, Blavier, McGuire, Jiang, Nemtchinov, Mills and Sander, 2001) or Harestua (Galle, Mellqvist, Arlander, Floisand, Chipperfield and Lee, 1999). In this case, one rotational axis is parallel to the Earth's axis. It enables a high tracking accuracy without a computer, since only one axis has to be driven at the Earth's rotation speed. To our knowledge, it is the only setup working without feedback on the Sun's position. On the other hand, equatorial mounts are large, need to be aligned accurately and their mechanical design is difficult. Most of the trackers used today are controlled by a computer enabling remote operation and automation. The computer first calculates the Sun position, moves the mirrors to point to the Sun and then controls these mirrors to optimize the signal on some kind of light sensor. For some trackers, the light sensor is attached to the moving part, whether it is a single mirror (Wiacek, Taylor, Strong, Saari, Kerzenmacher, Jones and Griffith, 2007) or amount of two mirrors (Neefs, de Maziere, Scolas, Hermans and Hawat, 2007). Compared to the solution presented below, the retroaction is simplified. The drawback is that the tracking is done some meters away from the spectrometer and is thus less accurate and stable.

The Alt-Azimuthal Tracker

The popular alt-azimuthal tracker design consists of two elliptical mirrors (M_1 and M_2) held in 45° relative to the vertical, facing each other. Both M_1 and M_2 rotate along the azimuthal axis and M_2 rotates as well around a horizontal axis (altitude direction). The mirror M_0, and possibly other fixed mirrors, direct the light beam into the spectrometer optical axis. A 4-quadrant photodiode is used as a position sensor for a closed-loop control of the mirrors position once their positioning towards the Sun has been set with enough accuracy, *i.e.*, once the Sun's image is visible by the photodiode. This alt-azimuthal setup is used with FTIR systems, e.g., in Park Falls, Wisconsin (Washenfelder, Toon, Blavier, Yang, Allen, Wennberg, Vay, Matross and Daube, 2006); it has been installed in Harestua to replace the equatorially mounted system (Merlaud, 2006). Compact versions have also been developed for field campaigns (Merlaud, 2004; Cordenier, 2004). A commercial version is sold by Bruker to be installed on their FTIR spectrometers (Geibel, Gerbig and Feist, 2010). A recent progress in the pointing accuracy has been reported (Gisi, Hase, Dohe and Blumenstock, 2011), replacing the traditional quadrant diode with a CCD camera, but the problems discussed hereafter remain the same.

Because developing a solar tracker is typically a master's thesis work (Merlaud, 2004; Cordenier, 2004), technical implementations are difficult to access in the literature. Some more information is available about the systems used in solar energy applications but their geometries differ.

Closed-Loop Solar Trackers

The closed loop solar tracker is a tracker with a closed-loop sensor, such as CCD camera or photo-detector, for sensing the position of the solar image on the receiver and sending a feedback signal to the controller if the solar image moves away from the receiver. Sun-tracking systems that employ closed-loop sensors are known as closed-loop Sun trackers. Over the past 30 years or so, the closed-loop tracking approach has been traditionally used in the active Sun-tracking scheme (Kalogirou, 1996; Berenguel, Rubio, Valverde, Lara, Arahal, Camacho and Lopez, 2004; Arbab, Jazi and Rezagholizadeh, 2009). For example, Kribus, Vishnevetsky, Yogev and Rubinov (2004) designed a closed-loop controller for heliostats which improved the pointing error of the solar image up to 0.1 mrad, with the aid of four CCD cameras set on the target. However, this method is rather expensive and complicated because it requires four CCD cameras and four radiometers to be placed on the target. Then the solar images captured by CCD cameras must be analyzed by a computer to generate the control correction feedback for correcting tracking errors.

In 2006, a Sun-tracking error monitoring system that uses a monolithic optoelectronic sensor for a concentrator photovoltaic system was presented by Luque-Heredia, Cervantes and Quémére (2006). According to the results from the case study, this monitoring system achieved a tracking accuracy of better than 0.1°. However, the criterion is that this tracking system requires full clear sky days to operate as the incidence light has to be above a certain threshold to ensure that the minimum required resolution is met. That same year, Aiuchi, Yoshida, Onozaki, Katayama, Nakamura and Nakamura (2006) developed a heliostat with an equatorial mount and a closed-loop photo-sensor control system. The experimental results showed that the tracking error of the heliostat was estimated to be 2 mrad during fine weather.

Nevertheless, this tracking method is not popular and only can be used for Sun-trackers with an equatorial mount configuration, which is not a common tracker mechanical structure and is complicated because the central of gravity for the solar collector is far off the pedestal. Furthermore, Chen, F., Feng, J. and Hong, Z. (2006) presented studies of digital and analogue sun sensors based on the optical vernier and optical nonlinear compensation measuring principle respectively. The proposed digital and analogue Sun sensors have accuracies of 0.02° and 0.2° correspondingly for the entire field of view of ±64° and ±62° respectively (Chen, Feng and Hong, 2006; Chen and Feng, 2007). The major disadvantage of these sensors is that the field of view, which is in the range of about ±64° for both elevation and azimuth directions, is rather small compared to the dynamic range of motion for a practical sun-tracker that is about ±70° and ±140° for elevation and azimuth directions, respectively. Besides that, it is just implemented at the testing stage in precise Sun sensors to measure the position of the sun and has not yet been applied in any closed-loop Sun-tracking system so far.

Open-Loop Solar Trackers

Although closed-loop sun-tracking system can produce a much better tracking accuracy, this type of system will lose its feedback signal when the sensor is shaded or when the sun is blocked by clouds. As an alternative method to overcome the limitation of closed loop sun-trackers, open-loop Sun trackers were introduced by using open-loop sensors that do not require any solar image as feedback.

The open-loop sensor will ensure that the solar collector is positioned at pre-calculated angles, which are obtained from a special formula or algorithm according to date, time and geographical information. In 2004, Abdallah and Nijmeh (2004) designed a two axes sun tracking system, which is operated by an open-loop control system. A programmable logic controller (PLC) was used to calculate the solar vector and to control the sun tracker so that it follows the sun's trajectory. In addition, Shanmugam and Christraj (2005) presented a computer program written in Visual Basic that is capable of determining the sun's position and thus drive a paraboloidal dish concentrator (PDS) along the East-West axis or North-South axis for receiving maximum solar radiation.

Hybrid Solar Trackers

In general, both sun-tracking approaches mentioned above have both strengths and drawbacks, so some hybrid sun-tracking systems have been developed to include both the open-loop and closed-loop sensors. Early in the 21st century, Nuwayhid, Mrad and Abu-Said (2001) adopted both the open-loop and closed-loop tracking schemes into a parabolic concentrator attached to a polar tracking system. In the open-loop scheme, a computer acts as controller to calculate two rotational angles, i.e., solar declination and hour angles, as well as to drive the concentrator along the declination and polar axes. In the closed-loop scheme, nine light-dependent resistors (LDR) are arranged in an array of a circular-shaped "iris" to facilitate sun-tracking with a high degree of accuracy. In 2006, Luque-Heredia, Gordillo and Rodriguez (2004) proposed a novel *PI* based hybrid sun-tracking algorithm for a concentrator photovoltaic system. In their design, the system can act in both open-loop and closed-loop mode. A mathematical model that involves a time and geographical coordinates function as well as a set of disturbances provides a feed forward open-loop estimation of the sun's position. To determine the sun's position with high precision, a feedback loop was introduced according to the error correction routine which is derived from the estimation of the error of the sun equations that are caused by external disturbances at the present stage based on its historical path. One year later, Rubio, Ortega, Gordillo and Lopez-Martinez (2007) fabricated and evaluated a new control strategy for a photovoltaic (PV) solar tracker that operated in two tracking modes, i.e., normal tracking mode and search mode. The normal tracking mode combines an open-loop tracking mode that is based on solar movement models and a closed-loop tracking mode which corresponds to the electro-optical controller to obtain a sun-tracking error that is smaller than a specified boundary value and enough for solar radiation to produce electrical energy. Search mode will be started when the sun-tracking error is large or no electrical energy is produced. The solar tracker will move according to a square spiral pattern in the azimuth-elevation plane to sense the sun's position until the tracking error is small enough.

Notes on Solar Trackers

Someone building a Sun tracker can quickly find ephemeris calculations in many programming languages, but other issues arise quickly. It is first necessary to characterize the field-of-view (FOV) of the 4-quadrant diode in the considered optical design. This serves two purposes: determining the accuracy needed for the ephemeris's algorithm and making sure this FOV is larger than the Sun's apparent diameter (9 mrad). This last point is important to track constantly the center of the Sun, which reduces the uncertainties in the air mass factor and avoids Doppler shifts on the edges of the Sun (Gisi, Hase, Dohe and Blumenstock, 2011). A second problem lies in the correction of the tracker orientation compared to the alt-azimuthal system in which the ephemeris is given, necessary for the calculated mode if the base of the solar tracker is not leveled. Thirdly, the relationship between the quadrant signal and the correction to apply on the mirrors positions depends on the tracker position itself (Gisi, Hase, Dohe and Blumenstock, 2011). Understanding this relationship is compulsory to achieve a smooth tracking.

As a matter of fact, the tracking accuracy requirement is very much reliant on the design as well as on the application of the sun-tracker. In this case, the longer the distance between the solar concentrator and the receiver the higher the tracking accuracy required will be because the solar image becomes more sensitive to the movement of the solar concentrator. As a result, a heliostat or off-axis sun-tracker normally requires much higher tracking accuracy compared to that of on-axis sun-tracker due to the fact that the distance between the heliostat and the target is normally much longer, especially for a central receiver system configuration. In this context, a tracking accuracy in the range of a few miliradians (mrad) is in fact sufficient for an on-axis sun-tracker to maintain its good performance when highly concentrated sunlight is involved (Chong, Wong, Siaw and Yew, 2010). Despite having many existing on-axis sun-tracking methods, the designs available to achieve a good tracking accuracy of a few mrad are complicated and expensive. It is worthwhile to note that conventional on-axis sun-tracking systems normally adopt two common configurations, which are azimuth-elevation and tilt-roll (polar tracking), limited by the available basic mathematical formulas of sun-tracking system. For azimuth-elevation tracking system, the Sun-tracking axes must be strictly aligned with both zenith and real north. For a tilt-roll tracking system, the sun-tracking axes must be strictly aligned with both latitude angle and real north. The major cause of Sun-tracking errors is how well the aforementioned alignment can be done and any installation or fabrication defects will result in low tracking accuracy. According to Chong and Wong (2009) for the azimuth-elevation tracking system, a misalignment of azimuth shaft relative to zenith axis of 0.4° can cause tracking error ranging from 6.45 to 6.52 mrad. In practice, most solar power plants all over the world use a large solar collector area to save on manufacturing cost and this has indirectly made the alignment work of the sun-tracking axes much more difficult. In this case, the alignment of the tracking axes involves an extensive amount of heavy-duty mechanical and civil works due to the requirement for thick shafts to support the movement of a large solar collector, which normally has a total collection area in the range of several tens of square meters to nearly a hundred square meters.

Under such tough conditions, a very precise alignment is really a great challenge to the manufacturer because a slight misalignment will result in significant sun-tracking errors. To overcome this problem, an unprecedented on-axis general sun-tracking formula has been proposed to allow the sun-tracker to track the sun in any two arbitrarily orientated tracking axes (Chong and Wong, 2009).

EQUATIONS FOR SOLAR TRACKING

The Sun's position in the sky varies both with the seasons (elevation) and time of day as the Sun moves across the sky. It is important to know the background of Sun-Earth geometrical relation. There are two motions should be considered.

- Seasons result from the yearly revolution of the Earth around the Sun (elliptical trajectory). It is responsible for the altitude variation of the Sun on the celestial sphere during one year.
- The earth rotates, spinning on its axis, thereby resulting in day and night. This is accounted for by the east-west daily path of the Sun.

The Sun travels through 360° east to west per day, but from the perspective of any fixed location the visible portion is 180° during an average 1/2 day period (more in spring and summer; less, in fall and winter). Local horizon effects reduce this somewhat, making the effective motion about 150°. A solar panel in a fixed orientation between the Sunrise and Sunset extremes will see a motion of 75° to either side, and thus will lose 75% of the energy in the morning and evening. Rotating the panels to the east and west can help recapture those losses. A tracker rotating in the east-west direction is known as a single-axis tracker.

The Sun also moves through 46° north and south during a year. The same set of panels set at the midpoint between the two local extremes will thus see the Sun move 23° on either side, causing losses of 8.3% A tracker that accounts for both the daily and seasonal motions is known as a dual-axis tracker. Generally speaking, the losses due to seasonal angle changes are complicated by changes in the length of the day, increasing collection in the summer in northern or southern latitudes. This biases collection toward the summer, so if the panels are tilted closer to the average summer angles, the total yearly losses are reduced compared to a system tilted at the spring/fall solstice angle (which is the same as the site's latitude).

Chen, Lim and Lim, (2006) was the pioneer group to derive a general sun-tracking formula for heliostats with arbitrarily oriented axes. The newly derived general formula by Chen, Lim and Lim, (2006) is limited to the case of off-axis sun tracker (heliostat) where the target is fixed on the earth surface and hence a heliostat normal vector must always bisect the angle between a sun vector and a target vector. As a complimentary to work of Chen, Lim and Lim, (2006), a general formula for the case of on-axis sun tracker, where the target is fixed along the optical axis of the reflector and therefore the reflector normal vector must be always parallel with the sun vector, was presented by Chong and Wong (2009). This formula is the most general form of Sun-tracking formula that embraces all the possible on-axis tracking methods. The use of azimuth-elevation tracking formula and tilt-roll tracking formula are the special case of it. The general Sun-tracking formula not only can provide a general mathematical solution, but more significantly it can improve the Sun-tracking accuracy by tackling the installation error of the solar collector. In this context, the precision of foundation alignment during the installation of solar collector becomes more tolerable because any imprecise alignment in the tracking axes can be easily compensated by changing the parameters' values in the general sun-tracking formulas. The integration of this formula in the algorithm of an open-loop sun-tracking system was also provided (Chong, Wong, Siaw, Yew, Ng, Liang, Lim and Lau, 2009).

The spinning-elevation tracking geometry included some mirror–pivot offset, orthogonal intersecting rotational axes and the elevation axis being parallel to the mirror surface plane. Guo, Wang, Zhang, Sun and Zhang (2011) analyzed the tracking accuracy of these standard spinning elevation tracking formulas to show that they are accurate with negligible tracking error. Hence, the mirror-surface-center normal obtained from these formulas is accurate for any dual-axis tracking heliostat. Then, the accurate mirror-surface-center normal information is used to determine general altitude–azimuth tracking angles for a heliostat with a mirror–pivot offset and other geometrical errors. The main geometrical errors in a typical altitude–azimuth tracking geometry are the azimuth axis tilt from the vertical, the non-orthogonality between the two heliostat rotational axes, the non-parallel degree between the mirror surface plane and the altitude axis, and the encoder reference errors. An actual heliostat in a solar field is used as an example to demonstrate use of the general altitude–azimuth tracking formulas, with the tracking angles for this heliostat on typical days graphically illustrated. The altitude–azimuth tracking angle formulas are further verified by an indoor laser-beam tracking test on a specially designed heliostat model.

Wei, Lu, Yu, Zhang and Wang (2011) derived the tracking and ray tracing equations for the target-aligned heliostat for solar tower power plants. Based on the equations, a new module for analysis of the target-aligned heliostat with an asymmetric surface has been developed and incorporated in the code HFLD. To validate the tracking and ray tracing equations, a target-aligned heliostat with a toroidal surface is designed and modeled. The image of the target-aligned heliostat is calculated by the modified code HFLD and compared with that calculated by the commercial software Zemax. It is shown that the calculated results coincide with each other very well. Therefore, the correctness of the tracking and ray tracing equations for the target-aligned heliostat is proved. Merlaud, De Mazière, Hermans and Cornet (2012) gathered the main geometrical formulas necessary for the use of a widely used kind of solar tracker, based on two 45° mirrors in alt-azimuthal set-up with a light sensor on the spectrometer, and illustrated them with a tracker developed by their group for atmospheric research.

Chong, Wong, Siaw, Yew, Ng, Liang, Lim and Lau (2009) introduced a novel sun-tracking system by integrating the general formula into the sun-tracking algorithm so that they could track the sun accurately and cost effectively, even if there is some misalignment from the ideal azimuth-elevation or tilt-roll configuration. In the new tracking system, any misalignment or defect can be rectified without the need for any drastic or labor intensive modifications to either the hardware or software components of the tracking system. In other words, even though the alignments of the azimuth-elevation axes with respect to the zenith-axis and real north are not properly done during the installation, the new Sun-tracking algorithm can still accommodate the misalignment by changing the values of the three orientation angles of the tracking axes in the tracking program. The advantage of the new tracking algorithm is that it can simplify the fabrication and installation work of solar collectors with higher tolerance in terms of the tracking axes alignment. This strategy has allowed great savings in terms of cost, time and effort by omitting more complicated solutions proposed by other researchers such as adding a closed-loop feedback controller or a flexible and complex mechanical structure to level out the sun-tracking error (Chen, Chong, Bligh, Chen, Yunus, Kannan, Lim, Lim, Alias, Bidin, Aliman, Salehan, Rezan, Tam and Tan, 2001).

Later on, Lim, Chong, Lim and Lai (2016) presented a latitude-orientated configuration of non-imaging focusing heliostat (LO-NIFH) where the sun tracking formulas of a non-imaging focusing heliostat (NIFH) are independent of the latitude. With the new configuration, the LO-NIFH introduces certain potential advantages in field applications; such as offering a standard in mechanical/optical designs, simplifying the mechatronic control schemes, offering high modularity and high scalability in both production and project implementation. One of the most important merits of latitude oriented configuration is that it makes a heliostat operating at a very narrow range of incident angles in a day; for instance, incident angles ranging from 0° to 52° for 14 h of tracking time per day. When a NIFH tracks the sun at a moderately small incident angle, it always makes a reasonably small focused spot (at the target) with a high optical efficiency above 88% at a *f/D* ratio of greater than 1. For an off-axis solar concentrator, such level of optical efficiency is considered very impressive as it is just 10% less than that of on-axis parabolic dish. We found that the astigmatic spread of a working LO-NIFH has little variation with time as compared to that of a spherical reflector operating in off-axis manner. Computer simulation of the focused spot images was conducted and compared with the spot images photographed at the site of a prototype LO-NIFH with an array of 5 × 5 mirror facets tracking at an incident angle of 23°.

Equations for Solar Ephemeris

In the first chapter, different acceptable formulas for determining the declination angle were noted. Moreover, in the item 1.6 of the first chapter, it was mentioned that, regions of the Earth are divided into certain time zones and in these time zones, noon does not necessarily correspond to the time when the Sun is highest in the sky. Similarly, Sun rise is defined as the stage when the Sun rises in one part of the time zone. However, due to the distance covered in a single time zone, the time at which the Sun actually clears the horizon in one part of the time zone may be quite different to the "defined" Sun rise (or what is officially recognized as the time of Sun rise). So, solar time is unique to each particular longitude. However, tracking algorithms require an accurate knowledge of the difference between solar time and the local clock time. On the other hand, the azimuth angle and the elevation angle are related to "solar time". Consequently, to calculate the Sun's position, first the local solar time is found and then the elevation and azimuth angles are calculated.

As mentioned before, equation (5), in the first chapter, is used to converts the local solar time (LST) into the number of degrees (the hour angle ω) which the Sun moves across the sky:

$$\omega = 15^{o}\left(LST - 12\right)$$

The hour angle expresses the time of day with respect to the solar noon. It is the angle between the plane of the meridian containing observer and meridian that touches the Earth-Sun line. It is zero at solar noon and increases by 15° every hour.

A solar time is a 24-hour clock with 12:00 as the exact time when the sun is at the highest point in the sky. The concept of solar time is to predict the direction of the sun's ray relative to a point on the earth. Solar time is location or longitudinal dependent. It is generally different from local clock time

(*LT*) (defined by politically time zones) because of the eccentricity of the Earth's orbit and because of human adjustments such as time zones and daylight saving. The conversion between solar time and local clock time requires knowledge of the location, the day of the year, and the standards to which local clocks are set. The conversion between solar time *LST* and local clock time (*LT*) (in 24-hour rather than AM/PM format) takes the form:

$$LT = LST - \frac{T_c}{60} - D \tag{1}$$

D is daylight saving time in (hr). Daylight saving time was initiated in the spring of 1918 to save fuel and promote other economies in a country at war (Jesperson and Fitz-Randolph, 1977). According to this concept, the standard time is advanced by 1 hour, usually from 2:00 Am on the last Sunday in April until 2:00 Am on the last Sunday in October. The T_c (min) being the time correction factor accounts for the variation of the LST within a given time zone due to the longitude variations within the time zone:

$$T_c = 4\left(Longitude - LSTM\right) + E \tag{2}$$

where Longitude is longitude of the site, LSTM is the local standard time meridian that is a reference meridian used for a particular time zone and is similar to the prime meridian, which is used for Greenwich mean time, and $E\left(min\right)$ is the equation of time (in minutes) which is an empirical equation that corrects for the eccentricity of the Earth's orbit and the Earth's axial tilt. The equation of time gives the difference between mean solar time and true solar time on a given date. The (LSTM) is calculated according to the equation:

$$LSTM = 15 * \Delta T_{cGMT} \tag{3}$$

where ΔT_{GMT} is the difference of the local time (LT) from Greenwich mean time (GMT) in hours. The factor of 4 minutes (in equation (2)) comes from the fact that the Earth rotates 1° every 4 minutes.

The level of accuracy required in determining the equation of time will depend on whether the designer is doing system performance or developing tracking equations. An approximation for calculating the equation of time in minutes is given by Woolf (1968) and claimed that is accurate to within about 30 seconds during daylight hours:

$$\begin{aligned} E = {} & 0.258\cos\left[\frac{2\pi\left(n-1\right)}{365.242}\right] - 7.416sin\left[\frac{2\pi\left(n-1\right)}{365.242}\right] \\ & -3.648cos\left[\frac{4\pi\left(n-1\right)}{365.242}\right] - 9.228sin\left[\frac{4\pi\left(n-1\right)}{365.242}\right] \end{aligned} \tag{5}$$

or given by Spencer (1971):

$$E = 229.2\left\{\begin{bmatrix} 0.000075 + 0.001868\cos\left[\frac{2\pi\left(n-1\right)}{365}\right] - 0.032077 sin\left[\frac{2\pi\left(n-1\right)}{365}\right] \\ -0.014615 cos\left[\frac{4\pi\left(n-1\right)}{365}\right] - 0.04089 sin\left[\frac{4\pi\left(n-1\right)}{365}\right] \end{bmatrix}\right\} \quad (6)$$

Since solar time is based on the Sun being due south at 12:00 noon on any specific day, the accumulated difference between mean solar time and true solar time can approach 17 minutes either ahead of or behind the mean, with an annual cycle.

To satisfy the control needs of concentrating collectors, a more accurate determination of the hour angle is often needed. An approximation of the equation of time claimed to have an average error of 0.63 seconds and a maximum absolute error of 2.0 seconds is proposed by Lamm (1981):

$$E = \sum_{k=0}^{5}\left\{A_k \cos\left[\frac{2\pi k n_1}{365.25}\right] + B_k sin\left[\frac{2\pi k n_1}{365.25}\right]\right\} \quad (7)$$

Here n_l *is* the number of days into a leap year cycle with n_l = 1 being January 1 of each leap year, and n_l =1461 corresponding to December 31 of the 4th year of the leap year cycle. The coefficients A_k and B_k are given in the Table 1. Arguments for the cosine and sine functions are in degrees. The resulting value is in minutes and is positive when the apparent solar time is ahead of mean solar time and negative when the apparent solar time is behind the mean solar time.

Iqbal (1983) introduced a value, called the fractional year (Ω):

$$\Omega\left(radians\right) = \frac{2\pi}{365}\left(n - 1 + \frac{t' - 12}{24}\right) \quad (8)$$

where t' is the Coordinated Universal Time (*UTC*) decimal time. The introduction of Ω in the equation of time E and declination angle δ should be accurate enough for most tracking purposes. Then the Spencer (1971) formulas for equation of time and declination angle could be modified to be:

Table 1. Coefficients for Equation (7)

k	$-10^{+4} \times A_k$ (hr)	$-10^{+4} \times B_k$ (hr)
0	- 2.0870	0
1	- 92.869	1222.9
2	522.58	1569.8
3	13.077	51.602
4	21.867	29.823
5	1.5100	2.3463

$$E\left(min\right) = 229.2\begin{bmatrix} 0.000075 + 0.001868\cos\left(\Omega\right) - 0.032077sin\left(\Omega\right) \\ -0.014615cos\left(2\Omega\right) - 0.04089sin\left(2\Omega\right) \end{bmatrix} \tag{9}$$

$$\delta\left(radians\right) = \begin{bmatrix} 0.006918 - 0.399912^{*}\cos\left(\Omega\right) + 0.070257^{*}\sin\left(\Omega\right) \\ -0.006758^{*}\cos\left(2^{*}\Omega\right) + 0.000907^{*}\sin\left(2^{*}\Omega\right) \\ -0.002697^{*}\cos\left(3^{*}\Omega\right) + 0.00148^{*}\sin\left(3^{*}\Omega\right) \end{bmatrix} \tag{10}$$

The declination is the equivalent of the latitude on the celestial sphere.

The offset T_{off} (in minutes) between the UTC time and the true solar time (*TST*) depends on the longitude (in degrees East) and is:

$$T_{off} = TST - 4 * Longitude) \tag{11}$$

The true solar time (*TST*, in minutes) is then:

$$TST = \text{hour}^{*}60 + \min + \frac{sec}{60} + T_{off} \tag{12}$$

where hour, min and sec are the components of the UTC time. The solar hour angle, in degrees, comes from the true solar time *TST* as:

$$\omega = \frac{TST}{4} - 180^{o} \tag{13}$$

For given latitude, the hour angle and the declination are converted to horizontal coordinates, i.e., solar zenith angle (θ_z) and azimuth (γ_s, from south positive eastwards) as:

$$\theta_z = arccos\left[sin\left(\varphi\right)sin\left(\delta\right) + cos\left(\varphi\right)cos\left(\delta\right)cos\left(\omega\right)\right] \tag{14}$$

$$\gamma_s = atan_2\left[sin\left(\omega\right)cos\left(\delta\right); cos\left(\omega\right)sin\left(\varphi\right)cos\left(\delta\right) - cos\left(\varphi\right)sin\left(\delta\right)\right] \tag{15}$$

Finally, the effect of refraction in arc-minutes can be approximated using Sæmundsson's formula (Meeus, 1998):

$$R' = \frac{1.02}{\tan\left(\alpha\right) + \frac{10.3}{\alpha + 5.1}} \tag{16}$$

where α is the un-refracted altitude in degrees.

The Derivation of General Formula

Prior to mathematical derivation, it is worthwhile to state that the task of the on-axis sun-tracking system is to aim a solar collector towards the sun by turning it about two perpendicular axes so that the sun ray is always normal relative to the collector surface (Blanco-Muriel, Alarcon-Padilla, Lopez-Moratalla and Lara-Coira,, 2001; Reda and Andreas, 2004; Sproul, 2007). Under this circumstance, the angles that are required to move the solar collector to this orientation from its initial orientation are known as sun-tracking angles. In the derivation of sun-tracking formulas, it is necessary to describe the sun's position vector and the collector's normal vector in the same coordinate reference frame, which is the collector-center frame. Nevertheless, the unit vector of the sun's position is usually described in the earth-center frame due to the sun's daily and yearly rotational movements relative to the earth. Thus, to derive the sun-tracking formula, it would be convenient to use the coordinate transformation method to transform the sun's position vector from earth center frame to earth-surface frame and then to collector center frame. By describing the sun's position vector in the collector-center frame, we can resolve it into solar azimuth and solar altitude angles relative to the solar collector and subsequently the amount of angles needed to move the solar collector can be determined easily.

According to the item 1.2 of the first chapter, the sun's position vector relative to the earth-center frame can be defined as shown in Figure 3, where CM, CE and CP represent three orthogonal axes from the center C of earth pointing towards the meridian, east and Polaris, respectively. The unified vector for the sun position $\boldsymbol{I}$ in the earth-center frame can be written in the form of direction cosines as follow:

$$\boldsymbol{I} = \begin{bmatrix} I_M \\ I_E \\ I_P \end{bmatrix} = \begin{bmatrix} cos\left(\delta\right) * cos\left(\omega\right) \\ -cos\left(\delta\right) * sin\left(\omega\right) \\ sin\left(\delta\right) \end{bmatrix} \tag{17}$$

Figure 4 depicts the coordinate system in the earth-surface frame that comprises of OZ, OE and ON axes, in which they point towards zenith, east and north, respectively. The coordinate transformation for the vector I from earth-center frame to earth-surface frame was presented in the first chapter. This transformation can be obtained through a rotation angle that is equivalent to the latitude angle, φ and it can be expressed as:

$$\left[\varphi\right] = \begin{bmatrix} \cos\left(\varphi\right) & 0 & \sin\left(\varphi\right) \\ 0 & 1 & 0 \\ -\sin\left(\varphi\right) & 0 & \cos\left(\varphi\right) \end{bmatrix} \tag{18}$$

Figure 3. The sun's position vector relative to the earth-center frame

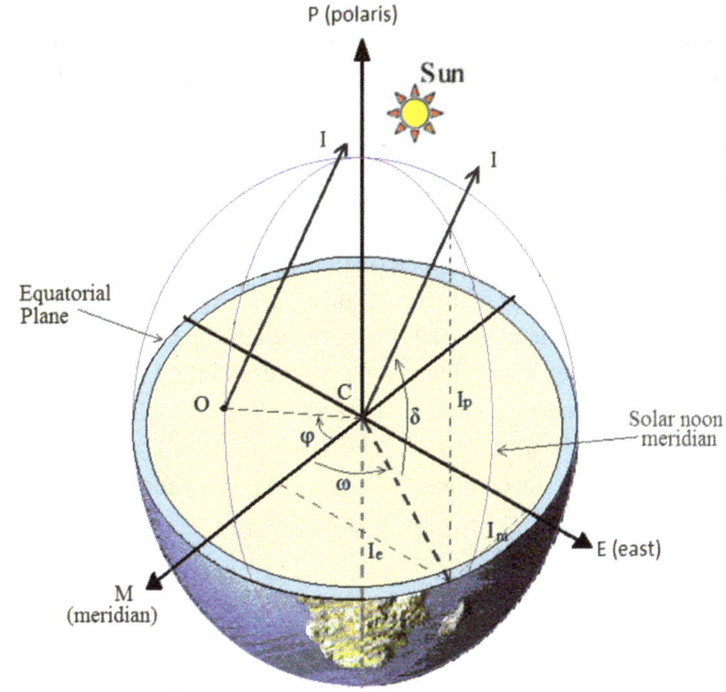

Figure 4. The coordinate system in the earth-surface frame that consists of OZ, OE and ON axes, in which they point towards zenith, east and north respectively

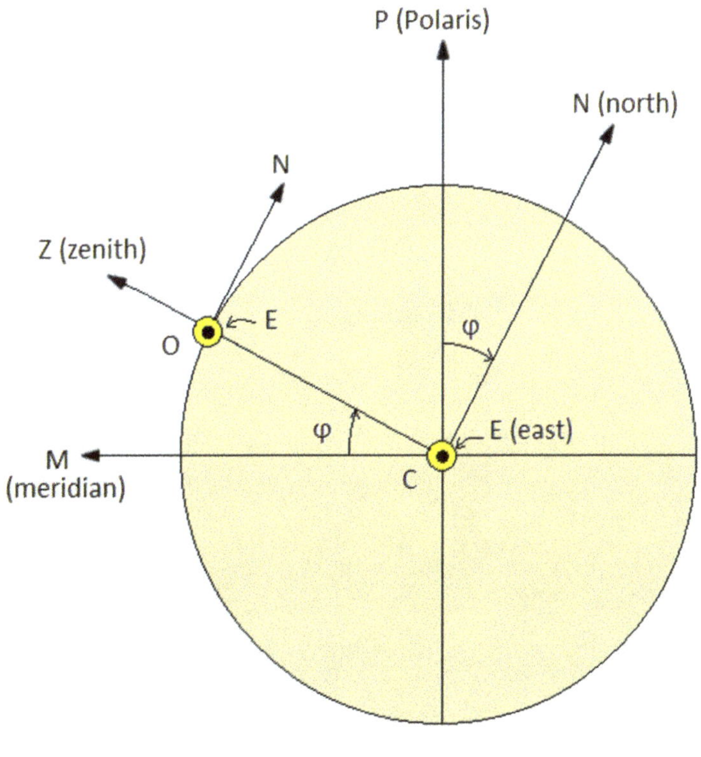

Now, let us consider a new coordinate system that is defined by three orthogonal coordinate axes in the collector- center frame as shown in Figure 5. For the collector-center frame, the origin O is defined at the center of the collector surface and it also coincides with the origin of earth-surface frame. OV is defined as vertical axis in this coordinate system and it is also parallel with first rotational axis of the solar collector. Meanwhile, OR is named as reference axis in which one of the tracking angle β is defined relative to this axis. The third orthogonal axis, OH, is named as horizontal axis and it is parallel with the initial position of the second rotational axis. The OR and OH axes form the level plane where the collector surface is driven relative to this plane. The simplest structure of solar collector that can be driven in two rotational axes: the first rotational axis that is parallel with OV and the second rotational axis that is known as EE0 dotted line (it can rotate around the first axis during the sun-tracking but must always remain perpendicular with the first axis). From the diagram, θ is the amount of rotational angle about EE0 axis measured from OV axis, whereas β is the amount of rotational angle about OV axis measured from OR axis. Furthermore, α is the solar altitude angle in the collector-center frame, which is expressed as α = 90o- θ.

In the collector-center frame, the sun position **I** can be written in the form of direction cosines as follow:

Figure 5. The orthogonal coordinate axes in the collector- center frame

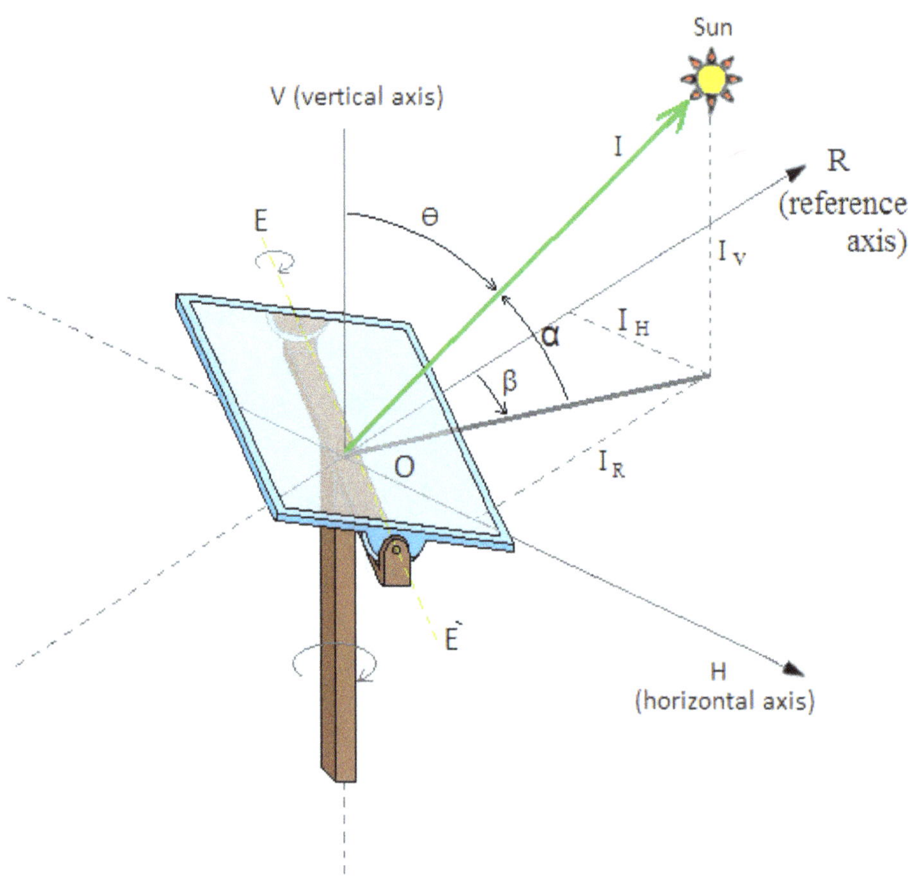

$$\boldsymbol{I} = \begin{bmatrix} I_V \\ I_H \\ I_R \end{bmatrix} = \begin{bmatrix} sin(\alpha) \\ cos(\alpha) * sin(\gamma_s) \\ cos(\alpha) * cos(\gamma_s) \end{bmatrix} \tag{19}$$

where α is elevation angle, γ_s is azimuth angle, ω is hour angle, δ is declination angle.

In an ideal azimuth-elevation system, OV, OH and OR axes of the collector-center frame are parallel with OZ, OE and ON axes of the earth-surface frame accordingly as shown in Figure 6. To generalize the mathematical formula from the specific azimuth-elevation system to any arbitrarily oriented sun-tracking system, the orientations of OV, OH and OR axes will be described by three tilted angles relative to the earth-surface frame. Three tilting angles have been introduced here because the two-axis mechanical drive can be arbitrarily oriented about any of the three principal axes of earth-surface frame: Φ is the rotational angle about zenith-axis if the other two angles are null, λ is the rotational angle about north-axis if the other two angles are null and ζ is the rotational angle about east-axis if the other two angles are null. On top of that, the combination of the above mentioned three angles can further generate more unrepeated orientations of the two tracking axes in earth-surface frame, which is very important in later consideration for improving sun-tracking accuracy of solar collector.

Figure 7 a–c shows the process of how the collector-center frame is tilted step-by-step relative to the earth-surface frame, where OV', OH' and OR' axes represent the intermediate position for OV, OH and OR axes, respectively. In Figure 7a, the first tilted angle, $+\phi$, is a rotational angle about the OZ axis in clockwise direction.

Figure 6. Ideal azimuth-elevation system

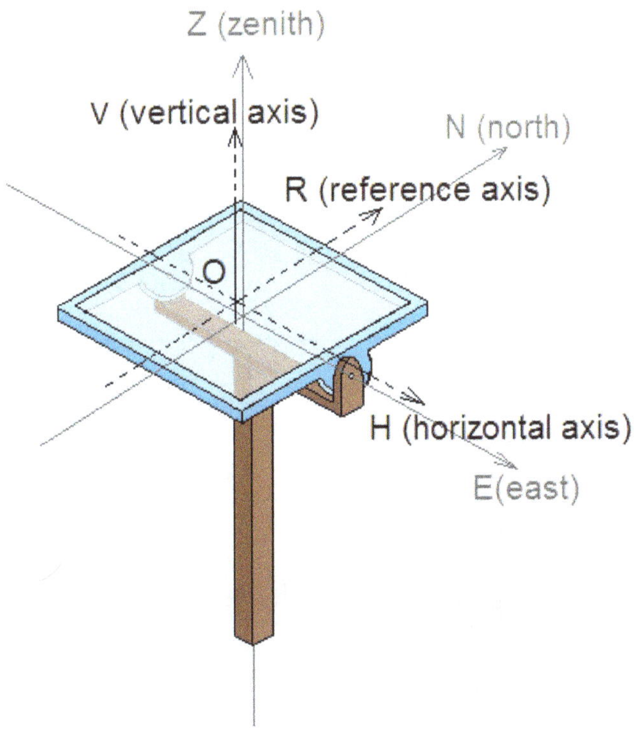

Figure 7. The diagram of the process of how the collector-center frame is tilted step-by-step relative to the earth-surface frame

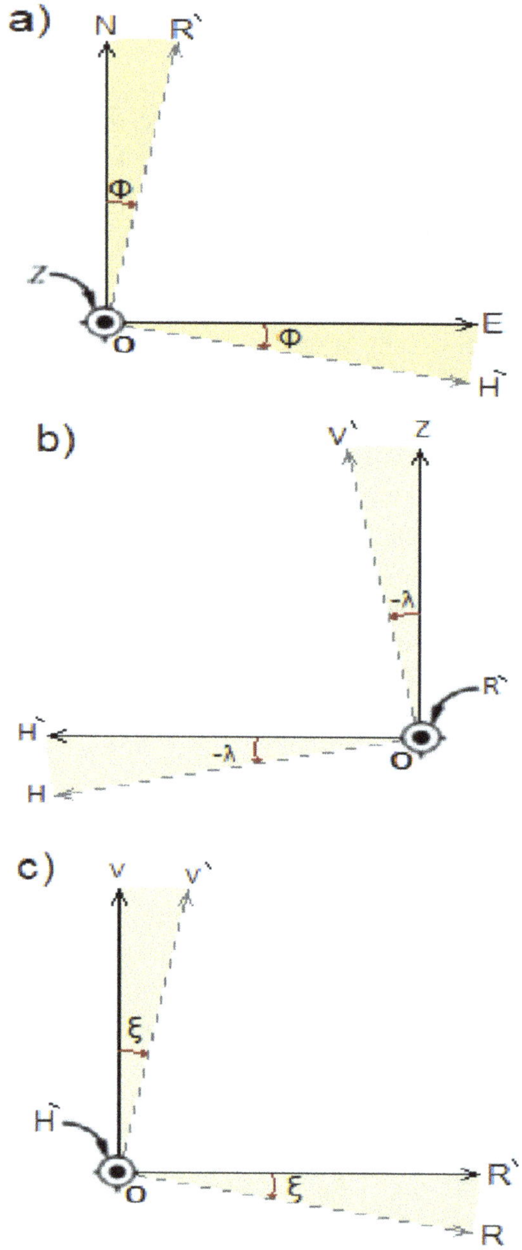

$$\begin{bmatrix} Z \\ H' \\ R' \end{bmatrix} = \begin{bmatrix} 1 & 0 & 0 \\ 0 & \cos(\phi) & -\sin(\phi) \\ 0 & \sin(\phi) & \cos(\phi) \end{bmatrix} \begin{bmatrix} Z \\ E \\ N \end{bmatrix} \quad (20)$$

In Figure 7b, the second tilted angle, -λ, is a rotational angle about OR' axis in counter-clockwise direction.

$$\begin{bmatrix} V' \\ H \\ R' \end{bmatrix} = \begin{bmatrix} \cos(\lambda) & -\sin(\lambda) & 0 \\ \sin(\lambda) & \cos(\lambda) & 0 \\ 0 & 0 & 1 \end{bmatrix} \begin{bmatrix} Z \\ H' \\ R' \end{bmatrix} \tag{21}$$

Lastly, in Figure 7c, the third tilted angle, ζ, is a rotational angle about OH axis in clockwise direction.

$$\begin{bmatrix} V \\ H \\ R \end{bmatrix} = \begin{bmatrix} \cos(\zeta) & 0 & \sin(\zeta) \\ 0 & 1 & 0 \\ -\sin(\zeta) & 0 & \cos(\zeta) \end{bmatrix} \begin{bmatrix} V' \\ H \\ R' \end{bmatrix} \tag{22}$$

Figure 8 shows the combination of the above three rotations in 3D view for the collector-center frame relative to the earth-surface frame, where the change of coordinate system for each axis follows the order: Z→ V' → V, E→ H' → H and N → R' → R.

Figure 8. The combination of the three rotations in 3D view from collector center frame to the earth-surface frame, where the change of coordinate system for each axis follows the order: Z→ V' → V, E→ H' → H and N → R' → R

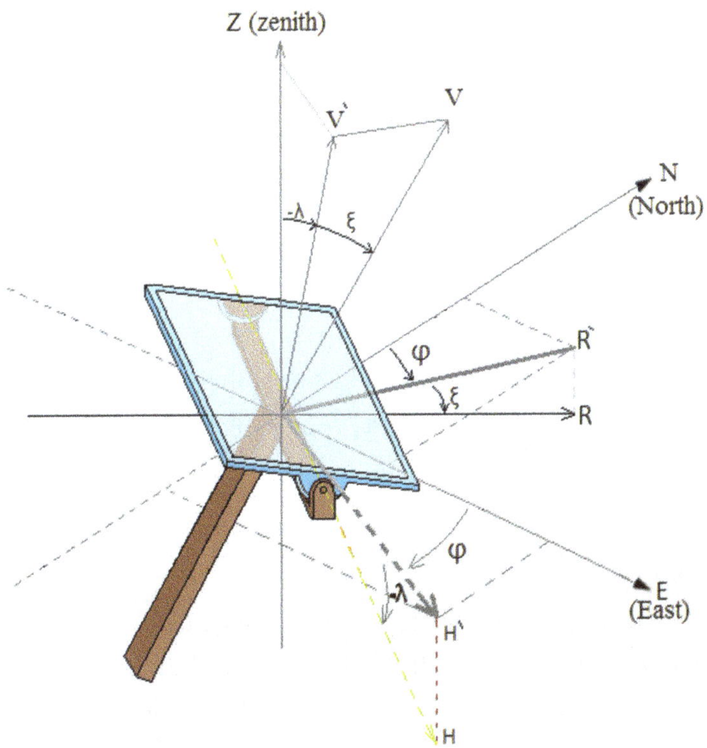

Similar to the latitude angle, in the direction representation of the three tilting angles, we define positive sign to the angles, i.e. ϕ, λ, ζ, for the rotation in the clockwise direction. In other words, clockwise and counter-clockwise rotations can also be named as positive and negative rotations, respectively. As shown in Figure 7a–c, the transformation matrices correspond to the three tilting angles ϕ, λ, and ζ can be obtained accordingly as follow:

$$\left[\phi\right]=\begin{bmatrix}1 & 0 & 0\\ 0 & \cos\left(\phi\right) & -\sin\left(\phi\right)\\ 0 & \sin\left(\phi\right) & \cos\left(\phi\right)\end{bmatrix} \tag{23}$$

$$\left[\lambda\right]=\begin{bmatrix}\cos\left(\lambda\right) & -\sin\left(\lambda\right) & 0\\ \sin\left(\lambda\right) & \cos\left(\lambda\right) & 0\\ 0 & 0 & 1\end{bmatrix} \tag{24}$$

$$\left[\zeta\right]=\begin{bmatrix}\cos\left(\zeta\right) & 0 & \sin\left(\zeta\right)\\ 0 & 1 & 0\\ -\sin\left(\zeta\right) & 0 & \cos\left(\zeta\right)\end{bmatrix} \tag{25}$$

The new set of coordinates, in the collector-center frame, can be interrelated with the earth-center frame based coordinate I through the process of four successive coordinate transformations. It will be first transformed from earth-center frame to earth-surface frame through transformation matrix [ϕ], then from earth-surface frame to collector-center frame through three subsequent coordinate transformation matrices that are [ϕ], [λ] and [ζ]. In mathematical expression, **I** can be obtained through multiplication of four successive rotational transformation matrices with the earth-center frame based coordinate and it is written as:

$$\begin{bmatrix}I_V\\ I_H\\ I_R\end{bmatrix}=\left[\zeta\right]\left[\lambda\right]\left[\phi\right]\left[\varphi\right]\begin{bmatrix}\cos\left(\delta\right)\cos\left(\omega\right)\\ -\cos\left(\delta\right)\sin\left(\omega\right)\\ \sin\left(\delta\right)\end{bmatrix}==>\begin{bmatrix}sin\left(\alpha\right)\\ cos\left(\alpha\right)*sin\left(\gamma_s\right)\\ cos\left(\alpha\right)*cos\left(\gamma_s\right)\end{bmatrix}=$$
$$=\begin{bmatrix}\cos\left(\zeta\right) & 0 & \sin\left(\zeta\right)\\ 0 & 1 & 0\\ -\sin\left(\zeta\right) & 0 & \cos\left(\zeta\right)\end{bmatrix}x\begin{bmatrix}\cos\left(\lambda\right) & -\sin\left(\lambda\right) & 0\\ \sin\left(\lambda\right) & \cos\left(\lambda\right) & 0\\ 0 & 0 & 1\end{bmatrix} \tag{26}$$
$$x\begin{bmatrix}1 & 0 & 0\\ 0 & \cos\left(\phi\right) & -\sin\left(\phi\right)\\ 0 & \sin\left(\phi\right) & \cos\left(\phi\right)\end{bmatrix}x\begin{bmatrix}\cos\left(\varphi\right) & 0 & \sin\left(\varphi\right)\\ 0 & 1 & 0\\ -\sin\left(\varphi\right) & 0 & \cos\left(\varphi\right)\end{bmatrix}x\begin{bmatrix}\cos\left(\delta\right)\cos\left(\omega\right)\\ -\cos\left(\delta\right)\sin\left(\omega\right)\\ \sin\left(\delta\right)\end{bmatrix}$$

Solving the above matrix equation for the solar altitude angle in the collector-center frame, α, we have:

$$\alpha = arcsin\left\{\begin{matrix} cos(\delta)cos(\omega)\begin{bmatrix}\cos\lambda\cos(\varphi)\cos(\zeta)-\cos(\zeta)\sin(\lambda)\sin(\phi)\sin(\varphi)\\ -\sin(\zeta)\cos(\phi)\sin(\varphi)\end{bmatrix} \\ -cos(\delta)\,sin(\omega)\left[\sin(\zeta)\sin(\phi)-\cos(\zeta)\sin(\lambda)\cos(\phi)\right] \\ +\sin(\delta)\begin{bmatrix}\cos(\zeta)\cos(\lambda)\sin(\varphi)+\\ \cos(\zeta)\sin(\lambda)\sin(\phi)\cos(\varphi)+\sin(\zeta)\cos(\phi)\cos(\varphi)\end{bmatrix}\end{matrix}\right\} \tag{27}$$

Thus, the first tracking angle along EE' axis is:

$$\theta = 90^o - arcsin\left\{\begin{matrix} cos(\delta)cos(\omega)\begin{bmatrix}\cos(\lambda)\cos(\varphi)\cos(\zeta)-\sin(\zeta)\cos(\phi)\sin(\varphi)\\ -\cos(\zeta)\sin(\lambda)\sin(\phi)\sin(\varphi)\end{bmatrix} \\ -cos(\delta)\,sin(\omega)\left[\sin(\zeta)\sin(\phi)-\cos(\zeta)\sin(\lambda)\cos(\phi)\right] \\ +\sin(\delta)\begin{bmatrix}\cos(\zeta)\cos(\lambda)\sin(\varphi)+\cos(\zeta)\sin(\lambda)\sin(\phi)\cos(\varphi)\\ +\sin(\zeta)\cos(\phi)\cos(\varphi)\end{bmatrix}\end{matrix}\right\} \tag{28}$$

Similarly, the other two remaining equations that can be extracted from the above matrix equation in the cosine format are as follow:

$$\sin(\gamma_s)=\frac{\left\{\begin{matrix} cos(\delta)cos(\omega)\left[\sin(\lambda)\cos(\varphi)+\cos(\lambda)\sin(\phi)\sin(\varphi)\right] \\ -cos\delta sin\omega\cos\lambda\cos\phi \\ +\sin(\delta)\left[\sin(\lambda)\sin(\varphi)-\cos(\lambda)\sin(\phi)\cos(\varphi)\right]\end{matrix}\right\}}{\cos(\alpha)} \tag{29}$$

$$\cos(\gamma_s)=\frac{\left\{\begin{matrix} cos(\delta)cos(\omega)\begin{bmatrix}\sin(\zeta)\cos(\lambda)\cos(\varphi)-\cos(\zeta)\cos(\phi)\sin(\varphi)\\ +\sin(\zeta)\sin(\lambda)\sin(\phi)\sin(\varphi)\end{bmatrix} \\ -cos(\delta)\,sin(\omega)\left[\sin(\zeta)\sin(\lambda)\cos(\phi)+\cos(\zeta)\sin(\phi)\right] \\ +\sin(\delta)\begin{bmatrix}-\sin(\zeta)\cos(\lambda)\sin(\varphi)+\cos(\zeta)\cos(\varphi)\cos(\phi)\\ -\sin(\zeta)\sin(\lambda)\sin(\phi)\cos(\varphi)\end{bmatrix}\end{matrix}\right\}}{\cos(\alpha)} \tag{30}$$

In fact, the second tracking angle along OV axis, γ_s, can be in any of the four trigonometric quadrants depending on location, time of day and the season. Since the arcsine and arccosine functions have two possible quadrants for their result, both equations of $\sin\gamma_s$ and $\cos(\gamma_s)$ require a test to ascertain a correct quadrant. Consequently:

$$\gamma_s = arcsin \frac{\begin{Bmatrix} cos(\delta)cos(\omega)\left[\sin(\lambda)\cos(\varphi)+\cos(\lambda)\sin(\phi)\sin(\varphi)\right] \\ -cos(\delta)sin(\omega)\cos(\lambda)\cos(\phi) \\ +\sin(\delta)\left[\sin(\lambda)\sin(\varphi)-\cos(\lambda)\sin(\phi)\cos(\varphi)\right] \end{Bmatrix}}{\cos(\alpha)} \tag{31}$$

when $cos(\gamma_s) \geq 0^o$ elsewhere:

$$\gamma_s = 180^o - arcsin \frac{\begin{Bmatrix} cos(\delta)cos(\omega)\left[\sin(\lambda)\cos(\varphi)+\cos(\lambda)\sin(\phi)\sin(\varphi)\right] \\ -cos(\delta)sin(\omega)\cos(\lambda)\cos(\phi) \\ +\sin(\delta)\left[\sin(\lambda)\sin(\varphi)-\cos(\lambda)\sin(\phi)\cos(\varphi)\right] \end{Bmatrix}}{\cos(\alpha)} \tag{32}$$

Special Cases of the General Formula

The derived general sun-tracking formula is the most general form of solution for various kinds of arbitrarily oriented on-axis sun tracker. In general, all the sun-tracking systems can be classified into two major groups: (i) full tracking system and (ii) partial tracking system (Kalogirou, 2004). In order to assess PV electricity cost ($/W) reduction, Solanki and Sangani (2007) designed and fabricated V-trough photovoltaic (PV) concentrator systems along with conventional 1-sun PV module for different types of tracking modes: seasonal, one axis north–south and two axes tracking. Three design models based on these tracking modes were used to develop the V-trough for a 2-sun concentration. Commercially available PV modules of different make and types were evaluated for their usability under 2-sun concentration. The V-trough concentrator system with geometric concentration ratio of 2 (2-sun) increases the output power by 44% as compared to PV flat-plate system for passively cooled modules. Design models with lower trough angles gave higher output power because of higher glass transmittivity. PV modules with lower series resistance gave higher gain in output power. The unit cost ($/W) for a V-trough concentrator, based on different design models, is compared with that of a PV flat plate system inclined at latitude angle (Mumbai, φ=19.12°)

In a full tracking system, the solar collector tracks the sun in two axes such that the sun vector is normal to the aperture as to achieve 100% energy collection efficiency. In contrast, a partial tracking system only tracks the sun in a single-axis such that the plane of the sun's motion is normal to the aperture as to reduce the cosine loss (Hein, Dimroth, Siefer and Bett, 2003). For the case of latitude-tilted tracking axis, the energy collection efficiency will be [cos (δ) x 100] % where δ is the declination angle of the sun that varies from -23.45° to 23.45°.

For the full tracking system, such as azimuth-elevation and tilt-roll tracking system, their tracking formulas can be derived from the general formula by setting different values to the parameters, such as

ϕ, λ and ζ. In the case of elevation-azimuth tracking system, the tracking formula can be obtained by setting the angles $\phi = 0^o$, $\lambda = 0^o$ and $\zeta = 0^o$ in the general formula. From that, the general formula can be simplified to:

$$\alpha = arcsin\left[cos\left(\delta\right)cos\left(\omega\right)\cos\left(\varphi\right) + \sin\left(\delta\right)\sin\left(\varphi\right)\right] \tag{33}$$

$$\gamma_s = arcsin\left[\frac{-cos\left(\delta\right)sin\left(\omega\right)}{\cos\left(\alpha\right)}\right] \tag{34}$$

when $cos\left(\gamma_s\right) \geq 0$ elsewhere:

$$\gamma_s = 180^o - arcsin\left[\frac{-cos\left(\delta\right)sin\left(\omega\right)}{\cos\left(\alpha\right)}\right] \tag{35}$$

$$\theta_i = 90^o - arcsin\left[cos\left(\delta\right)cos\left(\omega\right)\cos\left(\varphi\right) + \sin\left(\delta\right)\sin\left(\varphi\right)\right] \tag{36}$$

On the other hand, tilt-roll tracking formula can also be obtained by setting the angles $\phi = 180^o$, $\lambda = 0^o$ and $\zeta = \varphi - 90^o$ in the general formula. For this case, the general tracking formula can be then simplified to:

$$\alpha = arcsin\left[\sin\left(\delta\right)\right] ==> \alpha = \delta \tag{37}$$

$$\theta_i = 90^o - \delta \tag{38}$$

$$\gamma_s = arcsin\left[sin\left(\omega\right)\right] ==> \gamma_s = \omega \tag{39}$$

when $cos\left(\gamma_s\right) \geq 0$ or ($-90^o \leq \omega \leq 90^o$) elsewhere:

$$\gamma_s = 180^o - \omega \tag{40}$$

For the partial tracking system, a single-axis tracking formula can be obtained from the full tracking formula by setting one of its tracking angles, which is either θ or γ_s, as a constant value. For example, one of the most widely used partial tracking systems is to track the sun in latitude-tilted tracking axis. It is also called an equatorial mounting where the tracking axis is tilted from the horizon by the latitude angle towards south and it is parallel to the earth's rotational pole. The tracking formula of latitude-tilted tracking axis can be derived from tilt-roll tracking formula by putting the angle θ to be 90° and the solar collector only tracks the sun with the angle $\gamma_s = \omega$.

Methodology of Using General Formula

According to the general formula, the sun-tracking accuracy of the system is highly reliant on the precision of the input parameters of the sun-tracking algorithm: latitude angle φ, hour angle ω, declination angle δ, as well as the three orientation angles of the tracking axes of solar concentrator, i.e., ϕ, λ and ζ. Among these values, local latitude, φ, and longitude of the sun tracking system can be determined accurately with the latest technology such as a global positioning system (GPS). On the other hand, ω and δ are both local time dependent parameters (please refer to item 6.4.1 for the details). These variables can be computed accurately with the input from precise clock that is synchronized with the Internet time server. As for the three orientation angles ϕ, λ and ζ, their precision are very much reliant on the care paid during the on-site installation of solar collector, the alignment of tracking axes and the mechanical fabrication. The following mathematical derivation is attempted to obtain analytical solutions for the three orientation angles based on the daily sun-tracking error results induced by the misalignment of sun-tracking axes.

The time dependency of ω and δ can be found from the equations (27), (29) and (30). Therefore, the instantaneous sun-tracking angles of the collector only vary with the angles ω and δ. Given three different local times LT_1, LT_2 and LT_3 on the same day, the corresponding three hours angles ω_1, ω_2 and ω_3 as well as three declination angles δ_1, δ_2 and δ_3 can result in three elevation angles α_1, α_2 and α_3 and three azimuth angles γ_{s1}, γ_{s2} and γ_{s3} accordingly as expressed in the equations (27), (29) and (30). Considering three different local times, we can actually rewrite each of the Equations (27), (29) and (30) into three linear equations. By arranging the three linear equations in a matrix form, the equations (27), (29) and (30) can subsequently form the matrices:

$$\begin{bmatrix} \sin(\alpha_1) \\ \sin(\alpha_2) \\ \sin(\alpha_3) \end{bmatrix} = \begin{pmatrix} \cos\varphi cos\delta_1 cos\omega_1 + sin\varphi sin\delta_1 & -cos\delta_1 sin\omega_1 & -\sin\varphi cos\delta_1 cos\omega_1 + cos\varphi sin\delta_1 \\ \cos\varphi cos\delta_2 cos\omega_2 + sin\varphi sin\delta_2 & -cos\delta_2 sin\omega_2 & -\sin\varphi cos\delta_2 cos\omega_2 + cos\varphi sin\delta_2 \\ \cos\varphi cos\delta_3 cos\omega_3 + sin\varphi sin\delta_3 & -cos\delta_3 sin\omega_3 & -\sin\varphi cos\delta_3 cos\omega_3 + cos\varphi sin\delta_3 \end{pmatrix} x \begin{bmatrix} \cos(\lambda)\cos(\zeta) \\ -\cos(\zeta)\sin(\lambda)\cos(\phi) + \sin(\zeta)\sin(\phi) \\ \cos(\zeta)\sin(\lambda)\sin(\phi) + \sin(\zeta)\cos(\phi) \end{bmatrix} \tag{41}$$

$$\begin{bmatrix} \sin(\gamma_{s1})cos(\alpha_1) \\ \sin(\gamma_{s2})cos(\alpha_2) \\ \sin(\gamma_{s3})cos(\alpha_3) \end{bmatrix} = \begin{pmatrix} \cos\varphi cos\delta_1 cos\omega_1 + sin\varphi sin\delta_1 & -cos\delta_1 sin\omega_1 & -\sin\varphi cos\delta_1 cos\omega_1 + cos\varphi sin\delta_1 \\ \cos\varphi cos\delta_2 cos\omega_2 + sin\varphi sin\delta_2 & -cos\delta_2 sin\omega_2 & -\sin\varphi cos\delta_2 cos\omega_2 + cos\varphi sin\delta_2 \\ \cos\varphi cos\delta_3 cos\omega_3 + sin\varphi sin\delta_3 & -cos\delta_3 sin\omega_3 & -\sin\varphi cos\delta_3 cos\omega_3 + cos\varphi sin\delta_3 \end{pmatrix} x \begin{bmatrix} \sin\lambda \\ \cos\lambda\cos\phi \\ -\cos\lambda\sin\phi \end{bmatrix} \tag{42}$$

$$\begin{bmatrix} \cos(\gamma_{s1})cos(\alpha_1) \\ \cos(\gamma_{s2})cos(\alpha_2) \\ \cos(\gamma_{s3})cos(\alpha_3) \end{bmatrix} = \begin{pmatrix} \cos\varphi cos\delta_1 cos\omega_1 + sin\varphi sin\delta_1 & -cos\delta_1 sin\omega_1 & -\sin\varphi cos\delta_1 cos\omega_1 + cos\varphi sin\delta_1 \\ \cos\varphi cos\delta_2 cos\omega_2 + sin\varphi sin\delta_2 & -cos\delta_2 sin\omega_2 & -\sin\varphi cos\delta_2 cos\omega_2 + cos\varphi sin\delta_2 \\ \cos\varphi cos\delta_3 cos\omega_3 + sin\varphi sin\delta_3 & -cos\delta_3 sin\omega_3 & -\sin\varphi cos\delta_3 cos\omega_3 + cos\varphi sin\delta_3 \end{pmatrix} x \begin{bmatrix} -\cos(\lambda)\sin(\zeta) \\ \sin(\zeta)\sin(\lambda)\cos(\phi) + \cos(\zeta)\sin(\phi) \\ -\sin(\zeta)\sin(\lambda)\sin(\phi) + \cos(\zeta)\cos(\phi) \end{bmatrix} \tag{43}$$

where the angles ϕ, φ, λ and ζ are constants with respect to the local time.

In practice, we can measure the sun tracking angles, i.e., (α_1, α_2, α_3) and (γ_{s1}, γ_{s2}, γ_{s3}) during sun-tracking at three different local times via a recorded solar image of the target using a CCD camera. With the recorded data, we can compute the three arbitrary orientation angles, ϕ, λ and ζ, of the solar collector using the third-order determinants method to solve the three simultaneous equations (41)–(43). From the equation (42), the orientation angle λ can be determined as follows:

$$\lambda = arcsin\left[\frac{\left\|\begin{pmatrix} sin\gamma_{s1} cos\alpha_1 & -cos\delta_1 sin\omega_1 & -\sin\varphi cos\delta_1 cos\omega_1 + cos\varphi sin\delta_1 \\ sin\gamma_{s2} cos\alpha_2 & -cos\delta_2 sin\omega_2 & -\sin\varphi cos\delta_2 cos\omega_2 + cos\varphi sin\delta_2 \\ sin\gamma_{s3} cos\alpha_3 & -cos\delta_3 sin\omega_3 & -\sin\varphi cos\delta_3 cos\omega_3 + cos\varphi sin\delta_3 \end{pmatrix}\right\|}{\left\|\begin{pmatrix} \cos\varphi cos\delta_1 cos\omega_1 + sin\varphi sin\delta_1 & -cos\delta_1 sin\omega_1 & -\sin\varphi cos\delta_1 cos\omega_1 + cos\varphi sin\delta_1 \\ \cos\varphi cos\delta_2 cos\omega_2 + sin\varphi sin\delta_2 & -cos\delta_2 sin\omega_2 & -\sin\varphi cos\delta_2 cos\omega_2 + cos\varphi sin\delta_2 \\ \cos\varphi cos\delta_3 cos\omega_3 + sin\varphi sin\delta_3 & -cos\delta_3 sin\omega_3 & -\sin\varphi cos\delta_3 cos\omega_3 + cos\varphi sin\delta_3 \end{pmatrix}\right\|}\right] \tag{44}$$

Similarly, the other two remaining orientation angles, ϕ and ζ can be resolved from the equations (42) and (43) respectively as follows:

$$\phi = arcsin\left[\frac{-\frac{1}{\cos(\lambda)}\begin{vmatrix} \cos\varphi cos\delta_1 cos\omega_1 + sin\varphi sin\delta_1 & -cos\delta_1 sin\omega_1 & \sin\gamma_{s1} cos\alpha_1 \\ \cos\varphi cos\delta_2 cos\omega_2 + sin\varphi sin\delta_2 & -cos\delta_2 sin\omega_2 & \sin\gamma_{s2} cos\alpha_2 \\ \cos\varphi cos\delta_3 cos\omega_3 + sin\varphi sin\delta_3 & -cos\delta_3 sin\omega_3 & \sin\gamma_{s3} cos\alpha_3 \end{vmatrix}}{\begin{vmatrix} \cos\varphi cos\delta_1 cos\omega_1 + sin\varphi sin\delta_1 & -cos\delta_1 sin\omega_1 & -\sin\varphi cos\delta_1 cos\omega_1 + cos\varphi sin\delta_1 \\ \cos\varphi cos\delta_2 cos\omega_2 + sin\varphi sin\delta_2 & -cos\delta_2 sin\omega_2 & -\sin\varphi cos\delta_2 cos\omega_2 + cos\varphi sin\delta_2 \\ \cos\varphi cos\delta_3 cos\omega_3 + sin\varphi sin\delta_3 & -cos\delta_3 sin\omega_3 & -\sin\varphi cos\delta_3 cos\omega_3 + cos\varphi sin\delta_3 \end{vmatrix}}\right] \quad (45)$$

$$\zeta = arcsin\left[\frac{\frac{1}{\cos(\lambda)}\begin{vmatrix} \cos\gamma_{s1} cos\alpha_1 & -cos\delta_1 sin\omega_1 & -\sin\varphi cos\delta_1 cos\omega_1 + cos\varphi sin\delta_1 \\ \cos\gamma_{s2} cos\alpha_2 & -cos\delta_2 sin\omega_2 & -\sin\varphi cos\delta_2 cos\omega_2 + cos\varphi sin\delta_2 \\ \cos\gamma_{s3} cos\alpha_3 & -cos\delta_3 sin\omega_3 & -\sin\varphi cos\delta_3 cos\omega_3 + cos\varphi sin\delta_3 \end{vmatrix}}{\begin{vmatrix} \cos\varphi cos\delta_1 cos\omega_1 + sin\varphi sin\delta_1 & -cos\delta_1 sin\omega_1 & -\sin\varphi cos\delta_1 cos\omega_1 + cos\varphi sin\delta_1 \\ \cos\varphi cos\delta_2 cos\omega_2 + sin\varphi sin\delta_2 & -cos\delta_2 sin\omega_2 & -\sin\varphi cos\delta_2 cos\omega_2 + cos\varphi sin\delta_2 \\ \cos\varphi cos\delta_3 cos\omega_3 + sin\varphi sin\delta_3 & -cos\delta_3 sin\omega_3 & -\sin\varphi cos\delta_3 cos\omega_3 + cos\varphi sin\delta_3 \end{vmatrix}}\right] \quad (46)$$

SINGLE-AXIS SOLAR TRACKERS

Literature Survey

In early 1978 Edwards (1978) proposed a design study of a computer based system which controls a large number of paraboloidal collectors for Sun following operation. The system operates with the computer changing the speeds of each of the collector actuators in the field at regular intervals over the day. It is assumed that each collector requires individual attention in the calculation of appropriate speeds. Sources of following error within the system are evaluated, and for specific data rates over the communication link between the central controller and the field of collectors, the variables within the system are chosen to minimize the following error. Accurate sun following is shown to require a data output from the central controller of only 500 bit/sec for 10,000 collectors.

Lynch and Salameh (1990) described the design of a sun tracker which uses two electro-optic sensors and a small, low-cost electronic control circuit. One sensor is a four-cell pyramid which is mounted on the tracker plane. The second sensor is a sunlight beam sensor which is fixed facing south. The control circuit tracking resolution is within 0.1 degrees. Lynch and Salameh (1990) claimed that, this system minimizes wandering on partially overcast days and it will never make multiple revolutions or face down towards the ground. A very simple, reliable solar tracker is described (Poulek and Libra, 2000). The solar tracker is based on an arrangement of auxiliary bifacial solar cell connected directly to D.C. motor.

Auxiliary solar cells (panels) can both sense and provide energy for tracking. The tilt angle needs to be optimized to obtain higher power output, which is dependent on the latitude.

Clifford and Eastwood (2004) presented a low cost solar tracker suitable for use in equatorial regions around the world. The solar tracker is passively activated by aluminium/steel bimetallic strips and controlled by a viscous damper. Computer modeling predicts an increase in efficiency of up to 23% over fixed solar panels. Experimental testing shows excellent agreement with the computer model. In addition, further developments to the design are critically evaluated in terms of complexity and benefit. The performance of passive tracker using mechanical system was found to be comparable to active tracking system. Even so, passive tracker system has not yet been widely accepted by the consumer even though the cost is often less expensive (Clifford and Eastwood, 2004).

Mavromatakis and Franghiadakis (2008) presented a single-axis azimuthal tracker. The novel feature of this tracker is the ability to move the collector's plane in two directions through a special support structure. This structure consists of a sliding mechanism on the central axis and a curved window on the cylindrical surface coaxial to the central axis. Consequently, the proposed novel heliotrope behaves similarly to a two-axis tracker. It is shown that the new tracker system can be very efficient since its plane intercepts, at least, 98% of the insolation with respect to a two-axis tracker. Huang, Sun and Ho (2006) proposed a PV system design, called "near-maximum power-point-operation" (nMPPO) that can maintain the performance very close to PV system with MPPT (maximum-power-point tracking) but eliminate hardware of the MPPT. The concept of nMPPO is to match the design of battery bank voltage V_{set} with the MPP (maximum-power point) of the PV module based on an analysis using meteorological data. Three design methods were used by the authors to determine the optimal V_{set}. The analytical results show that nMPPO is feasible and the optimal V_{set} falls in the range 13.2–15.0 V for MSX60 PV module. The long-term performance simulation shows that the overall nMPPO efficiency η_{nMPPO} is higher than 94%. Two outdoor field tests were carried out by the authors to verify the design of nMPPO. The test results for a single PV module (60Wp) indicate that the nMPPO efficiency η_{nMPPO} is mostly higher than 93% at various PV temperature T_{pv}. Another long-term field test of 1 kWp PV array using nMPPO shows that the power generation using nMPPO is almost identical with MPPT at various weather conditions and T_{pv} variation from 24 °C to 70 °C.

In an attempt to solve the issues related to the limited space available in an urban area and low power output, Huang and Sun (2007) developed a one-axis three position Sun tracking PV module (1A-3P), which is a type of building-integrated photovoltaic system (BIPV) that can be directly mounted onto a wall on the roof of a building in an urban area. The nature of the design means that there is more space for the installation within the limited area available on the roof of a building to increase the power output. The new PV design (1A-3P) is with low concentration ratio reflector. Every PV module is designed with a low concentration ratio reflector and is mounted on an individual Sun tracking frame. The one axis tracking mechanism adjusts the PV position only at three fixed angles (three position tracking): morning, noon and afternoon. This (1A-3P) can be designed in a simple structure with low cost. The analytical results show that the optimal stopping angle β in the morning or afternoon is about 50° from the solar noon position and the optimal switching angle that controls the best time for changing the attitude of the PV module is half of the stopping angle, i.e. $\theta_H = \beta/2$, and both are independent of the latitude. The power generation increases by approximately 24.5% as compared to a fixed PV module for latitude $\phi <$ 50°. The analysis also shows that the effect of installation misalignment away from the true south direction is negligible (<2%) if the alignment error is less than 15°. An experiment performed by Huang and Sun (2007) indicates that the PV power generation can increase by about 23% using low concentration

(2*X*) reflectors. Hence, combining with the power output increase of 24.5%, by using one axis three position tracking, the total increase in power generation is about 56%. The economic analysis shows that the price reduction is between 20% and 30% for the various market prices of flat plate PV modules.

A simple single-axis tracking solar panel was designed using PIC microcontroller for controlling the mechanical movement based on the predetermined position of Sun (Chang, 2009a). The result shows that more than 20% daily power output is generated by tracking system in comparison to a fixed panel. The increment of power output of a solar tracking system was about 20 to 40% as compared to fixed solar panel. However, an external energy such as electrical and mechanical equipment is required to keep the solar panel position always perpendicular to the Sun irradiation and transfer it into useful form of energy. The gain in extraterrestrial radiation received by a single-axis tracked panel relative to a fixed panel was systematically analyzed over a specific period of time (Chang, 2009a). The dynamic angle that the tracked panel should rotate by in order to follow the Sun was derived through a series of spherical trigonometric procedures. The instantaneous incident angle of Sunlight upon the panel was then calculated, assuming that the panel would simultaneously follow the Sun's position. Thus, instantaneous increments of solar energy received by the tracked panel relative to the fixed panel were originally presented. The results show that the angle the tracked panel has to rotate by is 0° at solar noon, and increases towards dawn or dusk. The incident angle of Sunlight upon the tracked panel is always smaller than that upon the fixed panel, except at solar noon. As for panels installed with a yearly optimal tilt angle in Taipei, the gains are between 36.3% and 62.1% for four particular days of year, between 37.8% and 60.8% for the four seasons and 49.3% over the entire year. The amount of radiation collected by the tracked panel is enhanced as the maximum rotation angle is increased. The irradiation ratio of the tracked panel to the fixed panel is close to 1.5 for latitudes below 65° and gradually increases for latitudes above this. The yearly optimal tilt angle of a south-facing fixed panel is approximately equal to 0.9 multiplied by the latitude (i.e. $0.9 \times \varphi$) for latitudes below 65° and is about $56 + 0.4 \times (\varphi - 65)$ otherwise. On the other hand, Chang (2009b) found substantial gains of 51.4%, 28.5% and 18.7% from the extraterrestrial, predicted and observed radiations, respectively, by using a single-axis tracking system.

Huang, Ding and Huang (2011) mentioned that, they built and tested the 1 axis-3 position (1A-3P) Sun tracking PV to measure the daily and long-term power generation of the solar PV system. A comparative test using a fixed PV and a 1A-3P tracking PV was carried out with two identical stand-alone solar-powered LED lighting systems. The field test in the particular days shows that the 1A-3P tracking PV can generate 35.8% more electricity than the fixed PV in a partly-cloudy weather with daily-total solar irradiation $H_T = 11.7$ MJ/m² day, or 35.6% in clear weather with $H_T = 18.5$ MJ/m² day. This indicates that the 1A-3P tracking PV can perform very close to a dual-axis continuous tracking PV (Kacira, Simsek, Babur and Demirkol, 2004). The long-term outdoor test results showed that, the increase of daily power generation of 1A-3P tracking PV increases with increasing daily-total solar irradiation. The increase of monthly-total power generation for 1A-3P Sun tracking PV is between 18.5–28.0%. The total power generation increase in the test period from March 1, 2010 to March 31, 2011, is 23.6% in Taipei (an area of low solar energy resource). The long-term performance, of the proposed 1X-3P tracking PV, is shown very close to the 1-axis continuous tracking PV in Taiwan (Chang, 2009b). If the 1A-3P tracking PV is used in the area of high solar energy resource with yearly-average $H_T > 17$ MJ/m² day, the increase of total long-term power generation with respect to fixed PV will be higher than 37.5%. This is very close to that of dual-axis continuous tracking PV. The 1A-3P tracker can be easily mounted on the wall of a building. The cost of the whole tracker is about the same as the regular mounting cost of a conventional rooftop PV system. This means that there is no extra cost for 1A-3P PV mounted on buildings. The 1A-

3P PV is quite suitable for building-integrated applications. Moreover, Huang, Huang, Chen, Hsu and Li (2013) developed and tested a one-axis three-position (1A-3P) low- cost sun tracking PV as a BIPV. The comparative test of a 1A-3P tracking PV system started from 2010/3/01 was carried out. The single-day increase of energy generation is 39% in clear days. The overall increase of total energy generation from March 1, 2010 to May 31, 2012, is 24.2% in Taipei. The expected overall increase of total energy generation with respect to fixed PV is 37.5% in high solar radiation areas. The installation cost of 1A-3P Sun tracking PV is about the same as the regular mounting cost of a conventional rooftop PV system, but providing 25-37% higher PV energy generation efficiency. For cost reduction in structural design, the wall mounting of 1A-3P tracker may cause PV not facing south exactly. A long-term comparative field test was carried out to study the effect of misalignment using three identical 1A-3P tracking PV module facing south, southeast 45°, and southwest 45°. The measured energy loss is 4.42~6.82% for southeast orientation and 4.31~6.79% for southwest orientation. The test result shows that the energy loss of 1A-3P PV due to misalignment of south is negligible. Many samples of 1A-3P trackers were manufactured and installed for field test. The longest test is over 30 months with satisfactory performance.

Photovoltaic technology allows to, directly, convert solar energy into electrical energy with clear advantages: no environmental impact during operation, reliability and durability of the systems, reduced operating costs and maintenance, ability to both supply remote customers and simply connect to the electrical network. In order to analyze the increase of daily produced energy by using the Sun tracking system Lazaroiu, Longo, Roscia and Pagano (2015) evaluated the performance of two photovoltaic systems: one fixed and one equipped with a Sun tracker. The analysis accounts also the energy consumption of the Sun tracker. An analytical approach was proposed. To validate the results through experimental tests, two alternative low power PV systems were built. Each system consists of a PV source, a MPPT (Maximum Power Point Tracker) power converter and a 12 V–40 A h electrochemical battery, which is used as electric load. The Sun tracker system evidenced an important growth of power production during morning and evening. Lazaroiu, Longo, Roscia and Pagano (2015) presented that a single-axis Sun tracker using a closed-loop operation and two photo sensors increases 12-20% of the produced energy.

Koo, Hong, Lee and Kim (2016) mentioned that, to successfully implement the PV system in a real project, several impact factors should be simultaneously considered. Therefore, they aimed to develop an integrated multi-objective optimization (IMOO) model for determining the optimal solution in implementing the rooftop PV system. They followed in their study six steps: (i) establishment of database; (ii) generation of the installation scenarios in the rooftop PV system; (iii) energy simulation using the software program RETScreen; (iv) economic and environmental assessment from the life cycle perspective; (v) establishment of the IMOO process using a genetic algorithm; and (vi) systemization of the IMOO model using a Microsoft-Excel-based VBA. Two criteria were used to assess the robustness and reliability of the developed model. In terms of effectiveness, the optimal solution was determined from a total of 399,883,120 (=91×49×19×80×59) possible scenarios by comprehensively considering various factors. In terms of efficiency, it was concluded that the time required for determining the optimal solution was 150 s. The developed model makes it possible for final decision-maker such as construction managers or contractors to determine the optimal solution in implementing the rooftop PV system in the early design phase.

Farhat, Barambones and Sbita (2017) presented a photovoltaic (PV) system with a maximum power point tracking (MPPT) facility in order to maximize power extraction from the photovoltaic generator (PVG). The authors achieved their goal by using a sliding mode controller (SMC) that drives a boost converter connected between the PVG and the load. The system is modeled and tested under MATLAB/

SIMULINK environment. In simulation, the sliding mode controller offers fast and accurate convergence to the maximum power operating point that outperforms the well-known perturbation and observation method (PandO). The sliding mode controller performance is evaluated during steady-state, against load varying and panel partial shadow (PS) disturbances. To confirm their theoretical conclusion, a practical implementation of the maximum power point tracker based sliding mode controller on a hardware setup is performed on a dSPACE real time digital control platform. The data acquisition and the control system are conducted all around dSPACE 1104 controller board and its RTI environment. The experimental results demonstrate the validity of the proposed control scheme over a stand-alone real photovoltaic system. Thus, the control algorithm for the 1A-3P Sun tracker is used to rotate the PV modules at three fixed positions: 50° east in the morning, 0° at noon, and 50° west in the afternoon, regardless of the level of power generation able to be achieved by the PV modules. However, the efficiency of a 1A-3P Sun tracker that is installed in an urban area is affected by the location of the Sun, irradiation, and shadows from surrounding buildings and cloud cover since the PV modules are fixed at three different angles (Farhat, Barambones and Sbita, 2017).

Rotational Angle for the Optimum Tracking

The beam radiation on a tracking surface could be maximized by orienting the surface, within the constraints of the tracking apparatus, so that the solar radiation incidence angle is minimized. The incidence angle, θ, is the angle between a ray from the Sun and the normal to the surface. Chapter 1 provides expressions for the incidence angle in terms of the surface tilt and azimuth angles for a tilted surface with any orientation. Braun and Mitchell (1983) provide expressions for the incidence angle for optimally tracking one- and two-axis surfaces.

Although the rotation angle is an intermediate value for determining the incidence angle, it has applications of its own for the control of tracker movement and for modeling the solar radiation available for a collector. For a motorized tracker with fixed gearing, the tracker rotation is directly proportional to the number of motor revolutions; consequently, the calculated rotation angle can be used to determine the number of motor revolutions to move the tracker to its optimum position. When modeling collector solar radiation, the rotation angle can also be used to account for non-optimum tracking that may occur when the optimum rotation angle exceeds the rotation limits of the tracker.

When considering the rotation angle, R, as the angle of rotation of collector about axis when observed from the inclined end of axis, - 180° to +180°. Equals zero when the normal to the surface is in the vertical plane, clockwise is positive, the surface tilt, β, and the surface azimuth, γ, become functions of the axis tilt, β_a, the axis azimuth, γ_a, and the rotation angle, R. Figure 9 could be used to determine the relationship between these angles. For the analysis, γ_a is the azimuth of the tracker axis when observed from the inclined end of the tracker axis, and R, also observed from the inclined end of the tracker axis, is positive for clockwise rotation and negative for counterclockwise rotation. In Figure 9, the normal to the surface is the unit normal represented by the line OA. Line OB is the unit normal rotated angle R about the axis. The triangles formed by the unit normal OA, the unit normal OB and the vertical axis are used to derive equations for the surface tilt and azimuth.

Recognizing that triangles AOC and DOE are similar triangles whose respective sides are proportional, so:

$$\frac{OD}{OA}=\frac{OE}{OC}==>\frac{cos\left(\beta\right)}{cos\left(\beta_a\right)}=\frac{cos\left(R\right)}{1}==>cos\left(\beta\right)=cos\left(R\right)cos\left(\beta_a\right) \tag{47}$$

Thus, the surface tilt angle could be expressed from the equation (47) as:

$$\beta = arccos\left[cos\left(R\right)cos\left(\beta_a\right)\right] \tag{48}$$

The surface azimuth angle differs from the axis azimuth angle by the angle BED:

$$\gamma = \gamma_a + \widehat{BED} \tag{49}$$

It is seen form the triangle BED in the Figure 9 that:

$$\sin\left(\widehat{BED}\right)=\frac{\sin\left(R\right)}{\sin\left(\beta\right)}==>\widehat{BED}=\arcsin\left[\frac{\sin\left(R\right)}{\sin\left(\beta\right)}\right] \tag{50}$$

Here it should be mentioned that, the equation (50) is valid for $\beta \neq 0^o$. For horizontal surface $\beta = 0^o$ the surface azimuth could not be determined and γ is assigned any value because the surface is horizontal and assumed to have no azimuth response. The second condition of the equation (49) applicability is deduced from Figure 9: $-90^o \leq R \leq 90^o$. Consequently, the surface azimuth is expressed as:

$$\gamma = \gamma_a + \arcsin\left[\frac{\sin\left(R\right)}{\sin\left(\beta\right)}\right] \tag{51}$$

Equation (51) will not give the correct solution when R is outside the range of -90° to +90° because it does not distinguish between trigonometric quadrants when performing the arcsine operation. R can fall outside the range of -90° to +90° when the solar azimuth differs by more than 90° from the axis azimuth and the axis tilt is greater than zero. Extreme cases are midnight Sun conditions for northernmost locations, where R can range from -180° to +180° as the tracker follows a Sun that never sets.

For R values outside the range of -90° to +90°, the surface azimuth angle γ could be determined using the following equation (Marion and Dobos, 2013):

$$\gamma = \begin{cases} \gamma_a - \arcsin\left[\frac{\sin\left(R\right)}{\sin\left(\beta\right)}\right] - 180^o \; for -180^o \leq R \leq -90^o \\ \gamma_a - \arcsin\left[\frac{\sin\left(R\right)}{\sin\left(\beta\right)}\right] + 180^o \; for +90^o \leq R \leq +180^o \end{cases} \tag{52}$$

Figure 9. The geometry for one-axis tracking surface

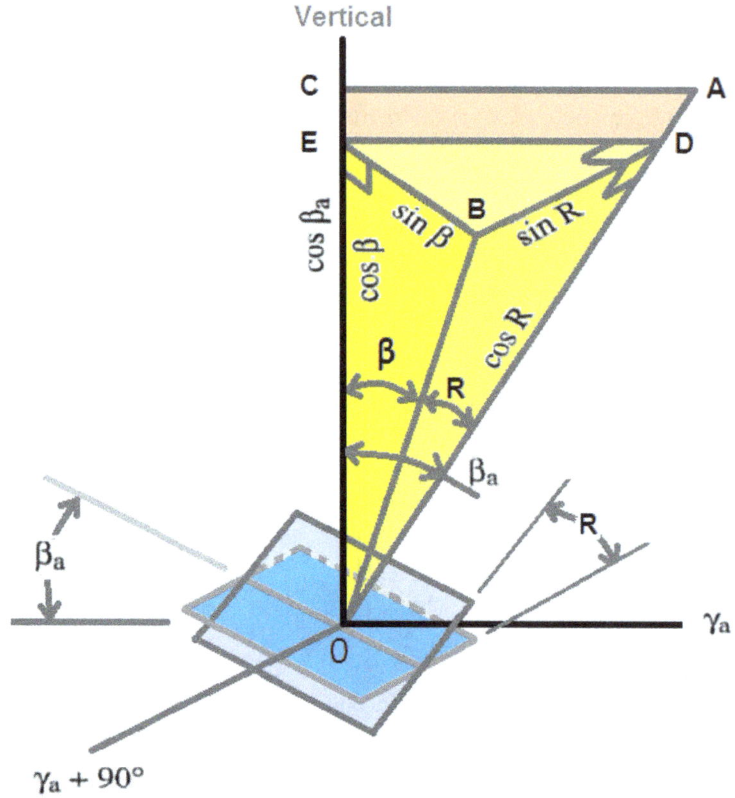

The equation (32) of the first chapter give the trigonometric relationship between the incidence angle θ_i, the zenith angle θ_z, the solar azimuth angle γ_s, the surface tilt angle β and the surface azimuth angle γ:

$$\cos\left(\theta_i\right) = \cos\left(\theta_z\right) * \cos\left(\beta\right) + \sin\left(\theta_z\right) * \sin\left(\beta\right) * \cos\left(\gamma_s - \gamma\right) \tag{53}$$

where the zenith angle θ_z can be determined using the equation (23) of the first chapter while the solar azimuth angle γ_s can be determined using the equation (21) and (22) of the first chapter.

To introduce angle R into equation (53) and remove β and γ from this equation, it is necessary to do the substitutions for $\cos\left(\beta\right)$, $\sin\left(\beta\right)$, and $\cos\left(\gamma_s - \gamma\right)$. From the equation (47) we have $cos\left(\beta\right) = cos\left(R\right) cos\left(\beta_a\right)$. Then, using the identity:

$$\sin^2\left(\beta\right) + \cos^2\left(\beta\right) = 1$$

we find:

$$\sin(\beta) = \sqrt{1-\cos^2(\beta)} = \sqrt{\left\{1-\left[cos(R)\, cos(\beta_a)\right]^2\right\}} \tag{54}$$

By supposing $a = \gamma_s - \gamma_a$ and $b = \widehat{BED}$ and taking into consideration, according to the equation (3), that:

$$a = \gamma_s - \gamma_a = \gamma_s - \left(\gamma + \widehat{BED}\right) ==> a + b = \gamma_s - \gamma \tag{55}$$

Thus, using the identity, $\cos(a-b) = cos(a) * cos(b) + sin(a) * sin(b)$,

$$\cos(\gamma_s - \gamma) = \cos(\gamma_s - \gamma_a)\cos\left(\widehat{BED}\right) + \sin(\gamma_s - \gamma_a)\sin\left(\widehat{BED}\right) \tag{56}$$

But, according to equations (50) and (54), we have:

$$\sin\widehat{BED} = \frac{\sin(R)}{\sin(\beta)} = \frac{\sin(R)}{\sqrt{1-\left[cos(R)\, cos(\beta_a)\right]^2}} \tag{57}$$

$$\begin{aligned}\cos\left(\widehat{BED}\right) &= \sqrt{1-\frac{\sin^2(R)}{1-\left[cos(R)\, cos(\beta_a)\right]^2}} = \sqrt{1-\frac{1-\cos^2(R)}{1-\left[cos(R)\, cos(\beta_a)\right]^2}} \\ &= \sqrt{\frac{\left[cos(R)\, cos(\beta_a)\right]^2+\cos^2(R)}{1-\left[cos(R)\, cos(\beta_a)\right]^2}} = \cos R\sqrt{\frac{-cos^2(\beta_a)+1}{1-\left[cos(R)\, cos(\beta_a)\right]^2}} \\ &= \frac{\cos R sin\beta_a}{\sqrt{1-\left[cos(R)\, cos(\beta_a)\right]^2}}\end{aligned} \tag{58}$$

When substituting (57) and (58) into (56) we obtain:

$$\cos(\gamma_s - \gamma) = \frac{\cos(R)\, sin(\beta_a)\cos(\gamma_s-\gamma_a) + \sin(R)\sin(\gamma_s-\gamma_a)}{\sqrt{\left[1-cos^2(R)\, cos^2(\beta_a)\right]}} \tag{59}$$

Then, by substituting $\cos(\beta)$, $\sin(\beta)$ and $\cos(\gamma_s - \gamma)$ using the equations (47), (54) and (59) respectively in the equation (53), we obtain:

$$\cos(\theta_i) = cos(R)\left[\cos(\theta_z) * cos(\beta_a) + \sin(\theta_z) * sin(\beta_a)\cos(\gamma_s - \gamma_a)\right] + \sin(R) * \sin(\theta_z) * \sin(\gamma_s - \gamma_a) \tag{60}$$

For optimum tracking, the value of *R* should give the minimum incidence angle, thereby maximizing the value of $\cos(\theta_i)$. This value of *R* is determined by differentiating equation (60) with respect to *R*, setting it equal to zero, and solving for *R*.

$$\frac{d\cos(\theta_i)}{dR} = -\sin(R)\left[\cos(\theta_z) * cos(\beta_a) + \sin(\theta_z) * sin(\beta_a)\cos(\gamma_s - \gamma_a)\right] + \cos(R) * \sin(\theta_z) * \sin(\gamma_s - \gamma_a) = 0 \tag{61}$$

Thus,

$$\tan(R) = \frac{\sin(\theta_z) * \sin(\gamma_s - \gamma_a)}{\left[\cos(\theta_z) * cos(\beta_a) + \sin(\theta_z) * sin(\beta_a)\cos(\gamma_s - \gamma_a)\right]} \tag{62}$$

So, with taking into consideration the above mentioned conditions, the rotational angle *R* is:

$$R = arctan\left\{\frac{\sin(\theta_z) * \sin(\gamma_s - \gamma_a)}{\left[\cos(\theta_z) * cos(\beta_a) + \sin(\theta_z) * sin(\beta_a)\cos(\gamma_s - \gamma_a)\right]}\right\} + \psi = arctan(X) + \psi \tag{63}$$

where,

$$X = \frac{\sin(\theta_z) * \sin(\gamma_s - \gamma_a)}{\left[\cos(\theta_z) * cos(\beta_a) + \sin(\theta_z) * sin(\beta_a)\cos(\gamma_s - \gamma_a)\right]} \tag{64}$$

$\psi = 0^o$ if X = 0, or if X > 0 and $(\gamma_s - \gamma_a) > 0^o$, or if X < 0 and $(\gamma_s - \gamma_a) < 0^o$

$\psi = 180^o$ if X < 0 and $(\gamma_s - \gamma_a) > 0^o$

$\psi = -180^o$, if X > 0 and $(\gamma_s - \gamma_a) < 0^o$.

The variable Ψ places *R* in the correct trigonometric quadrant. For determining which value of Ψ to use in equation (63), the difference $(\gamma_s - \gamma_a)$ is evaluated as the angular displacement between two vectors so that it falls within the range of -180° to +180°. For example, if γ_s =20° and γ_a = 210°, $(\gamma_s - \gamma_a)$ is evaluated as 20° + 360° - 210° = 170° and not 20° - 210° = -190°.

Procedure for Determining Incidence Angle

The use of the equations of items 6.5.2 to determine the beam incidence angle on a one-axis tracking surface might be considered a series of steps:

Step 1: Calculate γ_s using the equations (21) and (22) of the first chapter and Θ_z using the equation (23) of the first chapter.
Step 2: Calculate *R* using equation (63).
Step 3: Calculate β using equation (48).
Step 4: Calculate γ using either equation (51) or equation (52).
Step 5: Calculate Θ_i using equation (53).

For concentrating collectors, the procedure can be simplified slightly by using equation (60) to calculate the incidence angle after *R* has been determined in Step 2. The surface tilt, β, although necessary to model the diffuse radiation for a flat-plate collector, has no role in determining the beam radiation for a concentrator; consequently, steps 3–5 are replaced by the use of equation (60). The use of *R* permits better modeling of tracker performance because it can be compared to a tracker's design parameters to see if *R* is within the range of the tracker's physical rotation limits. If not, an additional step in the procedure can set *R* equal to the limit of the tracker rotation range before completing steps 3–5. A similar step could be performed using β, but *R* is more convenient because for a given tracker the physical limits of rotation, unlike the surface tilt, do not change if the axis tilt is changed.

Large one-axis trackers often have a horizontal axis because its simpler construction offers cost advantages. For a horizontal axis, the equations can be simplified because the range for *R* is -90° to +90°, and $sin(\beta_a)$ and $cos(\beta_a)$ can be evaluated for $\beta_a = 0^o$. This results in the following equations for *R*, β, and γ. After putting $cos(\beta_a) = 1$, we obtain from the equations (47) and (48):

$$\beta = |R| \tag{65}$$

After putting $cos(\beta_a) = 1$ and $sin(\beta_a) = 0$, we obtain from the equation (63):

$$R = arctan\left[Tan(\theta_z)*\sin(\gamma_s - \gamma_a)\right] \tag{66}$$

Finally, for $\beta \neq 0^o$, the equation (51) could be applied.

If β equals zero, γ can be assigned any value because the surface is horizontal. The tracker axis azimuth, γ_a, can be assigned from either end of the tracker.

Single Axis Solar Tracking Example

An algorithm for a one-axis maximum power generation system based on the single axis Sun tracker is recently developed that automatically identifies the optimal stopping angle for the PV modules, avoiding any shading and increasing the amount of power generated. The PV modules for the tracking system are automatically rotated from east to west once per half an hour during daylight hours in order to identify the optimal angle that generates the maximum amount of power. The range for each rotation is from 60° southeast to 60° southwest. A good quality timer IC is used in order to achieve precise timing. The proposed control algorithm 100 is called the "direct-search method". The impact on power generation when using the direct-search method compared with the previously designed mechanical one axis tracker is investigated. A disadvantage of the direct-search method, however, is that the high detection frequency and the range used for each rotation will cause the lifetime of the components to be reduced as well as increasing the cost of maintenance. In order to solve these problems, the detection frequency and range are optimized, and then the influence on the power generation of the optimization is also investigated. Therefore, by comparing with the reported research, the advantage of the proposed single axis tracker is real-timely calculating the instantaneous power corresponding to the current irradiation and automatically avoiding the environment interferences so that it has higher efficiency with low-cost, higher stability and reliability. The new MPPT technique, named nMPPO, is used to track the maximum power point (Huang, Sun and Ho, 2006).

The single axis sun tracker consists of a supporting stainless steel bar, a PV mounting frame, and a linear actuator, as shown in Figure 2. The structure of the tracker is designed to hold two PV modules (250 W_{peak} each). The linear actuator contains a DC motor that is used to rotate the PV mounting frame, and also includes a potentiometer that is used to detect the position of the linear actuator, thereby locating the current position of the PV modules on the sun tracker and allowing them to be rotated to the next desired angle.

In order to investigate the power generated by the sun tracker using the direct-search method, the sun tracker was connected to batteries and lamps to form a standalone solar PV system (Huang, Hsu, Wu and Ho, 2010). Each solar PV system contains two 250 W_{peak} PV modules connected in series, and two 38 Ah/12 V batteries also connected in series. The load is composed of four 12 V/50 W Halogen lamps connected in both series and parallel.

The controller first receives a signal to indicate the current angle of the PV modules, and then instructs the modules to rotate from 60° southeast to 60° southwest, which takes about 25 seconds for each rotation. The instantaneous power generated at different angles is recorded, meaning that the optimal angle for generating the maximum amount of power can be determined. Once identified, the PV modules are then rotated to the resulting angle. The controller is connected to the computer via an RS-485 interface. A solar charge and discharge controller is used to determine whether to deliver the generated power to the battery or the load, or to both at the same time.

As the PV modules begin to sense solar irradiation from sunlight, the PV voltage gradually increases. Once the PV voltage reaches a level greater than the battery voltage, the solar charge controller instructs the system to charge the battery using the generated current. At the same time, both the battery voltage and the current are measured. Once the battery voltage reaches the upper limit of the charging voltage, the charging current is reduced so as to protect the battery from potential damage caused by overcharging. Because the charge current is forcibly reduced when the battery voltage reaches the upper limit of the charging voltage, the generated power is also reduced (P=IV). Consequently, in order to retain the generated power, the system will begin delivering the generated power to the load and continue charging the battery until it again reaches the upper limit of the charging voltage.

The lead-acid battery which is widely used in stand-alone solar system is easily damaged by a poor charging control which causes overcharging. The battery charging control is thus usually designed to stop charging after the overcharge point. This will reduce the storage energy capacity and reduce the service time in electricity supply. The design of charging control system however requires a good understanding of the system dynamic behaviour of the battery first. Manwell and Mcgowan (1993) described a battery model developed for use in time series performance models of hybrid energy systems. The model is based on the approach of chemical kinetics and is specifically concerned with the apparent change in capacity as a function of charge and discharge rates. It assumes that the charge can be stored in two ways, either as immediately available or as chemically bound. Huang, Hsu, Wu and Ho (2010) derived experimentally a first-order system dynamics model of lead-acid battery at different operating points near the overcharge voltage. Then, a charging control system based on PI algorithm was developed using PWM charging technique. The feedback control system for battery charging after the overcharge point (14 V) was designed to compromise between the set-point response and the disturbance rejection. The experimental results show that the control system can suppress the battery voltage overshoot within 0.1 V when the solar irradiation is suddenly changed from 337 to 843 W/m^2. A long-term outdoor test for a solar LED lighting system showed that the battery voltage never exceeded 14.1 V for the set point 14 V and the control system can prevent the battery from overcharging. The test result also indicated that the control system is able to increase the charged energy by 78%, as compared to the case that the charging stops after the overcharge point (14 V).

As the level of solar irradiation from sunlight decreases, the PV voltage gradually decreases. Once the PV voltage level is lower than the battery voltage, the controller instructs the system to discharge using the pulse-width-modulation (PWM) technique to control the discharging current (Huang, Hsu, Wu and Ho, 2010). Once the voltage level in the battery reduces to below the over-discharge voltage level, the system will discontinue the discharge process to protect the battery from damage.

Since the sun tracker can be rotated, changing the orientation of the PV modules to determine the optimal stopping angle can prevent or reduce the effects of any shaded area, thereby increasing the ability to generate power (Ahmed and Salam, 2015). No matter whether the PV module is affected by shade, this rotation ability ensures that the optimal stopping angle can be identified. In addition, whether the battery voltage is in charging or discharging mode, any variation in battery voltage over a short period of time is very small. Consequently, the battery voltage can be regarded as a constant when considering short durations. This means that the generated instantaneous power is proportional to the generated instantaneous current.

Experimentally it was found that, at a mean daily total solar irradiation of H_t=10 MJ/m^2 day, the power generation obtained from the direct-search method based on the reduced detection range is almost the same as that obtained from the original direct-search method. This indicates that the reduction in the detection range does not significantly affect the amount of power that can be generated.

Finally, it should be mentioned that, the long-term experimental results indicate that the amount of power that can be generated using the direct-search method are 3.4% higher than that achieved using the mechanical control in spring/autumn, 5.4% higher in summer, and 8.3% higher in winter. This indicates that the direct-search method is more efficient in summer and winter because the weather conditions in Taiwan are more unstable during spring and autumn.

DUAL-AXIS SOLAR TRACKERS

Literature Survey

Augmentation in use of solar energy has been reiterated in the literature. The significant role of this gradual but continuous growth lies in its ubiquitously known ride through other conventional energy sources in terms of supplement to increasing energy demand. Whereas the only limitation with solar photovoltaic (PV) system is that it suffers from low efficiency energy conversion, therefore it becomes more important to draw maximum available power from solar PV. As the I-V characteristic of PV is nonlinear and alters with different values of insolation and temperature, it causes alteration in Maximum Power Point (MPP), also at the time of partial shading condition the point of maximum power deviates. In literature, good number of publications proclaimed for tracking the maximum power point for uniform insolation as well as for partial shading condition. All methods in the field of MPPT technique have their unique advantages and disadvantages which requisites a thorough and informative comparative analysis of different methods.

Using two axes Sun tracking system is inevitable in some solar system and it can be adapted to all solar systems to improve collection of solar radiation (Abdallah and Nijmeh, 2004). It was found experimentally that, the measured collected solar energy on the moving surface was significantly larger than that on a fixed surface. The two axes tracking surface showed a better performance with an increase in the collected energy of up to 41.34% compared with the fixed surface tilted at 32° towards the south (Abdallah and Nijmeh, 2004).

Seme and Stumberger (2011) determined the trajectories of a dual-axis sun tracking system for a photovoltaic system by an optimization procedure in order to maximize the electrical energy production within a photovoltaic system, by considering the tracking system consumption. The procedure used for determining the tilt angle and azimuth angle trajectories is described as a nonlinear and bounded optimization problem. Since an explicit form of the objective function is unavailable, a stochastic search algorithm called Differential Evolution was applied as the optimization tool. In order to evaluate the objective function, models for calculating the available solar radiation and tracking system consumption were applied together with the efficiencies of solar cells, a DC/DC converter and inverter. A new algorithm was introduced for the time dependent prediction of available solar radiation. It is based on the length of a sunbeam's path through the atmosphere and the statistical data of a pyranometer measured total and diffuse solar radiation at a given location on the Earth. The optimization bounds were given in the form of angular speed, lower and upper bounds for both angles and angle quantization. The results presented by Seme and Stumberger (2011) showed, that the optimal trajectories can help to increase the electrical energy production within photovoltaic systems by Sun tracking.

Quesada, Guillon, Rousse, Mehrtash, Dutil and Paradis (2015) developed tracking strategy for dual-axis Sun tracker in Montreal, Canada and in other high latitudes. They assessed a solar tracking photovoltaic panel hourly and seasonally in high latitudes. A theoretical method based on an isotropic sky model was formulated, implemented, and used in a case study analysis of a grid-connected photovoltaic system in Montreal, Canada. The results obtained, based on the definition of a critical hourly global solar radiation, were validated numerically and experimentally. The study confirmed that a zenith-set Sun tracking strategy for overcast or mostly cloudy days in summer is not advantageous.

Yilmaz, Ozcalik, Dogmus, Dincer, Akgol and Karaaslan (2015) mentioned that, In order to design new PV systems that will be installed to operate in more efficient and more feasible way, it is necessary to analyze parameters like solar radiation values, the angle of incidence of the genus, temperature etc. Therefore, in this study, theoretical works have been performed for solar radiation and angle of incidence values of any location, plus an experimental study was carried out on a system tracking the Sun in two axes and in a fixed system. The performed prototype is also adapted into a PV system with 4.6 kW power. Theoretical data are consistent with the data obtained from the PV system with 4.6 kW power. This study could be an important guide for the future PV power stations.

Fathabadi (2016) proposed a novel high accurate sensorless dual-axis solar tracker controlled by the maximum power point tracking unit available in almost all photovoltaic systems. The maximum power point tracking controller continuously calculates the maximum output power of the photovoltaic module/panel/array, and uses the altitude and azimuth angles deviations to track the Sun direction where the greatest value of the maximum output power is extracted. Unlike all other sensorless solar trackers, the proposed solar tracking system is a closed loop system which means it uses the actual direction of the Sun at any time to track the Sun direction, and this is the main contribution of Fathabadi (2016). The proposed solar tracker has the advantages of both sensor based and sensorless dual-axis solar trackers, but it does not have their disadvantages. Other sensorless solar trackers all are open loop, i.e., they use offline estimated data about the Sun path in the sky obtained from solar map equations, so low exactness, cloudy sky, and requiring new data for new location are their problems. A photovoltaic system has been built, and it is experimentally verified that the proposed solar tracking system tracks the Sun direction with the tracking error of 0.11° which is less than the tracking errors of other both sensor based and sensorless solar trackers. An increase of 28.8–43.6% depending on the seasons in the energy efficiency is the main advantage of utilizing the proposed solar tracking system.

Several studies show that from about 20% to 50% more solar energy can be recovered by using photovoltaic systems that track the Sun rather than systems set at a fixed angle. For overcast or cloudy days, recent studies propose the use of a set position in which each photovoltaic panel faces toward the zenith (horizontal position). Compared to a panel that follows the Sun's path, this approach claims that a horizontal panel increases the amount of solar radiation captured and subsequently the quantity of electricity produced.

In order to enhance power generation by focusing on photovoltaic component of the hybrid system, Sinha and Chandel (2016) analyzed fixed tilt and Sun tracking photovoltaic based micro wind hybrid power systems along with determining the optimum configurations for a 6 kW_p roof mounted micro wind based hybrid system using fixed and tracking photovoltaic systems to enhance the power generation potential in a low windy Indian hilly terrain with good solar resource. A comparative power generation analysis of different configurations of hybrid systems with fixed tilt, monthly optimum tilt, yearly optimum tilt and 6 different Sun tracking photovoltaic systems was carried out using Hybrid Optimization Model for Electric Renewables. Monthly and seasonal optimum tilt angles determined for the location vary between 0° and 60° with annual optimum tilt angle as 29.25°. The optimum configurations for all Sun tracking systems except for the two axis tracking system is found to be 7 kW_p photovoltaic system, one 5 kW_p wind turbine, 10 batteries and a 2 kW_p inverter. The optimum configuration for two axis tracking system and two types of fixed tilt systems, was found to be a 8 kW_p photovoltaic system, one 5 kW_p wind turbine, 10 batteries and a 2 kW_p inverter. The results showed that horizontal axis with monthly adjustment, horizontal axis with daily adjustment, horizontal axis with continuous adjustment and two axis tracking system could generate 4.88–26.29% more energy per year than the existing fixed tilt photovoltaic system. The cost of energy was found to be more for two axis tracking than all tracking systems and fixed tilted photovoltaic system. However, two axis tracking was found to be more advantageous than other systems despite higher costs to maximize power output of the hybrid systems. The methodology followed can be applied to improve the power generation potential of photovoltaic–micro wind based hybrid systems for any location worldwide.

Hu and Yao (2016) proposed an effective methodology for calculating the photovoltaic field output. A combination of two methods was first presented for optical performance calculation: point projection method for direction radiation, and Monte Carlo ray-tracing method for both diffuse radiation and albedo radiation. Based on the optical calculation, an accurate output of the photovoltaic field can be obtained through a cell-level simulation of PV system. Several simplified measures were taken to reduce the large amount of calculation work. The proposed methodology has been validated for accurate and fast calculation of field output. With the help of the developed code, Hu and Yao (2016) dealt with the performance comparison between four typical tracking strategies. Through the comparative analysis, the field output was proved to be related to the tracking strategy. For a regular photovoltaic field, the equatorial and elevation-rolling tracking showed the superior performance in annual field output to the azimuth-elevation and rolling-elevation tracking.

Skouri, Ben Haj Ali, Bouadila, Ben Salah and Ben Nasrallah (2016) mentioned that, in order to design a new solar parabolic concentrator that will be installed to operate in more efficient and more feasible way, it is necessary in the first step to analyze such parameters like solar irradiations values, and solar angles. Therefore, Skouri, Ben Haj Ali, Bouadila, Ben Salah and Ben Nasrallah (2016) studied solar angles using time and geographic parameters in Tunisia. An experimental measurement of solar irradiations has been done using a high precision metrological station to determine the diffuse, the direct and the global irradiations on the horizontal plane. Belkaid, Colak and Isik (2016) claimed that, even Perturb and Observe (PandO) and Incremental Conductance (INC) are widely used as Maximum Power Point Tracking (MPPT) techniques in Photovoltaic (PV) systems, they fail under rapidly varying of sunlight levels. Therefore, they proposed a new MPPT technique, which can make a distinction between perturbation in the reference voltage and sudden-changing of sunlight and thus optimize the PV system efficiency. This method consists on a modified INC algorithm, which is used to fine-tune the duty cycle of the DC/DC converter in order to avoid divergences of the maximum power point (MPP) when using basic INC under fast varying of luminosity levels. The proposed PV-MPPT system, which is composed by a step-up converter as the interface to feed the load, is tested by simulation within the Matlab/Simulink software by taking into account the luminosity, the temperature and the load variation. The simulation results are satisfactory and demonstrate that the improved INC technique can track the PV maximum power at diverse operating conditions with the most excellent performance, the energy conversion efficiency is increased by approximately 5%.

Under uniform conditions, there is only a single peak on the P-U curve (Farahat, Enany and Nasr, 2015); in this condition, the different conventional MPPT schemes can be used to extract the maximum power point (MPP). Within these methods, one can mention perturbation and observation (P&O) (Piegari and Rizzo, 2010). Fermia, Granozio, Petrone and Vitelli (2007) claimed that, P&O method is widely diffused because of its low-cost and ease of implementation. When atmospheric conditions are constant or change slowly, the PandO method oscillates close to MPP. However, when these change rapidly, this method fails to track MPP and gives rise to a waste of part of the available energy. Therefore, Fermia, Granozio, Petrone and Vitelli (2007) proposed an adaptive P&O method that has faster dynamics and improved stability compared to the traditional P&O. The proposed MPPT algorithm was set up and validated by means of numerical simulations and experimental tests, confirming the effectiveness of the method. Abdelsalam, Massoud, Ahmed and Enjeti (2011) proposed a high-performance adaptive perturb and observe MPPT technique for photovoltaic-based micro-grids. The incremental conductance is one of these methods (INC).The INC algorithm, one of the MPPT strategies, is widely used for its high tracking accuracy, good adaptability to rapidly changing atmospheric conditions, and easy implementation (Tian, Xia, Xu and Sun, 2014). Xu, Yang, Zhou, Li, Lei and Chen (2015) mentioned that, the incremental conductance (INC) method is one of the most widely applied MPPT methods. However, the choice of the step size still remains controversial. Xu, Yang, Zhou, Li, Lei and Chen (2015) presents an improved variable step size INC MPPT algorithm that uses four different step sizes. This method has the advantages of INC but with the ability to validly adjust the step size to adapt to changes of the PV's power curve. The presented algorithm also simultaneously achieves increased rapidity and accuracy when compared with the conventional fixed step size INC MPPT algorithm. In addition, the theoretical derivation and specific applications of the proposed algorithm were also presented. The proposed method was validated by simulation and experimental results. Kou, Xia and Ye (2015) built a new variable step-size function based on the logarithm form. Through modeling in Matlab/Simulink and experiment, both results showed that the proposed method can improve the MPPT response in both tracking speed and accuracy steadily compared with other published results. Alajmi, Ahmed, Finney and Williams (2011) proposed a new fuzzy-logic controller for maximum power point tracking of photovoltaic (PV) systems basing on a new version of hill climbing. In order to reduce the power losses in MPPT process that occur as a result of intermittent time based short circuit current measurements, Sher, Murtaza, Noman, Addoweesh and Chiaberge (2015) presented an improved fractional short circuit current (FSCC) maximum power point tracking technique in which an additional control loop is used to find the proper moment to measure the SCC. The proposed modification enables the conventional FSCC MPPT to decide intelligently about the measurement of SCC thus reduces the number of times the photovoltaic (PV) module is isolated from the load. Algazar, Al-Monie, El-Halim and Salem (2012), proposed an intelligent control method for the maximum power point tracking (MPPT) of a photovoltaic system under variable temperature and insolation conditions basing on a fuzzy logic controller applied to a DC–DC converter device. The different steps of the design of this controller are presented together with its simulation. However, under partial shading conditions (PSCs) (due to clouds, trees, buildings, and so on), the above mentioned conventional strategies cannot be used, because the P-U curve exhibits several peaks that cannot be distinguished. Ding, Wang, Zhai, XuZhang and Liu (2014) proposed an improved global maximum power point tracking method based on voltage interval for PV array under partially shaded conditions. Maximum Power Point Tracking (MPPT) is a technique employed to extract maximum power available from a photovoltaic (PV) module under varying atmospheric conditions. It traces the PV operating voltage corresponding to the maximum power point (MPP) and operates the panel at MPP. However, if a PV array is partially shaded, the conventional MPPT techniques track local MPP and fail to track global MPP. Also, if modules with different optimal currents are connected in series—parallel local MPPs occur in the *P-V* curves and conventional MPPT techniques fail to search global maxima. A lot of literature is available on global MPPT techniques to increase overall system efficiency. The power conditioning unit should, therefore, be capable of searching global maximum power point also. Therefore, Bhatnagar and Nema (2013) presented a number of conventional and global MPPT techniques and discussed them in detail on the basis of certain performance parameters. Verma, D., Nema, S., Shandilya, A. M. and Dash, S. K. (2015) presented, through an encyclopedic review of MPP Tracking (MPPT) technique which may overcome the distraction of researchers, a comprehensive analysis of maximum power point tracking techniques in solar photovoltaic systems under uniform insolation and partial shaded condition.

Partial shading is one of the unavoidable complications in the field of solar power generation. Although the most common approach in increasing a photovoltaic (PV) array's efficiency has always been to introduce a bypass diode to the said array, this poses another problem in the form of multi-peaks curves whenever the modules are partially shaded. To further complicate matters, most conventional Maximum Power Point Tracking methods develop errors under certain circumstances (for example, they detect the local Maximum Power Point (MPP) instead of the global MPP) and reduce the efficiency of PV systems even further. Presently, much research has been undertaken to improve upon them. To alleviate the above mentioned problem, a number of intelligent schemes have been proposed. Among them is particle swarm optimization (PSO) (Liu, Huang, Huang and Liang, 2012). Ishaque, Salam, Amjad and Mekhilef (2012) proposed an improved maximum power point tracking (MPPT) method for the photovoltaic (PV) system using a modified particle swarm optimization (PSO) algorithm. Later on, Ishaque and Salam (2013) proposed a deterministic particle swarm optimization to improve the maximum power point tracking (MPPT) capability for photovoltaic system under partial shading condition. The main idea was to remove the random number in the accelerations factor of the conventional PSO velocity equation. Seyedmahmoudian, Mekhilef, Rahmani, Yusof and Shojaei (2014) proposed maximum power point tracking of partial shaded photovoltaic array using an evolutionary algorithm technique, also known as a particle swarm optimization technique in MPP detection. Under partial shading conditions (e.g., due to buildings, trees, and clouds), multiple peaks may exist on the power-voltage (P-V) characteristic curve of photovoltaic (PV) array, leading to the conventional maximum power point tracking methods fail to extract the global maximum power point (GMPP). Tian, Xia, Sun, Xu and Zheng (2014) proposed a mathematical model of PV array under partial shading conditions with a voltage calculated principle. The non-linear characteristics of current-voltage (I-V) and P-V curves of PV array, as well as quickly develop the GMPP tracking controllers were studied on the proposed model. Besides, an adaptive random particle swarm optimization (ARPSO) algorithm is presented to accurately extract the GMPP under partial shading conditions. Five simulation cases with different partial shading patterns are used to evaluate the performance of the presented approach by comparing with the conventional PSO, perturbation and observation (P&O), incremental conductance (INC), and genetic algorithm (GA) methods. Simulation results show that the ARPSO algorithm can rapidly find the GMPP under different shading conditions compared with the conventional PSO algorithm. Furthermore, the presented algorithm can accurately extract the GMPP when the shading condition sharply changes, while the P&O and INC algorithms fail to track the GMPP, but only detect the rightmost MPP encountered either local or global and regardless of the course. Besides, the ARPSO can rapidly and accurately converge to the GMPP with smaller population size and higher convergence speed compared with the GA.

The cuckoo search (CS) exhibits several advantages which include fast convergence, higher efficiency using fewer tuning parameters. Ahmed and Salam (2014) outlined the concept of CS by highlighting the significance of the Lévy flight in influencing the algorithm's convergence. The main equations that govern the behavior of the search were also explained. To justify CS as a viable MPPT option, a comprehensive assessment was carried out against two well established methods, namely Perturbed and Observed (P&O) and Particle Swarm Optimization (PSO). The evaluations include (1) gradual irradiance and temperature changes, (2) step change in irradiance and (3) rapid change in both irradiance and temperature. These tests are carried out for both large and medium-sized PV systems. Furthermore, the ability of the algorithm to handle the partial shading condition was demonstrated. For the maximum utilization of solar energy, photovoltaic (PV) power generation systems are operated at the maximum power point (MPP) under varying atmospheric conditions, and MPP tracking (MPPT) is generally achieved using several conventional methods. One of these methods is artificial bee colony algorithms, proposed by Sundareswaran, Sankar and Nayak (2015).

The development of an effective maximum power point tracking (MPPT) algorithm is important in order to achieve maximum power operation in a photovoltaic system (PV). In this context, Kofinas, Dounis and Papadakis (2015) developed a direct neural control (DNC) scheme. The intelligent MPPT controller consists of a hybrid learning mechanism; an on-line learning rule based on gradient decent method and an off-line learning rule based on Big Bang–Big Crunch (BB–BC) algorithm. The effectiveness of the proposed system is tested under partial shading conditions by applying the cascaded converter topology. The feasibility of the DNC is evaluated by the simulation results and compared to the conventional perturbation and observation (P&O). The characteristics of a photovoltaic (PV) array are affected by temperature, solar insolation, and shading. In fact, under partially shaded conditions, the PV array characteristics get more complex with multiple maxima in the *P–V* and *I–V* characteristics. The two common algorithms, the Perturb and Observe (P&O) and the Incremental of Conductance (INC), fail to extract the maximum power of the PV panel if the PV generator is partially shaded. So, in these conditions, these techniques fail to extract the global maximum; however, they only detect the first maximum encountered either local or global and regardless of the course. Two common algorithms, the Perturb and Observe (P&O) and the Incremental of Conductance (IncCond), fail to extract the maximum power of the PV panel if the PV generator is partially shaded. So, in these conditions, these techniques fail to extract the global maximum; however, they only detect the first maximum encountered either local or global and regardless of the course. To resolve these problems, Shaiek, Y., Smida, M. B., Sakly, A. and Mimouni, M. F. (2013) studied and simulated under the same software a photovoltaic solar system composed of a solar panel, under Matlab/Simpower system software, a technique based on Genetic Algorithm (GA). Shaiek, Y., Smida, M. B., Sakly, A. and Mimouni, M. F. (2013)_showed that, the GA method has succeeded to overcome these difficulties and reach the global MPP.

Sundareswaran, Sankar and Nayak (2014) reported the development of a maximum power-point tracking (MPPT) method for photovoltaic (PV) systems under partially shaded conditions using firefly algorithm (FA). The major advantages of the proposed method are simple computational steps, faster convergence, and its implementation on a low-cost microcontroller.

Despite their global tracking ability, these intelligent schemes are typically complex and more time-consuming than the conventional methods.

Dual Axis Solar Tracking Example

A simple tracking system was constructed and tested (Yilmaz, Ozcalik, Dogmus, Dincer, Akgol and Karaaslan, 2015). Its precision was calculated in terms of the angle between the two LDR sensors. The optimal angle between these photo-resistors was evaluated numerically and experimentally. Despite the use of some approximations, such as the linearity of the used LDRs and the constant value of the diffuse radiation, a good agreement between these results was recorded. More precise results could be found if the system is working in close-loop operating-mode. This system was used for the command of photovoltaic panels. It demonstrates acceptable solar tracking results. Closed loop systems with photo sensors are traditionally used as the main method of the control of the sun tracking systems. The photo sensors are used to discriminate the sun's position and to send electrical signals proportional to the error to the controller, which actuates the motors to track the sun (Bentaher, Kaich, Ayadi, Ben Hmouda, Maalej and Lemmer, 2013).

In the design of this system, β is the slope of the surface, and γ is the surface azimuth angle. For two axes tracking, the surface positions are determined as follows: $\beta=\theta_z$ and $\gamma=\gamma_s$, where θ_z is the zenith angle of the sun and γ_s is the sun's azimuth angle. Two tracking motors are used, one for the joint rotating through the horizontal north south axis to control β and the other for the joint rotating through the vertical axis to control γ. Schematic views of the movement range and sun position sensor are shown in Figure 10. Established system has 180° horizontal and 90° vertical moving ability. For the horizontal axis, a DC motor with 0–180° moving ability was used.

Higher difference values of the light intensity that the sensors measured than the reference value trigger the movement of the DC motor in both directions. DC motor is driven by increasing the current via a transistor. Direction control is made by using four transistors. Q1 and Q4 are conducted when the motor turns to the east, Q2 and Q3 is conducted when it turns to the west. Our panel can rotate 1801 on the horizontal axis (i.e. east–west direction). In the horizontal axis, limit switch A is located in the west side and limit switch B is located in the east side. The panel is set as 0° when it is completely in the east and 180° when it is in completely west. When the panel goes to 180° while rotating from east to the west, limit switch A is opened and the engine is turned off. Similarly, when the panel goes to the 0° while rotating from west to the east, limit switch B is opened and the engine is turned off. The circuit is fed with 9 V and uses light sensors (LDR-A and LDR-B) which are identical. Two sensors are connected in series and the connected at the point X.

Rotation direction of the motor is controlled by using two differential amplifiers (Op-Amp-Ave Op-Amp-B). Reference voltage values are entered to the op-amps via voltage dividing resistors (POT-A, POT-B, R1, R2). Again with the voltage dividing resistors (POT-A, POT-B, R1, R2), reference voltage point-Y is connected to number-2 (minus) port of op-amp-A and point-Z is connected to number-5 (plus) port of op-amp-B. When the light sensors have the same amount of resistance values (i.e. when same rate of light is taken), point X becomes 4.5 V. In this case (sensors get the same amount of lights), point Y is set to be 4.73 and point Z is set to be 4.27 V. Additionally, op-amps give the same output (0 V) and since the potential difference at the motor terminals is zero, the panel does not rotate. Circuit scheme of the designed system for horizontal axis is shown in Figure 11.

When the sun goes to the west, LDR-A will get more light and its resistance will decrease. In this case, the voltage at point X will rise over 4.5 V and when it passes 4.73, op-amp-A and op-amp-B will give +9 V and 0 V, respectively. The engine will start to turn towards to the west. Thus, the resistance

Figure 10. Schematic view of the movement range and the sun position sensor

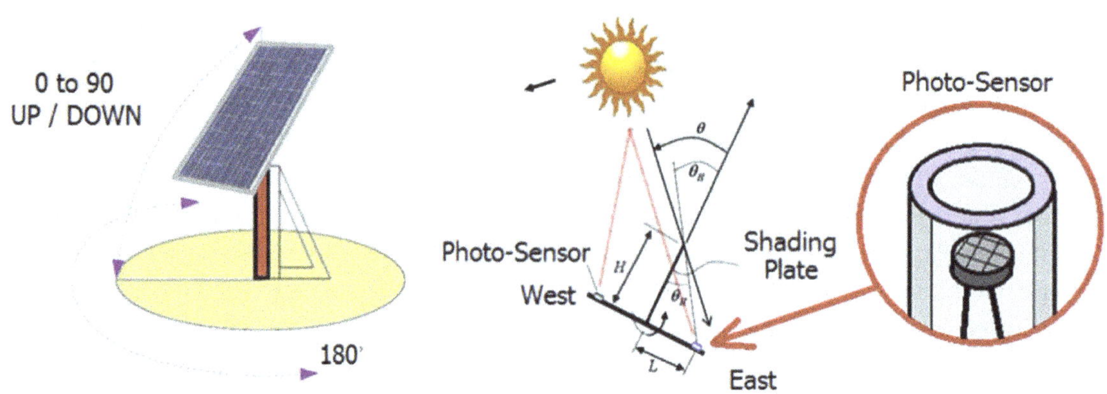

Figure 11. Circuit scheme of the designed system for horizontal axis

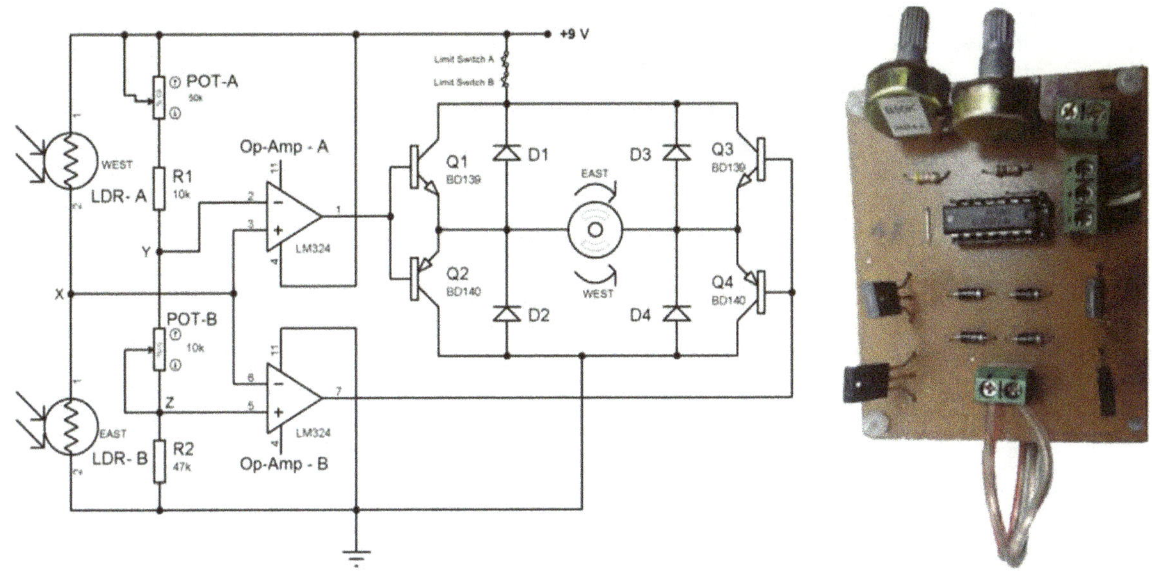

of LDR-B starts to drop and so does the voltage of point X. When the voltage at point X goes below to 4.73 V, op-amps gives the same output (0 V) and the motor stops. Therefore, when the sun goes to the west, the panels also turn to the west. When it opens the limit A switch, then the panel does not rotate anymore. Same process also occurs when the sun goes to the east. When the sun rises, LDR-B will get more sunlight and the resistance will fall. In this case, the voltage at the point X will drop below to 4.5 V and when it passes 4.27 V, op-amp-A and op-amp-B will give 0 V and +9 V, respectively. The motor will start to turn towards to the east. Thus the resistance of LDR-A will decrease and the voltage at point X will rise. When the voltage at point X goes above 4.27 V, op-amps give again the same output (0 V) and the motor stops. If this requirement does not take place, the panel continues to turn towards to the east until it opens limit B switch when the panel does not turn anymore. The algorithm created for the motors that are used separately for both axes is shown in Figure 12 and PV panel was ensured to operate its movements accordingly.

For vertical axis, a step motor with 0–90° moving ability was used. Different values which are higher than the reference value of the light intensity measured by the sensors triggers the step motor to rotate in both directions. Step motor is driven by increasing the current via MOSFETs. Direction control is performed by using 4-MOSFETs. When motor rotated downward, Q1 and Q2 are in transmission, when it goes upward Q3 and Q4 is in transmission stage. The panel can rotate on the vertical axis (i.e. up and down). In vertical axis, limit switch C is located in the up side and limits switch D is located in the downside. When the panel is fully below, it will be accepted as 0° and when it is in the top level, it will be accepted as 90°. When the panel rotates from bottom to the top, it opens limit switch C when it hits 90° and the motor stops. Similarly, when the panel rotates from top to the bottom, it opens limit switch D when it hits 0° and the engine stops again.

The circuit is fed with 5 V and the light sensors of LDR-C and LDR-D are identical. The rotation direction of the motor is controlled by using two differential amplifiers (op-amp-C and op-amp-D). Again with the voltage dividing resistors (POT-C, R3, R4), reference voltage values were entered to the

Figure 12. Designed algorithm to control PV panel movement

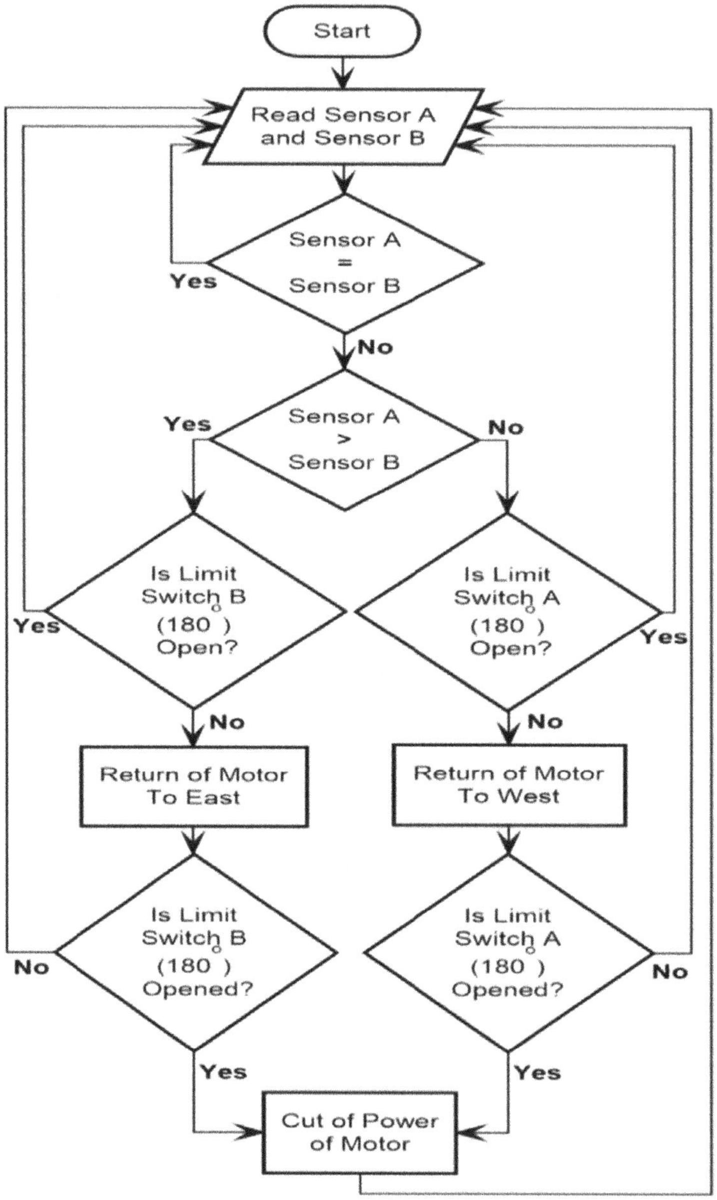

op-amps. Point Y1 is connected to the port-3 of op-amp-C (plus), point Z1 is connected to port-6 of op-amp-D (minus). Outputs of the op-amps are connected to RA0 and RA1 pins of PIC. In addition, by passing op-amp outputs through a NAND gate, they are connected to RA2 port of PIC. While light sensors have the same resistance values (i. e. when they receive same amount of light), point X1 becomes 2.5 V. In this position (when sensors get the same amount of light), point Y1 is set as 2.76 V and point Z1 is set as 2.24 V. In this case, op-amps give the same output (+5 V) (RA0=1 and RA1=1) and the output of the NAND gate becomes 0 (zero) and the signal of 0(zero) goes to RA2. When RA2=0, the motor will not rotate. Circuit scheme of the designed system for vertical axis is shown in Figure 13.

Figure 13. Circuit scheme of the designed system for vertical axis

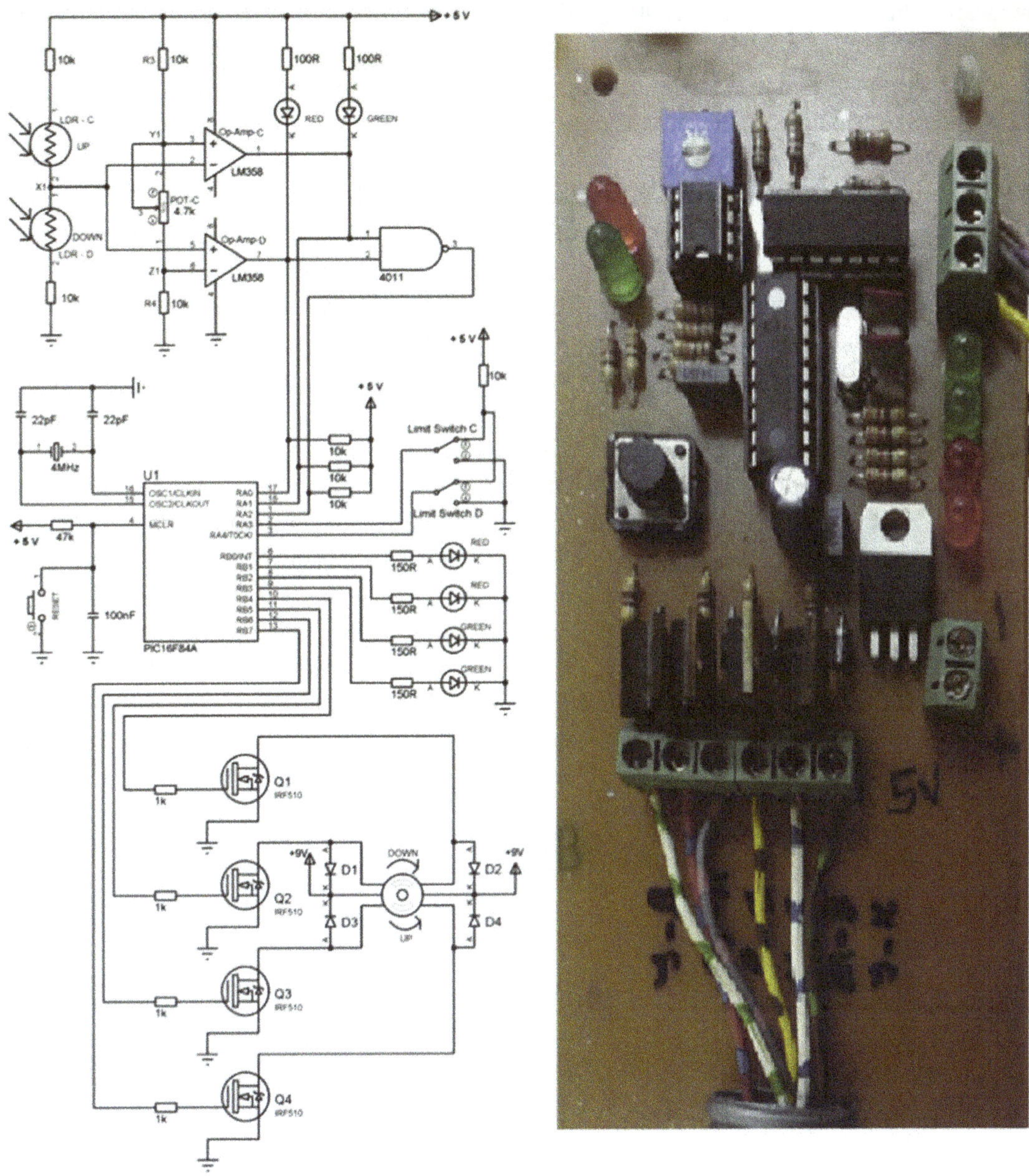

When the sun goes to the upward, LDR-C will have more light and its resistance will decrease. In this case, voltage at point X1 will rise and it starts to be more than 2.5 V and when it passes 2.76 V, op-amp-C and op-amp-D will give 0 V and +5 V, respectively. The engine will start to turn upward. Thus, the resistance of LDR-D starts to decrease and the voltage at point X1 will drop. When the voltage, at point X1, drops below to 2.76 V, op-amps will again give the same output (+5 V) and the motor stops. Thus, when the sun goes upward, the panel also turns to the upward. When the limit C switch is opened, the panel does not rotate anymore. When the sun goes down, LDR-D will have more light and its resistance will fall. In this case, the voltage at point X1 will drop and becomes lower than 2.5 V and when it passes 2.24 V, op-amp-C and op-amp-D will give +5 V and 0 V, respectively. The motor will start to turn downward. Thus the resistance of LDR-C will drop and the voltage at point X1 will rise. When the voltage at point X1 goes above 2.24 V, op-amps give again the same value (+5 V) and the motor stops. Thus, the more the sun goes down, the more the panels rotates to towards to downward. When the limit D switch is opened, the panel does not rotate anymore. A prototype has been created for the experiment and the average January energy amount produced by the fixed system and by the sun tracker system with 10W power can be seen in Figure 14. Sun azimuth angles can be calculated with Matlab m-file.

Sun tracker system produced maximum power as 8.93Wat 11:00 o'clock and the fixed system produced the maximum power as 7.67W at 11:00 o'clock. Sun tracker system produced 55.91Wh throughout the day and the fixed system produced 41.71Wh energy through the day. Sun tracker system produced 34.02% more energy than the fixed system does. Two-axis prototype solar tracking system with 10W power operated successfully and produced more energy comparing with the fixed system.

For sunny conditions, the system, so-called 2-axis tracking system, orients solar modules so that they are perpendicular to the direct rays from the sun produces the most solar energy for a given photovoltaic module area, i.e. 30–50% more solar energy than a PV system with a fixed tilt. However, the measurements show that such solar tracking dramatically decreases the solar energy output for cloudy conditions. The presented model allows generating 34.02% more energy than any conventional fixed system. Various sun tracking structures have been studied in the literature with energy generation percentages higher than the regular fixed ones. For instance, Rustemli, Dincadam and Demirtas (2010) used a similar model and obtained 29% more produced energy while Abu-Khader, Badran and Abdallah (2008) used north–south axes tracking method and achieved to obtain 30 to 45% more energy comparing to a fixed PV system. Two axes sun tracking structure was used by Abdallah and Nijmeh (2004) who obtained much better performance, 41.34% more energy production than a traditional fixed structure. With the

Figure 14. Measured average results for month of January

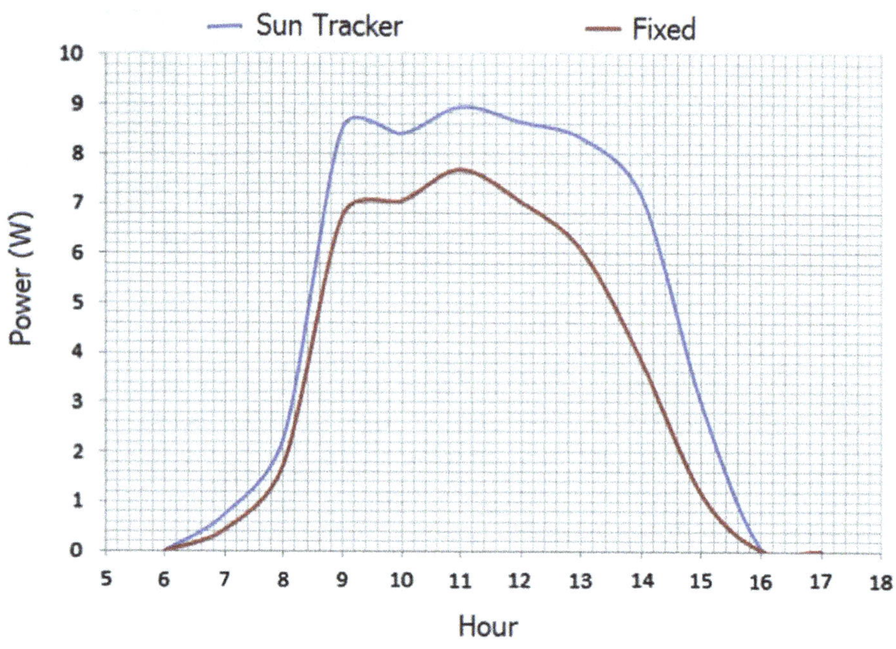

similar comparison, 24.5% more energy was produced with a one-axis sun tracking system used by Mousazadeh, Keyhani, Javadi, Mobli, Abrinia and Sharifi (2009).

Moreover, by taking the completed theoretical calculations into account, a similar system was established for Gaziantep province located in the south-east side of Turkey (Yilmaz, Ozcalik, Dogmus, Dincer, Akgol and Karaaslan, 2015). The data obtained from the prototype was verified by the high-power also as shown in Table 2.

ADVANTAGES AND DISADVANTAGES OF SOLAR COLLECTORS

The energy conversion efficiency of commercial solar PV is around 15-19%. Millions of dollars are spent in order to gain a small percentage in energy conversion efficiency. The power generation of solar PV can be increased using Sun tracking technology. However, the design of the Sun tracker is usually very complicated and expensive. Units are heavy and prone to breakdown and installation is difficult. Dual axis solar trackers have both a horizontal and a vertical axis and thus they can track the Sun's apparent motion virtually anywhere in the world. CSP applications using dual axis tracking include solar power towers and dish (Stirling engine) systems. Dual axis tracking is extremely important in solar tower applications due to the angle errors resulting from longer distances between the mirror and the central receiver located in the tower structure. Many traditional solar PV applications employ two axis trackers to position the solar panels perpendicular to the Sun's rays. This maximizes the total power output by keeping the panels in direct sunlight for the maximum number of hours per day.

Solar trackers are rising in popularity, but not everyone understands the complete benefits and potential drawbacks of the system. Solar panel tracking solutions are a more advanced technology for mounting photovoltaic panels. Stationary mounts, which hold panels in a fixed position, can have their productivity compromised when the Sun passes to a less-than-optimal angle. Compensating for this, solar trackers automatically move to "track" the progress of the Sun across the sky, thereby maximizing output. It's a fantastic system for energy output, but there are a few considerations to bear in mind before pursuing one for a particular jobsite.

Advantages

- Trackers generate more electricity than their stationary counterparts due to increased direct exposure to solar rays. This increase can be as much as 10 to 25% depending on the geographic location of the tracking system.
- There are many different kinds of solar trackers, such as single-axis and dual-axis trackers, all of which can be the perfect fit for a unique jobsite. Installation size, local weather, degree of latitude

Table 2. Comparison of a system capable of tracking the sun in fixed and double axis

System Type	Fixed System	Sun Tracker
Operation time	4113 h	4211 h
Electricity generation	7233 kW h	10096 kW h

and electrical requirements are all important considerations that can influence the type of solar tracker best suited for a specific solar installation.

- Solar trackers generate more electricity in roughly the same amount of space needed for fixed tilt systems, making them ideal for optimizing land usage.
- In certain states some utilities offer Time of Use (TOU) rate plans for solar power, which means the utility will purchase the power generated during the peak time of the day at a higher rate. In this case, it is beneficial to generate a greater amount of electricity during these peak times of day. Using a tracking system helps maximize the energy gains during these peak time periods.
- Advancements in technology and reliability in electronics and mechanics have drastically reduced long-term maintenance concerns for tracking systems.

Disadvantages

- Solar trackers are slightly more expensive than their stationary counterparts, due to the more complex technology and moving parts necessary for their operation. This is usually around a $0.08 – $0.10/W increase depending on the size and location of the project.
- Even with the advancements in reliability there is generally more maintenance required than a traditional fixed rack, though the quality of the solar tracker can play a role in how much and how often this maintenance is needed.
- Trackers are a more complex system than fixed racking. This means that typically more site preparation is needed, including additional trenching for wiring and some additional grading.
- Single-axis tracker projects also require an additional focus on company stability and bankability. When it comes to getting projects financed, these systems are more complex and thus are seen as a higher risk from a financier's viewpoint.
- Solar trackers are generally designed for climates with little to no snow making them a more viable solution in warmer climates. Fixed racking accommodates harsher environmental conditions more easily than tracking systems.
- Fixed racking systems offer more field adjustability than single-axis tracking systems. Fixed systems can generally accommodate up to 20% slopes in the E/W direction while tracking systems typically offer less of a slope accommodation usually around 10% in the N/S direction.

Overall, solar trackers are highly efficient installations, and are a great fit for both large and small project sites given the proper location and site conditions.

THE MAIN ELEMENTS OF SOLAR COLLECTORS

The main elements of a tracking system are as follows:

- Sun tracking algorithm: This algorithm calculates the solar azimuth and zenith angles of the Sun. These angles are then used to position the solar panel or reflector to point toward the Sun. Some algorithms are purely mathematical based on astronomical references while others utilize real-time light-intensity readings.

- Control unit: The control unit executes the Sun tracking algorithm and coordinates the movement of the positioning system.
- Positioning system: The positioning system moves the panel or reflector to face the Sun at the optimum angles. Some positioning systems are electrical and some are hydraulic. Electrical systems utilize encoders and variable frequency drives or linear actuators to monitor the current position of the panel and move to desired positions.
- Drive mechanism/transmission: The drive mechanisms include linear actuators, linear drives, hydraulic cylinders, swivel drives, worm gears, planetary gears, and threaded spindles.
- Sensing devices: For trackers that use light intensity in the tracking algorithm, pyranometers are needed to read the light intensity.

Ambient condition monitoring for pressure, temperature and humidity may also be needed to optimize efficiency and power output.

- Limit switches are used to control speed and prevent over travels. The mechanical over travel limits are used to prevent tracker damage.
- Elevation feedback is accomplished by either 1) a combination of limit switches and motor encoder counts, or 2) an inclinometer (a sensor that provides the tilt angle).
- An anemometer is used to measure wind speed. If the wind conditions are too strong, the panels are usually driven to a safe horizontal position and remain in the safety position until the wind speed falls below the set point.

Here, it should be noted that, the tracker control algorithms typically incorporate a control strategy that is a hybrid between open-loop and closed-loop control. The open-loop component is needed because the Sun can be obscured by clouds, eliminating or distorting the feedback signals. The closed-loop component is needed to eliminate errors that result from variability in installation, assembly, calibration, and encoder mounting. Closed loop systems track the Sun by relying on a set of lenses or sensors with a limited field of view, directed at the Sun, and are fully illuminated by Sunlight at all times.

As the Sun moves, it begins to shade one or more sensors, which the system detects and activates motors or actuators to move the device back into a position where all sensors are once again equally illuminated. Open loop systems track the Sun without physically following the Sun via sensors (although sensors may be used for calibration). These systems typically employ electronic logic which controls device motors or actuators to follow the Sun based on a mathematical formula.

CONCLUSION

Basing on the provided information in the first to four chapters where it was clearly demonstrated that, the maximum utilization of solar energy depends upon determining the exact location of the sun position. By proper calculation and computer programs, the solar path for any geographical location can be tracked. Thus, a clear idea about the prospects of solar energy for any location can be obtained and accordingly decisions and, measures can be adopted for harnessing solar energy in that area.

General formula for on-axis sun-tracking system has been derived using coordinate transformation method. The derived sun-tracking formula is the most general form of mathematical solution for various kinds of arbitrarily oriented on-axis sun tracker, where azimuth-elevation and tilt-roll tracking formulas are specific cases. The application of the general formula is to improve the sun tracking accuracy because the misalignment of solar collector from an ideal azimuth-elevation or tilt-roll tracking during the installation can be corrected by a straightforward application of the general formula. In the provided example of single-axis solar tracking, 0.4^o installation error of azimuth-elevation sun tracker has given a significant effect to the performance of high concentration solar modules.

On the example of single-axis Sun tracking system where direct-search method is proposed that automatically adjusts the orientation of PV modules to avoid the effects of shading caused by cloud cover and shadows from adjacent buildings to ensure that the maximum amount of power is generated wherever the system is installed. Three identical standalone PV systems using the direct-search method control algorithm, together with a reduced detection range and frequency are tested and then compared to the original single-axis mechanical control technique. The long-term experimental results indicate that the amount of power that can be generated using the direct-search method are 3.4% higher than that achieved using the mechanical control in spring/autumn, 5.4% higher in summer, and 8.3% higher in winter. This indicates that the direct-search method is more efficient in summer and winter because the weather conditions in Taiwan are more unstable during spring and autumn. However, the tracking system using the original direct-search method control algorithm based on a higher detection frequency and wider detection range has the potential to cause reliability issues with the sun tracker and result in a greater power loss due to the number of rotations required during detection operations. The results also indicate that the performance of the Sun tracking system based on the reduced detection range is very close to the same system when using the original direct-search method, but requires lower energy consumption and lower maintenance costs. Even when the number of detections is reduced to five per day, the amount of power generated is still the highest.

On the example of a photovoltaic dual axis sun-tracking system, the performances of dual axis tracking and fixed latitude tilt identical PV systems are theoretically and experimentally realized. The advantages of the proposed system over fixed systems are clearly discussed and compared with the results from other studies in literature. In order to verify the results obtained from the prototype, same system with the same working principle is applied to a high power plant and the results belong to the latter one were also support the concluded claim that the proposed sun tracking system provide more efficient results. As a result, the suggested model and the system can be easily used for solar power plants, solar home systems and so on for more efficient energy production.

The information of this chapter will be used in the chapters 7 and 8 where the technical and economic considerations are the heart of them respectively.

REFERENCES

Abdallah, S. (2004). The effect of using Sun tracking systems on the voltage-current characteristics and power generation of at plate photovoltaics. *Energy Conversion and Management*, *45*(11-12), 1671–1679. doi:10.1016/j.enconman.2003.10.006

Abdallah, S., & Nijmeh, S. (2004). Two axes sun tracking system with PLC control. *Energy Conversion and Management*, *45*(11-12), 1931–1939. doi:10.1016/j.enconman.2003.10.007

Abdelsalam, A. K., Massoud, A. M., Ahmed, S., & Enjeti, P. (2011). High-performance adaptive perturb and observe MPPT technique for photovoltaic-based microgrids. *IEEE Transactions on Power Electronics, 26*(4), 1010–1021. doi:10.1109/TPEL.2011.2106221

Abu-Khader, M. M., Badran, O. O., & Abdallah, S. (2008). Evaluating multi-axes sun tracking system at different modes of operation in Jordan. *Renewable & Sustainable Energy Reviews, 12*(3), 864–873. doi:10.1016/j.rser.2006.10.005

Ahmed, J., & Salam, Z. (2014). A maximum power point tracking (MPPT) for PV system using cuckoo search with partial shading capability. *Applied Energy, 119*(15), 118–130. doi:10.1016/j.apenergy.2013.12.062

Ahmed, J., & Salam, Z. (2015). An improved perturb and observe (P&O) maximum power point tracking (MPPT) algorithm for higher efficiency. *Applied Energy, 150*, 97–108. doi:10.1016/j.apenergy.2015.04.006

Aiuchi, K., Yoshida, K., Onozaki, M., Katayama, Y., Nakamura, M., & Nakamura, K. (2006). Sensor-controlled heliostat with an equatorial mount. *Solar Energy, 80*(9), 1089–1097. doi:10.1016/j.solener.2005.10.007

Alajmi, B. N., Ahmed, K. H., Finney, S. J., & Williams, B. W. (2011). Fuzzy-logic-control approach of a modified hill-climbing method for maximum power point in microgrid standalone photovoltaic system. *IEEE Transactions on Power Electronics, 26*(4), 1022–1030. doi:10.1109/TPEL.2010.2090903

Algazar, M. M., Al-Monie, H., El-Halim, H. A., & Salem, M. E. E. K. (2012). Maximum power point tracking using fuzzy logic control. *International Journal of Electrical Power & Energy Systems, 39*(1), 21–28. doi:10.1016/j.ijepes.2011.12.006

Arbab, H., Jazi, B., & Rezagholizadeh, M. (2009). A computer tracking system of solar dish with two-axis degree freedoms based on picture processing of bar shadow. *Renewable Energy, 34*(4), 1114–1118. doi:10.1016/j.renene.2008.06.017

Barret, B., De Mazière, M., & Demoulin, P. (2002). Retrieval and characterization of ozone profiles from solar infrared spectra at the Jungfraujoch. *Journal of Geophysical Research, 107*(D24), 4788. doi:10.1029/2001JD001298

Belkaid, A., Colak, I., & Isik, O. (2016). Photovoltaic maximum power point tracking under fast varying of solar radiation. *Applied Energy, 179*, 523–530. doi:10.1016/j.apenergy.2016.07.034

Benghanem, M. (2011). Optimization of tilt angle for solar panel: Case study for Madinah, Saudi Arabia. *Applied Energy, 88*(4), 1427–1433. doi:10.1016/j.apenergy.2010.10.001

Bentaher, H., Kaich, H., Ayadi, N., Ben Hmouda, M., Maalej, A., & Lemmer, U. (2013). A simple tracking system to monitor solar PV panels. *Energy Conversion and Management, 78*, 872–875. doi:10.1016/j.enconman.2013.09.042

Berenguel, M., Rubio, F. R., Valverde, A., Lara, P. J., Arahal, M. R., Camacho, E. F., & Lopez, M. (2004). An artificial vision-based control system for automatic heliostat positioning offset correction in a central receiver solar power plant. *Solar Energy, 76*(5), 563–575. doi:10.1016/j.solener.2003.12.006

Bhatnagar, A. P., & Nema, B. R. K. (2013). Conventional and global maximum power point tracking techniques in photovoltaic applications. *J. Renewable Sustainable Energy, 5*(3), 032701. doi:10.1063/1.4803524

Blanco-Muriel, M., Alarcon-Padilla, D. C., Lopez-Moratalla, T., & Lara-Coira, M. (2001). Computing the solar vector. *Solar Energy*, *70*(5), 431–441. doi:10.1016/S0038-092X(00)00156-0

Braun, J. A., & Mitchell, J. C. (1983). Solar Geometry for Fixed and Tracking Surfaces. *Solar Energy*, *31*(5), 439–444. doi:10.1016/0038-092X(83)90046-4

Cageao, R. P., Blavier, J., McGuire, J. P., Jiang, Y., Nemtchinov, V., Mills, F. P., & Sander, S. P. (2001). High-Resolution Fourier-transform ultraviolet-visible spectrometer for the measurement of atmospheric trace species: Application to OH. *Applied Optics*, *40*(12), 2024–2030. https://doi.org/10.1364/AO.40.002024 doi:10.1364/AO.40.002024 PMID:18357206

Chang, T. P. (2009a). The gain of single-axis tracked panel according to extraterrestrial radiation. *Applied Energy*, *86*(7-8), 1074–1079. doi:10.1016/j.apenergy.2008.08.002

Chang, T. P. (2009b). Output energy of a photovoltaic module mounted on a single-axis tracking system. *Applied Energy*, *86*(10), 2071–2078. doi:10.1016/j.apenergy.2009.02.006

Chen, F., & Feng, J. (2007). Analogue sun sensor based on the optical nonlinear compensation measuring principle. *Measurement Science & Technology*, *18*(7), 2111–2115. doi:10.1088/0957-0233/18/7/042

Chen, F., Feng, J., & Hong, Z. (2006). Digital sun sensor based on the optical vernier measuring principle. *Measurement Science & Technology*, *17*(9), 2494–2498. doi:10.1088/0957-0233/17/9/017

Chen, Y. T., Chong, K. K., Bligh, T. P., Chen, L. C., Yunus, J., Kannan, K. S., & Tan, K. K. et al. (2001). Non-imaging, focusing heliostat. *Solar Energy*, *71*(3), 155–164. doi:10.1016/S0038-092X(01)00041-X

Chen, Y. T., Lim, B. H., & Lim, C. S. (2006). General sun tracking formula for heliostats with arbitrarily oriented axes. *Journal of Solar Energy Engineering*, *128*(2), 245–250. doi:10.1115/1.2189868

Chong, K., & Wong, C. (2009). General formula for on-axis sun-tracking system and its application in improving tracking accuracy of solar collector. *Solar Energy*, *83*(3), 298–305. doi:10.1016/j.solener.2008.08.003

Chong, K. K., Wong, C. W., Siaw, F. L., & Yew, T. K. (2010). Optical characterization of non-imaging planar concentrator for the application in concentrator photovoltaic system. *J. Sol. Energy Eng. Journal of Solar Energy Engineering*, *132*(1), 011011, 011011–011019. doi:10.1115/1.4000355

Chong, K. K., Wong, C. W., Siaw, F. L., Yew, T. K., Ng, S. S., Liang, M. S., & Lau, S. L. et al. (2009). Integration of an on-axis general sun-tracking formula in the algorithm of an open-loop sun-tracking system. *Sensors (Basel, Switzerland)*, *9*(10), 7849–7865. doi:10.3390/s91007849 PMID:22408483

Clifford, M. J., & Eastwood, D. (2004). Design of a novel passive solar tracker. *Solar Energy*, *77*(3), 269–280. doi:10.1016/j.solener.2004.06.009

Cordenier, A. (2004). *Système Numérique de Régulation en Position de Miroirs Destinés Au Suivi Du Rayonnement Solaire* (M.Sc. Thesis). Ecole Centrale des Arts et Métiers, Brussels, Belgium.

De Mazière, M., Vigouroux, C., Gardiner, T., Coleman, M., Woods, P., Ellingsen, K., & Rockmann, A. et al. (2005). The exploitation of ground-based Fourier transform infrared observations for the evaluation of tropospheric trends of greenhouse gases over Europe. *Environmental Sciences*, *2*(2-3), 283–293. doi:10.1080/15693430500405179

De Miguel, A., Bilbao, J., & Diez, M. (1995). Solar radiation incident on tilted surfaces in Burgos, Spain: Isotropic models. *Energy Conversion and Management*, *36*(10), 945–951. doi:10.1016/0196-8904(94)00067-A

Ding, K., Wang, X., Zhai, Q. X., Xu, J. W., Zhang, J. W., & Liu, H. H. (2014). Improved global maximum power point tracking method based on voltage interval for PV array under partially shaded conditions. *J. Power Electron.*, *14*(4), 722–732. doi:10.6113/JPE.2014.14.4.722

Edwards, B. P. (1978). Computer based sun following system. *Solar Energy*, *21*(6), 491–498. doi:10.1016/0038-092X(78)90073-7

Farahat, M. A., Enany, M. A., & Nasr, A. (2015). Assessment of maximum power point tracking techniques for photovoltaic system applications. *J. Renewable Sustainable Energy*, *7*(4), 042702. doi:10.1063/1.4928680

Farhat, M., Barambones, O., & Sbita, L. (2017). A new maximum power point method based on a sliding mode approach for solar energy harvesting. *Applied Energy*, *185*, 1185–1198. doi:10.1016/j.apenergy.2016.03.055

Fathabadi, H. (2016). Novel high accurate sensorless dual-axis solar tracking system controlled by maximum power point tracking unit of photovoltaic systems. *Applied Energy*, *173*, 448–459. doi:10.1016/j.apenergy.2016.03.109

Fermia, N., Granozio, D., Petrone, G., & Vitelli, M. (2007). Predictive and adaptive MPPT perturb and observe method. *IEEE Transactions on Aerospace and Electronic Systems*, *43*(3), 935–950. doi:10.1109/TAES.2007.4383584

Galle, B., Mellqvist, J., Arlander, D. W., Floisand, I., Chipperfield, M. P., & Lee, A. M. (1999). Ground based FTIR measurements of stratospheric species from harestua, norway during Sesame and comparison with models. *Journal of Atmospheric Chemistry*, *32*(1), 147–164. doi:10.1023/A:1006137924562

Geibel, M. C., Gerbig, C., & Feist, D. G. (2010). A new fully automated FTIR system for total column measurements of greenhouse gases. *Atmos. Meas. Tech.*, *3*(5), 1363–1375. doi:10.5194/amt-3-1363-2010

Gisi, M., Hase, F., Dohe, S., & Blumenstock, T. (2011). Camtracker: A new camera controlled high precision solar tracker system for FTIR-spectrometers. *Atmos. Meas. Tech.*, *4*(1), 47–54. doi:10.5194/amt-4-47-2011

Gong, X., & Kulkarni, M. (2005). Design optimization of a large scale rooftop photovoltaic system. *Solar Energy*, *78*(3), 362–374. doi:10.1016/j.solener.2004.08.008

Guo, M., Wang, Z., Zhang, J., Sun, F., & Zhang, X. (2011). Accurate altitude-azimuth tracking angle formulas for a heliostat with mirror pivot offset and other fixed geometrical errors. *Solar Energy*, *85*(5), 1091–1100. doi:10.1016/j.solener.2011.03.003

Hein, M., Dimroth, F., Siefer, G., & Bett, A. W. (2003). Characterisation of a 300x photovoltaic concentrator system with one-axis tracking. *Solar Energy Materials and Solar Cells*, *75*(1-2), 277–283. doi:10.1016/S0927-0248(02)00170-8

Hu, Y., & Yao, Y. (2016). A methodology for calculating photovoltaic field output and effect of solar tracking strategy. *Energy Conversion and Management*, *126*, 278–289. doi:10.1016/j.enconman.2016.08.007

Huang, B. J., Ding, W. L., & Huang, Y. C. (2011). Long-term field test of solar PV power generation using one-axis 3-position Sun tracker. *Solar Energy*, *85*(9), 1935–1944. doi:10.1016/j.solener.2011.05.001

Huang, B. J., Hsu, P. C., Wu, M. S., & Ho, P. Y. (2010). System dynamic model and charging control of lead-acid battery for stand-alone solar PV system. *Solar Energy*, *84*(5), 822–830. doi:10.1016/j.solener.2010.02.007

Huang, B.-J., Huang, Y.-C., Chen, G.-Y., Hsu, P.-C., & Li, K. (2013). Improving Solar PV System Efficiency Using One-Axis 3-Position Sun Tracking. *Energy Procedia*, *33*, 280–287. doi:10.1016/j.egypro.2013.05.069

Huang, B. J., & Sun, F. S. (2007). Feasibility study of one axis three positions tracking solar PV with low concentration ratio reflector. *Energy Conversion and Management*, *48*(4), 1273–1280. doi:10.1016/j.enconman.2006.09.020

Huang, B. J., Sun, F. S., & Ho, R. W. (2006). Near-maximum-power-point-operation (nMPPO) design of photovoltaic power generation system. *Solar Energy*, *80*(8), 1003–1020. doi:10.1016/j.solener.2005.06.013

Iqbal, M. (1983). *An Introduction to Solar Radiation*. Waltham, MA: Academic Press.

Ishaque, K., & Salam, Z. (2013). A deterministic particle swarm optimization maximum power point tracker for photovoltaic system under partial shading condition. *IEEE Transactions on Industrial Electronics*, *60*(8), 3195–3206. doi:10.1109/TIE.2012.2200223

Ishaque, K., Salam, Z., Amjad, M., & Mekhilef, S. (2012). An improved particle swarm optimization (PSO)–based MPPT for PV with reduced steady-state oscillation. *IEEE Transactions on Power Electronics*, *27*(8), 3627–3638. doi:10.1109/TPEL.2012.2185713

Jesperson, J., & Fitz-Randolph, J. (1977). National Bureau of Standards Monograph. In From Sundials to Atomic Clocks (2nd ed.). Harcourts Brace & Company.

Kacira, M., Simsek, M., Babur, Y., & Demirkol, S. (2004). Determining optimum tilt angles and orientations of photovoltaic panels in Sanliurfa, Turkey. *Renewable Energy*, *29*(8), 1265–1275. doi:10.1016/j.renene.2003.12.014

Kalogirou, S. A. (1996). Design and construction of a one-axis sun-tracking system. *Solar Energy*, *57*(6), 465–469. doi:10.1016/S0038-092X(96)00135-1

Kalogirou, S. A. (2004). Solar thermal collectors and applications. *Progress in Energy and Combustion Science*, *30*(3), 231–295. doi:10.1016/j.pecs.2004.02.001

Kim, S. (2007). Robust maximum power point tracker using sliding mode controller for the three-phase grid-connected photovoltaic system. *Solar Energy*, *81*(3), 405–414. doi:10.1016/j.solener.2006.04.005

Kofinas, P., Dounis, A. I., Papadakis, G., & Assimakopoulos, M. N. (2015). An intelligent MPPT controller based on direct neural control for partially shaded PV system. *Energy and Building*, *90*(1), 51–64. doi:10.1016/j.enbuild.2014.12.055

Koo, C., Hong, T., Lee, M., & Kim, J. (2016). An integrated multi-objective optimization model for determining the optimal solution in implementing the rooftop photovoltaic system. *Renewable & Sustainable Energy Reviews*, *57*, 822–837. doi:10.1016/j.rser.2015.12.205

Kou, Y., Xia, Y., & Ye, Y. (2015). Fast variable step-size maximum power point tracking method for photovoltaic systems. *J. Renewable Sustainable Energy*, *7*(4), 043126. doi:10.1063/1.4928519

Koyuncu, B., & Balasubramanian, K. (1991). A Microprocessor Controlled Automatic Sun Tracker. *IEEE Transactions on Consumer Electronics*, *37*(4), 913–917. doi:10.1109/30.106958

Kribus, A., Vishnevetsky, I., Yogev, A., & Rubinov, T. (2004). Closed loop control of heliostats. *Energy*, *29*(5-6), 905–913. doi:10.1016/S0360-5442(03)00195-6

Lamm, L. O. (1981). A new analytic expression for the equation of time. *Solar Energy*, *26*(5), 465. doi:10.1016/0038-092X(81)90229-2

Lazaroiu, G. C., Longo, M., Roscia, M., & Pagano, M. (2015). Comparative analysis of fixed and Sun tracking low power PV systems considering energy consumption. *Energy Conversion and Management*, *92*, 143–148. doi:10.1016/j.enconman.2014.12.046

Lim, B. H., Chong, K. K., Lim, C. S., & Lai, A. C. (2016). Latitude-orientated mode of non-imaging focusing heliostat using spinning-elevation tracking method. *Solar Energy*, *135*, 253–264. doi:10.1016/j.solener.2016.05.038

Liu, Y. H., Huang, S. C., Huang, J. W., & Liang, W. C. (2012). A particle swarm optimization-based maximum power point tracking algorithm for PV systems operating under partially shaded conditions. *IEEE Transactions on Energy Conversion*, *27*(4), 1027–1035. doi:10.1109/TEC.2012.2219533

Lubitz, W. D. (2011). Effect of manual tilt adjustments on incident irradiance on fixed and tracking solar panels. *Applied Energy*, *88*(5), 1710–1719. doi:10.1016/j.apenergy.2010.11.008

Luque-Heredia, I., Cervantes, R., & Quéméré, G. (2006). A sun tracking error monitor for photovoltaic concentrators. *Proceedings of the IEEE 4th World Conference on Photovoltaic Energy Conversion*. doi:10.1109/WCPEC.2006.279553

Luque-Heredia, I., Gordillo, F., & Rodriguez, F. (2004). A *PI* based hybrid sun tracking algorithm for photovoltaic concentration. *Proceedings of the IEEE 19th European Photovoltaic Energy Conversion*.

Lynch, W. A., & Salameh, Z. M. (1990). Simple electro-optically controlled dual-axis sun tracker. *Solar Energy*, *45*(2), 65–69. doi:10.1016/0038-092X(90)90029-C

Maish, A. B. (1988). A self-aligning photovoltaic array tracking controller. *Proceedings of the 20th IEEE photovoltaic specialists conference*. doi:10.1109/PVSC.1988.105917

Manwell, J. F., & Mcgowan, J. G. (1993). Lead-Acid-Battery Storage Model for Hybrid Energy-Systems. *Solar Energy*, *50*(5), 399–405. doi:10.1016/0038-092X(93)90060-2

Marion, W. F., & Dobos, A. P. (2013). *Rotation angle for the optimum tracking of one-axis trackers*. Technical Report, NREL/TP-6A20-58891. Retrieved from http://www.osti.gov/bridge

Mavromatakis, F., & Franghiadakis, Y. (2008). A highly efficient novel azimuthal heliotrope. *Solar Energy*, *82*(4), 336–342. doi:10.1016/j.solener.2007.09.001

Meeus, J. (1998). *Astronomical Algorithms* (2nd ed.). London, UK: Atlantic Books.

Merlaud, A. (2004). *Development of Solar Tracker for Studies of Volcanic Gas Emissions* (Master's thesis). Ecole Nationale Supérieure de Physique de Grenoble, Grenoble, France.

Merlaud, A. (2006). *Development of a Solar Tracker for Monitoring of Atmospheric Gases at Harestua Observatory- Technical Report*. Chalmers Insitute of Technology.

Merlaud, A., De Mazière, M., Hermans, Ch., & Cornet, A. (2012). Equations for Solar Tracking. *Sensors (Basel)*, *12*(4), 4074–4090. doi:10.3390/s120404074 PMID:22666019

Mousazadeh, H., Keyhani, A., Javadi, A., Mobli, H., Abrinia, K., & Sharifi, A. (2009). A review of principle and Sun-tracking methods for maximizing solar systems output. *Renewable & Sustainable Energy Reviews*, *13*(8), 1800–1818. doi:10.1016/j.rser.2009.01.022

Nann, S. (1990). Potentials for tracking photovoltaic systems and V-troughs in moderate climates. *Solar Energy*, *45*(6), 385–393. doi:10.1016/0038-092X(90)90160-E

Neefs, E., de Maziere, M., Scolas, F., Hermans, C. & Hawat, T. (2007). BARCOS, an automation and remote control system for atmospheric observations with a Bruker interferometer. *Rev. Sci. Instrum.*, *78*(3), 035109:1–035109:8. DOI:10.1063/1.2437144

Neville, R. C. (1978). Solar energy collector orientation and tracking mode. *Solar Energy*, *20*(1), 7–11. doi:10.1016/0038-092X(78)90134-2

Nuwayhid, R. Y., Mrad, F., & Abu-Said, R. (2001). The realization of a simple solar tracking concentrator for the university research applications. *Renewable Energy*, *24*(2), 207–222. doi:10.1016/S0960-1481(00)00191-9

Piegari, L., & Rizzo, R. (2010). Adaptive perturb and observe algorithm for photovoltaic maximum power point tracking. *IET Renewable Power Generation*, *4*(4), 317–328. doi:10.1049/iet-rpg.2009.0006

Poulek, V., & Libra, M. (1998). New solar tracker. *Solar Energy Materials and Solar Cells*, *51*(2), 113–120. doi:10.1016/S0927-0248(97)00276-6

Poulek, V., & Libra, M. (2000). A very simple solar tracker for space and terrestrial applications. *Solar Energy Materials and Solar Cells*, *60*(2), 99–103. doi:10.1016/S0927-0248(99)00071-9

Quesada, G., Guillon, L., Rousse, D. R., Mehrtash, M., Dutil, Y., & Paradis, P. L. (2015). Tracking strategy for photovoltaic solar systems in high latitudes. *Energy Conversion and Management*, *103*, 147–156. doi:10.1016/j.enconman.2015.06.041

Reda, I., & Andreas, A. (2004). Solar position algorithm for solar radiation applications. *Solar Energy*, *76*(5), 577–589. doi:10.1016/j.solener.2003.12.003

Roth, P., Georgiev, A., & Boudinov, H. (2004). Design and construction of a system for sun-tracking. *Renewable Energy*, *29*(3), 393–402. doi:10.1016/S0960-1481(03)00196-4

Rubio, F. R., Ortega, M. G., Gordillo, F., & Lopez-Martinez, M. (2007). Application of new control strategy for Sun tracking. *Energy Conversion and Management*, *48*(7), 2174–2184. doi:10.1016/j.enconman.2006.12.020

Rumala, S. S. N. (1986). A shadow method for automatic tracking. *Solar Energy*, *37*(3), 245–247. doi:10.1016/0038-092X(86)90081-2

Rustemli, S., Dincadam, F., & Demirtas, M. (2010). Performance comparison of the sun tracking system and fixed system in the application of heating and lighting. *Arabian Journal Science Engineering*, *35*(2B), 171–183.

Seme, S., & Stumberger, G. (2011). A novel prediction algorithm for solar angles using solar radiation and Differential Evolution for dual-axis sun tracking purposes. *Solar Energy*, *85*(11), 2757–1770. doi:10.1016/j.solener.2011.08.031

Seyedmahmoudian, M., Mekhilef, S., Rahmani, R., Yusof, R., & Shojaei, A. A. (2014). Maximum power point tracking of partial shaded photovoltaic array using an evolutionary algorithm: A particle swarm optimization technique. *J. Renewable Sustainable Energy*, *6*(2), 023102. doi:10.1063/1.4868025

Shaiek, Y., Smida, M. B., Sakly, A., & Mimouni, M. F. (2013). Comparison between conventional methods and GA approach for maximum power point tracking of shaded solar PV generators. *Solar Energy*, *90*(4), 107–122. doi:10.1016/j.solener.2013.01.005

Shanmugam, S., & Christraj, W. (2005). The tracking of the sun for solar paraboloidal dish concentrators. *Journal of Solar Energy Engineering*, *127*(1), 156–160. doi:10.1115/1.1824103

Sharan, A. M., & Prateek, M. (2006). Automation of minimum torque-based accurate solar tracking systems using microprocessors. *Journal of the Indian Institute of Science*, *86*(5), 415–437.

Sher, H. A., Murtaza, A. F., Noman, A., Addoweesh, K. E., & Chiaberge, M. (2015). An intelligent control strategy of fractional short circuit current maximum power point tracking technique for photovoltaic applications. *J. Renewable Sustainable Energy*, *7*(1), 013114. doi:10.1063/1.4906982

Sinha, S., & Chandel, S. S. (2016). Analysis of fixed tilt and Sun tracking photovoltaic–micro wind based hybrid power systems. *Energy Conversion and Management*, *115*, 265–275. doi:10.1016/j.enconman.2016.02.056

Skouri, S., Ben Haj Ali, A., Bouadila, S., Ben Salah, M., & Ben Nasrallah, S. (2016). Design and construction of Sun tracking systems for solar parabolic concentrator displacement. *Renewable & Sustainable Energy Reviews*, *60*, 1419–1429. doi:10.1016/j.rser.2016.03.006

Solanki, C. S., & Sangani, C. S. (2007). Experimental evaluation of V-trough (2 suns) PV concentrator system using commercial PV modules. *Solar Energy Materials and Solar Cells*, *91*(6), 453–459. doi:10.1016/j.solmat.2006.10.012

Soulayman, S., & Hammoud, M. (2016). Optimum tilt angle of solar collectors for building applications in mid-latitude zone. *Energy Conversion and Management*, *124*, 20–28. doi:10.1016/j.enconman.2016.06.066

Spencer, J. W. (1971). Fourier series representation of the position of the Sun. *Search*, *2*(5), 172.

Sproul, A. B. (2007). Derivation of the solar geometric relationships using vector analysis. *Renewable Energy*, *32*(7), 1187–1205. doi:10.1016/j.renene.2006.05.001

Stanciu, C., & Stanciu, D. (2014). Optimum tilt angle for flat plate collectors all over the World – A declination dependence formula and comparisons of three solar radiation models. *Energy Conversion and Management*, *81*, 133–143. doi:10.1016/j.enconman.2014.02.016

Sundareswaran, K., Sankar, P., & Nayak, P. (2014). MPPT of PV systems under partial shaded conditions through a colony of flashing fireflies. *IEEE Transactions on Energy Conversion*, *29*(2), 463–472. doi:10.1109/TEC.2014.2298237

Sundareswaran, K., Sankar, P., Nayak, P., Simon, S. P., & Palani, S. (2015). Enhanced energy output from a PV system under partial shaded conditions through artificial bee colony. *IEEE Trans. Sustainable Energy*, *6*(1), 198–209. doi:10.1109/TSTE.2014.2363521

Tian, Y., Xia, B., Sun, W., Xu, Z., & Zheng, W. (2014). Modeling and global maximum power point tracking for photovoltaic system under partial shading conditions using modified particle swarm optimization algorithm. *J. Renewable Sustainable Energy*, *6*(6), 063117. doi:10.1063/1.4904436

Tian, Y., Xia, B. Z., Xu, Z. H., & Sun, W. (2014). Modified asymmetrical variable step size incremental conductance maximum power point tracking method for photovoltaic systems. *J. Power Electron.*, *14*(1), 156–164. doi:10.6113/JPE.2014.14.1.156

Tomson, T. (2008). Discrete two-positional tracking of solar collectors. *Renewable Energy*, *33*(3), 400–405. doi:10.1016/j.renene.2007.03.017

Van Roozendael, M., Peeters, P., Roscoe, H. K., Backer, H. D., Jones, A. E., Bartlett, L., & Simon, P. C. et al. (1998). Validation of ground-based visible measurements of total ozone by comparison with dobson and brewer spectrophotometers. *Journal of Atmospheric Chemistry*, *29*(1), 55–83. doi:10.1023/A:1005815902581

Verma, D., Nema, S., Shandilya, A. M., & Dash, S. K. (2015). Comprehensive analysis of maximum power point tracking techniques in solar photovoltaic systems under uniform insolation and partial shaded condition. *J. Renewable Sustainable Energy*, *7*(4), 042701. doi:10.1063/1.4926844

Wang, S., Pongetti, T. J., Sander, S. P., Spinei, E., Mount, G. H., Cede, A., & Herman, J. (2010). Direct Sun measurements of NO2 column abundances from Table Mountain, California: Intercomparison of low- and high-resolution spectrometers. *Journal of Geophysical Research. Atmospheres*, *115*(D13). doi:10.1029/2009JD013503

Washenfelder, R. A., Toon, G. C., Blavier, J., Yang, Z., Allen, N. T., Wennberg, P. O., & Daube, B. C. et al. (2006). Carbon dioxide column abundances at the Wisconsin Tall Tower site. *Journal of Geophysical Research, D, Atmospheres*, *111*(D22), D22305. doi:10.1029/2006JD007154

Wei, X., Lu, Z., Yu, W., Zhang, H., & Wang, Z. (2011). Tracking and ray tracing equations for the target-aligned heliostat for solar tower power plants. *Renewable Energy*, *36*(10), 2687–2693. doi:10.1016/j.renene.2011.02.022

Wiacek, A., Taylor, J. R., Strong, K., Saari, R., Kerzenmacher, T. E., Jones, N. B., & Griffith, D. W. T. (2007). Ground-based solar absorption FTIR spectroscopy: Characterization of retrievals and first results from a novel optical design instrument at a new NDACC complementary station. *Journal of Atmospheric and Oceanic Technology*, *24*(3), 432–448. doi:10.1175/JTECH1962.1

Woolf, H. M. (1968). *On the Computation of Solar Evaluation Angles and the Determination of Sunrise and Sunset Times*. National Aeronautics and Space Administration Report NASA TM-X -164, September.

Xu, Z., Yang, P., Zhou, D., Li, P., Lei, J., & Chen, Y. (2015). An Improved variable step size MPPT algorithm based on INC. *J. Power Electron.*, *15*(2), 487–496. doi:10.6113/JPE.2015.15.2.487

Yan, R., Saha, T. K., Meredith, P., & Goodwin, S. (2013). Analysis of yearlong performance of differently tilted photovoltaic systems in Brisbane, Australia. *Energy Conversion and Management*, *74*, 102–108. doi:10.1016/j.enconman.2013.05.007

Yilmaz, S., Ozcalik, H. R., Dogmus, O., Dincer, F., Akgol, O., & Karaaslan, M. (2015). Design of two axes Sun tracking controller with analytically solar radiation calculations. *Renewable & Sustainable Energy Reviews*, *43*, 997–1005. doi:10.1016/j.rser.2014.11.090

Chapter 7
Technical Consideration

ABSTRACT

The maximum output of the solar receiver is achieved when the solar receiver is perpendicular to the Sun's rays. Different attempts were made for making the solar receiver utilizing the maximum portion of incident solar radiation. The use of a dual-axis sun tracker versus a fixed-flat position is evidently profitable, but from economic point of view it is questionable. A mathematical conception has been developed and applied in this chapter to determine the energy gain resulted from different installations of PV systems. The experimental measurements and the model results show that, it is not economical to track the sun in hot and sunny regions because of the overheating effect on the PV panels' performance. The provided data, in literature, compare the performance of dual or single axis tracking with fixed solar receiver even the long term solar tracking is possible and effective with a negligible increase of the price of the unit of useful energy. This can be achieved by choosing the best monthly or even seasonally optimum tilts. The introduced concept of energy gain, see chapter 3, is calculated in this chapter all over the world and it was found that it is very useful in evaluating the performance of different types of tracking. This concept allows to evaluate the effectiveness of daily, weekly, fortnightly, monthly, seasonally, biannually and yearly adjustment of the solar receiver tilt angle in relation with the ideal instantaneous dual tracking.

INTRODUCTION

Optimization is a method of finding the conditions that give maximum or minimum value of the objective function to obtain the accurate results under specified conditions. It is useful in the design, construction and maintenance of any engineering system (Rao, 2010). For finding optimum tilt angle, solar radiation on the tilted surface is taken as an objective function which is solved using different optimization techniques like Genetic Algorithm (GA), Simulated Annealing (SA) technique and particle swarm optimization(PSO). Genetic algorithm (GA) is suitable for optimization problems which include complex nonlinear variables, a sit can find the global optimum solution with a high probability (Goldberg, 1989). A population of points is used for starting the GA instead of a single design point (Sivanandam and Deepa, 2008). It involves principles of natural genetics and natural selection. The natural genetics are reproduction, crossover and mutation which are used in the genetic search procedure (Beasley, Bull

DOI: 10.4018/978-1-5225-2950-7.ch007

and Martin, 1993). Talebizadeh, Mehrabian and Abdolzadeh (2011) used GA to calculate hourly, daily, monthly, seasonally and yearly optimum tilt angle for Iran and showed that the optimum hourly surface azimuth angle is not zero and optimum tilt angles of photovoltaic panels and solar collector are found to be the same. The solar energy gain at daily, monthly optimum tilt angle is found to be the same but energy gain is more at hourly tilt angle. Therefore, the solar tracker is beneficial for hourly variation of tilt angle. Čongradac, Prica, Paspalj, Bojanič and Čapko, (2012) used GA and fuzzy logic process to find out optimum blind tilt angle (angle rotated in anticlockwise and clockwise direction) for maintaining accurate brightness of the room. This process is useful in maintaining user's comfort and saving energy. Simulated Annealing (SA) derives its name from the simulation of thermal annealing of critically heated solid sand is used to find the global optimum with a high probability of objective functions which contain numerous local minima. Chen, Lee and Wu (2005) used SA for calculating the optimum installation angle for fixed solar-cell panels. Particle Swarm Optimization (PSO) is a stochastic technique for exploring the search space for optimization (Kennedy and Eberhart, 1995). It is achieved by particles in multidimensional space that have a position and a velocity (Beasley, Bull and Martin, 1993). In PSO particles are flying through hyper space and adjust their own best position i.e. the smallest objective function values. Chang (2010a) used the varying inertia weight methods (Shi and Eberhart, 1998; Shi and Eberhart, 1999; Eberhart and Shi, 2001; Ratnaweera, Halgamuge and Watson, 2004; Chatterjee and Siarry, 2006) and proposed a particle-swarm optimization method with nonlinear time-varying evolution (PSO—NTVE) to determine the tilt angle of PV modules for maximum output electrical energy of the modules in Taiwan. The yearly optimal angle for Taipei area is 18.16° and 17.3°, 16.15°, 15.79°, 15.17°, 17.16°, 15.94° for Taichung, Tainan, Kaosiung, Hengchung, Hualian, Taitung respectively. The PSO NTVE is faster than the other three PSO methods and the GA method for achieving an optimum solution. Therefore, these optimization techniques will give better results than the other methods (latitude based method, maximizing radiation by changing β from 0° to 90° at fixed steps).

Artificial Neural Network techniques (ANN) are excellent tools for research as these are able to solve nonlinear function approximation, data classification, clustering and simulation (Kalogirou, 2000). These are used in different fields of science and technology (Sumathi and Paneerselvam, 2010; Hontoria, Aguilera and Zufiria, 2002; Agarwal, 1997; Haykin, 1994) and also for the solar radiation predictions (Hontoria, Aguilera, Riesco and Zufiria, 2001). Chang (2009) used sequential neural-network approximation and orthogonal arrays (SNAOA) to determine the optimum tilt angle for maximum output power energy of PV modules in seven areas of Taiwan. The back propagation neural network is used to search optimum tilt angle design. The maximum and minimum values of optimum tilt angle and voltage are used as constraint to maximize power on PV module. The annual optimal angle is found to be 23.25° for Taipei site. Mehleri, Zervas, Sarimveis, Palyvos and Markatos (2010a) incorporates global solar irradiance on a horizontal surface, extraterrestrial radiation, solar zenith angle and solar incidence angle on a tilted plane as inputs to RBFNN for estimating global solar irradiance on inclined surfaces in Athens. The absolute fraction of variance R2) is 96%, proving the estimation by the model accurate. Mehleri, Zervas, Sarimveis, Palyvos and Markatos (2010b) used Radial Basis Function neural network (RBFNN) to predict solar radiation on tilted surface using tilt angle and orientation as input parameters. Fuzzy algorithm is used for training RBFNN and it is found that changing tilt angle seasonally improves performance and uniformity of PV array output power. Notton, Paoli, Vasileva, Nivet, Canaletti and Cristofari, 2012) developed three ANN models for Mediterranean site of Ajaccio, France, to estimate hourly global irradiation on the tilted plane. The first and second ANN models estimate hourly global irradiation on 45°, 60° tilted planes respectively whereas the third ANN model estimates both on 45°, 60°

tilted planes. The ANN models incorporate declination angle, hour, zenith angle, hourly extraterrestrial horizontal irradiation, hourly global irradiation as inputs and hourly global irradiation on 45°, 60° tilted planes as output parameters. The (R^2) in the first, second, third ANN models are 99.79%, 99.82% and 99.70% respectively. Celik and Muneer (2013) used generalized regression neural networks (GRNN) to predict solar radiation on the tilted surface for Iskenderun, Turkey. The GRNN utilize input parameters as global solar irradiation on horizontal surface, declination and hour angles. The R^2 and mean absolute percentage error are 98.7% and 14.9 Wh/m^2 respectively. Chatterjee and Keyhani (2012) used 14 inputs (latitude, ground reflectivity and 12 months irradiance value) to estimate optimum tilt and total irradiance on tilted surface by ANN. The activation function in hidden, output layers is hyperbolic tangents, and linear and Levenberg–Marquardt algorithm is used for training. The difference between predicted and analytically determined optimum tilt angle is found to be 3° showing accurate estimation by the ANN techniques. Therefore, the ANN techniques reduce the computation part of optimum tilt angle calculations, and estimate optimum angle and tilted surface solar radiation at different sites with better accuracy.

Many authors have determined optimum tilt angles analytically and experimentally by searching for the maximum total solar radiation falling on the PV modules for a number of locations. Asl-Soleimani, Farhangi and Zabihi (2001) determined optimum tilt angle for Tehran using an experimental setup of five 11-Wp PV mono-crystalline modules by measuring the maximum energy output for a period of one year and emphasized the need for validating the analytically determined optimal tilt angles experimentally also. Li, Lam and Chu (2008) proposed a sunshine hour data based model to calculate solar radiation on the tilted surface with different orientations. The maximum error in this model is found to be less than 5.2%, so it can be used to determine the optimum tilt angle for sites where only sunshine hour data are measured. Chang (2010b) used orthogonal array experiment technique and an ant direction hybrid differential evolution algorithm (ADHDEOA) for tilt angle of PV modules in Taipei area. An ant colony method is used to find the optimum tilt angle at which power is maximal. The annual optimal angle for Taipei area is determined (Taichung 17.3°, Tainan 16.15°, Kaosiung 15.79°, Hengchung 15.17°, Hualian 17.16°, Taitung 15.94°). The computer simulated optimum angles are also verified experimentally. The ADHDEOA is found to be better than GA method in computation and is much effective in optimum tilt angle determination. Beringer, Schilke, Lohse and Seckmeyer (2011) carried out the experimental measurements for monthly optimum tilt angles for cloudy mid-latitude location Hannover, Germany, using eight multi-crystalline silicium solar photovoltaic panels with south orientation mounted at different tilt angles from 0° to 70° in steps of 10°. The maximum power is found to be at 50–70° tilt angles in winter and 0–30° in summer months. The yearly performance difference is found to be less than 6% showing minor differences at various tilt angles. However, further follow up studies to determine optimal tilt angles for such locations with cloudy conditions can further be taken up using the optimization techniques. Pour, Beheshti and Rahnama (2011) used isotropic Liu and Jordan model (1963) to determine the optimum tilt angle and the non-isotropic Klein (1977) and Hay and Davies (1980) models, which consider the effects of the azimuth angle, are used to determine the optimum tilt and azimuth angles for Isfahan, Iran [lat 32° N, Long 51° E]. The optimum tilt angles determined using the Liu and Jordan model for south orientation shows that the yearly optimum tilt angle is nearly equal to the latitude of the site, whereas the results using the non-isotropic models i. e. the Klein (1977) and Hay and Davies (1980) models show that the panels at optimum tilt and azimuth angles receive higher solar energy compared with the fixed solar panel which indicates that the determination of both optimum tilt and azimuth angles is important for maximizing incident solar radiation at a location. A number of studies have been carried out to maximize power output and cooling load reduction of building integrated photovoltaic [BIPV] as such the optimum tilt angle determination for BIPV design and development is another important area of PV applications. Elhassan, Zain, Sopian and Awadalla (2011) determined optimum tilt angles for building integrated photovoltaic design and applications at Kuala Lumpur, Malaysia [Lat 2° 30′ N Long 112° 30′ E], by using a four PV module experimental setup inclined in the North, South, East and West directions to determine the optimum tilt angles for these directions as solar radiation received during the day varies for each direction. The optimum tilt angle for this location was found to be nearly equal to the latitude of the location. Sunderan, Ismail, Singh and Mohamed (2011) investigated that the optimum tilt angle and orientation enhance the power generation of Standalone Photovoltaic Electricity Generation Systems (SPVEGS) in Ipoh, Malaysia. The orientation of PV module is North from April to August and south for the other months. The energy gains of 6.4% and 6.1% are achieved for monthly and yearly optimum tilt angles respectively. Lucio, Valdés and Rodríguez (2012) have given an algorithm to evaluate optimum tilt angle for obtaining minimum loss of load probability and optimum design of stand-alone photovoltaic systems in Europe. Asowata, Swart and Pienaar (2012) used a photovoltaic experimental setup at Vaal Triangle, South Africa, lying in the southern hemisphere [Lat. 29° 00′ S and Long 24° 00′ E] to validate the optimum tilt angles 16°, 26° and 36° determined analytically for winters for this location. Sun, Lu and Yang (2012) determined the optimum tilt angles for the shading-type BIPV claddings at different orientations for Hong Kong [Lat 22.25°N Long 114.1667°S]. The maximum electricity generation per unit PV area is found when the PV modules are installed on south façades at the tilt angle of 10°, thus increasing the energy gain significantly. Bojić, Bigot, Miranville, Parvedy-Patou and Radulović (2012) determined optimum tilt angles for PV systems that are located in four towns (Les Avirons, Petite-France, Saint- Benoit, and Piton Saint-Leu) in Reunion Island, France. The Hooke– Jeeves algorithm is used to find optimum tilt angle at which electrical energy generation becomes maximum and found a discrepancy between latitude and optimum tiltangles, showing the need for determining optimum tilt angles for every location. Siraki and Pillay (2012) proposed a modified HDKR (Hay, Davies, Klucher, Reindl model) anisotropic sky model which also considers the effects of the adjacent buildings in urban locations to calculate the optimum angle. The optimum angles for the five different latitudes were determined and the tilt angle dependence on latitude, weather conditions, surrounding obstacles and optimum azimuth was found. The optimum tilt angle is found to be close to the location latitude for small latitude, while for higher latitude the optimum angle is found to be smaller than the location latitude. These studies indicate that for maximum solar radiation capturing both orientation and optimum tilt angle of PV array needs to be determined accurately for any location. Several authors have given correlations of optimum tilt angle (β_{opt}) in terms of different parameters for the calculation of optimum tilt angles. Hartley, Martinez-Lozano, Utrillas, Tena and Pedro (1999) compared Temps–Coulson's, Klucher's and Hay's models for the estimation of total radiation on vertical planes facing north, south, east and west to find the best model that can be used to calculate solar irradiance on tilted planes in Valencia Spain [Lat 39.5° N, Long 0.67°W]. Hay's model was found to be accurate in determining hourly variations of optimum tilt angles for south facing collector but the average monthly optimum tilt angle was not given accurately. Tang and Wu (2004) developed a method to estimate the optimal tilt angles of a solar collector based on the monthly horizontal radiation and prepared a map showing optimal tilt angles for the year for south facing collectors in China and used Collares-Perara and Rabl (Rabl, 1985) correlation to estimate diffuse solar radiation and Klein's method (Klein, 1977) for the calculation of R_b. It is found that Collares-Perara and Rabl correlations give accurate optimum tilt angle except for low clearness index sites. Fahl and Ganapathisubbu (2011) computed optimum tilt angles for Bangalore [Lat 12.97° N, Long 77.56° E] under various tracking conditions.

The optimum tilt angles are estimated between 15° and 17° for south facing collectors. The incident solar radiation is found to be the same with in variation of 73° in optimum tilt angle. The output of a monthly adjusted collector at optimum angle is found to be increased by 10% in comparison to that received at horizontal orientation. But the output is found to be 35% more for collectors with tracking system. Chang (2008) calculated the optimal tilt angles for six locations in Taiwan using the extraterrestrial, predicted global radiation model and ten year measured solar radiation data. The results indicate that the tilt angles calculated from extraterrestrial, predicted solar radiation are latitude dependent whereas those calculated using measured data vary from site to site and shows that the optimal tilt angle is accurately determined using measured data for a location with cloudy or pollutant environment. The method of maximizing extraterrestrial radiation for tilt angle calculation does not consider attenuation of the solar radiation so, it will not give accurate tilt angles. Shariah, Al-Akhras and Al-Omari (2002) used the annual solar fraction as an indicator to find the optimum tilt angles for a thermo-syphoning solar water heater in the northern and southern parts of Jordan. The optimum tilt angle varies from φ to $\varphi \pm 20°$ for high solar fraction system. Tang, Gao, Yu and Chen (2009) developed a mathematical procedure for determining optimum tilt angles based on maximum annual collectible radiation for evacuated glass tube collectors in China and found that the optimum tilt angles are found to be about 10° less than the latitude for sites with latitudes greater than 30°.

The basic component of PV systems comprises of the photovoltaic module or panel, mounting stands, other balance of system (BOS), and battery depending on the systems design such as standalone etc. The performance of the systems depends entirely on the following, mounting orientation such as fixed, single axis tracking and dual axis tracking system, the efficiency of module used, and solar radiation of the area of installation. According to Sungur (2009), the amount of electric energy produced from PV Systems is directly proportional to the intensity of sunlight which falls on the panel even though, the change observed in sun light does not occur linearly. Therefore, it is desired that solar panels be fixed in a way that they face the sun or have a system that tracks the sun for optimal performance. Solar radiation data is used in sizing and estimation of performance of photovoltaic system over time and it varies from one location to the other depending on the hemisphere. However, it is highly impracticable to purchase and install Pyranometers at all orientations to obtain required solar radiation data, thus many scholars have developed different mechanism used in estimating solar radiation of any location such as loss of load probability, observed time series, and bright sunshine hours with solar angles, respectively. (Egido and Lorenzo, 1992; Markvart, Fragaki and Ross, 2006; Li, Lam and Chu, 2008). Most PV studies are aimed at improving the efficiencies and reducing the cost of installation, so as to enable photovoltaic systems competes favorably with other conventional energy generation means.

The two main types of mounting systems are:

1. Sun tracking systems - An array mounting system, that automatically orients the array to the position of the sun. Here the PV module is installed on a tracker with an actuator to follow the sun. The movement of the tracker is produced by either active (uses electric motor and gear drives) or passive means. There are two types of sun trackers: one/single axis tracker that follows the sun from east to west during the day, and the two axis tracker which follows the sun from east to west during the day and from north to south, during the seasons of the year (azimuth and solar altitude). The sun tracking PV systems is costly because additional cost of trackers and requires more maintenance due to complexity in its design and operations.

2. Fixed systems - An array mounting system, that permanently secures module on a non-movable position, horizontally or at a specific tilt angle. The advantages of the fixed system over the tracking systems are; it's cost effective, easy to install and requires less maintenance.

A review of literatures on the two mounting orientations of photovoltaic systems design shows that generally the sun tracking systems increases yield over the years when compared with the fixed array design operating under the same conditions. Karimov, Saqib, Akhter, Ahmed, Chattha and Yousafzai (2005) described a simple photo-voltaic (PV) tracking system which has been designed and manufactured using a pyramidal stand as a base. A rotating unit consisting of two pairs of modules fixed at an angle of 170° between them was installed at the upper edge of the stand. The four modules and a DC motor were connected to a bridge circuit making the system sensitive to solar tracking. The PV tracker has a DC-DC (24/2 V) converter, DC-AC (24/220 V) inverter and a battery. The modules are able to provide a maximum power of ~100 W, which feeds the converter, inverter and DC motor. Wind resistance of the tracker is quite low. The total area of the four modules is 1.26 m ^{2}. The inclination angle between the modules and the horizontal plane is adjustable (34°±11°) and the tracking angle is 120°. The system can track solar motion with an error of ±10°. The designed PV tracking system, with modules fixed at an angle of 170° to feed the load as well as the DC motor, exhibited it to be an efficient energy-conversion system. The fabricated system offers low wind resistance. The cost analysis data revealed that the proposed design was very economical and cost effective.

The concept of energy gain is very useful in evaluating the performance of different types of tracking. This concept allows to evaluate the effectiveness of daily, weekly, fortnightly, monthly, seasonally, biannually and yearly adjustment of the solar receiver tilt angle in relation with the ideal instantaneous dual tracking, where the sun rays are kept permanently perpendicular to the flat-plate surface of the receiver. Thus, the incidence angle is kept to be zero all over the day. This evaluation determines the optimum tilt application over any period of consecutive days from technical point of view.

PHOTOVOLTAIC SYSTEMS TECHNOLOGY

Photovoltaic are notable renewable solar energy generation technology that has come to stay, it is believed by many scholar to be capable of playing a very important role in future global energy generation mix. This technology has history dated as back as 1839 when Henry Becquerel discovers the emission property of electron, subsequently in 1954 it was developed to terrestrial application in Bells Laboratory by Chapel et al. It works on the photovoltaic principle.

Solar energy in the form of direct electricity conversion (photovoltaics) is already very popular in countries such as the United States, Germany and Japan. The enormous potential of photovoltaic (PV) technology is also obvious and favorable in countries with high irradiation such as the Mediterranean region. In spite of very high solar energy sources potentials in hot countries like Sunbelt countries, e.g. Jordan, Syria and Egypt, these sources are utilized inefficiently. The most solar installations are specifically for water heating purpose. These countries have annual solar radiation estimated 2100 kWh/m^2 at horizontal surface with nearly 300 sunny days. This solar irradiation is much higher when compared to the sunniest area of Germany one of the world's solar photovoltaic largest market (Makrides, Zensser, Norton and Georghiou, 2010).

Photovoltaic technologies are classified into first generation PV systems (fully commercial) of Mono and poly crystalline which use the wafer based silicon (c-Si) technology, the second generation PV systems (early market deployment) of amorphous silicon a-Si, cadmium telluride (CdTe), copper indium selenide etc. that are based on thin film technology and the third generation PV systems which includes technologies of organic PV cells and concentrating PV (CPV) that are still under demonstration (Raugei and Frankl, 2009). A detailed description and review of these technologies were also performed and presented by Chaar, Lamont and Zein (2011). These technologies are however grouped as either standalone (uses energy storage) or on grid connected systems.

PHOTOVOLTAIC SYSTEM'S PERFORMANCE

Lubitz (2011) utilized hourly typical meteorological year (TMY3) data with the Perez radiation model to simulate solar radiation on fixed, azimuth tracking and two axis tracking surfaces at 217 geographically diverse temperate latitude sites across the contiguous United States of America. The optimum tilt angle for maximizing annual irradiation on a fixed south-facing panel varied from being equal to the latitude at low-latitude, high clearness sites, to up to 14° less than the latitude at a north-western coastal site with very low clearness index. Across the United States, the optimum tilt angle for an azimuth tracking panel was found to be on average 19° closer to vertical than the optimum tilt angle for a fixed, south-facing panel at the same site. Azimuth tracking increased annual solar irradiation incident on a surface by an average of 29% relative to a fixed south-facing surface at optimum tilt angle. Two axis tracking resulted in an average irradiation increase of 34% relative to the fixed surface. Introduction of manual surface tilt changes during the year produced a greater impact for non-tracking surfaces than it did for azimuth tracking surfaces. Even monthly tilt changes only resulted in an average annual irradiation increase of 5% for fixed panels and 1% for azimuth tracked surfaces, relative to using a single optimized tilt angle in each case. In practice, the decision whether to manually tilt panels requires balancing the added cost in labor and the panel support versus the extra energy generation and the cost value of that energy.

Mehrtash, Guillermo, Yvan and Daniel (2012) investigated the performance of photovoltaic (PV) systems with different types of solar trackers in Northern climates. To this end, four PV systems were simulated; horizontally fixed, inclined fixed, azimuth tracking, and a dual-axis tracking. The simulations have been carried out by use of PVSOL Pro for daily, monthly, and annual periods. The analyses have been done for climate conditions prevailing in Montreal, Canada. Annual analyses show an increase of solar irradiation upon a tilted system, azimuth tracker system, and dual axis tracker system as compared to the horizontal system. This yearly increase is 16.8%, 50.1%, and 55.7% respectively. The results from daily analyses show, as expected, that in clear days the dual axis tracker PV system provides the highest performance, but in overcast conditions all systems perform almost similarly and the optimum position is horizontal. The results indicated that a dual-axis tracking array is the optimum system if it goes to the horizontal position in overcast condition.

Cruz-Peragón, Casanova-Peláez, Díaz, López-García and Palomar (2011) showed an alterative method for determining the tracking energy advantage, defined as the additional electrical energy produced by two axes tracking systems respect to fixed devices, in order to analyze the economical profitability in Spain. For this purpose, 52 main cities of the Spain have been analyzed. The proposed methodology starts from irradiation data, combining diffuse models and daily-hourly relations. Different types of losses have been evaluated, and the electrical behavior of the systems has been incorporated. Final annual energetic results demonstrate that two axes devices show a relevant energy advantage (higher than 20%) for most of the Spain territory.

The results of these and other studies were present in Table 1. The reference efficiency is the result of the fixed system generated at different locations and is compared to the tracking systems (single axis or/and dual axis). It is clearly seen that both single and dual axis tracking systems, produce higher amount of energy when compared with the fixed systems, this efficiencies are between 10 to 55%. It is also observed that dual axis sun tracking systems produce much more energy than single axis systems as expected. The variation in efficiencies in Table 1 is due to locations of studies, tracking methods and different types of module used in the studies.

Table 1. Summary of percentage yield of sun tracking over fixed systems

Date	Tracking System Type		Location	Reference
	One Axis	Dual Axis		
2000	11 - 18%	30%	Egypt	(Helwa, Bahgat, El Shafee, and El Shenawy, 2000)
2004	-	41.34%	Jordan	(Abdallah and Nijmeh, 2004)
2004	15.69-37.53%	43.87%	Jordan	(Abdallah, 2004)
2004		62%	DamascusSyria	(Al-Mohamad, 2004)
2005	-	20%	Tajikistan	Karimov, Saqib, Akhter, Ahmed, Chattha and Yousafzai, 2005)
2006	46.46%	-	Greece	(Bakos, 2006)
2008	42.6%	-	Turkey	(Sungur, 2009)
2008	10 - 20%	-	Estonia	(Tomson, 2008)
2008	-	30 - 45%	Jordan	(Mazen, Abu-Khader, Omar, Badranb and Salah, 2008)
2009	-	29%	Turkey	(Rustemli, Dincadam and Demirtas, 2009)
2009	10 - 35%	25 - 45%	Turkey	(Sefa, Mehmet and Ilhami, 2009)
2011	29%	34%	USA	(Lubitz, 2011)
2011	-	10-40%	Spain	(Cruz-Peragón, Casanova-Peláez, Díaz, López-García and Palomar, 2011)
2012	-	30.79%	Turkey	Eke and Sentruk (2012)
2012		30%	Alepo Syria	Dakkak and Babeli (2012)
2012	15.6%	19.1%	DamascusSyria	(Soulayman and Sarsar, 2012)
2012	38%	-	Romania	(Tudorache, Oancea and Kreindler, 2012)
2012	33.3%	38.9%	Canada	(Mehrtash, Guillermo, Yvan and Daniel, 2012)
2013	-	7.1-10.5%	Bayan Lepas, Malaysia	(Lee, Rahim and Al-Turki, 2013)
2013	-	4.8-9.1%	Ipoh, Malaysia	(Lee, Rahim and Al-Turki, 2013)
2013	-	5.6-11.2%	Kuantan, Malaysia	(Lee, Rahim and Al-Turki, 2013)
2013	-	6.6-17.5%	Langkawi, Malaysia	(Lee, Rahim and Al-Turki, 2013)
2013	-	4.1-8.1%	Muadzam Shah, Malaysia	(Lee, Rahim and Al-Turki, 2013)
2013	-	1.5-6.4%	Senai, Malaysia	(Lee, Rahim and Al-Turki, 2013)

continued on following page

Table 1. Continued

Date	Tracking System Type		Location	Reference
	One Axis	Dual Axis		
2013	-	6.1-9.9%	Subang, Malaysia	(Lee, Rahim and Al-Turki, 2013)
2013		40.2%	Cyprus	(Okoye and Abbasoğlu, 2013)
2013	-	23%	Mexico	(Lee, Rahim and Al-Turki, 2013)
2013	-	23%	UK	(Lee, Rahim and Al-Turki, 2013)
2013	-	10-20%	Estonia	(Lee, Rahim and Al-Turki, 2013)
2013	-	72%	Mongolia	(Lee, Rahim and Al-Turki, 2013)
2013	-	24.5%	Taiwan	(Lee, Rahim and Al-Turki, 2013)
2016	-	40% (30%)	Berlin, Germany	(Sharaf Eldin, Abd-Elhady and Kandil, 2016)
2016	-	16.8% (6.8%)	Stuttgart, Germany	(Sharaf Eldin, Abd-Elhady and Kandil, 2016)
2016	-	10.61% (0.61%)	Cairo Egypt	(Sharaf Eldin, Abd-Elhady and Kandil, 2016)
2016	-	8.16% (-1.84%)	Aswan Egypt	(Sharaf Eldin, Abd-Elhady and Kandil, 2016)

LITERATURE SURVEY ON SOLAR TRACKING EFFECTIVENESS

General

Aberle (2009) pointed to the rapid progress that is being made with inorganic thin-film photovoltaic (PV) technologies, both in the laboratory and in industry. While amorphous silicon based PV modules have been around for more than 20 years, recent industrial developments include the first polycrystalline silicon thin-film solar cells on glass and the first tandem solar cells based on stacks of amorphous and microcrystalline silicon films ("micromorph cells"). Significant thin-film PV production levels are also being set up for cadmium telluride and copper indium diselenide. Even though, one of the biggest problems involved with solar energy today is the efficiency coefficient (Aberle, 2009). Even with the newest and best performers the highest lab recordings for panel efficiencies have been ~20% (CIS) and falls to ~13% when attempting commercial applications. Science has tried to combat these low efficiencies by using different materials and production techniques. The influence of the TiO_2/electrode interface, on electron transport properties at the interface and in TiO_2 porous film, in back contact dye-sensitized solar cells, was investigated. Analysis of dye-sensitized solar cells (DSCs) with Ti and TCO indicated that electron transport properties at TiO_2/Ti and TiO_2/TCO interfaces are similar despite the former's lack of a 'built-in potential'. The dependence of short circuit current density on TiO_2 thickness indicated that TiO_2 electron transport is not affected by 'built-in potential' or electrode structure. Electron transport thus appears similar in back contact dye-sensitized solar cells and DSCs. A back contact dye-sensitized solar cell fabricated with a Ti electrode and optimum TiO_2 porous film showed a conversion efficiency of 7.8% with a metal mask under an air mass of 1.5 sunlight (Fuke, Fukui, Islam, Komiya, Yamanaka, Harima and Han, 2009). It is believed that the successful implementation of the antenna solar energy conversion (ASEC) to the commercial level would revolutionize the solar energy utilization. Screening suitable materials and technologies by analyzing potential pros and cones all the way to commercial level would conserve much time and resources. Here carbon nanotube (CNT) known as the material of the 21st century is evaluated concerning many aspects to implement ASEC concept (Wijewardane, 2009). Thermo-mechanical and electrical properties of CNT as regards to ASEC, interaction of CNT and CNT arrays with optical spectrum and existing and future CNT technologies that have the potential to overcome possible future obstacles are discussed. The economic aspects which govern the future of ASEC concept, with CNT as a basic material are also discussed. Graphite sheets with carbon nanotubes and titanium oxide panels are a couple of examples, but even they are still in the preliminary testing stages (Wijewardane, 2009).

Another approach may combine current fossil fuel power technology with solar energy in an attempt to bridge the difference and drag the price /kWh down to a point where it is profitable. In this piece, the hybrid is a gas/solar station (Schwarzbozl, Buck, Sugarmen, Ring, Crespo, Altwegg and Enrile, 2006). The combination of high solar shares with high conversion efficiencies is one of the major advantages of solar gas turbine systems compared to other solar-fossil hybrid power plants. Pressurized air receivers are used in solar tower plants to heat the compressed air in the gas turbine to temperatures up to 1000 °C. Therefore solar shares in the design case of 40% up to 90% can be realized and annual solar shares up to 30% can be achieved in base load. Using modern gas turbine systems in recuperation or combined cycle mode leads to conversion efficiencies of the solar heat from around 40% up to more than 50%. This is an important step towards cost reduction of solar thermal power. Together with the advantages of hybrid power plants—variable solar share, fully dispatchable power, 24 h operation without storage—solar gas turbine systems are expected to have a high potential for market introduction in the mid-term view. Schwarzbozl, Buck, Sugarmen, Ring, Crespo, Altwegg and Enrile (2006) presented the design and performance assessment of several prototype plants in the power levels of 1 MW, 5 MW and 15 MW. Advanced software tools are used for design optimization and performance prediction of the solar tower gas turbine power plants. Ismail, Moghavvemi and Mahlia (2013) investigated the feasibility of using the micro-turbines as backup sources in the hybrid systems. A scenario depending on PV standalone system and other scenario depending on micro-turbine only were also studied. A simulation program was developed to optimize both the sizes of the PV system and the battery bank, and consequently determine the detailed specifications of the different components that make up the hybrid system. The optimization of the PV tilt angle that maximizes the annual energy production was also carried out. Powering a rural community using micro-Hybrid systems are defined as systems that utilize more than one energy source to supply a certain load. The implementation of a hybrid system that is based upon Photovoltaic (PV) to supply power to remote and isolated locations is considered a viable option. This is especially true for areas that receive sufficient amounts of annual solar radiation.

Solar Tracking Performance

The attractive point of having a tracking system is that energy gain and consequently energy output can be increased as a PV system can be kept perpendicular to the incoming solar radiation as the sun's location changes throughout the day. Moreover, tracking is very helpful on cloudy days. Eke and Sentruk (2012) in Turkey found that 30.79% power gain could be achieved when applying a solar tracker to a PV system in comparison with using a latitude tilt fixed PV system. Moreover, Dakkak and Babeli (2012) reported a 30% increase in the supplied energy to a utility from a tracking system. An earlier study by Huang and

Sun (2007) was able to prove that 56% power output increase could be reached by using one-axis three-position tracking system. Senpinar and Cebeci (2012) studied the effect of tracking on the performance of PV cells in Turkey. They concluded that the most advantage of a two-axis solar tracking array is its ability to maintain its position without being affected by clouds or other environmental conditions. Their system produced a 13.25% higher average output power than that of a fixed system. However, the same study questioned the economic benefit of the tracking system. Koussa, Haddadi, Saheb, Malek, Hadji (2012) studied the difference in efficiencies between single-axis tracking systems and two-axis tracking systems and found that the gain does not exceed 3%. The same study concluded that the use of a sun tracker system would only be useful on clear-sky days and unnecessary on cloudy days.

Alexandru and Tatu (2013) studied numerically and experimentally the idea of PV string control, such that the daily motion is transmitted to all the modules of the string simultaneously with a multi-parallelogram mechanism and a single motor. Alexandru and Tatu (2013) concluded that the energetic efficiency of the PV string control system was much higher than those in cases of a fixed system and tracking independently each module of the array using its own tracking system.

Another issue to be discussed is the parameters influencing the selection of a PV tracking system. For instance, temperature is a parameter that has a great effect on the PV module performance. The simulation of module temperature from Nominal Operation Cell Temperature (NOCT) is widely used to easily estimate module performance along the year. In this context, it is important to determine this parameter in a reliable way, as it is used to compare the performance of different module designs and can influence system predictions. At present there are several international standards that indicate the method to calculate NOCT in crystalline and thin-film terrestrial photovoltaic modules. Garcia and Balenzategui (2004) presented the results obtained when applying these standards to different types of PV modules, including glass–glass and glass–tedlar structures, crystalline and thin-film technologies, and some special module designs for building integration applications. NOCT values so calculated have been used to estimate the yearly module temperature and performance for different orientations and tilted angles, analyzing temperature influence in these estimations. Possible error sources that could bring about erroneous values of NOCT are also analyzed.

One of the main obstacles that face the operation of photovoltaic panels (PV), especially crystalline silicon panels, is overheating due to excessive solar radiation and high ambient temperatures. Overheating reduces the efficiency of the panels dramatically (Akbarzadeh and Wadowski, 1996). A research carried out by Moharram, Abd-Elhady, Kandil and El-Sherif (2013) concluded that the temperature coefficient of the PV panels that were used in their study is -0.5%/°C, which indicates that every 1 °C of temperature rise corresponds to a drop in the efficiency of 0.5%. However, the efficiency loss for thin-film modules is only about half of that of crystalline silicon technology (Rustemli, Dincer, Unal, Karaaslan and Sabah, 2013). Many researchers have shown that the operating temperature of the PV panel plays a central role in the PV conversion process. In this context, Ye, Nobre, Reindl, Luther and Reise (2013) analyzed the influencing factors, including rooftop material, ventilation, module framing and other environmental conditions on module temperature of selected photovoltaic (PV) systems in Singapore have been analyzed in detail. The variance of module temperature has turned out to be much larger than the variance of ambient temperature on the different project sites. Skoplaki and Palyvos (2009) presented a brief discussion regarding the operating temperature of one-sun commercial grade silicon-based solar cells/modules and its effect upon the electrical performance of photovoltaic installations. Mattei, Notton, Cristofari, Muselli and Poggi (2006) studied the performance of a photovoltaic module versus environmental variables such as solar irradiance, ambient temperature and wind speed.

Two types of simplified models were studied in this paper: a PV module temperature model and a PV module electrical efficiency model. These models have been validated utilizing experimental data from two experiments: a 850 Wp grid connected photovoltaic system and a p-Si module with eight temperature sensors integrated into the module. Both models have been coupled to determine the PV array output power versus the three meteorological parameters. This simple model using a simple energy balance and neglecting the radiation effects is in good agreement with the experimental data. It is clear that, overheating of the PV panel due to excessive exposure to solar irradiance and hot climate, as in Sunbelt countries, can affect the performance of the panel. Therefore, a tracking system could be unnecessary for Sunbelt countries.

Sun trackers can be categorized based on their electricity consumption (aka drive type) and their actuator movements. In the electricity consumption classification, the sun trackers can be divided into two main groups of passive and active sun trackers.

Comments on the Passive Sun Trackers

A passive sun tracker is a non-precision mechanism for reorienting photovoltaic panels. It is often based on a low boiling-point compressed gas fluid (e.g., Freon) that is driven from one side to another because of pressure created by the solar heat. The pressure causes the tracker to move. For example, Parmar, Parmar and Gautam (2015) discussed a passive solar tracking system that inserts a low-boiling-point compressed gas fluid to metal canisters that are fixed on both sides of the photovoltaic panel mount, and these canisters are connected to each other by a metal pipe. A shadow is installed on both canisters, and the shadow covers the outer half portion of the canisters. When not shaded, the canisters are heated by the sun and the liquid inside the non-shaded side is forced to move to the shaded side. While this phenomenon is occurring, the solar panel that is placed between two metal canisters will follow the sun. This tracking system uses gravity as its moving force. Parmar, Parmar and Gautam (2015) stated ''Zomework increases the electrical output of photovoltaic modules by 23.33% or more compared to modules on fixed mounts'' (p. 144).

Clifford and Eastwood (2004) used a different passive sun tracking based on bimetallic strips. This passive sun tracker places two bimetallic strips made of aluminum and steel on a wooden frame symmetrically on either side of a central horizontal axis. Then, the bimetallic strips are introduced to shade, so the strip that is further from the sun absorbs solar radiation, while the other strip remains shaded. This was the same design as the compressed gas tracker; however, a different material was applied to the compressed gas tracker. According to this paper ''this system also boasts the absence of complex fluid containment and accurate fitting of pistons'' (p. 271). As a result of this tracker, solar panel efficiency can be increased by 23%; however, the system in question requires manual reposition every morning. Therefore, the economic feasibility of such tracking is questionable.

Comments on the Active Sun Trackers

Based on actuator movements, sun trackers can be broadly classified into single- and dual-axis trackers. Sungur (2007) implemented an electromechanical control system of a photovoltaic panel tracking the sun on the axis it moves along according to its azimuth angle. In this experiment, the azimuth angle of the sun was measured by a PLC analog module, and the data collected were sent to the actuator motor, and then the actuator motor, which is on a single-axis system, moves the panel. According to measurements that were observed at 37.6 degrees latitude (Konya, Turkey), photovoltaic panels with a single-axis tracking system obtained 32.5% more energy compared to fixed-position PV panels.

Comparison between single- and dual-axis tracking systems has not led to one absolute superiority for one versus other because it depends on a variety of environmental factors (e.g., sky clarity and location) and technical factors (e.g., photovoltaic/concentrated photovoltaic type and axis tilt/rotation mechanism). While dual-axis tracking systems are more accurate and produce more energy compared to single-axis tracking systems, these systems are more complex and consume more energy; thus, they are more expensive (Cooke, 2011). Research by Afiqah, Syafawati, Idris, Haziah and Syahril (2015) proposed a dual-axis solar tracking system in which the solar panel moves according to control from timer system for two encoder motors, a PIC microcontroller and driver L293D. This active solar tracker enabled a solar panel to collect 12.93% higher sunlight compared to a fixed solar panel without a tracking system. Also, when using a tracking system, the output voltage was stable within a range of 18–20 V, while a solar panel without a tracking system was unstable. The average output voltage for a tracking system was 19.59 V as compared to 18.89 V for a non-tracking system. The tracking system was effective in avoiding the shading effect during sunrise and sunset. This study did not provide motor maintenance cost, although the use of the timer system reduced the need for real-time sun positioning sensors and their related operating and maintenance costs.

Here, it should be mentioned that, the power gain which could be achieved when applying a solar tracker to a PV system is mostly calculated basing on power output of a latitude tilt fixed PV system. Moreover, the value of this energy gain varies widely. For example, Senpinar and Cebeci (2012) found for Turkey that the tracking system produced a 13.25% higher average output power than that of a fixed system while Eke and Sentruk (2012) reported that, in Turkey 30.79% power gain could be achieved. Koussa, Haddadi, Saheb, Malek, Hadji (2012) found that the energy gain of the two-axis tracking systems exceeds that of the single-axis tracking systems by less than 3%. Thus, with taking into account that, there will always be higher capital costs in a tracking system due to installed motors and moving joints, the economic feasibility of solar trackers to PV systems in comparison with using the PV system with daily, monthly, seasonally and biannually tilt adjustment is not evident.

Comments on Aging and Degradation

The lifespan of photovoltaic panels according to some manufacturers is 20 years (Honsberg and Bowden, N. D.). The efficiency of the panels will decrease with time, and the quantification of power drop over time is known as the degradation rate. The median value of the degradation rate, in general, is 0.5% per year (Jordan and Kurtz, 2013). Many factors can contribute and cause this degradation: the environment (pollution is one major issue), discoloration of the encapsulation (EVA layer) or glass due to the ambient temperature, lamination defects, mechanical stress, and humidity cell contact breakdown (Kaplani, 2012). The different technologies used to manufacture photovoltaic panels can cause different types of degradation. Crystalline modules will suffer irreversible light-induced degradation due to defects activated by the initial light exposure (Livingonsolarpower 2013, June 10). Amorphous silicon cells may face a decrease in output power of 10–30% in the first six months of light exposure; then, it will stabilize (Livingonsolarpower 2013, June 10). Therefore, an empirical approach to measuring and comparing realistic power generation and associated benefits/costs should be provided. This kind of work was provided by Asiabanpour, Almusaied, Aslan, Mitchell, Leake, Lee, H., Fuentes, Rainosek, Hawkes and Bland (2016) using two similar solar panels where one is oriented by accurate manual sun tracking in 15-min time intervals and the other operated in a fixed horizontal position.

MAXIMAL POSSIBLE GAIN OF TRACKING

Theoretical Estimation

Let us consider the ideal case - the ideal instantaneous dual tracking. In this case the incidence angle satisfies the following equation:

$$\theta_i = 0^o \tag{1}$$

Thus, the daily extraterrestrial solar radiation $H_{0,d}\left(n\right)$ on a day number n, in this case, is:

$$H_{0,d}\left(n\right) = G_s\left(n\right).S_0\left(n\right) \tag{2}$$

where $S_0\left(n\right)$ is the maximum possible sunshine duration (day length) during the day number n.

When dividing the value of the equation (2) by the value of the daily extraterrestrial solar radiation on the horizontal plane, the energy gain factor of daily instantaneous dual tracking with relation to horizontal surface could be obtained as follows:

$$R = \left[\frac{H_{0,d}\left(n,\theta_i = 0^o\right)}{H_{0,d}\left(n,\theta_i variable\right)}\right] = \left\{\frac{\frac{2}{15}\arccos\left[-\tan\left(\varphi\right)\tan\left(\delta\right)\right]}{cos\left(\varphi\right)cos\left(\delta\right)sin\left(\omega_s\right) + \frac{\pi * \omega_s}{180}sin\left(\varphi\right)sin\left(\delta\right)}\right\} \tag{3}$$

On the other hand, the energy gain factor of daily instantaneous dual tracking with relation to daily optimum tilted surface could be obtained as:

$$R' = \left[\frac{H_{0,d}\left(n,\theta_i = 0^o\right)}{H_{0,d}\left(n,\beta_{opt,d}\right)}\right] = \left\{\frac{\frac{2}{15}\arccos\left[-\tan\left(\varphi\right)\tan\left(\delta\right)\right]}{cos\left(\varphi - \beta_{opt,d}\right)cos\left(\delta\right)sin\left(\omega_{ss}\right) + \frac{\pi * \omega_{ss}}{180^o}sin\left(\varphi - \beta_{opt,d}\right)sin\left(\delta\right)}\right\} \tag{4}$$

The results of R and R' calculation using the equations (3) and (4) for different are given in the Figures 1 to 3.

It is clear from Figures 1 to 3 that, the condition described in chapter 5 for optimum tilt determination which says that, the optimum tilt angle in the mid-latitude zone could be restricted to positive values in the NH and to negative values in the SH. This means that, in mid-latitude zone it is reasonable to orientate the solar receivers to the Equator as the losses in the energy gain are negligible (see the sum-

Figure 1. The daily energy gain of the daily instantaneous dual tracking with relation to horizontal (▲) and daily optimum tilted (x) surfaces for φ = 23.45°N

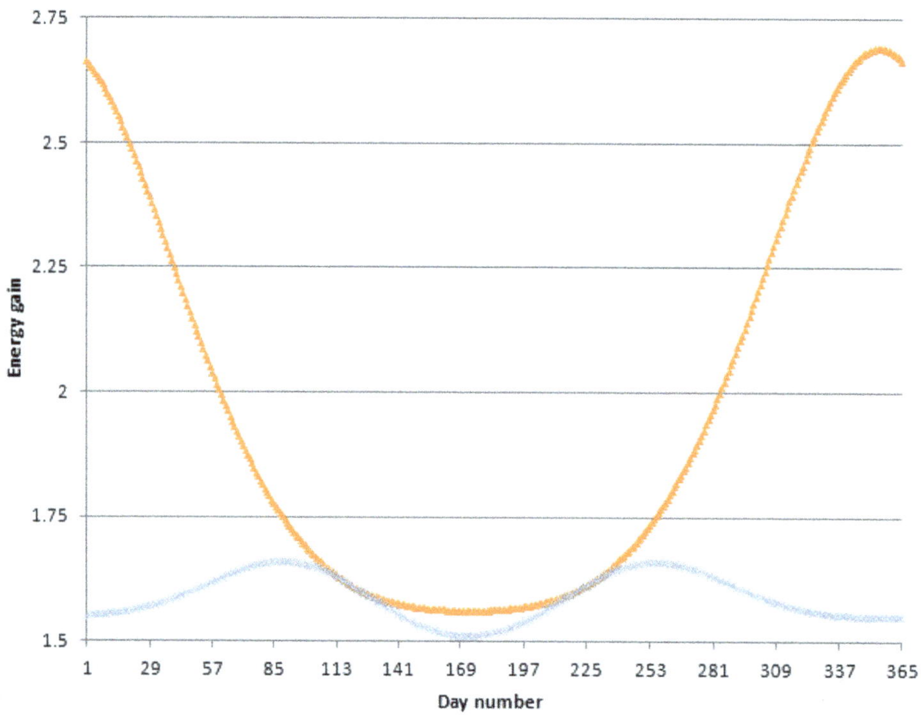

Figure 2. The daily energy gain of the daily instantaneous dual tracking with relation to horizontal (▲) and daily optimum tilted (x) surfaces for φ = 29.49°N

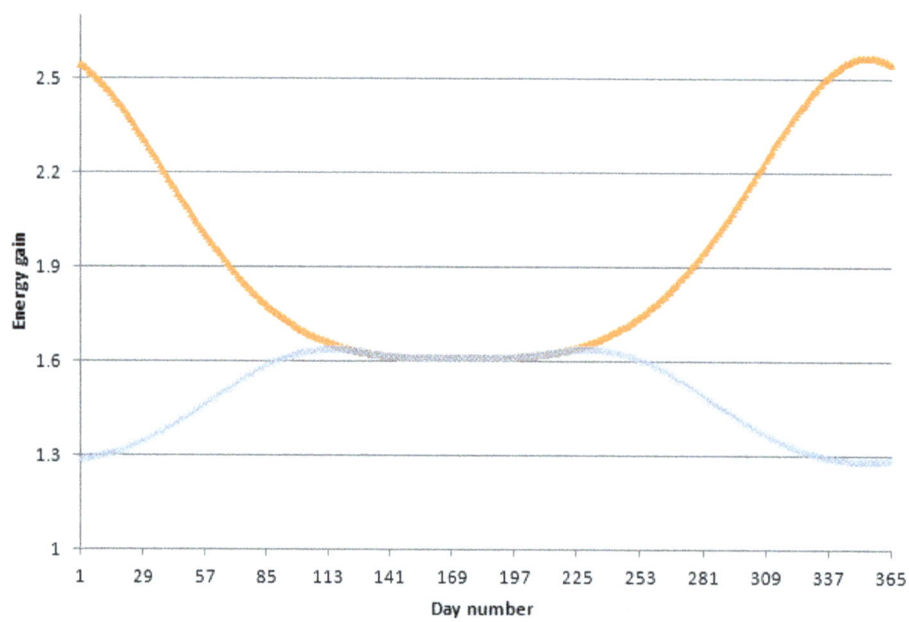

Figure 3. The daily energy gain of the daily instantaneous dual tracking with relation to horizontal (-) and daily optimum tilted (♦) surfaces for φ = 43.45°N

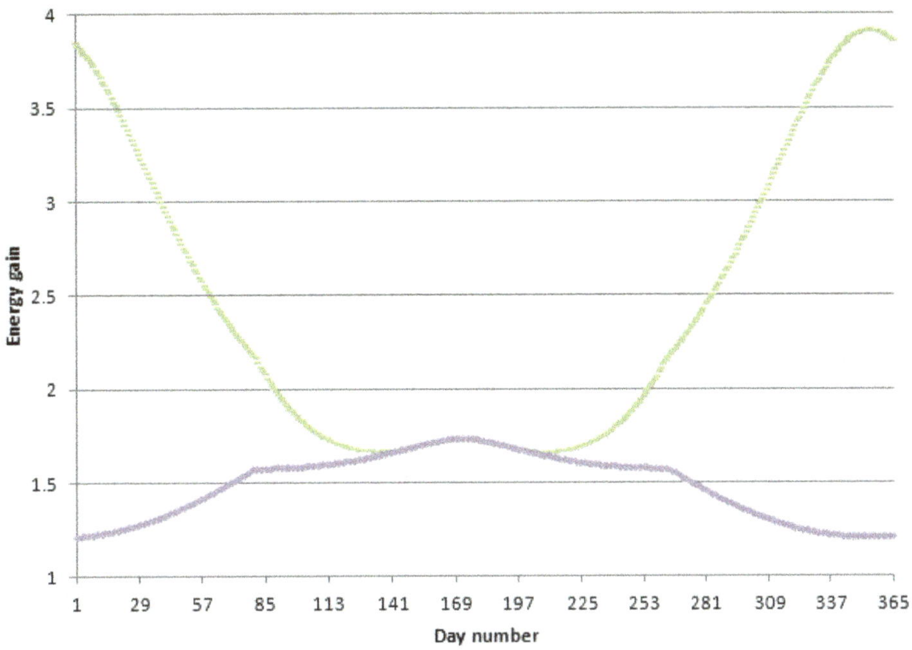

mer period in the Figures 1 and 2). The above mentioned losses are maximal at latitudes 23.45°N and 23.45°S. Figure 4 shows these losses for the latitude 23.45°N. It is seen from Figure 4 that these losses are restricted to the summer period and the daily values of these losses do not exceed 3%.

On the other hand, the instantaneous dual tracking is more effective than an installation with daily adjusting the tilt angle whatever the latitude is. Its effectiveness becomes higher with increasing the day length.

Real Estimation for PV Panel

Let us consider three cases:

1. Dual tracking case where the photovoltaic panel is placed perpendicular to the solar irradiance during tracking the sun.
2. Miss tracking case where the photovoltaic panel is parallel to the solar irradiance during miss tracking the sun. This case is considered in order to make the panel exposed to diffused solar irradiance only.
3. No tracking case (fixed tilted case) where the photovoltaic panel is installed at a fixed position during the no tracking case.

The mathematical model will be used to calculate the output power of photovoltaic (PV) panel in these three cases. The model comprises a set of equations that calculate the output power from a photo-

Figure 4. The daily energy gain of solar receiver with only positive daily optimum tilt (blue) and with positive and negative daily optimum tilt (red) at latitude φ=23.45°N

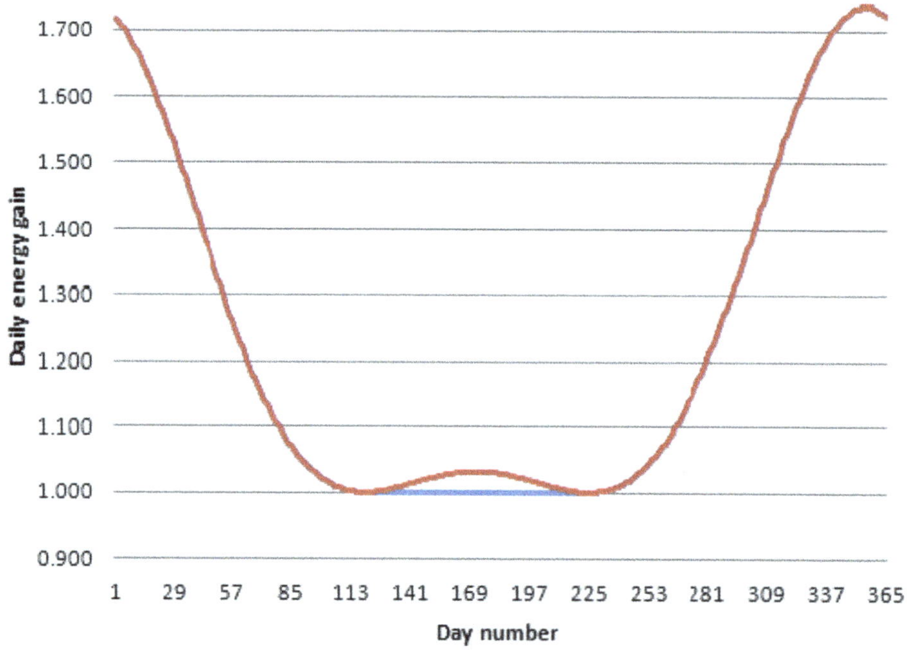

voltaic (PV) panel in any of the above stated conditions. The analysis starts by calculating the module temperature T_m using the empirical model developed by Garcia and Balenzatgui (2004) as follows:

$$T_m = T_{amb} + \frac{(NOST - 20)G}{800} \tag{5}$$

where T_{amb} is the ambient temperature, *NOCT* is the nominal operating cell temperature of the PV module, and *G* is the solar irradiance. The module temperature is based on the ambient conditions, i.e. solar irradiance, *G*, ambient temperature, T_{amb}, and the nominal operating cell temperature, *NOCT*. The *NOCT* is a constant value given by the manufacturer of the PV panel, and it is 45 °C for the panels used in the performed experiments. The solar irradiance and the ambient temperature are input variables that vary with time and location. The power output, P, from the PV panel depends on the module temperature, T_m, and it is calculated based on the model developed by Skoplaki and Palyvos (2009) as follows:

$$P = G\eta_{ref}A\left[1 - \beta_{ref}\left(T_m - 25\right)\right] \tag{6}$$

where η_{ref} is the photovoltaic panel at the reference temperature, T_{ref}, which is usually specified by the PV panel manufacturer, *A* is the photovoltaic panel's surface area and β_{ref} is the power temperature coefficient which is usually specified by the PV panel manufacturer. Thus, the power output from the

PV panel depends on the parameters shown in the equation (2). In the following calculations η_{ref} is taken to be 12% and β_{ref} is taken to be 0.005°C^{-1}.

It is clear from the equations (5) and (6) that, in order to determine the power output from a PV panel, it is required to know the ambient temperature, T_{amb}, as well as the solar irradiance, *G*, falling on the PV panel. The solar irradiance is measured between sunrise and sunset in each day while the ambient temperature is measured as an average temperature of the whole day.

The PV panel in the tracking case is assumed to be always perpendicular to the solar irradiance from sunrise till sunset. In this case, the total solar irradiance on the PV panel is considered in the calculations of the module power output. The solar irradiance reaching the PV panel, G_{panel}, is given by:

$$G_{panel} = G \tag{7}$$

where *G* is the total solar irradiance.

For the full miss tracking case, the panel is placed parallel to the solar irradiance, making the panel exposed to diffused solar irradiance only. In this case, the solar irradiance reaching the panel, G_{panel}, is equal to the diffused solar irradiance, G_d. The solar irradiance, G_d, is taken as 20% from the total solar irradiance in the full miss tracking case. This value represents the diffused solar irradiance, as proposed and verified by Safaripour and Mehrabian (2011). The miss tracking condition has been thought of as a method of avoiding overheating the PV panels installed in hot and sunny regions.

In the no tracking case, the PV panel is fixed at an angle β with the horizontal towards the Equator which is known as the tilt angle. The tilt angle β is taken to be equal to the latitude angle φ. The sun moves along the day from east to west making an inclination angle with the horizontal that varies along the day. The solar irradiance is calculated as a function of the total solar irradiance and the orientation of the PV panel, in case of no tracking the sun, as follows:

$$G_{panel} = G\cos\left(\theta_i\right) \tag{8}$$

where *G* is the total solar irradiance, and θ_i is the solar incidence angle, which is defined as the angle between the sun's rays and the normal on the PV panel surface (see chapter 1). θ_i is equal to zero in case of a dual tracking system such that equation (8) becomes equal to the equation (7). The angle of incidence in case of a PV panel located in the Northern Hemisphere facing south is calculated using the following equation (see chapter 1):

$$\cos\left(\theta_i\right) = \cos\left(\varphi - \beta\right)cos\left(\delta\right)cos\left(\omega\right) + sin\left(\varphi - \beta\right)sin\left(\delta\right) \tag{9}$$

where β is the tilt angle of the PV panel with the horizontal, δ is the solar declination angle, which is the angle between the sun-earth centerline and the projection of this line on the equatorial plane, ω is the solar hour angle of a point on the earth's surface, which is equal to the angle through which the earth would turn to bring the meridian of the point directly under the sun and φ is the local latitude angle defined as the angle between the line from the center of earth to the site of interest and the equatorial plane (see chapter 1 for the definitions of these angles and the equations which determine these angles). The hour angle can also be obtained from the apparent solar time (see chapter 6). The tilt angle of the PV panel could be adjusted to latitude angle φ of the site or to yearly optimum tilt angle, $\beta_{opt,y}$, of solar receiver at that site.

EXPERIMENTAL VERIFICATION IN THE MID-LATITUDE ZONE

Dual Tracking and Horizontal Fixed Case

An experimental setup was constructed to evaluate the energy gain of a dual tracking system in Texas State University (Asiabanpour, Almusaied, Aslan, Mitchell, Leake, Lee, Fuentes, Rainosek, Hawkes and Bland, 2016). The experimental setup consists of, see Figure 5, a mechanical system to hold and control the tilt and orientation of the photovoltaic panel, and an electrical system (i.e., wire-wound, adjustable

Figure 5. Mechanical system design, manufacturing, and assembly in the experimental setup

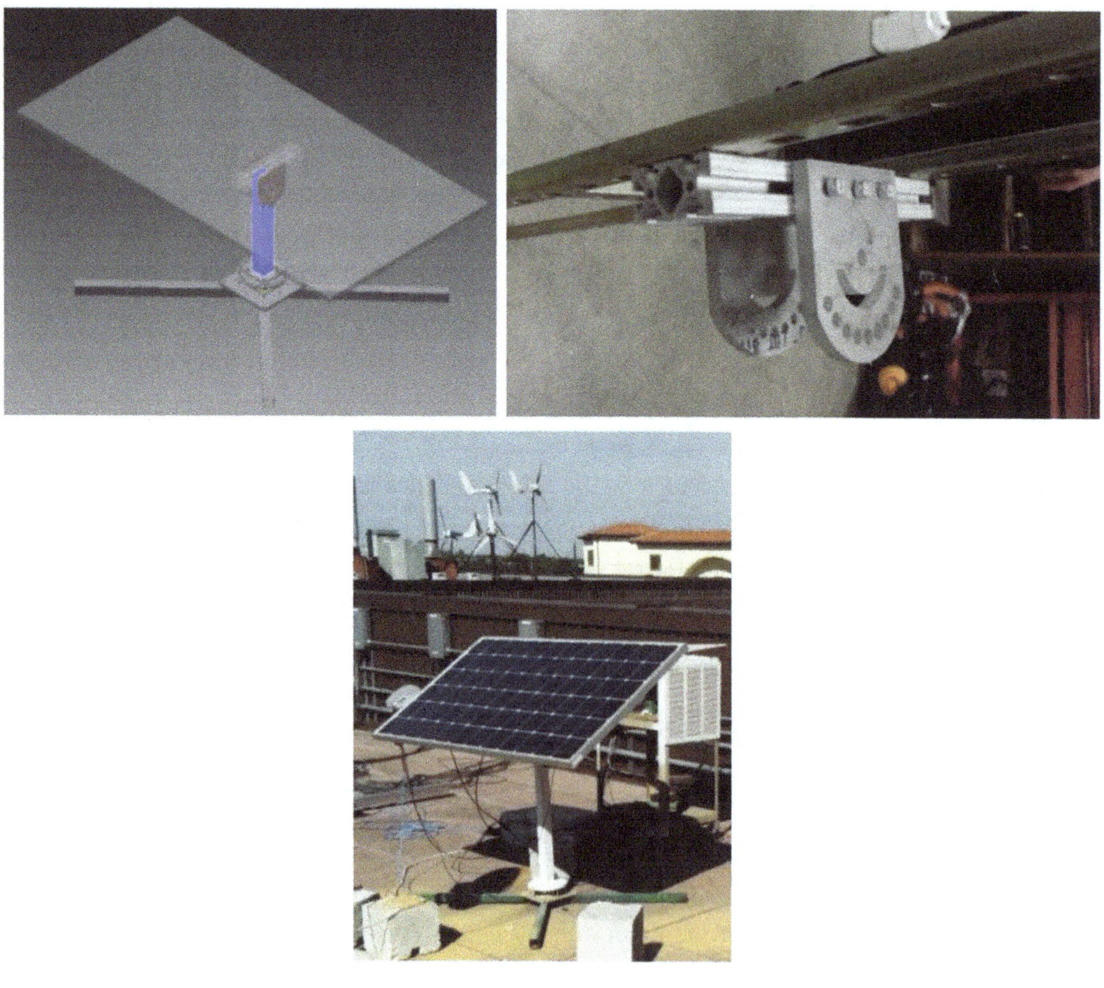

tube resistors) and a Web-based data acquisition system (Figure 6). The data acquisition system used in this research consists of an eGauge, DC current transducer, power injector, RS485 to Ethernet converter, Sunny sensor box, ambient temperature sensor, photovoltaic panel temperature sensor, router, Ethernet cable, and wires.

The experimental setup was comprised of two 190 Watt mono-crystalline photovoltaic panels that contain 72 cells each with the following dimensions (125 * 125 mm) and a weight of 15 kg (Solar Systems USA Online Solar Panels, 2016), rheostats, a manual dual-axis mechanical system, data acquisition system, and proper wiring. The specifications of the used photovoltaic 190 W_{peak} PV panel panels are given in Table 2. The power generated by these photovoltaic panels should be 90% for ten years and 85% for twenty-five years (Solar Systems USA Online Solar Panels 2016). The lifespan of the panels was assumed to be 25 years with a degradation rate of 0.5% based on the relevant literature (Jordan and Kurtz, 2013). A fixed-flat panel was mounted to the roof. The second panel was mounted to the manual dual-axis sun tracking system. Both panels were connected to the rheostats, as the loads to dissipate the generated electricity, and the data acquisition system. The experiment was conducted on February 16 starting at 10:50 a.m. and concluded at 4:00 p.m. The dual-axis tracker was controlled manually to attain the perpendicular condition. The perpendicular condition was sensed through the use of a shadow sensor, which consisted of a nail upright on cardboard. The shadow sensor was leveled to the panel. The manual sun tracker was adjusted every 15 min to retain perpendicular reception of the irradiance.

The tracking system is consisted of the following main components

Figure 6. Wiring diagram of the experimental setup

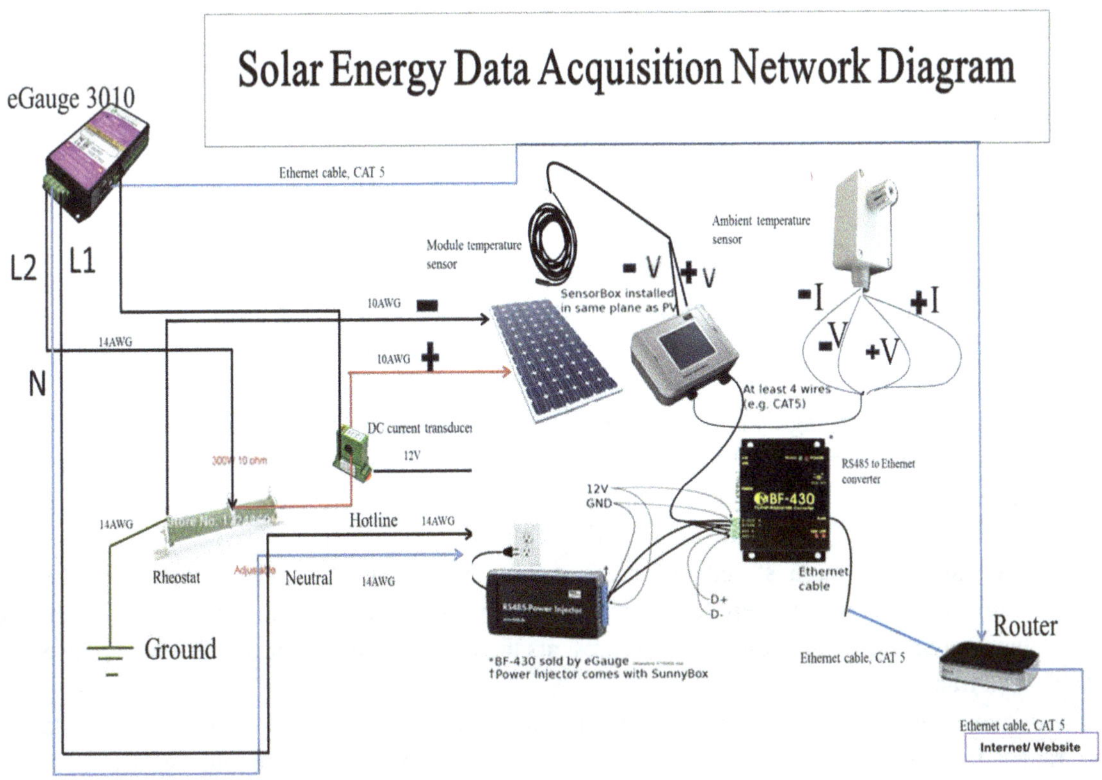

Table 2. Specifications of the used photovoltaic 190 W_{peak} PV panel panels

Specification	Value
Panel dimensions	1580 x 808 x 35mm
Maximum Power (P_{mp})	190 W
Voltage at maximum power (V_{mp})	37.08 V
Current at maximum power (I_{mp})	0.833 A
Short circuit current (I_{sc})	5.55 A
Open circuit voltage (V_{oc})	44.90 V
Type	Mono-crystalline silicon
Operating temperature	From -40°C to +85°C
Cell conversion efficiency (η_{ref})	17.04%
Power temperature coefficient (β_{ref})	-0.4459%/°C
Temperature coefficient of open circuit voltage	-0.3882%/°C
Temperature coefficient of short circuit current	-0.0625%/°C
Nominal operating cell temperature (NOCT)	47 ± 3°C
Standard testing conditions	1000 W/m² at 25 °C

- 6” actuator,
- 12” actuator,
- a dual-axis solar tracker controller, and
- framing brackets.

The nominal output of the system is 8–25 V (DC). The max load current is of 6 amps and the no-load current is of less than 1 amp. The solar tracker is equipped with a weather detector, which will alter the solar panels movements to yield the most efficient light energy capture, as the system will stop moving on a cloudy day and lie flat at night or on a rainy day. The system has a maximum load capacity of 1500 N (Eco-worthy Solar Tracking System N.D.). Based on this load capacity, up to 10 solar panels can be incorporated into the system, as each solar panel weighs 15 kgs (Solar Systems USA Online Solar Panels 2016). The life span of the sun tracking system was assumed to be more than 25 years, with a constant power consumption rate. The specifications of the actuators are given in Table 3.

The data acquisition system stored performance parameters for both panels used in the experiment. The unnatural spikes in the data were removed. The power output graphs of both panels were generated. The calculated area between the two curves represents the power saving achieved through the use of the dual-axis sun tracker.

It was found that the gross power improvement by using dual-axis sun tracking versus fixed-flat solar panel ranges from 31% in the midday to 294% in the early and late times of the intervals, see Figure 7.

Table 3. Actuator specifications as reported by Eco-worthy Solar Tracking System

Specification	North/South Actuator	East/West Actuator
Voltage	12 V	12 V
Load capacity	1500 N	1500 N
Speed	5.7 mm/s	5.7 mm/s
Stroke length	6 inches	12 inches
Minimum installation distance	10.24 inches	16.54 inches
Duty cycle	25%	25%
Waterproof	IP65	IP65

Figure 7. Fixed versus sun tracking systems power generation

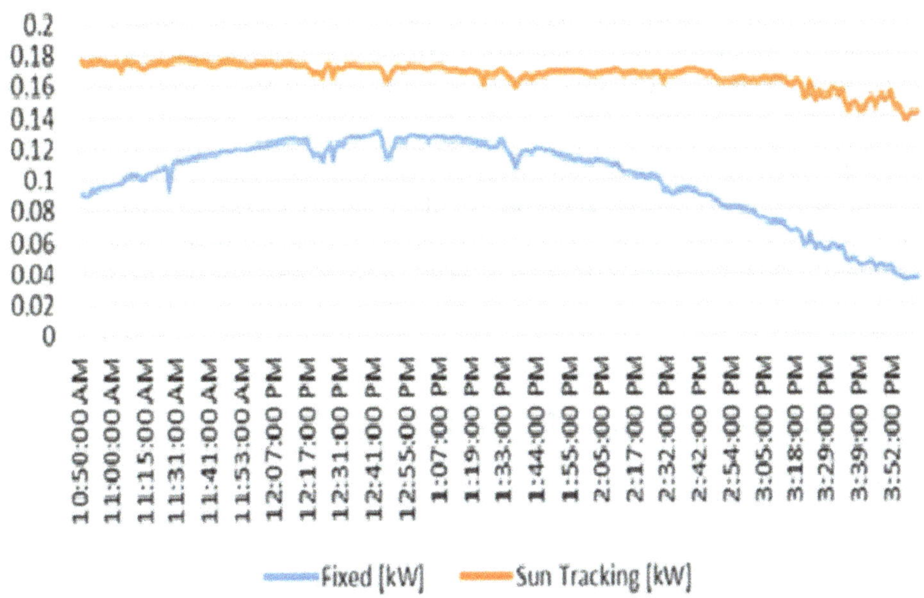

The overall average improvement for the entire interval was 82%. The influence of the panel temperature on its output was not considered.

To accurately calculate the power generation difference between the two systems, areas between the generated powers, by the two systems between two reading times by the automated system, were calculated as area for a trapezoid (Figure 8):

$$A_i = \frac{t_i\left(W_{1,i} + W_{2,i}\right)}{2} \tag{10}$$

where t_i is the time interval between two reading times. Thus, the average power saving is:

$$Average\,power\,saving = \frac{\sum_i A_i}{\sum_i t_i} \tag{11}$$

The average saving for one hour, x, based on a five-hour-interval data collection was found to be 0.0686 kWh (See Figure 8).

Taking into consideration that, the latitude of Texas state university is 29.8884° N, the maximal energy gain with relation to the horizontal plane was calculated using the equations (3) and (4) when the fixed system is horizontal or tilted by an optimum daily tilt angle or tilted by a latitude angle. As information is not provided with regard to solar radiation components or ambient temperature during the day of experiment (February 16), we will restrict comparison with maximal possible energy gain on that day only.

On February 16, the maximal daily energy gain of a dual tracking with relation to the fixed latitude tilted surface was found to be 69% and with relation to the surface with optimum daily tilt angle it was found to be 60%. These values are smaller than 82% cited in (Asiabanpour, Almusaied, Aslan, Mitchell, Leake, Lee, Fuentes, Rainosek, Hawkes and Bland, 2016). The final possible case is to suppose that the roof of the building is horizontal or approximately horizontal. When calculating the maximal daily energy gain of a dual tracking with relation to the horizontal plane it was found to be 116%. This value is higher than 82% cited in (Asiabanpour, Almusaied, Aslan, Mitchell, Leake, Lee, Fuentes, Rainosek, Hawkes and Bland, 2016). As the experiment started at 10:50 a.m. and concluded at 4:00 p.m. and the day length on February 16 is much longer than 5 hours, the difference between the experimental and theoretical results is due to omitted hours at the beginning of that day where the energy gain is much greater than its value at midday. Thus, it can be concluded that the agreement between the experiment and theory is good.

Dual Tracking and Latitude Tilted Fixed Case

An experimental setup was designed and built in Cairo, Egypt to evaluate the performance of dual solar tracking. A schematic diagram of the setup is shown in the Figure 9, and a photo of the setup is shown in

Figure 8. Power difference between two time readings

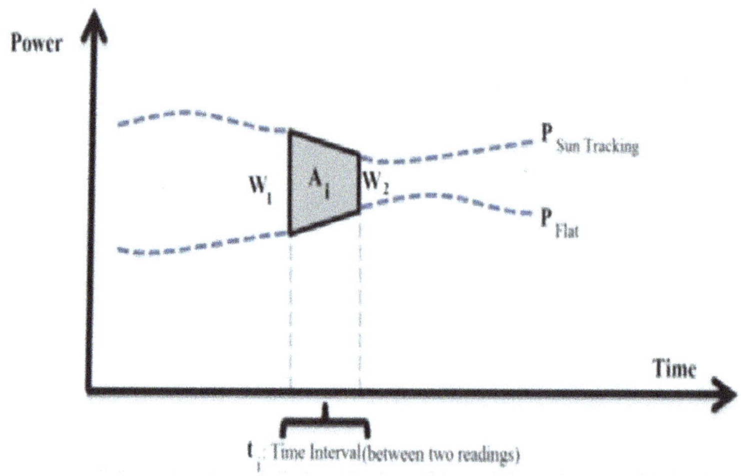

Figure 9. Schematic diagram of the experimental setup

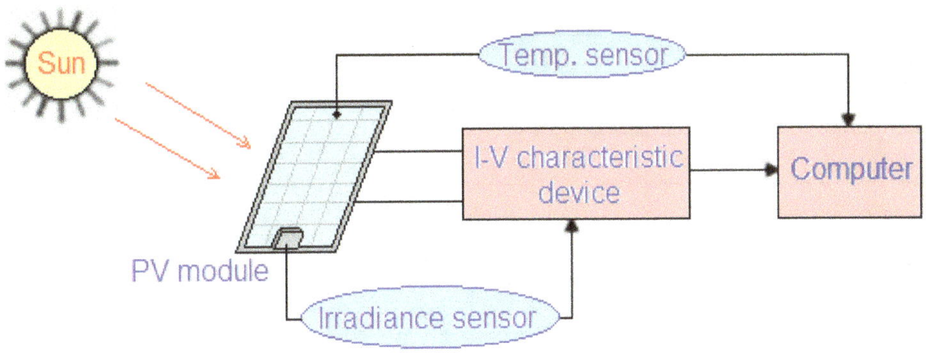

the Figure 10. The setup consists of; (1) a turn table to hold the PV modules and to allow the module to track or miss track the sun, (2) PV modules with the specification given in Table 4, and (3) the measuring system. The solar tracking system is based on azimuthal tracking, where there are two independent motions, one for the daily motion and is made by rotating the module around the polar axis, and the other motion is determined by the seasonal variation of the sun. Further details about the azimuthal tracking mechanism and other tracking mechanisms could be found in (Alexandru, 2013). The azimuthal tracking has been selected from a constructive point of view, i.e. easiness in construction. The PV panel position was manually adjusted every 15 min during the tracking case from sun rise till sunset, such that it can be assumed that the panel was always perpendicular to the solar irradiance. It can be concluded that the actuation errors, instrumentation errors as well as the transmission errors are minimized due to this manual tracking process. The measuring system consists of (a) an I-V characteristic device (PVPM 2540C, n.d) used to access the I-V characteristics of the PV module (Vatau, Musuroi, Barbulescu and Babescu, 2014), (b) a solar radiation sensor (SOZ-03, n.d) to measure the corresponding solar irradiance falling

Figure 10. The experimental setup consists of (1) the turn table, (2) the PV module, (3.a) the solar irradiance sensor and (3.b) the I-V characteristic device

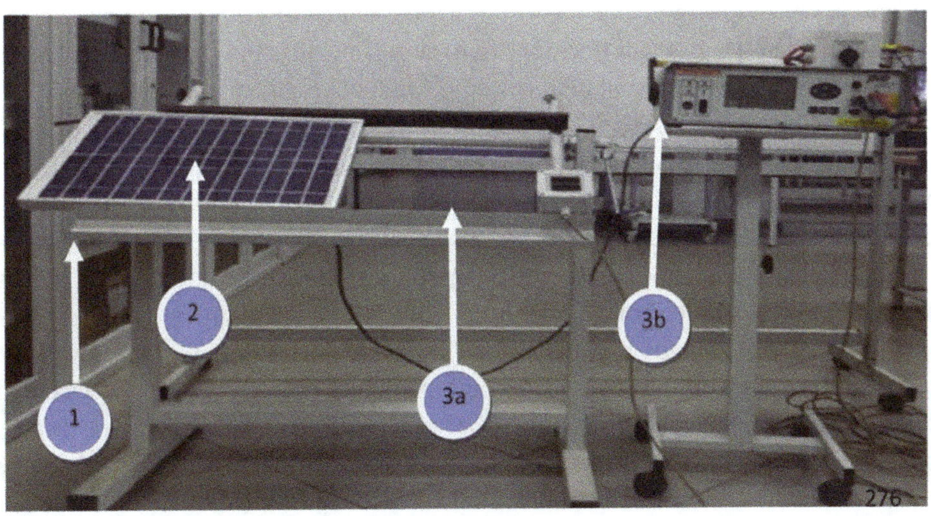

Table 4. Specifications of the tested photovoltaic 10 W_{peak} PV panel panels

Specification	Value
Panel dimensions	265 mm x 350 mm x 17 mm
Maximum Power (P_{mp})	10 W
Voltage at maximum power (V_{mp})	12 V
Current at maximum power (I_{mp})	0.833 A
Short circuit current (I_{sc})	0.874 A
Open circuit voltage (V_{oc})	14 V
Type	Mono-crystalline silicon
Operating temperature	From -40°C to +85°C
Module conversion efficiency (η_{ref})	12%
Power temperature coefficient (β_{ref})	-0.5%/°C
Nominal operating cell temperature (NOCT)	45°C
Standard testing conditions	1000 W/m² at 25 °C

on the PV modules, and (c) a temperature sensor to measure the module temperature. In the performed experiments in Cairo, Egypt, the tilt angle of the fixed mounted PV panel is adjusted to be tilted by 30° (Sharaf Eldin, Abd-Elhady and Kandil, 2016).

A set of experiments were performed in Cairo, Egypt, to compare the output power of the PV module in cases of tracking, no tracking, and full miss tracking of the sun. Three identical 10Wpeak PV panels (see Table 4) were placed in the sun at the same time, but at the three different positions mentioned above, from 8:00 am till 4:00 pm. The power output, solar irradiance and module temperature for each of the above mentioned cases were recorded every 15 min. The power output from the 10 W_{peak} PV panels installed in Cairo has been measured and the results are compared to the power output calculated based on the mathematical model, in cases of tracking, no tracking and miss tracking the sun. The solar irradiance has been measured for all cases while the average ambient temperature is taken as 45°C based on the Egyptian weather forecast on the day of the experiments.

It was found that, the solar irradiance in case of tracking the sun is higher than that in case of non-tracking the sun during the morning period, i.e. from 8.00 a.m. to 12 p.m. This is because the angle of incidence in case of tracking the sun is 0°, while the angle of incidence in case of non-tracking the sun varies with the relative motion of the sun to the earth. The solar irradiance in cases of tracking and non-tracking becomes almost the same at noon time. The difference between the latitude angle φ, i.e. 30° 6' in case of Cairo city, and the tilt angle of the PV panel, i.e. 30°, is very small, such that as we approach the noon time, ω = 0, the sun rays become perpendicular to the earth surface and the solar incidence angle in case of tracking and no tracking becomes almost the same, as can be determined from the equation (8). Therefore, the solar irradiances in cases of tracking and no tracking become the same at noon time. The solar irradiance in case of tracking becomes once more higher than the non-tracking case after the noon time, because of the increase in the angle of incidence of the sun rays in case of no tracking compared to that of the tracking case.

It was found, that the modeled and measured power outputs in case of tracking the sun are in good agreement where they are very close with an average difference of 5.1% and the time dependence of the both values are having the same trend. The power outputs of the no tracking case as modeled and measured are very close to one another, with an average difference of 2.8%. In case of miss tracking, the average difference between the model results and the experiments is a bit high, i.e. 12%. The large percentage difference is due to the low output power values, which are in the range of 1.5W-2.4W, such that a small difference results in a large percentage error. It can be concluded that, the applied mathematical model is capable of predicting the power output from PV panels as a function of the operating conditions. The operating conditions include tracking or not tracking the sun, as well as the ambient conditions, i.e. solar irradiance and ambient temperature. Therefore, the applied mathematical model could be used to assess the power outputs from PV panels in cases of low and high solar irradiance regions, as a method to determine the feasibility of using tracking systems instead of fixed systems in different ambient conditions.

Moreover, it was found that, the output power of the tracking case starts ahead of that of the no tracking panel, however, due to overheating because of the excess solar irradiance the power output of the panel in case of no tracking starts getting very close to that of the tracking case from 11:00 am until sunset. It can be concluded that tracking the sun in Cairo, i.e. a Sunbelt country, could be a waste of energy, and this can be observed through the modeled and the measured power outputs. The power outputs from the miss tracking case as modeled and measured are far from both tracking and no tracking cases. It is noticed that, miss tracking the sun has the least output power due to its dependence only on the diffused solar irradiance.

The generated electrical energy from the PV panel during the time of the performed experiment in Cairo is calculated by computing the area under the Power-Time curve. The energy gain from a tracking panel in Cairo compared to a no tracking one did not exceed 10.61%.

Rustemli, Dincer, Unal, Karaaslan and Sabah (2013) found that the energy consumed by the tracker varies between 2 and 3% of the collected energy of the tracking system. However, Mousazadeh, Keyhani, Javadi, Mobli, Abrinia and Sharifi (2009) noted that it is not recommended to use tracking system for small solar panels because of the high energy losses in the driving systems compared to the energy output from the panels. The energy consumed by the tracker could be estimated to be 10% of the generated electrical energy as proposed by Ahmad, Shafie and Kadir (2013) based on (i) the initial, (ii) the running and (iii) the estimated maintenance costs. Therefore, if the energy consumed by the tracker is taken into consideration, then the net gain in electrical energy from tracking the sun in Cairo will not exceed 0.61%. It can be concluded from the performed experiments that sun tracking in Cairo is not an attractive solution to improve the performance of PV panels.

The input data for the mathematical model in case of Aswan city located in south of Egypt, including solar irradiance and ambient temperature, were measured on the 15th of July 2013. The solar irradiance and the ambient temperature were obtained from the Egyptian Meteorological Authority (2014). The model results of the output power from the PV panel installed in Aswan city show that the power outputs in cases of tracking and no tracking the sun are very close to each other. The tracking case shows a higher output over the no tracking case from the morning until noontime. From the noon until sunset, the results are almost identical. The average energy gained from the tracking PV panel compared to that from the no tracking PV panel is higher by about 8.16%. This low energy gain from tracking is due to the overheating of the panel because of the increased solar irradiance and high ambient temperature leading to a reduction in the efficiency of the tracking PV panel. Moreover, according to Ahmad, Shafie and Kadir (2013) a tracker would consume 5.89% of the total generated power during a sunny day. Therefore, by deducting the tracking energy costs the gain from tracking over no tracking in Aswan would reach 2.27%. If the maintenance costs are included, the losses from the tracker will even increase more, i.e. become close to the 10%. This comparison clearly shows that tracking is not an economic option for PV panels in Aswan, Egypt. The miss tracking PV panel records the least out powers and this is due to low solar irradiance exposure.

Dual Tracking and Tilted Fixed Case

Two grid-connected PV systems with different mounting schemes were installed in central Iowa, USA; one roof-mounted stationary system and one pole-mounted dual-axis tracking system. Both systems were designed as "turn-key" installations for building energy generation applications. The orientation of the stationary system was selected to optimize for annual energy generation. All PV and Balance-Of-System (BOS) equipment used in the installations are "off-the-shelf" components and considered standard for residential and commercial applications (Warren, 2008). For example, flat-plate PV modules made of silicon nitride polycrystalline silicon (pc-Si) cells were used. The PV systems are equipped with extensive data acquisition (DAQ) systems capable of collecting accurate performance and meteorological data, archiving data in a central repository, and visually displaying real-time and historical data through an interactive online interface. Additionally, webcams are installed at each site to show real-time and historical streaming video and photographs that can be compared to the data presented in the interface.

The two systems have a few differences in equipment and installation. The systems have different inverters, number of modules per string, DC wiring lengths, and surrounding ground cover. Different inverters were necessary due to differing building AC line voltages; the stationary and dual-axis tracking systems are coupled to electrical systems having nominal 208 VAC and 120 VAC, respectively. Additionally, due to size constraints, the tracking system was limited to six modules wired in series where the stationary system consists of three strings of nine modules in series, each wired to a different but identical inverter. The DC wiring lengths for the stationary and dual-axis tracking system are 64 m (average of three subsystems) and 19 m, respectively.

The stationary system is located in Ames, Iowa, at a latitude of 42° 2' N and a longitude of 93° 48' W. The system has a total installed capacity of 4.59 kW_{peak} and total PV array area of 34.02 m^2. This system is designed as three side-by-side, identical, and independently operating sub-systems. Each subsystem consists of nine, 170 Watt modules operated at their peak power point and wired in series to a single inverter; all three inverters are identical. All modules are attached directly to a south-facing standing-seam white metal roof at a slope of 36° (oriented for maximum annual energy generation). The tracking system is located in Nevada, Iowa, 42° 1' N latitude and 93° 27' W longitude and was approximately 14 kilometers from the stationary system. The system has an installed capacity of 1.02 kW_{peak} and total PV array area of 7.56 m^2. This system uses six, 170 Watt modules wired in series to a single inverter. The pole-mounted array follows the sun throughout the day using an actively controlled dual-axis, azimuth drive solar tracker. The azimuth range of the tracker is 270°. The tracker is controlled by using an electronic transducer that is mounted on top of the array. The transducer has sensors on all sides and adjusts position until an equal amount of sun is sensed on all sides. Both systems use the same

PV modules made of silicon nitride polycrystalline silicon (pc-Si) cells (Warren, 2008). Electrical and mechanical specifications for these modules are given in Table 5.

The PV systems are equipped with extensive data acquisition systems (DAQ). Each DAQ is capable of collecting accurate data at a high sampling rate, archiving data in a central repository, and visually displaying real-time and historical data of both systems simultaneously through an online interactive interface. All data is measured at ten second intervals, and stored as one-minute averages. To adequately characterize the performance of the PV systems, both operating parameters of each array and meteorological conditions were monitored.

Both systems were operated and monitored simultaneously for one year. Experimental data was collected for performance parameters and meteorological conditions from September 2007 through August 2008. Data was sampled at 10 second intervals and stored as one-minute averages. The PV systems were new at the onset of data collection. During the test period, the systems were allowed to operate as real-world systems; modules were never cleaned of snow or soiling and the systems were not purposely taken off-line or interrupted for any reason.

The amount of solar energy that can be utilized by a PV system at any given time is dependent upon the beam radiation angle of incidence relative to the array surface. The inherent advantage of a tracking system when compared to a stationary system is its ability to orient the array perpendicular to the solar beam radiation to maximize the collection of available solar energy at all times. The annual average daily solar radiation seen by the stationary and dual axis tracking systems was found to be 4.37 and 5.95

Table 5. Specifications of the tested photovoltaic 10 W_{peak} PV panel panels

Specification	Value
Panel dimensions	1593 x 790 x 50 mm
Maximum Power (P_{mp})	170 W
Voltage at maximum power (V_{mp})	35.4 V
Current at maximum power (I_{mp})	4.8 A
Short circuit current (I_{sc})	5.0 A
Open circuit voltage (V_{oc})	44.2 V
Type	poly-crystalline silicon
Operating temperature	From -40°C to +85°C
Module conversion efficiency (η_{ref})	12%
Power temperature coefficient (β_{ref})	-(0.5 + 0.05) %/°C
Nominal operating cell temperature (NOCT)	47±2°C
Standard testing conditions	800 W/m^2 at 20 °C and wind speed 1m/s
Temperature coefficient of short circuit current	(0.065 + 0.015)%/°C
Temperature coefficient of open circuit voltage	-(160 + 20) mV/°C

kWh/m^2, respectively. On an annual basis the tracking system harvested 36% more solar energy than the stationary system (Warren, 2008).

EXPERIMENTAL VERIFICATION IN THE HIGH-LATITUDE ZONE

When applying the proposed algorithm for determining the PV panel performance in Stuttgart (φ =48.7758°N), Germany, the solar irradiance and the ambient temperature have been obtained from the German Meteorological Service (2014). The model results of the output powers from the PV panels installed in Stuttgart, Germany, show that, the tracking PV panel has higher power output throughout the whole day except for the noon period. Both tracking and no tracking PV panels have almost same output power, because at this time both of them have the same inclination angle with the sun. In Stuttgart, the effect of tracking is more emphasized, as the tracking PV panel has an energy gain of 16.8% over the no tracking PV panel. In this case, tracking the sun might be more beneficial as after deducting the tracker power consumption and the maintenance costs, which is approximately equal to 10% of the generated power, then the net gain from tracking will be 6.8%. As expected, the miss tracking PV panel records the least output power due to low exposure to solar irradiance.

When applying the proposed for determining the PV panel performance in Berlin (φ =52.52°N), Germany, the solar irradiance and the ambient temperature have been obtained from the German Meteorological Service (2014). This city is of relatively low ambient temperature and solar irradiance. The results of the model on the 15th of July 2013 show that, the tracking panel has the highest power output due to highest solar irradiance exposure and the absence of the overheating effect. The no tracking PV panel has a lower output power compared to the tracking panel due to low solar irradiance. This leads to a clear result that in Berlin the tracking PV panel makes an energy gain of 40% over the no tracking PV panel. The net gain of energy after deducting the tracker consumption, i.e. ~10%, will be 30%.

MAXIMAL POSSIBLE ENERGY GAIN OF LONG TERM TRACKING

The concept of energy gain is very useful in evaluating optimum tilt application over any period of consecutive days. In this context, it is reasonable to introduce the following functions:

- Energy gain factor of daily optimum tilted surface with relation to horizontal surface:

$$R = \left[\frac{H_t \left(\beta = \beta_{opt,d} \right)}{H} \right] \tag{12}$$

- Energy gain factor of monthly optimum tilted surface with relation to horizontal surface:

$$R_1 = \left[\frac{H_t \left(\beta = \beta_{opt,m} \right)}{H} \right] \tag{13}$$

- Energy gain factor of seasonally optimum tilted surface with relation to horizontal surface:

$$R_2 = \left[\frac{H_t\left(\beta = \beta_{opt,s}\right)}{H} \right] \tag{14}$$

- Energy gain factor of biannually optimum tilted surface with relation to horizontal surface (It is advised here to consider that the solar year starts on 22/3 and ends on 21/3. So, the first half year covers the period from 22/3 to 21/9 while the second half year covers the period from 22/9 to 21/3):

$$R_3 = \left[\frac{H_t\left(\beta = \beta_{opt,b}\right)}{H} \right] \tag{15}$$

- Energy gain factor of yearly optimum tilted surface with relation to horizontal surface:

$$R_4 = \left[\frac{H_t\left(\beta = \beta_{opt,y}\right)}{H} \right] \tag{16}$$

- Energy gain factor of latitude tilted surface with relation to horizontal surface:

$$R_5 = \left[\frac{H_t\left(\beta = \varphi\right)}{H} \right] \tag{17}$$

Based on the functions in the equations (12) to (17), one can get reliable information regarding the solar receiver installation from practical and economic points of view. So, these functions should be considered for any studied period: daily, weekly, fortnightly, monthly, seasonally, half-yearly and yearly.

As example, Figure 11 shows the daily energy gain (in times of solar radiation on horizontal plane) of *EF* solar collector for φ= 43.45 °N obtained for daily, monthly, seasonally and biannually solar collector tilt angle adjustments. It is seen from Figure 11 that: a) it is difficult to distinguish between the results obtained for daily and monthly adjustments which seems suggesting that the energy losses resulting from tilt angle adjustment on monthly basis is negligible; and b) even daily and monthly adjustments are slightly better than seasonally and half-yearly ones, all these four adjustments lead to comparable energy gain all over the year as the energy losses are not of practical importance. The daily yearly averages of R, R_1, R_2, R_3 are 1.7 and this means that it is reasonable, from practical and economical point of view, to adjust solar collector tilt twice a year.

When calculating the daily energy gain in the case of fixed installation over all the year with yearly optimum tilt angle and latitude tilted angle it was found that, even $\beta_{opt,y}$ is smaller than φ by few degrees (for φ= 43.45°N, $\beta_{opt,y}$ = 40.38°), the practical difference between the energy gain results is negligible

Figure 11. Calculated daily energy gain, on the base of daily (♦), monthly (■), seasonally (▲) and biannually (x) adjustments, for 43.45°N

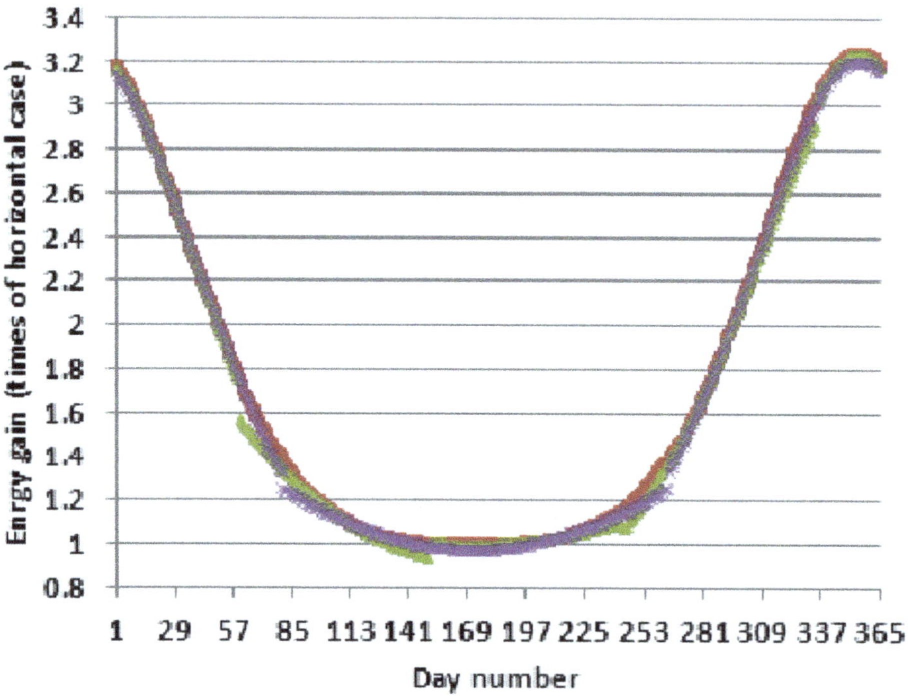

(see Figure 12). Moreover, the daily yearly averages of R_4 and $R_5 = 1.6$. This means that the second rule of thumb is valid for mid-latitude zone.

Here it should be noted that even optimum orienting solar collector towards Equator could provide an increase in the produced power in relation to other orientations, a lower power production ratio compared with that produced by the tracking systems is envisaged especially during morning and evening. The effect of tilt angle and system orientation is evaluated theoretically and experimentally (Lazaroiu, Longo, Roscia and Pagano, 2015) on the basis of two photovoltaic systems: one fixed and one equipped with a sun tracker. The study verified that the PV module equipped with a single axis sun tracker evidenced an important growth of power production during morning and evening.

Denoting that, $H\left(\beta = 0\right)$ is the average daily total solar radiation on the horizontal plane, $H\left(\beta = \beta_{opt,d}\right)$ is the average daily total solar radiation on a plane, tilted with an optimum daily tilt angle $\beta_{opt,d}$, $H\left(\beta = \varphi\right)$ is the average daily total solar radiation on a plane, tilted with latitude tilt angle, $H\left(\beta = \beta_{opt,m}\right)$ is the average daily total solar radiation on a plane, tilted with an optimum monthly tilt angle $\beta_{opt,m}$, $H\left(\beta = \beta_{opt,s}\right)$ is the average daily total solar radiation on a plane, tilted with an optimum seasonally tilt angle $\beta_{opt,s}$ and $H\left(\beta = \beta_{opt,y}\right)$ is the average daily total solar radiation on a plane, tilted with an optimum seasonally tilt angle $\beta_{opt,y}$. Then, when integrating the above mentioned losses all over the year for the following cases: $H\left(\beta = \beta_{opt,d}\right) / H\left(\beta = 0\right)$, $H\left(\beta = \beta_{opt,d}\right) / H\left(\beta = \varphi\right)$, $H\left(\beta = \beta_{opt,d}\right) / H\left(\beta = \beta_{opt,y}\right)$,

Figure 12. The calculated daily energy gain for $\beta = \beta_{opt,y}$ (x) and $\beta = \varphi$ (■) for 43.45°N

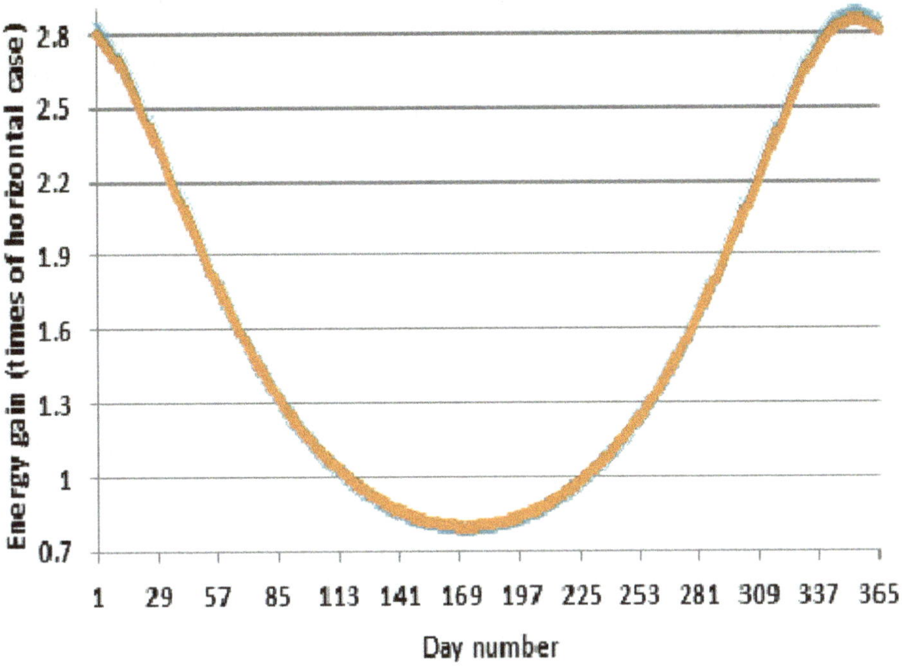

$H\left(\beta = \beta_{opt,d}\right) / H\left(\beta = \beta_{opt,s}\right)$ and $H\left(\beta = \beta_{opt,d}\right) / H\left(\beta = \beta_{opt,m}\right)$ it was found that the percentage of these losses is 0.4%, 0.5%, 0.5%, 0.4% and 0.1% respectively. This proves the validity of the proposed approximation. The results of calculations of $H\left(\beta = \beta_{opt,d}\right) / H\left(\beta = 0\right)$, $H\left(\beta = \beta_{opt,d}\right) / H\left(\beta = \varphi\right)$, $H\left(\beta = \beta_{opt,d}\right) / H\left(\beta = \beta_{opt,y}\right)$, $H\left(\beta = \beta_{opt,d}\right) / H\left(\beta = \beta_{opt,s}\right)$ and $H\left(\beta = \beta_{opt,d}\right) / H\left(\beta = \beta_{opt,m}\right)$ in tropical, mid-latitude and high-latitude zones are given in the appendixes A to C.

OPTIMUM ORIENTATION FOR VERTICAL SURFACES

For an EF vertical surface, solar radiation is decreased considerably in summer months and goes up to its maximum in winter months. For a west-facing vertical collector, monthly solar radiation in summer months is higher than an EF collector. Therefore, using a west-facing vertical collector is not an adequate choice for heating systems, however it can be used for cooling systems application. In order to determine the best orientation for vertical surfaces it is very important to investigate R_6 the energy gain factor of γ-orientated vertical surface with relation to the EF vertical surface:

$$R_6 = \left[\frac{H_t\left(\beta = 90,\ \gamma\right)}{H_t\left(\beta = 90,\ \gamma = 0\right)}\right] \tag{18}$$

The daily and monthly dependence of R_6 on γ for surfaces of different tilt angles are studied in chapter 5 for φ=43.45°N. The statement mentioned above regarding west-facing and EF vertical collector is clearly demonstrated on Figures 13 and 14. However, changing γ by ±20° has no important effect of on R_6.

The yearly energy gain R_6 for φ=43.45°N and for different γ is reported in Table 6 where it is clearly shown that vertical surface with γ €[50°, 70°] is more suitable for cooling systems application compared with an EF vertical collector.

CONCLUSION

The performance of the solar systems depends on many factors. One factor involves the light reception angles at the solar receiver in which the intensity of the received solar radiation from the Sun at the Earth is affected significantly by the diurnal and seasonal movement of the Earth. The maximum output of the solar receiver is achieved when the solar receiver is perpendicular to the Sun's rays. Different attempts were made for making the solar receiver utilizing the maximum portion of incident solar radiation without

Table 6. Yearly R_6 of vertical surface of different G with respect to EF surface for φ=43.45°N

γ(°)	10	20	30	40	50	60	70	80	90
R_6	1.01	1.04	1.09	1.13	1.16	1.17	1.16	1.12	1.07

Figure 13. Daily energy gain for β= 90° and different γ for 43.45°N. Curves (from upper to down on 169 day number) correspond γ= 10°, 20°, 30°, 40°, 50°, 60° and 90° respectively

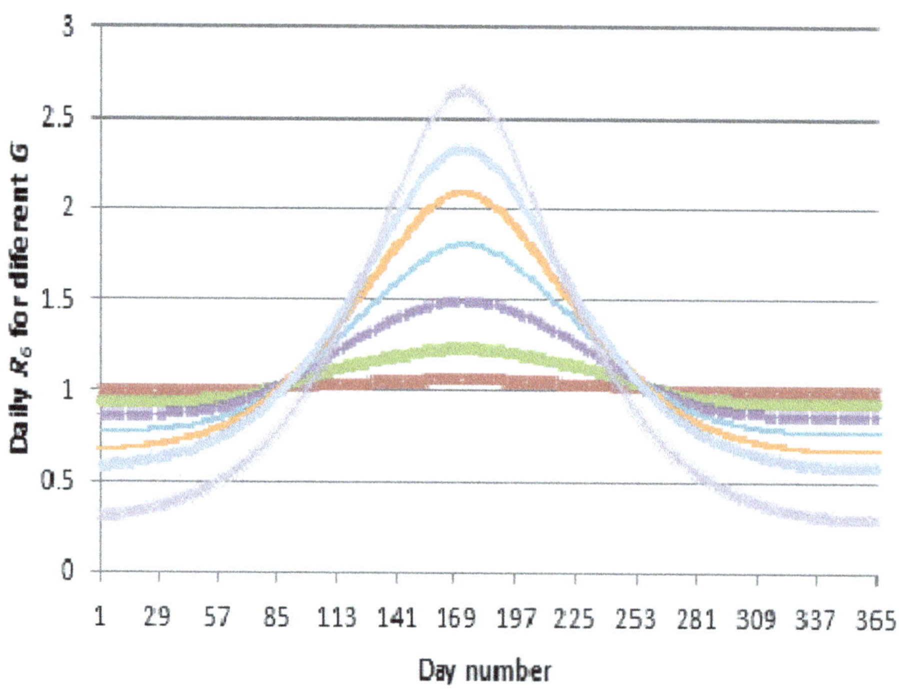

Figure 14. The calculated monthly energy gain for β= 90° and different γ for 43.45°N. Symbols ♦, ■, ▲ and x stand for γ= 20°, 50°, 70° and 90° respectively

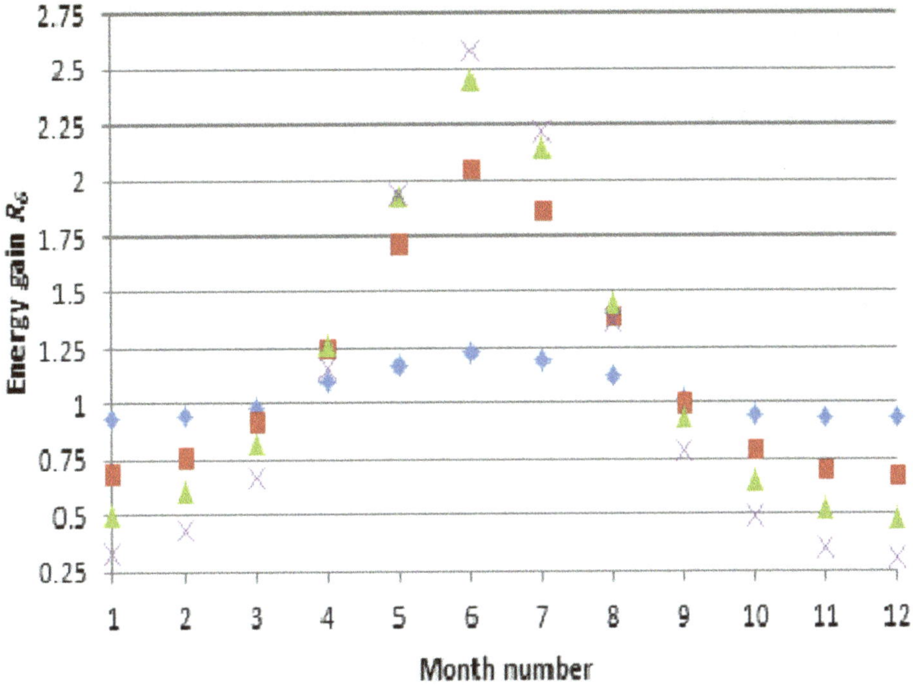

paying much money for production the unit of useful energy. From technical point of view, the benefit of using a dual-axis sun tracker versus a fixed-flat position is evident, but from economic point of view it is questionable. The provided data, in literature, compare the performance of dual or single axis tracking with fixed solar receiver even the long term solar tracking is possible and effective with a negligible increase of the price of the unit of useful energy. This can be achieved by choosing the best monthly or even seasonally optimum tilts. The introduced concept of energy gain, see chapter 3, calculated in this chapter all over the world and it was found that it is very useful in evaluating the performance of different types of tracking. This concept allows to evaluate the effectiveness of daily, weekly, fortnightly, monthly, seasonally, biannually and yearly adjustment of the solar receiver tilt angle in relation with the ideal instantaneous dual tracking, where the sun rays are kept permanently perpendicular to the flat-plate surface of the receiver. Thus, the incidence angle is kept to be zero all over the day. This evaluation determines the optimum tilt application over any period of consecutive days from technical point of view.

A mathematical conception has been developed and applied to determine the energy gain resulted from different installations. The performance of crystalline silicon PV panels as a function of tracking the Sun, and the operating conditions can be determined theoretically. The results of the applied model are in good agreement with the experimental measurements. The experimental measurements and the model results show that, it is not economical to track the sun in hot and sunny regions, such as Cairo and Aswan in Egypt, because of the overheating effect on the PV panels' performance. The power output from PV panels in case of solar tracking is sensitive to the ambient conditions, so it is very important to access the ambient conditions before deciding to use a solar tracking system.

It can be concluded that, the applied mathematical model is capable of predicting the power output from PV panels as a function of the operating conditions. The operating conditions include tracking or not tracking the sun, as well as the ambient conditions, i.e. solar irradiance and ambient temperature. Therefore, the applied mathematical model could be used to assess the power outputs from PV panels in cases of low and high solar irradiance regions, as a method to determine the feasibility of using tracking systems instead of fixed systems in different ambient conditions. It is clear that, overheating of the PV panel due to excessive exposure to solar irradiance and hot climate, as in Sunbelt countries, can affect the performance of the panel. Therefore, a tracking system could be unnecessary for Sunbelt countries.

REFERENCES

Abdallah, S. (2004). The effect of using Sun tracking systems on the voltage-current characteristics and power generation of at plate photovoltaics. *Energy Conversion and Management*, *45*(11-12), 1671–1679. doi:10.1016/j.enconman.2003.10.006

Abdallah, S., & Nijmeh, S. (2004). Two axes sun tracking system with PLC control. *Energy Conversion and Management*, *45*(11-12), 1931–1939. doi:10.1016/j.enconman.2003.10.007

Aberle, A. G. (2009). Thin-film solar cells. *Thin Solid Films*, *517*(17), 4706–4710. doi:10.1016/j.tsf.2009.03.056

Afiqah, N. N., Syafawati, N. A., Idris, S. H., Haziah, A.H. & Syahril, M. D. (2015). Development of dual axis solar tracking system performance at Ulu Pauh, Perlis. *Applied Mechanics and Materials, 793*(3), 328-332. Retrieved from www.scientific.net/AMM.793.328

Agarwal, M. (1997). A systematic classification of neural-network-based control. *IEEE Control Systems Magazine*, *17*(2), 75–93. doi:10.1109/37.581297

Akbarzadeh, A., & Wadowski, T. (1996). Heat pipe-based cooling systems for photovoltaic cells under concentrated solar radiation. *Applied Thermal Engineering*, *16*(1), 81–87. doi:10.1016/1359-4311(95)00012-3

Al-Mohamad, A. (2004). Efficiency improvements of photo-voltaic panels using a sun-tracking system. *Applied Energy*, *79*(3), 345–354. doi:10.1016/j.apenergy.2003.12.004

Alexandru, C. (2013). A novel open-loop tracking strategy for Photovoltaic systems. *Sci. World J., 2013*, Article ID 205396. 10.1155/2013/205396

Alexandru, C., & Tatu, N. I. (2013). Optimal design of the solar tracker used for a photovoltaic string. *J. Renew. Sustain. Energy*, *5*(2), 023133. doi:10.1063/1.4801452

Asiabanpour, B., Almusaied, Z., Aslan, S., Mitchell, M., Leake, E., Lee, H., Fuentes, J., Rainosek, K., Hawkes, N. & Bland, A. (2016). Fixed versus sun tracking solar panels: an economic analysis. *Clean Techn. Environ. Policy*, 1-9. doi:10.1007/s10098-016-1292-y

Asl-Soleimani, E., Farhangi, S., & Zabihi, M. S. (2001). The effect of tilt angle, air pollution on performance of photovoltaic systems in Tehran. *Renewable Energy*, *24*(3-4), 459–468. doi:10.1016/S0960-1481(01)00029-5

Asowata, O., Swart, J., & Pienaar, C. (2012). Optimum tilt angles for photovoltaic panels during winter months in the Vaal Triangle, South Africa. *Smart Grid and Renewable Energy*, *3*(2), 119–125. doi:10.4236/sgre.2012.32017

Bakos, G. C. (2006). Design and construction of two-axis sun tracking system for parabolic trough collector (ptc) efficiency improvement. *Renewable Energy*, *31*(15), 2411–2421. doi:10.1016/j.renene.2005.11.008

Beasley, D., Bull, D. R., & Martin, R. R. (1993). An overview of genetic algorithms: Part 1, fundamentals. *University Computing*, *15*(2), 58–69.

Beringer, S., Schilke, H., Lohse, I., & Seckmeyer, G. (2011). Case study showing that the tilt angle of photovoltaic plants is nearly irrelevant. *Solar Energy*, *85*(3), 470–476. doi:10.1016/j.solener.2010.12.014

Bojić, M., Bigot, D., Miranville, F., Parvedy-Patou, A., & Radulović, J. (2012). Optimizing performances of photovoltaic in Reunion Island—tilt angle. *Progress in Photovoltaics: Research and Applications*, *20*(8), 923–935. doi:10.1002/pip.1159

Celik, A. N., & Muneer, T. (2013). Neural network based method for conversion of solar radiation data. *Energy Conversion and Management*, *67*, 117–124. doi:10.1016/j.enconman.2012.11.010

Chaar, E. A., Lamont, L. A., & Zein, E. N. (2011). Review of photovoltaic technologies. *Renewable & Sustainable Energy Reviews*, *15*(5), 2165–2175. doi:10.1016/j.rser.2011.01.004

Chang, T. P. (2008). Study on the optimal tilt angle of solar collector according to different radiation types. *International Journal of Applied Science and Engineering*, *6*(2), 151–161.

Chang, Y. (2009). Optimal design of discrete-value tilt angle of PV using sequential neural-network approximation and orthogonal array. *Expert Systems with Applications*, *36*(3 part 2), 6010–6018. doi:10.1016/j.eswa.2008.06.105

Chang, Y. (2010a). Optimal the tilt angles for photovoltaic modules using PSO method with nonlinear time-varying evolution. *Energy*, *35*(5), 1954–1963. doi:10.1016/j.energy.2010.01.010

Chang, Y. (2010b). An ant direction hybrid differential evolution algorithm in determining the tilt angle for photovoltaic modules. *Expert Systems with Applications*, *37*(7), 5415–5422. doi:10.1016/j.eswa.2010.01.015

Chatterjee, A., & Keyhani, A. (2012). Neural network estimation of micro-grid maximum solar power. *IEEE Transactions on Smart Grid*, *3*(4), 1860–1866. doi:10.1109/TSG.2012.2198674

Chatterjee, A., & Siarry, P. (2006). Nonlinear inertia weight variation for dynamic adaptation in particle swarm optimization. *Computers & Operations Research*, *33*(3), 859–871. doi:10.1016/j.cor.2004.08.012

Chen, Y. M., Lee, C. H., & Wu, H. C. (2005). Calculation of the optimum installation angle for fixed solar-cell panels based on the genetic algorithm and the simulated annealing method. *IEEE Transactions on Energy Conversion*, *20*(2), 467–473. doi:10.1109/TEC.2004.832093

Clifford, M., & Eastwood, D. (2004). Design of a novel passive solar tracker. *Solar Energy*, *77*(3), 269–280. doi:10.1016/j.solener.2004.06.009

Compaan, A. D. (2006). Photovoltaic: Clean power for the 21st century. *Solar Energy Materials and Solar Cells*, *90*(15), 2170–2180. doi:10.1016/j.solmat.2006.02.017

Čongradac, V., Prica, M., Paspalj, M., Bojanič, D., & Čapko, D. (2012). Algorithm for blinds control based on the optimization of blind tilt angle using a genetic algorithm and fuzzy logic. *Solar Energy*, *86*(9), 2762–2770. doi:10.1016/j.solener.2012.06.016

Cooke, D. (2011). Single versus dual axis solar tracking. *AltEnergyMag*. Retrieved from http://www.altenergymag.com/content.php?post_type=1690

Cruz-Peragón, F., Casanova-Peláez, P. J., Díaz, F. A., López-García, R., & Palomar, J. M. (2011). An approach to evaluate the energy advantage of two axes solar tracking systems in Spain. *Applied Energy*, *88*(12), 5131–5142. doi:10.1016/j.apenergy.2011.07.018

Dakkak, M., & Babeli, A. (2012). Design and performance study of a PV tracking system (100W-24Vdc/220Vac). *Energy Procedia*, *19*, 91–95. doi:10.1016/j.egypro.2012.05.188

Duffie, J. A., & Beckman, W. A. (2013). *Solar engineering of thermal processes* (3rd ed.). New York, NY: Wiley & Sons; doi:10.1002/9781118671603

Eberhart, R. C., & Shi, Y. (2001). Tracking and optimizing dynamic systems with particle swarms. *IEEE International Conference on Evolutionary Computation*, 94–100. doi:10.1109/CEC.2001.934376

Egido, M. A., & Lorenzo, E. (1992). Review of sizing and a proposed new method. *Solar Energy Materials and Solar Cells*, *26*(1-2), 51–69. doi:10.1016/0927-0248(92)90125-9

Egyptian Meteorological Authority. (2014). Retrieved from http://ema.gov.eg

Eke, R., & Sentruk, A. (2012). Performance comparison of a double-axis sun tracking versus fixed PV system. *Solar Energy*, *86*(9), 2665–2672. doi:10.1016/j.solener.2012.06.006

Elhassan, Z. A. M., Zain, F. M., Sopian, K., & Awadalla, A. (2011). Optimum energy of photovoltaic module directed at optimum slope angle in Kuala Lumpur, Malaysia. *Research Journal of Applied Sciences*, *6*(2), 104–109. doi:10.3923/rjasci.2011.104.109

Englander, D. (2009). Solar's New Important Players: Chevron, Lockheed Martin. *Seeking Alpha*. Retrieved from http://seekingalpha.com/article/138253-solar-s-new-important-players-chevronlockheed-martin?source=email

EPIA. (2012). Global market outlook for photovoltaic until 2016. European Photovoltaic Industry Association.

Fahl, P. & Ganapathisubbu, S. (2011). Tracking benefits for solar collectors installed in Bangalore. *Journal of Renewable and Sustainable Energy, 3*(2). 10.1063/1.3558863

Fuke, N., Fukui, A., Islam, A., Komiya, R., Yamanaka, R., Harima, H., & Han, L. (2009). Influence of TiO2/electrode interface on electron transport properties in back contact dye-sensitized solar cells. *Solar Energy Materials and Solar Cells*, *93*(6-7), 720–724. doi:10.1016/j.solmat.2008.09.037

Garcia, V., & Balenzategui, J. L. (2004). Estimation of photovoltaic module yearly temperature and performance based on nominal operation cell temperature. *Renewable Energy, 29*(12), 1997–2010. doi:10.1016/j.renene.2004.03.010

Giakoumelos, E., Malamatenois, C., Hadjipaschalis, I., Kourtis, G., & Poullikkas, A. (2008). Energy policy deployment for the promotion of distributed generation technologies in the mediterranean region. *Proceedings of the international conference on deregulated electricity market issues in south-eastern Europe.*

Goldberg, D. E. (1989). *Genetic algorithms in search, optimization and machine learning*. Reading, MA: Addison Wesley.

Hartley, L. E., Martinez-Lozano, J. A., Utrillas, M. P., Tena, F., & Pedro, R. (1999). The optimization of the angle of inclination of a solar collector to maximize the incident solar radiation. *Renewable Energy, 6*(3), 180–298. doi:10.1016/S0960-1481(98)00763-0

Hay, J. E., & Davies, J. A. (1980). Calculation of the solar radiation on an inclined surface. In *Pros. First Canadian Solar Radiation Data Workshop*. Ministry of Supply and Services Canada.

Haykin, S. (1994). *Neural networks. A comprehensive foundation*. New York: Macmillan Publishing Company.

Helwa, N. H., Bahgat, A. B. G., El Shafee, A. M. R., & El Shenawy, E. T. (2000). Maximum collectable solar energy by different solar tracking systems. *Energy Sources, 22*(1), 23–34. doi:10.1080/00908310050014180

Higgens, J. M. (2009). Your Solar Powered Future: It's Closer Than You Thought. *The Futurist*, (May-June), 25–29.

Honsberg, C., & Bowden, S. (n.d.). *Degradation and failure modes*. Retrieved from http://pveducation.org/pvcdrom/modules/degradation-and-failuremodes

Hontoria, L., Aguilera, J., Riesco, J., & Zufiria, P. J. (2001). Recurrent neural supervised models for generating solar radiation. *Journal of Intelligent & Robotic Systems, 31*(1), 201–221. doi:10.1023/A:1012031827871

Hontoria, L., Aguilera, J., & Zufiria, P. J. (2002). Generation of hourly irradiation synthetic series using the neural network multilayer perceptron. *Solar Energy, 72*(5), 441–446. doi:10.1016/S0038-092X(02)00010-5

Huang, B. J., & Sun, F. S. (2007). Feasibility study of one axis three positions tracking solar PV with low concentration ratio reflector. *Energy Conversion and Management, 48*(4), 1273–1280. doi:10.1016/j.enconman.2006.09.020

IRENA. (2012). *Renewable energy technologies. Cost Analysis series, Power sector issue 4/5*. Solar Photovoltaic.

Ismail, M. S., Moghavvemi, M., & Mahlia, T. M. I. (2013). Design of an optimized photovoltaic and microturbine hybrid power system for a remote small community: Case study of Palestine. *Energy Conversion and Management, 75*, 271–281. doi:10.1016/j.enconman.2013.06.019

Jordan, D. C., & Kurtz, S. R. (2013). Photovoltaic degradation rates-an analytical review. *Progress in Photovoltaics: Research and Applications*, *21*(1), 12–29. doi:10.1002/pip.1182

Kalogirou, S. A. (2000). Applications of artificial neural-networks for energy systems. *Applied Energy*, *67*(1-2), 17–35. doi:10.1016/S0306-2619(00)00005-2

Kanellos, M. (2005). Google founders invest in solar energy. *CNET News*. Retrieved from http://news.cnet.com/8301-10784_3-5749586-7.html

Karimov, K. S., Saqib, M. A., Akhter, P., Ahmed, M. M., Chattha, J. A., & Yousafzai, S. A. (2005). A simple photo-voltaic tracking system. *Solar Energy Materials and Solar Cells*, *87*(1–4), 49–59. doi:10.1016/j.solmat.2004.08.010

Kennedy, J., & Eberhart, R. (1995). Particle swarm optimization. *International Joint Conference on Neural Networks*, 1942–1948.

Klein, S. A. (1977). Calculation of monthly average insolation on tilted surfaces. *Solar Energy*, *19*(4), 325–329. doi:10.1016/0038-092X(77)90001-9

Koussa, M., Haddadi, M., Saheb, D., Malek, A., & Hadji, S. (2012). Sun tracker systems effects on flat plate photovoltaic PV systems performance for different Sky States: A case of an arid and hot climate. *Energy Procedia*, *18*, 817–838. doi:10.1016/j.egypro.2012.05.097

LaMonica, M. (2009). California utility PG&E buys big into solar power. *CNET News*. Retrieved from http://news.cnet.com/8301-11128_3-10171036-54.html

Lazaroiu, G. C., Longo, M., Roscia, M., & Pagano, M. (2015). Comparative analysis of fixed and sun tracking low power PV systems considering energy consumption. *Energy Conversion and Management*, *92*, 143–148. doi:10.1016/j.enconman.2014.12.046

Lee, J. F., Rahim, N. A. & Al-Turki, Y. A. (2013). Performance of dual-axis solar tracker versus static solar system by segmented clearness index in Malaysia. *International Journal of Photoenergy, 2013*, Article ID 820714, 13 pages. 10.1155/2013/820714

Li, D. H. W., Lam, T. N. T., & Chu, V. W. C. (2008). Relationship between the total solar radiation on tilted surfaces and the sunshine hours in Hong Kong. *Solar Energy*, *82*(12), 1220–1228. doi:10.1016/j.solener.2008.06.002

Liu, B. Y. H., & Jordan, R. C. (1963). A rational procedure for predicting the long-term average performance of flat plate solar energy collectors. *Solar Energy*, *7*(2), 53–70. doi:10.1016/0038-092X(63)90006-9

Livingonsolarpower. (2013). *Solar PV power plants: major causes of performance degradation* [Blog Post]. Retrieved from https://livingonsolarpower.wordpress.com/2013/06/10/solar-pv-power-plants-majorcauses-of-performance-degradation/

Lubitz, W. D. (2011). Effect of manual tilt adjustments on incident irradiance on fixed and tracking solar panels. *Applied Energy*, *88*(5), 1710–1719. doi:10.1016/j.apenergy.2010.11.008

Lucio, J. H., Valdés, R., & Rodríguez, L. R. (2012). Loss-of-load probability model for standalone photovoltaic systems in Europe. *Solar Energy*, *86*(9), 2515–2535. doi:10.1016/j.solener.2012.05.021

Makrides, G., Zensser, B., Norton, M., & Georghiou, E. G. (2010). Potential of photovoltaic systems in countries with high solar irradiation. *Renewable & Sustainable Energy Reviews*, *14*(2), 754–762. doi:10.1016/j.rser.2009.07.021

Markvart, T., Fragaki, A., & Ross, J. N. (2006). PV system sizing using observed time series of solar radiation. *Solar Energy*, *80*(1), 46–50. doi:10.1016/j.solener.2005.08.011

Mattei, M., Notton, G., Cristofari, C., Muselli, M., & Poggi, P. (2006). Calculation of the polycrystalline PV module temperature using a simple method of energy balance. *Renewable Energy*, *31*(4), 553–567. doi:10.1016/j.renene.2005.03.010

Mazen, M., & Abu-Khader, O. (2008). Evaluating multi-axes sun tracking systems at different mode of operation in Jordan. *Renewable & Sustainable Energy Reviews*, *12*(3), 864–873. doi:10.1016/j.rser.2006.10.005

Mehleri, E. D., Zervas, P. L., Sarimveis, H., Palyvos, J. A., & Markatos, N. C. (2010a). A new neural network model for evaluating the performance of various hourly slope irradiation models: Implementation for the region Athens. *Renewable Energy*, *35*(7), 1357–1362. doi:10.1016/j.renene.2009.11.005

Mehleri, E. D., Zervas, P. L., Sarimveis, H., Palyvos, J. A., & Markatos, N. C. (2010b). Determination of the optimal tilt angle and orientation for solar photovoltaic arrays. *Renewable Energy*, *35*(11), 2468–2475. doi:10.1016/j.renene.2010.03.006

Mehrtash, M., Guillermo, Q., Yvan, D. L., & Daniel, R. (2012). *Performance evaluation of sun tracking photovoltaic systems in Canada*. 20thAnnual International Conference on Mechanical Engineering-ISME2012, Shiraz, Iran.

Moharram, K. A., Abd-Elhady, M. S., Kandil, H. A., & El-Sherif, H. (2013). Influence of cleaning using water and surfactants on the performance of photovoltaic panels. *Energy Conversion and Management*, *68*, 266–272. doi:10.1016/j.enconman.2013.01.022

Mousazadeh, H., Keyhani, A., Javadi, A., Mobli, H., Abrinia, K., & Sharifi, A. (2009). A review of principle and sun-tracking methods for maximizing solar systems output. *Renewable & Sustainable Energy Reviews*, *13*(8), 1800–1818. doi:10.1016/j.rser.2009.01.022

Nebraska Energy Office, State of Nebraska. (2009). *Electricity Rate Comparison by State*. Retrieved from http://www.neo.ne.gov/statshtml/115.htm

Notton, G., Paoli, C., Vasileva, S., Nivet, M. L., Canaletti, J., & Cristofari, C. (2012). Estimation of hourly global solar irradiation on tilted planes from horizontal one using artificial neural networks. *Energy*, *39*(1), 166–179. doi:10.1016/j.energy.2012.01.038

Okoye, C. O., & Abbasoğlu, S. (2013). Empirical investigation of fixed and dual axis sun tracking photovoltaic system installations in Turkish republic of Northern Cyprus. *Journal of Asian Scientific Research*, *3*(5), 440–453.

Park, C. S. (2002). *Contemporary Engineering Economics*. Prentice-Hall, Inc.

Parmar, N. J., Parmar, A. N., & Gautam, V. S. (2015). Passive solar tracking system. *Int. J. Emerg. Technol. Adv. Eng.*, *5*(1), 138–145. Retrieved from www.ijetae.com

Peak power measuring device and curve tracer for photovoltaic modules (PVPM). (n.d.). Retrieved from http://www.pvengineering.de/fileadmin/user_upload/download/Datasheets/

Pour, H. S. S., Beheshti, H. K., & Rahnama, M. (2011). The gain of the energy under the optimum angles of solar panels during a Year in Isfahan, Iran. *Energy Sources*, *33*(13), 1281–1290. doi:10.1080/15567036.2010.549923

Rabl, A. (1985). *Active solar collectors and their applications*. Oxford, UK: Oxford University Press.

Rao, S. S. (2010). *Engineering Optimization Theory and Practice*. New Age International Pvt. Ltd.

Ratnaweera, A., Halgamuge, S. K., & Watson, H. C. (2004). Self-organizing hierarchical particle swarm optimizer with time-varying acceleration coefficients. *IEEE Transactions on Evolutionary Computation*, *8*(3), 240–255. doi:10.1109/TEVC.2004.826071

Raugei, M., & Frankl, P. (2009). Life cycle impact and cost of photovoltaic systems; current state of the art and future outlooks. *Energy*, *34*(3), 392–399. doi:10.1016/j.energy.2009.01.001

Rustemli, S., Dincadam, F. & Demirtas, M. (2009). Performance comparison of the sun tracking and fixed system in the application of heating and lighting. *The Arabian Journal of Science and Engineering*, *35*(2B), 171-183.

Rustemli, S., Dincer, F., Unal, E., Karaaslan, M., & Sabah, C. (2013). The analysis on sun tracking and cooling systems for photovoltaic panels. *Renewable & Sustainable Energy Reviews*, *22*, 598–603. doi:10.1016/j.rser.2013.02.014

Safaripour, M. H., & Mehrabian, M. A. (2011). Predicting the direct, diffuse, and global solar radiation on a horizontal surface and comparing with real data. *Heat and Mass Transfer*, *47*(12), 1537–1551. doi:10.1007/s00231-011-0814-8

Sanden, B. A. (2005). The economic and institutional rationale of PV subsidies. *Solar Energy*, *78*(2), 137–146. doi:10.1016/j.solener.2004.03.019

Schwarzbozl, P., Buck, R., Sugarmen, C., Ring, A., Crespo, M. J. M., Altwegg, P., & Enrile, J. (2006). Solar gas turbine systems: Design, cost and perspectives. *Solar Energy*, *80*(10), 1231–1240. doi:10.1016/j.solener.2005.09.007

Sefa, I., Mehmet, D., & Ilhami, C. (2009). Application of one-axis sun tracking system. *Energy Conversion and Management*, *50*(11), 2709–2718. doi:10.1016/j.enconman.2009.06.018

Senpinar, A., & Cebeci, M. (2012). Evaluation of power output for fixed and two-axis tracking PV arrays. *Applied Energy*, *92*, 677–685. doi:10.1016/j.apenergy.2011.07.043

Sharaf Eldin, S. A., Abd-Elhady, M. S., & Kandil, H. A. (2016). Feasibility of solar tracking systems for PV panels in hot and cold regions. *Renewable Energy*, *85*, 228–233. doi:10.1016/j.renene.2015.06.051

Shariah, A. M., Al-Akhras, A., & Al-Omari, I. A. (2002). Optimizing the tilt angle of solar collectors. *Renewable Energy*, *26*(4), 587–598. doi:10.1016/S0960-1481(01)00106-9

Shi, Y., & Eberhart, R. C. (1998). A modified particle swarm optimizer. *IEEE International Conference on Evolutionary Computation*, 69–73.

Shi, Y., & Eberhart, R. C. (1999). Empirical study of particle swarm optimization. *IEEE International Conference on Evolutionary Computation*, 1945–1950.

Siraki, A. G., & Pillay, P. (2012). Study of optimum tilts angles for solar panels in different latitudes for urban applications. *Solar Energy*, *86*(6), 1920–1928. doi:10.1016/j.solener.2012.02.030

Sivanandam, S. N., & Deepa, S. N. (2008). *Introduction to genetic algorithms*. Berlin: Springer Verlag.

Skoplaki, E., & Palyvos, J. A. (2009). On the temperature dependence of photovoltaic module electrical performance: A review of efficiency/power correlations. *Solar Energy*, *83*(5), 614–624. doi:10.1016/j.solener.2008.10.008

Solar Tracker Market Size and Share | Industry Report, 2022. (2016). Retrieved from http://www.grandviewresearch.com/industry-analysis/solar-tracker-industry

Soulayman, S., & Sarsar, W. (2012). The effect of two-axis solar tracking on the photovoltaic solar panel. *2nd International Conference on Renewable Energy (ICRE'2012)*, Bejaia, Algeria.

Sumathi, S. & Paneerselvam, S. (2010). *Computational intelligence paradigms theory and applications using MATLAB*. CRC Press.

Sun, L., Lu, L., & Yang, H. (2012). Optimum design of shading-type building-integrated photovoltaic claddings with different surface azimuth angles. *Applied Energy*, *90*(1), 233–240. doi:10.1016/j.apenergy.2011.01.062

Sunderan, P., Ismail, A. M., Singh, B., & Mohamed, N. M. (2011). Optimum tilt angle and orientation of standalone photovoltaic electricity generation. *Journal of Applied Sciences*, *11*(7), 1219–1224. doi:10.3923/jas.2011.1219.1224

Sungur, C. (2007). Sun-tracking system with PLC control for photovoltaic panels. *Int. J. Green Energy*, *4*(6), 635–643. doi:10.1080/15435070701665404

Sungur, C. (2009). Multi-axes sun-tracking system with PLC control for photovoltaic panels in turkey. *Renewable Energy*, *34*(4), 1119–1125. doi:10.1016/j.renene.2008.06.020

Systems USA Online Solar Panels. (2016). *Pallet of eoplly 190 watt mono solar panel*. Retrieved from http://www.solarsystems-usa.net/solarpanels/eoplly/ep125m-72-190/

Talebizadeh, P., Mehrabian, M. A., & Abdolzadeh, M. (2011). Prediction of the optimum slope and surface azimuth angles using the Genetic Algorithm. *Energy and Building*, *43*(11), 2998–3005. doi:10.1016/j.enbuild.2011.07.013

Tang, R., Gao, W., Yu, Y., & Chen, H. (2009). Optimal tilt-angles of all-glass evacuated tube solar collectors. *Energy*, *34*(9), 1387–1395. doi:10.1016/j.energy.2009.06.014

Tang, R., & Wu, T. (2004). Optimal tilt-angles for solar collectors used in China. *Applied Energy*, *79*(3), 239–248. doi:10.1016/j.apenergy.2004.01.003

Tomson, T. (2008). Discrete two-positional tracking of solar collectors. *Renewable Energy*, *33*(3), 400–405. doi:10.1016/j.renene.2007.03.017

Tudorache, T., Oancea, C. D. & Kreindler, L. (2012). Performance evaluation of a solar tracking PV panel. *U.P.B. Sci. Bull., Series C, 74*(1), 1-9.

Vatau, D., Musuroi, S., Barbulescu, C., & Babescu, M. (2014). PV systems modelling and optimal control. *Energy Conversion and Management, 84*, 448–456. doi:10.1016/j.enconman.2014.04.032

Warren, R. D. (2008). A feasibility study of stationary and dual-axis tracking grid-connected photovoltaic systems in the Upper Midwest. *Graduate Theses and Dissertations.* Paper 11161.

Wijewardane, S. (2009). Potential applicability of CNT and CNT/composites to implement ASEC concept: A review article. *Solar Energy, 89*(8), 1379–1389. doi:10.1016/j.solener.2009.03.001

Ye, Z., Nobre, A., Reindl, T., Luther, J., & Reise, C. (2013). On PV module temperatures in tropical regions. *Solar Energy, 88*, 80–87. doi:10.1016/j.solener.2012.11.001

APPENDIX A

Energy Gain of Long Term Tracking in the Tropical Zone

Let us give the following abbreviations for simplicity:

$$R_{d,o} = H\left(\beta = \beta_{opt,d}\right) / H\left(\beta = 0\right) \tag{1}$$

$$R_{d,\varphi} = H\left(\beta = \beta_{opt,d}\right) / H\left(\beta = \varphi\right) \tag{2}$$

$$R_{d,y} = H\left(\beta = \beta_{opt,d}\right) / H\left(\beta = \beta_{opt,y}\right) \tag{3}$$

$$R_{d,s} = H\left(\beta = \beta_{opt,d}\right) / H\left(\beta = \beta_{opt,s}\right) \tag{4}$$

$$R_{d,m} = H\left(\beta = \beta_{opt,d}\right) / H\left(\beta = \beta_{opt,m}\right) \tag{5}$$

The day numbers given in all tables of these appendixes correspond the dates given in Table 7. Tables 8-13 show the energy gain in the tropical zone.

Table 7. The chosen dates

Day	Day/Month	Remark
1	1/1	-
10	10/1	-
17	17/1	Characteristic day
20	20/1	-
30	30/1	-
32	1/2	-
41	10/2	-
47	16/2	Characteristic day
51	20/2	-
59	28/2	-
60	1/3	-
69	10/3	-
75	16/3	Characteristic day
79	20/3	-
89	30/3	-
91	1/4	-
100	10/4	-
105	15/4	Characteristic day
110	20/4	-
120	30/4	-
121	1/5	-
130	10/5	-
135	15/5	Characteristic day
140	20/5	-
150	30/5	-
152	1/6	-
161	10/6	-
162	11/6	Characteristic day
171	20/6	-

continued in next column

Table 7. Continued

Day	Day/Month	Remark
181	30/6	-
182	1/7	-
191	10/7	-
198	17/7	Characteristic day
201	20/7	-
211	30/7	-
213	1/8	-
222	10/8	-
228	16/8	Characteristic day
232	20/8	-
242	30/8	-
244	1/9	-
253	10/9	-
258	15/9	Characteristic day
263	20/9	-
273	30/9	-
274	1/10	-
283	10/10	-
288	15/10	Characteristic day
293	20/10	-
303	30/10	-
305	1/11	-
314	10/11	-
318	14/11	Characteristic day
324	20/11	-
334	30/11	-
335	1/12	-
344	10/12	Characteristic day
355	20/12	-
365	31/12	-

Table 8. Energy gain for Equator

Day	$R_{d,o}$	$R_{d,\varphi}$	$R_{d,y}$	$R_{d,s}$	$R_{d,m}$
1	1.202	1.202	1.026	1.026	1.001
10	1.186	1.186	1.022	1.022	1.000
17	1.167	1.167	1.016	1.016	1.000
20	1.158	1.158	1.014	1.014	1.000
30	1.124	1.124	1.006	1.006	1.002
32	1.117	1.117	1.005	1.005	1.005
41	1.085	1.085	1.001	1.001	1.001
47	1.064	1.064	1.000	1.000	1.000
51	1.051	1.051	1.001	1.001	1.001
59	1.029	1.029	1.008	1.008	1.007
60	1.026	1.026	1.009	1.009	1.006
69	1.009	1.009	1.026	1.026	1.000
75	1.002	1.002	1.044	1.044	1.001
79	1.000	1.000	1.059	1.059	1.004
89	1.004	1.004	1.049	1.001	1.001
91	1.006	1.006	1.042	1.061	1.011
100	1.021	1.021	1.018	1.031	1.001
105	1.033	1.033	1.010	1.020	1.000
110	1.047	1.047	1.004	1.011	1.001
120	1.079	1.079	1.000	1.002	1.009
121	1.083	1.083	1.000	1.002	1.005
130	1.115	1.115	1.002	1.000	1.000
135	1.133	1.133	1.005	1.001	1.000
140	1.150	1.150	1.008	1.002	1.000
150	1.179	1.179	1.014	1.006	1.002
152	1.184	1.184	1.015	1.007	1.000
161	1.202	1.202	1.020	1.010	1.000
162	1.203	1.203	1.020	1.010	1.000
171	1.210	1.210	1.022	1.011	1.000

continued in next column

Table 8. Continued

Day	$R_{d,o}$	$R_{d,\varphi}$	$R_{d,y}$	$R_{d,s}$	$R_{d,m}$
181	1.206	1.206	1.021	1.010	1.000
182	1.204	1.204	1.020	1.025	1.001
191	1.189	1.189	1.017	1.021	1.000
198	1.172	1.172	1.012	1.016	1.000
201	1.163	1.163	1.011	1.014	1.000
211	1.130	1.130	1.004	1.007	1.002
213	1.123	1.123	1.003	1.005	1.007
222	1.090	1.090	1.000	1.001	1.001
228	1.069	1.069	1.000	1.000	1.000
232	1.056	1.056	1.002	1.001	1.000
242	1.028	1.028	1.012	1.009	1.008
244	1.023	1.023	1.016	1.012	1.005
253	1.007	1.007	1.038	1.032	1.000
258	1.002	1.002	1.056	1.048	1.001
263	1.000	1.000	1.077	1.068	1.005
273	1.005	1.005	1.049	1.001	1.013
274	1.006	1.006	1.046	1.062	1.012
283	1.022	1.022	1.021	1.032	1.002
288	1.034	1.034	1.012	1.020	1.000
293	1.048	1.048	1.006	1.012	1.001
303	1.081	1.081	1.000	1.002	1.008
305	1.088	1.088	1.000	1.001	1.004
314	1.120	1.120	1.002	1.000	1.000
318	1.134	1.134	1.003	1.001	1.000
324	1.154	1.154	1.007	1.002	1.000
334	1.183	1.183	1.012	1.006	1.002
335	1.185	1.185	1.013	1.006	1.000
344	1.202	1.202	1.016	1.009	1.000
355	1.210	1.210	1.018	1.010	1.000
365	1.204	1.204	1.017	1.009	1.000

Table 9. Energy gain for latitudes 5°S and 5°N

Day	5°S					5°N				
	$R_{d,o}$	$R_{d,\varphi}$	$R_{d,y}$	$R_{d,s}$	$R_{d,m}$	$R_{d,o}$	$R_{d,\varphi}$	$R_{d,y}$	$R_{d,s}$	$R_{d,m}$
1	1.147	1.211	1.020	1.020	1.001	1.271	1.194	1.034	1.034	1.001
10	1.133	1.193	1.015	1.015	1.000	1.251	1.178	1.029	1.029	1.000
17	1.118	1.173	1.011	1.011	1.000	1.228	1.161	1.023	1.023	1.000
20	1.111	1.164	1.009	1.009	1.000	1.217	1.152	1.020	1.020	1.000
30	1.083	1.128	1.003	1.003	1.002	1.176	1.120	1.011	1.011	1.002
32	1.077	1.121	1.002	1.002	1.006	1.167	1.113	1.009	1.009	1.005
41	1.052	1.087	1.000	1.000	1.001	1.127	1.082	1.003	1.003	1.001
47	1.036	1.065	1.001	1.001	1.000	1.101	1.062	1.000	1.000	1.000
51	1.027	1.052	1.004	1.004	1.001	1.085	1.050	1.000	1.000	1.001
59	1.012	1.029	1.014	1.014	1.008	1.054	1.028	1.004	1.004	1.007
60	1.010	1.027	1.015	1.015	1.002	1.051	1.026	1.004	1.004	1.012
69	1.001	1.009	1.037	1.037	1.000	1.025	1.009	1.018	1.018	1.002
75	1.000	1.002	1.095	1.003	1.003	1.012	1.002	1.032	1.032	1.000
79	1.002	1.000	1.075	1.001	1.001	1.006	1.000	1.045	1.045	1.001
89	1.015	1.004	1.036	1.003	1.003	1.000	1.004	1.088	1.088	1.013
91	1.019	1.006	1.031	1.059	1.011	1.000	1.006	1.055	1.064	1.011
100	1.043	1.020	1.011	1.030	1.001	1.007	1.021	1.026	1.033	1.001
105	1.060	1.032	1.005	1.019	1.000	1.014	1.033	1.016	1.021	1.000
110	1.079	1.046	1.001	1.011	1.001	1.024	1.048	1.008	1.012	1.001
120	1.121	1.077	1.000	1.002	1.008	1.048	1.081	1.001	1.002	1.009
121	1.125	1.081	1.001	1.001	1.004	1.051	1.085	1.001	1.002	1.005
130	1.165	1.111	1.005	1.000	1.000	1.076	1.119	1.001	1.000	1.001
135	1.187	1.128	1.009	1.001	1.000	1.090	1.137	1.002	1.001	1.000
140	1.208	1.144	1.012	1.002	1.000	1.104	1.155	1.004	1.002	1.000
150	1.243	1.172	1.020	1.006	1.002	1.128	1.186	1.010	1.006	1.003
152	1.249	1.177	1.021	1.007	1.000	1.133	1.192	1.011	1.007	1.000
161	1.270	1.193	1.026	1.009	1.000	1.147	1.210	1.014	1.010	1.000
162	1.272	1.195	1.027	1.009	1.000	1.148	1.211	1.015	1.011	1.000
171	1.280	1.201	1.029	1.011	1.000	1.154	1.219	1.016	1.012	1.000
181	1.275	1.197	1.027	1.010	1.000	1.150	1.214	1.015	1.011	1.000
182	1.274	1.196	1.027	1.033	1.001	1.149	1.213	1.015	1.019	1.001
191	1.255	1.182	1.023	1.028	1.000	1.137	1.197	1.012	1.015	1.000
198	1.234	1.165	1.018	1.023	1.000	1.122	1.178	1.008	1.011	1.000

continued on following page

Table 9. Continued

Day	5°S					5°N				
	$R_{d,o}$	$R_{d,\varphi}$	$R_{d,y}$	$R_{d,s}$	$R_{d,m}$	$R_{d,o}$	$R_{d,\varphi}$	$R_{d,y}$	$R_{d,s}$	$R_{d,m}$
201	1.223	1.157	1.016	1.020	1.000	1.115	1.169	1.007	1.009	1.000
211	1.183	1.125	1.008	1.011	1.002	1.088	1.134	1.002	1.003	1.002
213	1.174	1.119	1.007	1.009	1.006	1.082	1.127	1.001	1.002	1.007
222	1.134	1.088	1.001	1.003	1.001	1.057	1.093	1.000	1.000	1.001
228	1.108	1.068	1.000	1.000	1.000	1.040	1.071	1.002	1.001	1.000
232	1.091	1.055	1.000	1.000	1.000	1.031	1.057	1.005	1.003	1.000
242	1.053	1.027	1.007	1.005	1.008	1.011	1.028	1.019	1.015	1.008
244	1.047	1.023	1.010	1.007	1.010	1.008	1.024	1.024	1.019	1.002
253	1.022	1.007	1.028	1.022	1.001	1.001	1.007	1.051	1.044	1.001
258	1.012	1.002	1.042	1.036	1.000	1.000	1.002	1.102	1.003	1.003
263	1.005	1.000	1.061	1.053	1.002	1.003	1.000	1.077	1.000	1.000
273	1.000	1.005	1.063	1.000	1.000	1.018	1.005	1.038	1.004	1.004
274	1.000	1.006	1.059	1.065	1.012	1.020	1.006	1.035	1.059	1.011
283	1.007	1.022	1.029	1.033	1.002	1.044	1.021	1.014	1.030	1.002
288	1.015	1.034	1.018	1.021	1.000	1.061	1.033	1.007	1.019	1.000
293	1.024	1.049	1.010	1.012	1.001	1.080	1.047	1.002	1.011	1.001
303	1.049	1.083	1.001	1.002	1.008	1.122	1.078	1.000	1.002	1.008
305	1.054	1.090	1.001	1.001	1.004	1.131	1.085	1.000	1.001	1.004
314	1.080	1.124	1.000	1.000	1.000	1.171	1.116	1.004	1.000	1.000
318	1.091	1.139	1.001	1.001	1.000	1.188	1.130	1.006	1.001	1.000
324	1.108	1.160	1.004	1.002	1.000	1.213	1.149	1.011	1.002	1.000
334	1.131	1.190	1.008	1.006	1.003	1.247	1.176	1.017	1.006	1.002
335	1.133	1.192	1.009	1.007	1.000	1.250	1.178	1.018	1.006	1.000
344	1.147	1.210	1.012	1.009	1.000	1.271	1.194	1.022	1.009	1.000
354	1.154	1.219	1.013	1.011	1.000	1.280	1.201	1.024	1.010	1.000
365	1.149	1.212	1.012	1.010	1.000	1.273	1.195	1.022	1.009	1.000

Table 10. Energy gain for latitudes 10°S and 10°N

Day	10°S					10°N				
	$R_{d,o}$	$R_{d,\varphi}$	$R_{d,y}$	$R_{d,s}$	$R_{d,m}$	$R_{d,o}$	$R_{d,\varphi}$	$R_{d,y}$	$R_{d,s}$	$R_{d,m}$
1	1.104	1.218	1.014	1.014	1.001	1.357	1.186	1.043	1.032	1.001
10	1.092	1.199	1.011	1.011	1.000	1.332	1.171	1.037	1.027	1.000
17	1.080	1.179	1.007	1.007	1.000	1.305	1.155	1.031	1.021	1.000
20	1.074	1.169	1.005	1.005	1.000	1.292	1.146	1.027	1.019	1.000
30	1.052	1.131	1.001	1.001	1.002	1.242	1.116	1.017	1.010	1.002
32	1.047	1.124	1.001	1.001	1.006	1.231	1.109	1.014	1.008	1.005
41	1.028	1.089	1.001	1.001	1.001	1.182	1.080	1.006	1.002	1.001
47	1.017	1.067	1.004	1.004	1.000	1.149	1.061	1.002	1.000	1.000
51	1.011	1.053	1.008	1.008	1.001	1.128	1.049	1.000	1.000	1.001
59	1.002	1.029	1.021	1.021	1.008	1.089	1.028	1.001	1.004	1.007
60	1.002	1.027	1.023	1.023	1.000	1.085	1.026	1.002	1.005	1.012
69	1.001	1.009	1.107	1.005	1.005	1.049	1.009	1.011	1.018	1.002
75	1.006	1.002	1.077	1.001	1.001	1.030	1.002	1.023	1.032	1.000
79	1.011	1.000	1.059	1.000	1.000	1.020	1.000	1.033	1.045	1.001
89	1.035	1.004	1.026	1.007	1.007	1.004	1.004	1.070	1.088	1.012
91	1.041	1.006	1.021	1.056	1.010	1.002	1.006	1.079	1.001	1.001
100	1.074	1.020	1.006	1.028	1.001	1.000	1.021	1.036	1.041	1.004
105	1.096	1.032	1.002	1.017	1.000	1.003	1.033	1.023	1.027	1.001
110	1.121	1.045	1.000	1.010	1.001	1.009	1.048	1.013	1.016	1.000
120	1.173	1.075	1.002	1.002	1.008	1.025	1.083	1.003	1.004	1.005
121	1.179	1.078	1.003	1.001	1.004	1.027	1.087	1.002	1.003	1.005
130	1.228	1.108	1.009	1.000	1.000	1.046	1.122	1.000	1.000	1.001
135	1.255	1.124	1.013	1.001	1.000	1.057	1.141	1.001	1.000	1.000
140	1.280	1.139	1.018	1.002	1.000	1.068	1.160	1.002	1.001	1.000
150	1.323	1.166	1.027	1.006	1.002	1.088	1.192	1.006	1.004	1.003
152	1.331	1.170	1.028	1.006	1.000	1.091	1.198	1.007	1.005	1.000
161	1.356	1.185	1.034	1.009	1.000	1.103	1.217	1.010	1.008	1.000
162	1.358	1.187	1.034	1.009	1.000	1.104	1.219	1.010	1.008	1.000
171	1.368	1.193	1.036	1.010	1.000	1.109	1.227	1.012	1.009	1.000
181	1.361	1.189	1.035	1.010	1.000	1.106	1.222	1.011	1.009	1.000
182	1.360	1.188	1.035	1.033	1.001	1.105	1.220	1.011	1.014	1.001

continued on following page

Table 10. Continued

Day	10°S					10°N				
	$R_{d,o}$	$R_{d,\varphi}$	$R_{d,y}$	$R_{d,s}$	$R_{d,m}$	$R_{d,o}$	$R_{d,\varphi}$	$R_{d,y}$	$R_{d,s}$	$R_{d,m}$
191	1.338	1.174	1.030	1.028	1.000	1.095	1.203	1.008	1.010	1.000
198	1.312	1.159	1.024	1.023	1.000	1.083	1.184	1.005	1.007	1.000
201	1.299	1.151	1.022	1.020	1.000	1.077	1.174	1.004	1.006	1.000
211	1.250	1.121	1.013	1.012	1.002	1.055	1.138	1.000	1.001	1.002
213	1.240	1.115	1.011	1.010	1.006	1.051	1.130	1.000	1.001	1.007
222	1.191	1.085	1.004	1.003	1.001	1.031	1.095	1.001	1.000	1.001
228	1.158	1.066	1.001	1.001	1.000	1.020	1.072	1.005	1.003	1.000
232	1.137	1.054	1.000	1.000	1.000	1.013	1.058	1.009	1.007	1.001
242	1.088	1.027	1.003	1.004	1.007	1.002	1.029	1.027	1.023	1.009
244	1.079	1.023	1.005	1.006	1.012	1.001	1.024	1.033	1.028	1.000
253	1.044	1.007	1.019	1.020	1.002	1.001	1.007	1.110	1.003	1.003
258	1.029	1.002	1.031	1.033	1.000	1.006	1.002	1.083	1.000	1.000
263	1.017	1.000	1.047	1.049	1.001	1.014	1.000	1.061	1.000	1.000
273	1.003	1.005	1.091	1.094	1.013	1.039	1.005	1.027	1.009	1.009
274	1.002	1.006	1.096	1.074	1.000	1.042	1.006	1.025	1.057	1.011
283	1.001	1.022	1.039	1.039	1.004	1.076	1.021	1.008	1.029	1.002
288	1.004	1.034	1.026	1.026	1.001	1.098	1.032	1.003	1.018	1.000
293	1.009	1.049	1.016	1.016	1.000	1.122	1.046	1.001	1.010	1.001
303	1.026	1.084	1.004	1.004	1.005	1.175	1.076	1.001	1.002	1.007
305	1.030	1.092	1.002	1.002	1.005	1.186	1.083	1.002	1.001	1.004
314	1.049	1.127	1.000	1.000	1.000	1.236	1.112	1.008	1.000	1.000
318	1.058	1.142	1.000	1.000	1.000	1.257	1.125	1.011	1.001	1.000
324	1.071	1.165	1.002	1.002	1.000	1.286	1.143	1.016	1.002	1.000
334	1.090	1.196	1.005	1.005	1.003	1.328	1.169	1.023	1.005	1.002
344	1.104	1.218	1.008	1.008	1.000	1.357	1.186	1.029	1.008	1.000
354	1.109	1.227	1.009	1.009	1.000	1.368	1.193	1.031	1.009	1.000
365	1.105	1.220	1.008	1.125	1.000	1.359	1.187	1.029	1.008	1.000

Table 11. Energy gain for latitudes 15°S and 15°N

Day	15°S					15°N				
	$R_{d,o}$	$R_{d,\varphi}$	$R_{d,y}$	$R_{d,s}$	$R_{d,m}$	$R_{d,o}$	$R_{d,\varphi}$	$R_{d,y}$	$R_{d,s}$	$R_{d,m}$
1	1.070	1.225	1.010	1.010	1.001	1.464	1.178	1.053	1.030	1.001
10	1.060	1.205	1.007	1.007	1.000	1.434	1.164	1.047	1.025	1.000
17	1.050	1.183	1.004	1.004	1.000	1.401	1.149	1.040	1.020	1.000
20	1.045	1.173	1.003	1.003	1.000	1.385	1.141	1.036	1.018	1.000
30	1.029	1.134	1.000	1.000	1.003	1.324	1.112	1.024	1.010	1.002
32	1.025	1.126	1.000	1.000	1.005	1.311	1.106	1.021	1.008	1.005
41	1.012	1.090	1.002	1.002	1.001	1.250	1.078	1.011	1.002	1.001
47	1.005	1.067	1.007	1.007	1.000	1.210	1.059	1.005	1.000	1.000
51	1.002	1.053	1.013	1.013	1.002	1.184	1.048	1.002	1.000	1.001
59	1.000	1.030	1.143	1.013	1.000	1.135	1.027	1.000	1.003	1.007
60	1.000	1.027	1.137	1.011	1.013	1.130	1.025	1.000	1.004	1.012
69	1.008	1.009	1.087	1.001	1.002	1.083	1.009	1.006	1.017	1.002
75	1.019	1.002	1.060	1.000	1.000	1.057	1.002	1.014	1.031	1.000
79	1.029	1.000	1.045	1.002	1.001	1.043	1.000	1.023	1.043	1.001
89	1.063	1.004	1.017	1.014	1.012	1.016	1.004	1.054	1.085	1.012
91	1.072	1.006	1.013	1.053	1.010	1.012	1.006	1.062	1.003	1.003
100	1.116	1.020	1.003	1.027	1.001	1.002	1.021	1.105	1.000	1.000
105	1.144	1.031	1.000	1.016	1.000	1.000	1.034	1.133	1.003	1.003
110	1.175	1.044	1.000	1.009	1.001	1.001	1.048	1.020	1.023	1.000
120	1.240	1.073	1.005	1.001	1.008	1.010	1.084	1.006	1.007	1.002
121	1.247	1.076	1.006	1.001	1.004	1.011	1.088	1.005	1.006	1.005
130	1.308	1.104	1.015	1.000	1.000	1.024	1.124	1.000	1.001	1.001
135	1.340	1.120	1.020	1.001	1.000	1.033	1.144	1.000	1.000	1.000
140	1.371	1.134	1.025	1.002	1.000	1.041	1.164	1.001	1.000	1.000
150	1.424	1.159	1.035	1.005	1.002	1.057	1.198	1.003	1.002	1.003
152	1.432	1.163	1.037	1.006	1.000	1.060	1.204	1.004	1.003	1.000
161	1.463	1.178	1.043	1.009	1.000	1.069	1.224	1.007	1.005	1.000
162	1.466	1.179	1.043	1.009	1.000	1.070	1.225	1.007	1.005	1.000
171	1.478	1.185	1.046	1.010	1.000	1.074	1.234	1.008	1.006	1.000

continued on following page

Table 11. Continued

Day	15°S					15°N				
	$R_{d,o}$	$R_{d,\varphi}$	$R_{d,y}$	$R_{d,s}$	$R_{d,m}$	$R_{d,o}$	$R_{d,\varphi}$	$R_{d,y}$	$R_{d,s}$	$R_{d,m}$
181	1.470	1.181	1.044	1.009	1.000	1.071	1.228	1.007	1.006	1.000
182	1.468	1.180	1.044	1.031	1.001	1.071	1.227	1.007	1.009	1.001
191	1.441	1.167	1.039	1.027	1.000	1.062	1.209	1.005	1.007	1.000
198	1.410	1.153	1.033	1.022	1.000	1.053	1.189	1.002	1.004	1.000
201	1.394	1.145	1.030	1.020	1.000	1.048	1.179	1.002	1.003	1.000
211	1.334	1.117	1.019	1.011	1.002	1.031	1.141	1.000	1.000	1.002
213	1.321	1.111	1.017	1.010	1.006	1.028	1.133	1.000	1.000	1.048
222	1.261	1.083	1.008	1.003	1.001	1.014	1.096	1.003	1.002	1.001
228	1.221	1.064	1.003	1.001	1.000	1.007	1.073	1.009	1.007	1.000
232	1.195	1.053	1.001	1.000	1.000	1.003	1.059	1.015	1.011	1.001
242	1.133	1.027	1.001	1.004	1.007	1.000	1.029	1.154	1.012	1.012
244	1.122	1.022	1.002	1.005	1.012	1.001	1.024	1.141	1.009	1.012
253	1.077	1.007	1.012	1.019	1.002	1.010	1.007	1.090	1.001	1.002
258	1.056	1.002	1.021	1.032	1.000	1.020	1.002	1.066	1.000	1.000
263	1.038	1.000	1.035	1.048	1.001	1.033	1.000	1.047	1.003	1.001
273	1.013	1.005	1.072	1.091	1.013	1.069	1.005	1.018	1.017	1.013
274	1.011	1.006	1.077	1.091	1.003	1.073	1.006	1.016	1.055	1.011
283	1.001	1.022	1.124	1.051	1.000	1.118	1.021	1.004	1.027	1.001
288	1.000	1.035	1.035	1.035	1.003	1.146	1.032	1.001	1.017	1.000
293	1.001	1.050	1.023	1.023	1.000	1.177	1.045	1.000	1.010	1.001
303	1.010	1.085	1.007	1.007	1.002	1.243	1.074	1.004	1.002	1.007
305	1.013	1.093	1.005	1.005	1.005	1.256	1.080	1.005	1.001	1.003
314	1.027	1.130	1.001	1.001	1.000	1.317	1.109	1.012	1.000	1.000
318	1.033	1.146	1.000	1.000	1.000	1.342	1.121	1.016	1.001	1.000
324	1.044	1.169	1.000	1.000	1.000	1.379	1.138	1.022	1.002	1.000
334	1.059	1.202	1.003	1.003	1.003	1.430	1.162	1.031	1.005	1.002
335	1.060	1.205	1.003	1.003	1.000	1.434	1.164	1.031	1.005	1.000
344	1.070	1.224	1.005	1.005	1.000	1.464	1.178	1.037	1.008	1.000
354	1.074	1.234	1.006	1.111	1.000	1.478	1.185	1.039	1.009	1.000
365	1.071	1.227	1.005	1.107	1.000	1.467	1.180	1.037	1.008	1.000

Table 12. Energy gain for latitudes 20°S and 20°N

Day	20°S					20°N				
	$R_{d,o}$	$R_{d,\varphi}$	$R_{d,y}$	$R_{d,s}$	$R_{d,m}$	$R_{d,o}$	$R_{d,\varphi}$	$R_{d,y}$	$R_{d,s}$	$R_{d,m}$
1	1.043	1.231	1.006	1.006	1.002	1.601	1.171	1.073	1.029	1.001
10	1.036	1.210	1.004	1.004	1.000	1.564	1.158	1.065	1.024	1.000
17	1.028	1.187	1.002	1.002	1.000	1.523	1.143	1.057	1.020	1.000
20	1.025	1.176	1.001	1.001	1.000	1.503	1.136	1.053	1.017	1.000
30	1.013	1.136	1.000	1.000	1.003	1.428	1.108	1.037	1.009	1.002
32	1.010	1.128	1.001	1.001	1.002	1.412	1.102	1.034	1.008	1.004
41	1.003	1.091	1.006	1.006	1.000	1.337	1.075	1.020	1.002	1.001
47	1.000	1.068	1.013	1.013	1.002	1.288	1.058	1.012	1.000	1.000
51	1.000	1.053	1.165	1.019	1.000	1.255	1.047	1.008	1.000	1.001
59	1.005	1.030	1.116	1.006	1.001	1.194	1.027	1.001	1.003	1.006
60	1.007	1.027	1.111	1.005	1.013	1.187	1.025	1.001	1.004	1.012
69	1.023	1.009	1.067	1.000	1.002	1.128	1.009	1.001	1.016	1.002
75	1.040	1.002	1.044	1.002	1.000	1.094	1.002	1.006	1.030	1.000
79	1.055	1.000	1.032	1.005	1.001	1.074	1.000	1.012	1.042	1.001
89	1.102	1.004	1.010	1.023	1.012	1.036	1.004	1.035	1.083	1.012
91	1.113	1.006	1.007	1.051	1.010	1.030	1.006	1.041	1.007	1.007
100	1.170	1.020	1.000	1.025	1.001	1.010	1.021	1.076	1.000	1.000
105	1.205	1.030	1.000	1.016	1.000	1.004	1.034	1.100	1.001	1.001
110	1.244	1.043	1.002	1.009	1.001	1.001	1.049	1.126	1.003	1.003
120	1.325	1.071	1.010	1.001	1.007	1.002	1.085	1.010	1.012	1.000
121	1.333	1.074	1.011	1.001	1.004	1.002	1.089	1.009	1.010	1.006
130	1.407	1.101	1.022	1.000	1.000	1.010	1.126	1.002	1.002	1.001
135	1.447	1.115	1.028	1.001	1.000	1.016	1.147	1.000	1.001	1.000
140	1.485	1.129	1.035	1.002	1.000	1.022	1.167	1.000	1.000	1.000
150	1.550	1.153	1.046	1.005	1.002	1.033	1.202	1.002	1.001	1.003
152	1.561	1.157	1.048	1.006	1.000	1.036	1.208	1.002	1.001	1.000
161	1.599	1.170	1.054	1.008	1.000	1.043	1.230	1.004	1.003	1.000
162	1.602	1.171	1.055	1.008	1.000	1.044	1.231	1.004	1.003	1.000
171	1.617	1.177	1.058	1.009	1.000	1.047	1.240	1.005	1.004	1.000
181	1.608	1.173	1.056	1.009	1.000	1.045	1.235	1.005	1.003	1.000
182	1.605	1.172	1.056	1.030	1.001	1.044	1.233	1.004	1.006	1.001

continued on following page

Table 12. Continued

Day	20°S					20°N				
	$R_{d,o}$	$R_{d,\varphi}$	$R_{d,y}$	$R_{d,s}$	$R_{d,m}$	$R_{d,o}$	$R_{d,\varphi}$	$R_{d,y}$	$R_{d,s}$	$R_{d,m}$
191	1.572	1.160	1.050	1.026	1.000	1.038	1.214	1.003	1.004	1.000
198	1.533	1.147	1.043	1.021	1.000	1.030	1.193	1.001	1.002	1.000
201	1.514	1.140	1.040	1.019	1.000	1.027	1.182	1.000	1.001	1.000
211	1.441	1.113	1.027	1.011	1.002	1.015	1.143	1.000	1.000	1.003
213	1.425	1.107	1.025	1.009	1.005	1.012	1.135	1.001	1.000	1.024
222	1.351	1.080	1.014	1.003	1.001	1.004	1.097	1.006	1.005	1.011
228	1.301	1.063	1.007	1.001	1.000	1.001	1.074	1.014	1.011	1.001
232	1.268	1.051	1.004	1.000	1.000	1.000	1.059	1.221	1.023	1.004
242	1.192	1.026	1.000	1.003	1.007	1.006	1.029	1.154	1.007	1.019
244	1.178	1.022	1.000	1.005	1.012	1.009	1.024	1.142	1.004	1.012
253	1.119	1.007	1.006	1.018	1.002	1.027	1.007	1.094	1.000	1.002
258	1.092	1.002	1.013	1.030	1.000	1.042	1.002	1.072	1.002	1.000
263	1.068	1.000	1.023	1.046	1.001	1.060	1.000	1.053	1.007	1.001
273	1.032	1.005	1.055	1.089	1.012	1.109	1.005	1.026	1.025	1.013
274	1.029	1.006	1.059	1.111	1.007	1.115	1.006	1.024	1.052	1.010
283	1.010	1.022	1.100	1.065	1.000	1.172	1.020	1.012	1.026	1.001
288	1.004	1.035	1.127	1.046	1.001	1.208	1.031	1.009	1.016	1.000
293	1.001	1.050	1.158	1.031	1.003	1.247	1.044	1.008	1.009	1.001
303	1.002	1.086	1.012	1.012	1.000	1.328	1.072	1.012	1.002	1.007
305	1.003	1.094	1.009	1.009	1.005	1.344	1.078	1.013	1.001	1.003
314	1.011	1.131	1.002	1.002	1.001	1.419	1.105	1.020	1.000	1.000
318	1.016	1.148	1.001	1.001	1.000	1.450	1.117	1.024	1.001	1.000
324	1.023	1.172	1.000	1.000	1.000	1.495	1.133	1.030	1.002	1.000
334	1.035	1.207	1.001	1.001	1.003	1.558	1.155	1.039	1.005	1.002
335	1.036	1.210	1.001	1.001	1.000	1.563	1.157	1.040	1.005	1.000
344	1.043	1.230	1.003	1.088	1.000	1.600	1.171	1.046	1.008	1.000
354	1.047	1.240	1.004	1.093	1.000	1.617	1.177	1.049	1.009	1.000
365	1.044	1.233	1.003	1.089	1.000	1.604	1.172	1.047	1.008	1.000

Table 13. Energy gain for latitudes 23.45°S and 23.45°N

Day	23.45°S					23.45°N				
	$R_{d,o}$	$R_{d,\varphi}$	$R_{d,y}$	$R_{d,s}$	$R_{d,m}$	$R_{d,o}$	$R_{d,\varphi}$	$R_{d,y}$	$R_{d,s}$	$R_{d,m}$
1	1.030	1.233	1.004	1.004	1.002	1.717	1.166	1.075	1.028	1.001
10	1.023	1.212	1.002	1.002	1.000	1.674	1.154	1.068	1.024	1.000
17	1.017	1.189	1.001	1.001	1.000	1.626	1.140	1.059	1.019	1.000
20	1.014	1.178	1.000	1.000	1.000	1.603	1.133	1.055	1.017	1.000
30	1.006	1.137	1.001	1.001	1.003	1.515	1.106	1.040	1.009	1.002
32	1.004	1.128	1.002	1.002	1.001	1.496	1.100	1.037	1.008	1.004
41	1.000	1.090	1.009	1.009	1.001	1.410	1.074	1.022	1.002	1.001
47	1.001	1.067	1.170	1.021	1.001	1.353	1.057	1.014	1.000	1.000
51	1.003	1.053	1.145	1.014	1.000	1.315	1.046	1.009	1.000	1.001
59	1.014	1.029	1.101	1.003	1.003	1.244	1.027	1.002	1.003	1.006
60	1.015	1.027	1.095	1.003	1.013	1.236	1.025	1.002	1.004	1.012
69	1.039	1.009	1.056	1.000	1.002	1.166	1.009	1.001	1.015	1.002
75	1.061	1.002	1.035	1.004	1.000	1.126	1.002	1.005	1.029	1.000
79	1.079	1.000	1.024	1.009	1.001	1.102	1.000	1.010	1.040	1.001
89	1.135	1.004	1.006	1.030	1.012	1.055	1.004	1.032	1.081	1.012
91	1.149	1.006	1.004	1.049	1.009	1.047	1.006	1.038	1.011	1.011
100	1.215	1.020	1.000	1.024	1.001	1.021	1.021	1.071	1.001	1.001
105	1.257	1.030	1.001	1.015	1.000	1.011	1.033	1.093	1.000	1.000
110	1.302	1.042	1.004	1.008	1.001	1.005	1.048	1.119	1.001	1.001
120	1.396	1.070	1.015	1.001	1.007	1.000	1.084	1.013	1.015	1.000
130	1.492	1.099	1.028	1.000	1.000	1.004	1.126	1.003	1.004	1.001
135	1.538	1.113	1.035	1.001	1.000	1.008	1.147	1.001	1.002	1.000
140	1.582	1.126	1.042	1.002	1.000	1.012	1.168	1.000	1.000	1.000
150	1.658	1.149	1.054	1.005	1.002	1.021	1.204	1.001	1.000	1.003
152	1.671	1.153	1.056	1.006	1.000	1.023	1.211	1.001	1.001	1.000
161	1.715	1.166	1.063	1.008	1.000	1.029	1.233	1.003	1.002	1.000
162	1.719	1.167	1.064	1.008	1.000	1.030	1.234	1.003	1.002	1.000
171	1.736	1.172	1.066	1.009	1.000	1.032	1.243	1.004	1.003	1.000
181	1.725	1.169	1.065	1.009	1.000	1.031	1.238	1.003	1.002	1.000
182	1.722	1.168	1.064	1.029	1.001	1.030	1.236	1.003	1.004	1.001
191	1.683	1.156	1.058	1.025	1.000	1.025	1.217	1.001	1.002	1.000
198	1.638	1.143	1.051	1.021	1.000	1.019	1.195	1.000	1.001	1.000

continued on following page

Table 13. Continued

Day	23.45°S					23.45°N				
	$R_{d,o}$	$R_{d,\varphi}$	$R_{d,y}$	$R_{d,s}$	$R_{d,m}$	$R_{d,o}$	$R_{d,\varphi}$	$R_{d,y}$	$R_{d,s}$	$R_{d,m}$
201	1.616	1.136	1.047	1.019	1.000	1.016	1.184	1.000	1.000	1.000
211	1.530	1.111	1.034	1.011	1.001	1.007	1.144	1.001	1.001	1.003
213	1.512	1.105	1.031	1.009	1.005	1.005	1.135	1.002	1.001	1.019
222	1.426	1.079	1.019	1.003	1.001	1.000	1.097	1.009	1.007	1.007
228	1.368	1.062	1.011	1.001	1.000	1.000	1.073	1.199	1.025	1.002
232	1.330	1.051	1.007	1.000	1.000	1.002	1.059	1.172	1.017	1.007
242	1.241	1.026	1.001	1.003	1.006	1.014	1.028	1.112	1.003	1.025
244	1.225	1.022	1.000	1.005	1.011	1.018	1.024	1.101	1.002	1.013
253	1.156	1.007	1.003	1.018	1.002	1.043	1.007	1.060	1.001	1.002
258	1.124	1.002	1.008	1.029	1.000	1.062	1.002	1.042	1.004	1.000
263	1.095	1.000	1.017	1.045	1.001	1.086	1.000	1.027	1.011	1.001
273	1.049	1.005	1.045	1.087	1.012	1.144	1.005	1.007	1.033	1.012
274	1.046	1.006	1.048	1.126	1.011	1.151	1.006	1.006	1.050	1.010
283	1.020	1.022	1.085	1.076	1.001	1.219	1.020	1.000	1.025	1.001
288	1.010	1.034	1.110	1.055	1.000	1.260	1.031	1.000	1.015	1.000
293	1.004	1.049	1.138	1.038	1.001	1.305	1.043	1.002	1.009	1.001
303	1.000	1.086	1.015	1.015	1.000	1.399	1.071	1.011	1.001	1.007
305	1.000	1.094	1.012	1.012	1.006	1.419	1.077	1.013	1.001	1.003
314	1.005	1.132	1.004	1.004	1.001	1.505	1.103	1.024	1.000	1.000
318	1.008	1.149	1.002	1.002	1.000	1.541	1.114	1.029	1.001	1.000
324	1.013	1.173	1.000	1.000	1.000	1.593	1.130	1.036	1.002	1.000
334	1.023	1.209	1.000	1.000	1.003	1.667	1.152	1.046	1.005	1.002
335	1.023	1.212	1.001	1.001	1.000	1.673	1.153	1.047	1.005	1.000
344	1.030	1.233	1.002	1.078	1.000	1.716	1.166	1.054	1.007	1.000
354	1.032	1.243	1.002	1.083	1.000	1.736	1.172	1.057	1.008	1.000
365	1.030	1.236	1.002	1.079	1.000	1.721	1.168	1.054	1.007	1.000

APPENDIX B

Energy Gain of Long Term Tracking in the Mid-Latitude Zone

Tables 14-18 show energy gains in the mid-latitude zone.

Table 14. Energy gain for latitudes 25°N and 25°S

Day	25°S					25°N				
	$R_{d,o}$	$R_{d,\varphi}$	$R_{d,y}$	$R_{d,s}$	$R_{d,m}$	$R_{d,o}$	$R_{d,\varphi}$	$R_{d,y}$	$R_{d,s}$	$R_{d,m}$
1	1.024	1.236	1.024	1.024	1.024	1.776	1.163	1.081	1.029	1.001
10	1.019	1.214	1.019	1.019	1.019	1.730	1.151	1.073	1.025	1.000
17	1.013	1.191	1.013	1.013	1.013	1.679	1.137	1.064	1.020	1.000
20	1.011	1.179	1.011	1.011	1.011	1.653	1.130	1.060	1.018	1.000
30	1.003	1.138	1.003	1.003	1.003	1.560	1.104	1.044	1.010	1.002
32	1.002	1.129	1.002	1.002	1.002	1.540	1.099	1.041	1.009	1.004
41	1.000	1.091	1.202	1.000	1.004	1.447	1.073	1.026	1.003	1.001
47	1.002	1.068	1.163	1.001	1.001	1.386	1.056	1.017	1.000	1.000
51	1.006	1.053	1.139	1.003	1.000	1.346	1.046	1.011	1.000	1.001
59	1.018	1.030	1.096	1.013	1.004	1.270	1.026	1.003	1.002	1.006
60	1.021	1.027	1.091	1.015	1.013	1.261	1.024	1.003	1.003	1.012
69	1.047	1.009	1.052	1.038	1.002	1.186	1.009	1.000	1.014	1.002
75	1.071	1.002	1.033	1.060	1.000	1.142	1.002	1.003	1.027	1.000
79	1.091	1.000	1.022	1.078	1.001	1.117	1.000	1.007	1.038	1.001
89	1.153	1.004	1.005	1.135	1.012	1.065	1.004	1.027	1.078	1.011
91	1.167	1.006	1.003	1.048	1.010	1.056	1.006	1.033	1.014	1.012
100	1.239	1.019	1.000	1.024	1.001	1.027	1.021	1.064	1.002	1.002
105	1.283	1.030	1.002	1.014	1.000	1.015	1.034	1.086	1.000	1.000
110	1.331	1.042	1.005	1.008	1.001	1.007	1.049	1.110	1.001	1.001
120	1.432	1.069	1.016	1.001	1.007	1.000	1.085	1.166	1.009	1.010
121	1.442	1.072	1.017	1.001	1.004	1.000	1.089	1.172	1.010	1.000
130	1.534	1.097	1.030	1.000	1.000	1.002	1.127	1.002	1.024	1.002
135	1.584	1.111	1.037	1.001	1.000	1.005	1.148	1.005	1.033	1.005
140	1.631	1.124	1.044	1.002	1.000	1.009	1.169	1.009	1.043	1.009
150	1.713	1.146	1.056	1.005	1.002	1.017	1.206	1.017	1.060	1.017
152	1.726	1.150	1.058	1.006	1.000	1.018	1.213	1.018	1.063	1.018
161	1.774	1.163	1.065	1.008	1.000	1.024	1.235	1.024	1.074	1.024
162	1.778	1.164	1.066	1.008	1.000	1.025	1.237	1.025	1.075	1.025

continued on following page

Table 14. Continued

Day	25°S					25°N				
	$R_{d,o}$	$R_{d,\varphi}$	$R_{d,y}$	$R_{d,s}$	$R_{d,m}$	$R_{d,o}$	$R_{d,\varphi}$	$R_{d,y}$	$R_{d,s}$	$R_{d,m}$
171	1.797	1.169	1.069	1.009	1.000	1.027	1.245	1.027	1.079	1.027
181	1.785	1.166	1.067	1.008	1.000	1.025	1.240	1.025	1.076	1.025
182	1.782	1.165	1.067	1.029	1.001	1.025	1.238	1.025	1.122	1.025
191	1.740	1.154	1.060	1.025	1.000	1.020	1.219	1.020	1.110	1.020
198	1.691	1.141	1.053	1.021	1.000	1.015	1.196	1.015	1.095	1.015
201	1.667	1.134	1.049	1.019	1.000	1.012	1.185	1.012	1.088	1.012
211	1.576	1.109	1.036	1.011	1.001	1.004	1.145	1.004	1.063	1.004
213	1.556	1.103	1.033	1.009	1.005	1.003	1.136	1.003	1.057	1.003
222	1.464	1.078	1.020	1.003	1.001	1.000	1.098	1.000	1.035	1.000
228	1.402	1.061	1.012	1.001	1.000	1.001	1.074	1.188	1.021	1.001
232	1.362	1.050	1.008	1.000	1.000	1.004	1.059	1.162	1.014	1.000
242	1.266	1.026	1.001	1.003	1.006	1.019	1.029	1.104	1.002	1.004
244	1.248	1.021	1.000	1.005	1.011	1.024	1.024	1.094	1.001	1.012
253	1.175	1.007	1.002	1.017	1.002	1.052	1.007	1.055	1.001	1.002
258	1.140	1.002	1.007	1.029	1.000	1.073	1.002	1.037	1.006	1.000
263	1.108	1.000	1.015	1.044	1.001	1.098	1.000	1.023	1.013	1.001
273	1.059	1.005	1.042	1.086	1.013	1.162	1.005	1.006	1.038	1.013
274	1.055	1.006	1.045	1.013	1.013	1.170	1.006	1.005	1.048	1.010
283	1.026	1.022	1.081	1.002	1.002	1.242	1.020	1.000	1.023	1.001
288	1.015	1.035	1.105	1.000	1.000	1.287	1.031	1.001	1.014	1.000
293	1.007	1.050	1.132	1.001	1.001	1.335	1.043	1.003	1.008	1.001
303	1.000	1.086	1.194	1.010	1.010	1.435	1.070	1.013	1.001	1.007
305	1.000	1.094	1.000	1.000	1.000	1.456	1.076	1.015	1.001	1.003
314	1.003	1.133	1.003	1.003	1.003	1.548	1.101	1.027	1.000	1.000
318	1.005	1.150	1.005	1.005	1.005	1.588	1.112	1.032	1.001	1.000
324	1.010	1.175	1.010	1.010	1.010	1.644	1.128	1.039	1.002	1.000
334	1.018	1.211	1.018	1.018	1.018	1.722	1.149	1.050	1.005	1.002
335	1.019	1.214	1.019	1.019	1.019	1.729	1.151	1.051	1.006	1.000
344	1.024	1.235	1.024	1.024	1.024	1.776	1.163	1.058	1.008	1.000
354	1.027	1.245	1.027	1.027	1.027	1.797	1.169	1.061	1.009	1.000
365	1.025	1.238	1.025	1.025	1.025	1.781	1.165	1.058	1.008	1.000

Table 15. Energy gain for latitudes 30°N and 30°S

Day	30°S					30°N				
	$R_{d,o}$	$R_{d,\varphi}$	$R_{d,y}$	$R_{d,s}$	$R_{d,m}$	$R_{d,o}$	$R_{d,\varphi}$	$R_{d,y}$	$R_{d,s}$	$R_{d,m}$
1	1.011	1.240	1.011	1.011	1.011	2.007	1.156	1.098	1.028	1.001
10	1.008	1.217	1.008	1.008	1.008	1.947	1.144	1.090	1.024	1.000
17	1.004	1.193	1.004	1.004	1.004	1.881	1.131	1.080	1.019	1.000
20	1.003	1.181	1.003	1.003	1.003	1.849	1.125	1.075	1.017	1.000
30	1.000	1.138	1.000	1.000	1.000	1.730	1.101	1.058	1.010	1.001
32	1.000	1.129	1.219	1.000	1.008	1.704	1.095	1.054	1.008	1.004
41	1.004	1.091	1.165	1.002	1.001	1.588	1.071	1.037	1.003	1.001
47	1.011	1.068	1.131	1.007	1.000	1.511	1.055	1.026	1.001	1.000
51	1.019	1.053	1.110	1.013	1.001	1.460	1.045	1.019	1.000	1.001
59	1.039	1.030	1.073	1.031	1.009	1.365	1.026	1.008	1.002	1.005
60	1.043	1.027	1.069	1.034	1.013	1.354	1.024	1.007	1.003	1.012
69	1.080	1.009	1.037	1.067	1.002	1.260	1.009	1.000	1.013	1.002
75	1.113	1.002	1.021	1.096	1.000	1.205	1.002	1.001	1.026	1.000
79	1.139	1.000	1.013	1.120	1.001	1.172	1.000	1.003	1.037	1.001
89	1.218	1.004	1.001	1.192	1.012	1.104	1.004	1.017	1.075	1.011
91	1.236	1.005	1.000	1.046	1.009	1.093	1.006	1.022	1.023	1.012
100	1.326	1.019	1.002	1.022	1.001	1.052	1.021	1.048	1.006	1.002
105	1.382	1.029	1.006	1.014	1.000	1.035	1.034	1.066	1.002	1.000
110	1.442	1.041	1.011	1.008	1.001	1.022	1.049	1.087	1.000	1.001
120	1.568	1.067	1.025	1.001	1.007	1.006	1.085	1.136	1.004	1.010
121	1.581	1.070	1.027	1.001	1.003	1.005	1.089	1.141	1.005	1.001
130	1.698	1.094	1.042	1.000	1.000	1.000	1.127	1.191	1.016	1.001
135	1.761	1.107	1.050	1.001	1.000	1.000	1.149	1.000	1.024	1.000
140	1.821	1.119	1.058	1.002	1.000	1.002	1.171	1.002	1.032	1.002
150	1.925	1.140	1.072	1.005	1.002	1.006	1.209	1.006	1.048	1.006
152	1.943	1.144	1.074	1.005	1.000	1.007	1.216	1.007	1.051	1.007
161	2.004	1.156	1.082	1.007	1.000	1.011	1.239	1.011	1.061	1.011
162	2.009	1.156	1.083	1.008	1.000	1.011	1.241	1.011	1.062	1.011
171	2.034	1.161	1.086	1.009	1.000	1.013	1.250	1.013	1.066	1.013
181	2.018	1.158	1.084	1.008	1.000	1.012	1.244	1.012	1.063	1.012
182	2.014	1.157	1.083	1.028	1.001	1.012	1.243	1.012	1.103	1.012
191	1.960	1.147	1.076	1.024	1.000	1.008	1.222	1.008	1.091	1.008

continued on following page

Table 15. Continued

Day	30°S					30°N				
	$R_{d,o}$	$R_{d,\varphi}$	$R_{d,y}$	$R_{d,s}$	$R_{d,m}$	$R_{d,o}$	$R_{d,\varphi}$	$R_{d,y}$	$R_{d,s}$	$R_{d,m}$
198	1.898	1.135	1.068	1.020	1.000	1.005	1.199	1.005	1.077	1.005
201	1.867	1.129	1.064	1.018	1.000	1.003	1.188	1.003	1.071	1.003
211	1.750	1.105	1.049	1.010	1.001	1.000	1.145	1.000	1.048	1.000
213	1.725	1.100	1.046	1.009	1.005	1.000	1.137	1.000	1.043	1.000
222	1.609	1.075	1.030	1.003	1.001	1.003	1.098	1.194	1.023	1.002
228	1.531	1.059	1.021	1.001	1.000	1.009	1.074	1.157	1.013	1.000
232	1.481	1.049	1.015	1.000	1.000	1.015	1.059	1.134	1.007	1.001
242	1.361	1.025	1.004	1.003	1.006	1.041	1.029	1.083	1.000	1.010
244	1.339	1.021	1.003	1.004	1.011	1.048	1.024	1.074	1.000	1.012
253	1.246	1.007	1.000	1.016	1.002	1.087	1.007	1.040	1.005	1.002
258	1.201	1.002	1.002	1.027	1.000	1.115	1.002	1.026	1.012	1.000
263	1.161	1.000	1.008	1.042	1.001	1.148	1.000	1.014	1.023	1.001
273	1.096	1.005	1.028	1.083	1.013	1.230	1.005	1.002	1.053	1.013
274	1.091	1.006	1.030	1.022	1.013	1.239	1.006	1.001	1.046	1.009
283	1.050	1.022	1.060	1.006	1.002	1.331	1.020	1.001	1.022	1.001
288	1.034	1.035	1.081	1.002	1.000	1.387	1.030	1.003	1.013	1.000
293	1.021	1.050	1.105	1.000	1.001	1.447	1.042	1.008	1.007	1.001
303	1.005	1.086	1.159	1.004	1.010	1.573	1.068	1.020	1.001	1.007
305	1.003	1.094	1.170	1.006	1.001	1.599	1.073	1.023	1.001	1.003
314	1.000	1.133	1.225	1.018	1.001	1.716	1.098	1.036	1.000	1.000
318	1.000	1.151	1.000	1.000	1.000	1.766	1.108	1.042	1.001	1.000
324	1.002	1.176	1.002	1.002	1.002	1.837	1.122	1.051	1.002	1.000
334	1.007	1.214	1.007	1.007	1.007	1.937	1.143	1.063	1.005	1.002
335	1.007	1.217	1.007	1.007	1.007	1.946	1.144	1.064	1.005	1.000
344	1.011	1.240	1.011	1.011	1.011	2.006	1.156	1.071	1.007	1.000
354	1.013	1.250	1.013	1.013	1.013	2.034	1.161	1.074	1.008	1.000
365	1.012	1.242	1.012	1.012	1.012	2.013	1.157	1.072	1.008	1.000

Table 16. Energy gain for latitudes 35°N and 35°S

Day	35°S					35°N				
	$R_{d,o}$	$R_{d,\varphi}$	$R_{d,y}$	$R_{d,s}$	$R_{d,m}$	$R_{d,o}$	$R_{d,\varphi}$	$R_{d,y}$	$R_{d,s}$	$R_{d,m}$
1	1.004	1.243	1.004	1.004	1.004	2.319	1.148	1.121	1.027	1.001
10	1.002	1.219	1.002	1.002	1.002	2.239	1.138	1.111	1.023	1.000
17	1.000	1.194	1.000	1.000	1.000	2.152	1.126	1.101	1.018	1.000
20	1.000	1.182	1.000	1.000	1.000	2.110	1.119	1.096	1.016	1.000
30	1.003	1.138	1.192	1.001	1.000	1.954	1.097	1.076	1.009	1.001
32	1.004	1.129	1.181	1.002	1.008	1.921	1.092	1.071	1.008	1.004
41	1.015	1.091	1.133	1.010	1.001	1.771	1.068	1.052	1.003	1.001
47	1.028	1.068	1.104	1.020	1.000	1.672	1.053	1.039	1.001	1.000
51	1.039	1.053	1.086	1.030	1.001	1.608	1.043	1.030	1.000	1.001
59	1.069	1.030	1.054	1.056	1.009	1.488	1.025	1.016	1.002	1.005
60	1.074	1.027	1.050	1.060	1.013	1.474	1.023	1.014	1.003	1.012
69	1.124	1.009	1.024	1.105	1.002	1.354	1.008	1.004	1.012	1.002
75	1.167	1.002	1.012	1.143	1.000	1.284	1.002	1.000	1.024	1.000
79	1.200	1.000	1.006	1.174	1.001	1.243	1.000	1.000	1.035	1.001
89	1.301	1.003	1.000	1.265	1.012	1.155	1.004	1.009	1.073	1.011
91	1.324	1.005	1.000	1.044	1.009	1.141	1.006	1.012	1.036	1.012
100	1.439	1.019	1.006	1.021	1.001	1.086	1.021	1.032	1.014	1.002
105	1.510	1.028	1.012	1.013	1.000	1.063	1.034	1.047	1.006	1.000
110	1.585	1.040	1.019	1.007	1.001	1.044	1.049	1.064	1.002	1.001
120	1.746	1.065	1.037	1.001	1.007	1.018	1.085	1.106	1.001	1.010
121	1.763	1.067	1.039	1.001	1.003	1.016	1.089	1.111	1.001	1.004
130	1.913	1.090	1.056	1.000	1.000	1.005	1.127	1.154	1.008	1.000
135	1.995	1.103	1.066	1.001	1.000	1.001	1.149	1.179	1.014	1.000
140	2.073	1.114	1.074	1.002	1.000	1.000	1.171	1.203	1.021	1.002
150	2.210	1.134	1.090	1.004	1.002	1.001	1.211	1.001	1.035	1.001
152	2.234	1.137	1.092	1.005	1.000	1.001	1.218	1.001	1.037	1.001
161	2.316	1.148	1.101	1.007	1.000	1.003	1.242	1.003	1.046	1.003
162	2.322	1.149	1.102	1.007	1.000	1.004	1.244	1.004	1.047	1.004
171	2.355	1.153	1.105	1.008	1.000	1.005	1.254	1.005	1.051	1.005
181	2.334	1.151	1.103	1.007	1.000	1.004	1.248	1.004	1.048	1.004
182	2.329	1.150	1.102	1.026	1.001	1.004	1.246	1.004	1.084	1.004
191	2.257	1.140	1.095	1.023	1.000	1.002	1.225	1.002	1.073	1.002
198	2.174	1.129	1.086	1.019	1.000	1.000	1.200	1.000	1.061	1.000

continued on following page

Table 16. Continued

Day	35°S					35°N				
	$R_{d,o}$	$R_{d,\varphi}$	$R_{d,y}$	$R_{d,s}$	$R_{d,m}$	$R_{d,o}$	$R_{d,\varphi}$	$R_{d,y}$	$R_{d,s}$	$R_{d,m}$
201	2.133	1.123	1.081	1.017	1.000	1.000	1.189	1.000	1.055	1.000
211	1.981	1.100	1.064	1.010	1.001	1.002	1.145	1.222	1.034	1.000
213	1.948	1.096	1.060	1.009	1.004	1.003	1.137	1.211	1.030	1.009
222	1.798	1.073	1.043	1.003	1.001	1.013	1.098	1.159	1.014	1.002
228	1.699	1.057	1.032	1.001	1.000	1.024	1.074	1.127	1.006	1.000
232	1.634	1.047	1.025	1.000	1.000	1.034	1.059	1.107	1.003	1.001
242	1.483	1.025	1.010	1.002	1.006	1.071	1.029	1.063	1.000	1.010
244	1.454	1.021	1.007	1.004	1.011	1.081	1.024	1.055	1.001	1.012
253	1.337	1.007	1.001	1.015	1.002	1.133	1.007	1.027	1.011	1.002
258	1.280	1.002	1.000	1.026	1.000	1.170	1.002	1.015	1.021	1.000
263	1.229	1.000	1.003	1.041	1.001	1.213	1.000	1.007	1.035	1.001
273	1.145	1.005	1.017	1.081	1.012	1.317	1.005	1.000	1.071	1.013
274	1.138	1.006	1.019	1.034	1.013	1.329	1.006	1.000	1.044	1.009
283	1.084	1.022	1.043	1.013	1.002	1.444	1.019	1.004	1.021	1.001
288	1.061	1.035	1.061	1.006	1.000	1.515	1.029	1.008	1.013	1.000
293	1.043	1.050	1.081	1.001	1.001	1.591	1.041	1.015	1.007	1.001
303	1.017	1.086	1.128	1.001	1.010	1.752	1.066	1.030	1.001	1.006
305	1.014	1.094	1.138	1.002	1.003	1.786	1.071	1.034	1.001	1.003
314	1.003	1.133	1.186	1.010	1.000	1.936	1.094	1.049	1.000	1.000
318	1.001	1.151	1.207	1.015	1.000	2.001	1.103	1.056	1.001	1.000
324	1.000	1.177	1.239	1.023	1.003	2.094	1.117	1.065	1.002	1.000
334	1.001	1.216	1.001	1.001	1.001	2.227	1.136	1.079	1.005	1.002
335	1.001	1.219	1.001	1.001	1.001	2.238	1.137	1.080	1.005	1.000
344	1.004	1.243	1.004	1.004	1.004	2.318	1.148	1.088	1.007	1.000
354	1.005	1.254	1.005	1.005	1.005	2.356	1.153	1.091	1.008	1.000
365	1.004	1.245	1.004	1.004	1.004	2.327	1.150	1.088	1.007	1.000

Table 17. Energy gain for latitudes 40°N and 40°S

Day	40°S					40°N				
	$R_{d,o}$	$R_{d,\varphi}$	$R_{d,y}$	$R_{d,s}$	$R_{d,m}$	$R_{d,o}$	$R_{d,\varphi}$	$R_{d,y}$	$R_{d,s}$	$R_{d,m}$
1	1.000	1.245	1.000	1.000	1.000	2.760	1.141	1.150	1.025	1.001
10	1.000	1.220	1.240	1.000	1.002	2.648	1.131	1.139	1.021	1.000
17	1.001	1.194	1.212	1.000	1.000	2.528	1.119	1.127	1.017	1.000
20	1.003	1.182	1.199	1.001	1.000	2.470	1.114	1.122	1.015	1.000
30	1.012	1.138	1.153	1.007	1.003	2.259	1.092	1.099	1.009	1.001
32	1.015	1.129	1.144	1.009	1.008	2.215	1.088	1.094	1.008	1.004
41	1.034	1.091	1.103	1.025	1.001	2.015	1.066	1.072	1.003	1.001
47	1.053	1.068	1.077	1.041	1.000	1.886	1.051	1.056	1.001	1.000
51	1.069	1.053	1.062	1.055	1.001	1.803	1.042	1.046	1.000	1.001
59	1.109	1.030	1.036	1.090	1.009	1.647	1.025	1.028	1.002	1.005
60	1.115	1.027	1.033	1.096	1.013	1.629	1.023	1.026	1.002	1.011
69	1.181	1.009	1.013	1.154	1.002	1.476	1.008	1.010	1.012	1.002
75	1.236	1.002	1.004	1.203	1.000	1.387	1.002	1.003	1.023	1.000
79	1.279	1.000	1.001	1.242	1.001	1.333	1.000	1.001	1.034	1.001
89	1.408	1.003	1.002	1.357	1.012	1.221	1.004	1.003	1.071	1.011
91	1.438	1.005	1.003	1.042	1.009	1.203	1.006	1.004	1.053	1.012
100	1.584	1.018	1.014	1.020	1.001	1.132	1.021	1.018	1.024	1.002
105	1.675	1.028	1.022	1.012	1.000	1.101	1.034	1.030	1.014	1.000
110	1.773	1.039	1.032	1.007	1.001	1.075	1.049	1.044	1.006	1.001
120	1.982	1.062	1.053	1.001	1.006	1.038	1.085	1.079	1.000	1.010
121	2.004	1.065	1.056	1.001	1.003	1.035	1.089	1.083	1.000	1.008
130	2.203	1.086	1.076	1.000	1.000	1.015	1.127	1.120	1.003	1.001
135	2.313	1.098	1.086	1.001	1.000	1.009	1.149	1.141	1.007	1.000
140	2.419	1.109	1.096	1.002	1.000	1.004	1.171	1.162	1.012	1.001
150	2.608	1.127	1.113	1.004	1.001	1.000	1.211	1.201	1.023	1.005
152	2.640	1.130	1.116	1.005	1.000	1.000	1.218	1.208	1.026	1.000
161	2.755	1.140	1.126	1.007	1.000	1.000	1.244	1.000	1.034	1.000
162	2.764	1.141	1.126	1.007	1.000	1.000	1.246	1.000	1.035	1.000
171	2.810	1.145	1.130	1.007	1.000	1.001	1.256	1.001	1.038	1.001
181	2.781	1.143	1.128	1.007	1.000	1.000	1.250	1.000	1.036	1.000
182	2.774	1.142	1.127	1.025	1.001	1.000	1.248	1.000	1.063	1.000
191	2.672	1.133	1.119	1.021	1.000	1.000	1.225	1.267	1.000	1.002
198	2.558	1.122	1.109	1.018	1.000	1.001	1.201	1.239	1.042	1.000

continued on following page

Table 17. Continued

Day	40°S					40°N				
	$R_{d,o}$	$R_{d,\varphi}$	$R_{d,y}$	$R_{d,s}$	$R_{d,m}$	$R_{d,o}$	$R_{d,\varphi}$	$R_{d,y}$	$R_{d,s}$	$R_{d,m}$
201	2.502	1.117	1.104	1.016	1.000	1.002	1.189	1.225	1.037	1.000
211	2.294	1.096	1.084	1.009	1.001	1.010	1.145	1.177	1.020	1.003
213	2.250	1.091	1.080	1.008	1.004	1.012	1.137	1.167	1.017	1.009
222	2.051	1.070	1.060	1.003	1.001	1.029	1.098	1.123	1.006	1.002
228	1.920	1.055	1.047	1.001	1.000	1.047	1.074	1.095	1.001	1.000
232	1.836	1.046	1.038	1.000	1.000	1.062	1.059	1.078	1.000	1.001
242	1.640	1.024	1.019	1.002	1.005	1.112	1.029	1.042	1.004	1.010
244	1.604	1.020	1.016	1.004	1.010	1.124	1.024	1.036	1.006	1.012
253	1.455	1.007	1.004	1.015	1.002	1.192	1.007	1.014	1.022	1.002
258	1.382	1.002	1.001	1.025	1.000	1.240	1.002	1.006	1.036	1.000
263	1.316	1.000	1.000	1.039	1.001	1.295	1.000	1.002	1.054	1.001
273	1.208	1.005	1.008	1.079	1.012	1.428	1.005	1.001	1.099	1.013
274	1.199	1.006	1.009	1.051	1.013	1.443	1.006	1.002	1.041	1.009
283	1.129	1.022	1.028	1.023	1.002	1.591	1.019	1.010	1.020	1.001
288	1.099	1.035	1.042	1.013	1.000	1.682	1.029	1.018	1.012	1.000
293	1.073	1.050	1.058	1.006	1.001	1.780	1.039	1.026	1.006	1.001
303	1.037	1.086	1.098	1.000	1.010	1.991	1.063	1.046	1.001	1.006
305	1.031	1.094	1.106	1.000	1.008	2.034	1.068	1.050	1.000	1.002
314	1.013	1.133	1.148	1.004	1.001	2.234	1.090	1.068	1.000	1.000
318	1.008	1.151	1.166	1.007	1.000	2.321	1.099	1.076	1.001	1.000
324	1.003	1.177	1.194	1.014	1.001	2.447	1.112	1.087	1.002	1.000
334	1.000	1.216	1.236	1.025	1.005	2.631	1.129	1.102	1.004	1.002
335	1.000	1.220	1.239	1.026	1.000	2.646	1.130	1.103	1.005	1.000
344	1.000	1.245	1.000	1.000	1.000	2.758	1.141	1.112	1.006	1.000
354	1.001	1.256	1.001	1.001	1.001	2.811	1.145	1.116	1.007	1.000
365	1.000	1.247	1.000	1.000	1.000	2.770	1.142	1.113	1.007	1.000

Table 18. Energy gain for latitudes 43.45°N and 43.45°S

Day	43.45°S					43.45°N				
	$R_{d,o}$	$R_{d,\varphi}$	$R_{d,y}$	$R_{d,s}$	$R_{d,m}$	$R_{d,o}$	$R_{d,\varphi}$	$R_{d,y}$	$R_{d,s}$	$R_{d,m}$
1	1.018	1.147	1.124	1.012	1.003	3.182	1.136	1.224	1.024	1.001
10	1.038	1.107	1.088	1.028	1.003	3.036	1.126	1.210	1.020	1.000
17	1.062	1.079	1.063	1.048	1.000	2.881	1.115	1.195	1.017	1.000
20	1.075	1.067	1.053	1.059	1.000	2.807	1.110	1.187	1.015	1.000
30	1.130	1.035	1.025	1.108	1.007	2.538	1.090	1.158	1.009	1.001
32	1.144	1.029	1.020	1.120	1.009	2.483	1.085	1.152	1.007	1.003
41	1.219	1.011	1.005	1.187	1.003	2.235	1.064	1.122	1.003	1.001
47	1.283	1.003	1.001	1.244	1.000	2.076	1.050	1.101	1.001	1.000
51	1.333	1.001	1.000	1.288	1.001	1.975	1.041	1.087	1.000	1.001
59	1.450	1.001	1.004	1.392	1.007	1.786	1.025	1.061	1.002	1.005
60	1.466	1.002	1.005	1.407	1.009	1.764	1.023	1.057	1.002	1.011
69	1.629	1.012	1.019	1.027	1.003	1.581	1.008	1.032	1.011	1.002
75	1.753	1.022	1.031	1.016	1.000	1.475	1.002	1.018	1.022	1.000
79	1.843	1.029	1.041	1.010	1.000	1.411	1.000	1.011	1.033	1.001
89	2.089	1.051	1.066	1.002	1.003	1.277	1.004	1.001	1.069	1.011
91	2.141	1.056	1.072	1.002	1.005	1.255	1.006	1.000	1.091	1.012
100	2.386	1.077	1.096	1.000	1.001	1.170	1.021	1.004	1.051	1.002
105	2.525	1.089	1.109	1.000	1.000	1.133	1.033	1.009	1.034	1.000
110	2.662	1.099	1.121	1.001	1.000	1.102	1.048	1.018	1.021	1.001
120	2.917	1.118	1.142	1.003	1.001	1.056	1.084	1.041	1.005	1.010
121	2.940	1.119	1.144	1.004	1.001	1.052	1.088	1.044	1.004	1.009
130	3.118	1.131	1.157	1.005	1.000	1.027	1.127	1.070	1.000	1.001
135	3.188	1.136	1.162	1.006	1.000	1.017	1.149	1.086	1.000	1.000
140	3.233	1.139	1.165	1.007	1.000	1.010	1.171	1.102	1.002	1.001
150	3.239	1.139	1.166	1.007	1.000	1.003	1.211	1.133	1.008	1.005
152	3.227	1.138	1.165	1.007	1.000	1.002	1.218	1.138	1.010	1.001
161	3.120	1.132	1.157	1.022	1.000	1.000	1.243	1.158	1.016	1.000
162	3.103	1.131	1.156	1.021	1.000	1.000	1.245	1.159	1.016	1.000
171	2.920	1.118	1.142	1.017	1.000	1.000	1.256	1.167	1.000	1.000
181	2.665	1.099	1.121	1.011	1.001	1.000	1.249	1.252	1.017	1.000
182	2.638	1.097	1.119	1.010	1.001	1.000	1.247	1.250	1.051	1.003
191	2.389	1.077	1.096	1.005	1.002	1.001	1.225	1.228	1.042	1.001
198	2.198	1.061	1.077	1.002	1.000	1.004	1.200	1.203	1.032	1.000

continued on following page

Table 18. Continued

Day	43.45°S					43.45°N				
	$R_{d,o}$	$R_{d,\varphi}$	$R_{d,y}$	$R_{d,s}$	$R_{d,m}$	$R_{d,o}$	$R_{d,\varphi}$	$R_{d,y}$	$R_{d,s}$	$R_{d,m}$
201	2.118	1.054	1.069	1.001	1.000	1.006	1.188	1.190	1.027	1.000
211	1.869	1.032	1.043	1.001	1.002	1.019	1.145	1.147	1.013	1.005
213	1.823	1.028	1.038	1.001	1.004	1.022	1.136	1.138	1.011	1.009
222	1.631	1.012	1.019	1.008	1.005	1.045	1.097	1.099	1.003	1.002
228	1.519	1.005	1.009	1.018	1.001	1.067	1.073	1.075	1.000	1.000
232	1.452	1.001	1.004	1.026	1.000	1.086	1.059	1.060	1.000	1.001
242	1.309	1.002	1.000	1.059	1.006	1.146	1.028	1.029	1.008	1.010
244	1.285	1.003	1.001	1.068	1.009	1.161	1.024	1.024	1.011	1.012
253	1.193	1.016	1.009	1.055	1.004	1.243	1.007	1.008	1.032	1.002
258	1.152	1.027	1.018	1.038	1.001	1.299	1.002	1.002	1.048	1.000
263	1.118	1.040	1.029	1.024	1.000	1.365	1.000	1.000	1.068	1.001
273	1.066	1.074	1.059	1.007	1.006	1.524	1.005	1.004	1.119	1.012
274	1.062	1.078	1.062	1.005	1.007	1.542	1.006	1.005	1.039	1.008
283	1.033	1.115	1.095	1.000	1.003	1.718	1.019	1.018	1.019	1.001
288	1.022	1.137	1.115	1.000	1.001	1.829	1.028	1.027	1.011	1.000
293	1.014	1.159	1.135	1.002	1.000	1.947	1.039	1.038	1.006	1.001
303	1.004	1.201	1.173	1.008	1.003	2.205	1.062	1.060	1.001	1.006
305	1.003	1.208	1.180	1.010	1.004	2.259	1.066	1.065	1.000	1.002
314	1.000	1.237	1.206	1.017	1.000	2.507	1.087	1.086	1.000	1.000
318	1.000	1.246	1.214	1.019	1.000	2.618	1.096	1.094	1.001	1.000
324	1.000	1.254	1.222	1.021	1.000	2.778	1.108	1.106	1.002	1.000
334	1.000	1.253	1.220	1.021	1.000	3.014	1.125	1.123	1.004	1.001
335	1.000	1.251	1.219	1.020	1.000	3.034	1.126	1.124	1.004	1.000
344	1.028	1.124	1.104	1.020	1.007	3.180	1.135	1.133	1.006	1.000
354	1.058	1.082	1.066	1.045	1.000	3.249	1.140	1.138	1.007	1.000
365	1.111	1.044	1.032	1.091	1.004	3.196	1.136	1.134	1.006	1.000

APPENDIX C

Energy Gain of Long Term Tracking in the High Latitude Zone

Tables 19-24 show energy gains in the high latitude zone.

Table 19. Energy gain for latitudes 45°N and 45°S

Day	45°S					45°N				
	$R_{d,o}$	$R_{d,\varphi}$	$R_{d,y}$	$R_{d,s}$	$R_{d,m}$	$R_{d,o}$	$R_{d,\varphi}$	$R_{d,y}$	$R_{d,s}$	$R_{d,m}$
1	1.001	1.245	1.200	1.000	1.003	3.418	1.132	1.221	1.023	1.001
10	1.003	1.220	1.179	1.001	1.001	3.252	1.123	1.207	1.020	1.000
17	1.008	1.194	1.156	1.004	1.000	3.075	1.113	1.193	1.016	1.000
20	1.011	1.182	1.145	1.006	1.000	2.991	1.108	1.185	1.015	1.000
30	1.027	1.138	1.107	1.019	1.005	2.690	1.088	1.157	1.009	1.001
32	1.032	1.129	1.100	1.022	1.008	2.628	1.083	1.150	1.007	1.003
41	1.060	1.091	1.067	1.046	1.001	2.352	1.063	1.121	1.003	1.001
47	1.086	1.068	1.047	1.068	1.000	2.176	1.049	1.100	1.001	1.000
51	1.107	1.053	1.036	1.087	1.001	2.065	1.040	1.087	1.000	1.001
59	1.161	1.030	1.017	1.134	1.009	1.859	1.024	1.061	1.002	1.005
60	1.169	1.027	1.015	1.141	1.013	1.835	1.022	1.057	1.002	1.011
69	1.254	1.009	1.003	1.215	1.002	1.635	1.008	1.032	1.011	1.002
75	1.325	1.002	1.000	1.278	1.000	1.519	1.002	1.018	1.022	1.000
79	1.380	1.000	1.001	1.326	1.001	1.450	1.000	1.011	1.032	1.001
89	1.547	1.003	1.010	1.473	1.011	1.306	1.004	1.001	1.068	1.011
91	1.585	1.005	1.013	1.039	1.008	1.282	1.006	1.000	1.092	1.012
100	1.776	1.018	1.031	1.019	1.001	1.190	1.021	1.003	1.051	1.002
105	1.896	1.027	1.043	1.011	1.000	1.150	1.034	1.009	1.034	1.000
110	2.025	1.037	1.056	1.006	1.001	1.116	1.049	1.017	1.021	1.001
120	2.307	1.060	1.083	1.001	1.006	1.066	1.085	1.041	1.005	1.010
121	2.337	1.062	1.086	1.001	1.003	1.062	1.089	1.043	1.004	1.008
130	2.612	1.082	1.111	1.000	1.000	1.033	1.127	1.070	1.000	1.001
135	2.767	1.093	1.123	1.001	1.000	1.022	1.149	1.086	1.000	1.000
140	2.918	1.103	1.135	1.002	1.000	1.014	1.171	1.102	1.002	1.001
150	3.193	1.120	1.155	1.004	1.001	1.004	1.211	1.132	1.008	1.005
152	3.240	1.123	1.159	1.004	1.000	1.003	1.218	1.138	1.010	1.001
161	3.411	1.132	1.170	1.006	1.000	1.001	1.244	1.157	1.015	1.000

continued on following page

Table 19. Continued

Day	45°S					45°N				
	$R_{d,o}$	$R_{d,\varphi}$	$R_{d,y}$	$R_{d,s}$	$R_{d,m}$	$R_{d,o}$	$R_{d,\varphi}$	$R_{d,y}$	$R_{d,s}$	$R_{d,m}$
162	3.425	1.133	1.171	1.006	1.000	1.001	1.246	1.159	1.016	1.000
171	3.494	1.137	1.175	1.007	1.000	1.000	1.257	1.167	1.019	1.000
181	3.450	1.134	1.172	1.006	1.000	1.000	1.250	1.251	1.017	1.000
182	3.439	1.134	1.172	1.023	1.001	1.001	1.248	1.250	1.051	1.003
191	3.287	1.125	1.162	1.020	1.000	1.002	1.225	1.227	1.042	1.001
198	3.119	1.115	1.150	1.017	1.000	1.006	1.201	1.202	1.032	1.000
201	3.037	1.111	1.144	1.015	1.000	1.009	1.189	1.190	1.028	1.000
211	2.740	1.091	1.121	1.009	1.001	1.024	1.145	1.146	1.014	1.005
213	2.678	1.087	1.116	1.008	1.004	1.028	1.137	1.138	1.011	1.009
222	2.401	1.067	1.092	1.003	1.001	1.054	1.098	1.098	1.003	1.002
228	2.223	1.053	1.075	1.001	1.000	1.078	1.074	1.074	1.000	1.000
232	2.110	1.044	1.064	1.000	1.000	1.098	1.059	1.060	1.000	1.001
242	1.850	1.023	1.038	1.002	1.005	1.164	1.029	1.029	1.008	1.010
244	1.802	1.020	1.033	1.003	1.010	1.180	1.024	1.024	1.011	1.012
253	1.607	1.007	1.015	1.014	1.002	1.269	1.007	1.007	1.031	1.002
258	1.513	1.002	1.007	1.023	1.000	1.330	1.002	1.002	1.048	1.000
263	1.428	1.000	1.002	1.037	1.001	1.401	1.000	1.000	1.068	1.001
273	1.289	1.005	1.001	1.076	1.012	1.573	1.005	1.004	1.119	1.013
274	1.277	1.006	1.001	1.091	1.013	1.592	1.006	1.006	1.039	1.008
283	1.187	1.022	1.011	1.050	1.002	1.785	1.018	1.018	1.018	1.001
288	1.147	1.035	1.021	1.034	1.000	1.905	1.028	1.027	1.011	1.000
293	1.114	1.050	1.033	1.021	1.001	2.035	1.038	1.038	1.006	1.001
303	1.064	1.086	1.063	1.005	1.010	2.318	1.061	1.060	1.001	1.006
305	1.056	1.094	1.070	1.003	1.008	2.378	1.065	1.064	1.000	1.002
314	1.030	1.133	1.103	1.000	1.001	2.655	1.085	1.085	1.000	1.000
318	1.021	1.151	1.118	1.000	1.000	2.778	1.094	1.093	1.001	1.000
324	1.012	1.177	1.141	1.003	1.001	2.959	1.106	1.105	1.002	1.000
334	1.004	1.216	1.175	1.009	1.005	3.226	1.122	1.121	1.004	1.001
335	1.003	1.220	1.178	1.010	1.001	3.249	1.123	1.122	1.004	1.000
344	1.001	1.245	1.200	1.015	1.000	3.416	1.132	1.131	1.006	1.000
354	1.000	1.257	1.211	1.018	1.000	3.495	1.137	1.135	1.007	1.000
365	1.001	1.247	1.203	1.016	1.000	3.434	1.133	1.132	1.006	1.000

Table 20. Energy gain for latitudes 50°N and 50°S

Day	50°S					50°N				
	$R_{d,o}$	$R_{d,\varphi}$	$R_{d,y}$	$R_{d,s}$	$R_{d,m}$	$R_{d,o}$	$R_{d,\varphi}$	$R_{d,y}$	$R_{d,s}$	$R_{d,m}$
1	1.005	1.245	1.198	1.002	1.003	4.491	1.124	1.211	1.022	1.001
10	1.011	1.220	1.176	1.006	1.001	4.217	1.115	1.199	1.019	1.000
17	1.019	1.194	1.154	1.012	1.000	3.933	1.106	1.185	1.015	1.000
20	1.024	1.182	1.143	1.016	1.000	3.800	1.101	1.178	1.014	1.000
30	1.049	1.138	1.105	1.035	1.005	3.336	1.083	1.152	1.008	1.001
32	1.055	1.129	1.098	1.040	1.008	3.243	1.079	1.146	1.007	1.003
41	1.094	1.091	1.065	1.072	1.001	2.839	1.060	1.118	1.003	1.001
47	1.129	1.068	1.046	1.102	1.000	2.588	1.047	1.099	1.001	1.000
51	1.157	1.053	1.035	1.126	1.001	2.432	1.039	1.086	1.000	1.000
59	1.227	1.030	1.016	1.186	1.009	2.148	1.023	1.061	1.001	1.004
60	1.237	1.027	1.015	1.194	1.013	2.115	1.022	1.058	1.002	1.011
69	1.347	1.009	1.003	1.288	1.002	1.847	1.008	1.033	1.010	1.002
75	1.439	1.002	1.000	1.367	1.000	1.694	1.002	1.019	1.020	1.000
79	1.511	1.000	1.001	1.428	1.001	1.603	1.000	1.011	1.030	1.001
89	1.730	1.003	1.010	1.613	1.011	1.414	1.004	1.001	1.065	1.011
91	1.781	1.005	1.013	1.037	1.008	1.383	1.006	1.000	1.092	1.012
100	2.036	1.017	1.031	1.017	1.001	1.264	1.021	1.003	1.051	1.002
105	2.198	1.026	1.043	1.010	1.000	1.212	1.034	1.009	1.034	1.000
110	2.376	1.036	1.055	1.005	1.001	1.169	1.049	1.017	1.021	1.001
120	2.774	1.057	1.081	1.001	1.005	1.102	1.085	1.040	1.005	1.010
121	2.817	1.059	1.084	1.001	1.002	1.096	1.089	1.042	1.004	1.008
122	2.860	1.061	1.087	1.000	1.002	1.091	1.093	1.045	1.004	1.007
130	3.220	1.078	1.107	1.000	1.000	1.057	1.127	1.069	1.000	1.001
135	3.453	1.088	1.119	1.001	1.000	1.041	1.149	1.084	1.000	1.000
140	3.686	1.097	1.130	1.002	1.000	1.029	1.171	1.100	1.002	1.001
150	4.121	1.112	1.148	1.004	1.001	1.013	1.211	1.130	1.008	1.005
152	4.198	1.115	1.151	1.004	1.000	1.011	1.218	1.136	1.010	1.001
161	4.479	1.123	1.162	1.006	1.000	1.005	1.244	1.155	1.015	1.000
162	4.502	1.124	1.163	1.006	1.000	1.005	1.246	1.157	1.016	1.000
171	4.619	1.127	1.167	1.006	1.000	1.003	1.257	1.165	1.019	1.000
181	4.545	1.125	1.164	1.006	1.000	1.004	1.250	1.247	1.017	1.000
182	4.526	1.125	1.163	1.021	1.000	1.005	1.248	1.245	1.051	1.003

continued on following page

Table 20. Continued

Day	50°S					50°N				
	$R_{d,o}$	$R_{d,\varphi}$	$R_{d,y}$	$R_{d,s}$	$R_{d,m}$	$R_{d,o}$	$R_{d,\varphi}$	$R_{d,y}$	$R_{d,s}$	$R_{d,m}$
191	4.275	1.117	1.154	1.019	1.000	1.009	1.225	1.223	1.042	1.001
198	4.003	1.108	1.144	1.016	1.000	1.017	1.201	1.198	1.032	1.000
201	3.873	1.104	1.138	1.014	1.000	1.021	1.189	1.186	1.028	1.000
211	3.412	1.086	1.117	1.009	1.001	1.044	1.145	1.143	1.014	1.005
213	3.319	1.082	1.112	1.007	1.004	1.050	1.137	1.135	1.011	1.009
222	2.909	1.063	1.090	1.003	1.001	1.086	1.098	1.096	1.003	1.002
228	2.654	1.051	1.074	1.001	1.000	1.118	1.074	1.072	1.000	1.000
232	2.494	1.042	1.063	1.000	1.000	1.145	1.059	1.058	1.000	1.001
242	2.136	1.023	1.038	1.002	1.005	1.231	1.029	1.028	1.008	1.010
244	2.071	1.019	1.034	1.003	1.010	1.252	1.024	1.023	1.011	1.012
253	1.810	1.006	1.015	1.013	1.002	1.366	1.007	1.007	1.031	1.002
258	1.685	1.002	1.008	1.022	1.000	1.445	1.002	1.002	1.048	1.000
263	1.574	1.000	1.003	1.035	1.001	1.538	1.000	1.000	1.068	1.001
273	1.393	1.005	1.001	1.073	1.012	1.765	1.005	1.005	1.118	1.012
274	1.377	1.006	1.001	1.091	1.013	1.791	1.006	1.006	1.036	1.008
283	1.260	1.022	1.011	1.050	1.002	2.048	1.018	1.019	1.017	1.001
288	1.209	1.035	1.020	1.034	1.000	2.211	1.027	1.028	1.010	1.000
293	1.166	1.050	1.032	1.021	1.001	2.390	1.037	1.038	1.005	1.001
303	1.100	1.086	1.062	1.005	1.010	2.790	1.058	1.059	1.001	1.005
305	1.089	1.094	1.068	1.003	1.008	2.877	1.062	1.063	1.000	1.002
314	1.052	1.133	1.101	1.000	1.001	3.284	1.081	1.082	1.000	1.000
318	1.040	1.151	1.116	1.000	1.000	3.471	1.088	1.090	1.001	1.000
324	1.026	1.177	1.139	1.003	1.001	3.749	1.099	1.101	1.002	1.000
334	1.012	1.216	1.173	1.009	1.005	4.175	1.114	1.116	1.004	1.001
335	1.011	1.220	1.176	1.010	1.001	4.212	1.115	1.117	1.004	1.000
344	1.005	1.245	1.198	1.015	1.000	4.488	1.124	1.126	1.005	1.000
354	1.003	1.257	1.208	1.018	1.000	4.620	1.127	1.129	1.006	1.000
365	1.005	1.247	1.200	1.016	1.000	4.518	1.124	1.126	1.006	1.000

Table 21. Energy gain for latitudes 55°N and 55°S

Day	55°S					55°N				
	$R_{d,o}$	$R_{d,\varphi}$	$R_{d,y}$	$R_{d,s}$	$R_{d,m}$	$R_{d,o}$	$R_{d,\varphi}$	$R_{d,y}$	$R_{d,s}$	$R_{d,m}$
1	1.012	1.245	1.194	1.007	1.003	6.509	1.114	1.200	1.020	1.001
10	1.022	1.220	1.173	1.014	1.001	5.972	1.107	1.189	1.017	1.000
17	1.034	1.194	1.150	1.023	1.000	5.441	1.098	1.177	1.014	1.000
20	1.042	1.182	1.140	1.029	1.000	5.200	1.094	1.170	1.012	1.000
30	1.076	1.138	1.103	1.056	1.005	4.396	1.077	1.146	1.007	1.001
32	1.085	1.129	1.095	1.064	1.008	4.241	1.074	1.141	1.006	1.003
41	1.136	1.091	1.063	1.106	1.001	3.592	1.057	1.115	1.002	1.000
47	1.182	1.068	1.045	1.144	1.000	3.208	1.045	1.097	1.001	1.000
51	1.219	1.053	1.033	1.174	1.001	2.975	1.037	1.084	1.000	1.000
59	1.310	1.030	1.015	1.250	1.009	2.562	1.022	1.060	1.001	1.004
60	1.324	1.027	1.014	1.261	1.013	2.515	1.021	1.057	1.002	1.010
69	1.466	1.009	1.002	1.379	1.002	2.140	1.008	1.033	1.009	1.002
75	1.588	1.002	1.000	1.480	1.000	1.930	1.002	1.019	1.019	1.000
79	1.684	1.000	1.001	1.558	1.001	1.807	1.000	1.012	1.028	1.001
89	1.980	1.003	1.011	1.796	1.011	1.555	1.004	1.001	1.063	1.011
91	2.049	1.005	1.014	1.034	1.007	1.514	1.006	1.000	1.092	1.012
100	2.403	1.017	1.032	1.016	1.001	1.359	1.021	1.003	1.051	1.002
105	2.634	1.025	1.043	1.009	1.000	1.291	1.034	1.008	1.034	1.000
110	2.892	1.034	1.055	1.005	1.001	1.234	1.049	1.016	1.021	1.001
120	3.492	1.054	1.080	1.001	1.005	1.147	1.085	1.039	1.005	1.010
121	3.558	1.056	1.082	1.001	1.002	1.140	1.089	1.041	1.004	1.008
130	4.203	1.073	1.104	1.000	1.000	1.087	1.127	1.067	1.000	1.001
135	4.593	1.082	1.115	1.001	1.000	1.066	1.149	1.083	1.000	1.000
140	4.997	1.090	1.125	1.001	1.000	1.049	1.171	1.099	1.002	1.001
150	5.790	1.104	1.142	1.003	1.001	1.026	1.211	1.128	1.008	1.005
152	5.937	1.106	1.144	1.004	1.000	1.022	1.218	1.134	1.010	1.001
161	6.485	1.114	1.154	1.005	1.000	1.013	1.244	1.153	1.015	1.000
162	6.531	1.114	1.155	1.005	1.000	1.012	1.246	1.154	1.016	1.000
171	6.768	1.117	1.158	1.006	1.000	1.009	1.257	1.162	1.019	1.000
181	6.617	1.116	1.156	1.005	1.000	1.011	1.250	1.243	1.017	1.000
182	6.579	1.115	1.155	1.019	1.000	1.011	1.248	1.241	1.051	1.003
191	6.084	1.108	1.147	1.017	1.000	1.019	1.225	1.219	1.042	1.001
198	5.569	1.100	1.137	1.014	1.000	1.031	1.201	1.194	1.032	1.000
201	5.331	1.096	1.132	1.013	1.000	1.038	1.189	1.182	1.028	1.000

continued on following page

Table 21. Continued

Day	55°S					55°N				
	$R_{d,o}$	$R_{d,\varphi}$	$R_{d,y}$	$R_{d,s}$	$R_{d,m}$	$R_{d,o}$	$R_{d,\varphi}$	$R_{d,y}$	$R_{d,s}$	$R_{d,m}$
211	4.524	1.080	1.113	1.008	1.001	1.069	1.145	1.140	1.014	1.005
213	4.367	1.077	1.109	1.007	1.003	1.078	1.137	1.132	1.011	1.009
222	3.703	1.060	1.087	1.003	1.001	1.126	1.098	1.093	1.003	1.002
228	3.308	1.048	1.073	1.001	1.000	1.169	1.074	1.070	1.000	1.000
232	3.067	1.040	1.063	1.000	1.000	1.203	1.059	1.056	1.000	1.001
242	2.544	1.022	1.039	1.002	1.004	1.315	1.029	1.026	1.008	1.010
244	2.453	1.019	1.034	1.003	1.009	1.343	1.024	1.022	1.011	1.012
253	2.089	1.006	1.016	1.012	1.002	1.492	1.007	1.006	1.031	1.002
258	1.918	1.002	1.008	1.021	1.000	1.596	1.002	1.002	1.048	1.000
263	1.768	1.000	1.003	1.033	1.001	1.720	1.000	1.000	1.068	1.001
273	1.526	1.005	1.001	1.070	1.012	2.027	1.005	1.005	1.118	1.012
274	1.506	1.006	1.001	1.091	1.013	2.062	1.005	1.006	1.033	1.007
283	1.353	1.022	1.010	1.050	1.002	2.420	1.017	1.019	1.015	1.001
288	1.287	1.035	1.019	1.034	1.000	2.653	1.026	1.028	1.009	1.000
293	1.230	1.050	1.031	1.021	1.001	2.912	1.035	1.037	1.005	1.001
303	1.144	1.086	1.060	1.005	1.010	3.517	1.054	1.057	1.001	1.005
305	1.131	1.094	1.066	1.003	1.008	3.651	1.058	1.062	1.000	1.002
314	1.081	1.133	1.099	1.000	1.001	4.309	1.075	1.079	1.000	1.000
318	1.064	1.151	1.114	1.000	1.000	4.623	1.082	1.086	1.001	1.000
324	1.045	1.177	1.136	1.003	1.001	5.109	1.092	1.096	1.002	1.000
334	1.023	1.216	1.170	1.009	1.005	5.892	1.105	1.110	1.003	1.001
335	1.022	1.220	1.172	1.010	1.001	5.963	1.107	1.111	1.004	1.000
344	1.012	1.245	1.194	1.015	1.000	6.503	1.114	1.119	1.005	1.000
354	1.009	1.257	1.204	1.018	1.000	6.771	1.117	1.122	1.006	1.000
365	1.012	1.247	1.196	1.016	1.000	6.564	1.115	1.120	1.005	1.000

Table 22. Energy gain for latitudes 60°N and 60°S

Day	60°S					60°N				
	$R_{d,o}$	$R_{d,\varphi}$	$R_{d,y}$	$R_{d,s}$	$R_{d,m}$	$R_{d,o}$	$R_{d,\varphi}$	$R_{d,y}$	$R_{d,s}$	$R_{d,m}$
1	1.020	1.245	1.191	1.014	1.003	11.533	1.104	1.189	1.018	1.000
10	1.033	1.220	1.169	1.024	1.001	10.037	1.097	1.179	1.015	1.000
17	1.052	1.194	1.147	1.038	1.000	8.698	1.090	1.168	1.013	1.000
20	1.062	1.182	1.137	1.046	1.000	8.131	1.086	1.162	1.011	1.000
30	1.108	1.138	1.100	1.082	1.005	6.404	1.071	1.140	1.007	1.001
32	1.120	1.129	1.093	1.092	1.008	6.097	1.068	1.135	1.006	1.003
41	1.188	1.091	1.062	1.146	1.001	4.887	1.053	1.112	1.002	1.000
47	1.248	1.068	1.043	1.194	1.000	4.225	1.042	1.095	1.001	1.000
51	1.296	1.053	1.032	1.233	1.001	3.842	1.035	1.083	1.000	1.000
59	1.415	1.030	1.015	1.329	1.009	3.192	1.021	1.060	1.001	1.003
60	1.433	1.027	1.013	1.343	1.013	3.121	1.020	1.058	1.001	1.010
69	1.622	1.009	1.002	1.494	1.002	2.563	1.008	1.034	1.008	1.002
75	1.786	1.002	1.000	1.623	1.000	2.262	1.002	1.020	1.017	1.000
79	1.917	1.000	1.001	1.725	1.001	2.089	1.000	1.013	1.026	1.001
89	2.333	1.003	1.012	2.038	1.011	1.742	1.004	1.001	1.059	1.011
91	2.431	1.005	1.015	1.031	1.007	1.686	1.006	1.000	1.092	1.012
100	2.952	1.016	1.032	1.014	1.001	1.479	1.021	1.002	1.051	1.002
105	3.303	1.024	1.043	1.008	1.000	1.390	1.034	1.008	1.034	1.000
110	3.709	1.032	1.054	1.004	1.001	1.316	1.049	1.015	1.021	1.001
120	4.711	1.050	1.077	1.001	1.005	1.202	1.085	1.037	1.005	1.010
121	4.827	1.052	1.080	1.000	1.002	1.192	1.089	1.040	1.004	1.008
130	6.022	1.067	1.099	1.000	1.000	1.123	1.127	1.065	1.000	1.001
135	6.807	1.075	1.109	1.001	1.000	1.095	1.149	1.081	1.000	1.000
140	7.672	1.082	1.118	1.001	1.000	1.071	1.171	1.096	1.002	1.001
150	9.563	1.095	1.134	1.003	1.001	1.039	1.211	1.126	1.008	1.005
152	9.943	1.097	1.136	1.003	1.000	1.034	1.218	1.131	1.010	1.001
161	11.464	1.103	1.144	1.004	1.000	1.020	1.244	1.150	1.015	1.000
162	11.599	1.104	1.145	1.004	1.000	1.019	1.246	1.151	1.016	1.000
171	12.315	1.106	1.148	1.005	1.000	1.014	1.257	1.159	1.019	1.000
181	11.854	1.105	1.146	1.005	1.000	1.017	1.250	1.237	1.017	1.000
182	11.740	1.104	1.146	1.017	1.000	1.018	1.248	1.235	1.051	1.003
191	10.337	1.098	1.138	1.015	1.000	1.030	1.225	1.213	1.042	1.001
198	9.010	1.092	1.130	1.013	1.000	1.047	1.201	1.189	1.032	1.000

continued on following page

Table 22. Continued

Day	60°S					60°N				
	$R_{d,o}$	$R_{d,\varphi}$	$R_{d,y}$	$R_{d,s}$	$R_{d,m}$	$R_{d,o}$	$R_{d,\varphi}$	$R_{d,y}$	$R_{d,s}$	$R_{d,m}$
201	8.437	1.088	1.125	1.012	1.000	1.056	1.189	1.178	1.028	1.000
211	6.663	1.074	1.108	1.007	1.001	1.099	1.145	1.136	1.014	1.005
213	6.345	1.071	1.104	1.006	1.003	1.110	1.137	1.128	1.011	1.009
222	5.085	1.055	1.084	1.003	1.001	1.174	1.098	1.090	1.003	1.002
228	4.394	1.045	1.071	1.001	1.000	1.230	1.074	1.067	1.000	1.000
232	3.992	1.038	1.061	1.000	1.000	1.275	1.059	1.054	1.000	1.001
242	3.165	1.021	1.039	1.001	1.004	1.422	1.029	1.025	1.008	1.010
244	3.026	1.018	1.034	1.002	1.009	1.458	1.024	1.020	1.011	1.012
253	2.489	1.006	1.017	1.010	1.002	1.656	1.007	1.005	1.031	1.002
258	2.245	1.002	1.009	1.019	1.000	1.798	1.002	1.001	1.048	1.000
263	2.034	1.000	1.003	1.030	1.001	1.967	1.000	1.000	1.068	1.001
273	1.703	1.005	1.000	1.066	1.012	2.400	1.004	1.006	1.117	1.012
274	1.676	1.006	1.001	1.091	1.013	2.451	1.005	1.007	1.030	1.007
283	1.472	1.022	1.009	1.050	1.002	2.976	1.017	1.020	1.013	1.001
288	1.384	1.035	1.018	1.034	1.000	3.332	1.024	1.028	1.008	1.000
293	1.311	1.050	1.029	1.021	1.001	3.741	1.033	1.037	1.004	1.001
303	1.198	1.086	1.058	1.005	1.010	4.754	1.051	1.056	1.001	1.004
305	1.180	1.094	1.064	1.003	1.008	4.993	1.054	1.060	1.000	1.002
314	1.115	1.133	1.096	1.000	1.001	6.229	1.069	1.076	1.000	1.000
318	1.093	1.151	1.111	1.000	1.000	6.869	1.076	1.083	1.001	1.000
324	1.066	1.177	1.133	1.003	1.001	7.923	1.084	1.092	1.001	1.000
334	1.036	1.216	1.166	1.009	1.005	9.825	1.096	1.104	1.003	1.001
335	1.034	1.220	1.169	1.010	1.001	10.014	1.097	1.105	1.003	1.000
344	1.020	1.245	1.190	1.015	1.000	11.516	1.104	1.112	1.004	1.000
354	1.014	1.257	1.201	1.018	1.000	12.322	1.107	1.115	1.005	1.000
365	1.018	1.247	1.193	1.016	1.000	11.695	1.104	1.113	1.004	1.000

Table 23. Energy gain for latitudes 65°N and 65°S

Day	65°S					65°N				
	$R_{d,o}$	$R_{d,\varphi}$	$R_{d,y}$	$R_{d,s}$	$R_{d,m}$	$R_{d,o}$	$R_{d,\varphi}$	$R_{d,y}$	$R_{d,s}$	$R_{d,m}$
1	1.020	1.245	1.186	1.020	1.003	42.597	1.092	1.175	1.015	1.000
10	1.040	1.220	1.165	1.034	1.001	28.342	1.086	1.166	1.013	1.000
17	1.066	1.194	1.143	1.053	1.000	20.274	1.080	1.157	1.011	1.000
20	1.079	1.182	1.133	1.064	1.000	17.643	1.077	1.152	1.010	1.000
30	1.142	1.138	1.097	1.112	1.005	11.463	1.064	1.133	1.006	1.001
32	1.158	1.129	1.090	1.124	1.008	10.582	1.061	1.128	1.005	1.002
41	1.247	1.091	1.059	1.193	1.001	7.560	1.048	1.107	1.002	1.000
47	1.326	1.068	1.041	1.255	1.000	6.161	1.039	1.092	1.001	1.000
51	1.389	1.053	1.030	1.304	1.001	5.415	1.032	1.081	1.000	1.000
59	1.548	1.030	1.013	1.427	1.009	4.249	1.020	1.060	1.001	1.003
60	1.571	1.027	1.012	1.445	1.013	4.128	1.019	1.057	1.001	1.009
69	1.828	1.009	1.002	1.640	1.002	3.219	1.007	1.035	1.007	1.002
75	2.058	1.002	1.000	1.809	1.000	2.756	1.002	1.021	1.015	1.000
79	2.246	1.000	1.001	1.945	1.001	2.497	1.000	1.014	1.024	1.001
89	2.863	1.003	1.013	2.371	1.011	1.995	1.004	1.001	1.055	1.010
91	3.014	1.005	1.016	1.027	1.006	1.917	1.006	1.001	1.092	1.012
100	3.844	1.015	1.033	1.012	1.001	1.634	1.021	1.002	1.051	1.002
105	4.440	1.022	1.043	1.007	1.000	1.514	1.034	1.007	1.034	1.000
110	5.166	1.030	1.054	1.004	1.001	1.415	1.049	1.014	1.021	1.001
120	7.173	1.046	1.075	1.000	1.004	1.265	1.085	1.036	1.005	1.010
121	7.428	1.047	1.077	1.000	1.002	1.253	1.089	1.038	1.004	1.008
130	10.375	1.061	1.094	1.000	1.000	1.162	1.127	1.063	1.000	1.001
135	12.706	1.068	1.103	1.000	1.000	1.124	1.149	1.079	1.000	1.000
140	15.763	1.074	1.111	1.001	1.000	1.093	1.171	1.094	1.002	1.001
150	25.124	1.084	1.125	1.002	1.001	1.048	1.211	1.123	1.008	1.005
152	27.668	1.086	1.127	1.003	1.000	1.042	1.218	1.128	1.010	1.001

continued on following page

Table 23. Continued

Day	65°S					65°N				
	$R_{d,o}$	$R_{d,\varphi}$	$R_{d,y}$	$R_{d,s}$	$R_{d,m}$	$R_{d,o}$	$R_{d,\varphi}$	$R_{d,y}$	$R_{d,s}$	$R_{d,m}$
161	41.752	1.092	1.134	1.004	1.000	1.021	1.244	1.147	1.015	1.000
162	43.436	1.092	1.134	1.004	1.000	1.019	1.246	1.148	1.016	1.000
171	54.193	1.094	1.137	1.004	1.000	1.012	1.257	1.156	1.019	1.000
181	46.883	1.093	1.136	1.004	1.000	1.016	1.250	1.230	1.017	1.000
182	45.292	1.092	1.135	1.015	1.000	1.018	1.248	1.228	1.051	1.003
191	30.640	1.087	1.129	1.013	1.000	1.035	1.225	1.207	1.042	1.001
198	21.892	1.081	1.121	1.011	1.000	1.059	1.201	1.183	1.032	1.000
201	19.013	1.078	1.117	1.010	1.000	1.072	1.189	1.172	1.028	1.000
211	12.252	1.066	1.102	1.007	1.001	1.130	1.145	1.131	1.014	1.005
213	11.291	1.064	1.098	1.006	1.002	1.145	1.137	1.123	1.011	1.009
222	8.012	1.050	1.081	1.002	1.001	1.228	1.098	1.086	1.003	1.002
228	6.503	1.041	1.069	1.001	1.000	1.303	1.074	1.064	1.000	1.000
232	5.702	1.035	1.060	1.000	1.000	1.362	1.059	1.051	1.000	1.001
242	4.203	1.020	1.039	1.001	1.003	1.556	1.029	1.023	1.008	1.010
244	3.969	1.017	1.035	1.002	1.009	1.605	1.024	1.019	1.011	1.012
253	3.104	1.006	1.018	1.009	1.002	1.876	1.007	1.005	1.031	1.002
258	2.731	1.002	1.010	1.016	1.000	2.074	1.002	1.001	1.048	1.000
263	2.416	1.000	1.004	1.027	1.001	2.318	1.000	1.000	1.068	1.001
273	1.941	1.005	1.000	1.061	1.012	2.966	1.004	1.007	1.116	1.012
274	1.903	1.006	1.001	1.091	1.013	3.044	1.005	1.008	1.026	1.006
283	1.624	1.022	1.009	1.050	1.002	3.885	1.016	1.020	1.012	1.001
288	1.506	1.035	1.017	1.034	1.000	4.490	1.023	1.029	1.007	1.000
293	1.408	1.050	1.028	1.021	1.001	5.227	1.030	1.037	1.004	1.001
303	1.260	1.086	1.055	1.005	1.010	7.268	1.046	1.054	1.000	1.004
305	1.237	1.094	1.062	1.003	1.008	7.799	1.049	1.058	1.000	1.001
314	1.151	1.133	1.093	1.000	1.001	10.956	1.063	1.072	1.000	1.000
318	1.121	1.151	1.107	1.000	1.000	12.907	1.068	1.078	1.000	1.000
324	1.085	1.177	1.129	1.003	1.001	16.764	1.075	1.086	1.001	1.000
334	1.044	1.216	1.162	1.009	1.005	26.846	1.085	1.097	1.003	1.001
335	1.040	1.220	1.165	1.010	1.001	28.172	1.086	1.098	1.003	1.000
344	1.020	1.245	1.186	1.015	1.000	42.386	1.092	1.104	1.004	1.000
354	1.012	1.257	1.196	1.018	1.000	54.320	1.094	1.107	1.004	1.000
365	1.018	1.247	1.188	1.016	1.000	44.680	1.092	1.104	1.004	1.000

Table 24. Energy gain for latitudes 66.45°N and 66.45°S

Day	66.45°S					66.45°N				
	$R_{d,o}$	$R_{d,\varphi}$	$R_{d,y}$	$R_{d,s}$	$R_{d,m}$	$R_{d,o}$	$R_{d,\varphi}$	$R_{d,y}$	$R_{d,s}$	$R_{d,m}$
1	1.016	1.244	1.184	1.021	1.003	163.207	1.088	1.171	1.015	1.000
10	1.039	1.219	1.163	1.037	1.001	56.940	1.083	1.163	1.013	1.000
17	1.067	1.194	1.142	1.058	1.000	32.081	1.077	1.154	1.011	1.000
20	1.083	1.181	1.131	1.069	1.000	26.107	1.074	1.149	1.010	1.000
30	1.151	1.138	1.096	1.121	1.005	14.713	1.062	1.131	1.006	1.001
32	1.168	1.129	1.089	1.134	1.008	13.328	1.060	1.126	1.005	1.002
41	1.265	1.091	1.058	1.209	1.001	8.945	1.047	1.106	1.002	1.000
47	1.350	1.067	1.040	1.275	1.000	7.085	1.038	1.091	1.001	1.000
51	1.419	1.053	1.030	1.328	1.001	6.135	1.032	1.081	1.000	1.000
59	1.592	1.029	1.013	1.461	1.009	4.700	1.020	1.060	1.001	1.003
60	1.618	1.027	1.011	1.480	1.013	4.554	1.018	1.058	1.001	1.009
69	1.901	1.009	1.001	1.691	1.002	3.480	1.007	1.035	1.007	1.002
75	2.156	1.002	1.000	1.876	1.000	2.947	1.002	1.022	1.015	1.000
79	2.367	1.000	1.002	2.025	1.001	2.652	1.000	1.014	1.023	1.001
89	3.069	1.003	1.013	2.496	1.011	2.086	1.004	1.002	1.053	1.010
91	3.243	1.005	1.016	1.026	1.006	1.999	1.006	1.001	1.091	1.012
100	4.215	1.015	1.033	1.011	1.001	1.687	1.021	1.002	1.051	1.002
105	4.931	1.022	1.043	1.007	1.000	1.556	1.033	1.007	1.034	1.000
110	5.824	1.029	1.053	1.004	1.001	1.448	1.048	1.014	1.021	1.001
120	8.421	1.044	1.074	1.000	1.004	1.284	1.084	1.035	1.005	1.010
121	8.765	1.046	1.076	1.000	1.002	1.271	1.088	1.038	1.004	1.009
130	13.011	1.059	1.093	1.000	1.000	1.173	1.127	1.062	1.000	1.001
135	16.756	1.065	1.101	1.000	1.000	1.131	1.149	1.078	1.000	1.000
140	22.292	1.071	1.109	1.001	1.000	1.097	1.171	1.093	1.002	1.001
150	45.576	1.081	1.122	1.002	1.001	1.048	1.211	1.122	1.008	1.005
152	54.360	1.083	1.124	1.002	1.000	1.041	1.218	1.127	1.010	1.001
161	151.82	1.088	1.131	1.003	1.000	1.017	1.243	1.145	1.016	1.000
162	175.81	1.089	1.131	1.003	1.000	1.015	1.245	1.147	1.016	1.000
171	791.31	1.091	1.134	1.004	1.000	1.006	1.256	1.155	1.019	1.000
181	246.96	1.089	1.132	1.004	1.000	1.012	1.249	1.228	1.017	1.000
182	209.45	1.089	1.132	1.014	1.000	1.013	1.247	1.226	1.051	1.003
191	66.671	1.084	1.126	1.013	1.000	1.034	1.225	1.205	1.042	1.001

continued on following page

Table 24. Continued

Day	66.45°S					66.45°N				
	$R_{d,o}$	$R_{d,\varphi}$	$R_{d,y}$	$R_{d,s}$	$R_{d,m}$	$R_{d,o}$	$R_{d,\varphi}$	$R_{d,y}$	$R_{d,s}$	$R_{d,m}$
198	36.185	1.079	1.119	1.011	1.000	1.060	1.200	1.181	1.032	1.000
201	29.119	1.076	1.115	1.010	1.000	1.074	1.188	1.170	1.027	1.000
211	15.997	1.064	1.100	1.006	1.001	1.138	1.145	1.130	1.013	1.005
213	14.438	1.062	1.097	1.005	1.002	1.154	1.136	1.121	1.011	1.009
222	9.566	1.049	1.080	1.002	1.001	1.245	1.097	1.085	1.003	1.002
228	7.531	1.040	1.068	1.001	1.000	1.325	1.073	1.063	1.000	1.000
232	6.498	1.034	1.060	1.000	1.000	1.390	1.059	1.050	1.000	1.001
242	4.645	1.019	1.039	1.001	1.003	1.602	1.028	1.022	1.008	1.010
244	4.363	1.017	1.035	1.002	1.009	1.655	1.024	1.018	1.011	1.012
253	3.347	1.006	1.018	1.009	1.002	1.954	1.007	1.004	1.032	1.002
258	2.918	1.002	1.010	1.016	1.000	2.175	1.002	1.001	1.048	1.000
263	2.559	1.000	1.004	1.027	1.001	2.448	1.000	1.000	1.068	1.001
273	2.026	1.005	1.000	1.060	1.011	3.188	1.004	1.007	1.117	1.012
274	1.984	1.006	1.000	1.091	1.013	3.278	1.005	1.008	1.025	1.006
283	1.676	1.022	1.008	1.051	1.002	4.264	1.016	1.021	1.011	1.001
288	1.547	1.034	1.016	1.034	1.000	4.991	1.023	1.029	1.006	1.000
293	1.440	1.049	1.027	1.021	1.001	5.899	1.030	1.037	1.003	1.001
303	1.279	1.086	1.055	1.005	1.010	8.547	1.045	1.054	1.000	1.004
305	1.254	1.094	1.061	1.004	1.007	9.272	1.048	1.057	1.000	1.001
314	1.160	1.133	1.092	1.000	1.001	13.910	1.061	1.071	1.000	1.000
318	1.128	1.150	1.106	1.000	1.000	17.098	1.066	1.077	1.000	1.000
324	1.089	1.176	1.128	1.003	1.001	24.281	1.073	1.084	1.001	1.000
334	1.043	1.216	1.160	1.009	1.005	51.368	1.082	1.095	1.002	1.001
335	1.039	1.219	1.163	1.010	1.001	56.279	1.083	1.095	1.003	1.000
344	1.016	1.244	1.184	1.015	1.000	160.252	1.088	1.101	1.003	1.000
354	1.006	1.256	1.194	1.018	1.000	818.337	1.091	1.104	1.004	1.000
365	1.014	1.247	1.186	1.016	1.000	197.338	1.089	1.102	1.004	1.000

Chapter 8
Economic Consideration

ABSTRACT

After giving a survey on tracking market and introducing the base elements of economic analysis, several examples were studied in this chapter in order to evaluate the economic feasibility of dual tracking systems in comparison with horizontally installed fixed panels and with latitude tilted fixed panels. It was found that tracking is feasible in relation with these two cases at high latitudes and it questionable at sunny belt region. Anyway, the diffusion of photovoltaic systems is hindered until today by high investment costs. Trackers are more expensive because now you have moving parts. Instead of something that is just sitting on the ground you now have a motor that moves the panels. O&M costs will be higher, as well. The motor needs to be maintained throughout the life of the tracker. However, PV power generation is justified for special purposes. It is clearly demonstrated that, the small scale applications such as telecommunication systems, rural electrification, cathode protection and water lifting are economically feasible. Moreover, the comparison of the effectiveness of tracking in relation to monthly adjusted tilt of PV panels where the simplicity and high energy gain is not considered. This will be done in the near future.

INTRODUCTION

Electricity is the fuel that the world craves; the only question is where will it come from? Currently three main resources satiate most of the world's electricity hunger: Coal (25%), natural gas (21%), and oil (34%) (IEA, 2006). These three source materials have two things in common; they are all burned to release the energy contained within and combustion releases pollutants. In recent years, there has been a big push towards renewable energy and energy efficient technologies like wind power, solar collectors, and LED lights. This last year the United States set aside $2.7 billion for communities to develop and complete projects designed to increase energy efficiency (USDOE, 2009).

Current renewable energy systems that are utilized in buildings include geothermal, wind, and solar energy systems. Although geothermal systems can reduce the consumption of energy, these systems do not generate energy, and thus, still require auxiliary power to operate. Additionally, geothermal energy systems can be difficult to integrate into existing buildings and can require significant amounts of space for a well field. Furthermore, these systems require routine maintenance due to moving parts and circulating fluid of the system.

DOI: 10.4018/978-1-5225-2950-7.ch008

Wind energy systems also have vast potential for local building energy generation, yet have some inherent disadvantages in comparison to other forms of renewable energy sources. Wind turbines and towers are large and require extensive capital and engineering to design and install. These systems have moving parts and are prone to regular maintenance requirements. In addition, many cities have regulations that restrict towers within city limits. Moreover, wind resources can be greatly minimized due to surrounding structures/ground cover to the site, which degrades their effectiveness in urban areas. Wind energy systems are most conducive to rural settings.

Solar energy is the most important renewable source of energy. It has the potential to be one of the key alternative clean and renewable sources to supply the increasing demand. In 2009, the world's consumption of energy was 11,164.3 million tons of oil equivalent (*mtoe*). Comparing this figure with the amount of received solar radiation during the same year, we find that the input solar radiation is 11,300 times greater than the world's total primary energy consumption (Shepherd and Shepherd, 2014). Solar energy is clean and friendly to the environment, and it will not be affected by fluctuations in the market, such as oil. The Solar Energy Industry Association (SEIA) report for 2014 shows a rapid growth in solar industry. Photovoltaic panels of 6201 MW have been installed plus 767 MW of concentrating solar power. This produces 20 Giga Watts (GW) of total installed capacity, which is enough to power four million houses in the USA. The report also shows that the industry has 175,000 workers, more than Google, Apple, Facebook, and Twitter combined*. Solar energy is now estimated for one-third of the USA new generating capacity in 2014, surpassing both wind energy and coal for the second year in a row (Resch, 2016; Solar Energy Industries Association, 2016).

Solar energy systems could show the most promise for existing and new building integration for electrical energy generation applications. PV systems can often be installed directly to a building without further strengthening the existing structure. These systems have no moving parts and maintenance requirements are relatively minimal. Additionally, in comparison to wind, PV systems can be simpler to design. Solar energy systems are currently more expensive on a peak Watt basis to implement in comparison to wind and geothermal systems. However, government incentives, technological advances, and the potential for PV equipment to replace building materials increase economic feasibility. Modern technology provides the means to have a clean and inexpensive form of energy through photovoltaic systems, which convert the solar energy into electrical energy. Recently, it has been a concern weather to install fixed photovoltaic systems or is it better to have a photovoltaic tracking system? A photovoltaic tracking system is a system that tracks the sun from sunrise till sunset for maximum energy gain (Bentaher, Kaich, Ayadi, Ben Hmouda, Maalej and Lemmer, 2014). In fact, such idea is to be viewed from different aspects in this study, as the feasibility of such approach must be investigated. For instance, on what scale a tracker would be efficient? Also, what are the parameters affecting the installation of a tracking system?

Estimating the economic internal rate of return on an investment in a PV system allows for a direct comparison to other types of financial investments with associated expectations of return. Finally, approximating the cost of energy generated by a PV system over its assumed life is useful for comparing options of other alternative sources of energy such as energy supplied by a local utility. It can be concluded from the literature on the solar tracking that, a PV tracking system could improve the efficiency and power yield of the modules. Moreover, it could have a negative influence in case of a hot country like Sunbelt countries, e.g. Syria and Egypt, due to excessive heating. One of the objectives of this chapter is to investigate the economic feasibility of the tracking systems in different regions of the world, i.e. hot and cold regions.

PRODUCTION OF SOLAR ELECTRICITY

Conventional methods for energy generation using fossil-fuel based sources in many countries are negatively impacting the economy, national security, environment, natural resources, and public health. Additionally, negative impacts resulting from the use of fossil-fuels are intensifying due to the country's increasing demand of energy. Thus, the increasing demand of energy must be met while decreasing the use of fossil-fuels. Conventional energy generation methods must be reformed by using sources that are renewable, domestic, distributed, and environmentally friendly.

The use of photovoltaic (PV) technology for generating electrical energy is one approach that could serve as a solution to help alleviate some of the nation's energy-related problems. However, the use of PV is not currently widespread in a big part of available countries. Several barriers to the widespread use of PV have been identified in a number of studies. One can mention some of them such as a lack of familiarity, awareness, and understanding of PV technology. A different survey conducted in the United Kingdom (U.K.) found that 88 percent of respondents would consider the use of integrated photovoltaic building products (BIPV) given greater evidence for the performance and reliability of these products (Fanney and Dougherty, 2001). This U.K. study also showed that 49 percent of those surveyed would consider the use of BIPV technology only after they had witnessed the actual use of it in demonstration sites (Fanney and Dougherty, 2001). Studies show that consumers and design professionals are often reluctant to adopt a particular technology without first observing the application of that technology. Although this survey was conducted in the U.K. for BIPV, it is expected that the acceptance of non-building integrated PV in the US would closely follow these trends (Fanney and Dougherty, 2001).

Electricity production from solar radiations depends on the choice of technologies and available solar radiation of the area under consideration. The implementation of solar power plants in regions of high insolation is a promising option for an environmentally compatible electricity supply strategy. In 1997, approximately 80% of the solar generated electricity is provided by solar thermal power plants, while 20% is supplied by photovoltaic systems. Decision-makers have the choice among the following solar technologies: (1) parabolic trough; (2) central receiver; (3) paraboloidal dish; (4) solar chimney; (5) solar pond; (6) photovoltaic cells. Trieb, Ole and Helmut (1997) compared available solar electricity technologies from the point of view of system analysis, taking into consideration their performance, costs and environmental impact. Trieb, Ole and Helmut (1997) showed that the different approaches cover a wide range from units of a few Watts to utiilty-size plants and from isolated to grid-connected systems. They also showed that there is still a need for intense efforts to integrate those technologies into the existing electricity supply scheme. They concluded that, if a continuous development and market introduction is achieved, solar power plants will contribute substantially to the reduction of global CO_2-emissions.

Trieb, Ole and Helmut (1997) presented detailed description of the various technologies that can be employed to convert solar radiation into electricity in their study, they emphasize that electricity are generated from solar via solar thermal plants and photovoltaic systems. However, no matter the method adopted and the technology used, available evident still shows that solar energy is inexhaustible, environmentally friendly and is capable of supplying the energy needs of the entire universe. Brief descriptions of various ways of converting solar radiations to electricity using some of technologies list by Trieb, Ole and Helmut (1997) are as follows;

1. Solar thermal systems employ solar collectors (flat plate or concentrating) and working fluids to convert solar radiation into heat energy; the heated transferred fluid is then used to generate steam that drives the heat engine, thus producing electricity.
2. Solar chemical systems employ solar reactors to produce chemicals which can be converted directly or indirectly to electricity via electrolysis and heat engines respectively.
3. Solar photovoltaic system is the only technology that converts solar energy from the sun directly to electricity. It uses solar cells made of silicon material which can be combined in parallel and series to form module.

Moreover, Trieb, Ole and Helmut (1997) presented a practical tool, for decision-makers, that facilitates a first estimate of the performance and costs of such plants under local conditions.

PHOTOVOLTAIC SYSTEMS MARKET

Apparently there is need to encourage the residential sector of the economy in hot countries like Sunbelt countries, e.g. Jordan, Syria and Egypt, which is the highest consumer of the generated capacity to be energy producer, feed in tariff system mechanism would be a viable tool as it has been successfully implemented in many neighboring European countries. Comprehensive detailed comparison of PV promoting scheme of different European and Mediterranean countries have been presented by Giakoumelos, Malamatenois, C., Hadjipaschalis, I., Kourtis, G. and Poullikkas (2008).

One of the most important issues surrounding the application of PV for building energy generation is that of economic feasibility. The economics of PV systems can be estimated in terms of many different measures. However, the basic economic premise for an investment in PV is comparing a known initial investment and estimated future operating/maintenance expenses with projected future savings in energy costs generated by the system (Duffie and Beckman, 2013). The economics of a grid-connected PV system can vary significantly depending on the solar resource, site, performance of the system, interconnection agreement type, operation and maintenance costs, financial incentives, costs of energy, and initial equipment and installation costs of the system.

Economic results for both the stationary and tracking systems include life cycle costs (LCC), payback period, internal rate of return (IRR), and average incremental cost of solar energy. Each economic parameter allows for comparisons to be made to other investments or alternative sources of energy and can offer insight to the attractiveness of an investment in PV. A LCC analysis is a method of analyzing the initial and future annual expenses and savings associated with a system over the life of the equipment; this method normalizes the annual cash flow to an overall net present value (NP) assuming a particular discount rate (Park, 2002). The life cycle cost can be defined as “the total discounted dollar cost of owning, operating, maintaining, and disposing of a system” over a given period of time. Payback period can be useful for assessing the amount of time required for the PV system to pay itself back or “break-even.”

Despite high initial capital cost, that is reflected in cost per kWh of the energy produced, and the low conversion efficiency of the photovoltaic systems. There has tremendous increase in the use of photovoltaic technology globally, from 1.8 GW in 2000 to 67.4 GW in 2011 as installed capacity (EPIA, 2012; IRENA, 2012). This number is driven by attractive policy incentives of feed in tariffs and tax breaks. Even with this increase, Mediterranean region shares the smaller part of installation with the rest of the world in spite of installing some PV farms by the leader countries in this field.

To increase the efficiency of PV technologies and lower the cost of its production, research has been intensified on development and utilization of recent technologies in the development of high efficiently low cost PV cells and providing incentives to customers that wish to install. A comprehensive analysis of current state of the Art of and future outlook of PV system technologies with regard to costs, market penetration and environment were presented by (Raugei and Frankl, 2009). Compaan (2006) explained that the cost of manufacturing of PV solar module has significantly diminished over the years with the advancement of technology. However, according to available scenarios, if economic incentives are supported long enough into the next ten to twenty years, PV looks set for a rosy future, and is likely to play a significant role in the future energy mix, while at the same time contributing to reduce the environmental impact of electricity supply.

LITERATURE SURVEY ON SOLAR TRACKING COST

General

The current cost of photovoltaic (PV) panels is too expensive to implement for most residential uses, but to lower the cost more participation is required. Subsidies may be needed to get consumer buy-in, taxes on current energy production methods to fund these subsidies, and time to allow advocacy groups to become more powerful (Sande'n, 2005). In terms of cost and performance, infant technologies, such as solar photovoltaics (PV), are normally inferior to entrenched technologies. It is a Catch-22 situation since the diffusion on larger markets that would be needed to reduce cost is hindered by the high cost. Therefore it would make sense to subsidize PV to increase sales, which would increase experience and induce investments in larger factories, which in turn would drive down costs and the subsidies needed. The total cost of such a scheme does not have to be prohibitive if cost reductions with increased volumes are large enough. Over the last 20 years the cost of PV modules was reduced by 18–23% per doubling of cumulative production (a progress ratio of 0.77–0.82). For a progress ratio of 0.80 and an annual growth rate of 30%, the modeled annual subsidy peaks at $14 US billion, which correspond to an additional electricity tax of no more than 0.1 US cents/kW h in the Organisation for Economic Co-operation and Development (OECD) countries. A market support program also creates institutional learning and increases the political power of the proponents of PV. The current federal German support program is a product of learning and network formation in earlier market stimulation and research, development and demonstration (RDD) programs of smaller scale. Hence, the current support program is now likely to create not only economic virtuous circles that reduce costs, but also institutional virtuous circles that work for the survival and expansion of the program itself. As the PV industry grows, care should be taken to maintain variety to reduce the risk of a premature lock-in of an inferior design. To maintain variety in the market place may prove costly when the market grows but variety creation at the level of RDD investments is fairly cheap. To increase the world expenditure on RDD of renewable energy technology by a factor of 10 would not cost more than $1 US/ton C or 0.02 US cent/kW h of electricity.

Hybrid systems are defined as systems that utilize more than one energy source to supply a certain load. The implementation of a hybrid system that is based upon Photovoltaic (PV) to supply power to remote and isolated locations is considered a viable option. This is especially true for areas that receive sufficient amounts of annual solar radiation. While analysis of hybrid systems that depend on diesel generators as backup sources can be found in many previous research works, detailed techno economic analysis of hybrid systems that depend on micro-turbines as backup sources are less addressed. The combination of current fossil fuel power technology with solar energy in the form of a gas/solar hybrid station is an attempt to bridge the difference and drag the price /kWh down to a point where it is profitable. The detailed cost assumptions for the solarized gas turbine, the solar tower plant and further equipment as well as for operation and maintenance are presented (Schwarzbozl, Buck, Sugarmen, Ring, Crespo, Altwegg and Enrile, 2006). Intensive performance and economic analysis of the prototype plants for different locations and capacity factors are shown. The cost reduction potential through automation and remote operation is revealed. A techno-economic analysis of a complete hybrid system that comprises of Photovoltaic (PV) panels, a battery system, and a micro-turbine as a backup power source for a remote community is presented in the paper of Ismail, Moghavvemi and Mahlia (2013) which investigated the feasibility of using the micro-turbines as backup sources in the hybrid systems. The comparison between different scenarios with regards to the cost of energy and pollutant emissions was also conducted. The effect of the variation of some parameters on the cost of energy was duly evaluated. Powering a rural community using micro-turbine alone indicates lower values of cost of energy (COE) production compared to the hybrid system in which a combination of PV panels, battery bank and micro-turbine has been used. The difference is very small and taking into account the environmental effect of the micro-turbine surely will make the hybrid system with limited running hours of the micro-turbine more attractive. Furthermore, as it is obvious from the sensitivity analysis, any reduction in the price of the PV panels or any increase in the natural gas price will make the hybrid system economically and environmentally more attractive.

Despite low demand, manufacturers and power companies such as Google (Kanellos, 2005), Lockheed and Martin (Englander, 2009), and PGandE (LaMonica, 2009) are investing in PV technology. As demand for energy is ever increasing it is possible that by the year 2050 we'll need to increase energy production by 46% or so to meet it. Coal alone cannot meet this increase in demand (Higgens, 2009). The possibility of solar power is as endless as the applications for energy production.

The major problem hampering the expansion of solar power is its cost per kWh. For solar energy to truly flourish, the generating costs of solar electricity must be $0.07- $0.14 per kWh (Schwarzbozl, Buck, Sugarmen, Ring, Crespo, Altwegg and Enrile, 2006) given the current production cost of solar panels. Today's market value for energy in Nebraska is about $0.07 kWh compared to the national average of roughly $0.10 per kWh (Nebraska Energy Office, 2009). To determine if solar energy is economically feasible in Nebraska, the cost associated with PV collectors (including composition, installation, and maintenance) and average daily receipt of solar radiation has to be considered. For example, current mass produced silicon has a stable average solar to electrical conversion efficiency of 6% or less (Aberle, 2009). Making PV cells with different materials can have a very large impact on the efficiency values of the panel. Solar energy is increasingly becoming a significant component in the energy profiles of several tropical nations. Agee, Obok-obok and deLazzer (2007) discussed trends in solar tracking technologies and analyzed the cost of acquisition, domains of application, maintenance costs and efficiency improvements. They concluded that hydraulic-based tracking systems are suitable for low capacity installations with low pay loads while polar axis tracking systems offer a performance nearly equal to that of two-axis tracking systems, at the cost of single axis trackers.

Solar Tracking Market

Many approaches and technologies have been used to create reliable and cost-effective sun trackers. The advancement in the efficiency of the photovoltaic panels, plus the development of enhanced manufacturing methods in the photovoltaic industry, reduced the energy payback time of the panels to 3–5 years (Rustemli and Dincer, 2011). This energy payback time is reliant on multiple factors such as the irradiance level at the designated site. The cost of the photovoltaic panels is estimated to be $2.5 per watt peak and ambitiously forecasted to be $1 per watt peak by 2020 (Rustemli and Dincer, 2011).

The global market of solar trackers is anticipated to grow to reach around 6.83 billion by 2022. According to the industry report on solar tracker market size and share (STMSS) the environmental awareness, governmental regulations and incentives plus many other factors lead to increase in photovoltaic size from 100.5 GW in 2012 to over 182.5 GW in 2014 (STMSS, 2016). Due to their low cost, single-axis trackers had a major share of the market in 2014. Residential and commercial areas had a high increase in purchases of these trackers. Dual-axis trackers are projected to be the fastest growing product as major companies prefer these types of trackers (STMSS, 2016).

However, the cost of energy generated from a PV tracking system is higher than the energy produced from a fixed system because of the running cost and the initial cost of the tracking system, which makes the economic advantage of a tracking system questionable (Ismail, Moghavvemi and Mahlia, 2013). Huang and Sun (2007) suggested that having additional PV modules connected in series or in parallel on the same tracker would be economic. Alexandru and Tatu (2013) concluded that there will always be higher capital costs in a tracking system due to installed motors and moving joints, but a 20% reduction in the capital cost can be achieved in high powers. For instance, the same motor may be used for a larger number of PV panels, as suggested by Dakkak and Babeli (2012). Huang and Sun (2007) concluded that, for a PV module size of 200 W_{peak}, the price reduction in the capital cost is 26.3% for the PV module sale price of 3.69 USD/W_{pea}k. A price reduction of 19.9% is still possible even if the market price of the flat plate PV module will be down to 3 USD/W_{peak} in the future. The price reduction will not change too much if the size of the flat-plate PV module is larger than 160 W_{peak}. For smaller flat-plate PV module sizes, lower than 100 W_{peak}, the price reduction of the energy generated from a PV module with one axis three position tracking and 2X concentration reflector becomes not significant if the market price of the flat plate PV module is less than 3 USD/W_{peak}. However, they could not conclude weather this price reduction was due to the use of a tracking system or due to the use of the concentrators.

Sun trackers can be categorized based on their electricity consumption (aka drive type) and their actuator movements. In the electricity consumption classification, the sun trackers can be divided into two main groups of passive and active sun trackers. The economic feasibility of passive trackers is questionable because of different restraints in its use as mentioned by Clifford and Eastwood (2004). Koussa, Haddadi, Saheb, Malek, Hadji (2012) found that the energy gain of the two-axis tracking systems exceeds that of the single-axis tracking systems by less than 3%. However, with taking into account that, there will always be higher capital costs in a tracking system due to installed motors and moving joints, the economic feasibility of solar trackers to PV systems in comparison with using the PV system with daily, monthly, seasonally and biannually tilt adjustment is not evident.

ECONOMIC ANALYSIS

Techno-economic analysis is the area of engineering where engineering judgment and experience are utilized in the application of scientific principles and techniques to problems of project cost control, profitability analysis, planning, scheduling and optimization of operational research etc. Therefore, it is important to cover a wide range of topics such as time value of money, maintenance, organizational structures, integral projects control, quality and resource management, life of cycle and risk analysis etc.

Close estimation becomes crucial and critical with the advancement of technology and society to remain competitive. An estimate based on our design may be too high to sustain, whereas that based on under design may be successful for a while but again it is not possible to sustain. An effective economic analysis can be made by the knowledge of cost analysis, which can be done by the aid of cash flow diagrams and some other methods.

Cost Analysis

Let us consider the main elements of cost analysis, namely:

- Capital recovery factor
- Unacost
- Sinking fund factor

Capital Recovery Factor

For the capital recovery factor, let *P* be the present amount invested at zero (*N*=0) time at the interest of *i* per year and S_N future value at the end of *N* years. Then, at the end of one year the value of *P* will be:

$$S_1 = P + iP = P(1+i) \tag{1}$$

At the end of the second year *P* becomes:

$$S_2 = S_1 + iS_1 = P + iP + i(P + iP) = P(1+i)^2 \tag{2}$$

At the end of the third year and N[th] year we have:

$$S_3 = P(1+i)^3 \tag{3}$$

$$S_N = P(1+i)^N \tag{4}$$

Assuming S_N to be S, then S_N can be written as:

$$S = P(1+i)^N \tag{5}$$

where *i* is the rate of interest and *N* is the number of years. Here, *S*>*P* for *i*>0, compound interest law:

$$S = PF_{PS} \tag{6}$$

where F_{PS} is more completely designed as $F_{PS,iN}$, and known as the compound interest factor or future value factor which converts *P* into *S*:

$$F_{PS,i,N} = \left(1+i\right)^{N} \tag{7}$$

Thus,

Future value = (present value) (compound interest factor)

If one year is divided into *p* equal units of period, then *N* becomes *Np* and *i* becomes *i/p* which is the rate of return per unit period. The substitution of these values in the equation (6) leads to:

$$S = P\left(1+\frac{i}{p}\right)^{Np} = P\left[\left(1+\frac{i}{p}\right)^{p}\right]^{N} \tag{8}$$

The expression $\left(1+i/p\right)^{p}$ could be written as:

$$\left(1+\frac{i}{p}\right)^{p} = 1 + effective\, rate\, of\, return \tag{9}$$

Thus,

$$effective\, rate\, of\, return = \left(1+\frac{i}{p}\right)^{p} - 1 ==> effective\, rate\, of\, return \begin{Bmatrix} = i\, for\, p = 1 \\ > i\, for\, p > 1 \end{Bmatrix}$$

The future value *S* of initial investment *P* are related to each other as:

$$P = S\left(1+i\right)^{-N} \tag{10}$$

Thus, the equations (5) and (10) can be combined as:

$$A_{t2} = A_{t1}\left(1+i\right)^{N} \tag{11}$$

Equation (16) should be read as follows:

Amount at time 2= Amount at time 1 (Compound interest operator)

Here N is positive with the calendar and negative against the calendar. Equation (16) is referred to as the time-value conversion relationship.

Unacost

For the unacost, it is to mention that, in solving engineering problems it is convenient to diagram expenditures and receipts as vertical lines positioned along a horizontal line representing time. Let us consider a uniform end-of-year annual amount *R* (unacost) for a period of *N* years and let *P* be a single present value at initial time. Then, using equation (10), we get:

$$P = R\left[\frac{1}{1+i} + \frac{1}{\left(1+i\right)^2} + \ldots + \frac{1}{\left(1+i\right)^N}\right] = R\sum_{k=1}^{k=N}\frac{1}{\left(1+i\right)^k} \tag{12}$$

Equation (12) is a geometric series which has $1/\left(1+i\right)$ as the first term and a ratio of *N* successive terms. The summation of geometric series can be evaluated as:

$$P = R\sum_{k=1}^{k=N}\frac{1}{\left(1+i\right)^k} = R\frac{\left(1+i\right)^N - 1}{i\left(1+i\right)^N} = RF_{RP,i,N} \tag{13}$$

Thus, the unacost present value factor $F_{RP,i,N}$ is:

$$F_{RP,i,N} = \frac{\left(1+i\right)^N - 1}{i\left(1+i\right)^N} \tag{14}$$

$F_{RP,i,N}$ is referred to as the equal-payment series present value factor or annuity present value factor. Equation (13) could be read as:

Present value = (unacost) (unacost present value factor)

Equation (13) could be written as:

$$R = P\frac{i\left(1+i\right)^N}{\left(1+i\right)^N - 1} = PF_{PR,i,N} \tag{15}$$

where $F_{PR,i,N}$ is capital recovery factor (CRF). Thus,

Unacost = (Present value) (Capital recovery factor)

$$F_{PR,i,N} = \frac{i\left(1+i\right)^{N}}{\left(1+i\right)^{N}-1} \tag{16}$$

Sinking Fund Factor

The future value, S, at the end of N years can also be converted into a uniform end-of-year annual amount R.

$$R = P\frac{i\left(1+i\right)^{N}}{\left(1+i\right)^{N}-1} = S\left(1+i\right)^{-N}\frac{i\left(1+i\right)^{N}}{\left(1+i\right)^{N}-1} = S\frac{i}{\left(1+i\right)^{N}-1} = SF_{SR,i,N} \tag{17}$$

where $F_{SR,iN}$ is the sinking fund factor (SFF):

$$F_{SR,i,N} = \frac{i}{\left(1+i\right)^{N}-1} \tag{18}$$

Thus, the future amount is but:

$$S = R\frac{\left(1+i\right)^{N}-1}{i} = RF_{RS.i,N} \tag{19}$$

$F_{RS.i,N}$ is the equal payment series future value factor.

$$F_{RS.i,N} = \frac{\left(1+i\right)^{N}-1}{i} \tag{20}$$

The equations (17) and (19) could be read as:

Unacost = (Future amount) (Sinking fund factor)

Future amount= (Unacost) (Equal payment series future value factor)

Cash-Flow Diagrams

A cash flow diagram is simply a graphical representation of cash flows drawn on a time scale.

Net cash flow = receipts – disbursements

Cost Comparisons With Equal Duration

A uniform expense is referred to as a uniform end-of-year cost.

Cost Comparisons With Unequal Duration

If two energy efficient systems have different duration of lives, a fair comparison can be made only on the basis of equal duration. One of the methods for comparison is to compare single present value of costs on the basis of a common denominator of their service lives. Let us now consider two methods in this context.

The First Method

This method is known as cost comparison by capitalized cost. If P_N is equivalent present value of a system lasting *N* years, then the capitalized cost is the present value on an infinite time basis for a system costing P_N and lasting *N* years.

The present value will be:

$$K = P_N \sum_{x=0}^{x=\infty} \frac{1}{\left(1+i\right)^{xN}} = P_N \left[1 + \frac{1}{\left(1+i\right)^N} + \frac{1}{\left(1+i\right)^{2N}} + \ldots\right] \tag{21}$$

The equation (21) is a geometric series with the first term as 1 and the ratio of the consecutive terms as $1 / \left(1+i\right)^N$. Its summation is given by:

$$\sum_{x=0}^{x=\infty} \frac{1}{\left(1+i\right)^{xN}} = 1 \frac{1 + \left[\frac{1}{\left(1+i\right)^N}\right]^\infty}{1 - \frac{1}{\left(1+i\right)^N}} = \frac{\left(1+i\right)^N}{\left(1+i\right)^N - 1} \tag{22}$$

Thus, the equation (21) becomes:

$$K = P_N F_{PK,i,N} \tag{23}$$

where *K* is the capitalized cost and $F_{PK,i,N}$ is known as the capitalized cost factor which is the factor that converts a present value to capitalized cost and is given by:

$$F_{PK,i,N} = \frac{\left(1+i\right)^{N}}{\left(1+i\right)^{N} - 1} \tag{24}$$

The equation (23) could be read as:

$$\textbf{Capitalized cost} = \left(\textbf{Present value basis } \boldsymbol{N} \textbf{ years duration}\right)\left(\textbf{Capitalized cost factor}\right)$$

Taking the equation (16) into consideration, the equation (24) could be written as:

$$F_{PR,i,N} = iF_{PK,i,N} \tag{25}$$

Thus,

$$\textbf{Capital recovery factor} = \left(\textbf{Rate of return}\right)\left(\textbf{Capitalized cost factor}\right)$$

On the other hand, taking into consideration the equations (15) and (23), we found that *R* and *K* are related to each other as:

$$R = iK \tag{26}$$

Thus,

$$\textbf{Unacost} = \left(\textbf{Rate of return}\right)\left(\textbf{Capitalized cost factor}\right)$$

The Second Method

This method is known as the cost comparison by cost ratio. As a matter of fact, it is possible to convert a present value P_{N1}, of N_1 years duration to an equivalent present value P_{N2}, of N_2 years duration. The use of equation (23) leads to:

$$K = P_{N1}F_{PK,i,N1} = P_{N2}F_{PK,i,N2} \tag{27}$$

which in its turn leads to:

$$P_{N1} = P_{N2}\frac{F_{PK,i,N2}}{F_{PK,i,N1}} \tag{28}$$

The values of various conversion factors with number of years for a given rate of interest are given in the Tables 1 to 10.

Table 1. The values of various conversion factors with number of years N for i=0.03

N	$F_{PS,iN}$	$F_{SP,iN}$	$F_{RP,iN}$	$F_{PR,iN}$	$F_{RS,iN}$	$F_{SR,iN}$	$F_{PK,iN}$
1	1.03	0.970874	0.970874	1.03	1	1	34.33333
2	1.0609	0.942596	1.91347	0.522611	2.03	0.492611	17.42036
3	1.092727	0.915142	2.828611	0.35353	3.0909	0.32353	11.78435
4	1.125509	0.888487	3.717098	0.269027	4.183627	0.239027	8.967568
5	1.159274	0.862609	4.579707	0.218355	5.309136	0.188355	7.278486
6	1.194052	0.837484	5.417191	0.184598	6.46841	0.154598	6.15325
7	1.229874	0.813092	6.230283	0.160506	7.662462	0.130506	5.350212
8	1.26677	0.789409	7.019692	0.142456	8.892336	0.112456	4.748546
9	1.304773	0.766417	7.786109	0.128434	10.15911	0.098434	4.281129
10	1.343916	0.744094	8.530203	0.117231	11.46388	0.087231	3.907684
11	1.384234	0.722421	9.252624	0.108077	12.8078	0.078077	3.602582
12	1.425761	0.70138	9.954004	0.100462	14.19203	0.070462	3.348736
13	1.468534	0.680951	10.63496	0.09403	15.61779	0.06403	3.134318
14	1.51259	0.661118	11.29607	0.088526	17.08632	0.058526	2.950878
15	1.557967	0.641862	11.93794	0.083767	18.59891	0.053767	2.792219
16	1.604706	0.623167	12.5611	0.079611	20.15688	0.049611	2.653695
17	1.652848	0.605016	13.16612	0.075953	21.76159	0.045953	2.531751
18	1.702433	0.587395	13.75351	0.072709	23.41444	0.042709	2.423623
19	1.753506	0.570286	14.3238	0.069814	25.11687	0.039814	2.327129
20	1.806111	0.553676	14.87747	0.067216	26.87037	0.037216	2.240524
21	1.860295	0.537549	15.41502	0.064872	28.67649	0.034872	2.162393
22	1.916103	0.521893	15.93692	0.062747	30.53678	0.032747	2.09158
23	1.973587	0.506692	16.44361	0.060814	32.45288	0.030814	2.02713
24	2.032794	0.491934	16.93554	0.059047	34.42647	0.029047	1.968247
25	2.093778	0.477606	17.41315	0.057428	36.45926	0.027428	1.914262

Payback Time/Payment Time/Payback Period

Profitability is a measure of the total income for the project compared to the total outlay. Money going into the project is taken to be negative and money coming back from the project is taken to be positive. Payout time is one of the criteria for profitability.

The payback period N_I is the number of years necessary to exactly recover the initial investment P. The payback period N_I could be calculated by summing the annual cash-flow values and using the following relation:

Table 2. The values of various conversion factors with number of years N for i=0.05

N	$F_{PS,iN}$	$F_{SP,iN}$	$F_{RP,iN}$	$F_{PR,iN}$	$F_{RS,iN}$	$F_{SR,iN}$	$F_{PK,iN}$
1	1.05	0.952381	0.952381	1.05	1	1	21
2	1.1025	0.907029	1.85941	0.537805	2.05	0.487805	10.7561
3	1.157625	0.863838	2.723248	0.367209	3.1525	0.317209	7.344171
4	1.215506	0.822702	3.545951	0.282012	4.310125	0.232012	5.640237
5	1.276282	0.783526	4.329477	0.230975	5.525631	0.180975	4.619496
6	1.340096	0.746215	5.075692	0.197017	6.801913	0.147017	3.940349
7	1.4071	0.710681	5.786373	0.17282	8.142008	0.12282	3.456396
8	1.477455	0.676839	6.463213	0.154722	9.549109	0.104722	3.094436
9	1.551328	0.644609	7.107822	0.14069	11.02656	0.09069	2.813802
10	1.628895	0.613913	7.721735	0.129505	12.57789	0.079505	2.590091
11	1.710339	0.584679	8.306414	0.120389	14.20679	0.070389	2.407778
12	1.795856	0.556837	8.863252	0.112825	15.91713	0.062825	2.256508
13	1.885649	0.530321	9.393573	0.106456	17.71298	0.056456	2.129115
14	1.979932	0.505068	9.898641	0.101024	19.59863	0.051024	2.020479
15	2.078928	0.481017	10.37966	0.096342	21.57856	0.046342	1.926846
16	2.182875	0.458112	10.83777	0.09227	23.65749	0.04227	1.845398
17	2.292018	0.436297	11.27407	0.088699	25.84037	0.038699	1.773983
18	2.406619	0.415521	11.68959	0.085546	28.13238	0.035546	1.710924
19	2.52695	0.395734	12.08532	0.082745	30.539	0.032745	1.6549
20	2.653298	0.376889	12.46221	0.080243	33.06595	0.030243	1.604852
21	2.785963	0.358942	12.82115	0.077996	35.71925	0.027996	1.559922
22	2.925261	0.34185	13.163	0.075971	38.50521	0.025971	1.51941
23	3.071524	0.325571	13.48857	0.074137	41.43048	0.024137	1.482736
24	3.2251	0.310068	13.79864	0.072471	44.502	0.022471	1.449418
25	3.386355	0.295303	14.09394	0.070952	47.7271	0.020952	1.419049

$$-P+\sum_{t=1}^{t=N_1} CF_t\left(F_{SP,i\%,t}\right)=0 \quad (29)$$

where CF_t is the net cash-flow at the end of the year *t*. If cash-flow is the same each year, the equation (29) could be written as:

$$-P+CF_1\left(F_{SP,i\%,N_1}\right)=0 \quad (30)$$

Table 3. The values of various conversion factors with number of years N for i=0.07

N	$F_{PS,iN}$	$F_{SP,iN}$	$F_{RP,iN}$	$F_{PR,iN}$	$F_{RS,iN}$	$F_{SR,iN}$	$F_{PK,iN}$
1	1.07	0.934579	0.934579	1.07	1	1	15.28571
2	1.1449	0.873439	1.808018	0.553092	2.07	0.483092	7.901311
3	1.225043	0.816298	2.624316	0.381052	3.2149	0.311052	5.443595
4	1.310796	0.762895	3.387211	0.295228	4.439943	0.225228	4.217545
5	1.402552	0.712986	4.100197	0.243891	5.750739	0.173891	3.484153
6	1.50073	0.666342	4.76654	0.209796	7.153291	0.139796	2.997083
7	1.605781	0.62275	5.389289	0.185553	8.654021	0.115553	2.65076
8	1.718186	0.582009	5.971299	0.167468	10.2598	0.097468	2.392397
9	1.838459	0.543934	6.515232	0.153486	11.97799	0.083486	2.192664
10	1.967151	0.508349	7.023582	0.142378	13.81645	0.072378	2.033964
11	2.104852	0.475093	7.498674	0.133357	15.7836	0.063357	1.905099
12	2.252192	0.444012	7.942686	0.125902	17.88845	0.055902	1.7986
13	2.409845	0.414964	8.357651	0.119651	20.14064	0.049651	1.709298
14	2.578534	0.387817	8.745468	0.114345	22.55049	0.044345	1.633499
15	2.759032	0.362446	9.107914	0.109795	25.12902	0.039795	1.568495
16	2.952164	0.338735	9.446649	0.105858	27.88805	0.035858	1.512252
17	3.158815	0.316574	9.763223	0.102425	30.84022	0.032425	1.463217
18	3.379932	0.295864	10.05909	0.099413	33.99903	0.029413	1.42018
19	3.616528	0.276508	10.3356	0.096753	37.37896	0.026753	1.382186
20	3.869684	0.258419	10.59401	0.094393	40.99549	0.024393	1.34847
21	4.140562	0.241513	10.83553	0.092289	44.86518	0.022289	1.318414
22	4.430402	0.225713	11.06124	0.090406	49.00574	0.020406	1.291511
23	4.74053	0.210947	11.27219	0.088714	53.43614	0.018714	1.267342
24	5.072367	0.197147	11.46933	0.087189	58.17667	0.017189	1.245557
25	5.427433	0.184249	11.65358	0.085811	63.24904	0.015811	1.225865

i.e. after N_1 years, the cash-flow will recover the investment and a return of I present. If the expected retention period N (life) of the asset/project is less than N_1 years$\left(N < N_1\right)$, then investment is not advisable.

Considering i to be zero, equation (29) becomes:

$$-P + \sum_{t=1}^{t=N_1} CF_t = 0 \tag{31}$$

and if CF_t values are equal, then:

Table 4. The values of various conversion factors with number of years N for i=0.09

N	$F_{PS,iN}$	$F_{SP,iN}$	$F_{RP,iN}$	$F_{PR,iN}$	$F_{RS,iN}$	$F_{SR,iN}$	$F_{PK,iN}$
1	1.09	0.917431	0.917431	1.09	1	1	12.11111
2	1.1881	0.84168	1.759111	0.568469	2.09	0.478469	6.316321
3	1.295029	0.772183	2.531295	0.395055	3.2781	0.305055	4.389497
4	1.411582	0.708425	3.23972	0.308669	4.573129	0.218669	3.429652
5	1.538624	0.649931	3.889651	0.257092	5.984711	0.167092	2.856583
6	1.6771	0.596267	4.485919	0.22292	7.523335	0.13292	2.476886
7	1.828039	0.547034	5.032953	0.198691	9.200435	0.108691	2.207672
8	1.992563	0.501866	5.534819	0.180674	11.02847	0.090674	2.007493
9	2.171893	0.460428	5.995247	0.166799	13.02104	0.076799	1.85332
10	2.367364	0.422411	6.417658	0.15582	15.19293	0.06582	1.731334
11	2.580426	0.387533	6.805191	0.146947	17.56029	0.056947	1.632741
12	2.812665	0.355535	7.160725	0.139651	20.14072	0.049651	1.551674
13	3.065805	0.326179	7.486904	0.133567	22.95338	0.043567	1.484073
14	3.341727	0.299246	7.78615	0.128433	26.01919	0.038433	1.427035
15	3.642482	0.274538	8.060688	0.124059	29.36092	0.034059	1.378432
16	3.970306	0.25187	8.312558	0.1203	33.0034	0.0303	1.336666
17	4.327633	0.231073	8.543631	0.117046	36.9737	0.027046	1.300514
18	4.71712	0.211994	8.755625	0.114212	41.30134	0.024212	1.269025
19	5.141661	0.19449	8.950115	0.11173	46.01846	0.02173	1.241449
20	5.604411	0.178431	9.128546	0.109546	51.16012	0.019546	1.217183
21	6.108808	0.163698	9.292244	0.107617	56.76453	0.017617	1.19574
22	6.6586	0.150182	9.442425	0.105905	62.87334	0.015905	1.176722
23	7.257874	0.137781	9.580207	0.104382	69.53194	0.014382	1.159799
24	7.911083	0.126405	9.706612	0.103023	76.78981	0.013023	1.144695
25	8.623081	0.115968	9.82258	0.101806	84.7009	0.011806	1.131181

$$N_1 = \frac{P}{CF} \tag{32}$$

There is a little value in techno-economic study for N_1 computed from equations (31) and (32). When $i\%>0$ is used to estimate N_1, the results incorporate the risk considered in the project undertaken.

Benefit-Cost Analysis

Let us define first the following:

Table 5. The values of various conversion factors with number of years N for i=0.11

N	$F_{PS,iN}$	$F_{SP,iN}$	$F_{RP,iN}$	$F_{PR,iN}$	$F_{RS,iN}$	$F_{SR,iN}$	$F_{PK,iN}$
1	1.11	0.900901	0.900901	1.11	1	1	10.09091
2	1.2321	0.811622	1.712523	0.583934	2.11	0.473934	5.308488
3	1.367631	0.731191	2.443715	0.409213	3.3421	0.299213	3.720119
4	1.51807	0.658731	3.102446	0.322326	4.709731	0.212326	2.93024
5	1.685058	0.593451	3.695897	0.27057	6.227801	0.16057	2.45973
6	1.870415	0.534641	4.230538	0.236377	7.91286	0.126377	2.148878
7	2.07616	0.481658	4.712196	0.212215	9.783274	0.102215	1.92923
8	2.304538	0.433926	5.146123	0.194321	11.85943	0.084321	1.766555
9	2.558037	0.390925	5.537048	0.180602	14.16397	0.070602	1.641833
10	2.839421	0.352184	5.889232	0.169801	16.72201	0.059801	1.543649
11	3.151757	0.317283	6.206515	0.161121	19.56143	0.051121	1.464736
12	3.498451	0.285841	6.492356	0.154027	22.71319	0.044027	1.400248
13	3.88328	0.257514	6.74987	0.148151	26.21164	0.038151	1.346827
14	4.310441	0.231995	6.981865	0.143228	30.09492	0.033228	1.302075
15	4.784589	0.209004	7.19087	0.139065	34.40536	0.029065	1.264229
16	5.310894	0.188292	7.379162	0.135517	39.18995	0.025517	1.23197
17	5.895093	0.169633	7.548794	0.132471	44.50084	0.022471	1.204286
18	6.543553	0.152822	7.701617	0.129843	50.39594	0.019843	1.18039
19	7.263344	0.137678	7.839294	0.127563	56.93949	0.017563	1.159659
20	8.062312	0.124034	7.963328	0.125576	64.20283	0.015576	1.141597
21	8.949166	0.111742	8.07507	0.123838	72.26514	0.013838	1.125799
22	9.933574	0.100669	8.175739	0.122313	81.21431	0.012313	1.111937
23	11.02627	0.090693	8.266432	0.120971	91.14788	0.010971	1.099738
24	12.23916	0.081705	8.348137	0.119787	102.1742	0.009787	1.088975
25	13.58546	0.073608	8.421745	0.11874	114.4133	0.00874	1.079457

- Benefits (*B*) are the advantages to the owner.
- Disbenefits (*D*) are the involved disadvantages to the owner. These disadvantages can be occurred when the project under consideration involves them.
- Costs are the anticipated expenditures for construction, operation, maintenance etc.
- Owner is the one who incurs the costs as the government.

Benefit-cost ratio (*B/C* ratio) is a tool to select the right project based on advantage versus disadvantage analysis. A project is considered to be attractive when the benefits from its execution exceed its associated costs.

The conventional *B/C* ratio is calculated as:

Table 6. The values of various conversion factors with number of years N for i=0.13

N	$F_{PS,iN}$	$F_{SP,iN}$	$F_{RP,iN}$	$F_{PR,iN}$	$F_{RS,iN}$	$F_{SR,iN}$	$F_{PK,iN}$
1	1.13	0.884956	0.884956	1.13	1	1	8.692308
2	1.2769	0.783147	1.668102	0.599484	2.13	0.469484	4.611412
3	1.442897	0.69305	2.361153	0.423522	3.4069	0.293522	3.257861
4	1.630474	0.613319	2.974471	0.336194	4.849797	0.206194	2.586109
5	1.842435	0.54276	3.517231	0.284315	6.480271	0.154315	2.187035
6	2.081952	0.480319	3.99755	0.250153	8.322706	0.120153	1.924256
7	2.352605	0.425061	4.42261	0.226111	10.40466	0.096111	1.739314
8	2.658444	0.37616	4.79877	0.208387	12.75726	0.078387	1.602975
9	3.004042	0.332885	5.131655	0.194869	15.41571	0.064869	1.498992
10	3.394567	0.294588	5.426243	0.18429	18.41975	0.05429	1.417612
11	3.835861	0.260698	5.686941	0.175841	21.81432	0.045841	1.352627
12	4.334523	0.230706	5.917647	0.168986	25.65018	0.038986	1.299893
13	4.898011	0.204165	6.121812	0.16335	29.9847	0.03335	1.256541
14	5.534753	0.180677	6.302488	0.158667	34.88271	0.028667	1.220519
15	6.25427	0.159891	6.462379	0.154742	40.41746	0.024742	1.190321
16	7.067326	0.141496	6.603875	0.151426	46.67173	0.021426	1.164817
17	7.986078	0.125218	6.729093	0.148608	53.73906	0.018608	1.143142
18	9.024268	0.110812	6.839905	0.146201	61.72514	0.016201	1.124622
19	10.19742	0.098064	6.937969	0.144134	70.74941	0.014134	1.108726
20	11.52309	0.086782	7.024752	0.142354	80.94683	0.012354	1.095029
21	13.02109	0.076798	7.10155	0.140814	92.46992	0.010814	1.083187
22	14.71383	0.067963	7.169513	0.139479	105.491	0.009479	1.072919
23	16.62663	0.060144	7.229658	0.138319	120.2048	0.008319	1.063993
24	18.78809	0.053225	7.282883	0.137308	136.8315	0.007308	1.056217
25	21.23054	0.047102	7.329985	0.136426	155.6196	0.006426	1.04943

$$\frac{B}{C} = \frac{\left(Benefits - Disbenefits\right)}{Cost} = \frac{\left(B - D\right)}{C} \tag{33}$$

The modified B/C ratio, which is gaining support includes operation and maintenance (O&M) costs in the numerator and treats them in a manner similar to disbenefits, and is given by:

$$\frac{B}{C} = \frac{\left(Benefits - Disbenefits - \text{O \& M Cost}\right)}{Initial\,investment} \tag{34}$$

The salvage value can also be considered in the denominator.

Table 7. The values of various conversion factors with number of years N for i=0.15

N	$F_{PS,iN}$	$F_{SP,iN}$	$F_{RP,iN}$	$F_{PR,iN}$	$F_{RS,iN}$	$F_{SR,iN}$	$F_{PK,iN}$
1	1.15	0.869565	0.869565	1.15	1	1	7.666667
2	1.3225	0.756144	1.625709	0.615116	2.15	0.465116	4.100775
3	1.520875	0.657516	2.283225	0.437977	3.4725	0.287977	2.919846
4	1.749006	0.571753	2.854978	0.350265	4.993375	0.200265	2.335102
5	2.011357	0.497177	3.352155	0.298316	6.742381	0.148316	1.98877
6	2.313061	0.432328	3.784483	0.264237	8.753738	0.114237	1.761579
7	2.66002	0.375937	4.16042	0.24036	11.0668	0.09036	1.602402
8	3.059023	0.326902	4.487322	0.22285	13.72682	0.07285	1.485667
9	3.517876	0.284262	4.771584	0.209574	16.78584	0.059574	1.39716
10	4.045558	0.247185	5.018769	0.199252	20.30372	0.049252	1.328347
11	4.652391	0.214943	5.233712	0.191069	24.34928	0.041069	1.273793
12	5.35025	0.186907	5.420619	0.184481	29.00167	0.034481	1.229872
13	6.152788	0.162528	5.583147	0.17911	34.35192	0.02911	1.19407
14	7.075706	0.141329	5.724476	0.174688	40.50471	0.024688	1.16459
15	8.137062	0.122894	5.84737	0.171017	47.58041	0.021017	1.140114
16	9.357621	0.106865	5.954235	0.167948	55.71747	0.017948	1.119651
17	10.76126	0.092926	6.047161	0.165367	65.07509	0.015367	1.102446
18	12.37545	0.080805	6.127966	0.163186	75.83636	0.013186	1.087909
19	14.23177	0.070265	6.198231	0.161336	88.21181	0.011336	1.075576
20	16.36654	0.0611	6.259331	0.159761	102.4436	0.009761	1.065076
21	18.82152	0.053131	6.312462	0.158417	118.8101	0.008417	1.056112
22	21.64475	0.046201	6.358663	0.157266	137.6316	0.007266	1.048438
23	24.89146	0.040174	6.398837	0.156278	159.2764	0.006278	1.041856
24	28.62518	0.034934	6.433771	0.15543	184.1678	0.00543	1.036199
25	32.91895	0.030378	6.464149	0.154699	212.793	0.004699	1.031329

The B/C ratio influences the decision on the project approval:

$$if \frac{B}{C}\begin{bmatrix} > 1, accept\, the\, project \\ < 1\ , reject\, the\, project \end{bmatrix} \tag{35}$$

Thus, in case of mutually exclusive projects, B/C ratio gives a method to compare them against each other.

Table 8. The values of various conversion factors with number of years N for i=0.17

N	$F_{PS,iN}$	$F_{SP,iN}$	$F_{RP,iN}$	$F_{PR,iN}$	$F_{RS,iN}$	$F_{SR,iN}$	$F_{PK,iN}$
1	1.17	0.854701	0.854701	1.17	1	1	6.882353
2	1.3689	0.730514	1.585214	0.630829	2.17	0.460829	3.710762
3	1.601613	0.624371	2.209585	0.452574	3.5389	0.282574	2.662198
4	1.873887	0.53365	2.743235	0.364533	5.140513	0.194533	2.144312
5	2.192448	0.456111	3.199346	0.312564	7.0144	0.142564	1.838611
6	2.565164	0.389839	3.589185	0.278615	9.206848	0.108615	1.638911
7	3.001242	0.333195	3.92238	0.254947	11.77201	0.084947	1.49969
8	3.511453	0.284782	4.207163	0.23769	14.77325	0.06769	1.398176
9	4.1084	0.243404	4.450566	0.224691	18.28471	0.054691	1.321709
10	4.806828	0.208037	4.658604	0.214657	22.39311	0.044657	1.262686
11	5.623989	0.17781	4.836413	0.206765	27.19994	0.036765	1.216263
12	6.580067	0.151974	4.988387	0.200466	32.82393	0.030466	1.179209
13	7.698679	0.129892	5.11828	0.195378	39.40399	0.025378	1.149283
14	9.007454	0.111019	5.229299	0.19123	47.10267	0.02123	1.124884
15	10.53872	0.094888	5.324187	0.187822	56.11013	0.017822	1.104836
16	12.3303	0.081101	5.405288	0.185004	66.64885	0.015004	1.088259
17	14.42646	0.069317	5.474605	0.182662	78.97915	0.012662	1.07448
18	16.87895	0.059245	5.533851	0.180706	93.40561	0.010706	1.062976
19	19.74838	0.050637	5.584488	0.179067	110.2846	0.009067	1.053338
20	23.1056	0.04328	5.627767	0.17769	130.0329	0.00769	1.045237
21	27.03355	0.036991	5.664758	0.17653	153.1385	0.00653	1.038412
22	31.62925	0.031616	5.696375	0.17555	180.1721	0.00555	1.032649
23	37.00623	0.027022	5.723397	0.174721	211.8013	0.004721	1.027773
24	43.29729	0.023096	5.746493	0.174019	248.8076	0.004019	1.023642
25	50.65783	0.01974	5.766234	0.173423	292.1049	0.003423	1.020138

Effect of Depreciation (Humphreys, 1991; Blank, Tarquin and Antony, 1989)

Let us define the following terms to be used:

- **Initial cost (C_i):** Also referred as first cost or initial value or single amount. It is the installed cost of the system. The cost includes the purchase price, delivery, installation fee and other depreciable direct cost incurred to ready the asset for use.
- **Depreciation (C_d):** An expenditure that decreases in value with time. This must be apportioned over its lifetime. The term used to describe this loss is known as depreciation.

Table 9. The values of various conversion factors with number of years N for i=0.19

N	$F_{PS,iN}$	$F_{SP,iN}$	$F_{RP,iN}$	$F_{PR,iN}$	$F_{RS,iN}$	$F_{SR,iN}$	$F_{PK,iN}$
1	1.19	0.840336	0.840336	1.19	1	1	6.263158
2	1.4161	0.706165	1.546501	0.646621	2.19	0.456621	3.403268
3	1.685159	0.593416	2.139917	0.467308	3.6061	0.277308	2.459515
4	2.005339	0.498669	2.638586	0.378991	5.291259	0.188991	1.994689
5	2.386354	0.419049	3.057635	0.32705	7.296598	0.13705	1.721317
6	2.839761	0.352142	3.409777	0.293274	9.682952	0.103274	1.543549
7	3.379315	0.295918	3.705695	0.269855	12.52271	0.079855	1.420289
8	4.021385	0.248671	3.954366	0.252885	15.90203	0.062885	1.330974
9	4.785449	0.208967	4.163332	0.240192	19.92341	0.050192	1.264169
10	5.694684	0.175602	4.338935	0.230471	24.70886	0.040471	1.213007
11	6.776674	0.147565	4.4865	0.222891	30.40355	0.032891	1.17311
12	8.064242	0.124004	4.610504	0.216896	37.18022	0.026896	1.141558
13	9.596448	0.104205	4.714709	0.212102	45.24446	0.022102	1.116327
14	11.41977	0.087567	4.802277	0.208235	54.84091	0.018235	1.095971
15	13.58953	0.073586	4.875863	0.205092	66.26068	0.015092	1.079431
16	16.17154	0.061837	4.9377	0.202523	79.85021	0.012523	1.065913
17	19.24413	0.051964	4.989664	0.200414	96.02175	0.010414	1.054812
18	22.90052	0.043667	5.033331	0.198676	115.2659	0.008676	1.045661
19	27.25162	0.036695	5.070026	0.197238	138.1664	0.007238	1.038093
20	32.42942	0.030836	5.100862	0.196045	165.418	0.006045	1.031817
21	38.59101	0.025913	5.126775	0.195054	197.8474	0.005054	1.026602
22	45.92331	0.021775	5.14855	0.194229	236.4385	0.004229	1.02226
23	54.64873	0.018299	5.166849	0.193542	282.3618	0.003542	1.01864
24	65.03199	0.015377	5.182226	0.192967	337.0105	0.002967	1.015617
25	77.38807	0.012922	5.195148	0.192487	402.0425	0.002487	1.013091

- **Salvage value (C_{sal}):** It is the expected market value at the end of useful life of asset. It is negative if dismantling cost or carrying away cost are anticipated. It can be zero also. For example, the window glass has zero salvage value.

$$C_{sal} = C_i - C_d \tag{36}$$

- **Book value (*B*):** It represents the remaining un-depreciated investment on corporate books. It can be obtained after the total amount of annual depreciation charges to date has been subtracted from the first cost (present value/initial cost). The book value is usually determined at the end of each year.

Table 10. The values of various conversion factors with number of years N for i=0.20

N	$F_{PS,iN}$	$F_{SP,iN}$	$F_{RP,iN}$	$F_{PR,iN}$	$F_{RS,iN}$	$F_{SR,iN}$	$F_{PK,iN}$
1	1.2	0.833333	0.833333	1.2	1	1	6
2	1.44	0.694444	1.527778	0.654545	2.2	0.454545	3.272727
3	1.728	0.578704	2.106481	0.474725	3.64	0.274725	2.373626
4	2.0736	0.482253	2.588735	0.386289	5.368	0.186289	1.931446
5	2.48832	0.401878	2.990612	0.33438	7.4416	0.13438	1.671899
6	2.985984	0.334898	3.32551	0.300706	9.92992	0.100706	1.503529
7	3.583181	0.279082	3.604592	0.277424	12.9159	0.077424	1.38712
8	4.299817	0.232568	3.83716	0.260609	16.49908	0.060609	1.303047
9	5.15978	0.193807	4.030967	0.248079	20.7989	0.048079	1.240397
10	6.191736	0.161506	4.192472	0.238523	25.95868	0.038523	1.192614
11	7.430084	0.134588	4.32706	0.231104	32.15042	0.031104	1.155519
12	8.9161	0.112157	4.439217	0.225265	39.5805	0.025265	1.126325
13	10.69932	0.093464	4.532681	0.22062	48.4966	0.02062	1.1031
14	12.83918	0.077887	4.610567	0.216893	59.19592	0.016893	1.084465
15	15.40702	0.064905	4.675473	0.213882	72.03511	0.013882	1.069411
16	18.48843	0.054088	4.729561	0.211436	87.44213	0.011436	1.057181
17	22.18611	0.045073	4.774634	0.20944	105.9306	0.00944	1.047201
18	26.62333	0.037561	4.812195	0.207805	128.1167	0.007805	1.039027
19	31.948	0.031301	4.843496	0.206462	154.74	0.006462	1.032312
20	38.3376	0.026084	4.86958	0.205357	186.688	0.005357	1.026783
21	46.00512	0.021737	4.891316	0.204444	225.0256	0.004444	1.02222
22	55.20614	0.018114	4.90943	0.20369	271.0307	0.00369	1.018448
23	66.24737	0.015095	4.924525	0.203065	326.2369	0.003065	1.015326
24	79.49685	0.012579	4.937104	0.202548	392.4842	0.002548	1.012739
25	95.39622	0.010483	4.947587	0.202119	471.9811	0.002119	1.010594

- **Depreciation rate (D_t):** It is the fraction of the first cost removed through depreciation from corporate book. This rate may be the same, i.e., straight-line (SL) rate or different for each year of the recovery period. Mathematically, it can be written for straight-line (SL) rate depreciation as follows:

$$D_t = \frac{C_i - C_{sal}}{N} \tag{37}$$

The book value at N[th] year can be expressed as:

$$B_N = C_i - ND_t \tag{38}$$

- **Fractional depreciation rate (*1/N*):** It is the ratio of depreciation rate to depreciation.
- **Recovery period (N):** It is the life of the asset (in years) for depreciation and tax purpose. It is also referred as expected life of asset in years.
- **Market value:** It is the actual amount that could be obtained after selling the asset in the open market. For example,
    - The market value of a commercial building tends to increase with period in the open market but the book value will decrease as depreciation charges are taken to account and
    - An electronic equipment (computer system) may have a market value much lower than book value due the rapid change of technology.

If C_i is an initial cost of the asset and C_{sal} is the salvage value after *N* years, then total depreciation or depreciable first cost is given by:

$$C_d = C_i - C_{asl} \tag{39}$$

Let $D_{f1}, D_{f2}, D_{f3}, D_{f4}, D_{f5}, \ldots, D_{fN-1}, D_{fN}$ are the fractional depreciation for each year, then depreciation for m[th] year will be:

$$D_m = D_{fm} C_d \tag{40}$$

The book value of the asset at end of the mth year can be obtained by subtracting the accumulated depreciation expense to that time from the original value of the asset. i.e.:

$$\textbf{Book value} = \textbf{Initial cost}\left(\textbf{first year}\right) - \textbf{Accumulated cost}$$

or:

$$B_m = C_i - C_d \sum_{i=1}^{i=N} D_{fi} \tag{41}$$

Subtracting and adding C_d in the right side of the equation (41), we get:

$$B_m = C_i - C_d + C_d - C_d \sum_{i=1}^{i=N} D_{fi} = C_{asl} + C_d \left(1 - \sum_{i=1}^{i=N} D_{fi}\right) \tag{42}$$

The book value may bear no relation to the resale value.

If fractional depreciation is same for all years, i.e.:

$$D_{f1} = D_{f2} = D_{f3} = D_{f4} = D_{f5} = \ldots = D_{fN-1} = D_{fN} = \frac{1}{N} \tag{43}$$

Then, depreciation for m[th] year is:

$$D_m = \frac{C_d}{N} \tag{44}$$

The accumulated depreciation up to m[th] year becomes:

$$\sum_{i=1}^{i=m} D_m = \frac{mC_d}{N} \tag{45}$$

The book value at the end of the mth year will be:

$$B_m = C_i - C_d \sum_{i=1}^{i=m} D_{fi} = C_i + C_d \frac{m}{N} = C_{asl} + C_d \left(1 - \sum_{i=1}^{i=m} D_{fi}\right) = C_{asl} + C_d \frac{\left(N - m\right)}{N} \tag{46}$$

The depreciation remaining for future years (from m[th] year to N[th] year) is:

$$C_d \sum_{i=m}^{i=N} D_{fi} = C_d \sum_{i=m}^{i=N} \frac{1}{N} = C_d \frac{\left(N - m\right)}{N} \tag{47}$$

The above result can be summarized as follows:

Book value = Salvage value + Future depreciation

The present value of the fractional depreciation is:

$$\begin{aligned} F_{SLP,i,N} &= \frac{1}{N}\left[\frac{1}{\left(1+i\right)} + \frac{1}{\left(1+i\right)^2} + \frac{1}{\left(1+i\right)^3} + \ldots + \frac{1}{\left(1+i\right)^N}\right] \\ &= \frac{1}{N}\frac{\left(1+i\right)^N - 1}{i\left(1+i\right)^N} = \frac{1}{N}F_{RP,i,N} \end{aligned} \tag{48}$$

Suppose now an annual deposit (D_a) is made at the end of each year to a sinking fund (SF) to restore the depreciable value at the end of N years. The annual deposit (D_a) can be expressed for a known first cost C_i and salvage value C_{sal} as:

$$D_a = C_d F_{SP,i,N} F_{PR,i,N} = C_d \frac{i}{\left(1+i\right)^N - 1} \tag{49}$$

The depreciation for any year is the sinking fund increase for that year, which is the deposit for year plus an interest earned by the fund for the year.

Cost Comparisons After Taxes

This can be done in two cases:

- Without depreciation
- With depreciation

Let us first deal the case without depreciation. If *i* the rate of return before taxes and *t* the tax rate, then the rate of return *r* after taxes will be:

$$r = i\left(1-t\right) \tag{50}$$

In this case, an investment compounds at a rate *i* and first cost *P* can be expressed in terms of unacost R as:

$$\begin{aligned} P &= \left[\frac{R\left(1-t\right)}{\left(1+r\right)} + \frac{R\left(1-t\right)}{\left(1+r\right)^2} + \frac{R\left(1-t\right)}{\left(1+r\right)^3} + \ldots + \frac{R\left(1-t\right)}{\left(1+r\right)^N}\right] \\ &= R\left(1-t\right)\frac{\left(1+r\right)^N - 1}{r\left(1+r\right)^N} = R\left(1-t\right)F_{RP,r,N} \end{aligned} \tag{51}$$

From the equation (51), an expression for unacost after taxes can be expressed as:

$$R = \frac{P}{\left(1-t\right)}\frac{r\left(1+r\right)^N}{\left(1+r\right)^N - 1} = \frac{P}{\left(1-t\right)}F_{PR,r,N} \tag{52}$$

Taking into consideration that:

$$S = P\left(1+r\right)^N \tag{53}$$

The equation (52) could be written as:

$$R = \frac{S}{(1-t)} \frac{r}{(1+r)^N - 1} = \frac{S}{(1-t)} F_{SR,r,N} \tag{54}$$

Let us deal now the case with depreciation. Consider C_i as the initial cost of an article that lasts N years with salvage value of C_{sal}. The depreciable cost will be given by the equation (39). There is no tax consideration at the time of purchase of an article.

Let $D_{f1}, D_{f2}, D_{f3}, D_{f4}, D_{f5}, \ldots, D_{fN-1}, D_{fN}$ are the fractional depreciation for each year, then:

$$C_d = \sum_{i=1}^{i=N} C_d D_{fi} \tag{55}$$

If $D_{f1} = D_{f2} = D_{f3} = D_{f4} = D_{f5} = \ldots, D_{fN-1} = D_{fN} = \frac{1}{N}$, then $C_d = C_d$.

Now, the taxable base is reduced $C_d D_{f1}$ and a saving or reduction in taxes amounting to $C_d D_{f1}$ is realized. In this case, by assuming $D_{f1} = D_{f2} = D_{f3} = D_{f4} = D_{f5} = \ldots, D_{fN-1} = D_{fN} = \frac{1}{N}$, the present value is:

$$\begin{aligned} P &= C_d - C_d D_f t \left[\frac{1}{(1+r)} + \frac{1}{(1+r)^2} + \frac{1}{(1+r)^3} + \ldots + \frac{1}{(1+r)^N} \right] \\ &= C_d - C_d t \frac{1}{N} \left[\frac{1}{(1+r)} + \frac{1}{(1+r)^2} + \frac{1}{(1+r)^3} + \ldots + \frac{1}{(1+r)^N} \right] \\ &= C_d - C_d t F_{SLP,r,N} = C_d \left(1 - t F_{SLP,r,N}\right) \end{aligned} \tag{56}$$

After knowing an expression for present value *P* one can write the expressions for unacost *R* and capitalized cost *K* as follows:

$$R = P F_{PR,r,N} = C_d \left(1 - t F_{SLP,r,N}\right) F_{PR,r,N} \tag{57}$$

$$K = P F_{PK,r,N} = C_d \left(1 - t F_{SLP,r,N}\right) F_{PK,r,N} \tag{58}$$

Here, it is important to note that an expression for conversion factor from straight-line depreciation to the present value with tax is given by the equation (48).

ECONOMIC CONSIDERATION

Solar panels are typically mounted at a fixed angle. Such systems have few parts, so are less costly than those with trackers and have fewer operations and maintenance (O&M) considerations. On the other hand, single-axis systems track the Sun east to west as it moves across the sky, allowing them to increase system performance by 20% or more over fixed systems in areas of high insolation. Dual-axis trackers angle through both the x and y axes, typically generating about 8 -10% more energy than single-axis types, depending on location. However, a major consideration when using trackers is land use. They generally take up much more land than fixed systems because their movement can create shadows which can affect neighbouring panels, so they must be spaced appropriately. A typical fixed-tilt system usually uses about 1.6 to 2.4 ha of land per MW, while a single-axis tracker can use anywhere between 1.8 and 3.0 ha/MW.

Commercial scale projects Since a solar power plant does not use any fuel, the cost consists mostly of capital cost with minor operational and maintenance cost. If the lifetime of the plant and the interest rate is known, then the cost per kWh can be calculated. This is called the leveled cost. The first step in the calculation is to determine the investment for the production of 1 kWh in a year. One of the famous commercial scale projects is the Andasol-1 project. Tables 11 and 12 give the main information of this project. For example, the fact sheet of the Andasol-1 project shows a total investment of 310 million euros for a production of 179 GWh a year. Since 179 GWh is 179 million kWh, the investment

Table 11. The main information on the Andasol-1 project

Item Number	Information Kind	Description
1.	Status	Operational
2.	Country	Spain
3.	City	Aldeire
4.	Region	Granada
5.	Lat/Long Location	37°13′ 50.83″ North, 3°4′ 14.08″ West
6.	Land Area	200 hectares
7.	Solar Resource	2,136 kWh/m^2/yr
8.	Source of Solar Resource	Meteorological Station
9.	Electricity Generation	158,000 MWh/yr (Expected/Planned)
10.	Contact(s)	Manuel Cortés; María Sánchez
11.	Company	ACS/Cobra Group
12.	Break Ground	July 3, 2006
13.	Start Production	November 26, 2008
14.	Construction Job-Years	600
15.	Annual O&M Jobs	40
16.	PPA/Tariff Date	September 15, 2008
17.	PPA/Tariff Type	Real Decreto 661/2007
18.	PPA/Tariff Rate	27.0 Euro cents per kWh
19.	PPA/Tariff Period	25 years
20.	Project Type	Commercial

Table 12. The main plant configuration of the Andasol-1 project

	Technology	Parabolic Trough
Solar Field	Solar-Field Aperture Area	510,120 m^2
	Number of Solar Collector Assemblies (SCAs)	624
	Number of Loops	156
	Number of SCAs per Loop	4
	SCA Aperture Area	817 m^2
	SCA Length	144 m
	Number of Modules per SCA	12
	SCA Manufacturer (Model)	UTE CT Andasol-1 (SKAL-ET)
	Mirror Manufacturer (Model)	Flabeg (RP3)
	Number of Heat Collector Elements (HCEs)	11,232
	HCE Manufacturer (Model)	Schott (PTR70)
	Number of HCEs	11,232
	HCE Manufacturer (Model)	Solel (UVAC 2008)
	Heat-Transfer Fluid Type	Dowtherm A
	Solar-Field Inlet Temperature	293°C
	Solar-Field Outlet Temperature	393°C
	Solar-Field Temperature Difference	100°C
Power Block	Turbine Capacity (Gross)	50.0 MW
	Turbine Capacity (Net)	49.9 MW
	Turbine Manufacturer	Siemens (Germany)
	Output Type	Steam Rankine
	Power Cycle Pressure	100.0 bar
	Cooling Method	Wet cooling
	Cooling Method Description	Cooling towers
	Turbine Efficiency	38.1% @ full load
	Annual Solar-to-Electricity Efficiency (Gross)	16%
	Fossil Backup Type	HTF heater
	Backup Percentage	12%
Thermal Storage	Storage Type	2-tank indirect
	Storage Capacity	7.5 hour(s)
	Thermal Storage Description	28,500 tons of molten salt. 60% sodium nitrate, 40% potassium nitrate. 1,010 MWh. Tanks are 14 m high and 36 m in diameter

per kWh a year production is 310/179 = 1.73 euro. Another example is Cloncurry solar power station in Australia. It produces 30 million kWh a year for an investment of 31 million Australian dollars. So, this price is 1.03 Australian dollars for the production of 1 kWh in a year. This is significantly cheaper than Andasol-1, which can partially be explained by the higher radiation in Cloncurry over Spain. The investment per kWh cost for one year should not be confused with the cost per kWh over the complete lifetime of such a plant.

In most cases the capacity is specified for a power plant (for instance Andasol-1 has a capacity of 50MW). This number is not suitable for comparison, because the capacity factor can differ. If a solar power plant has heat storage, then it can also produce output after sunset, but that will not change the capacity factor, it simply displaces the output. The average capacity factor for a solar power plant, which is a function of tracking, shading and location, is about 20%, meaning that a 50MW capacity power plant will typically provide a yearly output of 50 MW x 24 hrs x 365 days x 20% = 87,600 MWh/year, or 87.6 GWh/yr.

Although the investment for one kWh year production is suitable for comparing the price of different solar power plants, it doesn't give the price per kWh yet. The way of financing has a great influence on the final price. If the technology is proven, an interest rate of 7% should be possible. However, for a new technology investors want a much higher rate to compensate for the higher risk. This has a significant negative effect on the price per kWh. Independent of the way of financing, there is always a linear relation between the investment per kWh production in a year and the price for 1 kWh (before adding operational and maintenance cost). In other words, if by enhancements of the technology the investments drop by 20%, then the price per kWh also drops by 20%.

If a way of financing is assumed where the money is borrowed and repaid every year, in such way that the debt and interest decreases, then the equation (14) can be used to calculate the division factor $F_{RP.i,N}$. For a lifetime of 25 years and an interest rate of 7%, the division number is $11.65358 \approx 11.65$ (see Table 3). For example, the investment of Andasol-1 was 1.73 euro, divided by 11.65 results in a price of 0.15 euro per kWh. If one cent operation and maintenance cost is added, then the leveled cost is 0.16 euro.

Other ways of financing, different way of debt repayment, different lifetime expectation, different interest rate, may lead to a significantly different number. If the cost per kWh may follow the inflation, then the inflation rate can be added to the interest rate. If an investor puts his money on the bank for 7%, then he is not compensated for inflation. However, if the cost per kWh is raised with inflation, then he is compensated and he can add 2% (a normal inflation rate) to his return. The Andasol-1 plant has a guaranteed feed-in tariff of 0.21 euro for 25 years. If this number is fixed, it should be realized that after 25 years with 2% inflation, 0.21 euro will have a value comparable with 0.13 euro now.

Finally, there is some gap between the first investment and the first production of electricity. This increases the investment with the interest over the period that the plant is not active yet. The modular solar dish (but also solar photovoltaic and wind power) have the advantage that electricity production starts after first construction.

Given the fact that solar thermal power is reliable, can deliver peak load and does not cause pollution, a price of US$0.10 per kWh starts to become competitive. Although a price of US$0.06 has been claimed with some operational cost a simple target is 1 dollar (or lower) investment for 1 kWh production in a year. Thus, energy price is one of the most influential factors in all case studies, yet it is among the most volatile indexes in the world market scale. Any deviation in energy price may significantly change the financial feasibility of any project.

Experimental Dual Tracking Verse Horizontal Fixed

An experimental setup was built in the Texas State University for evaluating the economic feasibility of PV dual tracking generating method verse a horizontally installed PV system of the same PV panels (Asiabanpour, Almusaied, Aslan, Mitchell, Leake, Lee, Fuentes, Rainosek, Hawkes and Bland, 2016). The experimental setup was comprised of two 190 Watt mono-crystalline photovoltaic panels that contain 72 cells each with the following dimensions (125 * 125 mm) and a weight of 15 kg (Solar Systems USA Online Solar Panels, 2016), rheostats, a manual dual-axis mechanical system, data acquisition system, and proper wiring. The power generated by these photovoltaic panels should be 90% for ten years and 85% for twenty-five years (Solar Systems USA Online Solar Panels 2016). The lifespan of the panels was assumed to be 25 years with a degradation rate of 0.5% based on the relevant literature (Jordan and Kurtz, 2013). A fixed-flat panel was mounted to the roof. The second panel was mounted to the manual dual-axis sun tracking system (Asiabanpour, Almusaied, Aslan, Mitchell, Leake, Lee, Fuentes, Rainosek, Hawkes and Bland, 2016).

To evaluate the monetary savings due to the use of the sun tracking mechanism, first the average power saving per hour should be calculated. For this purpose let us consider that, the average saving for one hour is x. Basing on a five-hour-interval data collection x was found to be 0.0686 kWh (see chapter 7). The extra power gained by a system of one PV panel using the dual-axis sun tracking system versus fixed horizontal format and considering a value, y, of the degradation rate and historic local number of sunny hours per year, j, could be calculated as follows:

$$Average\,power\,saving\,in\,year\,i = \text{j.x.}\left(1-y\right)^{i-1};1 \leq i \leq N \tag{59}$$

where N is the life cycle of the tracking system (in years). Then, the extra power gained by a 1-, 4-, and 9-panel array using the dual-axis sun tracking system versus fixed horizontal format and considering 0.5% degradation rate and historic local number of sunny hours per year (i.e., 2645 h in Austin Texas (Average Annual Sunshine in American Cities, 2016)) could be calculated. A summary of two scenarios is given in the comparison study (see Table 13). Then, current local energy price rates and predicted inflation market price should be applied in the calculation (i.e., the energy market rate is 11.3 cents per kWh in Texas with a historic average inflation rate of 1.81% (The price of electricity in your State, 2011). Then,

$$Benefit\,year\,m = q\left(E_i - E_c\right)\left(1+f\right)^{m-1} \tag{60}$$

where E_i is the average power saving in year i, E_c is the sun tracking power consumption, q is the energy rate in base year, and f average energy inflation rate.

Taking into consideration that, the initial dual-axis sun tracking cost capable of moving up to 10 panels = $1000 and considering all power-saving benefits and investment cost, the present value for each year was calculated and shown in Tables 14, 15 and 16. Figures 1 to 3 demonstrate these calculations.

Taking into consideration that the life cycle of the system is 25 years, the breakeven point (BP) for each scenario could be calculated from the following equation:

Table 13. kWh extra power gained by using dual-axis sun tracking system against fixed horizontal system using 1, 4, and 9 panels

Year Number	1-Panel System	4-Panel System	9-Panel System
1	181.4	725.8	1633.0
2	180.5	722.2	1624.
3	179.6	718.5	1616.7
4	178.7	715.0	1608.7
5	177.8	711.4	1600.6
6	177.0	707.8	1592.6
7	176.1	704.3	1584.6
8	175.2	700.8	1576.7
9	174.3	697.3	1568.8
10	173.4	693.8	1561.0
11	172.6	690.3	1553.2
12	171.7	686.9	1545.4
13	170.9	683.4	1537.7
14	170.0	680.0	1530.0
15	169.2	676.6	1522.4
16	168.3	673.2	1514.7
17	167.5	669.9	1507.2
18	166.6	666.5	1499.6
19	165.8	663.2	1492.1
20	165.0	659.9	1484.1
21	164.1	656.6	1477.3
22	163.3	653.3	1469.9
23	162.5	650.0	1462.5
24	161.7	646.8	1455.2
25	160.9	643.5	1447.9

$$NP = \left[1000 * integer\left(\frac{25+N}{N}\right)\right] + \sum_{m=1}^{m=N} P\left[\begin{pmatrix}\text{Average power saving in year m} \\ -\text{Sun tracking power consumption}\end{pmatrix} * \text{Energy rate}*\left(1+\text{Energy inflation rate}\right)^{m-1}\right] \quad (61)$$

where P is the present value, NP is net present value, and N is the first year that NP equation sign is turned from negative to positive.

To evaluate the impact of the energy price change in the breakeven points, the model will be solved for two extreme scenarios, when the energy price inflates twice and deflates equal to the historical energy inflation rate. Table 17 illustrates sensitivity analysis of the breakeven points for three-panel configurations

Table 14. Overall saving for dual-axis sun tracking versus fixed for a 1-panel configuration

Year Number	Benefit/Cost ($)	Present Value ($)	Sum of Present Values ($)
0	-1000.0	-1000.0	-1000.0
1	17.5	17.4	-982.6
2	17.7	17.4	-965.3
3	18.0	17.4	-947.8
4	18.2	17.5	-930.4
5	18.4	17.5	-912.9
6	18.6	17.5	-895.3
7	18.8	17.6	-877.8
8	19.1	17.6	-860.1
9	19.3	17.6	-842.5
10	19.5	17.7	-824.8
11	19.8	17.7	-807.1
12	20.0	17.8	-789.3
13	20.3	17.8	-771.5
14	20.5	17.8	-753.7
15	20.7	17.9	-735.8
16	21.0	17.9	-717.9
17	21.2	17.9	-700.0
18	21.5	18.0	-682.0
19	21.8	18.0	-664.0
20	22.0	18.1	-645.9
21	22.3	18.1	-627.9
22	22.6	18.1	-609.7
23	22.8	18.2	-591.6
24	23.1	18.2	-573.4

and three energy inflation/deflation scenarios. As shown in Table 17, the increase in the energy price will positively impact the usage of the solar energy as it makes them to reach the breakeven point faster.

The power gained was calculated to include a degradation rate of .5% per year for 25 years. A market-available dual-axis sun tracking system, with an initial cost of $1000, along with its specifications was used to calculate the power consumption and benefit. The breakeven point was found for three different scenarios. The first scenario with a system capacity for 1 panel had no breakeven point (NP never turns positive). The second with a system capacity for 4 panels had a breakeven point of 13 years. The third system with a capacity of 9 panels had a breakeven point of 6 years. Finally, sensitivity analysis of the breakeven points for three-panel configurations and three energy inflation/deflation scenarios illustrated that higher energy prices make the sun tracking systems more economical.

Table 15. Overall saving for dual-axis sun tracking versus fixed for a 4-panel configuration

Year Number	Benefit/Cost ($)	Present Value ($)	Sum of Present Values ($)
0	-1000.0	-1000.0	-1000.0
1	79.0	78.3	-921.7
2	80.1	78.5	-843.3
3	81.1	78.7	-764.6
4	82.1	78.9	-685.7
5	83.2	79.1	-606.5
6	84.2	79.4	-527.2
7	85.3	79.6	-447.6
8	86.4	79.8	-367.8
9	87.	80.0	-287.8
10	88.6	80.2	-207.6
11	89.8	80.5	-127.1
12	90.9	80.7	-46.4
13	92.1	80.9	34.5

Table 16. Overall saving for dual-axis sun tracking versus fixed for a 9-panel configuration

Year Number	Benefit/Cost ($)	Present Value ($)	Sum of Present Values ($)
0	-1000.0	-1000.0	-1000.0
1	181.6	179.8	-820.2
2	183.9	180.3	-640.0
3	186.3	180.8	-459.2
4	188.7	181.3	-277.8
5	191.1	181.8	-96.0
6	193.6	182.4	86.4

Dual Tracking Verse Latitude Tilted Fixed

The electrical energy output from a 1 kW_{peak} PV system installed in Stuttgart, Germany, has been calculated over the year 2013, based on the applied mathematical model (see chapter 7). It has been found that the difference in the electrical energy output in case of tracking and no tracking is equal to 350 kWh/year. Therefore, the net gain in electricity will equal to 315 kWh/year, based on a 10% energy lost in the tracker. The net gain in electricity corresponds to 95 USD/year, based on a feed-in-tariff of 0.3 USD/kWh (Jäger-Waldau, Szabo', Scarlat and Monforti-Ferrario, 2011; Krozer, 2013) for electricity generated from renewable energy in Europe. The price of a single axis tracker supporting a 1 kW_{peak} PV panel systems is about 200 USD, therefore, the revenue time for the tracker based on the electrical energy saving is about 2.1 years (Sharaf Eldin, Abd-Elhady and Kandil, 2016).

Figure 1. Breakeven point for 1- panel configuration (13 years)

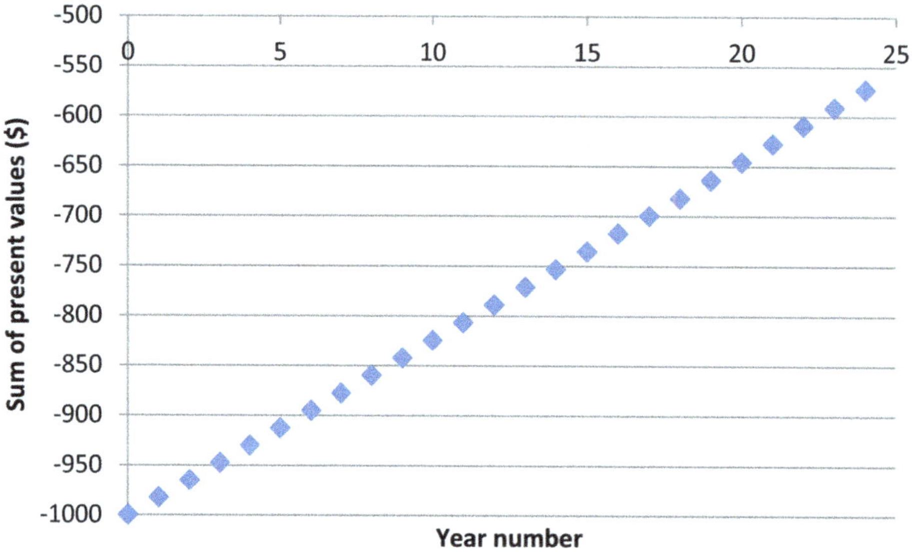

Figure 3. Breakeven point for 9- panel configuration (13 years)

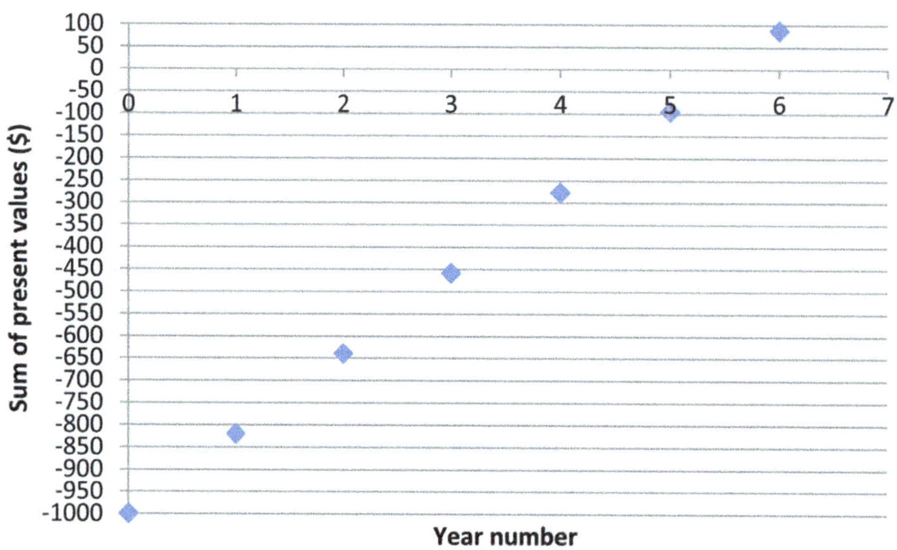

The electrical energy output from a 1 kW_{peak} PV system installed in Berlin, Germany, has been also calculated over the year 2013, based on the applied mathematical model and the meteorological data of Berlin (see chapter 7). It has been found that the net gain in electricity due to tracking is equal to 365 kWh/year. This value corresponds to a money gain of 110 USD/year, based on a feed-in-tariff of 0.3 USD/kWh. Therefore, the revenue time for the tracker based on the electrical energy gain and the price of the tracker, i.e. 200 USD, is about 1.8 years. It can be concluded that PV panel solar tracking is a

Table 17. Sensitivity analysis of the breakeven points for three-panel configurations and three energy inflation/deflation scenarios

Panel Configuration	Energy Price Inflation (3.62%)	Historical Energy Price Inflation (1.81%)	Energy Price Deflation (-1.81%)
1-Panel configuration	Never	Never	Never
4-Panel configuration	12 years	13 years	17 years
9-Panel configuration	6 years	6 years	7 years

Figure 2. Breakeven point for 4- panel configuration (13 years)

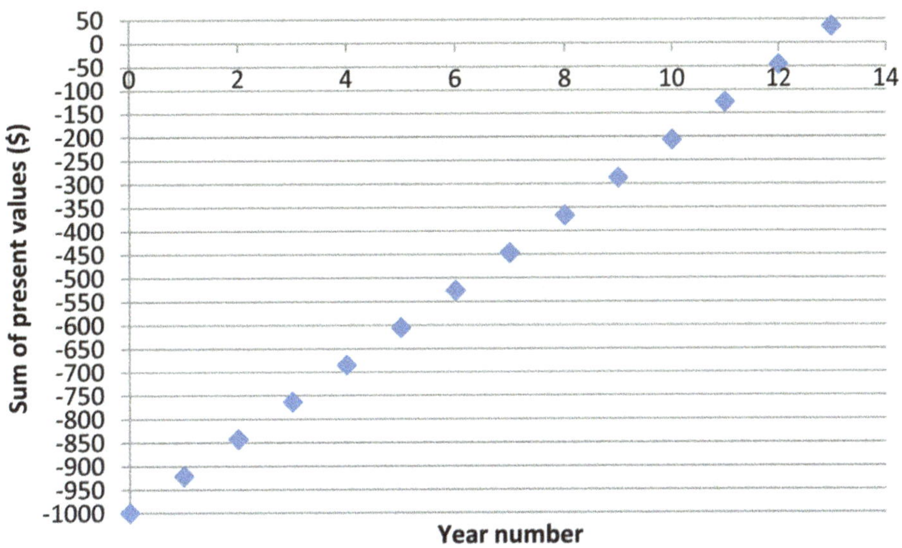

very attractive technique to increase the power output in cold and cloudy countries. It can be concluded from the performed cost analysis for the cities of Stuttgart and Berlin that, it is highly recommended and beneficial to use a solar tracking system for PV panels.

PV Systems Verse Traditional Sources of Energy

Two grid-connected PV systems with different mounting schemes were installed in central Iowa, USA; one roof-mounted stationary system and one pole-mounted dual-axis tracking system (Warren, 2008). The technical description of the studied systems is given in chapter 7.

The economic analysis, for both systems, focuses on measures of life cycle costs (LCC), payback period, internal rate of return (IRR), and average incremental cost of solar energy. For the LCC analysis, three discount rates are assumed; the first representing general inflation, the second essentially representing a risk-free investment and the last indicative of long-term liquidity and risk one may be subjected to in competitive market conditions. These specific discount factors will allow the systems to be compared to a wide range of alternative investment risk levels and also compared to each other directly. The LCC analysis is performed over an assumed useful system life period of twenty-five years. Payback period is defined as the number of years that it takes for the accumulated annual savings to equal to or become greater than the accumulated annual expenses of the system. Internal rate of return estimated in this research is defined as the discount rate at which the net present value (NP) of the investment in PV over the assumed life of the system is zero. The average incremental cost of solar energy generated by the systems were also estimated considering all expenses related to the PV systems and estimated lifetime energy generation. Finally, the potential economic feasibility of the systems was assessed assuming more favorable economic conditions in that the incremental energy cost of utility supplied energy was varied and initial rebates and feed-in tariffs were considered.

All initial and future expenses and revenues were quantified including: initial costs, operating and maintenance costs, energy savings, and end-of-life salvage value. However, the costs of the data acquisition (DAQ) system and other expenses considered "non-typical" for residential and commercial applications were omitted from the analysis. Initial costs and first-year energy savings for stationary and dual-axis tracking systems were attained experimentally. All future expenses and revenues for both PV systems were estimated over their assumed useful life on an annual basis. Incentives applicable to this project include sales tax exemption and net-metering. The economic analysis neglects all tax considerations; the PV systems evaluated in this research were not financed, and no initial or operating costs will be claimed as business expenses.

One of the most important issues surrounding the application of PV for building energy generation is that of economic feasibility. The economics of PV systems can be estimated in terms of many different measures. However, the basic economic premise for an investment in PV is comparing a known initial investment and estimated future operating/maintenance expenses with projected future savings in energy costs generated by the system (Duffie and Beckman, 2013). The economics of a grid-connected PV system can vary significantly depending on the solar resource, site, performance of the system, interconnection agreement type, operation and maintenance costs, financial incentives, costs of energy, and initial equipment and installation costs of the system.

Economic results for both the stationary and tracking systems include LCC, payback period, IRR, and average incremental cost of solar energy. Each economic parameter allows for comparisons to be made to other investments or alternative sources of energy and can offer insight to the attractiveness of an investment in PV. A LCC analysis is a method of analyzing the initial and future annual expenses and savings associated with a system over the life of the equipment; this method normalizes the annual cash flow to an overall net present value (NP) assuming a particular discount rate (Park, 2002). The life-cycle cost can be defined as "the total discounted dollar cost of owning, operating, maintaining, and disposing of a system" over a given period of time (Fuller and Petersen, 1995). Payback period can be useful for assessing the amount of time required for the PV system to pay itself back or "break-even". Estimating the economic internal rate of return on an investment in a PV system allows for a direct comparison to other types of financial investments with associated expectations of return. Finally, approximating the cost of energy generated by a PV system over its assumed life is useful for comparing options of other alternative sources of energy such as energy supplied by a local utility.

Warren (2008) found that, grid-connected PV systems, used for building energy generation in the Upper Midwest, are not economically feasible. Poor economic results for the stationary and dual-axis tracking systems are primarily due to high initial costs of PV systems, relatively low incremental costs of utility supplied electrical energy, and insufficient financial incentives for the implementation and/or operation of PV systems. However, due to the volatile nature and unpredictability of future energy costs, actual economics of the systems may vary significantly from the provided estimations. Additionally, the performance and economics of grid-connected PV systems are very sensitive to the specific site and applicable circumstances (e.g., solar resource, incremental electric rates, available incentives, etc.).

The cost effectiveness of photovoltaic panels for use by the University of Nebraska-Lincoln as a means of electricity generation was investigated (Schwarz, 2010) and it was found that, photovoltaic panels are not cost effective for the University of Nebraska – Lincoln campus. Current electrical rates are too inexpensive, due mostly to a cheap source of coal, oil, and natural gas. Also, solar panels themselves are too expensive, attributed to the lack of demand, high costs of production and stricter commercial building codes.

PV Economic Feasibility Applications

The diffusion of photovoltaic systems is hindered until today by high investment costs. However, PV power generation is justified only for special purposes. Anyway, as a consequence of cost reduction and improved component reliability, it is ever more frequently taken into account to install PV systems for small scale power generation in remote areas, where the supply of electric energy by means of the utility network or other conventional systems is difficult or costly. Applications include telecommunication systems, rural electrification, cathode protection and water lifting.

In order to evaluate the economic convenience of PV systems in any country, we can compare them with a diesel generator. The analysis is based on the following data:

- Yearly required energy: 500, 1000, 2000 or 10000 kWh.
- Corresponding diesel generator power: 0.75, 1.5, 3 and 15kW.
- Generator cost in the four cases: 320\$, 320\$, 320\$ and 800\$ respectively (no diesel reduction for low power requirements, as minimum available size is 2 to 3 kW).
- Generator average efficiency: 20%.
- Generator useful life: 10 years.
- Generator maintenance cost: major overhaul needed every 2 years with costs at 30% of initial cost.
- Oil price: 0.5\$/liter.
- PV array required power (assuming an irradiation on the module plane of 2000kWh/m^2 per year): 0.25, 0.5, 1 and 5 kW$_{peak}$ respectively for a yearly required energy of 500, 1000, 2000 and 10000 kWh.
- PV module useful life: 25 years.
- Battery required in the four cases: 5, 10, 20 and 100kWh respectively.
- Battery cost: 400\$/kWh
- Battery useful life: 10years
- Conversion system required in the 4 cases: 1.25, 2.5, 5 and 25kVA respectively.
- Conversion system useful life: 20 years.
- PV system yearly maintenance cost 1.5% of the initial cost.
- Net discount rate: 5%.

Figure 4 compares the cost of the energy output from a diesel generator with the cost of the energy output from a PV system. Figure 4 shows that, at the present cost of photovoltaic modules (0.5\$/W), PV power generation is profitable only for very low yearly energy needs. This is also true if PV systems

Figure 4. The produced energy price verse produced power price using PV and diesel generators. (♦) stands for diesel generator of 10000kWh/year, (■) stands for diesel generator of 2000kWh/year, (▲) stands for diesel generator of 1000kWh/year, (ж) stands for PV system and (x) stands for diesel generator of 500kWh/year

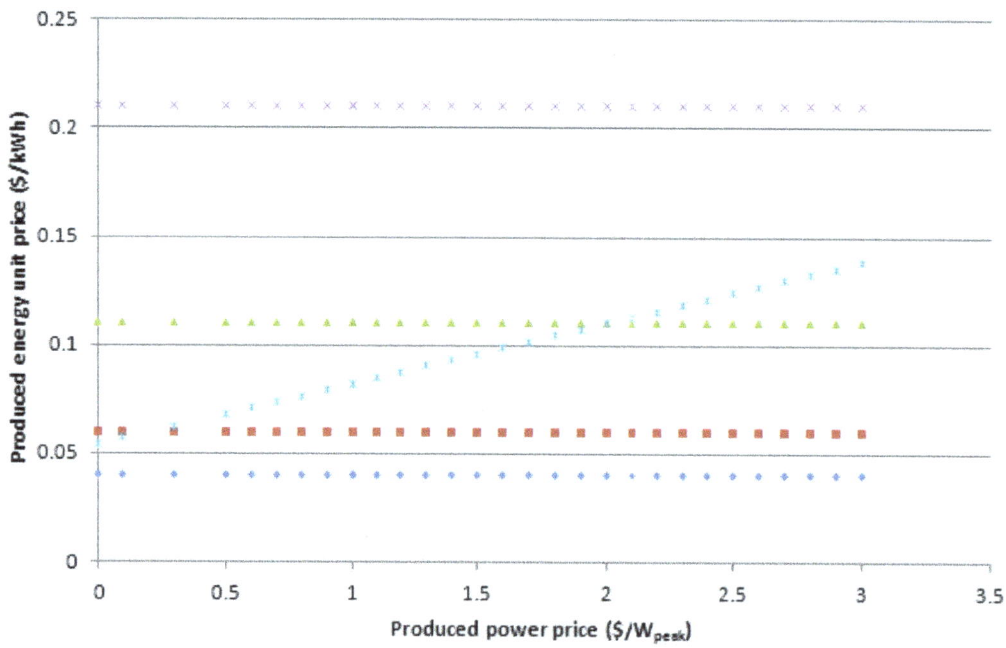

are used for water lifting for drinking or irrigation purposes. In this case, solar energy is converted to electricity using photovoltaic panels and then mechanical energy by means of a pump.

The comparison procedure is similar to that carried out for PV power generation: the result is that PV water lifting is already economically convenient when the daily energy needed, expressed in m^4 (i.e. volume of lifted water expressed in m^3 multiplied by the head expressed in m), is a few hundreds m^4. That is the case of small scale applications only.

Here, it should be pointed out that, even if PV systems are usually more costly than diesel generators, the latter generally suffer from unmanned operation, whilst the fixed panel installation, having no moving parts, are the best choice wherever there is a need for electric energy and there is no personnel to run the plant.

Moreover, the economic analysis has taken into account mono and poly-crystalline silicon module technologies and costs. It should be superseded if the amorphous, thin films and organic PV cells technology went out of the present research and development stage.

CONCLUSION

Tracking technology has made some drastic improvements. Equipment and O&M costs have come down and standards are higher. Now there are a lot more projects in the field that use tracker technology, it has become a much more bankable aspect of a solar project, and developers are more comfortable using it.

However, solar tracking systems would probably increase the efficiency of a PV module in comparison with latitude tilted PV module. Thus, per panel you get more kilowatt hours. The big question is when and where this can be arrived. There are many factors that affect the performance of PV panels, especially crystalline silicon panels, e.g. overheating due to excessive exposure to solar irradiance in a hot climate as in Sunbelt countries. So, it could be the case that a tracking system is not necessary for a Sunbelt country. For example, in comparison with latitude tilted PV systems, the gain in electrical energy from dual tracking the sun is about 39% in case of a cold city as Berlin, Germany, while this gain does not exceed 8% in case of a hot city as Aswan, Egypt, due to overheating of the PV panels. However, if the energy needed for running the tracking system, which ranges from 5% to 10% of the energy generated, is included in this analysis then tracking the sun will not be feasible in hot countries. The issue with dual axis is basically that the added generation from upgrading from single to dual-axis doesn't economically pan out in terms of the extra materials and costs. There are certainly players in the market that are still doing dual axis and most CPV companies still in the game are typically using dual-axis tracking, but single-axis is more preferred. Dual-axis players have found success in smaller-scale markets. Much of that success comes from consumers that may not be as economically driven, such as corporates that have a goal to offset as much consumption as possible, or those that favour energy output rather than the lowest leveled cost of energy. However, there is a little bit of a trade-off. Trackers are more expensive because now you have moving parts. Instead of something that is just sitting on the ground you now have a motor that moves the panels. O&M costs will be higher, as well. The motor needs to be maintained throughout the life of the tracker.

After giving a survey on tracking market and introducing the base elements of economic analysis, several examples were studied in order to evaluate the economic feasibility of dual tracking systems in comparison with horizontally installed fixed panels and with latitude tilted fixed panels were it was found that tracking is feasible in relation with these two cases at high latitudes and it questionable at sunny belt region. Moreover, the comparison of the effectiveness of tracking in relation to monthly adjusted tilt of PV panels where the simplicity and high energy gain is not considered. This will be done in the near future. Anyway, the diffusion of photovoltaic systems is hindered until today by high investment costs. However, PV power generation is justified for special purposes. It is clearly demonstrated that, the small scale applications such as telecommunication systems, rural electrification, cathode protection and water lifting are economically feasible.

REFERENCES

Aberle, A. G. (2009). Thin-film solar cells. *Thin Solid Films*, *517*(17), 4706–4710. doi:10.1016/j.tsf.2009.03.056

Agee, J. T., Obok-Obok, A. & de Lazzer, M. (2007). Solar tracker technologies: market trends and field applications. *Advanced Materials Research, 18-19*, 339-344. Retrieved from www.scientific.net/AMR.18-19.339

Alexandru, C., & Tatu, N. I. (2013). Optimal design of the solar tracker used for a photovoltaic string. *J. Renew. Sustain. Energy*, *5*(2), 023133. doi:10.1063/1.4801452

Asiabanpour, B., Almusaied, Z., Aslan, S., Mitchell, M., Leake, E., Lee, H., Fuentes, J., Rainosek, K., Hawkes, N. & Bland, A. (2016). Fixed versus sun tracking solar panels: an economic analysis. *Clean Techn. Environ. Policy*, 1-9. doi:10.1007/s10098-016-1292-y

Average Annual Sunshine in American Cities. (2016). Retrieved from https://www.currentresults.com/Weather/US/average-annual-sunshineby-city.php

Bentaher, H., Kaich, H., Ayadi, N., Ben Hmouda, M., Maalej, A., & Lemmer, U. (2014). A simple tracking system to monitor solar PV panels. *Energy Conversion and Management*, *78*, 872–875. doi:10.1016/j.enconman.2013.09.042

Blank, L. T., & Tarquin, A. J. (2012). *Engineering Economy* (7th ed.). New York: McGraw-Hill Int.

Clifford, M., & Eastwood, D. (2004). Design of a novel passive solar tracker. *Solar Energy*, *77*(3), 269–280. doi:10.1016/j.solener.2004.06.009

Compaan, A. D. (2006). Photovoltaic: Clean power for the 21st century. *Solar Energy Materials and Solar Cells*, *90*(15), 2170–2180. doi:10.1016/j.solmat.2006.02.017

Concentrating Solar Power Projects - Andasol-1. (n.d.). Retrieved from https://www.nrel.gov/csp/solarpaces/project_detail.cfm/projectID=3

Dakkak, M., & Babeli, A. (2012). Design and performance study of a PV tracking system (100W-24Vdc/220Vac). *Energy Procedia*, *19*, 91–95. doi:10.1016/j.egypro.2012.05.188

Duffie, J. A., & Beckman, W. A. (2013). *Solar engineering of thermal processes* (3rd ed.). New York, NY: Wiley & Sons. doi:10.1002/9781118671603

Englander, D. (2009). Solar's New Important Players: Chevron, Lockheed Martin. *Seeking Alpha*. Retrieved from http://seekingalpha.com/article/138253-solar-s-new-important-players-chevronlockheed-martin?source=email

EPIA. (2012). Global market outlook for photovoltaic until 2016. European Photovoltaic Industry Association.

Fanney, A. H., & Dougherty, B. P. (2001). Building Integrated Photovoltaic Test Facility. *Journal of Solar Energy Engineering, Transactions of the ASME*, *123*(3), 194-199. doi:10.1115/1.1385823

Fuller, S. K., & Petersen, S. R. (1995). *Life-Cycle Costing Manual for the Federal Energy Management Program. Nist Handbook, 135*. Gaithersburg, MD: U.S. Dept. of Commerce, Technology Administration, National Institute of Standards and Technology.

Giakoumelos, E., Malamatenois, C., Hadjipaschalis, I., Kourtis, G., & Poullikkas, A. (2008). Energy policy deployment for the promotion of distributed generation technologies in the mediterranean region. *Proceedings of the international conference on deregulated electricity market issues in south-eastern Europe.*

Higgens, J. M. (2009). Your Solar Powered Future: It's Closer Than You Thought. *The Futurist*, (May-June), 25–29.

Huang, B. J., & Sun, F. S. (2007). Feasibility study of one axis three positions tracking solar PV with low concentration ratio reflector. *Energy Conversion and Management*, *48*(4), 1273–1280. doi:10.1016/j.enconman.2006.09.020

Humphreys, K. K. (1991). Jelen's cost and optimization engineering (3rd ed.). New York: McGraw-Hill, Inc. Retrieved from http://www.library.trisakti.ac.id/opac/index.php/collection/detail/0000000000000059903

International Energy Agency (IEA). (2006). *Key World Energy Statistics*. United Nations. Retrieved from http://www.iea.org/Textbase/nppdf/free/2006/Key2006.pdf

IRENA. (2012). *Renewable energy technologies. Cost Analysis series, Power sector issue 4/5*. Solar Photovoltaic.

Ismail, M. S., Moghavvemi, M., & Mahlia, T. M. I. (2013). Design of an optimized photovoltaic and microturbine hybrid power system for a remote small community: Case study of Palestine. *Energy Conversion and Management*, *75*, 271–281. doi:10.1016/j.enconman.2013.06.019

Jäger-Waldau, A., Szabo, M., Scarlat, N., & Monforti-Ferrario, F. (2011). Renewable electricity in Europe. *Renewable & Sustainable Energy Reviews*, *15*(8), 3703–3716. doi:10.1016/j.rser.2011.07.015

Jordan, D. C., & Kurtz, S. R. (2013). Photovoltaic degradation rates-an analytical review. *Progress in Photovoltaics: Research and Applications*, *21*(1), 12–29. doi:10.1002/pip.1182

Kanellos, M. (2005). Google founders invest in solar energy. *CNET News*. Retrieved from http://news.cnet.com/8301-10784_3-5749586-7.html

Koussa, M., Haddadi, M., Saheb, D., Malek, A., & Hadji, S. (2012). Sun tracker systems effects on flat plate photovoltaic PV systems performance for different Sky States: A case of an arid and hot climate. *Energy Procedia*, *18*, 817–838. doi:10.1016/j.egypro.2012.05.097

Krozer, Y. (2013). Cost and benefit of renewable energy in the European Union. *Renewable Energy*, *50*, 68–73. doi:10.1016/j.renene.2012.06.014

LaMonica, M. (2009). California utility PG&E buys big into solar power. *CNET News*. Retrieved from http://news.cnet.com/8301-11128_3-10171036-54.html

Lee, J. F., Rahim, N. A. & Al-Turki, Y. A. (2013). Performance of dual-axis solar tracker versus static solar system by segmented clearness index in Malaysia. *International Journal of Photoenergy, 2013*, Article ID 820714, 13 pages. 10.1155/2013/820714

Livingonsolarpower. (2013). *Solar PV power plants: major causes of performance degradation* [Blog Post]. Retrieved from https://livingonsolarpower.wordpress.com/2013/06/10/solar-pv-power-plants-majorcauses-of-performance-degradation/

Nebraska Energy Office, State of Nebraska. (2009). *Electricity Rate Comparison by State*. Retrieved from http://www.neo.ne.gov/statshtml/115.htm

Park, C. S. (2002). *Contemporary Engineering Economics*. Prentice-Hall, Inc.

Rustemli, S. & Dincer, F. (2011). Economic analysis and modeling process of photovoltaic power systems. *Przeglad Elektrotechniczny (Electrical Review)*, *87*(9A), 243–247.

Rustemli, S., Dincer, F., Unal, E., Karaaslan, M., & Sabah, C. (2013). The analysis on sun tracking and cooling systems for photovoltaic panels. *Renewable & Sustainable Energy Reviews*, *22*, 598–603. doi:10.1016/j.rser.2013.02.014

Sanden, B. A. (2005). The economic and institutional rationale of PV subsidies. *Solar Energy*, *78*(2), 137–146. doi:10.1016/j.solener.2004.03.019

Schwarz, C. (2010). The economic feasibility of solar panels for the University of Nebraska- Lincoln. *Environmental Studies, Undergraduate Student Theses.* Paper 19. Retrieved from http://digitalcommons.unl.edu/envstudtheses/19

Schwarzbozl, P., Buck, R., Sugarmen, C., Ring, A., Crespo, M. J. M., Altwegg, P., & Enrile, J. (2006). Solar gas turbine systems: Design, cost and perspectives. *Solar Energy*, *80*(10), 1231–1240. doi:10.1016/j.solener.2005.09.007

Sharaf Eldin, S. A., Abd-Elhady, M. S., & Kandil, H. A. (2016). Feasibility of solar tracking systems for PV panels in hot and cold regions. *Renewable Energy*, *85*, 228–233. doi:10.1016/j.renene.2015.06.051

Shepherd, W., & Shepherd, D. W. (2014). Energy studies. *Imperial College Press.*

Solar Energy Industries Association. (2016). *US solar market insight.* SEIA. Retrieved from http://www.seia.org/research-resources/us-solar-marketinsight

Solar Tracker Market Size and Share | Industry Report, 2022. (2016). Retrieved from http://www.grandviewresearch.com/industry-analysis/solar-tracker-industry

Systems USA Online Solar Panels. (2016). *Pallet of eoplly 190 watt mono solar panel.* Retrieved from http://www.solarsystems-usa.net/solarpanels/eoplly/ep125m-72-190/

The Price of Electricity in Your State. (2011). Retrieved from http://www.npr.org/sections/money/2011/10/27/141766341/the-price-of-electricityin-your-state

Trieb, F., Ole, L., & Helmut, K. (1997). Solar electricity generation—a comparative view of technologies, costs and environmental impact. *Solar Energy*, *59*(1-3), 89–99. doi:10.1016/S0038-092X(97)80946-2

United States Department of Energy (USDOE). (2009). *Energy Efficiency and Conservation Block Grant Program.* Retrieved from http://www.eecbg.energy.gov/

Warren, R. D. (2008). A feasibility study of stationary and dual-axis tracking grid-connected photovoltaic systems in the Upper Midwest. *Graduate Theses and Dissertations.* Paper 11161.

Nomenclature

Below is a partial listing of symbols. Those that are used infrequently or in limited parts of the book are defined locally and do not appear on this list. In some cases references to the section where a symbol is defined or where there might be confusion about its significance are given in parentheses.

A The photovoltaic panel's surface area
A_t Amount at time t
a_0 Absorption coefficient (mm)
AM Air mass
a_w Absorptivity of the water vapor
a, b Coefficients in empirical relationships
B Benefits, book value
BP The breakeven point
C Cost, initial investment
C_d Depreciation
C_i Initial cost
C_{sal} Salvage value
CF_t The net cash-flow at the end of the year *t*
*C*1, *C*2 Planck's first and second radiation constants
COE Cost of energy
Cp Specific heat
D Disbenefits, Diameter (defined locally), down-payment fraction
D_A Aerosol scatter diffuse component
D_a Annual deposit
D_{fm} The fractional depreciation for m[th] year
D_R Rayleigh scatter diffuse component
D_t Depreciation rate
DAQ data acquisition
d Market discount rate
E Equation of time
E_b Black body emitted energy
E_c The sun tracking power consumption
E_i The average power saving in year *i*

E_λ Spectral radiation
F_{c-s} View factor from the sky to the collector
$F_{PK,i,N}$ The capitalized cost factor
$F_{PS,iN}$, The compound interest factor or future value factor
$F_{RP,i,N}$ The equal-payment series present value factor or annuity present value factor
$F_{RS.i,N}$ The equal payment series future value factor
$F_{SR,iN}$ Sinking fund factor (SFF)
F_{sg} Angle factor between the surface and the ground
F_{ss} Angle factor between the surface and the sky
f Average energy inflation rate
G Solar irradiance W/m^2
G_0 Solar irradiance on a horizontal surface W/m^2
$G_{0,\beta}$ Solar irradiance on a Equator facing tilted surface W/m^2
G_b Beam irradiance W/m^2
G_c Clear sky global irradiance incident on a horizontal surface on the Earth's surface W/m^2
G_{cb} Clear sky beam irradiance incident on a surface on the Earth's surface W/m^2
G_{cd} Clear sky diffuse irradiance incident on a horizontal surface on the Earth's surface W/m^2
G_{cnb} Clear sky beam irradiance incident normally on a surface on the Earth's surface W/m^2
G_{cnd} Clear sky diffuse irradiance incident normally on a surface on the Earth's surface W/m^2
G_{cr} Clear sky reflected irradiance incident on a surface on the Earth's surface W/m^2
G_{ct} Global solar irradiance incident on a tilted surface on the Earth's surface W/m^2
G_d Diffuse irradiance W/m^2
G_s Normal extraterrestrial solar irradiance W/m^2
$G_{s,cd}$ Characteristic day extraterrestrial solar irradiance W/m^2
$G_{s,m}$ Daily average monthly extraterrestrial solar irradiance W/m^2
GW Giga Watts
GMT Greenwich mean time
$H_{0,b}$ Biannually extraterrestrial solar radiation on a horizontal plane J/m^2 or MJ/m^2 or kWh/m^2
$H_{0,\beta b}$ Biannually extraterrestrial solar radiation on a tilted plane J/m^2 or MJ/m^2 or kWh/m^2
$H_{0,\beta d}$ Daily extraterrestrial solar radiation on a tilted plane J/m^2 or MJ/m^2 or kWh/m^2
$H_{0,\beta f}$ Fortnightly extraterrestrial solar radiation on a tilted plane J/m^2 or MJ/m^2 or kWh/m^2
$H_{0,\beta m}$ Monthly extraterrestrial solar radiation on a tilted plane J/m^2 or MJ/m^2 or kWh/m^2
$H_{0,\beta s}$ Seasonally extraterrestrial solar radiation on a tilted plane J/m^2 or MJ/m^2 or kWh/m^2
$H_{0,\beta y}$ yearly extraterrestrial solar radiation on a tilted plane J/m^2 or MJ/m^2 or kWh/m^2
$H_{0,\beta w}$ Weekly extraterrestrial solar radiation on a tilted plane J/m^2 or MJ/m^2 or kWh/m^2
$H_{0,cd}$ Characteristic day solar radiation on a horizontal plane J/m^2 or MJ/m^2 or kWh/m^2
$H_{0,d}$ Daily extraterrestrial solar radiation on a horizontal plane J/m^2 or MJ/m^2 or kWh/m^2
$H_{0,f}$ Fortnightly extraterrestrial solar radiation on a horizontal plane J/m^2 or MJ/m^2 or kWh/m^2
$H_{0,m}$ Monthly extraterrestrial solar radiation on a horizontal plane J/m^2 or MJ/m^2 or kWh/m^2

$H_{0,s}$ Seasonally extraterrestrial solar radiation on a horizontal plane J/m^2 or MJ/m^2 or kWh/m^2
$H_{0,w}$ Weekly extraterrestrial solar radiation on a horizontal plane J/m^2 or MJ/m^2 or kWh/m^2
$H_{0,y}$ Yearly extraterrestrial solar radiation on a horizontal plane J/m^2 or MJ/m^2 or kWh/m^2
$H\left(N_1, N_2\right)$ Solar radiation over a period J/m^2 or MJ/m^2 or kWh/m^2
h_{atm} Atmospheric height
H_b Beam solar energy on a horizontal plane J/m^2 or MJ/m^2 or kWh/m^2
$H_{b,m,\beta}$ Monthly beam radiation on a tilted surface J/m^2 or MJ/m^2 or kWh/m^2
$H_{c,m}$ Monthly average daily clear sky global solar radiation on a horizontal surface J/m^2 or MJ/m^2 or kWh/m^2
$H_{d,d}$ Daily diffuse solar radiation J/m^2 or MJ/m^2 or kWh/m^2
$H_{d,m,\beta}$ Monthly diffuse component on a tilted surface J/m^2 or MJ/m^2 or kWh/m^2
$H_{r,m,\beta}$ Monthly ground reflected component on a tilted surface J/m^2 or MJ/m^2 or kWh/m^2
$H_{t,d}$ Daily solar radiation on a horizontal surface J/m^2 or MJ/m^2 or kWh/m^2
$H_{t,m,\beta}$ Monthly total incident radiation on a tilted surface J/m^2 or MJ/m^2 or kWh/m^2
I_0 Hourly extraterrestrial solar radiation on a horizontal surface J/m^2 or MJ/m^2 or kWh/m^2
I_b Hourly beam solar radiation on a horizontal surface J/m^2 or MJ/m^2 or kWh/m^2
$I_{b,\beta}$ Hourly beam solar radiation on a tilted surface J/m^2 or MJ/m^2 or kWh/m^2
I_d Hourly diffuse solar radiation on a horizontal surface J/m^2 or MJ/m^2 or kWh/m^2
$I_{d,\beta}$ Hourly diffuse solar radiation on a tilted surface J/m^2 or MJ/m^2 or kWh/m^2
I_{mp} Current at maximum power
I_{sc} Short circuit current
I_t Hourly total solar radiation on a horizontal surface J/m^2 or MJ/m^2 or kWh/m^2
$I_{t,\beta}$ Hourly total solar radiation on a tilted surface J/m^2 or MJ/m^2 or kWh/m^2
I_s Solar constant
IRR Internal rate of return
i Inflation rate, the annual interest
K The present value
K_d Daily diffuse transmittance index
$K_{d,m}$ Monthly average diffuse clearness index
$K_{t,d}$ Daily clearness index
$K_{t,m}$ Monthly average clearness index
k_b Hourly beam transmittance index
k_d Hourly diffuse transmittance index
k_t Hourly clearness index
L Shadow length
L_p Mast length
LCC Life cycle costs
LST Local solar time

LSTM Local standard time Meridian
LT Local time
MBE Mean bias error
MPE Mean Percentage Error
mtoe Million tons of oil equivalent
N_I The payback period
NH Northern Hemisphere
NMBE Normalized mean bias error
NOCT The nominal operating cell temperature of the PV module
NRMSE Normalized root mean square error
NSE Nash–Sutcliffe equation
n Day of year, index of refraction
n_1 First day number of the period in the year
n_2 Last day number of the period in the year
Np Payback time
NP Net present value
O&M Operation and maintenance
P Power, the present amount invested at zero (*N*=0) time
P_{mp} Maximum Power
q The energy rate in base year
R Energy gain factor of daily tracking with relation to horizontal surface, unacost
R' Energy gain factor of daily tracking with relation to optimum tilted surface
R_1 Energy gain factor of monthly optimum tilted surface with relation to horizontal surface
R_2 Energy gain factor of seasonally optimum tilted surface with relation to horizontal one
R_3 Energy gain factor of half-yearly optimum tilted surface with relation to horizontal one
R_4 Energy gain factor of yearly optimum tilted surface with relation to horizontal one
R_5 Energy gain factor of latitude tilted surface with relation to horizontal one
R_6 Energy gain factor of vertical surface with relation to EF vertical one
$\boldsymbol{R_{b,m}}$ Ratio of mean daily beam radiation on the tilted surface to that on a horizontal surface
RDD Research, development and demonstration
$\boldsymbol{R_{d,o}}$ Daily energy gain of daily optimum tilted surface with relation to horizontal surface
$\boldsymbol{R_{d,m}}$ Daily energy gain of daily optimum tilted surface with relation to monthly optimum tilted surface
$\boldsymbol{R_{d,s}}$ Daily energy gain of daily optimum tilted surface with relation to seasonally optimum tilted one
$\boldsymbol{R_{d,y}}$ Daily energy gain of daily optimum tilted surface with relation to yearly optimum tilted one
$\boldsymbol{R_{d,\varphi}}$ Daily energy gain of daily optimum tilted surface with relation to latitude tilted one
R_E Earth's radius
R_H Relative humidity
RMSE Root mean square error
R_s Actual Sun-Earth distance
R_{sa} mean Sun-Earth distance
r coefficient of correlation; The rate of return
r_d Ratio of diffuse radiation in an hour to diffuse in a day

r_S Sun radius
r_t Ratio of total radiation in an hour to total in a day
s Monthly average sunshine duration on the horizontal surface
S' Maximum sunshine duration on the tilted surface on the Earth's surface
S_0 Daylight length on the horizontal surface on the Earth's surface
S_N Future value at the end of N years
SH Southern Hemisphere
T Temperature
T_0 Transmittance of the ozone layer
T_a Transmittance for aerosols
T_{amb} Ambient temperature,
T_c Time correction factor
T_m The module temperature
T_r Transmittance of the atmosphere due to Rayleigh scattering
t Time
t_a Average aerosol optical depth
t Time, the tax rate
t_{sr} Sunrise time
t_{ss} Sunset time
V_{mp} Voltage at maximum power
V_{oc} Open circuit voltage
Greek
α Solar altitude (elevation) angle (o)
αn_{oon} Noon solar altitude (elevation) angle (o)
β Slope (tilt) (°)
β_A Ångström's turbidity coefficient
β_{opt} Optimum tilt (°)
$\beta_{opt,d}$ Daily optimum tilt (°)
$\beta_{opt,w}$ Weekly optimum tilt (°)
$\beta_{opt,f}$ Fortnightly optimum tilt (°)
$\beta_{opt,m}$ Monthly optimum tilt (°)
$\beta_{opt,s}$ Seasonally optimum tilt (°)
$\beta_{opt,b}$ Biannually optimum tilt (°)
$\beta_{opt,y}$ Yearly optimum tilt (°)
β_{ref} The power temperature coefficient
γ Surface azimuth angle (°)
γ_s Solar azimuth angle (°)
γ_{st} Azimuth angle of Sun's position on the inclined surface (°)
δ Declination, thickness (defined locally), dispersion (°)
δ_{cd} Characteristic day declination (°)
δ_m Average monthly daily declination (°)

ΔT_{GMT} The difference of the local time (LT) from Greenwich mean time (GMT) in hours
ε Sky clearness
η_{ref} Solar cell conversion efficiency
Θ_i Angle (defined locally), angle between surface normal and incident radiation (°)
Θ_z Zenith angle (°)
$\theta_{z.noon}$ Noon zenith angle (°)
λ Wavelength
λ_L Linke turbidity coefficient
λ_{max} Wavelength displacement
ρ_g Ground albedo
σ Stefan-Boltzmann constant, standard deviation
τ_b Transmittance to beam solar radiation
τ_d Transmittance to diffuse solar radiation
τ_a Aerosols scattering transmittance
τ_g Gas scattering transmittance
τ_O Ozone scattering transmittance
τ_r Rayleigh scattering transmittance
τ_w Water scattering transmittance
Φ The utilizability
φ Latitude, angle (defined locally)
Φ_h Hourly utilizability
ψ Mast's azimuth angle
ω Hour angle
ω_r Sunrise hour angle on the horizontal plane
ω_s Sunset hour angle on the horizontal plane
ω_{sr} Sunrise hour angle on the tilted plane
ω_{ss} Sunset hour angle on the tilted plane

Index

G

I

L

M

N

P

R

S

T

W

www.ingramcontent.com/pod-product-compliance
Lightning Source LLC
LaVergne TN
LVHW060251011026
849161LV00011B/2690